Insect Ultrastructure

Volume 2

Insect Ultrastructure

Volume 2

Edited by
Robert C. King
Northwestern University
Evanston, Illinois, USA

and

Hiromu Akai
Sericultural Experiment Station
Yatabe, Ibaraki, Japan

Plenum Press • New York and London

Library of Congress Cataloging in Publication Data

Main entry under title:

Insect ultrastructure.

Includes bibliographies and indexes.
1. Insects—Anatomy. 2. Insects—Cytology. 3. Ultrastructure (Biology) I. King, Robert C. II. Akai, Hiromu, 1930-
QL494.I49   1982   595.7′04   82-5268

ISBN 978-1-4612-9685-0   ISBN 978-1-4613-2715-8 (eBook)
DOI 10.1007/978-1-4613-2715-8

© 1984 Plenum Press, New York

Softcover reprint of the hardcover 1st edition 1984

A Division of Plenum Publishing Corporation
233 Spring Street, New York, N.Y. 10013

All rights reserved

No part of this book may be reproduced, stored in a retrieval system, or transmitted, in any form or by any means, electronic, mechanical, photocopying, microfilming, recording, or otherwise, without written permission from the Publisher

Contributors

Hiromu Akai — The Sericultural Experiment Station
Yatabe, Ibaraki 305, Japan

Christiane Ballan-Dufrancais — Laboratory of Fundamental and Applied Histophysiology
Pierre and Marie Curie University
75005 Paris, France

Stanley D. Carlson — Department of Entomology
Neuroscience Training Program
University of Wisconsin
Madison, Wisconsin 53706, USA

Che Chi — Biomedical Division
Nicolet Instrument Corporation
Madison, Wisconsin 53711, USA

Imré Foldi — Department of General and Applied Entomology
National Museum of Natural History
75005 Paris, France

Elisabeth Gateff — Institute of Genetics
Johannes Gutenberg University
Mainz 6500, Federal Republic of Germany

George M. Happ — Department of Zoology
University of Vermont
Burlington, Vermont 05405, USA

Roberta T. Hess — Department of Entomological Sciences
University of California
Berkeley, California 94720, USA

ERWIN HUEBNER	Department of Zoology University of Manitoba Winnipeg, Manitoba R3T 2N2, Canada
THOMAS A. KEIL	Max Planck Institute of Behavioral Sciences D-8131 Seewiesen, Federal Republic of Germany
MICHAEL LOCKE	Cell Science Laboratories Department of Zoology University of Western Ontario London, Ontario N6A 5B7, Canada
ROGER MARTOJA	Laboratory of Fundamental and Applied Histophysiology Pierre and Marie Curie University 75005 Paris, France
KEIICHIRO MIYA	Faculty of Agriculture Iwate University Morioka 020, Japan
ROSE M. RIZKI	Division of Biological Sciences University of Michigan Ann Arbor, Michigan 48109, USA
TAHIR M. RIZKI	Division of Biological Sciences University of Michigan Ann Arbor, Michigan 48109, USA
RICHARD L. SAINT MARIE	Department of Anatomy School of Medicine Boston University Boston, Massachusetts 02118, USA
BONNIE JOY SEDLAK	Developmental Biology Center University of California Irvine, California 92717, USA
ROSHANA SHRESTHA	Department of Zoology Triburan University Kathmandu, Nepal
DAVID SPENCER SMITH	Hope Entomological Collections and Department of Zoology Oxford University Oxford OX1 3PS, England

Contributors

VEIKKO SORSA   Department of Genetics
                            University of Helsinki
                            SF-00100 Helsinki 10, Finland

R. ALEXANDER STEINBRECHT   Max Planck Institute of Behavioral Sciences
                            D-8131 Seewiesen, Federal Republic of Germany

YOSHINORI TANADA   Department of Entomological Sciences
                            University of California
                            Berkeley, California 94720, USA

YOSHIO WAKU   Biology Laboratory
                            Kyoto Technical University
                            Matsugasaki, Sakyo-ku
                            Kyoto 606, Japan

Preface

Cell biology is moving at breakneck speed, and many of the results from studies on insects have helped in understanding some of the central problems of biology. The time is therefore ripe to provide the scientific community with a series of up-to-date, well illustrated reviews of selected aspects of the submicroscopic cytology of insects. The topics we have included fall into four general groups: seven chapters deal with gametogenesis, four concern developing somatic cells, seventeen chapters describe specialized tissues and organs, and three chapters cover cells in pathological states. These accounts are illustrated with over 600 electron micrographs.

The more than 1100 pages in the two volumes of *Insect Ultrastructure* represent the combined labors of 49 dedicated contributors from 11 countries. These authors have digested and critically summarized a very large body of information, and some measure of this effort can be gained from consulting the bibliographies that close each of the 31 chapters. These contain 2400 publications authored by 1500 different scientists. However, before we congratulate ourselves on the advanced state of our knowledge, it is worth remembering that representatives of less than 0.01% of the known species of insects have been examined with the electron microscope.

The Conference on the Ultrastructure and Functioning of Insect Cells, held in Sapporo, Japan, in August of 1982, provided an excellent opportunity for an international group of entomologists to exchange information and ideas concerning insect cytology. Many of the chapters that follow arose as a result of discussions held during the Sapporo meetings. We deeply appreciate the efforts of our many Japanese colleagues who contributed to the success of this meeting, especially Dr. Seijiro Morohoshi (President, Tokyo University of Agriculture and Technology), Dr. Mikio Arei (President, Hokkaido University), Dr. Toshio Iizuka (Sericulture Laboratory, Hokkaido University), Dr. Kazuo Hazama (Director, Sericultural Experiment Station, Yatabe), and Mr. Genkichi Hara (Managing Director, The Kajima Foundation).

Finally, we regret to report that one of the contributors to Volume 1 is no longer with us. Dr. Annette Szöllösi died in France on August 24, 1982. Her research has exerted a profound effect on all students of insect spermatogenesis, and her memory is held in affection and admiration by us all.

R. C. King and H. Akai
March 1984

Contents

I. THE ULTRASTRUCTURE OF DEVELOPING CELLS

Chapter 1
The Ultrastructure and Development of the Telotrophic Ovary
Erwin Huebner

1. Introduction	3
1.1. Groups Having Telotrophic Ovaries	4
2. Adult Ovary—General Aspects	4
3. Ovarian and Ovariole Sheaths, Tunica Propria, and Terminal Filaments	5
4. Inner Sheath or Interstitial Cells	8
5. Tropharium—General Features and Nurse Cell Functions	9
5.1. Nurse Cell Nuclear Morphology	11
5.2. Nurse Cell Cytoplasm	12
6. Tropharium—Arrangement of Nurse Cell Compartments	12
7. Trophic Cords or Trophic Tubes	16
8. Nurse Cell–Oocyte Interaction: Transport Regulation and Mechanisms	19
9. Oocytes: General Features, Nuclear Morphology, Cortex Morphology	20
10. Follicle Cells	24
11. Adult Ovary Summary	28
12. Ovarian Development: General Aspects	28
13. Postembryonic Early or Larval Ovary Stages: Mitotic Growth and Establishment of Somatic Tissue–Germ Cell Interaction	30
14. Postembryonic Late Ovary Stages: Larval–Adult Transformation, Nurse Cell, and Oocyte Differentiation	33
15. Summary	38
References	40

Chapter 2
Early Embryogenesis of Bombyx mori
Keiichiro Miya

1. Introduction	49
2. The Structure of Mature Eggs from the Oviduct	50
2.1. The Micropylar Region	50
2.2 Other Egg Regions	55
3. Maturation and Fertilization	59
3.1. Structure of Sperm in the Spermatheca	59
3.2. Changes of the Vitelline Membrane and Periplasm Accompanying the Entry of Sperm	60
3.3. The Sperm and Egg Nuclei	61
4. Cleavage and Blastoderm Formation	63
4.1. The Cleavage Stage	63
4.2. The Formation of Blastoderm Cells	65
4.3. The Vitellophages	67
5. Formation of the Germband and the Embryonic Envelopes	69
5.1. The Germband	69
5.2. The Serosa and the Amnion	69
6. Summary	70
References	71

Chapter 3
Electron Microscopic Mapping and Ultrastructure of Drosophila Polytene Chromosomes
Veikko Sorsa

1. Introduction	75
2. Light Microscopic Maps of the Salivary Gland Chromosomes of *Drosophila melanogaster*	76
3. Electron Microscopic Methods for the Mapping of Polytene Chromosomes	77
3.1. Thin-Sectioning of Squashed Chromosomes	77
3.2. Electron Microscopy of Whole-Mounted Polytene Chromosomes	78
4. Electron Microscopic Maps of the Salivary Gland Chromosomes of *Drosophila melanogaster*	78
5. Ultrastructural Studies on *Drosophila* Polytene Chromosomes	79
5.1. Ultrastructure of Interbands	81
5.2. Ultrastructure of Bands	83
5.3. Functional Organization of Polytene Chromosomes	86
6. Local Variation of the Degree of Polyteny	90
7. Summary	91
References	103

II. THE ULTRASTRUCTURE OF THE DEVELOPMENT, DIFFERENTIATION, AND FUNCTIONING OF SPECIALIZED TISSUES AND ORGANS

Chapter 4
The Structure of Insect Muscles
David Spencer Smith

1. Introduction 111
2. Contractile System 114
   2.1. Myofilament Organization and Disposition 114
   2.2. Z-Band Structure and Supercontraction 119
   2.3. Paramyosin and Connecting Protein 124
3. Muscle Insertions 125
4. Membrane Systems 129
   4.1. General Features 129
   4.2. Synchronous Muscles 131
   4.3. Asynchronous Muscles 137
5. Tracheation 142
6. Summary 145
   References 146

Chapter 5
The Structure and Development of the Vacuolar System in the Fat Body of Insects
Michael Locke

1. Introduction 151
   1.1. The Fat Body 151
   1.2. The Ultrastructure of Fat Body Cells 152
   1.3. The Fat Body Vacuolar System 152
   1.4. Kinds of Fat Body Vacuole 154
2. Posttransition Compartments 158
   2.1. Tyrosine Storage Vacuoles and Provacuoles 158
   2.2. Urate Storage Vacuoles 160
   2.3. Urate Granules 162
   2.4. Protein Storage Granules 167
   2.5. Secretory Vesicles 171
   2.6. Vacuoles Containing Symbionts 173
   2.7. Vacuoles Associated with Glycogen 176
   2.8. Autophagic Vacuoles 177
   2.9. Multivesicular Bodies 182
   2.10. Lamellar Bodies 184
   2.11. Basal Lamina Phagocytic Vacuoles 184

3. Pretransition Compartments 186
   3.1. Distensions of the RER 189
   3.2. Peroxisomes 190
   3.3. Concretions in the RER 190
4. Discussion 191
5. Summary and Conclusions 192
   References 194

Chapter 6
The Ultrastructure of the Digestive and Excretory Organs
Roger Martoja and Christiane Ballan-Dufrançais

1. Introduction 199
2. The Midgut: General Organization 201
   2.1. Structure of the Wall 201
   2.2. Striated Border 202
   2.3. Basement Membrane (Basal Lamina) and Related
        Structures 204
3. Types of Epithelial Cells in the Midgut 205
   3.1. Columnar Cells 205
   3.2. Goblet Cells 208
   3.3. Peculiar Secretory Cell Types: The Goblet Cells of
        Thysanura and the Bottle-Shaped Cells of Odonata 209
   3.4. Endocrine Cells 211
   3.5. Storage Products of the Columnar Cells 211
   3.6. Secretion of the Peritrophic Membrane 219
4. Changes in Fine Structure 220
   4.1. Functional Subdivisions of the Gut and Cellular
        Cycles 220
   4.2. Effects of Unsuitable Diets and Starvation 221
   4.3. Blood-Sucking Insects 222
5. Cell Degeneration and Renewal 223
   5.1. Cell Degeneration 223
   5.2. Cell Differentiation 223
   5.3. Differentiation of Microvilli in Embryonic Cells 225
   5.4. Peculiar Aspects of Metamorphosis in Some
        Holometabola 226
6. The Proctodeal Fermentation Chambers 228
7. Filter Chambers and Related Organs 229
   7.1. Midgut Filter Chambers 229
   7.2. The Composite Segment of Termite Intestines 232
8. The Salivary Glands 232
   8.1. The Tubular Glands of Adult Flies 233

	8.2. The Tubular Salivary Glands of Adult Lepidoptera	234
	8.3. The Tubular Salivary Glands of Mosquitos	234
	8.4. The Acinar Glands of Dictyoptera and Orthoptera	235
	8.5. The Sac-Shaped and Tubular Salivary Glands of Lower Diptera	239
9.	The Malpighian Tubules	240
	9.1. The General Organization of the Malpighian Tubules	240
	9.2. Types of Cells	240
	9.3. Compounds Deposited in the Lumen of the Tubules	245
10.	The Proctodeum	245
	10.1. General Organization of the Epithelium	247
	10.2. Types of Epithelial Cells	247
11.	The Pericardial Cells	255
12.	The Other Excretory Organs	257
13.	Concluding Remarks	258
	References	261

Chapter 7
The Ultrastructure of Interacting Endocrine and Target Cells
Bonnie Joy Sedlak

1.	Introduction	269
2.	The Insect Endocrine System	270
3.	The Ultrastructure of the Endocrine Glands of *Manduca*	272
4.	The Corpora Allata	272
	4.1. Synthesis of Juvenile Hormone	275
	4.2. Secretion of Juvenile Hormone	277
	4.3. Axons of PTTH Release	279
	4.4. Degradation of Juvenile Hormone	281
5.	Prothoracic Glands	281
	5.1. Synthesis of Ecdysone	282
	5.2. Secretion of Ecdysone	283
	5.3. Stimulation of Prothoracic Glands by PTTH	285
	5.4. Stimulation of Prothoracic Glands by 20HE	287
6.	Summary of Gland Cell Structure	287
7.	Target Tissue: Epidermal Cells and Cuticle Secretion	288
	7.1. Hormonal Influences on the Epidermis	289
	7.2. Effects of Ecdysteroids on the Epidermis	291
	7.3. Effects of Juvenile Hormone on the Epidermis	295
8.	Summary of Target Tissue Responses to Hormones	296
9.	Conclusions	296
	References	296

Chapter 8
The Fine Structure of Insect Glands Secreting Waxy Substances
Yoshio Waku and Imiré Foldi

1. The Waxy Secretions of Insects	303
1.1. Chemical Properties	303
1.2. Site of Synthesis of Waxy Secretions	304
1.3. The Function of Insect Waxes	305
1.4. The Economical Importance of Insect Waxes	306
2. Classification of Wax Glands	306
2.1. Classification of Epidermal Glands	306
2.2. Simple Wax Glands	306
2.3. Complex Wax Glands	311
2.4. The Mechanism of Pattern Formation by the Products of Wax Glands	316
3. Development of Wax Glands	318
4. Taxonomic Importance of Wax Glands	320
References	321

Chapter 9
The Ultrastructure and Functions of the Silk Gland Cells of Bombyx mori
Hiromu Akai

1. Introduction	323
2. The General Cytology of the Silk Gland	324
3. The Ultrastructure and Functioning of the Silk Gland Cells	327
3.1. Posterior Silk Gland	327
3.2. Middle Silk Gland	342
3.3. Anterior Silk Gland	345
3.4. Filippi's Gland	347
3.5. Liquid Silk and the Cocoon Filament	348
4. The Hormonal Control of Silk Gland Development	351
5. Mutations Influencing Silk Production	352
6. Summary	356
References	357

Chapter 10
Structure and Development of Male Accessory Glands in Insects
George M. Happ

1. Introduction	365
2. Accessory Glands and Their Secretions	366
3. Muscular Coats and Basement Membranes	369
4. The Secretory Epithelia	369
4.1. Intercellular Junctions	371
4.2. Absorption of Precursors	373
4.3. Biosynthetic Machinery	373
4.4. Export of Product	377
5. Semisolid Secretory Products	381
5.1. Organized Secretion Masses	381
5.2. Secretion into Separate Compartments Arranged in Series	382
5.3. Secretion in Parallel to a Common Lumen	383
6. Formation of the Spermatophore	385
7. Evacuation of the Spermatophore	386
8. Development of Accessory Glands	387
9. Endocrine Control of Accessory Gland Development	390
10. Summary and Prospects	391
References	392

Chapter 11
The Photoreceptor Cells
Stanley D. Carlson, Richard L. Saint Marie, and Che Chi

1. Introduction	397
2. Historical Record	398
3. Ommatidial Morphology	399
4. Photoreceptor Fine Structure	401
5. Ultrastructure of the Cell Soma	403
5.1. Rhabdom	403
5.2. Nucleus	409
5.3. Endoplasmic Reticulum	409
5.4. Screening Pigment Granules	410

	5.5. Mitochondria	411
	5.6. Golgi Material	413
	5.7. Ciliary Structures	413
	5.8. Other Intracellular Structures	414
6.	Cells Associated with Photoreceptor Cells	414
7.	The Axon	417
8.	The Synapse	422
	8.1. Chemical Synapses	422
	8.2. Electrical Synapses	427
9.	Concluding Remarks	427
	References	428

Chapter 12
The Glial Cells of Insects
Richard L. Saint Marie, Stanley D. Carlson, and Che Chi

1.	Introduction	435
	1.1. Historical Concepts of Neuroglia	435
	1.2. Insect Neuroglia: General Concepts	436
	1.3. The Insect Eye	436
2.	Ultrastructure of Insect Neuroglia	438
	2.1. The Nucleus	439
	2.2. Cytoplasmic Organelles	439
	2.3. Elaborations of the Plasma Membrane	445
	2.4. Interglial Junctions	447
3.	Specialized Supporting Cells of the Insect Eye	453
	3.1. Pigmented Glia	453
	3.2. Tracheal Cells	457
4.	Neuro-glial Junctions	458
	4.1. Desmosomes	458
	4.2. Septate Junctions	458
	4.3. Scalariform Junctions	459
	4.4. Tight Junctions	461
	4.5. Gap Junctions	461
5.	Structure-Function Relationships	462
	5.1. Structural Support	462
	5.2. The Blood-Brain Barrier and Ionic Homeostasis	463
	5.3. Metabolic Commerce	466
	5.4. Phagocytosis	467
	5.5. Glia in Neurogenesis	468
	5.6. Compartmentalization of Neuropil	469
6.	Concluding Remarks	470
	References	471

Chapter 13
Mechanosensitive and Olfactory Sensilla of Insects
Thomas A. Keil and R. Alexander Steinbrecht

1. Introduction	477
2. General Organization and Morphogenesis of Sensilla	478
3. Fine Structure of the Cellular Components of Sensilla	481
3.1. The Sensory Neurons	481
3.2. The Auxiliary Cells	483
3.3. Membrane Contact Structures	486
3.4. The Receptorlymph Spaces	487
3.5. Glia Cells and Neuron–Glia Interrelations	489
4. Fine Structure of Mechanosensitive Sensilla	490
4.1. General Remarks	490
4.2. Cuticular Parts	491
4.3. Cellular Parts	491
4.4. Stimulus Uptake and Transmission	493
4.5. The Macrochaetae of *Calliphora*	493
4.6. Filiform and Clavate Hairs on the Cerci of *Acheta*	494
5. Fine Structure of Olfactory Sensilla	499
5.1. General Remarks	499
5.2. Single-Walled Olfactory Sensilla with Pore Tubules	501
5.3. Double-Walled Olfactory Sensilla with Spoke Channels	507
6. Final Remarks	509
References	510

Chapter 14
The Cytopathology of Baculovirus Infections in Insects
Yoshinori Tanada and Roberta T. Hess

1. Introduction	517
2. Mode of Virus Infection	519
3. Gross Pathology of Nuclear Polyhedroses of Lepidoptera	525
4. Gross Pathology of Nuclear Polyhedroses of Hymenoptera	526
5. Gross Pathology of Nuclear Polyhedroses of Diptera	526
6. Gross Pathology of Granuloses	527
7. Gross Pathology of Baculovirus of *Oryctes*	528
8. Gross Pathology of Parasitoid Baculovirus	528
9. Cytopathology of Occluded NPV in Lepidoptera	529

10. Cytopathology of Occluded NPV in Hymenoptera 539
11. Cytopathology of Occluded NPV in Diptera 541
12. Cytopathology of Granuloses 542
13. Cytopathology of Nonoccluded NPV in
    Oryctes rhinoceros 547
14. Cytopathology of Nonoccluded NPV in *Gyrinus natator* ... 548
15. Cytopathology of Nonoccluded NPV of Parasitoids 550
16. Summary 550
    References 552

III. THE ULTRASTRUCTURE OF CELLS IN PATHOLOGICAL STATES

Chapter 15
Comparative Ultrastructure of Wild-Type and Tumorous Cells of Drosophila
Elisabeth Gateff, Roshana Shrestha, and Hiromu Akai

1. Introduction 559
2. Comparative Ultrastructure of Wild-Type and Tumorous
   Imaginal Discs 559
3. Comparative Ultrastructure of Wild-Type and Malignant
   Optic Neuroblasts and Ganglion Mother Cells 564
   3.1. Ultrastructure of the Neuroblasts and Ganglion Mother
        Cells in the Mature Wild-Type Larval Brain 564
   3.2. Ultrastructure of the Neuroblasts and Ganglion Mother
        Cells in the Mature $l(2)gl^4$ Larval Brain 564
4. Comparative Ultrastructure of Wild-Type and Tumorous
   Blood Cells 567
   4.1. Ultrastructure of Blood Cell Precursors in the Hematopoietic Organs and of the Free Blood Cells in the Hemolymph
        of Wild-Type Larvae 568
   4.2. Ultrastructure of Tumorous Blood Cells 570
5. Discussion 575
   References 577

Chapter 16
The Cellular Defense System of Drosophila melanogaster
Tahir M. Rizki and Rose M. Rizki

1. Introduction 579
2. Distribution of Hemocytes in the Larva 580

3. Classification of the Larval Hemocytes 581
   3.1. The Plasmatocyte and Its Variants 583
   3.2. The Crystal Cell 585
4. Crystal Cells in Other *Drosophila* Species 585
5. Phagocytosis 588
   5.1. Cell Disruption as a Source of Humoral Factors 589
   5.2. Absence of Opsonization in Phagocytosis 589
   5.3. A Mutation Affecting Phagocytosis 592
6. Encapsulation 592
   6.1. Capsule Formation in Melanotic Tumor Mutants 592
   6.2. Basement Membrane as a Factor for Recognition
        of Self 593
   6.3. Ultrastructural Examination of Basement
        Membrane 595
7. Differentiation of Competent Lamellocytes 596
8. Contribution of Crystal Cells to Melanization Reactions ... 599
9. Wound Healing 600
10. Genetic Dissection of the Cellular Defense System 600
11. Summary and Concluding Remarks 601
    References 602

Author Index 605
Subject Index 617

The Ultrastructure of Developing Cells

I

The Ultrastructure and Development of the Telotrophic Ovary

ERWIN HUEBNER

1. Introduction

The production of a viable insect oocyte is achieved by an intimate interaction between the oocyte and various cell types during differentiation within the ovary. Functions such as organelle production, transport of cytoplasmic constituents, transport of yolk precursors, regulation of oocyte growth, production of egg envelopes, and others are carried out by these accessory cells. Meroistic ovaries exemplify the most complex forms of interaction within the insects (Gross, 1901; Telfer, 1975). Gross (1903) subdivided the meroistic into the polytrophic and telotrophic (sometimes referred to as acrotrophic) types.

Since the early histological studies by investigators such as Wielowiejski (1886), Korschelt (1887, 1889), Henking (1892), Preusse (1895), Gross (1901, 1903), Köhler (1907), and Hegner (1914) (see also Telfer, 1975), there have been numerous light microscopic studies and, more recently, electron microscopic studies on meroistic ovaries. We have learned how remarkable these interacting cell systems are and the unique opportunities they provide for experimentation on basic cellular developmental problems such as regulation of oocyte differentiation, RNA synthesis, transport, and storage, cell communication, protein transport, and pinocytosis, to cite a few. Thus, a detailed view of the structure of meroistic ovaries is not only useful for understanding insect reproduction and evolution, but also for a variety of other disciplines focusing on basic cellular and biochemical mechanisms (see Garel, 1982; Huebner, 1984).

ERWIN HUEBNER • Department of Zoology, University of Manitoba, Winnipeg, Manitoba R3T 2N2, Canada.

The polytrophic ovary type has been far more extensively studied with a huge body of elegant and sophisticated data accumulated (see Bonhag, 1958; King, 1970; Mahowald, 1972; Telfer, 1975; Berry, 1982; King et al., 1982). Telotrophic ovaries are histologically and ultrastructurally far more complex, and it is only within the last 10 years that ultrastructural studies have attempted to resolve the cellular organization.

There have been a number of excellent reviews of insect oogenesis generally (Bonhag, 1958; Engelmann, 1970; King, 1970; Mahowald, 1972; Telfer, 1975; King et al., 1982; King and Büning, 1984) some of which include details on the telotrophic ovary. This review will focus on the ultrastructural organization of the telotrophic ovary and builds on the foundation established by these earlier reviews. Hopefully, what emerges will be a detailed description of the cellular components of the telotrophic ovary and some of the variety in organization and cytoarchitecture one sees by surveying a broad spectrum of insects possessing telotrophic ovaries. The development of the ovary will also be discussed.

1.1. Groups Having Telotrophic Ovaries

The telotrophic ovary is characteristic of the Hemiptera, polyphagic Coleoptera, and Megaloptera (Sialidae) including the closely related Raphidioptera (Gross, 1903; Bonhag, 1958; Matsuzaki and Ando, 1977; Büning, 1979a,b,c, 1980). Büning (1979a,b) has examined 32 species from 23 families and verified that all polyphagic Coleoptera are telotrophic. Matsuzaki and Ando (1977) confirmed, as Gross (1903) suggested, but Bonhag (1958) doubted, that the megalopteran *Sialis mitsuhashii* has a typical telotrophic ovary. Confirmation of this in a second megalopteran species as well as in the closely related Raphididae comes from Büning's work (1979c, 1980). Studies of a broad spectrum of hemipterans have established they are telotrophic (Bonhag, 1955a,b, 1958; Davis, 1956; Masner, 1966, 1968; Brunt, 1971; Huebner and Anderson, 1972a,b,c; Schreiner, 1977a,b; Couchman and King, 1979; Książkiewicz, 1980; Büning, 1981a,b).

Ovary type and variation in complexity should be related to the phylogenetic organization of insects (Heming, 1977; Kristensen, 1981).

Heming suggested in his book review that the telotrophic ovary type arose twice independently; however, he has the Megaloptera listed as polytrophic. Kristensen (1981) groups the Raphidoptera with the Neuroptera and Megaloptera into the superorder Neuropterida. They do not share the same ovary type (see King et al., 1982). It is now apparent (Matsuzaki and Ando, 1977; Büning, 1979c, 1981b; Książkiewicz, 1980) that the telotrophic ovary has a polyphyletic origin having three distinct subtypes each characteristic for one of the three groups having them. The cellular differences between the three subtypes will be discussed in the detailed description of ovarioles in later sections.

2. Adult Ovary—General Aspects

In telotrophic ovaries, the number of ovarioles ranges from just a few in aphids (Couchman and King, 1979, Büning, 1981a) to relatively many in the megalopteran *Sialis* (Matsuzaki and Ando, 1977) and certain Coleoptera such as *Lytta nuttalli* (Gerber *et al.*, 1971) and the Meloini Coleoptera with about 1000 ovarioles (Jeannel and Paulian, 1944). Hemipterans such as *Rhodnius prolixus* and *Oncopeltus fasciatus* typically have seven ovarioles per ovary.

In the following sections, *R. prolixus* provides a central example for a description of the various layers and parts of the ovariole both to provide the basic hemipteran structure and also for comparative discussion to include the variations in structure present in other telotrophic systems.

3. Ovarian and Ovariole Sheaths, Tunica Propria, and Terminal Filaments

All the ovarioles collectively are enclosed in an ovarian sheath which has a rich tracheal supply and forms a continuous layer, the peritoneal sheath. Within this ovarian connective tissue sheath, each ovariole is invested by a double cellular layer, the ovariole sheath (Figures 1, 2). Some confusion in terminology and whether or not these sheaths were syncytial or cellular existed in the earlier literature (see Bonhag and Wick, 1953; Masner, 1966; Brunt, 1971; Huebner and Anderson, 1972a; Schreiner, 1977a). In *Rhodnius* as in *Dysdercus fasciatus* (Brunt, 1971) and *O. fasciatus* (Schreiner, 1977a), the outer ovariole sheath (or outer epithelial sheath) is cellular and consists of muscle cells, tracheoles, and connective tissue cells (Figures 1, 2). The muscle fibers of the outer ovariole sheath are organized in a regular branching network which undergoes a regular systematic contraction causing a regular differential rolling and "massaging" of the ovariole (Huebner, unpublished data). The fiber organization is readily seen in living preparations using interference or polarizing microscopy. Thin basal laminae provide an upper and lower boundary. A regular geometric substructure characterizes the outer sheath basal laminae in *Rhodnius* (Figures 3, 4). Between the lower basal lamina of the outer sheath and the basal lamina (or tunica propria) of the ovariole are scattered pleomorphic cells referred to as the inner envelope (Figure 2) (Bonhag and Wick, 1953; Brunt, 1971; Schreiner, 1977a). These cells resemble the plasmatocytes described by Wigglesworth (1979) on the epidermal basal lamina. In living preparations, one often sees their long microvilli move. Irregular cell shape and pseudopodia were described for these cells in *Oncopeltus* (Schreiner, 1977a). Thin sections (Figure 2) show the inner ovariole envelope cells contain distended RER and well-developed Golgi. Their ultrastructure, coupled with the observation that their distribution is related to tunica propria thickness (Schreiner, 1977a), suggests they are responsible at least in part for synthesis of the tunica propria. The tropharium tip and previtellogenic area where the tunica propria is thickest in *Rhodnius* has the largest number of inner envelope cells. There is a parallel between the description of these cells and tunica propria formation and the description of Wigglesworth (1979) of epidermal basement lamina formation.

Removal of the ovariole sheaths exposes the smooth tunica propria that covers the ovariole proper (Figure 1). The lanceolate trophic chamber containing the nurse cells and the vitellarium with various sized follicles are now clearly

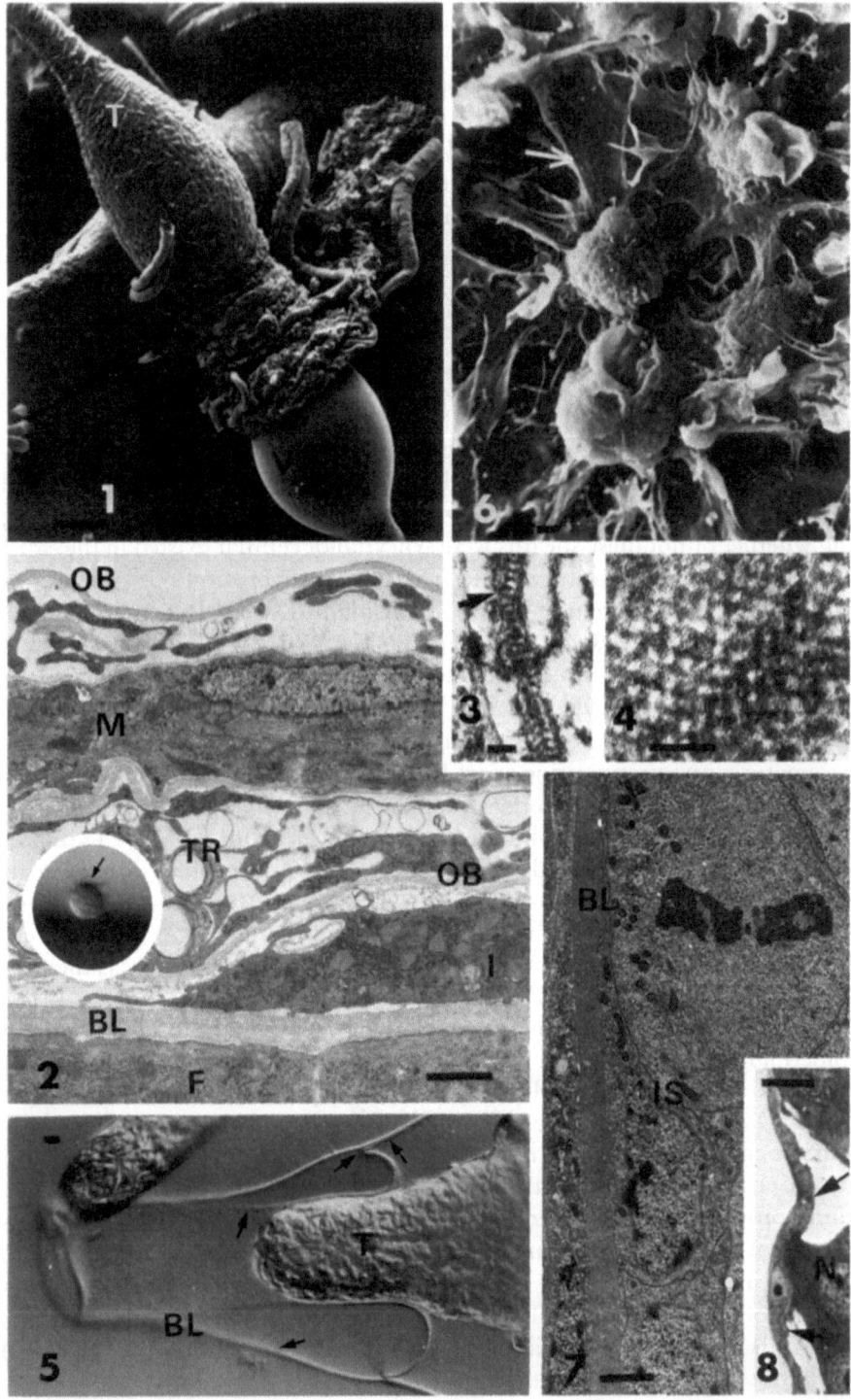

discerned. The basal laminae of the sheaths and the tunica propria are permeable to hemolymph proteins, such as vitellins, that must reach oocytes for vitellogenesis. However, since this only involves certain regions of the tunica propria, the consideration of variations in permeabilities over the various ovariole regions as well as biochemical composition has been inadequately addressed. The tunica propria in *Rhodnius* is thinnest over vitellogenic and chorionating eggs and thickest over the tropharium. Trypsin digestion experiments (Watson and Huebner, unpublished data) clearly confirm the tropharium tunica propria is more substantial. Elastase digestion (Types I and III) did not remove the tunica propria (Watson and Huebner, unpublished data), unlike the effect on the Malpighian tubule basal lamina (Bradley, 1981). However, it did result in a characteristic splitting of the tunica propria at the tropharium tip (Figure 5). Part of the tunica propria remains adherent to the nurse cells while the outer part expands out. Interestingly, in the adult hemipteran ovary, it is only in this region of the tropharium where mitotic activity occurs (Huebner, 1981a).

Anteriorly, the tunica propria of hemipterans and coleopterans continues over the tropharium tip forming a layer, the transverse septum, between the terminal filament and tropharium. The Megaloptera differ significantly here. There is no transverse septum, with the tunica propria simply continuing over the terminal filament (Matsuzaki and Ando, 1977, Büning, 1979a,b, 1980). The terminal filaments are thin cellular strands that join to form a single muscular strand, which in *Rhodnius* attaches to the internal dorsal anterior lip of the thoracic cuticle (Huebner, unpublished data). The terminal filaments are highly birefringent and have the ovariole sheaths attached and often extended over them basally. Many earlier studies on Hemiptera and Coleoptera indicated terminal filaments were syncytial but electron microscopy has indicated they are cellular (see Matsuzaki and Ando, 1977).

←

Figure 1. A scanning electron micrograph (SEM) of a *Rhodnius* ovariole with the ovariole sheaths pulled back over the vitellarium revealing the smooth tunica propria. T, tropharium; V, vitellarium. From Huebner (1981a) courtesy of Longman Group. (Bar = 0.1 μm.)

Figure 2. A thin section of the ovariole sheaths in the area around a previtellogenic follicle. Note outer ovariole sheath with basal laminae (OB), muscle cells (M), tracheoles (TR), and inner ovariole sheath cell (I) resting on the tunica propria (BL). (Inset) An interference light micrograph (LM) of ovariole with outer ovariole sheath removed; note rounded inner ovariole sheath cell (arrow). (Bar = 1 μm.)

Figures 3 and 4. Cross-sectional and oblique thin sections of tannic acid-fixed outer ovariole sheath basal lamina. Note regular substructure. (Bars = 0.1 μm.)

Figure 5. A Nomarski differential interference LM of an elastase-treated ovariole tip. Note the tunica propria (BL) is double-layered (arrows) over the tropharium tip. (Bar = 10 μm.)

Figure 6. An SEM of a tropharium denuded of its tunica propria and placed in cation-free saline. Note the inner sheath cells have rounded somas with flattened cell extremities (arrow) contacting adjacent cells. (Bar = 1 μm.)

Figure 7. A thin section of the inner sheath cells (IS) in a late fifth-instar ovariole. Note mitotic figure and differentiating thin mononucleate cells under the basal lamina (BL). From Huebner and Anderson (1972c) courtesy of Wistar Press. (Bar = 1 μm.)

Figure 8. An epoxy semithin toluidine blue-stained section of a tropharium briefly placed in cation-free saline. Note the inner sheath (arrows) forms a continuous layer. The nurse cells have separated from them. (Bar = 10 μm.)

4. Inner Sheath or Interstitial Cells

Beneath the tunica propria we encounter both somatic and germ cells that constitute the ovariole proper. With the exception of some aphids (Wightman, 1973; Couchman and King, 1979), in most hemipteran and megalopteran telotrophic ovaries, all the germ tissue is encased in at least one layer of mesodermally derived somatic cells. These somatic cells are specialized into various cell types including the inner sheath cells (Huebner and Anderson, 1972a; Schreiner, 1977a; Büning, 1979c; Huebner, 1984) or interstitial cells in the coleopterans (Schlottman and Bohnag, 1956; Büning, 1979a,b) associated with nurse cells or trophocytes, the prefollicular tissue around early previtellogenic oocytes and the follicle cells around growing oocytes, the interfollicular plug cells and pedicel cells adjacent to follicles.

The inner sheath cells refer to the single layer of flattened pleiomorphic cells that form a continuous layer covering the entire nurse cell surface, thus providing an epithelial cell barrier between nurse cells and the tunica propria (Figures 6-8). These, or an equivalent cell type, have been described for all three types of telotrophic ovaries (Huebner and Anderson, 1972a,c; Matsuzaki and Ando, 1977; Schreiner, 1977a; Büning, 1979b,c; Huebner, 1981a, 1984).

In *Rhodnius* as in some species, they are found exclusively on the outer surface of the nurse cells although they may have slender extensions toward the tropharium interior. The projections extending between nurse cells are more extensive in *Oncopeltus* than *Rhodnius* which prompted Schreiner (1977a) to suggest a nutrient function and that they be referred to as *peripheral trophocytes*. Since the term *trophocyte* is used synonymously with nurse cells in many papers and these flat cells are of mesodermal origin, I prefer the term *inner sheath*. In *Rhodnius*, they form a continuous layer and their ultrastructure does not provide support for the notion that they may be synthetically active. They could alternatively provide a physiological barrier or play a role in nurse cell differentiation.

Exposure of *Rhodnius* ovarioles to salines of normal osmolarity but altered cation composition can cause nurse cell clusters to separate from the inner sheath and thereby reveal that this inner layer is indeed continuous and also physiologically responds differently than nurse cells (Figure 8) (Watson and Huebner, unpublished data). Scanning electron microscopy shows these cells as a layer of interlocking irregular cells. Upon drastic treatment intended to dissociate cells after basal lamina removal, these cells retract much of their cytoplasm into a central soma and become stellate while still maintaining some contacts with neighboring cells (Figure 6) until they are physically washed off. Based on their tenacity to maintain a layer and their dramatic changes with experimental manipulation, we should determine their function and importance for nurse cells. Even after their removal, the tropharium maintains an approximately normal shape (Figure 12) so I would suggest their role is not merely a mechanical one.

The equivalent cells in the polyphagic Coleoptera do not appear to completely encase the nurse cells (Huebner, 1984) and some are also dispersed in between them in small groups referred to as *interstitial cells* (Büning, 1972, 1979b). Neither the inner sheath or interstitial cells of the Coleoptera nor the inner sheath cells of the Hemiptera and Megaloptera have an ultrastructure that suggests they are active in RNA or protein synthesis. Autoradiographic study supports the notion of synthetic inactivity (Büning, 1972). Büning (1979b) discusses their possible roles in the structure and mechanical stability of the tropharium and in differentiation, as postulated by Kloc and Matuszewski (1977).

Thus, the precise role(s) of the mesodermal cells in the telotrophic tropharium is still unresolved. Recent results indicate they may play an important role in the electrophysiology of meioistic ovaries generally (Huebner, 1984; Woodruff et al., 1984). Nurse cells or trophocytes comprise the major portion of the tropharium.

5. Tropharium—General Features and Nurse Cell Functions

The trophic tissue and its organization is at the heart of what makes the telotrophic ovary so unique and remarkable. As discussed in the ovarian development section, nurse cells and oocytes are sister cells both originating, by mitosis, from common undifferentiated germ cells. Incomplete cytokinesis during ovarian development gives rise to a cytoarchitecturally complex arrangement of sibling nurse cell clusters connected to oocytes. Studies on the polytrophic ovary have clearly shown the precise organization of nurse cells and their bridges with other nurse cells and the oocyte (King et al., 1982). Although a general view of the organization of telotrophic nurse cells has existed for many years (Bonhag, 1958; Masner, 1966; Büning, 1972; Huebner and Anderson, 1972a,b,c), the complexity made unraveling cell interactions difficult. Recent electron microscopic, developmental, and microinjection studies have furthered our understanding of nurse cell–nurse cell and nurse cell–oocyte interactions in all three major telotropic ovariole types (Huebner, 1977, 1981a, 1982, 1984; Kloc and Matuszewski, 1977; Büning, 1978, 1981b; Lutz and Huebner, 1978, 1980, 1981).

The primary and highly amplified function of the trophic tissue is to produce and transport the spectrum of RNAs (Bier, 1965, 1967; Zinsmeister and Davenport, 1971; Büning, 1972; Mahowald, 1972; Mays, 1972; Duspiva et al., 1973; Davenport, 1974, 1976; Winter, 1974, 1975; Telfer, 1975; Choi and Nagl, 1977b; Schmidt and Jäckle, 1978; Capco and Jeffrey, 1979; Telfer et al., 1981; Berry, 1982) and proteins (Vanderberg, 1963; Telfer, 1975; Telfer et al., 1981) to the oocyte for use in its differentiation and/or storage for embryonic development. Evidence that ribosomal RNA and messenger RNA are transported from the tropharium to oocytes comes from many species and is based on a wide spectrum of experimental approaches including electron microscopy, autoradiography, ligaturing, and biochemistry (MacGregor and Stebbings, 1970; Büning, 1972; Mays, 1972; Ullmann, 1973; Telfer, 1975; Davenport, 1976; Capco and Jeffrey, 1979; Huebner, 1981a). Analysis of the functional aspects of nurse

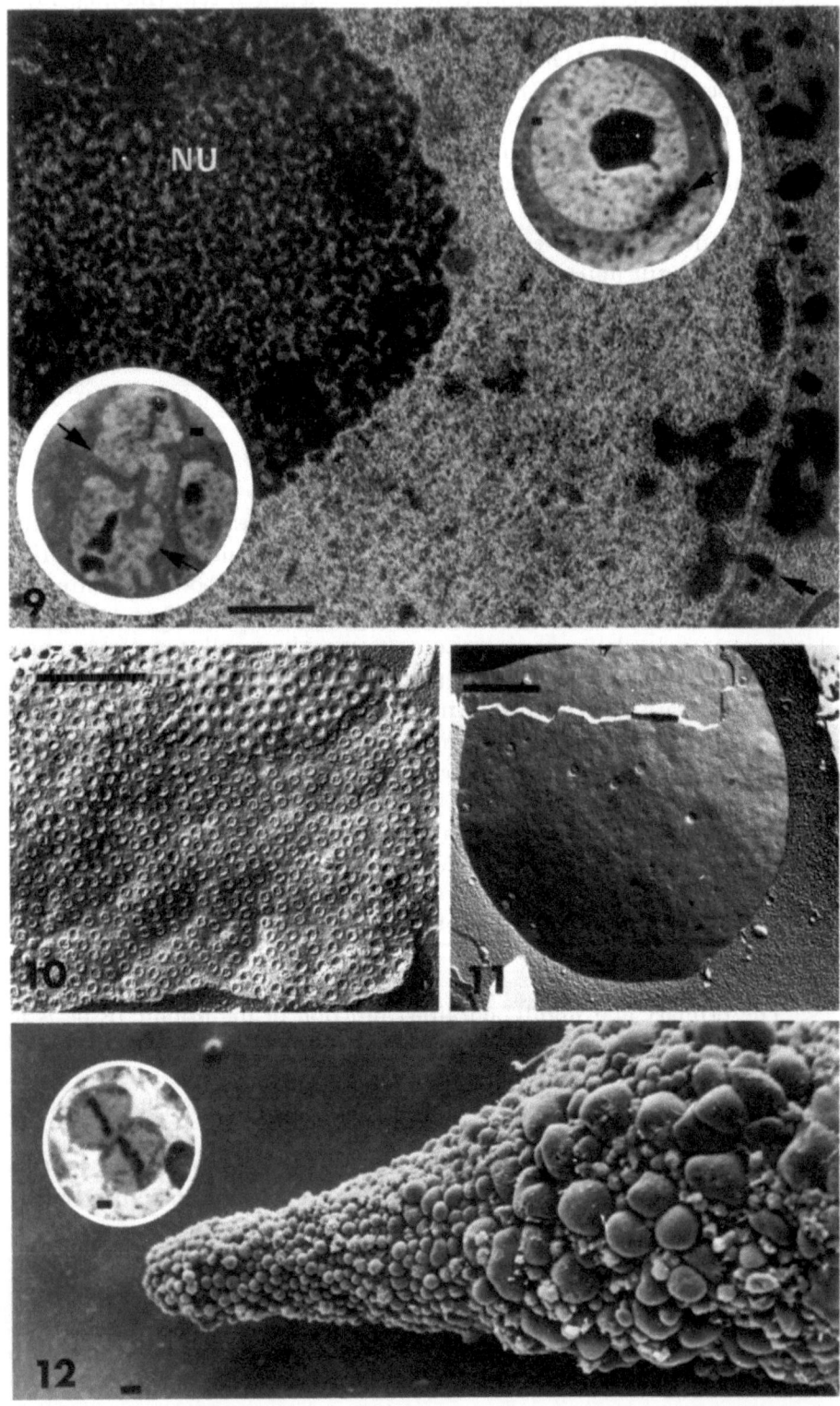

cells (see reviews of Telfer, 1975; Berry, 1982) impresses one with the fact that in order to achieve synthesis of these various biochemical species, nurse cells in general must share certain common subcellular features that comprise the synthetic components.

5.1. Nurse Cell Nuclear Morphology

For convenience, I first consider the nurse cell nuclei as the primary synthetic machinery and then will examine how their cytoplasmic compartments are organized. Nurse cell nuclei can vary from spherical to irregular shape (Figure 9 and insets), are large, and usually have prominent nucleoli (Bonhag, 1955a; Anderson and Beams, 1956; Huebner and Anderson, 1972c; Büning, 1979a,b). Büning (1979a,b) has presented the most comprehensive comparative study of the variation seen in the Coleoptera. In the Hemiptera, there is a zonation in nuclear structure from the tropharium apex to the base (Schrader and Leuchtenberger, 1952; Bonhag, 1955a; Eschenberg and Dunlap, 1966; Brunt, 1971; Huebner and Anderson, 1972c; Matsuzaki, 1975; Sahai, 1975; Schreiner, 1977a) but in the Coleoptera and Megaloptera, their morphology is similar throughout the tropharium (Davis, 1956; Schlottman and Bonhag, 1956; Aggarwal, 1967, Büning, 1972, 1979a,b, 1980; Ullmann, 1973; Kurihara, 1975; Matsuzaki and Ando, 1977) with no discrete zones.

Mitotic activity of adult nurse cells is not seen in the Coleoptera or Megaloptera (Büning, 1979b,c, 1981b) but is a common feature of the hemipteran tropharium apical region (Figure 12, inset) (Payne, 1912; Bonhag, 1955a; Huebner and Anderson, 1972c; Telfer, 1975; Huebner, 1981a). As nurse cell nuclei differentiate to become active in RNA production, they have generally been shown to increase in size and DNA content (see Telfer, 1975). Earlier studies (e.g., Schrader and Leuchtenberger, 1952) suggested this increase was due to nuclear fusion. However, more recent studies have the general consensus that individual nuclei enlarge (Buning, 1972, 1979a, b; Huebner and Anderson, 1972c; May, 1972; Ullmann, 1973; Kloc, 1980). A fundamental question concerns whether or not the extra copies of DNA for amplified RNA synthesis arise by selective DNA amplification or endopolyploidization (see Telfer, 1975; Choi and Nagl, 1977a,b; Kloc, 1980). Results on the hemipterans *Oncopeltus* and *Gerris najas* and a variety of coleopteran species (Cave, 1975; Choi and Nagl, 1977a,b;

Figure 9. (Upper inset) An LM of an epoxy section of a *Rhodnius* nurse cell nucleus. Note spherical shape, prominent nucleolus, and perinuclear material (arrow). From Huebner and Anderson (1972b). (Bar = 1 μm.) (Lower inset) The irregular nurse cell nucleus shape (arrows) in *Oncopeltus*. (Bar = 1 μm.) The electron micrograph shows a portion of a *Rhodnius* nurse cell nucleus. Note the large nucleolus (NU) and nuclear envelope-associated material (arrow). From Huebner and Anderson (1972c) courtesy of Wistar Press. (Bar = 10 μm.)

Figure 10. A freeze-fracture replica revealing the high nuclear pore density of the nurse cell nucleus. From Huebner (1981a) courtesy of Longman Group. (Bar = 1 μm.)

Figure 11. A freeze-fracture replica of the nucleus of a small early previtellogenic oocyte. Note nuclear pores are scarce. (Bar = 1 μm.)

Figure 12. SEM illustrating a basement membrane-denuded anterior tropharium. Note it tapers to a small tip consisting of small round cells. (Bar = 10 μm.) The LM inset shows mitotic cells that reside in the tropharium tip. (Bar = 10 μm.)

Kloc and Matuszewski, 1977; Büning, 1979b; Kloc, 1980) indicate telotrophic nurse cells increase their activity and DNA content by an endopolyploidization (or endomitosis) without DNA amplification. There is, however, evidence of an extrachromosomal DNA that is a marker for postembryonic oocyte differentiation in the ovary of the beetle *Creophilus maxillosus* (Kloc and Matuszewski, 1977; Kloc, 1980). This will be discussed in the ovarian development section.

To facilitate nucleocytoplasmic flow of the magnitude required for these large polyploid nurse cell nuclei, one expects numerous well-developed nuclear pores. One of the first electron microscopic studies showing nucleocytoplasmic transfer and pores was done on the nurse cells of *Rhodnius* (Anderson and Beams, 1956). These pores and adjacent cytoplasmic nuage material (Figure 9, inset) have been verified (Huebner and Anderson, 1972c; Lutz and Huebner, 1980; Huebner, 1981a). Freeze-fracture preparations reveal the dense packing of nuclear pores in *Rhodnius* nurse cells (Figure 10). The oocyte nucleus, which is inactive, by contrast has very few pores (Figure 11) as do the prefollicular, and early previtellogenic follicle cells. Later stages of follicle cells have a large pore density also reflecting their increased RNA synthetic activity in preparation for and during chorion synthesis. Evidence for a correlation between nuclear pore packing and synthetic activity has been reported in sperm development (Fawcett and Chemes, 1979).

5.2. Nurse Cell Cytoplasm

As expected, a prominent feature is the abundance of free ribosomes with only sparse amounts of RER and Golgi (MacGregor and Stebbings, 1970; Huebner and Anderson, 1972c; Büning, 1979a,b; Huebner, 1981a). Microtubules are usually not found in regions close to the nuclei but have been reported in *Raphidia flavipes* (Büning, 1980). Mitochondria are numerous and well developed, and lipid droplets are seen in some species (Bonhag, 1958; Huebner and Anderson, 1972c; Telfer, 1975). Whether or not components such as mitochondria, lipids, and glycogen are transported from nurse cell cytoplasm to oocytes, as often suggested, is unclear (Telfer, 1975). Histochemical and biochemical studies of polytrophic oocytes indicate lipid and glycogen are made in the oocyte (Engels, 1970; Telfer, 1975; Wiemerslage, 1976). In telotrophic ovarioles, mitochondria are commonly seen in the cords connecting nurse tissue with the oocytes. However, video time-lapse viewing of *Rhodnius* did not reveal any mitochondria movement (Inoué and Huebner, unpublished data). Without labeling studies, the possibility that nurse cell cytoplasm generates mitochondria for oocytes cannot at present be ruled out. Besides the RNA and possibly mitochondria, some nonyolk proteins from nurse cell cytoplasm are also transported (Vanderberg, 1963; Büning, 1972; Mays, 1972; Telfer, 1975; Korfsmeier, 1980; Telfer *et al.*, 1981).

6. Tropharium—Arrangement of Nurse Cell Compartments

While a similarity exists in the nuclear and cytoplasmic structure across a broad spectrum of telotrophic ovaries, major differences and variety exist in the structural arrangement of the nurse cell compartments. In the Hemiptera, nurse cells are organized around a central nucleus-free area, the trophic core (Bonhag, 1958; Telfer, 1975) while in the Coleoptera, the tropharium is relatively homogeneous with no core development (Büning, 1972; Telfer, 1975). In the Megaloptera, tapetal cells, which appear to be equivalent to nurse cells, surround a central core that harbors many nuclei (Matsuzaki and Ando, 1977; Büning, 1979c, 1980). Not only are there significant differences between the three major telotrophic types, but there are also differences suggesting evolutionary trends within the polyphagic Coleoptera (Büning, 1979a,b, 1981b) and Hemiptera (Książkiewicz, 1980).

The hemipteran telotrophic tropharium usually has a mitotic tip region (Figure 12), zones of differentiating nurse cell nuclei (see above) and a basal area where nurse cell nuclei degenerate (Cone and Eschenberg, 1966; Eschenberg and Dunlap, 1966; Huebner and Anderson, 1972c). Recent microinjection experiments using intracellular tracers suggest that the *Rhodnius* tropharium consists of two parts, the apical mitotic cells and the larger broader syncytial part with numerous nuclei contained in extensions off a common central core (Huebner, 1981a,b). Bonhag (1955a) noted that in *Oncopeltus*, the apical mitotic zone cell boundaries were very distinct unlike the rest of the tropharium. The differences, noted earlier, in the tunica propria structure over the tip versus the rest of the tropharium may be important to the functional differences of the two parts. Since the *Rhodnius* syncytial tropharium is similar to other species (Bonhag, 1955a; Wick and Bonhag, 1955; Masner, 1966, 1968; Brunt, 1971; Schreiner, 1977a), the following summary provides the typical hemipteran situation (see also Huebner, 1981a). The differences in the aphids and coccids will be noted later.

Individual elongate lateral projections from the fibrous trophic core often contain nuclei (5-10) (Figure 13), ultrastructurally in a similar state. These cytoplasmic projections can best be seen in ovarioles with the tunica propria removed and exposed to cell dissociation buffers (Figures 12, 13) (Watson and Huebner, unpublished data). Scanning electron microscopy of such denuded preparations reveals rounded projections like "elongated grapes" off a stem, the trophic core (Figure 13, inset). The narrower stalks that connect into the trophic core are fibrous in the light microscopic preparations and contain microtubules as seen in electron microscopy. The trophic core is devoid of nuclei and intervening membranes (Figure 14) and contains ribosomes, mitochondria, and an impressive array of microtubules (Figure 15) (Huebner and Anderson, 1970, 1972c; MacGregor and Stebbings, 1970; Hyams and Stebbings, 1977a, 1979a,b; Huebner, 1981a). Products from the hundreds of nurse cell nuclei flow into this common cytoplasmic core and subsequently flow selectively into growing oocytes attached to the tropharium base by trophic cords (see below).

Unlike most Hemiptera, however, the coccids and aphids have small nurse cell numbers and their telotrophic nature has been questioned in the past (Shinji, 1919; Toth, 1933; Oseto and Helms, 1971; Elliot *et al.*, 1975; Couchman and King, 1979; Książkiewicz, 1980; Büning, 1981a). Recent EM work on the scale insect *Aspidiotus hederae* revealed its atypical, but nevertheless authentic, telotrophic nature (Książkiewicz, 1980). The ovariole of this coccid consists of three nurse cells and one oocyte which presumably constitute a sibling cluster

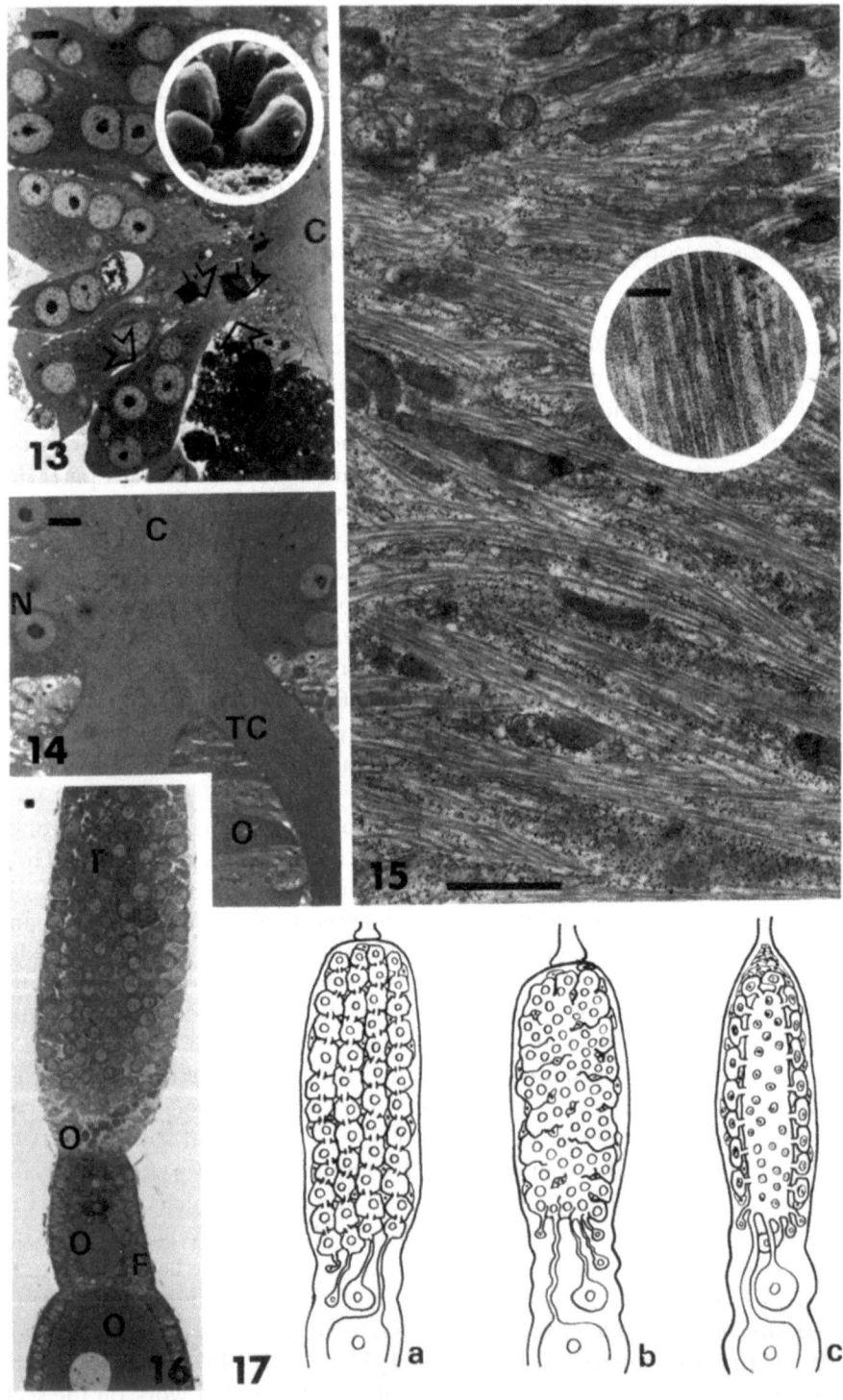

(see also Büning, 1981a). The nurse cells were like other telotrophic ovaries connected to the oocyte by a trophic cord which contained microtubules (Ksiazkiewicz, 1980). Książkiewicz suggested the individual sister clusters are more pronounced in the homopteran scale insects than other Heteroptera and the evolutionary trend toward formation of syncytial trophic chambers with reduction of nurse cell individuality is also characteristic of the Hemiptera as Büning (1979b) had suggested for the polyphagic Coleoptera.

Aphids also have a small number of nurse cells. Recent results of Couchman and King (1979) and Büning (1981a) on a number of aphid species prove the nurse cells connect directly to the oocyte via a microtubule-containing trophic cord. Büning showed each of the five ovarioles has a constant number of nurse cells and oocytes around a core. Thus, each ovariole is 1 germ cell cluster with 2^n oocytes and 2^{n+1} nurse cells, where $n = 4$ or 5.

The tropharium of polyphagic Coleoptera appears homogeneously cellular with no mitotic activity (Figures 16, 17). The elegant ultrastructural studies by Büning (1972, 1978, 1979a,b) on many species have shown that the nurse cell nuclei are housed in cytoplasmic compartments all interconnected by bridges. Thus, in general terms, the coleopteran tropharium has far more membranes than the hemipteran one. Once the intercellular bridges are fully differentiated, fusomal material is no longer present and cytoplasmic transport is possible. By

Figure 13. (Inset) SEM view of the "fingerlike" projection of the tropharium. (Bar = 10 μm.) The light micrograph is a toluidine blue-stained 1-μm section of a basement membrane-denuded tropharium. Fingerlike projections (arrows) containing numerous nurse cell nuclei branch from the nuclei-free trophic core (C). (Bar = 10 μm.)

Figure 14. A thick section of the basal *Rhodnius* tropharium showing the nuclei (N) in lateral projections, the fibrous trophic core (C), two large trophic cords (TC), and a small oocyte (O). From Huebner (1981a) courtesy of Longman Group. (Bar = 10 μm.)

Figure 15. Thin sections showing the rich array of microtubules, ribosomes, and mitochondria in the trophic core. This tropharium has been exposed to cation-free saline. (Bars = 1 μm and 0.1 μm.)

Figure 16. Thick section of an ovariole of the coleopteran *Zygogramma exclamationis* (sunflower beetle) showing the cellular tropharium (T) with no core, small oocytes (O) at the base and larger ones with follicle cells (F) lower. The tropharium has cytoplasmic continuity between adjacent nuclei containing compartments. (Bar = 10 μm.)

Figure 17. Schematic diagrams illustrating the tropharium organization of the polyphagic Coleoptera (a, b) and the Megaloptera-Raphidioptera (c). The coleopteran schematic is based on Büning (1979b, Fig. 23a, c; 1981b) and the work of Kloc and Matuszewski (1977). The megalopteran schematic is based on Matsuzaki and Ando (1977) and Büning (1979c, 1980, 1982b). (a) The primary coleopteran situation characterized by sibling cell clusters connected by intercellular bridges. (b) The secondary or more advanced stage where membranes are reduced and a more open syncytium exists. In both (a) and (b), oocytes connect to the tropharium base by small cords. (c) The *Sialis* megalopteran tropharium with a common nuclei-containing central syncytium with no intervening membranes. The peripheral cells or tapetal cells at the edge connect to the central syncytium by bridges. Spaces between these cells and the central syncytium are rare. The spaces seen in this diagram are there only to clearly show the peripheral cells are separate compartments attached by bridges. Apically, the tunica propria does not form a transverse septum. In all three types there are small somatic interstitial cells associated with some of the nurse cells. Although not shown in the diagram, the interstitial cells also contact each other. The trophic cords connect to the nurse cell or central syncytium by a persistent intercellular bridge and by the elongated growth of oocyte membrane.

examining many species, Büning (1979b) has proposed three basic polyphagic ovary subtypes based on the degree of compartment fusion and membrane reduction. In the more primitive situation (Figure 17a), intercellular bridges are present in the adult and connect the compartments of a sibling cell cluster (see also Kloc and Matuszewski, 1977), while other species in the intermediate or transitionary stage retain no intercellular bridges (Figure 16) and membranes between compartments in the same cluster are reduced. In the most evolved situation or second stage, cell membranes are significantly reduced between all nurse cells, creating less nurse cell individuality and a highly syncytial tropharium (Figure 17b) (Büning, 1979b, 1981a). The nurse cell clusters can vary in their geometry from a more irregular pattern to a very linear arrangement (Kloc and Matuszewski, 1977; Büning, 1979b).

The megalopteran (Sialidae) and raphidiopteran tropharia consist of mononucleated compartments arranged as a single layer around a multinucleated syncytial core (Figure 17c) (Matsuzaki and Ando, 1977; Büning, 1979c, 1980; Huebner, 1984). The mononucleated compartments were termed *peripheral cells* by Matsuzaki and Ando and *tapetal cells* by Büning. Since they are derived from germ cells and connected to the syncytial core by intercellular bridges, one assumes they are specialized nurse cells. There are no intervening membranes between the nuclei of the syncytial core.

Clearly, by using a distinguishing feature of the telotrophic ovary, the tropharium, the general view possible when Bonhag wrote his 1958 review and Telfer his 1975 review can now be updated to include three distinct types of telotrophic ovaries with evolutionary trends toward the reductions of membranes and nurse cell compartments. These recent data support a polyphyletic origin of the telotrophic ovary.

7. Trophic Cords or Trophic Tubes

Besides the tropharium, a central characteristic of the telotrophic ovary is the presence of trophic cords or tubes (sometimes referred to as nutritive tubes). Their presence has now been confirmed in all three telotrophic ovary types (Telfer, 1975; Matsuzaki and Ando, 1977; Büning, 1979a; Huebner, 1981a, 1984). Ultrastructural analysis of developing ovaries indicates that part of the cords are derived from highly elongated intercellular bridges arising from incomplete cytokinesis (Huebner and Anderson, 1972c; Büning, 1978; Lutz and Huebner, 1980, 1981). Cords are most conspicuous and best studied in the Hemiptera (Huebner, 1970; MacGregor and Stebbings, 1970; Hyams and Stebbings, 1977b, 1979a,b; Bennett and Stebbings, 1979). Even in the Homoptera where they tend to be smaller (8 μm for *Bothrogonia japonica*), they can reach diameters of 25 μm (Matsuzaki, 1975; Matsuzaki and Ando, 1977). In the polyphagic Coleoptera, they are very small and inconspicuous (Figure 22) and, as a result, often overlooked (Bryan, 1954; Aggarwal, 1967; Telfer, 1975). EM study has verified their existence in most if not all polyphagic Coleoptera (Büning, 1972, 1978, 1979a,b; Kloc and Matuszewski, 1977). Cord diameters range from 1 μm in early previtellogenesis to 8 μm during the main growth phase and in some species can reach lengths of 5-10 mm (Büning, 1979a,b). Unlike hemipteran cords, the coleopteran ones are more tortuous in their path to the oocyte. Microtubules are usually sparce. Cords are present in the Megaloptera and as in the Coleoptera usually do not have abundant microtubules (Matsuzaki and Ando, 1977).

Both from the perspective of size and ultrastructural complexity, the hemipteran trophic cords are the best developed and provide a good example of how the tropharium interacts with oocytes. In *Rhodnius*, trophic cords of varying diameters branch from the central base of the trophic core (Figures 14, 20, 21) and connect to all stages of oocytes up to midvitellogenesis (Huebner, 1981a). Transmission and scanning electron microscopy show the cord plasma membrane is relatively smooth and continuous with the oolemma (Figure 18). Late previtellogenic and early vitellogenic oocytes have microvilli on the oolemma. The microvilli continue a short distance onto the cord where an abrupt transition to the smooth surface occurs. The cords pass through pre-follicular tissue as well as through small follicles where the cord abuts the oolemma of the oocyte (Huebner, 1981a). Procion yellow, fluorescein, and lucifer yellow microinjection indicate little if any junctional contact exists between the oolemma and cords passing by to supply larger oocytes (Huebner, 1981a; Figure 21). EM studies show hemipteran trophic cords contain one of the richest arrays of microtubules reported for any tissue in the animal kingdom (Hamon and Folliot, 1969; Brunt, 1970; Huebner and Anderson, 1970; MacGregor and Stebbings, 1970). Single trophic cords may contain 30 to 50 thousand microtubules (MacGregor and Stebbings, 1970; Hyams and Stebbings, 1977a; Huebner, 1981a).

The microtubules of trophic cords are very stable upon isolation (Figure 20) (Hyams and Stebbings, 1979b) and are resistant to cold treatment and colchicine *in vitro* (Hyams and Stebbings, 1977b, 1979a,b; Huebner, 1981a). Upon vinblastine treatment, the microtubule protein is reorganized into crystals or aggregates (Huebner and Anderson, 1970; Stebbings, 1971, 1975; Huebner, 1981a). Despite the descriptions of microtubules in a variety of species and data on spacing distribution as well as biochemistry (Brunt, 1970; Huebner and Anderson, 1970; MacGregor and Stebbings, 1970; Hyams and Stebbings, 1977a,b, 1979a,b; Schreiner, 1977a; Bennett and Stebbings, 1979; Huebner, 1981a), the precise role of microtubules in cytoplasmic transport is still uncertain (Hyams and Stebbings, 1979b). Hyams and Stebbings (1977a) suggested thay may act as a structural sieve to prevent nuclei and other larger cell inclusions from entering cords (see Huebner, 1981a, for further discussion). Since the cords of most Coleoptera lack microtubules, they are either not essential for cytoplasmic transport or different mechanisms operate in the two systems.

After cytoplasmic flow to oocytes is no longer necessary, cords close and microtubules become packed together in a crystallinelike array in the so-called redundant cords (Bennett and Stebbings, 1979; Hyams and Stebbings, 1979a). Cords near the oocyte retract anteriorly and are absorbed rapidly (Hyams and Stebbings, 1979a; Huebner, 1981a). Thus, unlike the case in polytrophic ovaries, there is no last surge of cytoplasm into the oocyte as they lose nurse cell contact.

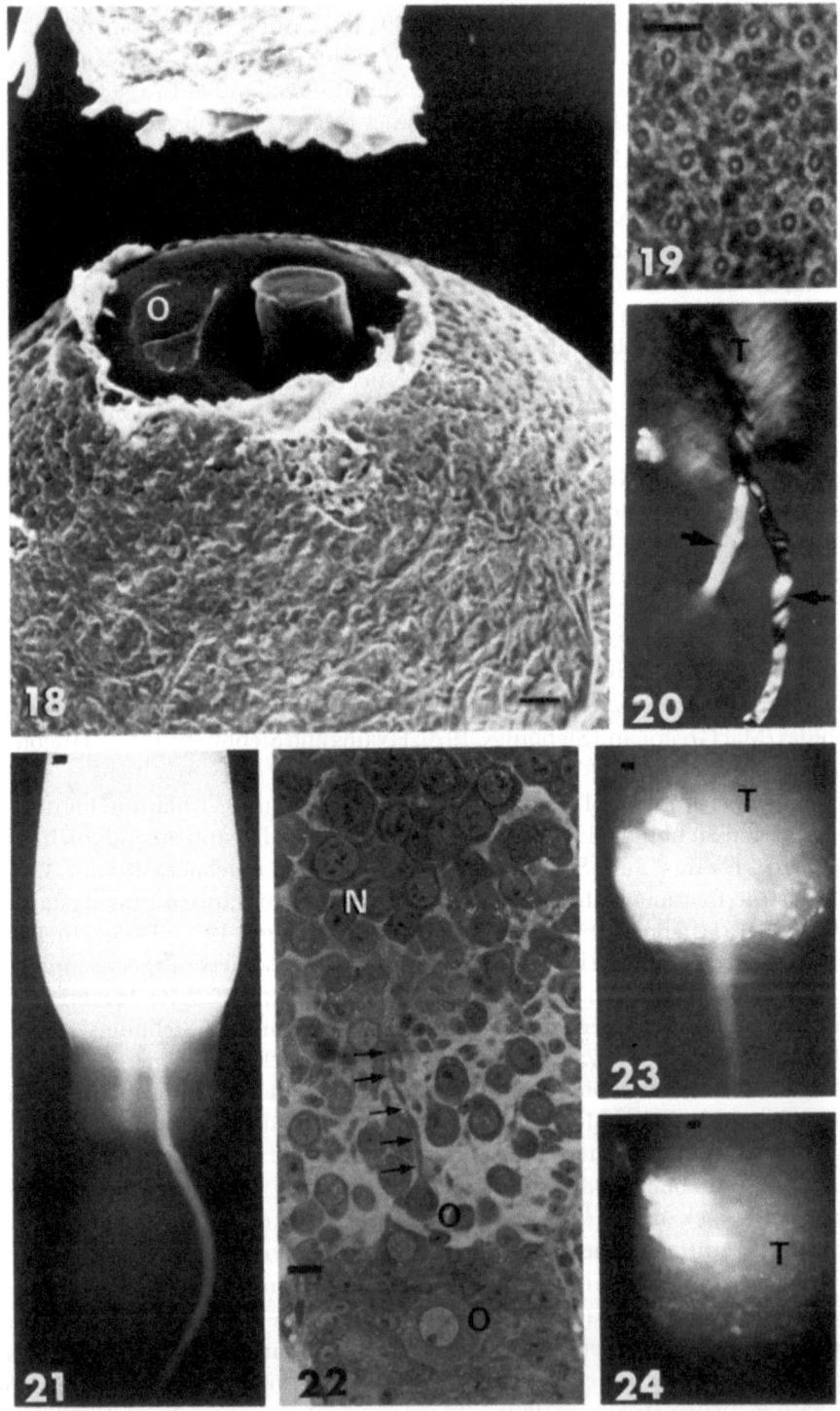

In *Notonecta glauca* and *N. maculata*, connection is lost at the onset of vitellogenesis with a decrease in size. This decrease is due to the reduction of transported materials that normally occupy 75% of the cord volume (Bennett and Stebbings, 1979). In *Rhodnius*, the connection is maintained into vitellogenesis with a reduction in diameter also resulting upon closure (Huebner, 1981a). *Rhodnius* has highly regulated ovarioles with only one terminal vitellogenic oocyte and the penultimate oocyte remaining in previtellogenesis (Huebner, 1981a, 1983). The relationship between the reduction of cord diameter as flow reduces and stops to the terminal oocyte and the increase in the diameter to the penultimate oocyte may be significant for intraovariole oocyte growth regulation.

8. Nurse Cell–Oocyte Interaction: Transport Regulation and Mechanisms

The interaction between the tropharium and the oocyte does not merely involve the oocyte as a passive recipient of the various components discussed earlier. Intraovariole feedback and an active role in the transport mechanism are two possibilities. Little is known about the first, but the elegant studies of Woodruff and Telfer over the last few years on the polytrophic ovaries of cecropia are noteworthy here. In cecropia, they revealed a 10-mV electrical gradient and polarized transport between nurse cells and oocytes (Woodruff and Telfer, 1973, 1974; Woodruff, 1979). Subsequently, Jaffe and Woodruff (1979) and Woodruff *et al.*, 1984), using the vibrating probe technique verified a steady flow of current in the follicles (see also Jaffe and Nuccitelli, 1977; Jaffe, 1981; Berry, 1982). Microinjection experiments using the positively and negatively charged probes, fluorescein-labeled lysozyme (FLY) and methyl carboxylated fluorescein-labeled lysozyme (McFLY), revealed differential protein movement was via an intra-

---

Figure 18. A microdissected *Rhodnius* ovariole revealing a prominent severed trophic cord connected to the anterior end of an oocyte (O). From Huebner (1981a) courtesy of Longman Group. (Bar = 10 μm.)

Figure 19. A thin section of a trophic cord showing abundant microtubules. From Huebner (1981a) courtesy of Longman Group. (Bar = 0.1 μm.)

Figure 20. Polarizing LM of a Triton X-100-extracted ovariole in a microtubule-stabilizing buffer showing the birefringence of the microtubule-rich tropharium (T) core and the cords (arrows) to oocytes. (Bar = 10 μm.)

Figure 21. A fluorescence LM of an ovariole with the tropharium microinjected with fluorescein. Note the dye has passed down the cords toward the oocytes. From Huebner (1981a) courtesy of Longman Group. (Bar = 10 μm.)

Figure 22. The narrow, wavy trophic cord (arrows) of the sunflower beetle (*Zygogramma exclamationis*) is evident in this light micrograph. N, nurse cells; O, oocytes. (Bar = 10 μm.)

Figures 23 and 24. Fluorescence LMs illustrating the distribution of the microinjected fluorescent proteins McFLY and FLY, respectively. McFLY disperses within the tropharium (T) and down trophic cords, while FLY remains concentrated near the injection site. (Bar = 10 μm.)

cellular electrophoresis mechanism (Woodruff and Telfer, 1980; Woodruff et al., 1984).

This experimental approach has now been extended to the telotrophic ovarioles of *Rhodnius* (Telfer et al., 1981). A similar electrical gradient of 9-10 mV was found if juvenile hormone was present. The tropharium was more negative than the oocytes. Microinjection of FLY and McFLY (Figures 23, 24) as in cecropia also suggested an electrophoresis mechanism. Application of the vibrating probe technique to *Rhodnius* ovarioles also reveals steady current flows leaving previtellogenic and early vitellogenic follicles and entering the lateral basal sides of the tropharium (see Huebner, 1984). Some current leaving the larger follicles also flows posteriorly and enters basally near the pedicel end (Sigurdson and Huebner, unpublished data).

Recent results of Dittmann et al. (1981) using the vibrating probe on the ovaries of *Dysdercus intermedius* are important to nurse cell oocyte transport and oocyte organization. They found two systems of steady extracellular current. One, similar to *Rhodnius*, was over the subgerminal tropharium and previtellogenic follicles with current leaving the follicles and entering the tropharium. The second system was around vitellogenic follicles with current entering laterally and leaving the interfollicular surfaces. This second current pattern was not seen in *Rhodnius*. This difference may be due to the fact that *Dysdercus* has many developing follicles, whereas *Rhodnius* has only one vitellogenic follicle. The tropharium of the coleopteran, *Ips*, has current leaving the lateral sides (Huebner, 1984). The is the reverse of the hemipterans and interestingly coincides with the presence of an incomplete inner sheath. For a comparative view of ovarian currents see Huebner (1984) and Woodruff et al. (1984).

With the greater geometric separation between oocytes and nurse cells in the telotrophic as compared to the polytrophic ovaries, the telotrophic ovary should be a particularly valuable system to study intracellular transport and localization. As the ultrastructural organization of the tropharium and cords is becoming better known, further experimental approaches will follow.

9. Oocytes: General Features, Nuclear Morphology, Cortex Morphology

In *Rhodnius*, as in a number of other telotrophic species (see development section), oocytes become distinguishable from nurse cells during the last larval instar (Lutz and Huebner, 1980, 1981). Initially, nuclei in the basal compartments of the syncytial tropharium remain small with condensing chromosomes attached to the inner nuclear envelope, and synaptonemal complexes indicative of meiosis become visible (Figure 25). The oocytes remain arrested in late meiotic prophase throughout adult oogenesis.

All the oocytes are located at the base of the tropharium and are embedded in the prefollicular tissue. Initiation of growth results in oocyte enlargement due to cytoplasmic flow through the trophic cords during previtellogenesis. Concomitant with this growth, the prefollicular cells differentiate into a follicular epithelium that becomes associated with the oolemma via gap junctions (Huebner, 1981b; see also Caveney and Berdan, 1982). The formed follicle then goes through the previtellogenic, vitellogenic, and postvitellogenic or chorionating growth phases.

The oocyte nucleus or germinal vesicle is centrally located during previtellogenesis but by vitellogenesis characteristically comes to lie just beneath the lateral oolemma (Figure 26) (Huebner and Anderson, 1972b). Examination of many histological and EM studies shows this is a common feature of most insects. The oocyte nucleus position during mid to late oogenesis marks the dorsal-ventral axis (Mulnard, 1954; Kleine-Schonnefeld and Engels, 1981). Little is known about how the migration and positioning of the nucleus are controlled. Recent work on the polytrophic ovary of *Musca domestica* shows all the oocyte nuclei in each ovary are specifically oriented to an imaginary center that is eccentric near the medial part of the abdomen (Kleine-Schonnefeld and Engels, 1981). These results suggest something specific must regulate the oocyte nucleus position. Whether or not a comparable situation exists in telotrophic or other polytrophic ovaries has not been studied.

During oocyte differentiation, the oocyte nucleus grows, forming a large germinal vesicle with a variety of inclusions (Büning, 1972; Huebner and Anderson, 1972b; Kurihara, 1975, 1976; Choi and Nagl, 1976, 1977a; Matuszewski and Kloc, 1976; Gruzova, 1979, 1980; Kloc, 1980; Magakyan *et al.*, 1980). The variety of inclusions reported in these papers include the karyosome, endobodies, nucleoli, nucleoluslike protein bodies, and DNA-positive ring bodies. In the lady beetle, various ringlike nucleolar bodies are induced during oocyte atresia (Kurihara, 1975, 1976). Recent findings in the beetle *Creophilus maxillosus* draw attention to nucleolar function and the role of extrachromosomal DNA in oocyte determination during development (Kloc, 1980). In adult telotrophic ovaries, however, oocyte nuclei are generally reported as inactive or have a low level of synthesis compared to nurse cells (Vanderberg, 1963; Bier, 1965, 1967; Winter, 1974; Telfer, 1975; Berry, 1982). Davenport (1976), using *Oncopeltus* ovarioles that were ligatured at the trophic cord, showed a very low ^{3}H incorporation into high-molecular-weight heterodispersed oocyte RNA and concluded the oocyte nucleus may participate in maternal RNA synthesis. But since mitochondria are also present, this finding does not establish that synthesis occurs in the oocyte nucleus. Capco and Jeffrey (1979) were able to confirm in *Oncopeltus* that the early oocyte nucleus was inactive in mRNA production and that the nurse cells were the major source. Autoradiographic studies (Büning, 1972; Mays, 1972; Ullmann, 1973) showed some silver grains over oocyte nuclei, but less than over nurse cells. The general consensus is that the germinal vesicle is not primarily involved in the production of oocyte RNA (Capco and Jeffrey, 1979; Berry, 1982). Oocytes may synthesize short-lived mRNA after vitellogenesis, but this is not stable (Winter *et al.*, 1977). During most of oogenesis, the RNA-producing functions are usurped by the nurse cells.

Along with accumulating various RNAs and organelles, the oocyte must also differentiate an oolemma and cortex capable of binding, incorporating, and

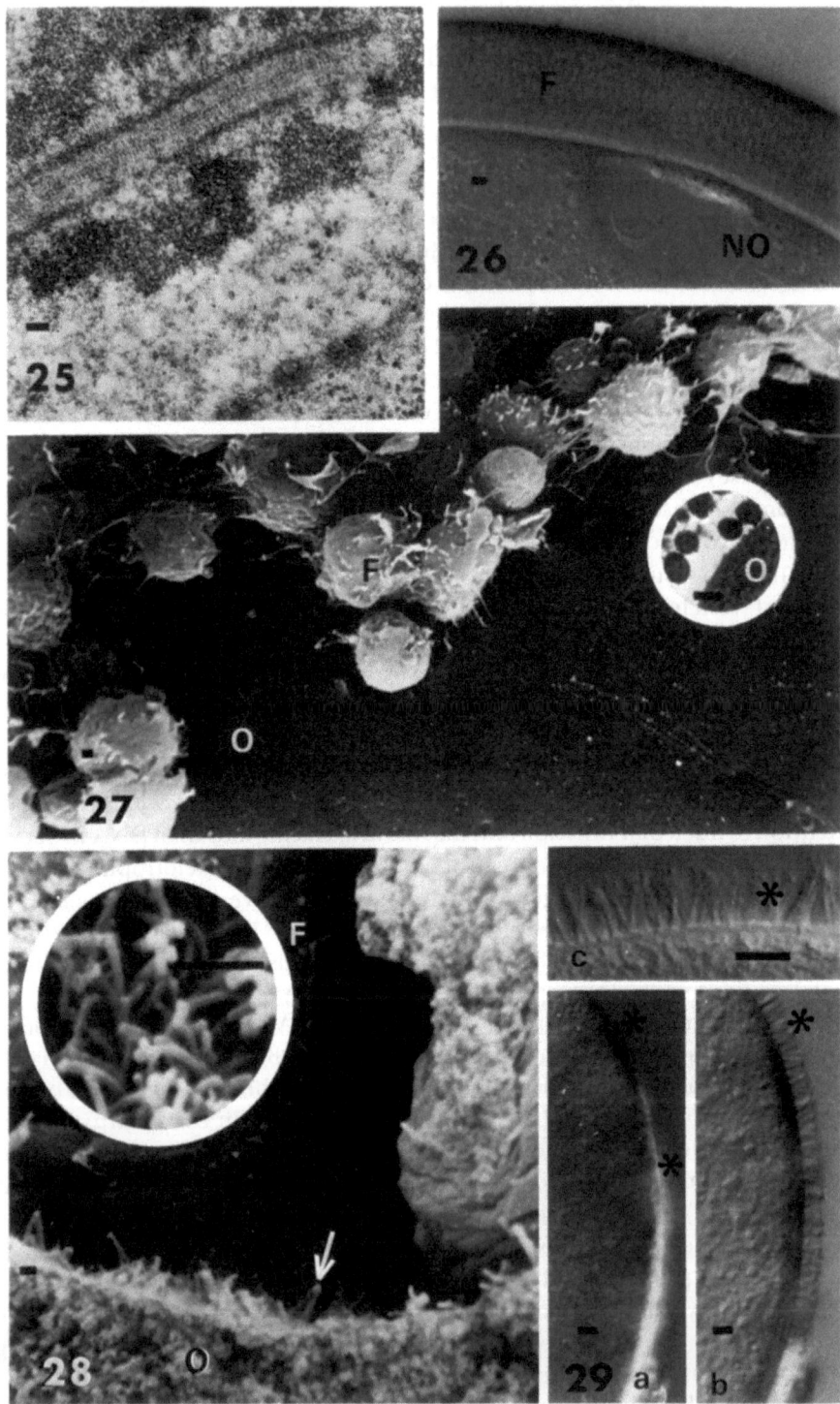

transporting yolk proteins and metabolites. The uptake of yolk precursors via pinocytosis in the oolemma has been reported in a large number of telotrophic ovaries (Kessel and Beams, 1963; Aggarwal, 1968; de Loof *et al.*, 1972; Huebner and Anderson, 1972b; Schreiner, 1977a,b; and many others). The ultrastructural details of pinocytosis and vitellogenesis are reviewed in Telfer *et al.* (1982) so only a few aspects on the oolemma are included here. Figures 27 and 28 demonstrate that initially the oolemma is fairly smooth but by late previtellogenesis numerous filament-containing microvilli appear. The oolemma is in junctional contact with the surrounding follicle in *Rhodnius* (Huebner, 1981b; see also Caveney and Berdan, 1982; Telfer *et al.*, 1982). Presumably, the oolemma's structure is mediated or influenced by follicle cells. One finds the greatest degree of microvillar development once the perivitelline space forms (Figure 28; Huebner and Anderson, 1972b; Matsuzaki and Ando, 1977; Huebner, 1981b). Experimentally, one can induce the formation of a brush border of microvilli on mid to late previtellogenic denuded follicle cell-free oocytes (Figure 29; Huebner, unpublished data). These results suggest the oolemma morphology is influenced by follicle cells. Formation of microvilli is known to be a dynamic process dependent on both ionic and pH conditions (Carron and Longo, 1982). de Loof *et al.* (1972) using *Leptinotarsa decemlineata* suggest oocyte microvilli are capable of movement and may play a role in the vitellogenic process. In the rotation of certain polytrophic follicles, the oocyte microvilli have also been implicated (Fux *et al.*, 1978). In telotrophic ovaries, the function of the oocyte microvilli is not fully known, but they provide for oocyte–follicle cell contact and presumably also for membrane turnover. For further details on vitellogenic oocyte growth and the structure of yolk, the reader is referred to other reviews (Telfer, 1965; Engelmann, 1979; Hagedorn and Kunkel, 1979; Telfer *et al.*, 1982). Figure 30 summarizes the three-dimensional structure of the syncytial tropharium and its association with growing oocytes. This reconstruction was based on thin sections and microinjection experiments and was subsequently verified by observations on basement membrane-free, cell-dissociated preparations with the scanning electron microscope.

⬅——————————————————————————————————————

Figure 25. A thin section of the nucleus of a small early previtellogenic oocyte in meiotic prophase with a synaptonemal complex evident. From Huebner and Anderson (1972b) courtesy of Wistar Press. (Bar = 0.1 μm.)

Figure 26. Differential interference LM of late previtellogenic oocyte with germinal vesicle (NO) in the cortex close to the follicular epithelium (F). From Huebner (1981b) courtesy of Academic Press. (Bar = 10 μm.)

Figure 27. An SEM and LM showing early previtellogenic oocytes (O) from a basement membrane-denuded ovariole placed in cell dissociation buffer. Many follicle cells (F) were washed off revealing the smooth oolemma (O). (Bars = 0.1 μm, 10 μm.)

Figure 28. An SEM of a fractured early vitellogenic follicle showing oocyte (O) cortex with microvilli (arrow) and patent follicle epithelium (F). (Bar = 1 μm.) (Inset) Higher-magnification surface view of oocyte microvilli in a vitellogenic follicle. (Bar = 1 μm.)

Figure 29. A Nomarski differential interference LM of the previtellogenic oocyte surface shortly after denuding of follicle cells (a) and 10 min in cation-free saline (b, c). Note the appearance of regular microvilli (∗). (Bars = 10 μm.)

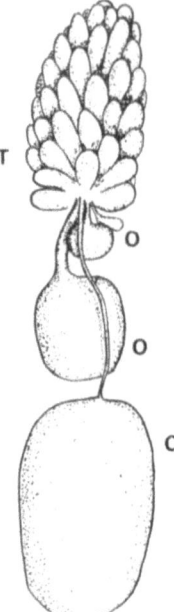

Figure 30. Line drawing summarizing the three-dimensional structure of the germ cell component of the *Rhodnius prolixus* ovariole. Note the tropharium (T) with lateral extensions and the trophic cords connecting to oocytes (O).

10. Follicle Cells

Oocytes of most telotrophic ovaries undergo their entire differentiation in intimate association with follicle cells. In certain parthenogenetic aphids, oocytes are only enclosed in follicles for a short period (Couchman and King, 1979). By mitosis, prefollicular cells generate a layer of follicle cells which undergo a sequence of morphological and physiological changes during oogenesis. Once follicles reach mid to late previtellogenesis, mitosis is rarely seen except in the area where the interfollicular tissue forms (Figure 26). To date, no intercellular bridges have been described between adjacent follicle cells in telotrophic ovaries as have been described in polytrophic ovaries (Meola *et al.*, 1977; Ramamurty and Engels, 1977; Fiil, 1978; King *et al.* 1978). Follicle cells of most hemipteran telotrophic ovaries are binucleate (Bonhag, 1958; Brunt, 1971; Huebner and Anderson, 1972a; Schreiner, 1977a) although certain aphids have mononucleate ones (Couchman and King, 1979). A number of Coleoptera and the Megaloptera also have mononucleate follicle cells (Büning, 1972, 1979c; Figure 32). In cases where DNA incorporation has been determined, follicle cells were found to be polyploid (Ullmann, 1973; Davenport, 1975a,b; see also Koeppe and Wellman, 1980; Koeppe *et al.*, 1980).

Davenport (1975a) working with *Oncopeltus* follicle cells stated "In spite of intimate association between follicle cells and oocytes and of the dramatic morphological changes that occur during oocyte development the role of the epithelium is far from clear." Both he and Kafatos (1976) point out, as have others, two clear roles are in chorion biosynthesis and yolk uptake.

Follicle cells are often viewed as one cell type when in fact examination of them over the entire oocyte surface reveals both functional and structural subtypes. In *Rhodnius*, for example, the cells on the anterior surface remain closely apposed and columnar throughout vitellogenesis while those on the lateral surfaces are short and undergo significant alteration to create spaces for vitellins to reach the oolemma (Huebner and Anderson, 1972a). The transition between the two areas is abrupt (Figure 31). These two groups of follicle cells are different despite the fact that both associate with the oocyte and with each other by gap junctions (Huebner and Caveney, 1984; Sigurdson and Huebner, unpublished data). Analysis of the morphogenesis of the various parts of the regionally complex *Rhodnius* chorion allowed Beament to identify a large number of different follicle cell types (Beament, 1946a,b,c). In the polytrophic ovary at least 10 distinct subpopulations of follicle cell types have been determined (King and Koch, 1963; Margaritis et al., 1979, 1980). With further work on telotrophic follicles, I am sure a parallel situation will emerge.

Follicle cells generally have a number of proven as well as postulated functions. Follicle cells of the telotrophic ovary are coupled to the oocyte via gap junctions (Huebner, 1981b). Coupling was previously described for the panoistic and polytrophic ovary types (Telfer et al., 1981; Caveney and Berdan, 1982). Follicle cell-oocyte coupling provides for a number of possible functions, such as have only recently been elucidated for granulosa cell-oocyte interaction in mammals. In mammals, up to 85% of the oocyte's metabolites are obtained from granulosa cells, with a dramatic reduction in denuded oocytes (Heller et al., 1981). Brower and Schultz (1982) showed that the degree of metabolic cooperation is proportional to surface area and directly influences oocyte growth rate. Eppig (1979) previously showed oocytes grow only when gap junctions are present. The possibility of metabolic cooperation has not been addressed in any insect follicle. Recent discovery of gap junctions in various insect eggs and the absence of junctions upon atresia in *Rhodnius* eggs (Huebner, 1981b) suggest the possibility of metabolic cooperation is worth investigating. In the highly specialized polytrophic paedogenic dipteran *Heteropezea*, Went and Junquera (1981) were able to obtain normal development without the follicular epithelium, but in cecropia incorporation of amino acids and proteins was only minimal without follicle cells (Anderson, 1971).

Besides the prospect of metabolic coupling, mammalian follicle cells have been implicated in meiotic arrest and regulation of the oocyte (Dekel et al., 1981). Since telotrophic oocytes also exhibit meiotic arrest, this possibility also deserves attention. However, Eppig (1982) has shown in the mouse that resumption of meiosis was not initiated by reduction of cell coupling. Hormonal regulation of oocytes via follicle cell junctions is nevertheless a strong possibility (Brown et al., 1979; Burghardt and Anderson, 1981).

A third possible follicle cell function, namely in vitellogenesis, has been widely studied in many systems including the telotrophic one (see de Loof and Lagasse, 1970; Abu-Hakima and Davey, 1977a; Huebner and Injeyan, 1980, 1981; Davey, 1981; Telfer et al., 1982). The follicular layer constitutes a permeability barrier that can regulate the flow of vitellins to the oolemma. Closely apposed

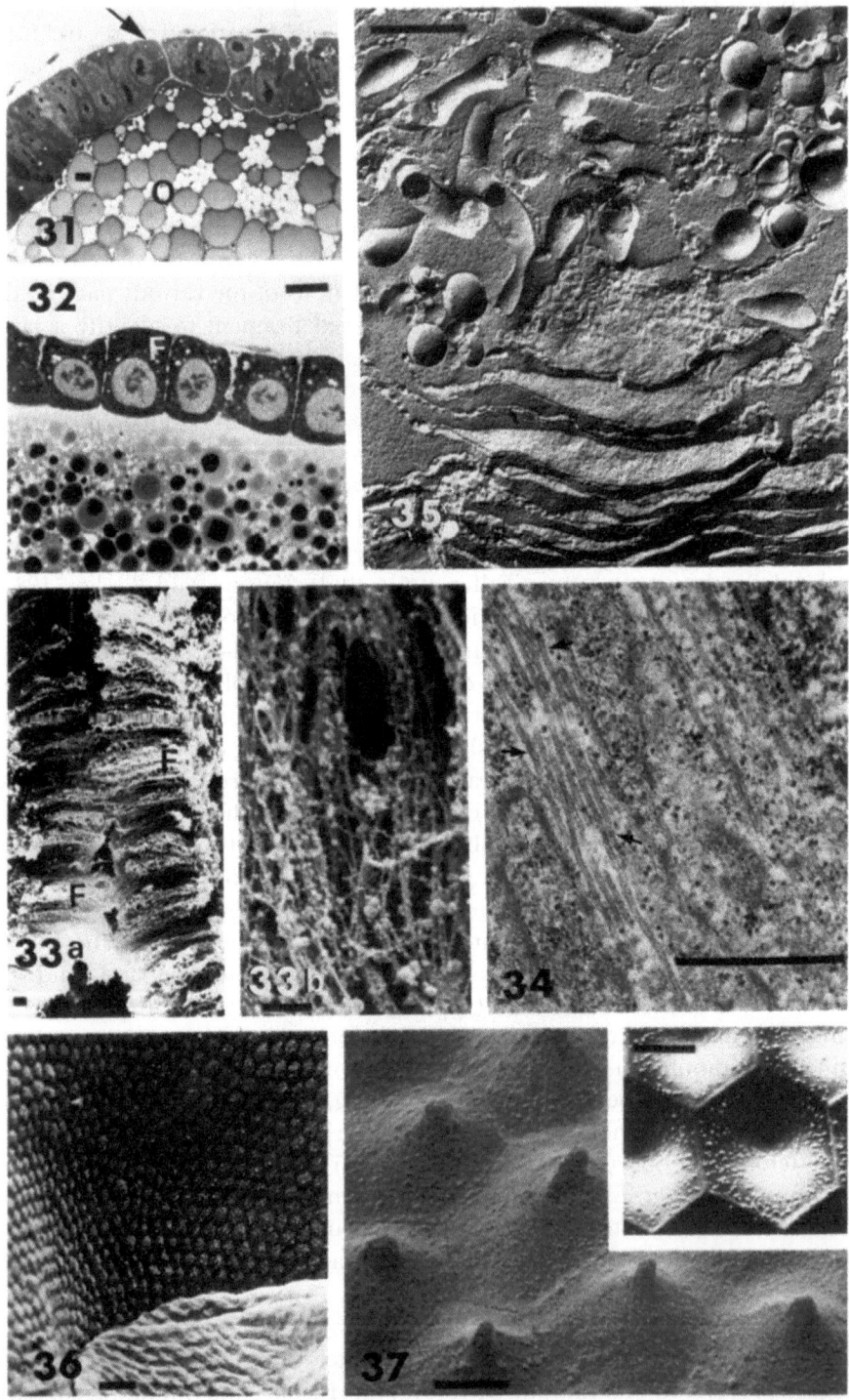

cells impede passage, whereas reduced cell-to-cell contacts alter cell shape and volume and create extracellular channels for easy access to the oolemma (Huebner and Injeyan, 1980, 1981). These spaces have been reported to a broad spectrum of telotrophic follicles (Masner, 1966, 1968; Brunt, 1971; Ullmann, 1973; Matsuzaki and Ando, 1977; Kelly and Telfer, 1979; Huebner, 1984; to cite a few).

Of the telotrophic follicles, *Rhodnius* and *Oncopeltus* are among the best studied and recently reviewed (Schreiner, 1977a,b; Huebner and Injeyan, 1980, 1981; Davey, 1981; Telfer *et al.*, 1982, for details). An important element in the follicular changes during vitellogenesis is the cytoskeleton (Huebner, 1976; Abu-Hakima and Davey, 1977b). Exposure of basement membrane-denuded *Rhodnius* follicles to buffers with altered cation composition shows the follicle cells can alter their shape and that they have a dynamic cytoskeleton (Watson and Huebner, unpublished data). Isolation of follicle cell cytoskeletons (Figure 33) and thin sectioning reveal they have well-developed cytoskeletal elements (Figure 34). Besides the presumed role of the cytoskeleton in follicle cell shape change, it may be of general importance to the overall mechanical role of the follicle in maintaining oocyte shape. A mechanical role was suggested for the midge follicle by Tucker and Meats (1976) and by Went (1978).

A fourth follicle cell function often suggested in earlier histological studies of telotrophic ovaries was the production of RNA for oocytes. EM, biochemical, and autoradiographic studies have ruled out this as a serious possibility (Mays, 1972; Duspiva *et al.*, 1973; Davenport, 1974, 1976; Telfer, 1975; Capco and Jeffrey, 1979).

Production of ovarian ecdysteroids is a function of panoistic and polytrophic follicles (Laguex *et al.*, 1977; Hetru *et al.*, 1978; Bulliere *et al.*, 1979; Hagedorn, 1980). Evidence that ecdysone is produced by telotrophic ovaries, presumably by follicle cells, has been presented for *Tenebrio molitor* (Laverdure *et al.*, 1977) and *Rhodnius* (Ruegg *et al.*, 1981, 1982). Correlations between follicle cell ultrastructure and ecdysone synthesis and release have not been

←―――――――――――――――――――――――――――――

Figure 31. LM of a *Rhodnius* vitellogenic follicle showing the distinction between the apical and lateral follicle cells (arrow). (Bar = 10 µm.)

Figure 32. LM illustrating the mononucleate follicle cells (F) of the vitellogenic oocyte of the sunflower beetle. (Bar = 10 µm.)

Figure 33. SEMs revealing the cytoskeletal framework of *Rhodnius* follicle cells (F) extracted with Triton X-100. (Bars = 1 µm.)

Figure 34. A thin section of a previtellogenic follicle cell showing the microtubules (arrows) of the cytoskeleton. From Huebner and Anderson (1972a) courtesy of Wistar Press. (Bar = 1 µm.)

Figure 35. The RER and elongate mitochondria of follicle cells are revealed in this freeze-fracture replica. (Bar = 1 µm.)

Figure 36. An SEM of an early stage of chorion production. Note the follicle cells are closely apposed. The oocyte has been removed and the thin endochorion peeled back. (Bar = 100 µm.)

Figure 37. Illustrated here is a late chorion stage, showing the conical end and central projection of the follicle cells. (Inset) The chorion surface from which the follicle cells were separated. Note the chorion sculpturing mirrors the follicle cells. (Bars = 10 µm.)

made. Follicle cells have a rich array of organelles such as RER, Golgi, and mitochondria (Figure 35).

A final phase of follicle cell function in most telotrophic ovaries is the synthesis, release, and sculpturing of the chorion (Figures 36, 37). The most detailed and elegant treatment of the role of follicle cells in synthesizing the vitelline membrane and chorion is in the polytrophic ovaries (King and Koch, 1963; King and Aggarwal, 1965; Kafatos, 1975, 1976; Margaritis et al., 1979, 1980; Mazur et al., 1982). Studies on *Leptinotarsa* ovaries by de Loof (1971) showed the follicle cells secrete the egg envelopes, and in *Rhodnius* Beament (1946a,b,c) and Huebner and Anderson (1972a) showed their involvement. The thorough classical work of Beament provides the best example of chorion cytochemistry and morphogenesis from any telotrophic ovary. His publications present the details of the five endochorion and two exochorion layers, the specializations in the cap, the neck, and how the micropyles form.

Transmission EM, freeze-fracture, and SEM confirm Beament's elegant work. Follicle cell nuclei become rich in pores and the cytoplasm in RER and secretory vesicles during chorion synthesis. The endo- and exochorion layers are deposited sequentially with the sculpturing of the general and specialized regions by follicle cell shape and contact (Figures 36, 37; Huebner, unpublished data). Separation of the follicular epithelium while the chorion is sculptured, dramatically reveals their shape (Figure 37 inset).

11. Adult Ovary Summary

The foregoing discussion has presented a brief selective description of the cellular organization of the "telotrophic ovariole" noting some of the diversity. Figure 38 summarizes this for *Rhodnius*. The telotrophic ovariole constitutes a regulated interacting cell system involving both germ cell and somatic cell derivatives. It has accentuated regional compartmentalization and cytoplasmic transport to an extreme, thereby providing a good system to study basic cell mechanisms (see also Telfer, 1975).

The focus has been on the cell interactions (NC-oocyte, FC-FC, FC-oocyte) that make the system work. Because of the complex syncytial nature, analysis of ovarian development is valuable to our understanding of its organization.

12. Ovarian Development: General Aspects

Relatively few studies have examined telotrophic ovary development (Wick and Bonhag, 1955; Masner, 1966, 1968; Huebner, 1970; Huebner and Anderson, 1972c; Büning, 1972, 1978, 1979c, 1981a,b; Ullmann, 1973; Choi and Nagl, 1976, 1977a,b; Kloc and Matuszewski, 1977; Furtado, 1979; Kloc, 1980; Lutz and Huebner, 1980, 1981; Huebner, 1982). It is generally accepted that the germ line gives rise to oocytes and nurse cells, with the rest of the ovarian cells originating from mesoderm (Bonhag, 1958; Telfer, 1975). A major stumbling block in

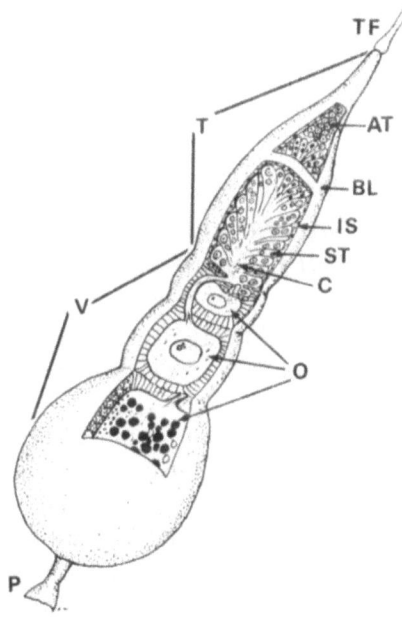

Figure 38. A diagram of the intact hemipteran telotrophic ovariole including both somatic cells and germ cells. The ovariole is bounded by the tunica propria (BL), suspended anteriorly by the terminal filament (TF). The tropharium has the cellular apical tip (AT) and the larger syncytial portion (ST) arranged around the common trophic core (C). Between the nurse cells and the BL is a thin layer of interstitial cells (IS). Trophic cords branch off the trophic core base and are confluent with the oocytes (O), which are surrounded by prefollicular or follicle cells. The ovary is divided into a tropharium (T) and vitellarium (V).

unraveling the organization and interrelationship between nurse cells and oocytes in telotrophic ovaries has been the difficulty in sibling tracing (Telfer, 1975). Because of the large germ cell number and complexity in the adult, clearly, study of the ontogeny of the telotrophic ovary provides valuable insight into its organization and evolution (Huebner and Anderson, 1972c; Büning, 1979b,c, 1981b). The detailed study of Wick and Bonhag (1955) provided the only thorough histological study of the hemipteran since those from the early 1900s. A few histological studies have been published more recently (Masner, 1966; Furtado, 1979) and provide general information, but resolution of sibling cell interactions and cellular details requires the resolution of EM. One of the first EM developmental studies on telotrophic ovaries of *Rhodnius* suggested some developmental similarities between telotrophic and polytrophic ovaries (Huebner, 1970; Huebner and Anderson, 1972c). At that time, we suggested in *Rhodnius* the developing ovary consisted of groups of interconnected sister cells or syncytial groups and proposed by transformation of bridge complexes with membrane fusion the adult condition develops. This suggested that one could view the telotrophic ovary similar to a "highly fused" polytrophic ovary. Analysis of bridges is an important aspect of telotrophic ovary development. The description of intercellular bridges between germ cells and subsequent morphogenesis has since also been shown in coleopteran and megalopteran telotrophic ovaries (Büning, 1972, 1978, 1979a,b, 1980, 1981b; Kloc and Matuszewski, 1977). Since this early ultrastructural study of the *Rhodnius* fifth-instar ovary, it has been further studied in great depth from the first instar through to the adult (Lutz and Huebner, 1980, 1981; Huebner, 1982, in preparation). Büning (1981a) has studied development of aphid ovaries and *Pyrrhocoris apterus* as well. Detailed studies also exist on the Coleoptera (Büning, 1972, 1978, 1981b; Choi and Nagl, 1977a,b; Kloc and Matuszewski, 1977; Kloc, 1980) and the Megaloptera (Büning, 1979c). These studies on six or seven different telotrophic species have attempted to clarify and resolve some of the questions still unanswered (Telfer, 1975).

13. Postembryonic Early or Larval Ovary Stages: Mitotic Growth and Establishment of Somatic Tissue-Germ Cell Interaction

The following discussion uses *Rhodnius* as a primary example but also draws heavily on the excellent work of Büning and Kloc and Matuszewski on other species.

Mellanby (1936) studied embryonic development in *Rhodnius*. By day 11, germ cells are established as two continuous strands running through abdominal segments 6–8 near where spiracular invaginations occur. By day 19, distinct gonads, each with about eight follicles, have formed with two types having significantly different lengths. Mellanby postulates the longer ones probably are ovaries and the shorter ones testes, and they remain in this condition until the insects become adults. Embryonic gonad development has been studied in the beetle *Leptinotarsa* by Richard-Mercier (1972, 1977). More recent work on *Rhodnius* showed in newly hatched first-instar nymphs the reverse is true, namely the shorter gonads are ovaries (Huebner, 1982, in preparation) and significant changes in growth and differentiation prior to the adult stage (Lutz and Huebner, 1980, 1981; Huebner, in preparation). The only other histological study of ovarian development in postembryonic *Rhodnius* is the unpublished thesis of Case (1970). He was unable to identify the sex of gonads until the third instar. Due to the difficulties resulting from the resolution of the techniques he used, a number of misinterpretations concerning the timing of oocyte differentiation in later instars resulted (see Huebner and Anderson, 1972c; Lutz and Huebner, 1980, 1981). Careful examination of gonads from first-instar nymphs with either phase-contrast or interference microscopy allows for clear identification of ovaries (Figures 39, 40). This is conclusively confirmed by light microscopic sections (Figure 41) and coincides nicely with the excellent work of Wick and Bonhag (1955) on *Oncopeltus*. Postembryonic ovary development in *Rhodnius* and *Oncopeltus* consists of five larval stages (Figure 39). At hatching, the first-instar nymph already has seven ovarioles present in each ovary (Figure 40). Each ovariole contains a group of 16–30 germ cells organized around a membranous entanglement of processes (Figures 41, 45; Huebner, 1982). A more extensive version of this tortuous trophic core or membrane labyrinth has been described previously in the fifth instar of *Rhodnius* (Huebner and Anderson, 1972c) and in aphid ovaries (Büning, 1981a). Developing coleopteran ovaries do not have a central membrane core or labyrinth (Büning, 1972). Basally, the germ cell group abuts onto a small area of somatic cells (Figures 40, 41) (the presumptive prefollicular and predicel cells). Thee mesodermal cells plus the germ cell cluster are enclosed in a thin basal lamina or tunica propria (Figure 43). At the apical area, where the tunica propria forms the transverse septum, a cylinder of flat waferlike cells is attached (Figures 40, 41). These are the presumptive terminal filaments which join together to form a single muscle strand attached to the anterior thoracic cuticle.

Although much of the germ cell surface rests directly on the tunica propria, thin cells from the presumptive prefollicular area are beginning to cover the germ cells and create a layer, the future inner sheath, between the germ cells and the tunica propria (Figure 43). TEM shows the germ cells have prominent nuclei with nucleoli, many free ribosomes, a few Golgi complexes, and sparse RER (Figure 45). Broad, often branching cytoplasmic channels connect the germ cell compartments into the tangled early instar core (Figures 41, 42, 45, 48). Analysis of serial sections indicates many, if not all, germ cells are interconnected (see also Büning, 1981b; Huebner, 1982), and further evidence that germ cells are already interconnected in the first-instar ovaries comes from cell dissociation experiments (Huebner, 1982). Presumably, these interconnections result from incomplete cytokinesis during embryonic development. Trypsinization and cell dissociation of first-instar ovaries removes the sheaths and somatic cells, leaving denuded germ cell clusters (Figure 42). The bridges interconnecting individual compartments via the core can be seen in live preparations with a through-focus series using an interference microscope. Whether or not all or only most of the germ cells in the intact ovary are interconnected is still unclear. If a few isolated cells were separate, they would have dissociated during treatment and thus be lost. Resolution of this question awaits completion of serial reconstructions. The presence of a few isolated cells would provide an easy explanation for the existence of the mitotic cell tip in the adult tropharium. However, separate cells are not essential since one could have a secondary separation of part of a sibling cluster later in postembryonic development. Precedent for secondary isolation of parts of a sibling cluster exists in the polytrophic ovaries of *Anisolabis maritima* (Yamauchi and Yoshitake, 1982). Büning (1981b) has presented an attractive model based on sister cell clusters that proposes the hemipteran germ cells are all interconnected and originate from a single germ cell. Büning considers the organization of germ cells around the core, which is a complex intercellular bridge system, as a highly modified rosette. Determination of whether or not the germ cells of *Rhodnius* arise from a single cell will require analysis of the embryonic gonad.

During the first three instars, there is mitotic activity of the nuclei within the cluster but the basic morphology remains similar except for growth and complete envelopment of the germ cells by somatic cells (Figures 39, 44). *Oncopeltus* is similar with only a 3-fold increase in ovarian volume while there was a 20-fold increase in body weight (Wick and Bonhag, 1955).

Feeding by the third-instar nymph activates considerable differentiation of the somatic cells, so that by the molt to the fourth instar, distinct larval prefollicular and pedicel cells have formed. The terminal filaments have become well developed, as have the ovariole sheaths and tracheoles (Figures 46, 47). The germarium enlarges due to mitotic activity (Figure 47). The ultrastructure of fourth-instar germ cells is generally similar to that of earlier stages but nuclear

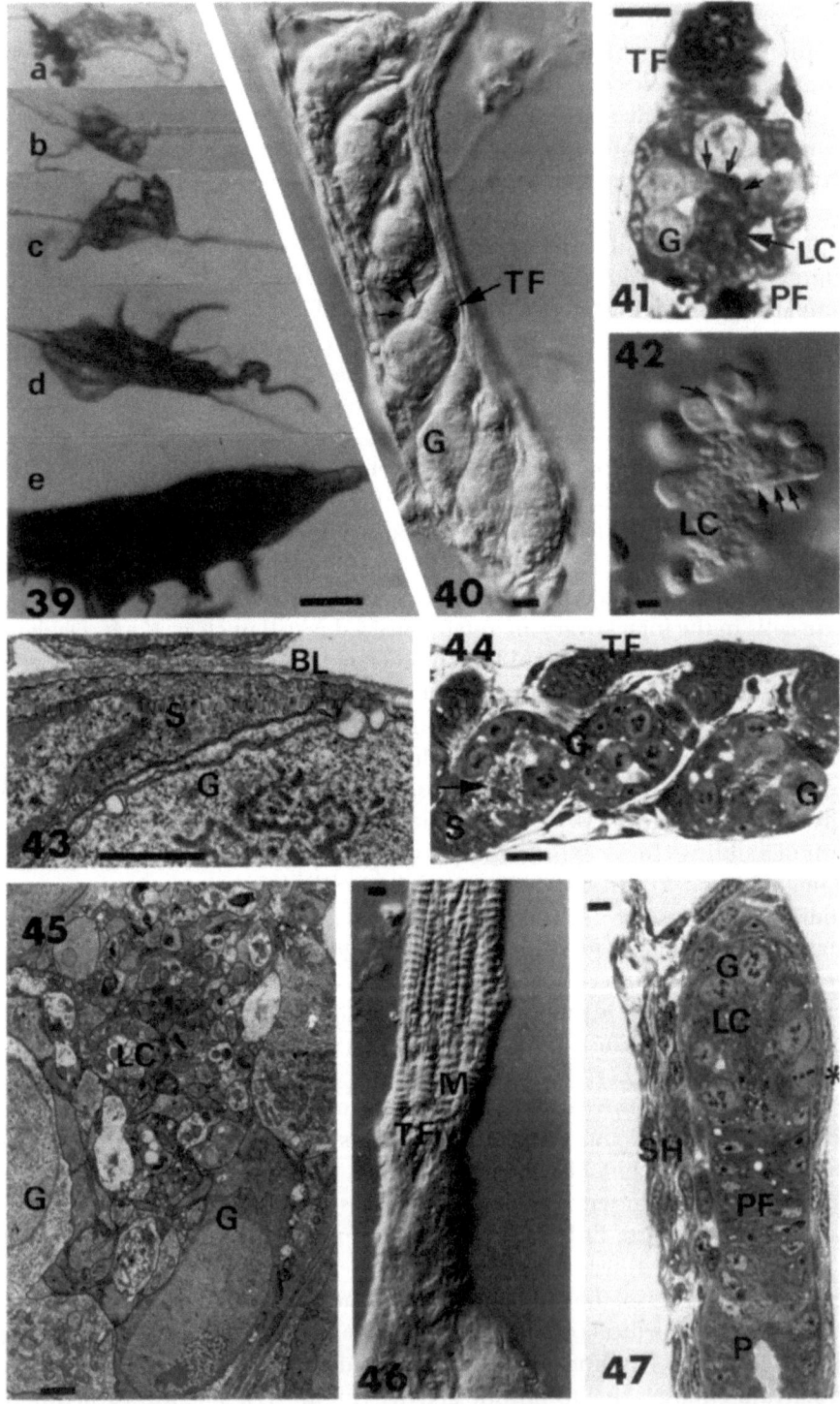

nuages and nucleoli are more prominent. The trophic core is well developed and consists of a tangled network of intercellular bridges and cell processes. The increasing number of germ cells remain interconnected by prominent intercellular bridges that arise from incomplete cytokinesis. Thus, a progressively larger, more complex syncytium of germ cells is generated. Developing telotrophic germ cell syncytia consist of either a number of sister cell clusters (Büning, 1972; Huebner and Anderson, 1972c; Kloc and Matuszewski, 1977) and/or as indicated above and recently proposed by Büning (1981b) in some instances a single large sister cell cluster.

14. Postembryonic Late Ovary Stages: Larval-Adult Transformation, Nurse Cell, and Oocyte Differentiation

Although a few reports (Wick and Bonhag, 1955; Choi and Nagl, 1977a,b; Robert, 1979) indicate that in some species oocyte determination occurs in the fourth or penultimate larval instar, in most other species the differentiation from

Figure 39. LMs showing *Rhodnius* ovaries at different developmental stages. From a to e are whole ovaries from first- to fifth-instar nymphs, respectively. Note there is relatively little ovarian growth from the first to third nymph, a slight increase by the fourth nymph, and a significant increase by the fifth nymph. (Bar = 100 μm.)

Figure 40. A Nomarski differential interference LM of a live, slightly spread, first-instar ovary. Note all seven ovarioles are present. Each has a central pad of germ cells (G) apically, a prominent cylinder of terminal filament cells (TF), and basally a poorly defined area of presumptive prefollicular tissue. All seven ovarioles join anteriorly to the anterior ovarian strand and basally to a thin larval oviduct (not shown). Around the individual ovarioles the somatic cells (arrows) that will form the ovariole sheath can be seen. (Bar = 10 μm.)

Figure 41. LM showing the group of germ cells (G) around a nuclei-free larval trophic core (LC) in a first-instar ovariole. Note the germ cells connected to the core by processes (arrows), the terminal filament cells (TF), and prefollicular area (PF). (Bar = 10 μm.)

Figure 42. Nomarski differential interference LM showing a live first-instar ovariole with the somatic cells removed by cell dissociation methods leaving only the germ cells. All the germ cell compartments remain connected to the core (LC) and the connections (arrows) of each can be seen in a through-focus series. (Bar = 10 μm.)

Figure 43. A thin section showing part of a thin somatic cell (S), the presumptive inner sheath, beginning to form a layer between the tunica propria (BL) and the germ cells (G). (Bar = 1 μm.)

Figure 44. LM showing that the ovary of a second-instar *Rhodnius* nymph has undergone little cellular change since the first instar. The germ cells (G) around the larval core (arrows) and the apical terminal filaments (TF) can be seen. (Bar = 10 μm.)

Figure 45. Low-magnification EM of a first-instar ovariole revealing that the larval core (LC) consists of a tangled network of cell processes and extensions with germ cells (G) around the periphery. (Bar = 1 μm.)

Figure 46. By the fourth-instar, the ovariole orientation changes and ovarian sheath and terminal filaments (TF) are well developed. This interference LM shows the terminal filaments are attached to long slender striated muscle cells (M) that form the single long terminal strand that suspends the ovary anteriorly. (Bar = 10 μm.)

Figure 47. LM of a fourth-instar ovariole. Note further mitotic growth (∗) of germ cells (G) around the larval core (LC). The posterior mesodermal tissue has differentiated into presumptive prefollicular (PF) tissue, and the pedicels (P) and the sheaths (SH) are well developed. (Bar = 10 μm.)

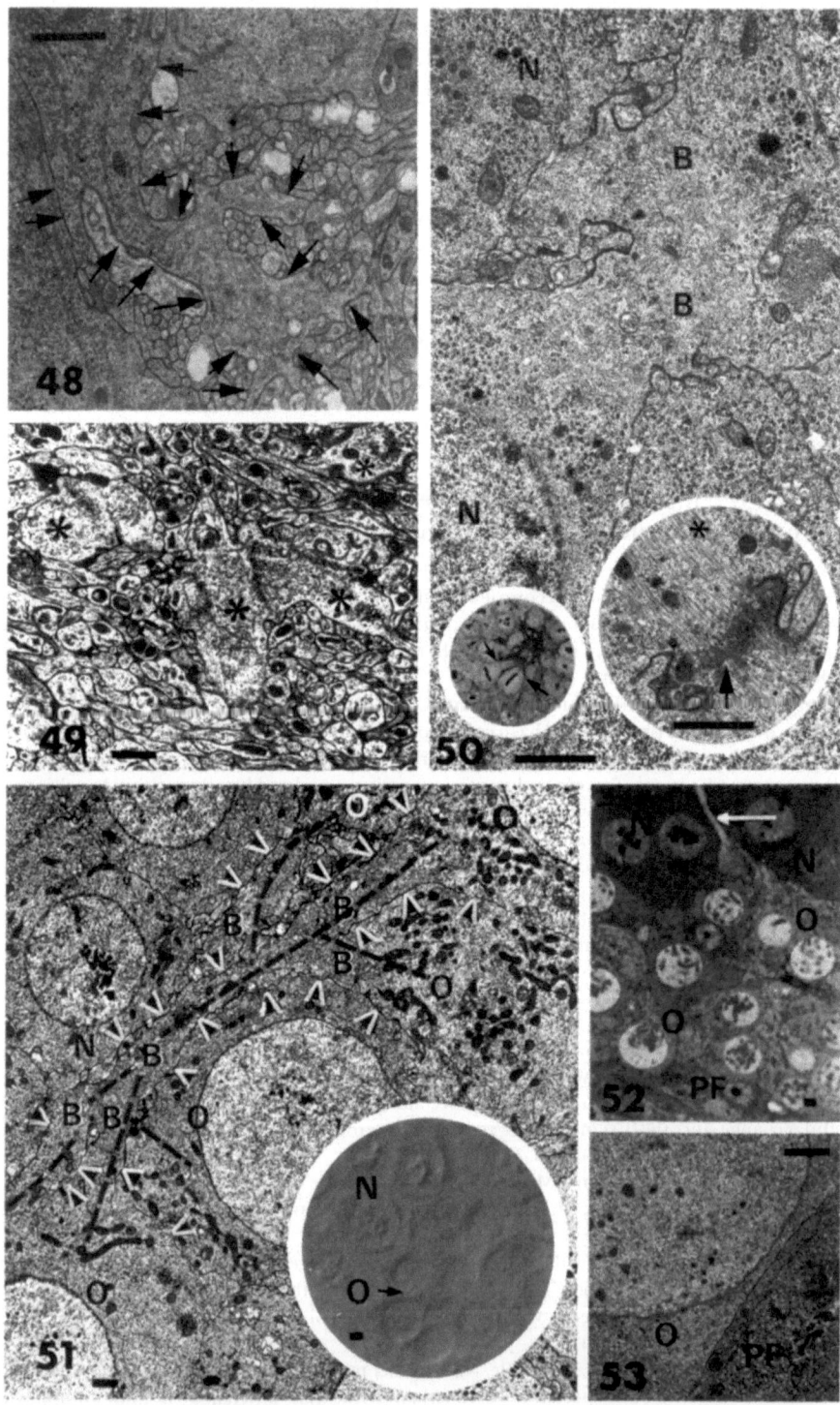

the larval to adult condition is triggered in the pupal coleopteran stage or last larval hemipteran instar (Masner, 1966; Vernier, 1970; Büning, 1972; Huebner and Anderson, 1972c; Huebner, 1977; Furtado, 1979; Lutz and Huebner, 1980).

The major growth and differentiation occurs during the larval–adult transformation when an altered hormonal milieu is present (Baehr et al., 1978; Furtado, 1979; Lutz, 1979). Furtado (1979) showed the thoracic glands were essential for meiosis to be initiated and oocyte determination to occur. This is supported by the transplant experiments of Lutz (1979) which suggested that ecdysone was the key factor in triggering ovary differentiation, with altered JH levels controlling whether larval germ cells or adult oocytes and nurse cells result. In *Pyrrhocoris apterus*, Masner (1969) indicated a 2-day absence of JH was necessary for proper prefollicular tissue development. Allatectomy of fourth-instar *Rhodnius* results in precocious oocyte and nurse cell differentiation (Huebner, unpublished data). Endocrine control of coleopteran ovary differentiation has not been clearly resolved (Laverdure and Huet, 1965; Laverdure, 1970, 1971, 1975; Alzouma, 1977).

Initiation of adult ovary differentiation begins with a period of mitotic growth (days 1· to 9 in *Panstrongylus megistus* and *Rhodnius*; Furtado, 1979; Lutz and Huebner, 1980). This mitosis, coupled with incomplete cytokinesis, generates an enlarged syncytium with many intercellular bridges (Figure 50). Numerous intercellular bridges also characterize the coleopteran ovary during these stages, although a trophic core is not present (Büning, 1972, 1979a). Following the mitotic phase, nuclei in most of the compartments begin to develop the ultrastructural features characteristic of nurse cells (see Lutz and Huebner, 1981). *Rhodnius* nurse cell nuclei increase approximately 10-fold in volume (Lutz and Huebner, 1980), and in *Oncopeltus* this increase has been shown to include a 32- to 64-fold increase in the haploid amount of DNA (Cave,

---

Figure 48. Thin section showing a portion of the third-instar larval core. The arrows outline the continuity of the germ cell compartment with the core and show part of the branching bridge network that in other sections connects to other germ cell compartments. (Bar = 1 μm.)

Figure 49. Thin section showing a bridge network (∗) in the membranous trophic core of the fifth-instar ovary. From Huebner and Anderson (1972c) courtesy of Wistar Press. (Bar = 1 μm.)

Figure 50. The left inset LM shows mitotic spindles in a projection from the core. The right thin section inset shows a mitotic bridge with the midbody (arrow) and spindle remnant microtubules (∗). From Lutz and Huebner (1981) courtesy of Longman Group. (Bar = 1 μm.) Figure 50 shows a bridge (B) system between differentiating nurse cells from a fifth-instar ovary (N). Spindle remnants are gone and fusomal material (which later disappears) occupies the bridges. From Lutz and Huebner (1981) courtesy of Longman Group. (Bar = 1 μm.)

Figure 51. (Inset) A Nomarski differential interference LM of the basal oocyte (O) clusters and nurse cells (N). From Lutz and Huebner (1980) courtesy of Longman Group. (Bar = 1 μm.) The thin section shows a bridge (B) complex interconnecting at least six oocytes (O). The arrowheads mark membranes and the dashed line the bridge confluencies. From Lutz and Huebner (1981) courtesy of Longman Group. (Bar = 1 μm.)

Figure 52. LM from a late fifth-instar ovary showing the elongated intercellular connection (arrow) between the nurse cells (N) and oocytes (O). These become the trophic cords of the adult. PF, prefollicular cells. From Lutz and Huebner (1981) courtesy of Longman Group. (Bar = 1 μm.)

Figure 53. As this thin section of a late fifth-instar ovary shows, the larval mesodermal cells have differentiated into prefollicular cells (PF) in contact with oocytes (O). From Lutz and Huebner (1981) courtesy of Longman Group. (Bar = 1 μm.)

1975). Nucleoli become prominent and all the adult characteristics develop well before the adult molt. Condensing chromosomes of meiotic prophase are seen around day 9-10 in nuclei in the basal compartments adjacent to prefollicular tissue in *Rhodnius*.

EM analysis of a development time course series in *Rhodnius* demonstrated intercellular bridges initially have the spindle remnants which are temporarily replaced by fusomal material that disappears leaving a network of ribosome-filled bridges (Figures 50, 51; Lutz and Huebner, 1981). Bridges interconnect adjacent nurse cell compartments as well as more distant cells via enlongated ones in the trophic core (Figures 49, 50) (Huebner, 1970; Lutz and Huebner, 1981). Besides nurse cell-nurse cell interconnections, the oocyte compartments at the base are also connected to each other (Figure 51) as well as the nurse cells. These data indicate that the earlier hypothesis of Huebner and Anderson (1972c) is still a reasonable one (Figure 54) with some modification. If there is more than one sibling cluster, then a number of basal compartments, rather than only one per cluster, become oocytes or, perhaps as proposed by Büning (1981b), we have a single, but large, sibling cluster where most of the basal compartments become oocytes.

The final phase of hemipteran ovary development involves a reduction of cell membranes, presumably by fusion and breakdown, transforming the larval membrane-rich core to the microtubule-filled membrane-free adult core (Huebner and Anderson, 1972c; Lutz and Huebner, 1981). The bridges that connected oocytes to nurse cell compartments become part of the elongating trophic cords (Figure 52) with the dense region of the original bridge still evident as part of the cord membrane (Lutz and Huebner, 1981).

Megalopteran ovary development also begins with the generation of a syncytium by incomplete cytokinesis (Büning, 1979c). The central region of the germarium is a membrane labyrinth in which intercellular bridges are entangled. This central membrane labyrinth is then transformed into the mem-

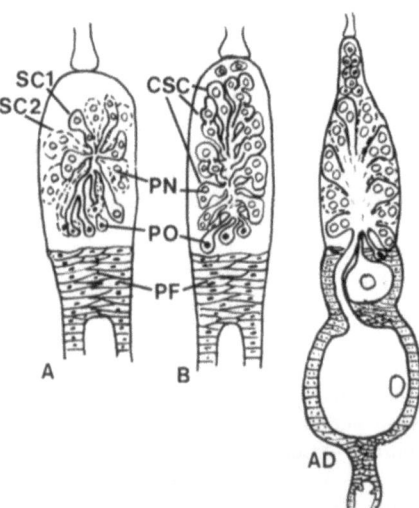

Figure 54. (A, B) Two possible interpretations of the germ cell arrangements in fifth-instar ovaries that could give rise to the arrangement in the adult (diagram AD). (A) illustrates the hypothesis that germ cells of the fifth-instar ovary are an entanglement of individual sibling cell clusters (SC1, SC2) that fuse during larval-adult transformation creating the common syncytium represented in the adult, AD. (B) presents a modified view (based on Büning, 1981b) that the germ cells are all one common sibling cell cluster (CSC) presumably originating from a single cell. The transformation involves reduction of membranes, bridges, and restructuring of an already complete syncytial network. PN, presumptive nurse cells; PO, presumptive oocytes; PF, presumptive prefollicular cells. (A) and (AD) are based on Figure 33 in Huebner and Anderson (1972c).

brane-free multinucleate central syncytium characteristic of the adult. The oocytes differentiate at the base of the central syncytium and remain connected to it by bridges that are part of the trophic cords (Büning, 1979c).

Büning (1978) clearly showed that the trophic cords of the polyphagic coleopteran *Bruchidius obtectus* also are highly modified intercellular bridges. Numerous intercellular bridges interconnect the germ cells of the *Bruchidius* ovary. Initially, the bridges contain fusomal material which subsequently disappears (Büning, 1978). Bridges interconnect adjacent compartments, and a tangled membranous core with bridges does not form during coleopteran ovary development (Büning, 1972, 1978). As discussed earlier in the adult section, there is some variety in the degree of membrane reduction in the ovaries of polyphagic coleopterans, with some species retaining more membranes and intercellular bridges, while the more advanced ones reduce membranes, lose bridges, and form a more open syncytium (Büning, 1979b). The membrane reduction process is very rapid (Büning, 1979b; Lutz and Huebner, 1981).

The work of Kloc and Matuszewski (1977) on the beetle *Creophilus maxillosus* showed that (like the more primitive situation in Büning's three types) sibling clusters initially are independent units in the form of strings of cells with the nuclei in each group in synchrony. Each string of germ cells (termed *chordoblasts* by them) is the descendent of a single cell. Because of the linear geometry in this species, they were able to trace mitotic growth in individual sibling clusters. Furthermore, they discovered that in the pupal stage the nucleus of a germ cell in a forming cluster undergoes a DNA amplification creating an extrachromosomal DNA body. With each subsequent linear division, this nuclear DNA body is segregated in a nonrandom fashion and always ends up in the terminal or basal cell of the string. After the final division, only the oocyte nucleus contains the extrachromosomal DNA and becomes arrested in meiosis. This provides a marker for studying oocyte determination similar to the classic work of the polytrophic *Dytiscus cybister* ovary (see Kloc and Matuszewski, 1977, for further discussion). Kloc (1980) has shown that the extra DNA contains amplified copies of the nucleolar organizer region and that it is active in RNA synthesis at the same time the DNA body is being synthesized and accumulated. As soon as the developing terminal oogonial cell becomes a meiotically arrested oocyte RNA synthesis ceases and the nucleolar material remains partly attached to the extra DNA body.

The significance, beyond being a marker, for determination of oocyte differentiation is unclear (Kloc and Matuszewski, 1977). Kloc (1980) speculates that the persistence of the association of the DNA body with nucleolar material may constitute a mechanism for favoring the preferential segregation of those nucleoli to accompany the extra DNA into the terminal oogonial cells and are eventually associated with the extra DNA in the oocyte nucleus.

In some Coleoptera, meiosis is initiated in all germ cells (Büning, 1972), but by the completion of oogenesis meiotic chromosomes are only evident in oocytes, while the nurse cells become polyploid and produce multiple nucleoli. Meiosis was only found in developing oocytes in *Rhodnius*. No chromosomal condensation and synapsis was evident in nurse cells (Lutz and Huebner, 1980). Choi and

Nagl (1977b) showed that the development of endoploidy in *Gerris* nurse cells coincided during development with the onset of ribonucleoprotein production so characteristic of the adult ovary.

Despite the fact that markers exist for oocytes and morphological studies catalog the appearance of distinguishing features and the location of the compartments destined to form oocytes, we do not know what determines that only the basal compartments always become oocytes (Kloc and Matuszewski, 1977).

Once oocytes differentiate and are distinguishable from nurse cells, no new oocytes form (Büning, 1972; Huebner and Anderson, 1972a,b,c; Mays, 1972). Concomitant with oocyte differentiation, the larval prefollicular cells develop into spindle-shaped adult prefollicular cells (Figure 53). To generate increased prefollicular tissue and follicles by the adult stage, mitotic activity becomes intense (Lutz and Huebner, 1980, 1981). The contact between presumptive oocyte compartments and prefollicular tissue has prompted the suggestion that this may play a decisive role in oocyte determination (Masner, 1966, 1968, 1969; Huebner and Anderson, 1972c; Kloc and Matuszewski, 1977; Büning, 1979a,b; Lutz and Huebner, 1980, 1981). The discussion of control of oocyte differentiation in telotrophic ovaries at present is primarily speculation with relatively little experimental data available as exists for the polytrophic system (see King *et al.*, 1982). It would be of great interest to determine if the prefollicular contact is merely coincidental or not and in the case of *Creophilus*, how the extra DNA marker becomes differentially segregated.

The recent elucidation of the electrophysiological properties of insect meroistic ovaries (see Telfer *et al.*, 1981) provides a possible new approach to this old problem. Beyond the role of the currents in intercellular electrophoretic transport and nurse cell regulation of the oocyte, studies of the onset of development of nurse cell–oocyte electrical gradients could provide clues about the control of differentiation as well. Are the gradients a consequence of differentiation and a means whereby oocyte growth is facilitated in the adult or could they also have played a role in the activation either of the nuclei in the basal compartments to become oocytes or the anterior ones to become nurse cells? Knowledge of when the gradient is first established and how it is generated is needed.

15. Summary

The telotrophic ovaries of the Hemiptera, polyphagic Coleoptera, the Megaloptera and Raphidioptera all have an anterior syncytial trophic chamber connected by modified intercellular bridges, the trophic cords, to the oocytes. The organization of the trophic tissue varies considerably between them. The hemipteran tropharium has a tip of mitotic cells and a larger syncytial portion organized as nuclei-containing cytoplasmic extensions off a microtubule-rich common trophic core. Prominent trophic cords also packed with microtubules connect to oocytes, providing the channel for cytoplasmic transfer from nurse cells to oocytes. Electrophysiological studies have suggested that primarily electronegative molecules flow in a polarized fashion to the oocyte due to a potential gradient. Polyphagic coleopteran tropharia lack a trophic core and show no mitosis in the adult condition. Nurse cell compartments of sibling cell clusters are interconnected by intercellular bridges in the most primitive situation. Reduction of membranes and fusions between clusters forms a more open syncytium in the more advanced situation. Trophic cords are generally small and inconspicuous. The Megaloptera and Raphidioptera have similar tropharia consisting of a membrane-free multinucleate central syncytium with a single layer of nurse cells around the periphery. The peripheral nurse cells are each connected to the central syncytium by an intercellular bridge. Oocytes are connected basally to the syncytium by small trophic cords. In all three types of telotrophic ovaries, there is species variation in tropharial ultrastructure. The prefollicular tissue and the envelopment of the oocyte by follicle cells is generally similar in all, but variation in whether or not follicle cells are mononucleate or binucleate occurs.

The nurse cells are in syncytial association with each other and the oocytes. This association arises during development by mitosis of germ cells with incomplete cytokinesis, thereby creating cytoarchitecturally complex sibling cell clusters. In the case of the Hemiptera and Megaloptera, some of the syncytial interaction is via a membranous larval trophic core of tangled intercellular bridges and cell processes. During transformation to the adult condition, membrane fusion and reduction occur, forming a central membrane-free area. In the polyphagic Coleoptera, although some membrane fusion and reduction between adjacent nurse cell compartments can occur, no core is present and the trophic tissue appears very "cellular" throughout development and in the adult. The intercellular bridges that connected the basal compartments adjacent to the prefollicular tissue in all three telotrophic ovary types are part of the elongated trophic cords of the adult. Only the cells in the basal compartments of the developing ovaries become oocytes, and they are all established by the adult stage. Some aspects of which factors may control and/or influence oocyte determination are discussed.

Both in the adult as well as in the developing ovary, there are many unresolved questions. Recent EM and physiological studies have helped clarify our understanding of the organization of the telotrophic ovary. Now we can probe further into the mechanisms that make it operate. By exploiting this marvellous syncytial and cellular system that nature has evolved, we may gain insights into how intracellular cytoplasmic transfer and localized distributions are achieved, the role of microtubules, the regulation of nuclear function, and protein transport across epithelia, to cite a few.

ACKNOWLEDGMENTS

I am grateful to my graduate students Douglas Lutz, Donald Lococo, Andrew Watson, and Wade Sigurdson for their help and comments during the course of the *Rhodnius* research discussed here. I am grateful to Dr. Jürgen Büning for the information presented at the 1981 International Development Conference and his insight into the sibling cell associations in telotrophic ovaries. Discussion and collaboration with Drs. William Telfer and Richard Woodruff on meroistic ovary electrophysiology over the past few years have been most helpful. Appreciation is also expressed to Drs. Robert King and Hiromu Akai for their comments and the opportunity to write this review. This chapter contains previously unpublished results of research conducted at the University of Manitoba and was supported by research grants from the Canadian Natural Sciences and Engineering Research Council and the University of Manitoba Research Board.

References

Abu-Hakima, R., and Davey, K. G., 1977a, The action of juvenile hormone on the follicle cells of *Rhodnius prolixus:* The importance of volume change, *J. Exp. Biol.* **69**:33-44.

Abu-Hakima, R., and Davey, K. G., 1977b, The action of juvenile hormone on the follicle cells of *Rhodnius prolixus:* The effect of colchicine and cytochalasin B, *Gen. Comp. Endocrinol.* **32**:360-370.

Aggarwal, S. K., 1967, Morphological and histochemical studies on oogenesis in *Callosobruchus analis* Fabr. (Bruchidae-Coleoptera), *J. Morphol.* **122**:19-33.

Aggarwal, S. K., 1968, Pinocytosis during vitellogenesis in the mealworm, *Tenebrio molitor* L. (Coleoptera, Tenebrionidae), *Acta Entomol. Bohemoslov.* **65**:272-275.

Alzouma, I., 1977, Étude de la différencitation des trophocytes et des ovocytes au cours du développement ovarien de *Tenebrio molitor* (Coleoptera: Tenebrionidae), Thèse de 3e cycle, Université de Paris-Sud, Orsay, France.

Anderson, E., and Beams, H. W., 1956, Evidence from electron micrographs for the passage of material through pores of the nuclear membrane, *J. Biophys. Biochem. Cytol.* **2**(Suppl.): 439-443.

Anderson, L. M., 1971, Protein synthesis and uptake by isolated *Cecropia* oocytes, *J. Cell Sci.* **8**:735-750.

Baehr, J. C., Porcheron, P., and Dray, F., 1978, Radioimmunological measurement of ecdysteroids and juvenile hormones in last 2 larval instar of *Rhodnius prolixus*, *C. R. Acad. Sci. Ser. D* **287**:523-526.

Beament, J. W. L., 1946a, The waterproofing process in eggs of *Rhodnius prolixus* Stähl, *Proc. R. Soc. London Ser. B* **133**:407-418.

Beament, J. W. L., 1946b, The formation and structure of the chorion of the egg in a hemipteran, *Rhodnius prolixus*, *Q. J. Microsc. Sci.* **87**:393-439.

Beament, J. W. L., 1946c, The formation and structure of the micropylar complex in the egg-shell of *Rhodnius prolixus* Stähl (Heteroptera Reduviidae), *J. Exp. Biol.* **23**:213-233.

Bennett, C. E., and Stebbings, H., 1979, Redundant nutritive tubes in insect ovarioles: The fate of an extensive microtubule transport system, *Cell Biol. Int. Rep.* **3**:577-583.

Berry, D. J., 1982, Maternal direction of oogenesis and early embryogenesis in insects, *Annu. Rev. Entomol.* **27**:205-227.

Bier, K., 1965, Zur Funktion der Nährzellen im meroistischen Insectenovar unter besonderer Berücksichtigung der Oogenese adephager Coleopteren, *Zool. Jahrb. Abt. Allg. Zool. Physiol. Tiere* **71**:371-384.

Bier, K., 1967, Oogenese, das Wachstum von Riesenzellen, *Naturwissenschaften* **54**:189-195.

Bonhag, P. F., 1955a, Histochemical studies of the ovarian nurse tissues and oocytes of the milkweed bug, *Oncopeltus fasciatus* (Dallas). I. Cytology, nucleic acids and carbohydrates, *J. Morphol.* **96**:381-439.

Bonhag, P. F., 1955b, Histochemical studies of the ovarian nurse cell tissues and oocytes of the milkweed bug, *Oncopeltus fasciatus* (Dallas). II. Sudanophilia, phospholipids and cholesterol, *J. Morphol.* **97**:283-311.

Bonhag, P. F., 1958, Ovarian structure and vitellogenesis in insects, *Annu. Rev. Entomol.* **3**:137-160.

Bonhag, P. F., and Wick, J. R., 1953, The functional anatomy of the male and female reproductive systems of the milkweed bug, *Oncopeltus fasciatus* (Dallas) (Heteroptera: Lygaerdae), *J. Morphol.* **95**:177-230.

Bradley, T., 1981, Elastase fully removes the basal lamella of insect Malpighian tubules, *J. Cell Biol.* **91**:153a.

Brower, P. T., and Schultz, R. M., 1982, Intercellular communication between granulosa cells and mouse oocytes: Existence and possible nutritional role during oocyte growth, *Dev. Biol.* **90**:144-153.

Brown, C. L., Wiley, H. S., and Dumont, J. N., 1979, Oocyte-follicle cell gap junctions in *Xenopus laevis* and the effects of gonadotropin on their permeability, *Science* **203**:182-183.

Brunt, A., 1970, Extensive system of mirotubules in the ovariole of *Dysdercus fasciatus* Signoret (Heteroptera: Pyrrhocoridae), *Nature (London)* **118**:80-81.

Brunt, A., 1971, The histology of the first batch of eggs and their associated tissue in the ovariole of *Dysdercus fasciatus* Signoret (Heteroptera: Pyrrhocoridae) as seen with the light microscope, *J. Morphol.* **134**:105-130.

Bryan, J. H. D., 1954, Cytological and cytochemical studies of oogenesis of *Popilius disjunctus* Illiger (Coleoptera-Polyphaga), *Biol. Bull.* **107**:64-79.

Bulliere, D., Bullière, F., and de Reggi, M., 1979, Ecdysteroid titres during ovarian and embryonic development in *Blaberus craniifer*, *Wilhelm Roux Arch. Dev. Biol.* **186**:103-114.

Büning, J., 1972, Untersuchungen am Ovar von *Bruchidius obtectus* Say (Coleoptera-Polyphaga) zur Klärung des Oocytenwachstums in der Prävitellogenese, *Z. Zellforsch. Mikrosk. Anat.* **128**:241-282.

Büning, J., 1978, Development of telotrophic-meroistic ovarioles of polyphage beetles with special reference to the formation of nutritive cords, *J. Morphol.* **156**:237-256.

Büning, J., 1979a, The trophic tissue of telotrophic ovarioles in polyphage Coleoptera, *Zoomorphologie* **93**:33-50.

Büning, J., 1979b, The telotrophic nature of ovarioles of polyphage Coleoptera, *Zoomorphologie* **93**:51-57.

Büning, J., 1979c, The telotrophic-meroistic ovary of Megaloptera. I. The ontogenetic development, *J. Morphol.* **162**:37-66.

Büning, J., 1980, The ovary of *Raphidia flavipes* is telotrophic and of the *Sialis* type (Insecta, Raphidioptera), *Zoomorphologie* **95**:127-131.

Büning, J., 1981a. The development of the aphid ovary, *IX Congr. Int. Soc. Dev. Biol.* Basle (Abstract).

Büning, J., 1981b, Cluster formation in telotrophic meroistic ovarioles, *IX Congr. Int. Soc. Dev. Biol.* Basle (Abstract).

Burghardt, R. C., and Anderson, E., 1981, Hormonal modulation of gap junctions in rat ovarian follicles, *Cell Tissue Res.* **214**:181-193.

Capco, D., and Jeffrey, W., 1979, Origin and spatial distribution of maternal messenger RNA during oogenesis of an insect, *Oncopeltus fasciatus*, *J. Cell Sci.* **39**:63-76.

Carron, C. P., and Longo, F. J., 1982, Relation of cytoplasmic alkalinization to microvillar elongation and microfilament formation in the sea urchin egg, *Dev. Biol.* **89**:128-137.

Case, D., 1970, Postembryonic development of the ovary of *Rhodnius prolixus* Stål, M.Sc. thesis, Department of Zoology, McGill University, Montreal, Quebec, Canada.

Cave, M. D., 1975, Absence of r-DNA amplification in the meroistic (telotrophic) ovary of the large milkweek bug, *Oncopeltus fasciatus* (Dallas), *J. Cell Biol.* **66**:461-469.

Caveney, S., and Berdan, R., 1982, Selectivity in junctional coupling between cells of insect tissues. In *Insect Ultrastructure*, vol. 1, edited by R. C. King and H. Akai, pp. 434-465, Plenum Press, New York.

Choi, W. C., and Nagl, W., 1976, Electron microscopic study of the differentiation and development of trophocytes and oocytes in *Gerris najas*, *Cytobios* **17**:47-62.

Choi, W. C., and Nagl, W., 1977a, Electron microscopic study on the development and functional morphology of the ovarian nutritive tissue in *Gerris najas* (Heteroptera), *Biol. Zentralbl.* **96**:513-522.

Choi, W. C., and Nagl, W., 1977b, Patterns of DNA and RNA synthesis during the development of ovarian nurse cells in *Gerris najas* (Heteroptera), *Dev. Biol.* **61**:262-272.

Cone, M. V., and Eschenberg, K. M., 1966, Histochemical localization of acid phosphatase in the ovary of *Gerris remigis* Say (Hemiptera), *J. Exp. Zool.* **161**:337-352.

Couchman, J. R., and King, P. E., 1979, Germarial structure and oögenesis in *Brevicoryne brassicae* (L.) (Hemiptera: Aphididae), *Int. J. Insect Morphol. Embryol.* **8**:1-10.

Davenport, R., 1974, Synthesis and intercellular transport of ribosomal RNA in the ovary of the milkweed bug, *Oncopeltus fasciatus, J. Insect Physiol.* **20**:1949-1956.

Davenport, R., 1975a, RNA synthesis in the follicular epithelium of the milkweed bug ovary, *J. Insect Physiol.* **21**:1175-1178.

Davenport, R., 1975b, Changes in polyribosome populations during follicle cell maturation in the ovary of the bug *Oncopeltus fasciatus, J. Insect Physiol.* **21**:1721-1723.

Davenport, R., 1976, Transport of ribosomal RNA into the oocytes of the milkweed bug, *Oncopeltus fasciatus, J. Insect Physiol.* **22**:925-926.

Davey, K. G., 1981, Hormonal control of vitellogenin uptake in *Rhodnius prolixus* Stål, *Am. Zool.* **21**:701-705.

Davis, N. T., 1956, The morphology and functional anatomy of the male and female reproductive systems in *Cimex lecularius* L. (Heteroptera: Cimicidae), *Ann. Entomol. Soc. Am.* **49**:466-493.

Dekel, N., Lawrence, T. S., Gilula, N. B., and Beers, W. H., 1981, Modulation of cell-to-cell communication in the cumulus-oocyte complex and the regulation of oocyte maturation by LH, *Dev. Biol.* **86**:356-362.

de Loof, A., 1971, Synthesis and deposition of oocyte envelopes in the Colorado beetle, *Leptinotarsa decemlineata* Say, *Z. Zellforsch. Mikrosk. Anat.* **115**:351-360.

de Loof, A., and Lagasse, A., 1970, The ultrastructure of the follicle cells of the ovary of the Colorado beetle in relation to yolk formation, *J. Insect Physiol.* **16**:211-220.

de Loof, A., Lagasse, A., and Bohyn, W., 1972, Proteid yolk formation in the Colorado beetle with special reference to the mechanism of the selective uptake of haemolymph proteins, *K. Ned. Akad. Wet. Sect. C* **75**:125-143.

Dittmann, F., Ehni, R., and Engels, W., 1981, Bioelectric aspects of the hemipteran telotrophic ovariole of *Dysdercus intermedius, Wilhelm Roux Arch. Dev. Biol.* **190**:221-225.

Duspiva, F., Scheller, K., Weiss, D., and Winter, H., 1973, Ribonucleinsauresynthese in der telotrophmeroistischen Ovariole von *Dysdercus intermedius* (Heteroptera, Pyrrhoc.), *Wilhelm Roux Arch. Dev. Biol.* **172**:83-130.

Elliot, J. H., McDonald, F. J. D., and Vesk, M., 1975, Germarial structure and function in a parthenogenetic aphid, *Aphis craccivora* Koch (Hemiptera: Aphididae), *Int. J. Insect Morphol. Embryol.* **4**:341-347.

Englemann, F., 1970, *The Physiology of Insect Reproduction*, Pergamon Press, Elmsford, N.Y.

Englemann, F., 1979, Insect vitellogenin: Identification, biosynthesis, and role in vitellogenesis, *Adv. Insect Physiol.* **14**:49-108.

Engels, W., 1970, Kälte-wirkungen auf die Glykogenspeicherung in Eizellen von *Musca domestica, Arch. Entwicklungsmech. Org.* **166**:89-104.

Eppig, J., 1979, A comparison between oocyte growth in co-culture with granulosa cells and oocytes with granulosa cell-oocyte junctional contact maintained *in vitro, J. Exp. Zool.* **209**:345-353.

Eppig, J., 1982, The relationship between cumulus cell-oocyte coupling, oocyte meiotic maturation and cumulus expansion, *Dev. Biol.* **89**:268-277.

Eschenberg, K. M., and Dunlap, H. L., 1966, The histology and histochemistry of oogenesis in the water strider, *Gerris remigis* Say, *J. Morphol.* **118**:297-316.

Fawcett, D. W., and Chemes, H. E., 1979, Changes in distribution of nuclear pores during differentiation of male germ cells, *Tissue Cell* **11**:147-162.

Fiil, A., 1978, Follicle cell bridges in the mosquito ovary: Syncytia formation and bridge morphology, *J. Cell Sci.* **31**:137-143.

Furtado, A., 1979, The hormonal control of mitosis and meiosis during oogenesis in a blood-sucking bug *Panstrongylus megistus, J. Insect Physiol.* **25**:561-570.

Fux, T., Went, D. F., and Camenzird, R., 1978, Movement pattern and ultrastructure of rotating follicles of the paedogenetic gall midge, *Heteropeza pygmaea* Winnertz (Diptera: Cecidomyiidae), *Int. J. Insect Morphol. Embryol.* **7**:415-426.

Garel, J.-P., 1982, The silkworm, a model for molecular and cellular biologists, *Trends Biochem. Sci.* **7**:105-108.

Gerber, G. H., Church, N. S., and Rempel, J. G., 1971, The anatomy, histology, and physiology of the reproductive systems of *Lytta nuttalli* Say (Coleoptera: Meloidae). I. The internal genitalia, *Can. J. Zool.* **49**:523-533.

Gross, J., 1901, Untersuchungen über das Ovarium der Hemipteren, zugleich ein Beitrag zur Amitosenfrage, *Z. Wiss. Zool.* **69**:139-201.

Gross, J., 1903, Untersuchungen über die Histologie des Insectenovariums, *Zool. Jahrb. Abt. Anat. Ontog. Tiere* **18**:71-186.

Gruzova, M. N., 1979, Nuclear structures in the telotrophic ovarioles of the darkling beetle Tenebrionidae: Polyphaga. 3. The nucleus of oocytes: Electron microscopic data, *Ontogenez* **10**:332-339.

Gruzova, M. N., 1980, Nuclear structures in telotrophic ovarioles of darkling beetles Tenebrionidae: Polyphaga. 3. Nucleus of oocytes, electron microscopic data, *Sov. J. Dev. Biol.* **10**:299-309.

Hagedorn, H. H., 1980, Ecdysone, a gonadal hormone in insects. In *Advances in Invertebrate Reproduction*, vol. II, edited by W. H. Clarke, Jr., and T. S. Adams, pp. 97-107, Elsevier/North-Holland, Amsterdam.

Hagedorn, H. H., and Kunkel, J. G., 1979, Vitellogenin and vitellin in insects, *Annu. Rev. Entomol.* **24**:475-505.

Hamon, C., and Folliot, R., 1969, Ultrastructure des cordons trophiques de l'ovaire de divers Homoptères Auchènorhynches, *C.R. Acad. Sci. Ser. D* **268**:577-580.

Hegner, R. W., 1914, Studies on germ cells, *J. Morphol.* **25**:375-509.

Heller, D. T., Cahill, D. M., and Schultz, R. M., 1981, Biochemical studies of mammalian oogenesis: Metabolic cooperativity between granulosa cells and growing mouse oocytes, *Dev. Biol.* **84**:455-464.

Heming, B. S., 1977, Book review—Matsuda R., 1976, *Quaest. Entomol.* **13**:75-81.

Henking, H., 1892, Untersuchungen über die ersten Entwicklungsvorgänge in den Eirn der Insekten, *Z. Wiss. Zool.* **54**:1-49.

Hetru, C., Lagueux, M., Bang, L., and Hoffman, J. A., 1978, Adult ovaries of *Locusta migratoria* contain the sequence of biosynthetic intermediates for ecdysone, *Life Sci.* **22**:2141-2154.

Huebner, E., 1970, Cytology of the female reproductive system of *Rhodnius prolixus* Stål: A normal and an experimental analysis, Ph.D. thesis, Department of Zoology, University of Massachusetts, Amherst.

Huebner, E., 1976, Experimental modulation of the follicular epithelium in *Rhodnius* oocytes by juvenile hormone and other agents, *J. Cell Biol.* **70**:251A.

Huebner, E., 1977, Cell interaction in the *Rhodnius prolixus* follicle, *Am. Zool.* **17**:944 (abstract).

Huebner, E., 1981a, Nurse cell-oocyte interaction in the telotrophic ovaries of an insect, *Rhodnius prolixus*, *Tissue Cell* **13**:105-125.

Huebner, E., 1981b, Oocyte-follicle cell interaction during normal oogenesis and atresia in an insect, *J. Ultrastruct. Res.* **74**:95-104.

Huebner, E., 1982, Ultrastructure and development of the telotrophic ovary. In *The Ultrastructure and Functioning of Insect Cells*, edited by H. Akai, R. C. King, and S. Morohoshi, pp. 9-12, Society for Insect Cells, Tokyo.

Huebner, E., 1983, Oostatic hormone-antigonadotropin and reproduction. In *Insect Endocrinology*, pp. 319-329, edited by H. Laufer and R. G. H. Downer, Alan R. Liss, Inc. New York.

Huebner, E., 1984, Developmental cell interactions in female insect reproductive organs. In *Advances in Invertebrate Reproduction*, Vol. 3, edited by W. Engels, Elsevier/North Holland Biomedical Publishers, Amsterdam, in press.

Huebner, E., and Anderson, E., 1970, The effects of vinblastine sulfate on the microtubular organization of the ovary of *Rhodnius prolixus*, *J. Cell Biol.* **46**:191-198.

Huebner, E., and Anderson, E., 1972a, A cytological study of the ovary of *Rhodnius prolixus*. I. The ontogeny of the follicular epithelium, *J. Morphol.* **136**:459-493.

Huebner, E., and Anderson, E., 1972b, A cytological study of the ovary of *Rhodnius prolixus*. II. Oocyte differentiation, *J. Morphol.* **137**:385-416.
Huebner, E., and Anderson, E., 1972c, A cytological study of the ovary of *Rhodnius prolixus*. III. Cytoarchitecture and development of the trophic chamber, *J. Morphol.* **138**:1-40.
Huebner, E., and Caveney, S., 1984, Invertebrate cell junctions. In *Cell Receptors and Cell Communication in Invertebrates*, edited by A. H. Greenberg, in press, Marcel Dekker Publishers, New York.
Huebner, E., and Injeyan, H. S., 1980, Patency of the follicular epithelium in *Rhodnius prolixus:* A re-examination of the hormone response and technique refinement, *Can. J. Zool.* **58**:1617-1625.
Huebner, E., and Injeyan, H., 1981, Follicular modulation during oocyte development in an insect: Formation and modification of septate and gap junctions, *Dev. Biol.* **83**:101-113.
Hyams, J. S., and Stebbings, H., 1977a, The distribution and function of microtubules in nutritive tubes, *Tissue Cell* **9**:537-545.
Hyams, J. S., and Stebbings, H., 1977b, The isolation and characterization of a cytoplasmic microtubule transport system, *J. Cell Biol.* **75**:271a.
Hyams, J. S., and Stebbings, H., 1979a, The formation and breakdown of nutritive tubes-Massive microtubular organelles associated with cytoplasmic transport, *J. Ultrastruct. Res.* **68**:46-57.
Hyams, J. S., and Stebbings, H., 1979b, The mechanism of microtubule associated cytoplasmic transport, *Cell Tissue Res.* **196**:103-116.
Jaffe, L. F., 1981, The role of ionic currents in establishing developmental pattern, *Philos. Trans. R. Soc. London Ser. B* **295**:553-566.
Jaffe, L. F., and Nuccitelli, R., 1977, Electrical controls of development, *Annu. Rev. Biophys. Bioeng.* **6**:445-476.
Jaffe, L. F., and Woodruff, R. I., 1979, Large electrical currents traverse developing cecropia follicles, *Proc. Natl. Acad. Sci. USA* **76**:1328-1332.
Jeannel, R., and Paulian, R., 1944, Morphologie abdominale des Coléoptères et systématiques de l'ordre, *Rev. Fr. Entomol.* **11**:65-110.
Kafatos, F. C., 1975, The insect chorion: Programmed expression of specific genes during differentiation. In *Control Mechanisms in Development*, edited by R. H. Meints and E. Davies, pp. 103-121, Plenum Press, New York.
Kafatos, F. C., 1976, Sequential cell polymorphism, a fundamental concept in developmental biology, *Adv. Insect Physiol.* **12**:1-15.
Kelly, T. J., and Telfer, W. H., 1979, The function of the follicular epithelium in vitellogenic *Oncopeltus* follicles, *Tissue Cell* **11**:663-672.
Kessel, R. G., and Beams, H. W., 1963, Micropinocytosis and yolk formation in oocytes of the small milkweed bug, *Exp. Cell Res.* **30**:440-443.
King, R. C., 1970, *Ovarian Development in Drosophila melanogaster*, Academic Press, New York.
King, R. C., and Aggarwal, S. K., 1965, Oogenesis in *Hyalophora cecropia*, *Growth* **29**:17-83.
King, R. C., and Büning, J., 1984, The origin and functioning of insect oocytes and nurse cells. In *Comprehensive Insect Physiology, Biochemistry and Pharmacology*, Vol. 1, *Embryogenesis and Reproduction*, edited by G. A. Kerkut and L. I. Gilbert, in press, Pergamon Press, Oxford.
King, R. C., and Koch, E. A., 1963, Studies on the ovarian follicle cells of *Drosophila*, *Q. J. Microsc. Sci.* **104**:297-320.
King, R. C., Bahns, M., Horowitz, R., and Larramendi, P., 1978, A mutation that affects female and male germ cells differentially in *Drosophila melanogaster* Meigen (Diptera: Drosophilidae), *Int. J. Insect Morphol. Embryol.* **7**:359-375.
King, R. C., Cassidy, J. D., and Rousset, A., 1982, The formation of clones of interconnected cells during gametogenesis in insects. In *Insect Ultrastructure*, vol. 1, edited by R. C. King and H. Akai, pp. 3-60, Plenum Press, New York.
Kleine-Schonnefeld, H., and Engels, W., 1981, Symmetrical pattern of follicle arrangement in the ovary of *Musca domestica* (Insecta, Diptera), *Zoomorphologie* **98**:185-190.
Kloc, M., 1980, Extrachromosomal DNA and its activity in RNA synthesis in oogonia and oocytes in the pupal ovary of *Creophilus maxillosus* (Staphylinidae, Coleoptera-Polyphaga), *Eur. J. Cell Biol.* **21**:328-334.
Kloc, M., and Matuszewski, B., 1977, Extrachromosomal DNA and the origin of oocytes in the

telotrophic-meroistic ovary of *Creophilus maxillosus* (L.) (Staphylinidae, Coleoptera-Polyphaga), *Wilhelm Roux Arch. Dev. Biol.* **183**:351-368.

Koeppe, J. K., and Wellman, S. E., 1980, Ovarian maturation in *Leucophaea maderae:* Juvenile hormone regulation of thymidine uptake into follicle cell DNA, *J. Insect Physiol.* **26**:219-228.

Koeppe, J. K., Hobson, K., and Wellman, S. E., 1980, Juvenile hormone regulation of structural changes and DNA synthesis in the follicular epithelium of *Leucophaea maderae, J. Insect Physiol.* **26**:229-240.

Köhler, A., 1907, Untersuchungen über das Ovarium der Hemipteren, *Z. Wiss. Zool. Abt. A* **87**:337-381.

Korfsmeier, K. H., 1980, Lysosomal enzymes and yolk platelets, *Eur. J. Cell Biol.* **22**:208.

Korschelt, E., 1887, Über einige interessante Vörgange bei der Bildung der Insekteneier, *Z. Wiss. Zool. Abt. A* **45**:327-397.

Korschelt, E., 1889, Über die Entstehung und Bedeutung der verschiedenen Zellenelemente des Insektenovariums, *Z. Wiss. Zool. Abt. A* **43**:537-720.

Kristensen, N. P., 1981, Phylogeny of insect orders, *Annu.Rev. Entomol.* **26**:135-157.

Książkiewicz, M., 1980, Ultrastructure of the trophic chamber and nutritive cord of *Aspidiotus hederae* (Homoptera, Coccoidea), *Cell Tissue Res.* **213**:149-157.

Kurihara, M., 1975, Anatomical and histological studies on the germinal vesicle in degenerating oocyte of starved females of the lady beetle, *Epilachna vigintioctomaculata* Motschulsky (Coleoptera, Coccinellidae), *Kont yû Tokyo* **43**:91-105.

Kurihara, M., 1976, Origin and formative process of two nuclear structures in degenerating oocyte of the starved lady beetle *Epilachna vigintioctomaculata* Motschulsky (Coleoptera: Coccinellidae), *J. Fac. Agric. Iwate Univ.* **13**:145-159.

Laguex, M., Hirn, M., and Hoffmann, J. A., 1977, Ecdysone during ovarian development in *Locusta migratoria, J. Insect Physiol.* **23**:109-119.

Laverdure, A. M., 1970, Action de l'ecdysone et de l'ester méthylique du farnesol sur l'ovaire nymphal de *Tenebrio molitor* (Coléoptère) cultivé *in vitro, Ann. Endocrinol.* **31**:516-521.

Laverdure, A. M., 1971, Etude des conditions hormonales nécessaires à l'évolution de l'ovaire chez la nymph de *Tenebrio molitor* (Coléoptère), *Gen. Comp. Endocrinol.* **17**:467-478.

Laverdure, A. M., 1975, Action de l'ecdystérone sur l'évolution de l'ovaire nymphal de *Tenebrio molitor* en culture *in vitro, C. R. Acad. Sci. Ser. D* **281**:1745-1748.

Laverdure, A. M., and Huet, C., 1965, Rôle du complexe céphalique et des glandes prothoraciques sur le développement de l'ovariole de la nymphe de *Tenebrio molitor* (Coléoptère), *C. R. Acad. Sci.* **159**:60-63.

Laverdure, A. M., Laguex, M., and Hoffmann, J. A., 1977, Ecdysone et développement ovarien chez la femelle adulte de *Tenebrio molitor, Bull. Soc. Zool. Fr.* **102**:311.

Lutz, D. A., 1979, Structural and physiological aspects of 5th instar ovarian development in *Rhodnius prolixus* (Insecta: Hemiptera), M.Sc. thesis, Department of Zoology, University of Manitoba, Winnipeg, Canada.

Lutz, D., and Huebner, E., 1978, Structural and physiological aspects of postembryonic ovarian development in *Rhodnius prolixus, J. Cell Biol.* **79**(2, part 2):G1026 (abstract).

Lutz, D., and Huebner, E., 1980, Development and cellular differentiation of an insect telotrophic ovary (*Rhodnius prolixus*), *Tissue Cell* **12**:773-794.

Lutz, D., and Huebner, E., 1981, Development of nurse cell-oocyte interactions in the insect telotrophic ovary of *Rhodnius, Tissue Cell* **13**:321-335.

MacGregor, H. C., and Stebbings, H., 1970, A massive system of microtubules associated with cytoplasmic movement in telotrophic ovarioles, *J. Cell Sci.* **6**:431-449.

Magakyan, Y. A., Makaryan, S. R., Akopyan, L. A., and Petrosyan, A. V., 1980, The oocyte nucleus in the oogenesis of cochineal functional morphology of chromosomo nuclear apparatus, *Biol. Zh. Arm.* **32**:1129-1134.

Mahowald, A. P., 1972, Oogenesis. In *Developmental Systems: Insects*, vol. 1, edited by S. J. Counce and C. H. Waddington, pp. 1-48, Academic Press, New York.

Margaritis, L. H., Petri, W. H., and Wyman, A. R., 1979, Structural and image analysis of a crystalline layer from dipteran eggshell, *Cell Biol. Int. Rep.* **3**:61-66.

Margaritis, L. H., Kafatos, F. C., and Petri, W. H., 1980, The eggshell of *Drosophila melanogaster*. I. Fine structure of the layers and regions of the wild-type eggshell, *J. Cell Sci.* **43:**1-35.

Masner, P., 1966, The structure, function and imaginal development of the female inner reproductive organs of *Adephocoris lineolatus* (Goeze) (Heteroptera: Miridae), *Acta Entomol. Bohemoslov.* **63:**177-199.

Masner, P., 1968, The inductors of differentiation of prefollicular tissue and the follicular epithelium in ovarioles of *Pyrrhocoris apterus*, *J. Embryol. Exp. Morphol.* **20:**1-13.

Masner, P., 1969, The effect of substances with juvenile hormone activity on morphogenesis and function of gonads in *Pyrrhocoris apterus* (Heteroptera), *Acta Entomol. Bohemoslov.* **66:**81-86.

Matsuzaki, M., 1975, Ultrastructural changes in developing oocytes, nurse cells, and follicular cells during oogenesis in the telotrophic ovarioles of *Bothrogonia japonica* Ishihara (Homoptera) (Telligellidae), *Kontyû Tokyo* **43:**75-90.

Matsuzaki, M., and Ando, H., 1977, Ovarian structures of the adult alderfly, *Sialis mitsuhashii* Okamoto (Megaloptera: Sialidae), *Int. J. Insect Morphol. Embryol.* **6:**17-29.

Matuszewski, B., and Kloc, M., 1976, Gene amplification in oocytes of the rove beetle *Creophilus maxillosus* (Staphylinidae, Coleoptera-Polyphaga), *Experientia* **32:**34-36.

Mays, U., 1972, Stofftransport im Ovar von *Pyrrhocoris apterus* L.: Autoradiographische Untersuchungen zum Stofftransport von den Nährzellen zur Oocyte der Feuerwanze *Pyrrhocoris apterus* L. (Heteroptera), *Z. Zellforsch. Mikrosk. Anat.* **123:**395-410.

Mazur, G. D., Regier, J. C., and Kafatos, F. C., 1982, Order and defects in the silkworm chorion, a biological analogue of a cholesteric liquid crystal. In *Insect Ultrastructure*, vol. 1, edited by R. C. King and H. Akai, pp. 150-185, Plenum Press, New York.

Mellanby, H., 1936, The later embryology of *Rhodnius prolixus*, *Q. J. Microsc. Sci.* **79:**1-42.

Meola, S. M., Mollenhauer, H. H., and Thompson, J. M., 1977, Cytoplasmic bridges within the follicular epithelium of the ovarioles of two diptera *Aedes aegypti* and *Stomoxys calctrans*, *J. Morphol.* **153:**81-85.

Mulnard, J., 1954, Etude morphologique et cytochimique de l'oogénèse chez *Acanthoscelides obtectus* Say (Bruchidae-Coléoptère), *Arch. Biol.* **65:**135-216.

Oseto, C. Y., and Helms, T. J., 1971, Embryonic and postparturienic reproductive-system development in *Schizaphis graminum* (Hemiptera) (Homoptera: Aphididae), *Ann. Entomol. Soc. Am.* **64:**603-608.

Payne, F., 1912, I. A further study of the chromosomes of the Reduviidae. II. The nucleolus in the young oocytes and origin of the ova in *Gelastocoris*, *J. Morphol.* **23:**331-347.

Preusse, F., 1895, Über die amitotische Kerntheilung in den Ovarien der Hemipteren, *Z. Wiss. Zool. Abt. A* **59:**305-349.

Ramamurty, P. S., and Engles, W., 1977, Occurrence of intercellular bridges between follicle epithelial cells in the ovary of *Apis mellifica* queens, *J. Cell Sci.* **24:**195-202.

Richard-Mercier, N., 1972, Embryogenèse et differenciation sexuelle de la gonade du Doryphore *Leptinotarsa decemlineata* Say, *Ann. Embryol. Morphol.* **5:**191-201.

Richard-Mercier, N., 1977, Organogenèse d'ovaries et de testicules steriles apres cauterisation des cellules polaires de l'embryon du Dorphore (*Leptinotarsa decemlineata* Say), *Wilhelm Roux Arch. Dev. Biol.* **183:**171-176.

Robert, A., 1979, Les premiers stades de l'ovogenese et les variations de la neurosecretion cerebrale chez deux especes sympatriques *Roscius elongatus*, Stål et *R. brazzavilliensis*, Robert (Heteroptera: Pyrrhocoridae), *Int. J. Insect Morphol. Embryol.* **8:**11-31.

Ruegg, R. P., Kriger, F. L., Davey, K. G., and Steel, C. G. H., 1981, Ovarian ecdysone elicits release of a myotropic ovulation hormone in *Rhodnius* (Insecta: Hemiptera), *Int. J. Invertebr. Reprod.* **3:**357-361.

Ruegg, R. P., Orchard, I., and Davey, K. G., 1982, 2O-Hydroxy-ecdysone as a modulator of electrical activity in neurosecretory cells of *Rhodnius prolixus*, *J. Insect Physiol.* **28:**243-248.

Sahai, Y. N., 1975, The ovarian development in *Dysdercus similis* (Freeman), *Zool. Jahrb. Abt. Anat. Ontog. Tiere* **94:**453-473.

Schlottman, L., and Bonhag, P. F., 1956, Histology of the ovary of the adult mealworm *Tenebrio molitor* L. (Coleoptera, Tenebrionidae), *Univ. Calif. Publ. Entomol.* **11:**351-394.

Schmidt, O., and Jäckle, H., 1978, RNA synthesized during oogenesis and early embryogenesis in an insect egg (*Euscelis plebejus*), *Wilhelm Roux Arch. Dev. Biol.* **184:**143-153.

Schrader, F., and Leuchtenberger, C., 1952, The origin of certain nutritive substances in the eggs of Hemiptera, *Exp. Cell Res.* **3**:136-146.

Schreiner, B., 1977a, Vitellogenesis in the milkweed bug, *Oncopeltus fasciatus:* A light and electron microscopic investigation, *J. Morphol.* **151**:35-80.

Schreiner, B., 1977b, The effect of the hormone(s) from the corpus allatum complex on the ovarian tissue of *Oncopeltus fasciatus:* A light and electron microscopic investigation, *J. Morphol.* **155**:81-110.

Shinji, G. O., 1919, Embryology of coccids, with special reference to the formation of the ovary, origin and differentiation of the germ cells, germ layers, rudiments of the midgut and intracellular symbiotic organisms, *J. Morphol.* **33**:73-167.

Stebbings, H., 1971, Influence of vinblastine sulfate on deployment of microtubules and ribosomes in telotrophic ovarioles, *J. Cell Sci.* **8**:111-125.

Stebbings, H., 1975. The role of microtubules in the assembly of vinblastine-induced crystals, *Cell Tissue Res.* **159**:141-145.

Telfer, W. H., 1965, The mechanism and control of yolk formation, *Annu. Rev. Entomol.* **10**:161-184.

Telfer, W. H., 1975, Development and physiology of the oocyte-nurse cell syncytium, *Adv. Insect Physiol.* **11**:223-319.

Telfer, W. H., Woodruff, R. I., and Huebner, E., 1981, Electrical polarity and cellular differentiation in meroistic ovaries, *Am. Zool.* **21**:675-686.

Telfer, W., Huebner, E., and Smith, D. S., 1982, The cell biology of vitellogenic follicles in *Hyalophora* and *Rhodnius*. In *Insect Ultrastructure*, vol. 1, edited by R. C. King and H. Akai, pp. 118-149, Plenum Press, New York.

Toth, L., 1933, Über die frühembryonale Entwicklung der viviparen Aphiden, *Z. Morphol. Oekol. Tiere* **27**:692-731.

Tucker, J. B., and Meats, M., 1976, Microtubules and control of insect egg shape, *J. Cell Biol.* **71**:207-217.

Ullmann, S. L., 1973, Oogenesis in *Tenebrio molitor:* Histological and autoradiographical observations on pupal and adult ovaries, *J. Embryol. Exp. Morphol.* **30**:179-217.

Vanderberg, J. P., 1963, Synthesis and transfer of DNA, RNA, and protein during vitellogenesis in *Rhodnius prolixus* (Hemiptera), *Biol. Bull.* **125**:556-575.

Vernier, J.-M., 1970, Anatomie et histologie des ovaries et de l'appareil génital de *Sitophilus granarius* [Col. Curculionidae], *Ann. Soc. Entomol. Fr.* **6**:243-265.

Went, D. F., 1978, Oocyte maturation without follicular epithelium alters egg shape in a dipteran species, *J. Exp. Zool.* **205**:149-155.

Went, D. F., and Junquera, 1981, Embryonic development of insect eggs formed without follicular epithelium, *Dev. Biol.* **86**:100-110.

Wick, J. R., and Bonhag, P. F., 1955, Postembryonic development of the ovaries of *Oncopeltus fasciatus* (Dallas), *J. Morphol.* **96**:31-59.

Wielowiejski, H. R. V., 1886, Zur Morphologie des Insectenovariums, *Zool. Anz.* **9**:132-139.

Wiemerslage, L. J., 1976, Lipid droplet formation during vitellogenesis in the cecropia moth, *J. Insect Physiol.* **22**:41-50.

Wigglesworth, V. B., 1979, Secretory activities of plasmatocytes and oenocytoids during the moulting cycle in an insect (*Rhodnius*), *Tissue Cell* **11**:69-78.

Wightman, J. A., 1973, Ovariole microstructure and vitellogenesis in *Lygocoris pabulinus* (L.) and other mirids (Hemiptera: Miridae), *J. Entomol.* **48**:103-115.

Winter, H., 1974, Ribonucleoprotein-Partikel aus dem telotroph-meroistischen Ovar von *Dysdercus intermedius* Dist. (Heteroptera, Pyrrhoc.) und ihr Verhalten im zellfreien Proteinsynthesesystem, *Wilhelm Roux Arch. Dev. Biol.* **175**:103-127.

Winter, H., 1975, Charakteristika mütterlicher RNP-Partikel aus dem Ei der Baumwollwanze, *Dysdercus intermedius* Dist. (Heteroptera, Pyrrhoc.), *Verh. Dtsch. Zool. Ges.* **67**:201-204.

Winter, H., Wiemann-Weiss, D., and Duspiva, F., 1977, Endogene synthese kurzlebiger Messenger-RNS in der Oocyte von *Dysdercus intermedius* Dist. nach Abschluss der Vitellogenese, *Wilhelm Roux Arch. Dev. Biol.* **182**:39-58.

Woodruff, R. I., 1979, Electrotonic junctions in cecropia moth ovaries, *Dev. Biol.* **69**:281-295.

Woodruff, R. I., and Telfer, W. H., 1973, Polarized intercellular bridges in ovarian follicles of the cecropia moth, *J. Cell Biol.* **58**:172-188.

Woodruff, R. I., and Telfer, W. H., 1974, Electrical properties of ovarian cells linked by intercellular bridges, *Ann. N.Y. Acad. Sci.* **238**:408-419.

Woodruff, R. I., and Telfer, W. H., 1980, Electrophoresis of proteins in intercellular bridges, *Nature (London)* **286**:84-86.

Woodruff, R. I., Huebner, E., and Telfer, W. H., 1984, The origin of electrical currents in insect ovarioles, abstract. In *Advances in Invertebrate Reproduction*, Vol. 3, edited by W. Engels, in press, Elsevier/North Holland and Biomedical Publishers, Amsterdam.

Yamauchi, H., and Yoshitake, N., 1982, Origin and differentiation of the oocyte-nurse cell complex in the germarium of the earwig, *Anisolabis maritima* (Dermaptera: Labiduridae), *Int. J. Insect Morphol. Embryol.* **11**:293-305.

Zinsmeister, P. P., and Davenport, D., 1971, An autoradiographic and cytochemical study of cell interactions during oogenesis in the milkweed bug, *Oncopeltus fasciatus*, *Exp. Cell Res.* **67**:273-278.

2

Early Embryogenesis of *Bombyx mori*

KEIICHIRO MIYA

1. Introduction

The first detailed studies on the embryogenesis of the domesticated silkmoth, *Bombyx mori* L., were carried out by Toyama (1902, 1909). He reported that the development of fertilized eggs becomes arrested at an early stage in which the germ layers are differentiated and a caudal segment has formed. After hibernation, the eggs resume embryogenesis and, under natural conditions in Japan, hatch between the end of April and the beginning of May.

Subsequent embryological studies of the silkmoth have been made chiefly for the solution of practical problems, such as the storage of the early embryos and the initiation, continuation, and termination of the diapause. An excellent review of this work was published by Takami (1969). Eggs of the silkmoth also have been used as convenient materials for studying several basic embryological problems, i.e., fertilization, cell determination and differentiation, embryonic induction, and the defective development of embryonic lethals, and such investigations have been reviewed by Kuwana and Takami (1957) and Tazima (1964).

With progress in the techniques of electron microscopy, the ultrastructure of the egg and the changes occurring during early and late embryogenesis have also been examined. Akai (1957) was the first to describe the ultrastructure of the *Bombyx* chorion, and Miya (1959, 1960) reported the changes in egg structure that accompanied fertilization, the cleavage stages, and the formation of the serosa in diapausing eggs. However, these early studies were not satisfactory because of fixation artifacts. Subsequently, Takei and Nagashima

---

KEIICHIRO MIYA • Faculty of Agriculture, Iwate University, Morioka 020, Japan.

(1975a,b) compared the early embryonic development of diapause and nondiapause eggs, and Takei and Yoshitake (1975) also examined the same process in pigmented, nondiapause eggs. Unfortunately, similar problems with the fixation procedures plagued these efforts. Since the *Bombyx* egg is enclosed by a thick, tough chorion, which protects it and the developing embryo from the dangers of desiccation and drowning (Mazur *et al.*, 1982), fixatives for electron microscopy, especially osmium tetroxide, hardly penetrate an intact egg. Consequently, appropriate pretreatments, such as cutting the egg or removing the chorion, are required before fixation.

Okada (1970) examined the ultrastructure of cells from embryos in pre- to postdiapause stages after dissecting off the egg shell before fixation. Takesue *et al.* (1976) compared the yolk granules of diapause and nondiapause eggs during early developmental stages. In this study, the chorion was removed before fixation. Using the same method, Miya *et al.* (1972) observed ultrastructural changes of yolk cells from embryos in diapause to postdiapause stages; subsequently, Miya (1975, 1976, 1978) described the embryonic development of gonads, alimentary canal, and Malpighian tubules, the architecture of the newly laid egg, and the ultrastructural changes that accompanied sperm entry. With the scanning electron microscope, Sakaguchi *et al.* (1973), Kanda *et al.* (1974), and Ohtsuki *et al.* (1977) studied the superficial structure of the chorion, and Keino and Takesue (1978) examined the surface structure of dechorionated embryos from oviposition to the stage of germband formation.

In this chapter, the term *early embryogenesis* covers the changes that occur in the mature egg as it passes through the oviduct, the changes coupled with sperm entry, and the development of the embryo during the prediapause stages.

2. The Structure of Mature Eggs from the Oviduct

2.1. The Micropylar Region

2.1.1. The Micropyle and the Vitelline Membrane

The mature egg of *B. mori* is a spheroid, approximately 1.3 mm long, 1.0 mm wide, and 0.6 mm thick. The outermost egg membrane, the chorion, is a thick, tough, and elastic membrane, formed by follicular cells during the latter part of oogenesis. The first electron microscopic observation of the chorion of the *Bombyx* egg was carried out by Akai (1957) using the replica method, and subsequently several authors examined it with the scanning electron microscope (Sakaguchi *et al.*, 1973; Kanda *et al.*, 1974; Ohtsuki *et al.*, 1977) and the transmission electron microscope (Matsuzaki, 1968; Akutsu and Yoshitake, 1974; Miya, 1978; Goldsmith *et al.*, 1978). The silkmoth chorion recently has been reexamined in great detail and the results are reviewed in the first volume of this treatise (Mazur *et al.*, 1982).

The surface of the micropylar region shows a petallike pattern, and at its center the micropylar canals open (Figure 1). The micropylar canals are normally 3 or 4 in number, but in some cases 6 to 8 canals are recognized (Kanda et al., 1974; Ohtsuki, 1979). The chorion of the micropylar region represents a conspicuous structure in thin sections. The outer lamellar layer is vague (Akai, 1958), and the inner trabecular layer becomes thinner and is not detected under the micropylar canals (Miya, 1978). As shown in Figure 2, the lamellae of the middle lamellar layer display an irregular arrangement similar to those observed in the abnormal chorions produced by mutant alleles of the *Gray egg* gene (Sakaguchi et al., 1973; Akutsu and Yoshitake, 1974).

Each micropylar canal is divided into two parts: a tube penetrating the chorion, about 1 μm in diameter, and a protrusion connecting the tube to the vitelline membrane, 16-18 μm long (Akai, 1958). In the egg of a noctuid moth, *Amathes c-nigrum*, Salkeld (1973) designated the former as "ectomicropyle" and the latter "entomicropyle." In the *Bombyx* egg, the entomicropyle is an elastic tube enclosed by a membrane of chorionic substance, 0.1 μm thick. The entomicropyle penetrates the outer layer of the vitelline membrane. The micropylar canals are filled with fine particles (Figure 3).

In most insect eggs, the vitelline membrane has been described as a thin, noncellular membrane (Johannsen and Butt, 1941), and a similar description was given for *Bombyx* (Takami, 1969; Ohtsuki, 1979). However, recent electron microscopic observations reveal that the vitelline membrane is not a simple thin membrane, but shows a more complex structure (Matsuzaki, 1968; Akutsu and Yoshitake, 1977), and its ultrastructure changes remarkably with the entry of sperm (Miya, 1978). The vitelline membrane is composed of an electron-dense outer layer, 0.2-0.5 μm thick, and an inner layer containing abundant, irregularly shaped, electron-dense granules of various size (Akutsu and Yoshitake, 1977; Miya, 1978). In the micropylar region, both outer and inner layers increase the thickness, especially the outer layer which reaches a thickness of 2.0-2.5 μm in the region adjacent to the entomicropyle. In this area a wide space occurs in the outer layer, from the end of which a somewhat crooked, tubular channel, about 0.6 μm in diameter and enclosed by an electron-dense thin membrane, penetrates the inner layer to reach the surface of the oocyte (Figure 4). Furthermore, a slender cytoplasmic process containing many fine vesicles protrudes from the oocyte into the lumen of the tubular channel (Figure 5).

Certain structural adaptations of the female reproductive organs and the egg have been evolved to facilitate sperm penetration and to ensure fertilization. Examples in *Musca domestica* are: the micropylar cap substance which serves as a protective medium for sperm before their entry into the egg, the breakdown of the plasma membrane prior to sperm entry into the fertilization chamber, and the canals that pass through the anterior vitelline membrane facilitate the rapid entry of sperm into the egg (Degrugillier and Leopold, 1976). In *Bombyx*, besides the micropylar canals running through the chorion, the egg provides tubular channels that penetrate the vitelline membrane. Furthermore, the sperm undergo a structural change in the spermatheca, as will be described later.

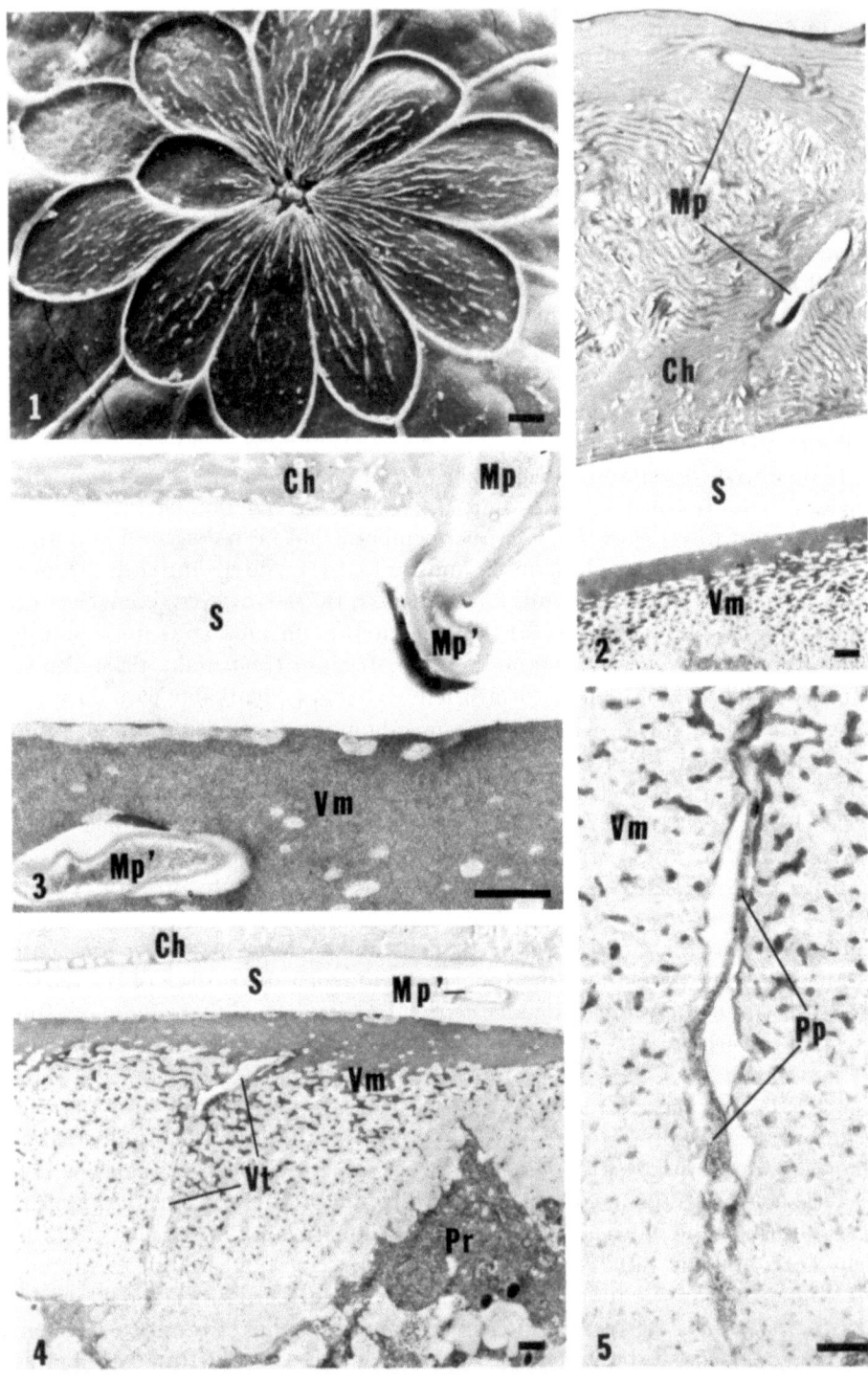

2.1.2. The Periplasm

As in most insect eggs (see review by Zissler and Sander, 1982), the ooplasm of the *Bombyx* egg is divided into a superficial layer, the periplasm, which is free of yolk spheres, and the endoplasm, which occupies the central part of the egg and is rich in yolk organelles. The periplasm is not a homogeneous layer of uniform thickness, but is rather a complex meshlike structure that connects with the endoplasm. The periplasm lying under the micropylar canals possesses remarkable structural differences, compared with the other egg regions. This periplasm consists of a relatively thicker layer of ooplasm which terminates in a series of small conelike cytoplasmic processess, from which short microvilluslike processes protrude into the inner layer of the vitelline membrane. Stacks of ER develop parallel to the egg surface (Figure 6). In the cytoplasmic processes, there are many vesicles of various size and also aggregates of minute vesicles, 80 nm in diameter, are found in the conelike processes. These minute vesicles contain fine dense particles (Figure 8). The scarcity of mitochondria is one of the characteristics of the region where lamellar stacks of RER develop (the ER zone).

Another ooplasmic region encloses the ER zone. Here the vesicles containing electron-dense granules are abundant and numerous slender mitochondria are present. This region is designated the *granule zone* (Figure 7). In the ER zone adjacent to the granule zone, organelles composed of abundant, small, tightly packed vesicles exist and whorls of ER radiate from them (Figure 8). The organelles are designated here as *ER-associated rosettes* and similar structures have been reported in *Smittia* (Zissler and Sander, 1982; their Figures 19, 20). The above-mentioned structures of periplasm in the micropylar region apparently play a role in sperm reception and the activation of the egg cell.

⬅————————————————————————————

Figures 1–5. Scanning (SEM) and transmission electron micrographs (TEM) of the micropylar region in the anterior pole of mature *Bombyx* eggs. The eggs in Figures 1, 3, 4, and 5 are from the oviduct. (Bars = 1 μm.)

Figure 1. SEM of the outer surface of the micropylar region showing six openings of micropylar canals in the center of a petallike pattern (reproduced from Ohtsuki, 1979).

Figure 2. TEM of a section through the micropylar region of a newly laid egg. Chorion of the micropylar region lacks the outer lamellar and the inner trabecular layer. The lamellae of the middle layer are arranged irregularly. The lumen of the micropylar canal is empty in contrast with that in Figure 3. Ch, chorion; Vm, vitelline membrane.

Figure 3. A higher-power EM of the micropylar canal. The micropyle (Mp) penetrates the chorion and protrudes inward as a tube enclosed by a thin chorionic substance. This tube, the entomicropyle (Mp′), runs through the outer layer of the vitelline membrane. The micropylar canal is filled with numerous fine particles.

Figures 4 and 5. EMs of a tubular channel (Vt) penetrating the inner layer of the vitelline membrane. In Figure 4 a segment of entomicropyle (Mp′) is present in the space (S) between the chorion and the vitelline membrane. Figure 5 shows an enlarged view of the tubular channel, into which a slender projection (Pp) extends from the surface of the oocyte. Pr, periplasm. The spaces (S) in Figures 2–4 may be artifacts of cytological processing.

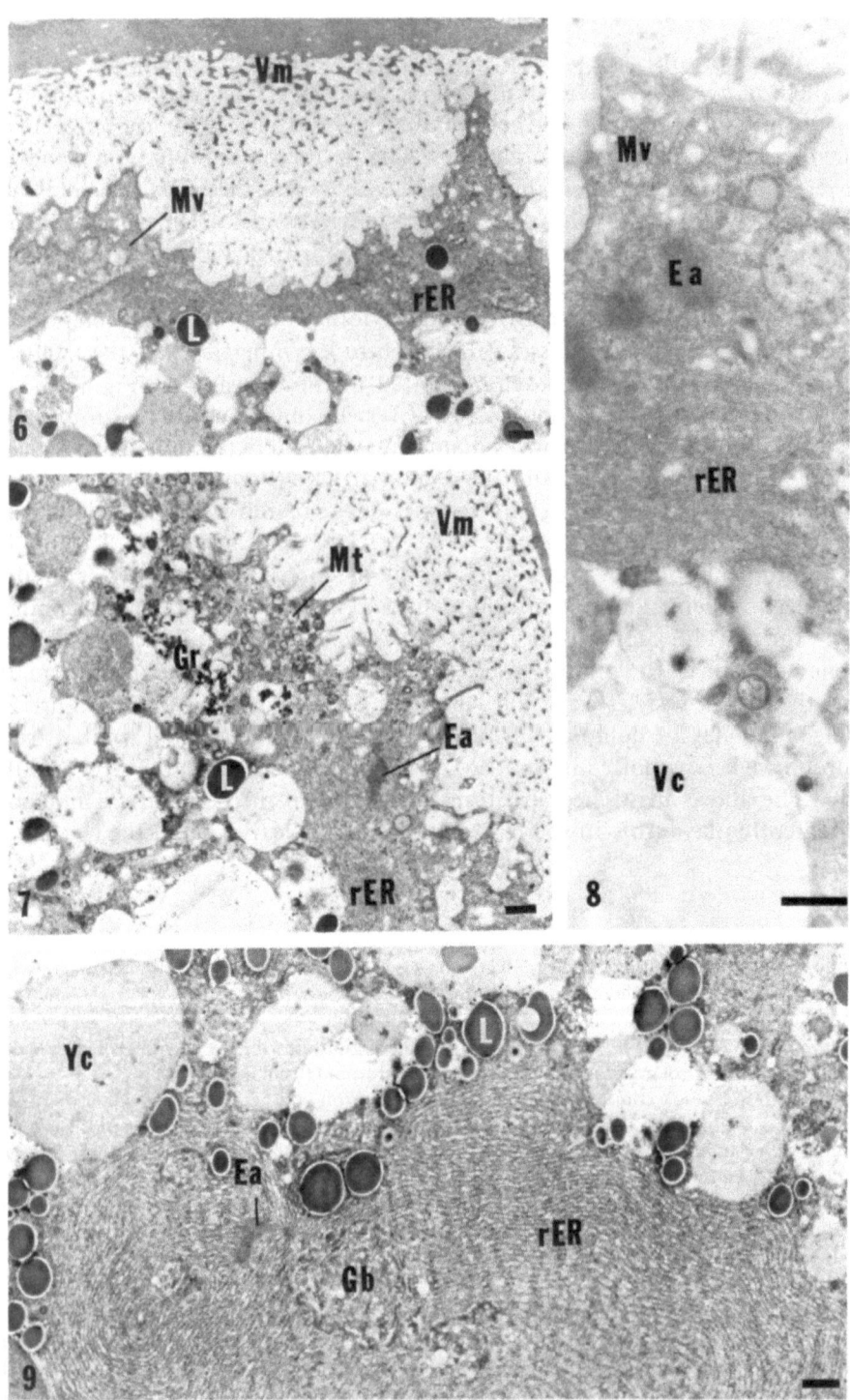

2.2. Other Egg Regions

2.2.1. The Anterior Polar Plasm

The island of anterior polar plasm in which the maternal chromosomes reside contains large whorls of ER and encloses the spindle arrested in metaphase of the first meiotic division. Many well-developed Golgi bodies reside among cisterns of ER (Figure 9). The structure and behavior of the chromosomes during the first meiotic metaphase have been described in detail by Rasmussen (1977).

2.2.2. The Periplasm

The periplasm in the *Bombyx* egg is not a homogeneous layer, but rather a cytoskeletal network, and in this matrix various organelles and inclusions are distributed. Numerous free ribosomes exist in the matrix. The cisterns and vesicles of ER are well developed, and sometimes small stacks form concentric whorls (Figure 10). In several places, larger whorls of ER attach to the network of periplasm (Figure 11). These structures coincide in position with the pyronine-positive granules found in the newly laid egg (Suzuki, 1969). The distribution of these organelles is limited to the ventral and lateral sides of egg, i.e., the presumptive embryonic region (Kobayashi and Miya, unpublished work).

Mitochondria are the most numerous organelles in the periplasm, and they are distributed evenly, aside from being absent within the stacks of RER. Golgi bodies are limited to the anterior polar plasm. The whorls of RER and the annulate lamellae are rare, although they were abundant during earlier stages of oogenesis. Multivesicular bodies are frequently observed, and their vesicles often contain electron-dense granules. These vesicles sometimes unite with c-yolk spheres as described later (Section 2.2.3).

Another interesting organelle is the "small yolk sphere," an electron-dense

---

Figures 6-9. EMs of the ooplasm in the anterior polar region of mature eggs from various positions in the oviduct. (Bars = 1 μm.)

Figure 6. A portion of the longitudinal section of periplasm under the micropylar canals. The periplasm protrudes in the inner layer of the vitelline membrane (Vm) as a series of small conelike processes which contain aggregates of fine vesicles (Mv). At the base of the conelike processes, stacks of ER run parallel to the egg surface.

Figure 7. The granule zone (Gr) contains many vesicles with electron-dense granules, and it connects the ER zone to the periplasm. Short rod-shaped mitochondria are abundant. In the periphery of the ER zone, dense organelles composed of numerous fine vesicles (Ea) interconnect the cisternae of ER.

Figure 8. An enlarged segment of the ER zone, showing the RER, dense organelles connecting the cisternae of ER (Ea), and aggregates of fine vesicles (Mv). Vc, vacuole.

Figure 9. A portion of the anterior polar plasm showing a large whorl-shaped stack of cisternae of RER. These develop around the spindle during the first meiotic metaphase. Many Golgi bodies (Gb) are located among the cisternae of ER. Yc, c-yolk sphere.

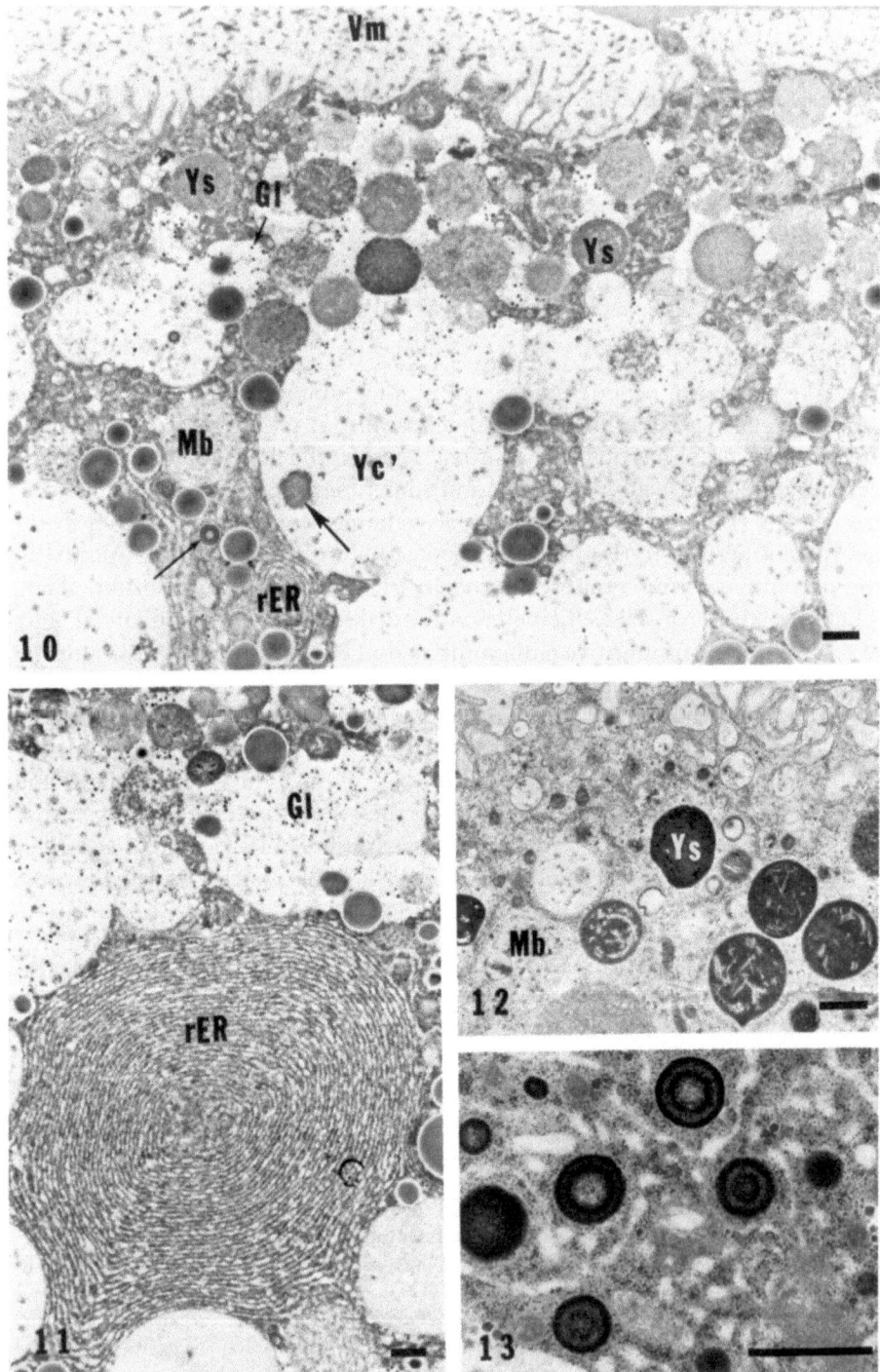

body containing needlelike or tubular inclusions (Figure 12). Small yolk spheres stain deep blue with toluidine blue in thick sections, and these are probably the same as the "dense bodies" of Miya et al. (1972), the "yolk granules" of Okada (1970), and the "small yolk granules" of Ohtsuki (1965). There are also round granules about 0.5 µm in diameter, which are characterized by a thick dense cortex (Figure 10, 13). These granules resemble the granules found in the epithelial cells of Malpighian tubules of pupae (Waku, 1974) and in the epithelial cells of the pupal midgut (Waku and Sumimoto, 1974). Similar granules were detected in the epithelial cells of the embryonic midgut, just after differentiation of cylinder and goblet cells (Miya, 1976), and these granules are also found in the germband cells. Their function remains unknown.

2.2.3. The Yolk System

The yolk system is composed of the ooplasmic reticulum, lipid droplets, glycogen granules, and proteid yolk spheres. In the ooplasmic network, there are areas rich in free ribosomes and areas without them. In the ribosome-rich area are scattered ER and mitochondria, and the ribosome-poor areas contain clusters of glycogen granules and lipid droplets. Proteid yolk spheres are inserted within the ooplasmic network. There are two types: homogeneous a-yolk spheres and b-yolk spheres which contain numerous electron-dense granules (Miya et al., 1972). Another special layer of ooplasm, the "subcortical layer," lies adjacent to the periplasm prior to the germband stage. This layer is thicker and more distinct in eggs fixed after removing the chorion. The subcortical layer contains "c-yolk spheres." These have been called "refractile bodies" by Akutsu and Yoshitake (1977) and "yolk granules 2" by Takesue et al. (1976). c-yolk spheres are of medium electron density and contain electron-dense globules. Besides the typical c-yolk spheres, there are spheres composed of many vesicles enclosing either electron-dense granules or glycogen granules. These spheres may result from the fusion of multivesicular bodies and glycogen granules. The subcortical layer also contains populations of large vacuoles.

Figures 10–13. EMs of the periplasm and the subcortical layer of mature eggs from the oviduct. (Bars = 1 µm.)
Figure 10. A portion of the ventral periplasm and the subcortical layer. The periplasm is composed of a reticulate matrix in which ER, numerous free ribosomes, lipid droplets, and mitochondria are scattered. Note the round granule with the thick cortex (small arrow). The cytoplasmic network contains many small yolk spheres (Ys), multivesicular bodies (Mb), lacunae containing glycogen granules (Gl), and modified c-yolk spheres (Yc′), which often contain a dense globule (large arrow). Vm, vitelline membrane.
Figure 11. A larger whorl-shaped arrangement of cisternae of ER within the network of periplasm.
Figure 12. Small yolk spheres of various morphologies.
Figure 13. Round granules shown at a higher magnification.

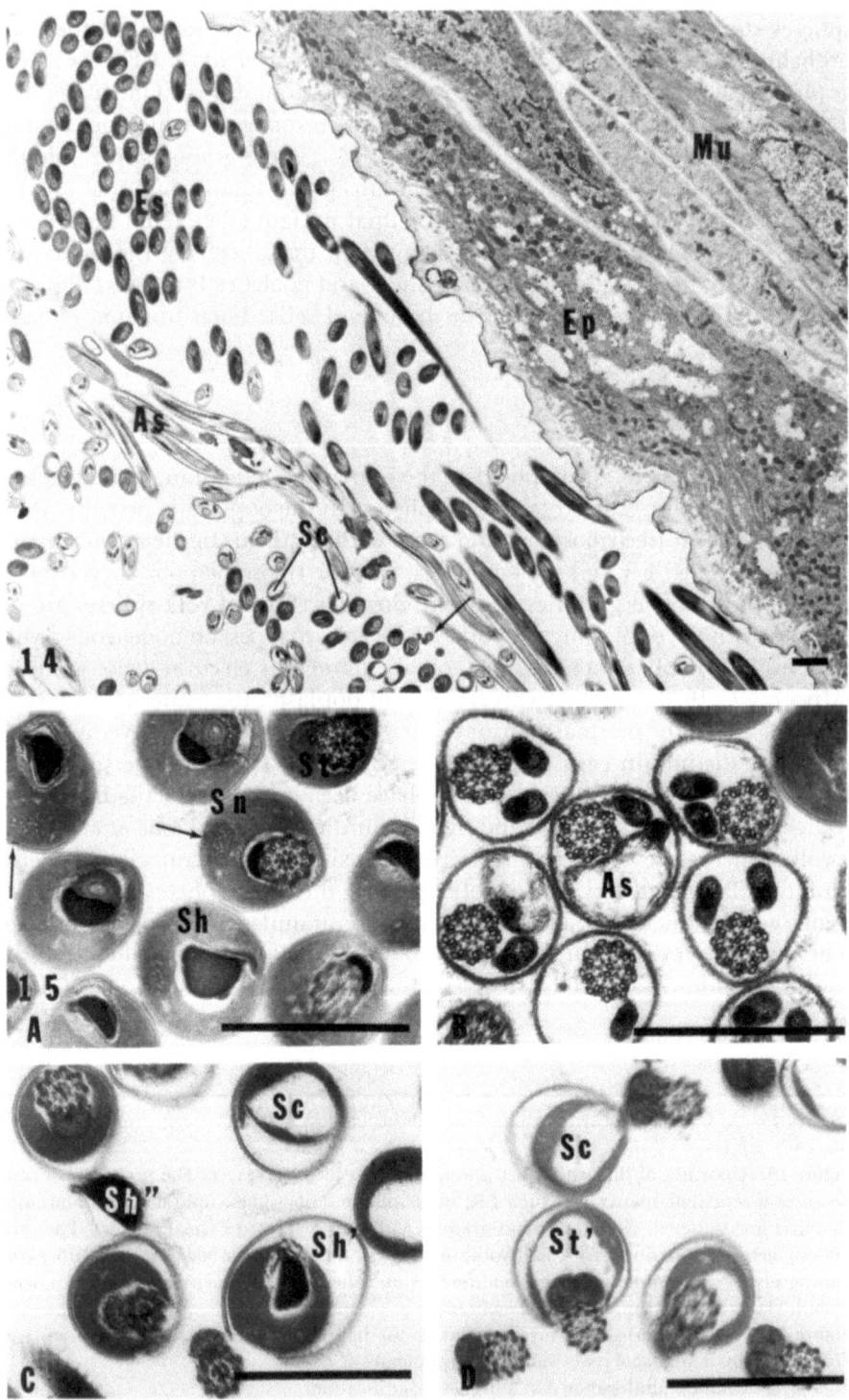

3. Maturation and Fertilization

3.1. Structure of Sperm in the Spermatheca

As the *Bombyx* egg descends through the oviduct, the chromosomes are arrested in metaphase of the first meiosis. Fertilization accomplishes the activation of the egg and transportation of the male genetic information to the egg. Sperm must first be activated so they will swim toward and enter the egg, and the egg must accept them before development can be initiated.

Ômura (1936, 1938a,b) studied in detail the behavior of sperm in male and female reproductive organs. The sperm migrating into the vas efferens from the testicular follicle pass via the vas deferens and seminal vesicles into the ejaculatory duct. After being ejaculated into the bursa copulatrix of the female, the sperm receive a secretion from the lower part of ejaculatory duct, which causes them to begin moving into the spermatheca.

Normal lepidopteran males produce two kinds of sperm: nucleate (eupyrene) and anucleate (apyrene). Katsuno (1977a,b,c) observed the behavior of eupyrene and apyrene sperm from their emergence from the testicular follicles to their ejaculation into the spermatheca during copulation. According to his results, eupyrene sperm migrate into the vas efferens in bundles, whereas apyrene sperm enter the vas efferens separately and earlier than eupyrene sperm bundles. They then enter into the bursa copulatrix, where the eupyrene sperm bundles may be dissolved by an action of apyrene sperm, and both sperm types migrate into the spermatheca (Figure 14). The spermatheca consists of two parts—a small lobe and a large lobe—and a long tubular spermathecal gland is attached to the large lobe. Eupyrene sperm mainly migrate from the bursa copulatrix into the large lobe, while apyrene sperm mainly migrate into the small lobe through the large lobe, and a large number of them degenerate there at the beginning of oviposition (Katsuno, 1977c). Thus, apyrene sperm play an important role in facilitating the passage of the eupyrene sperm bundles through the basement membrane of the testicular follicle and the separation of bundles in the bursa copulatrix, and only eupyrene sperm fertilize the egg.

Sperm length in *Bombyx* has been measured by several authors. Eupyrene sperm bundles are about 700 μm long (Ômura, 1942; Katsuno, 1978), and lengths of about 620 μm and 275 μm were obtained for individual eupyrene and apyrene sperm, respectively, in the spermatheca (Katsuno, 1978). The head

←

Figures 14 and 15. EMs of the spermatheca of an ovipositing moth and the sperm in its cavity. (Bars = 1 μm.)

Figure 14. In the spermatheca there are eupyrene sperm (Es), apyrene sperm (As), and sperm shedding their coats (arrow). Each kind of sperm is distributed separately in the spermatheca. Ep, epithelial cell; Mu, muscle layer; Sc, shed sperm coat.

Figure 15. (A–D) Higher-power EMs of cross-sectioned sperm. As apyrene sperm tail; Sc, shed coat; Sh, head; Sh′, shedding head; Sh″, shed head; Sn, midpiece; St, tail; St′, shedding sperm tail of eupyrene sperm. Arrows show an electron-dense junction in the cell membrane of eupyrene sperm.

of each eupyrene sperm contains the nucleus. The head is about 0.7 μm in diameter and somewhat convex on one side, while the tail is 0.61–0.64 μm in diameter. The eupyrene sperm is covered with a cell membrane about 28 nm thick, and the space between the cell membrane and the sperm nucleus or the axial filament–mitochondrial derivative complex is occupied by a thick envelope containing fine, electron-dense particles (Figure 15A). In cross-section, the sperm nucleus shows a semispherical or triangular contour. At the anterior end of the sperm, there is a conical acrosome with a distinct membrane. On the other hand, apyrene sperm lack such an inner dense thick membrane, and an axial filament separates from a pair of mitochondrial elements (Figure 15B).

After migration of the sperm into the spermatheca, some eupyrene sperm shed their coats, which consist of the cell membrane and an inner envelope of electron-dense material, in the lower part of the large lobe and the upper part of the spermathecal duct. Figures 15C and D illustrate cross-sections of sperm during this shedding process. A very interesting structure of the eupyrene sperm is an electron-dense junction in the cell membrane (Figure 15A, arrows). Here the membrane splits open longitudinally, allowing the sperm to escape. The cell membrane then closes and appears in electron micrographs as a shed coat (Figures 15C, D). The cause and the significance of this shedding phenomenon remain unsolved, but a few shed sperm occur in the vestibulum near the end of the oviduct, and the sperm penetrating the egg has no cell membrane. These observations suggest that shedding of the cell membrane may be a prerequisite for the entry of sperm into the egg. Degrugillier and Leopold (1976) concluded for the housefly that the loss of the plasma membrane prior to the entry of sperm into the fertilization chamber was one of the structural adaptations to facilitate sperm penetration and to ensure the fertilization during periods of rapid oviposition.

3.2. Changes of the Vitelline Membrane and Periplasm Accompanying the Entry of Sperm

In insect eggs, several changes take place as the sperm enters the ooplasm: an active streaming motion by the egg, an accumulation of cytoplasm by the sperm, a distinctive basophilic trail following the migrating sperm, and an increase in the activity of those enzymes involved in nucleic acid synthesis (Counce, 1973). Rempel and Church (1965) reported that the vitelline membrane of *Lytta viridana* became porous less than 5 min after laying. Then the pores became progressively smaller during first $\frac{1}{2}$ hr and finally closed after 1 to $1\frac{1}{2}$ hr, forming a membrane of uniform thickness. Presumably, these changes in the vitelline membrane were associated with sperm entry. Subsequently, Gerrity *et al.* (1967) studied the ultrastructure of the vitelline membrane of the same beetle. It was found to consist of a complex, three-dimensional membrane system that showed a solid structure after penetration of sperm because of a condensation of the membranes which occurred first at the anterior end and proceeded posteriorly.

In the housefly egg, entry of a sperm causes an immediate response during which vacuolated areas form in the periplasm directly beneath the micropyle (Degrugillier and Leopold, 1976). Within these vacuolated areas, the oolemma unfolds and separates from the vitelline membrane. The nucleus and tail of the sperm were found in this vacuolated area.

In the *Bombyx* egg, several eupyrene sperm enter the egg, passing through the micropylar canals and the tubular channels penetrating the inner layer of the vitelline membrane (Figures 16, 17). With the entry of sperm, various changes are induced in the vitelline membrane and the periplasm of the micropylar region. At first, numerous large vesicles and aggregates of smaller vesicles included in the periplasm are discharged into the inner layer of the vitelline membrane, and the parallel stacks of ER change to vesicular and moniliform configurations (Figures 18, 19). The electron density of the outer layer and of the irregular-shaped granules in the inner layer of the vitelline membrane decreases, and fibrous material is formed in the space between the chorion and the oolemma. Then a fertilization membrane is produced, and the entomicropyles degenerate and the tubular channels which allow passage of sperm through the inner layer of the vitelline membrane disappear (Figures 20, 21). The contour of periplasm changes remarkably to produce many blunt projections. The changes in the vitelline membrane and the stacks of cisternae of RER begin in the micropylar region and proceed backward until reaching the posterior pole.

3.3. The Sperm and Egg Nuclei

When a sperm enters the egg, the nucleus immediately separates from the tail and migrates inwards enclosed by an envelope which originates from the periplasm of the micropylar region (Figures 22, 23). This envelope is star-shaped, and it contains vesicular and irregularly shaped cisternae of RER. Other organelles are absent, except for a few mitochondria scattered at the periphery. Next the nucleus swells to form a spherical male pronucleus.

In the newly laid egg, the chromosomes are enclosed in an anterior island of polar plasm. The chromosomes, which were arrested in metaphase, resume the first meiotic division. The first polar nucleus is produced, and the second maturation division follows, generating the female pronucleus about 100 min after the egg is laid (Tazima and Ônuma, 1967; Ohtsuki and Murakami, 1968). The whorls of RER in the anterior polar plasm change into vesicular and irregularly shaped configurations, similar to those in the periplasm of the micropylar region (Figure 24). Finally, the female pronucleus migrates toward the male pronucleus, while the three polar nuclei remain in the anterior polar plasm (Figure 25).

About 2 hr after the egg is laid, the male and female pronuclei unite at a definite position (12% of egg length from the anterior pole; Takami, 1969). However, fusion of male and female pronuclei does not take place in the *Bombyx* egg (Kawamura, 1978). As in the case of *Drosophila* (Sonnenblick,

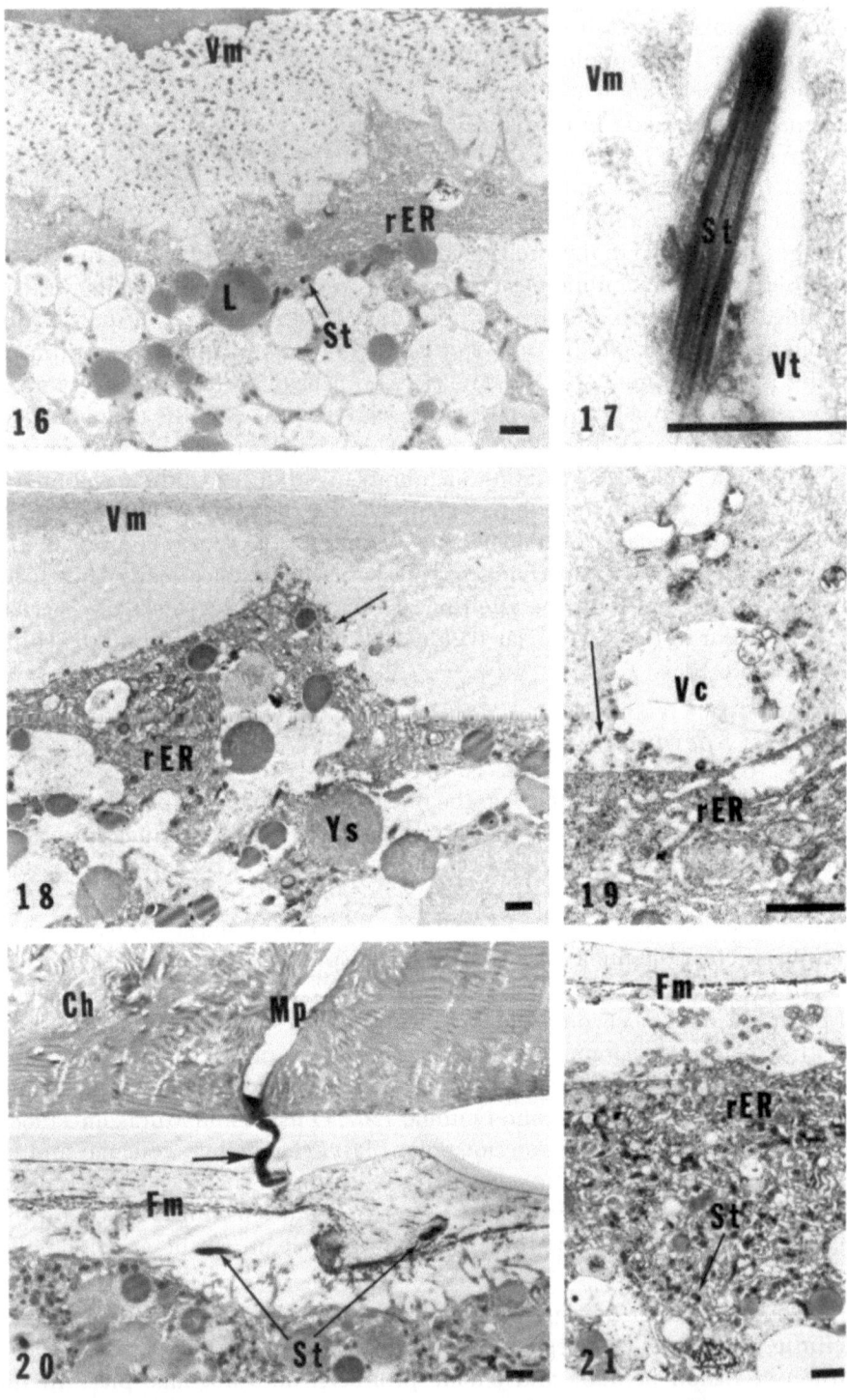

1950), gonomery occurs. During this process, a spindle is formed in each of the pronuclei, and at metaphase and anaphase the paternal and maternal chromosomes lie in a group on a separate spindle. Karyogamy occurs at telophase, and the maternal and paternal chromosomes mix for the first time at the subsequent interphase.

4. Cleavage and Blastoderm Formation

4.1. The Cleavage Stage

Synchronous mitotic divisions begin about 3 hr after egg laying to produce many cleavage nuclei, and these subsequently migrate toward the periphery. These nuclei are surrounded with ooplasm originating from the endoplasm. At interphase, the cleavage nuclei are enclosed with a typical double membrane containing pores, and the chromatin appears in sections as small electron-dense masses in the nucleoplasm. However, nucleoli have not yet formed (Figure 26). Each nucleus is surrounded by a layer of ooplasm. This contains vesicles and layers of ER and a few peripheral mitochondria. Dividing cleavage nuclei are surrounded by layers of ER (Figure 27). Subsequently, halos of microtubules appear around the cleavage nuclei, and Takesue and Keino (1980) suggest that these microtubules are responsible for the movement of cleavage nuclei toward the egg surface. Wolf (1969) reports that migrating cleavage nuclei in the fly, *Wachtliella persicariae*, are surrounded by a complex multilayered membrane, but such complex nuclear membranes are not found in the cleavage nuclei of *Bombyx* eggs.

During the cleavage stages, the ooplasm changes in contour, i.e., the microvilli seen projecting from the oolemma just after the egg is laid disappear gradually and are replaced by knoblike projections (Keino and Takesue, 1978).

◀——————————————————————

Figures 16–21. EMs of the micropylar region of the fertilized, laid egg showing structural changes with the lapse of time—Figures 16, 17: immediately after egg laying; Figures 18, 19: 5 min after laying; Figures 20, 21: 30 min after laying. (Bars = 1 μm.)

Figure 16. The vitelline membrane (Vm) and the periplasm do not yet show any changes. A cross-sectioned sperm tail (St) lies beneath a stack of cisternae of ER. L, lipid droplet.

Figure 17. A semilongitudinal section of a sperm tail passing through the tubular channel (Vt) and penetrating the inner layer of the vitelline membrane.

Figure 18. The parallel arrangement of cisternae of ER is disturbed and diverse organelles (arrow) are discharged into the vitelline membrane, resulting in the decrease of electron density.

Figure 19. Higher-power EM showing a large vacuole (Vc) and many minute vesicles (arrow) being discharged into the vitelline membrane from the periplasm.

Figure 20. The electron-dense material in the vitelline membrane disappears and a fertilization membrane (Fm) is formed. Fragments of sperm tail (St) are observed both within and outside of this membrane. The lumen of the micropyle (Mp) is empty and the entomicropyle (large arrow) degenerates. Ch, chorion.

Figure 21. A portion of the periplasm showing the irregular arrangement of cisternae of ER and a sectioned sperm tail.

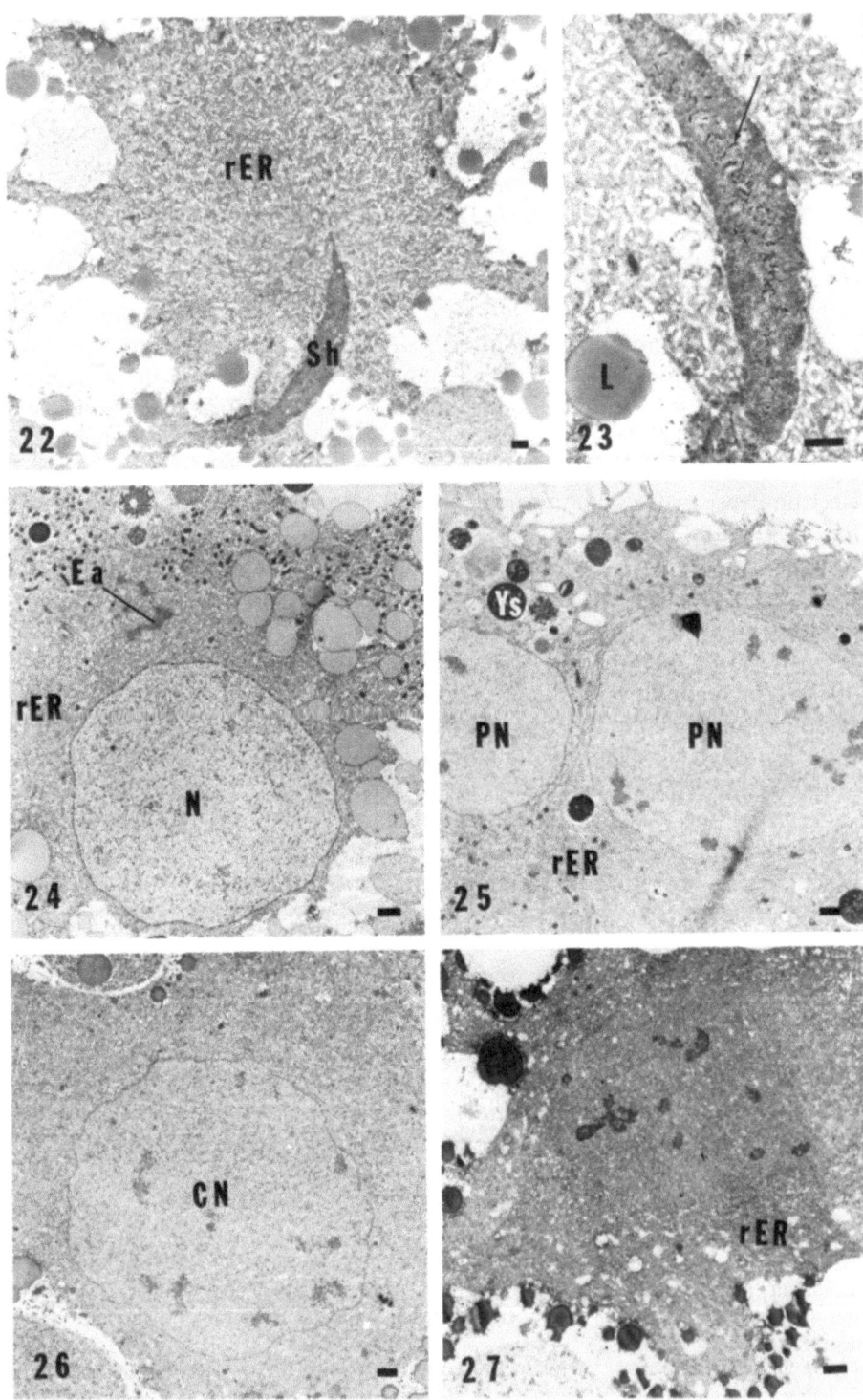

4.2. The Formation of Blastoderm Cells

The cleavage nuclei enter the periplasm after 10 synchronous mitoses according to Ohtsuki (1979). However, the penetration into the periplasm of all the cleavage nuclei does not occur at the same time, but begins first at the anterior half of the egg and then proceeds toward the posterior. Before the penetration of the cleavage nuclei, larger microprojections of the periplasm develop than were present at the cleavage stage, but the periplasm does not show any characteristic change in the distribution of organelles. Then the periplasm intermingles with the coat of cytoplasm that accompanies the cleavage nuclei (Figure 29).

In most insects, when the syncytial blastoderm develops to the cellular blastoderm, the oolemma invaginates between the nuclei entering the periplasm to make partitions. Each infolding then branches at both sides, parallel to the egg surface. The T-shaped infolds form the inner plasmalemma of the blastoderm cells, and therefore the blastoderm and the inner yolk system are separated (e.g., *W. persicariae*, Wolf, 1969; *Melosoma saliceti*, Biliński, 1978; *Drosophila*, Fullilove and Jacobson, 1980). However, Takesue et al. (1977, 1980) report that the formation of cell membranes by blastoderm cells is carried out by *Bombyx* in a manner different from other insects. The cleavage nuclei that penetrate into the periplasm project like knobs, and then they are pinched off from the yolk system to produce the blastoderm cells.

The blastoderm cells of *Bombyx* contain various organelles and inclusions previously described in the periplasm and the coats of cytoplasm that enveloped the nuclei. There are numerous ribosomes, abundant slender mitochondria with dense matrices, a small amount of ER, a few microtubules and Golgi bodies, small vesicles, lipid droplets, small yolk spheres, and glycogen granules. There are also some inclusions from the subcortical layer (c-yolk spheres and large vacuoles) (Figures 28, 31). There are two types of blastoderm cells. One, found in the presumptive germband, consists of small, cuboidal cells attached compactly to one another. The other, found in the extraembryonic region,

---

Figures 22–27. (Bars = 1 μm.)

Figure 22. EM of the cytaster enclosing a sperm nucleus (Sh) in the fertilized egg 1 hr after laying. Irregular cisternae of ER are the predominant organelles, and mitochondria are restricted to the periphery. The sperm nucleus is swelling as it transforms into the male pronucleus.

Figure 23. An enlarged micrograph of the sperm nucleus, showing the unfolding of the chromatin (arrow). L, lipid droplet.

Figure 24. The female pronucleus in the fertilized egg 2 hr after laying. The cytoplasmic island containing the nucleus (N) is filled with numerous vesicular and irregularly shaped cisternae of RER. Dense organelles similar to those displayed in Figure 7 (Ea) exist.

Figure 25. Two polar nuclei (PN) in the fertilized egg 4 hr after laying. The nuclei are surrounded by multilamellar cisternae of ER. Ys, small yolk sphere.

Figures 26 and 27. Cleavage nuclei from a late cleavage stage embryo.

Figure 26. An interphase nucleus (CN) migrating to the egg periphery.

Figure 27. A nucleus in mitosis remaining in the yolk. Lamellar and moniliform cisternae of ER develop well, but mitochondria are rare.

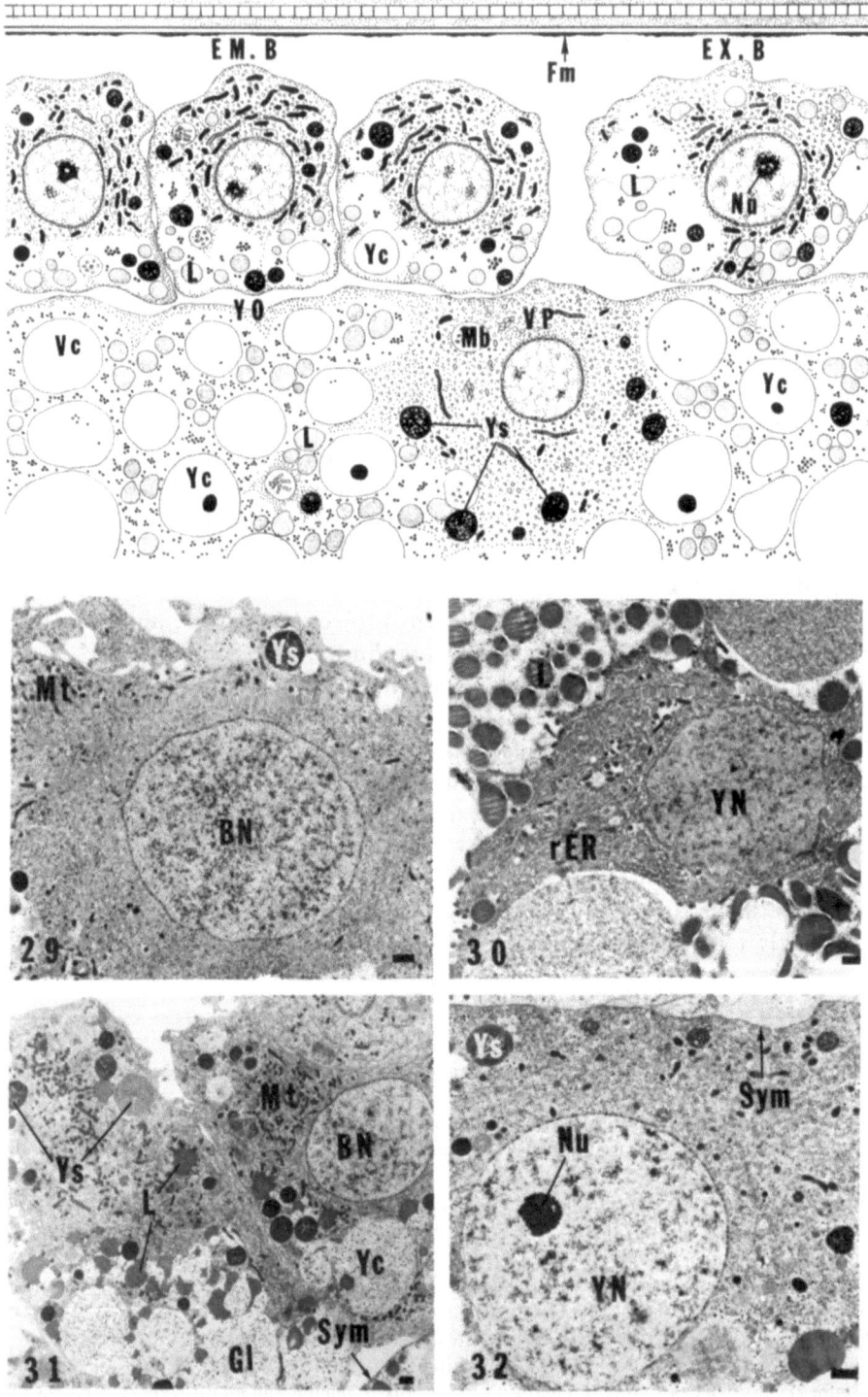

consists of larger and flattened, amoeboid cells with cytoplasmic processes and a multitude of inclusions. Nucleoli appear just after the completion of cell membranes by the blastoderm cells, and each blastoderm cell produces one or two large compact nucleoli.

4.3. The Vitellophages

At the early cleavage stage, all cleavage nuclei and their cytoplasmic envelopes have the ultrastructure described in Section 4.1. With migration of the cleavage nuclei toward the egg surface, some conspicuous changes in structure take place. The nuclei remaining in the yolk have a cytoplasmic coat containing many ribosomes, short rod-shaped mitochondria, and stacks of ER (Figure 30). In the migrating nuclei, the cytoplasmic envelopes resemble those of the nuclei penetrating the periplasm (Figure 32). These characteristic features remain until the early germband stage.

As the differentiation of vitellophages proceeds, a nucleolus appears in the nucleus, the conspicuous cisternae of ER become scattered, and the mitochondria increase in number and elongate. Berg and Gassner (1978) divided the vitellophages of *Pectinophora gossypiella* into three types on the basis of their ultrastructure, and concluded that these changes represented stages in the maturation of the vitellophages. Changes of ER in *Bombyx* vitellophages indicate a similar sequence of developmental stages.

At the cellular blastoderm stage, when the secondary yolk membrane is formed between the blastoderm cells and the yolk system, some cleavage nuclei migrating toward the egg periphery are prevented from entering the periplasm, and they remain attached to the secondary yolk membrane (Figure 32). These cells also become vitellophages.

⬅───────────────────────────────

Figure 28. A drawing of a cortical portion of the egg at the late blastoderm stage. In the extraembryonic region, large cells exist freely in the space between the fertilization membrane (Fm) and the yolk system (YO) to form the extraembryonic blastoderm (EX. B), while in the embryonic region smaller cells attach to one another to make a cell layer, the embryonic blastoderm (EM. B). Other nuclei exist in the cytoplasmic islands both in the periphery and in the interior of the yolk system. These become vitellophages (VP). Mb, multivesicular body; L, lipid droplet; Nu, nucleolus; Vc, vacuole; Yc, c-yolk sphere; Ys, small yolk sphere.
Figures 29 and 30. EMs of nuclei at an early blastoderm stage.
Figure 29. A nucleus (BN) penetrating into the periplasm. The plasmalemma is not completed yet. Mitochondria (Mt) are distributed chiefly in the periphery, and most of them seem to originate from the periplasm. Ys, small yolk sphere.
Figure 30. A nucleus (YN) remaining in the interior of yolk system. The ooplasm enclosing the nucleus contains a well-developed, lamellar ER.
Figures 31 and 32. EMs of cells and nuclei (BN, YN) at a late blastoderm stage.
Figure 31. Embryonic blastoderm cells. With the formation of cell membrane, the cells have incorporated various inclusions, such as small yolk spheres (Ys), lipid droplets (L), glycogen granules (Gl), and c-yolk spheres (Yc). Mt, mitochondria; Sym, secondary yolk membrane.
Figure 32. A nucleus of a vitellophage lying in the periphery of the yolk system. A nucleolus (Nu) has formed.

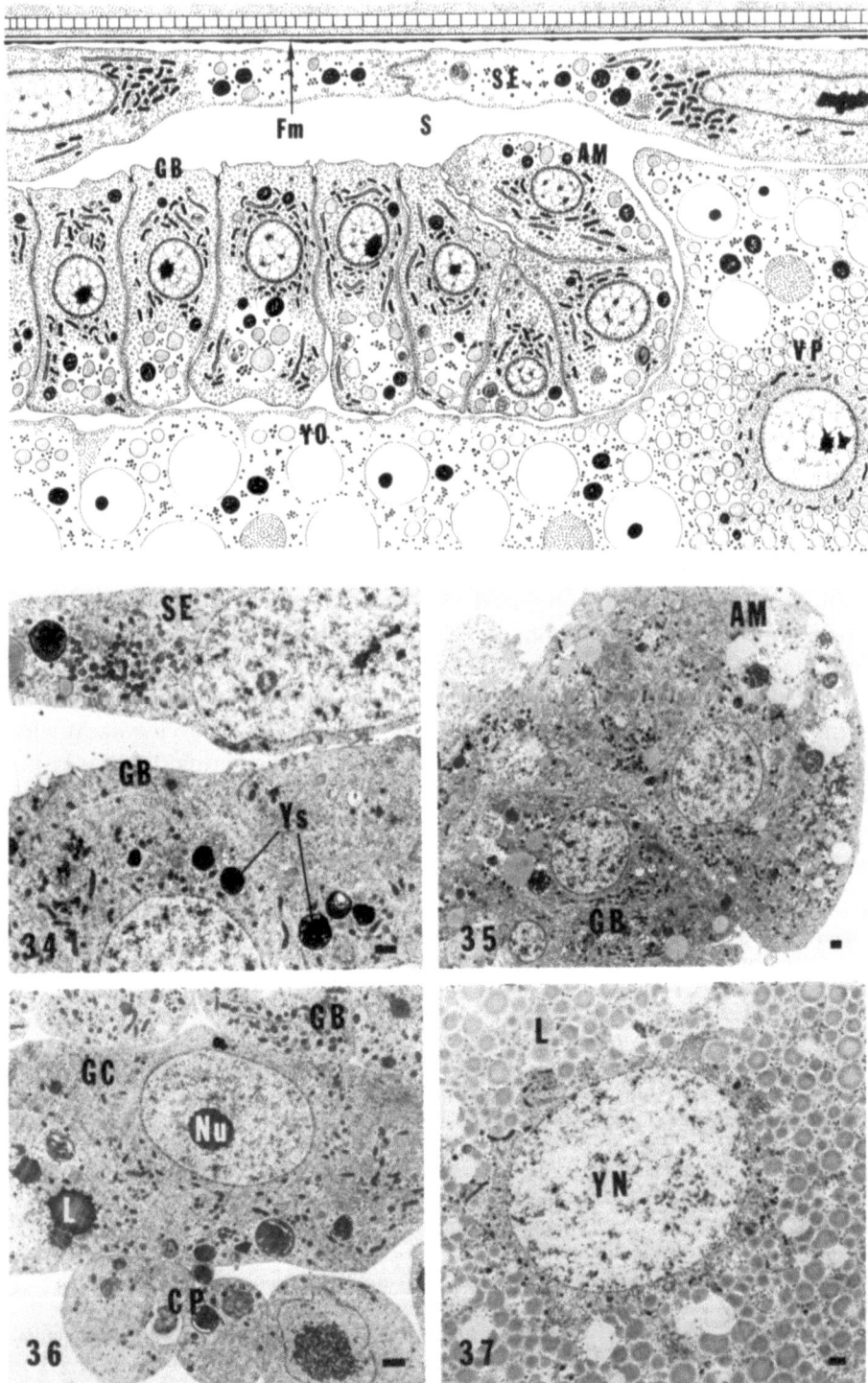

5. Formation of the Germband and the Embryonic Envelopes

5.1. The Germband

The blastoderm cells in the embryonic region divide asynchronously several times to form a germband, and the germband cells change gradually from a cuboidal to a cylindrical shape. The germband cells are surrounded by a relatively smooth plasmalemma except for a few cytoplasmic microprojections on the outside, and they are connected to adjacent cells by desmosomes (Figures 33, 34).

Certain organelles and inclusions in the cytoplasm of germband cells show a polarized distribution. Mitochondria are clustered around the nucleus, and microtubules run parallel to the long axis along the lateral sides of cells. The area adjacent to the yolk system contains lipid droplets and a wide lacuna packed with glycogen deposits (Figures 33–35).

A group of cells attaches to the inside of a specific region of the germband at this stage. The cytoplasm of these cells appears denser than that of germband cells, because the ribosomes are more concentrated. From their position, these cells are judged to be primordial germ cells (Figure 36).

During the late germband stage, the secondary yolk membrane forms infoldings, and the yolk system is divided into many masses, each enclosing one or several nuclei and yolk organelles. This process is called *yolk segmentation*, and prior to the yolk segmentation numerous lipid droplets assemble around the nucleus of vitellophages (Figure 37).

5.2. The Serosa and the Amnion

In most insects, two embryonic envelopes, an outer serosa and an inner amnion, are formed during the course of early embryonic development. The serosa is a derivative of the extraembryonic blastoderm and the amnion is derived from the margin of the germband. The most typical steps in the development of these envelopes are the appearance of an amniotic fold in the periphery of the germband, and the extension and the closure of the lips of the midventral

←

Figure 33. A drawing of a cortical portion of the egg at the late germband stage. The cells of the germband (GB) change from a cuboidal to a cylindrical shape. This causes the germband to contract toward the ventral side of the egg and at the same time to sink slightly into the yolk system (YO). As a result, a space (S) forms between the fertilization membrane (Fm) and the germband. Extraembryonic blastoderm cells migrate into this space and produce a flat layer of cells, the serosa (SE). Next the peripheral cells of the germband elongate and migrate over the surface of the germband to make another embryonic membrane, the amnion (AM). Numerous lipid droplets aggregate around the nucleus of a vitellophage (VP) prior to the yolk segmentation.

Figures 34–37. EMs of cortical portions of the egg at the late germband stage. Figure 34 illustrates cells of the germband (GB) and serosa (SE), Figure 35 germband and amnion (AM) cells, Figure 36 germband cells and a primoridal germ cell (GC), and Figure 37 a vitellophage. YN, yolk nucleus; Nu, nucleolus; L, lipid droplet; Ys, small yolk sphere; CP, cytoplasmic processes from a primordial germ cell.

side of the germband. In the *Bombyx* egg, these typical steps have been reported predominantly (Takami, 1969; Ohtsuki, 1979), but the serosa may occur prior to the amnion. In the space formed by the slight sinking of the germband into the yolk system, many extraembryonic blastoderm cells migrate and then become flattened to produce a continuous cellular membrane, the serosa (Figures 33, 34). Various inclusions, especially c-yolk spheres, contained in the presumptive serosa cells at the late blastoderm stage, disappear gradually with the flattening of the cells. About 36 hr after egg laying, a feltlike serosal cuticle begins to be formed on the surface of the serosa, and at the same time numerous pigment granules appear in the cytoplasm of the serosa in the diapause egg (Miya *et al.*, 1972).

After the completion of the serosa, the amnion is formed as another embryonic envelope. In the *Bombyx* egg, this envelope does not form as an amniotic fold, but originates from the peripheral cells of the germband. These cells are somewhat larger than the other germband cells and contain more cytoplasmic vesicles. They evaginate and then become flattened as they cover the ventral surface of the germband (Figures 33, 35).

6. Summary

In the domesticated silkmoth, *B. mori*, there are several structural adaptations of both egg and sperm for rapid egg laying, as in the housefly. Tubular channels are provided in the inner layer of the vitelline membrane to allow entry of sperm into the egg, while each sperm sheds its cell membrane before migration to the vestibulum. The periplasm in the micropylar region has various characteristic organelles, which may be responsible for the changes in the vitelline membrane and periplasm that accompany the entry of a sperm. Fusion of the male and female pronuclei does not take place. Rather gonomery occurs, as in the case of *Drosophila*.

During cleavage, halos of microtubules appear around the cleavage nuclei and may be responsible for their movement toward the egg surface. The penetration of all the cleavage nuclei into the periplasm does not occur at the same time, but begins first at the anterior half of the egg, and the cellular blastoderm forms shortly thereafter. The formation of cell membranes by blastoderm cells is carried out by knoblike projections of the cleavage nuclei that penetrate into the periplasm. The formation of the serosa and amnion does not occur at the same time. The serosa is formed at first by migration of the extraembryonic blastoderm cells into the space between the fertilization membrane and the germband, and the amnion appears later as an evagination of the flattened peripheral cells of the germband.

ACKNOWLEDGMENTS

My special thanks are due to the editors, Drs. Robert C. King and Hiromu Akai, who encouraged me and gave much valuable advice during preparation of the manuscript. I am grateful to Mr. Ichiro Tanimura, Laboratory of Electron Microscopy, Iwate University, for preparing the samples and for assistance with electron microscopic observations. Thanks are also due to Dr. Yoshiki Ohtsuki, Sericultural Experiment Station, who provided Figure 1. My research described herein was supported by grants from the Ministry of Education (Nos. 148078 and 444016).

References

Akai, H., 1957, Studies on the electron microscopic structure of egg shell in *Bombyx mori* L., *J. Sericult. Sci. Jpn.* **26**:335-340 (Japanese).

Akai, H., 1958, Observations of the micropylar region of the chorion in the silkworm, *Bombyx mori* L., *Acta Sericol.* **26**:19-21 (Japanese).

Akutsu, S., and Yoshitake, N., 1974, Electron microscopic observations on th grey eggshell of the silkworm, *Bombyx mori* L., *J. Sericult. Sci. Jpn.* **43**:461-466 (Japanese).

Akutsu, S., and Yoshitake, N., 1977, Electron microscopic observation on the vitelline membrane formation in the silkworm, *Bombyx mori* L., *J. Sericult. Sci. Jpn.* **46**:509-514 (Japanese).

Berg, G. J., and Gassner, G., 1978, Fine structure of the blastoderm embryo of the pink bollworm, *Pectinophora gossypiella* (Saunders) (Lepidoptera: Gelechiidae), *Int. J. Insect Morphol. Embryol.* **7**:81-105.

Biliński, S., 1978, Electron microscope studies on *Melosoma* embryo during the formation of the cellular blastoderm, *Folia Biol.* (*Krakow*) **26**:189-194.

Counce, S. J., 1973, The causal analysis of insect embryogenesis. In *Developmental Systems: Insects*, vol. 2, edited by S. J. Counce and C. H. Waddington, pp. 1-156, Academic Press, New York.

Degrugillier, M. E., and Leopold, R. A., 1976, Ultrastructure of sperm penetration of housefly egg, *J. Ultrastruct. Res.* **56**:312-325.

Fullilove, S. L., and Jacobson, A. G., 1980, Embryonic development—Descriptive. In *The Genetics and Biology of Drosophila*, vol. 2c, edited by M. Ashburner and T. R. F. Wright, pp. 105-227, Academic Press, New York.

Gerrity, R. G., Rempel, J. G., Sweeny, P. R., and Church, N. S., 1967, The embryology of *Lytta viridana* Le Conte (Coleoptera: Meloidae). II. The structure of the vitelline membrane, *Can. J. Zool.* **45**:497-503.

Goldsmith, M. R., Paule, M., Weare, B., and Clermont-Rattner, E., 1978, Development of the chorion in *Bombyx mori*, *J. Cell Biol.* **79**(2, part 2):32A.

Johannsen, O. A., and Butt, F. H., 1941, *Embryology of Insects and Myriapods*, McGraw-Hill, New York.

Kanda, T., Matsumura, H., and Ohtsuki, Y., 1974, Surface structure of silkworm (*Bombyx mori*) egg. I. Fine structure of micropylar apparatus, *J. Sericult. Sci. Jpn.* **43**:379-383 (Japanese with an English summary).

Katsuno, S., 1977a, Studies on eupyrene and apyrene spermatozoa in the silkworm, *Bombyx mori* L. (Lepidoptera: Bombycidae). II. The intratesticular behavior of the spermatozoa after emergence, *Appl. Entomol. Zool.* **12**:236-240.

Katsuno, S., 1977b, Studies on eupyrene and apyrene spermatozoa in the silkworm, *Bombyx mori* L. (Lepidoptera: Bombycidae). III. The post-testicular behavior of the spermatozoa at various stages from the pupa to adult, *Appl. Entomol. Zool.* **12**:241-247.

Katsuno, S., 1977c, Studies on eupyrene and apyrene spermatozoa in the silkworm, *Bombyx mori* L. (Lepidoptera: Bombycidae). IV. The behavior of the spermatozoa in the internal reproductive organs of female adults, *Appl. Entomol. Zool.* **12**:352-359.

Katsuno, S., 1978, Studies on eupyrene and apyrene spermatozoa in the silkworm, *Bombyx mori* L. (Lepidoptera: Bombycidae). VIII. The length of spermatozoa, *Appl. Entomol. Zool.* **13**:127-129.

Kawamura, N., 1978, The early embryonic mitosis in normal and cooled eggs of the silkworm, *Bombyx mori*, *J. Morphol.* **158**:57-72.

Keino, H., and Takesue, S., 1978, Scanning electron microscopic observations on the early embryogenesis in the silkworm, *Bombyx mori*, *Zool. Mag.* **87**:342 (Japanese).

Kuwana, J., and Takami, T., 1957, Insecta. In *Embryology of Invertebrata*, edited by M. Kume and K. Dan, pp. 287-343, Baihukan, Tokyo (Japanese).

Matsuzaki, M., 1968, Electron microscopic observations on the chorion formation of the silkworm, *Bombyx mori* L., *J. Sericult. Sci. Jpn.* **37**:483-490 (Japanese with an English summary).

Mazur, G. D., Regier, J. C., and Kafatos, F. C., 1982, Order and defects in the silkmoth chorion, a biological analogue of a cholesteric liquid crystal. In *Insect Ultrastructure*, vol. 1, edited by R. C. King and H. Akai, pp. 150-185, Plenum Press, New York.

Miya, K., 1959, Electron microscopic studies on the embryonic development in the silkworm, *Bombyx mori*. I. Observations on the hibernating egg, *J. Sericult. Sic. Jpn.* **28**:163-164 (Japanese).

Miya, K., 1960, Electron microscopic studies on the embryonic development in the silkworm, *Bombyx mori*. II. Fine structure of the serosa, *J. Sericult. Sci. Jpn.* **29**:273 (Japanese).

Miya, K., 1975, Ultrastructural changes of embryonic cells during organogenesis in the silkworm, *Bombyx mori*. I. The gonad, *J. Fac. Agric. Iwate Univ.* **12**:329-363.

Miya, K., 1976, Ultrastructural changes of embryonic cells during organogenesis in the silkworm, *Bombyx mori*. II. The alimentary canal and the Malpighian tubules, *J. Fac. Agric. Iwate Univ.* **13**:95-122.

Miya, K., 1978, Electron microscope studies on the early embryonic development of the silkworm, *Bombyx mori*. I. Architecture of the newly laid egg and the changes by sperm entry, *J. Fac. Agric. Iwate Univ.* **14**:11-35.

Miya, K., Kurihara, M., and Tanimura, I., 1972, Changes of fine structures of the serosa cell and the yolk cell during diapause and post-diapause development in the silkworm, *Bombyx mori* L., *J. Fac. Agric. Iwate Univ.* **11**:51-87.

Ohtsuki, Y., 1965, Studies on the yolk formation in the silkworm, *Bombyx mori* L.: An analysis of its mechanism revealed by the comparison between normal and abnormally small egg, *Bull. Fac. Text. Fibers Kyoto Ind. Arts Text. Fibers* **4**:314-344 (Japanese with an English summary).

Ohtsuki, Y., 1979, Silkmoth eggs. In *A General Textbook of Sericulture*, edited by the Sericultural Society of Japan, pp. 156-173, Nihon Sanshi Shinbun-sha, Tokyo (Japanese).

Ohtsuki, Y., and Murakami, A., 1968, Nuclear division in the early embryonic development of the silkworm, *Bombyx mori* L., *Zool. Mag.* **77**:383-387 (Japanese with an English summary).

Ohtsuki, Y., Kanda, T., and Matsumura, H., 1977, Surface structure of silkworm (*Bombyx mori*) egg. II. Characteristics of different areas of egg surface, *J. Sericult. Sci. Jpn.* **46**:45-50 (Japanese with an English summary).

Okada, M., 1970, Electron microscope studies on diapause embryos of the silkworm, *Bombyx mori* L., *Sci. Rep. Tokyo Kyoiku Daigaku Sect. B* **14**:95-111.

Ômura, S., 1936, Studies of the reproductive system of the male of *Bombyx mori*. I. Structure of testis and the intratesticular behavior of the spermatozoa, *J. Fac. Agric. Hokkaido Imp. Univ.* **38**:151-181.

Ômura, S., 1938a, Structure and function of the female genital system of *Bombyx mori* with special reference of the mechanism of fertilization, *J. Fac. Agric. Hokkaido Imp. Univ.* **40**:111-128.

Ômura, S., 1938b, Studies on the reproductive system of the male of *Bombyx mori*. II. Post-testicular organs and post-testicular behavior of the spermatozoa, *J. Fac. Agric. Hokkaido Imp. Univ.* **40**:129-170.

Ômura, S., 1942, The length of spermatozoa in the silkworm, *Bombyx mori* L., *J. Sericult. Sci. Jpn.* **13**:30-38 (Japanese).

Rasmussen, S. W., 1977, The transformation of the synaptonemal complex into the "elimination chromatin" in *Bombyx mori* oocytes, *Chromosoma* **60**:205-221.

Rempel, J. G., and Church, N. S., 1965, The embryology of *Lytta viridana* Le Conte (Coleoptera: Meloidae). I. Maturation, fertilization, and cleavage, *Can. J. Zool.* **43**:915-925.

Sakaguchi, B., Chikushi, H., and Doira, H., 1973, Observations of the egg shell structure controlled by gene action in *Bombyx mori*, *J. Fac. Agric. Kyushu Univ.* **18**:53-62.

Salkeld, E. H., 1973, The chorionic architecture and shell structure of *Amathes c-nigrum* (Lepidoptera: Noctuidae), *Can. Entomol.* **105**:1-10.
Sonnenblick, B. P., 1950, The early embryology of *Drosophila melanogaster*. In *The Biology of Drosophila*, edited by M. Demerec, pp. 62-167, Wiley, New York.
Suzuki, K., 1969, Distribution and change of pyronin-positive granules during the early development in the silkworm, *Bombyx mori* L., B.A. thesis, Fac. Agric. Iwate Univ., pp. 1-30 (Japanese).
Takami, T., 1969, *A General Textbook of the Silkworm Egg*, Zenkoku Sanshu Kyokai, Tokyo (Japanese).
Takei, R., and Nagashima, E., 1975a, Electron microscope investigation on the early developmental stages of diapause and non-diapause eggs in the silkworm, *Bombyx mori* L., *J. Sericult. Sci. Jpn.* **44**:118-124 (Japanese).
Takei, R., and Nagashima, E., 1975b, Electron microscope investigation on the yolk granules in the silkworm egg, *J. Sericult. Sci. Jpn.* **44**:161-164 (Japanese).
Takei, R., and Yoshitake, N., 1975, Electron microscope investigation on the early developmental stages of the pigmented non-diapause egg in the silkworm, *Bombyx mori* L., *J. Sericult. Sci. Jpn.* **44**:169-175 (Japanese).
Takesue, S., and Keino, H., 1980, The mechanism of nuclear migration before blastoderm formation in the silkworm egg (*Bombyx mori* L.), *Zool. Mag.* **89**:408 (Japanese).
Takesue, S., Keino, H., and Endo, K., 1976, Studies on the yolk granules of the silkworm, *Bombyx mori* L.: The morphology of diapause and non-diapause eggs during early developmental stages, *Wilhelm Roux Arch. Dev. Biol.* **180**:93-105.
Takesue, S., Onitake, K., and Keino, H., 1977, Studies on the blastoderm and germband formation during early embryogenesis in the silkworm, *Bombyx mori* L., *Zool. Mag.* **86**:345 (Japanese).
Takesue, S., Keino, H., and Onitake, K., 1980, Blastoderm formation in the silkworm egg (*Bombyx mori* L.), *J. Embryol. Exp. Morphol.* **60**:117-124.
Tazima, Y., 1964, *The Genetics of the Silkworm*, Logos Press, London.
Tazima, Y., and Ônuma, A., 1967, Experimental induction of androgenesis, gynogenesis, and polyploidy in *Bombyx mori* by treatment with CO_2 gas, *J. Sericult. Sci. Jpn.* **36**:286-292.
Toyama, K., 1902, Contributions to the study of silkworm. I. On the embryology of the silkworm, *Bull. Coll. Agric. Tokyo Imp. Univ.* **5**:73-118.
Toyama, K., 1909, *A Textbook of the Silkworm Egg*, Maruyamasha, Tokyo (Japanese).
Waku, Y., 1974, Ultrastructure of Malpighian tubule cells in the silkworm, *Bombyx mori* L. with special regard to metamorphosis, *Zool. Mag.* **83**:152-162 (Japanese with an English summary).
Waku, Y., and Sumimoto, K., 1974, Metamorphosis of midgut epithelial cells in the silkworm (*Bombyx mori* L.) with special regard to the calcium salt deposits in the cytoplasm. II. Electron microscopy, *Tissue Cell* **6**:127-136.
Wolf, R., 1969, Kinematik und Feinstruktur plasmatischer Faktorenbereiche des Eies von *Wachtliella persicariae* L. (Diptera). I. Das Verhalten ooplasmatischer Teilsysteme im normalen Ei, *Wilhelm Roux Arch. Dev. Biol.* **162**:121-160.
Zissler, D., and Sander, K., 1982, The cytoplasmic architecture of the insect egg cell. In *Insect Ultrastructure*, vol. 1, edited by R. C. King and H. Akai, pp. 189-221, Plenum Press, New York.

3

Electron Microscopic Mapping and Ultrastructure of *Drosophila* Polytene Chromosomes

VEIKKO SORSA

1. Introduction

The number of molecular studies involving cloned segments of DNA has been growing rapidly. This development holds a promise of rapid progress in the study of individual genes of *Drosophila melanogaster*. The exact sequences of DNA of several genes evidently will be available in the near future (cf. Spradling and Rubin, 1981). Consequently, the cytogenetic localization of cloned sequences on the chromosomes is also going to meet new challenges. More exact cytological methods have to be developed to demonstrate the normal banding pattern of polytene chromosomes and to recognize minute rearrangements in it. The quality and usability of all new methods should be tested against the revised light microscopic reference maps of Bridges (cf. Lindsley and Grell, 1968). An urgent task, however, is the reliable verification and possible revision of light microscopic maps by means of high-resolution electron microscopy. EM division maps for the salivary gland chromosome *2L* of *D. melanogaster* are included in this review.

The EM maps are based on EM analyses of the banding pattern carried out using thin-sectioned salivary gland chromosomes (cf. Saura and Sorsa, 1979a,b,c,d; Saura, 1980, 1983). The first part of the present chapter contains a brief review of the mapping of the salivary gland chromosomes of *D. melanogaster* and of the development of EM methods suitable for the studies of polytene chromosomes. The latter part attempts to review some recent results and aspects

---

VEIKKO SORSA • Department of Genetics, University of Helsinki, SF-00100 Helsinki 10, Finland.

of ultrastructural studies on bands and interbands of polytene chromosomes in general, and in particular, of studies on certain genetically interesting regions. Structural and functional models combining the results of molecular studies and cytogenetics are also presented. For a more comprehensive introduction to the topic, the reader is referred to the reviews of Beermann (1972), Lefevre (1974a,b), and Spradling and Rubin (1981).

2. Light Microscopic Maps of the Salivary Gland Chromosomes of Drosophila melanogaster

Painter's first report on December 22, 1933, in *Science* already included the first map for the salivary gland X chromosome of *D. melanogaster*. The first map of the whole salivary gland genome of *D. melanogaster* was published only a year later by Painter (1934). The map showed the approximate location of about 30 genes in the polytene chromosomes. Shortly after Painter's first report, C. B. Bridges also started the mapping of the salivary gland chromosomes of *D. melanogaster*. The original camera lucida drawings of Bridges, which were first shown in the annual exhibition of the Carnegie Institution of Washington in December 1934, were published in 1935 with a key to the banding of chromosomes (Bridges, 1935).

Bridges' system for cataloguing the salivary gland chromosome bands of *D. melanogaster* divides the whole genome into 102 numbered divisions. Each division is divided further into six subdivisions designated by capital letters (A through F). The total number of bands drawn in the original maps of Bridges (1935) is 3540. That is about 35 bands per division and about 5.8 bands per subdivision. With the exception of divisions beginning from the end of a chromosome, each division was planned to begin with a prominent band, called the *main band* by Bridges (1935). The subdivisions also were constructed to start with a sharp band.

A revision of chromosome maps soon became necessary because the improvement of preparation methods of polytene chromosomes showed that there were numerous new bands. C. B. Bridges (1938) started the revision from the X chromosome map. This was followed by the revision of the map of *2R* (Bridges and Bridges, 1939). The work was continued by P. N. Bridges (1941a,b, 1942). The numbering of individual bands was added into the revised maps of X and *2R*, while in the later maps of *2L*, *3L*, and *3R*, only the total number of lines was given per subdivision. The total number of bands in the divisions 1-100 of the revised maps of chromosomes *1*, *2*, and *3* was 5012. This means more than 50 bands per division. Slizynski (1944) made a cytological revision of the Bridges' map for chromosome *4*, but the number of bands in the revised map is obviously much too high (see e.g. King, 1975; Lefevre, 1976).

As C. B. Bridges envisioned in 1938, the original maps (Bridges, 1935) continued to serve *Drosophila* cytologists as a guide map. King (1965, 1970b, 1975) published new revisions of the cytological reference maps redrawn on the basis of original maps of Bridges (1935) and including the newest results of cytogenetic localization of genes. The latest revision of King (1975) included the new guide map for chromosome *4* prepared by B. Hochman. More recently, Lefevre (1976) compiled an excellent photographic representation of all the salivary gland chromosomes of *D. melanogaster* to be used together with the original guide maps of Bridges (1935).

In exact studies of specific bands or regions, the revised maps—as suggested by C. B. Bridges (1938)—will provide the necessary reference to the normal banding pattern. Unfortunately, the later revisions of chromosome arms *3L* (Bridges, 1941a), *3R* (Bridges, 1941b), and *2L* (Bridges, 1942) are more schematically drawn and difficulties may appear when using them for the direct comparison to the salivary gland chromosomes in the preparations or in photomicrographs.

A special character of the revised maps of Bridges (see Lindsley and Grell, 1968) is the high number of double bands, which is extremely difficult, in many cases impossible to verify by means of light microscopy. In addition, there are hundreds of faint bands depicted in the revised maps of Bridges, which hardly can ever be recognized in the normal light microscopic preparations. The resolution limits of light microscopy make it also difficult to localize exactly small deletions or other rearrangements in the polytene chromosomes.

3. Electron Microscopic Methods for the Mapping of Polytene Chromosomes

The polytene chromosomes were first subjected to thin-section EM over 30 years ago (see Palay and Claude, 1949; Yasuzumi *et al.*, 1951). An early effort to use a total mount method for the EM of polytene chromosomes was made by Herskowitz (1952). Improvements of both the fixation and embedding methods of biological material for the EM made it possible to recognize and study the ultrastructure of Balbiani rings in the salivary gland chromosomes of *Chironomus* (see Beermann and Bahr, 1954). Thin-section EM of smeared and methacrylate-embedded salivary gland chromosomes of *D. melanogaster* made it possible to compare the same regions of polytene chromosomes in the light and electron microscopes (Gay, 1955), and to demonstrate interband fibrils between the bands (Swift, 1962). However, the banding pattern depicted in the reference maps of Bridges was extremely difficult to identify in the electron micrographs of smeared and thin-sectioned salivary gland chromosomes.

3.1. Thin-Sectioning of Squashed Chromosomes

Comparison of the reference maps and the electron micrographs became much easier when the method originally developed in the early 1960s for studying the longitudinal sections of synaptonemal complexes in the SMCs of ferns was applied to the EM of salivary gland chromosomes (Sorsa and Sorsa, 1967a,b). Berendes (1968) improved the method by designing a block holder for the final trimming of chromosomes in transmitted light. A good correlation of the electron micrographs and the revised reference maps of Bridges was obtained by using an immediate fixation of salivary gland chromosomes with cold acetic alcohols (see Sorsa, 1969b; Sorsa and Sorsa, 1967a), while the fixation with glutaraldehyde seemed to reduce the number of bands from that already depicted in the reference maps (Berendes, 1968, 1970). Thus, the acetic acid-methanol (AM) fixation was mainly used for the localization of the *white* gene in the X chromosome of *D. melanogaster* (Sorsa *et al.*, 1973).

An extremely clear banding pattern is also obtained after immediate fixation of salivary gland chromosomes with a FAR (4% formaldehyde in Ringer) solution (see Sorsa and Saura, 1980a,b; Saura, 1980). Both the AM and FAR fixation methods are used for the EM analyses of the banding pattern in the salivary gland chromosomes, which is going on in our laboratory for the EM mapping of the whole genome of *D. melanogaster* (see Saura and Sorsa, 1979a,b,c,d; Saura, 1980; Sorsa and Saura, 1980a,b; Sorsa *et al.*, 1983). The first part of the EM map compiled on the basis of EM analyses included the Bridges' divisions 1-5 of the X chromosome (Sorsa, 1982b). The second part of the EM map comprising the whole chromosome limb 2L of *D. melanogaster* is included in this review (see Figure 10).

3.2. Electron Microscopy of Whole-Mounted Polytene Chromosomes

The identification of salivary gland chromosome bands, which is necessary for mapping, was impossible from the early whole mounts of Rae (1966). Ris (1961) introduced a whole-mount method for the EM of squashed chromosomes, which was later modified also to the EM of polytene chromosomes (Ris, 1976). Recently, Laird (1980) and Laird *et al.* (1980, 1981) have shown that the banding pattern depicted in the original reference maps of Bridges (1935) can be recognized with high-voltage EM in whole-mounted, squashed polytene chromosomes prepared by using the method of Ris (1976). The capability of Ris' method for the demonstration of all bands depicted in the revised reference maps of Bridges has not yet been proved. A good correlation with the reference maps and whole-mounted salivary gland chromosomes can be obtained if the AM-fixed polytene chromosomes are spread for the EM instead of squashing them (Burkholder, 1976; Alanen and Sorsa, 1978; Alanen, 1981; Kalisch and Hägele, 1982). For instance, a comparison of the banding pattern in the *zeste-white* region of chromosome X shows that essentially all the bands found in the thin-section EM of the same region (see Sorsa, 1979, 1982a) can also be recognized in the whole-mounted spread chromosomes (Alanen, 1981).

4. Electron Microscopic Maps of the Salivary Gland Chromosomes of Drosophila melanogaster

The use of thin-section EM of acid-fixed and squashed salivary gland chromosomes for the EM mapping of the 3R chromosome of *D. melanogaster* was demonstrated by Sorsa (1969b). The first detailed EM map was published by

Berendes (1970) for the distal divisions of the X chromosome. The map was based on thin sections of glutaraldehyde-fixed salivary gland chromosomes. According to the EM map of Berendes (1970), the total number of bands in the region *1A-4E* of the X chromosome is only about 67% of that depicted by Bridges (1938) for the same region. A similar reduction of band number was found in the same region by Lossinsky and Lefever (1978) by using thin-section EM of X chromosomes fixed with a combination of acetic and lactic acids. Even more drastic reduction of band number was found in the X chromosome region *4F-10F* also studied by Lossinsky and Lefever (1978), and in the distal divisions of *3R* studied by Kerkis *et al.* (1977).

However, extensive analyses of thousands of electron micrographs taken from thin-sectioned salivary gland chromosomes of *D. melanogaster* have indisputably shown that practically all Bridges' bands and a certain number of new, mainly faint bands can be found in good-quality electron micrographs (see Saura and Sorsa, 1979a,b,c,d; Sorsa, 1979, 1982a; Saura, 1980; Sorsa and Saura, 1980a,b; Sorsa *et al.*, 1983). An essential difference between the EM and the revised reference maps of Bridges is the number of double bands. The problem has been considered by Sorsa and Saura (1980a) and Sorsa (1982c). It should be emphasized that many of Bridges' doublets seem to be pairs of more or less equal but in the EM clearly separate bands with a distinct interband area in between them. Usually, quite a few Bridges' doublets can never be seen as double structures in thin-section EM. It should also be kept in mind that the method used by Bridges to handle the polytene chromosomes was particularly designed for unraveling the possible double structure in bands and for demonstrating the maximum number of bands. As already shown by the total length of salivary gland chromosomes in the camera lucida drawings used for the construction of revised reference maps, the chromosomes cannot be stretched that far after the fixations used in EM. It means that at least some of the bands drawn by Bridges with double lines cannot be verified as doublets in the EM simply for methodological reasons (see Figure 1).

5. Ultrastructural Studies on Drosophila Polytene Chromosomes

In the development of EM methods suitable for localization of submicroscopic rearrangements in the salivary gland chromosomes, two most essential requirements are: The method should reveal exactly identifiable banding pattern and at the same time it should be able to uncover most detailed ultrastructure of bands and interbands. In the early 1960s, thin-section EM of suitably fixed and squashed salivary gland chromosomes seemed to be the only possible way to fulfill these requirements. Application of the method in cytogenetic analyses of the *white* locus (Sorsa *et al.*, 1973; Sorsa 1974b, 1979; Green, Beermann, and Sorsa, unpublished analyses of deletions going leftwards from *white*) has shown the possibilities of thin-section EM in the localization of genes. Thin-section EM of squashed salivary gland chromosomes was also used in the cytogenetic analysis of the band complex *10A1-2* (Zhimulev *et al.*, 1981a,b). By using the

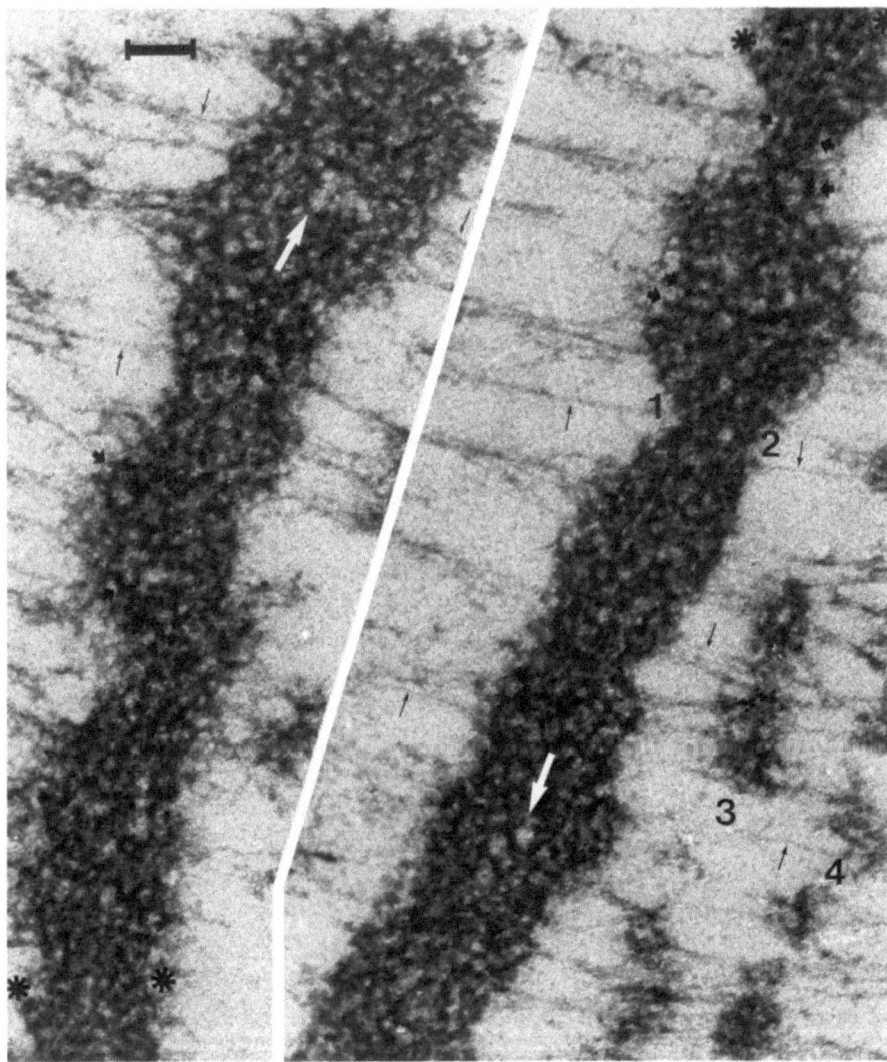

Figure 1. A high-magnification electron micrograph of Bridges' doublet *26D1-2*, which is interpreted as a single band in the EM map (Figure 10), because the double organization of the band is observable only locally (white arrows). The tiny bands *26D3* and *4* are shown only in the right panel. The points where the panels overlap are marked with asterisks. The band chromatin is composed mostly of about 10- to 15-nm-thick fibers (some of them are indicated with thick black arrows). The chromatin fibers in the band material obviously represent tightly packed nucleosome strands, also called chromatosome fibers (see McGhee *et al.*, 1980). The interband fibrils are only about 5 nm thick (some individual fibrils are indicated with thin black arrows). Thicker fibers in interbands are obviously bundles of 5-nm strands. The axial length of each interband between the bands *26D2* and *3*, *D3* and *4* is only about 100 nm. The thin-sectioned salivary gland chromosome (*2L*) is derived from the *D. melanogaster* stock $gt^1 w^a$ (obtained from B. Hochman). The chromosomes were squashed in 45% acetic acid after fixation with FAR solution. The squash preparations were stained in UAM (3% uranyl acetate in absolute methanol) during the dehydration, before embedding. The contrast was increased by staining the sections in Pb citrate (see Saura and Sorsa, 1979a). (Bar = 0.1 μm.)

same method, the beaded structure of band chromatin (Sorsa, 1976, 1982a) and the different organization of fibers in the interbands and bands (Sorsa, 1982a,b,d) have been demonstrated. The main drawback of the thin-section method is the long and laborious preparation procedure of the chromosomes. The recent development of whole-mount spreading methods (e.g., Alanen, 1981; Kalisch and Hägele, 1982) promises an easier and quicker way to a reasonably good resolution and identifiability of bands in the EM of polytene chromosomes.

5.1. Ultrastructure of Interbands

Differentiation of interphase chromosomes into the chromomeres and interchromomeres appears as a regular banding pattern in the polytenized chromosomes. EM of thin-sectioned salivary gland chromosomes shows that there is an essential structural difference in the chromatin fibers of interbands and bands (see Sorsa, 1982a,b,d). The continuity of the DNA molecule through the entire chromatids of *Drosophila* was quite convincingly demonstrated by Kavenoff and Zimm (1973; Kavenoff *et al.*, 1974). It implies that DNA is responsible for the integrity of chromatids also in the interband regions of polytene chromosomes.

From the results of labeling experiments with [^{3}H]thymidine on the stretched interbands, Steffensen (1963) concluded that DNA is probably absent from the interbands of polytene chromosomes. However, the central axis of the induced lampbrush stage of polytene chromosomes, which is proposed to be composed of interloop fibers, shows a continuous labeling with [^{3}H]thymidine (see Sorsa and Virrankoski-Castrodeza, 1972). The continuity of DNA through the interbands was also demonstrated by staining the RNA-depleted polytene chromosomes with acridine orange (Wolstenholme, 1966). Even stronger evidence for the presence of DNA and its role in the continuity of chromatids in interbands was obtained from the results of DNase digestion of isolated, unfixed polytene chromosomes (Lezzi, 1965).

Several arguments have been presented in favor of the hypothesis of interband location of genes in the polytene chromosomes (see Crick, 1971; Speiser, 1974; Zhimulev and Belyaeva, 1975). The view was recently supported by the localization of RNA polymerase B in the interbands (Jamrich *et al.*, 1977; Sass, 1982). The finding of uridine labeling (Semeshin *et al.*, 1979) and of RNP particles (Skaer, 1977; Mott *et al.*, 1980) in the area of interbands has been interpreted as evidence for transcriptional activity in the interbands of polytene chromosomes. The low DNA content of interbands (Beermann, 1972) argued against the interband location of genes. The structural organization of interband fibers in whole mounts (Ris, 1976; Laird, 1980) indicated, however, that DNA content is much higher in interbands than previously thought and thus the location of transcriptional units in the interbands is quite possible.

There are, however, some other observations that argue against the concept of interband genes. According to the results of mutation saturation experiments (see Hochman, 1971, 1974; Judd *et al.*, 1972; Kaufman *et al.*, 1975), the number of genes is closely equal to the number of bands in a given salivary gland

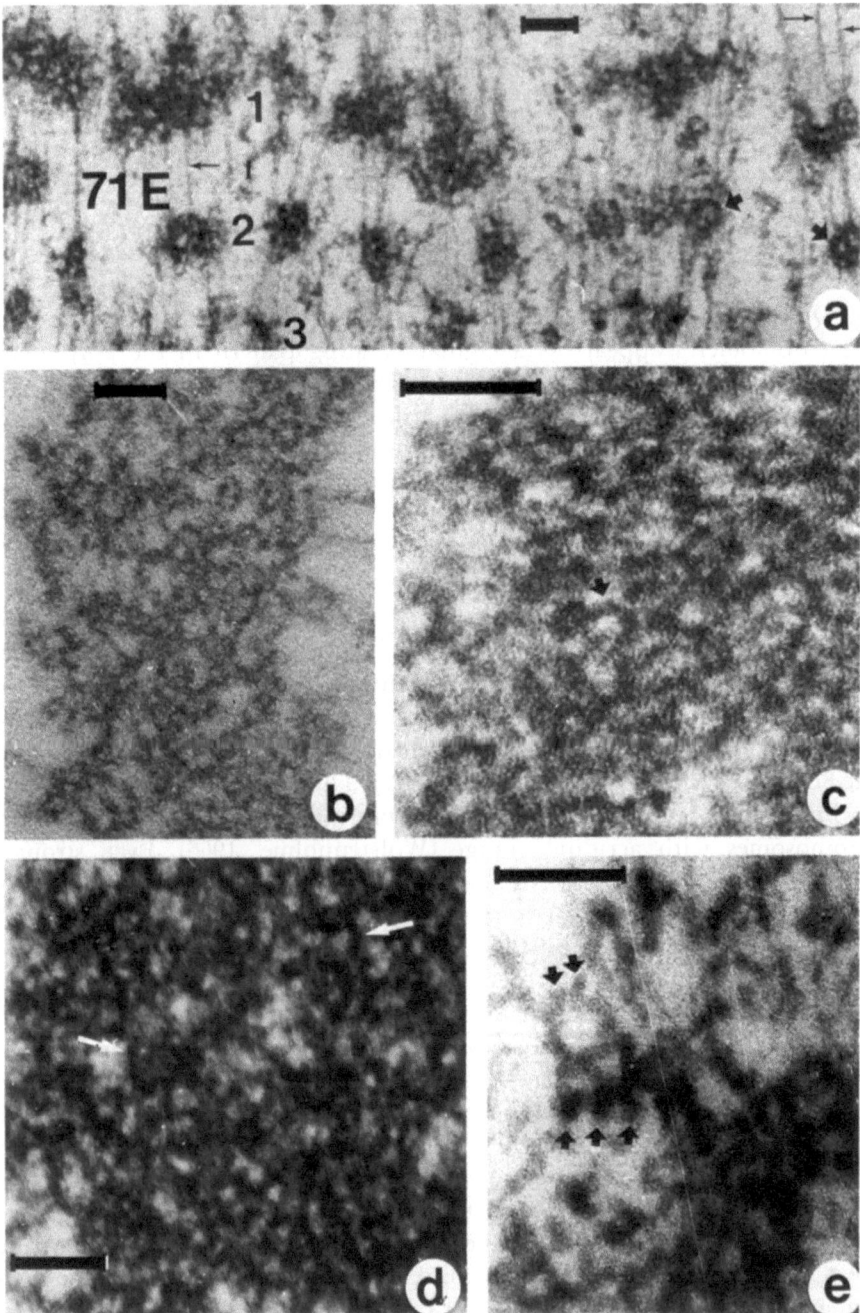

Figure 2. High-resolution electron micrographs of interband and band chromatin in the salivary gland chromosomes of *D. melanogaster*. (a) Well-separated groups of very small chromomeres in the subdivision *71E* of chromosome *3L*. Some of the individual chromomeric loops are indicated by thick arrows. As was drawn by Bridges (1942), the "doublet" *71E1-2* is composed of two bands of different size. In electron micrographs, the bands *71E1* and *E2* appear as two separate bands with a

chromosome region. This implies that the total number of recognizable gene loci is about 5000-6000 in the salivary gland chromosomes of *D. melanogaster*. If all the interbands represent active genes in the polytene chromosomes, their number alone is about equal to the number of all detectable gene loci, and a question arises: Why are so many active genes needed in the cells of salivary glands? The dimensions of fibrils within interbands in the high-resolution electron micrographs of thin-sectioned salivary gland chromosomes indicate that their average DNA content is very low per chromatid (Sorsa, 1982a,b,d), only about 0.3-0.5 kb. Such a segment is obviously too short to contain eukaryotic genes, especially if the regulatory DNA is also located in the interbands (see Nordheim *et al.*, 1981). The results of direct localizations of genes to specific bands of the salivary gland chromosomes of *D. melanogaster* (which are discussed more in Section 5.3) constitute strong evidence against the interband gene concept.

The absence of histone H1 from the interbands of polytene chromosomes (Jamrich *et al.*, 1977) and the absence of nucleosome-sized beads from the interband fibrils (Figures 1 and 2) both indicate that the basic structure of chromatin is different in the interbands and bands. The localization of left-handed methylated form of DNA in the interbands of salivary gland chromosomes of *D. melanogaster* (Nordheim *et al.*, 1981) may partly explain the structural difference found in the EM of thin-sectioned polytene chromosomes.

5.2. Ultrastructure of Bands

The diameter of chromatin fibers in the bands of polytene chromosomes varies from ca. 12 to ca. 15 nm in thin sections. Thicker helical structures reported from whole mounts may also exist, particularly in heavy bands (see Sorsa, 1976, 1982a,c), though it is difficult to demonstrate them in thin sections. A lampbrush type of organization with chromomeric loops and a central core formed by thin axial fibrils can be induced in the polytene chromosomes (see Sorsa *et al.*, 1970; Sorsa and Sorsa, 1970; Derksen and Sorsa, 1972; Sorsa and Virrankoski-Castrodeza, 1972, 1976; Sorsa, 1974b). In the thin-section EM of fixed and squashed polytene chromosomes, the open chromomere loops can be found only in the puffed bands (see Sorsa, 1969a) and in very small chromomeres (Figures 2 and 3). Rosettelike organization of a certain type of small chromomeres was demonstrated by Sorsa and Sorsa (1968); see also Sorsa, 1974b, 1979, 1982a). Obviously, the manner of further coiling or folding of the chromatosome fiber in bands is dependent on the length of the chromomeric loop as illustrated in Figures 3 and 4. Usually, the fiber of chromomeric loops seems to be quite

---

clear interband between them. Some individual interband fibrils are indicated with thin arrows. For possible organization of chromomeres, see Figure 3. (b, c) FAR-fixed and UAM-stained chromatin fibers in the heavy bands of division *21* of *2L*. (d) UAM + Pb citrate-stained chromatin in the heavy band complex *22A1-2* of *2L*. (e) AM-fixed and UAM-stained chromatin in the border area of a heavy band. The band chromatin is obviously composed of chromatosome fibers. Some individual fibers are indicated with arrows in (c), (d), and (e). Larger regular helices are difficult to recognize in thin sections (see Figures 4 and 5). (Bars = 0.1 μm.)

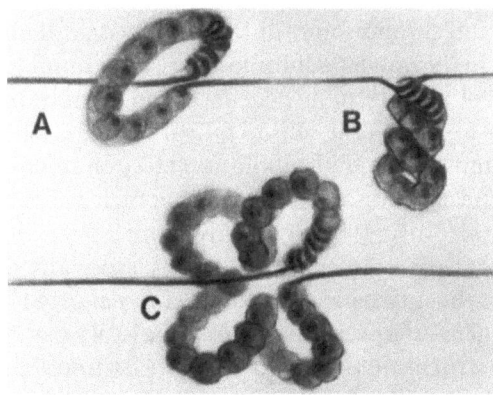

Figure 3. Models proposing possible structures of small chromomeres in the polytene chromosomes. (A) An open loop organization. (B) A closed loop containing about 12–15 nucleosomes (Sorsa and Sorsa, 1967b; Sorsa, 1973, 1976). (C) Rosette organization of chromomeric fiber (M. Sorsa and V. Sorsa, 1968; Sorsa, 1974b, 1979, 1982a).

irregularly kinked in larger bands, and flattened against the axial parts of chromatids (see Comings and Okada, 1974). The flattening of chromomeric loops also appears in the whole-mounted squashes of polytene chromosomes (see Laird *et al.*, 1980). Some evidence has been presented by Sorsa and Sorsa (1967b), Sorsa (1976), and Alanen (1981) for the formation of plectonemically coiled hairpin loops in certain types of chromomeres.

Although the flattened loop model of chromomeres may give an oversimplified view of the structural organization of chromatids in band regions, the model has been used to illustrate the location of genes (Figures 6 and 7), and the structural variation of double bands (Figure 5).

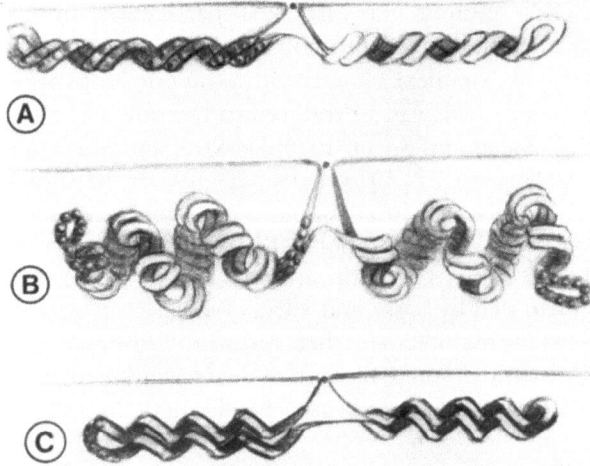

Figure 4. Closed-loop models illustrating the organization in larger chromomeres (Comings and Okada, 1974; Sorsa, 1976; Sass, 1980; Alanen, 1981). (A) A plectonemically coiled closed loop of about 100 nucleosome units. (B) Additional coiling in a closed loop composed of about 300 nucleosomes, corresponding to about 60 kb of DNA. (C) Paranemic coils in closed loops of a flattened chromomere. The closed-loop models are consistent with the EM observations that the chromomeric fibers tend to exist as pairs in polytene chromosomes (Sorsa and Sorsa, 1968; M. Sorsa, 1969a; V. Sorsa, 1976, 1982c; Alanen, 1981).

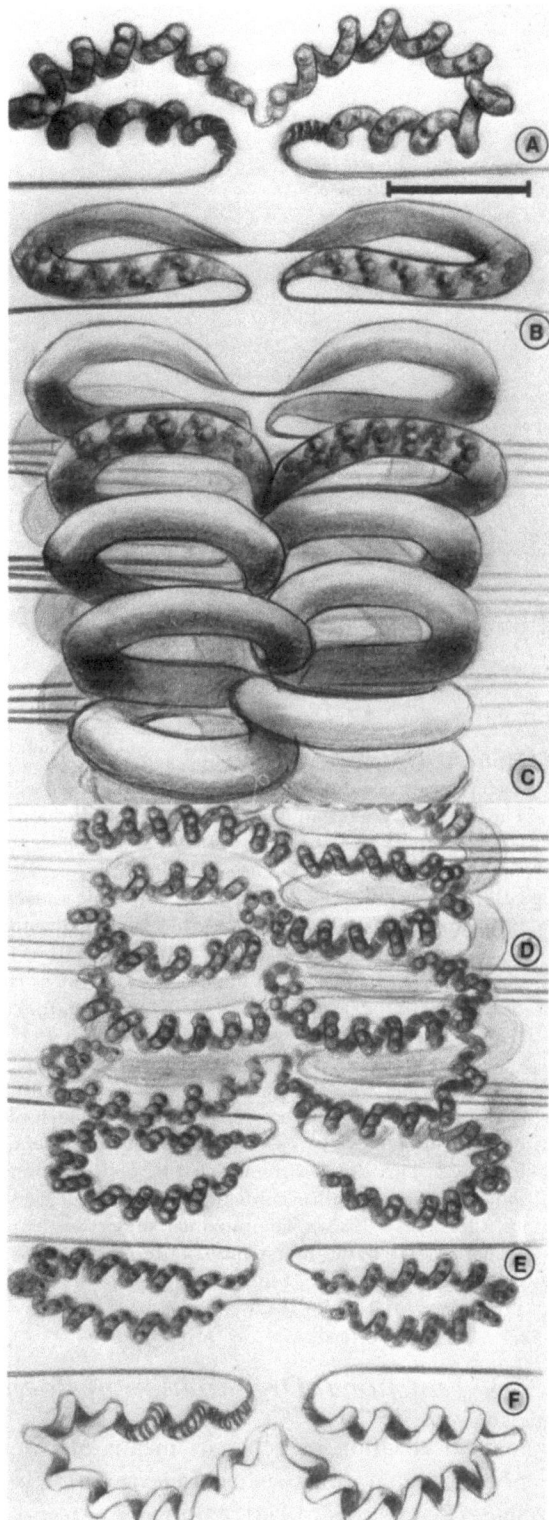

Figure 5. A schematic representation of the assumed structural organization in a double band as it usually appears in thin sections of polytene chromosomes. Individual chromatids consist of tandemly duplicated chromomere loops. The chromomeres are depicted as flattened, open loops formed by sparsely coiled helices of a nucleosome fiber. The length of a single loop is about 15 kb of DNA. The DNA content of a tandem loop is about 10 μm, representing an average-sized chromomere in *D. melanogaster* (Sorsa, 1982a). The double structure is clearest in the surface of polytene chromosomes (b, e). Proposed organization of chromomeres is illustrated also with separate chromatids (a, f). In the middle of the band the double structure of chromomeres is less distinct because of complicated overlapping of fixed chromatin (c, d). See the band structure in Figure 1. (Bar = 0.1 μm.)

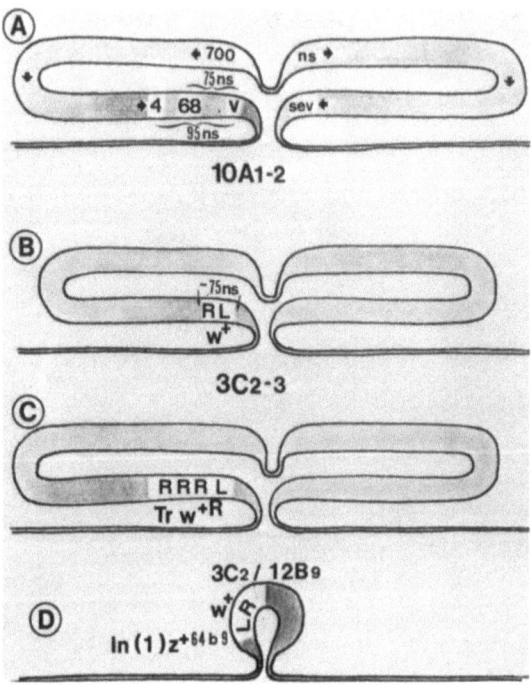

Figure 6. The arrangements of the genes detected in the double bands *10A1-2* and *3C2-3* diagrammed as cytological maps of the chromomeres (see Sorsa *et al.*, 1973; Zhimulev *et al.*, 1981a,b). The structural organizations of huge chromomeric loops of *10A1-2* and *3C2-3* are simplified. The DNA content of each doublet is probably about 200 kb per chromatid (see Gehring and Paro, 1980; Zhimulev *et al.*, 1981b). (A) The cytological location of genetic markers *vermilion* (*v*), l(1)L68 (*68*), *l(1)BP4* (*4*), and seven less [(*sev*) a locus controlling visual sensitivity to ultraviolet light (Harris *et al.*, 1976)] in the double chromomere *10A1-2* is diagrammed according to the analyses of Zhimulev *et al.* (1981a,b). The distance of the vermilion gene from the nearest left-side marker is about 70 nucleosome (ns) units, which implies that the *v* locus is close to the interband between *9F13* and *10A1*. Thus, the area including all the genes found from the distal part of band *10A1* is confined to about 120-150 ns units, while the whole doublet *10A1-2* contains about 1000 ns units. Correspondingly, the *sev* locus has been localized close to the proximal end of *10A2* (see Zhimulev *et al.*, 1981b). (B) The *white*-locus area (w^+) in the distal of Bridges' doublet *3C2-3* (see Sorsa *et al.*, 1973) consists of about 70-80 ns units per chromatid, which is less than 1/10th of the total length of the doublet chromomere *3C2-3*. (C) The triplication of $white^{+R}$ markedly increases the material in the band *3C2*. This can be demonstrated by both light and electron microscopy (Sorsa, 1979) and indicates that the length of chromomeric loops is reflected in the axial length of the band (see Laird *et al.*, 1981; Sorsa, 1982a). The proximal sublocus w^{+R} probably includes the sites w-*eosin*$^+$ and w-*spotted*$^+$. (D) In the inversion $In(1)zeste^{+64b9}$, the DNA of the distal part of band *3C2*, which still includes a normal *white*$^+$ function, is joined to the inverted fraction of band *12B9* (see Sorsa *et al.*, 1973; Sorsa, 1974b, 1979).

5.3. Functional Organization of Polytene Chromosomes

Localization of genes by means of the light microscope both from the normal preparations (see e.g. Lefevre, 1974a,b) and from the induced lampbrush stage (Bencze *et al.*, 1978) of polytene chromosomes has strongly suggested that the genes are in chromomeres. A considerable amount of evidence, derived

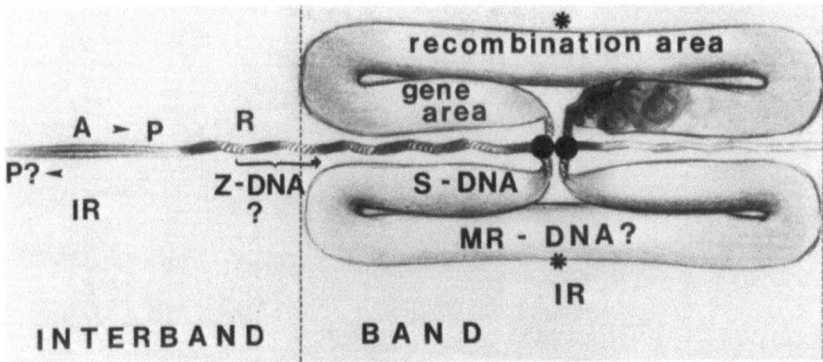

Figure 7. A general model illustrating the structural and functional organization of the chromomere unit of polytene interphase chromosomes. The model attempts to combine ideas of previous models (see Paul, 1972; Sorsa, 1975) with the results of recent cytogenetic and ultrastructural analyses of polytene chromosomes. The replication of the loop has temporarily terminated at the regulatory region of the transcriptional unit (R). The regulatory region is proposed to be a zone of Z-DNA (Nordheim *et al.*, 1981), which is thus located outside of the duplicated chromomeric loop in the interband area. According to the general model of Paul (1972), the promoter site (P) and the address locus (A) are located upstream from the regulatory zone, which implies that they may be in the interband, too, or in the adjacent band. The regulatory region may also contain sites for specific proteins (Paul, 1972). The actual genes representing single-sequence (S) DNA have been localized in the band material close to the interband (see Figure 6). The middle part of the chromomere loop is assumed to contain repetitive DNA sequences (MR-DNA), which participate in the synapsis and recombination between the homologs (see King, 1970a; Comings and Okada, 1974; Lefevre, 1974a; Sorsa, 1975). The location of inverted repeats (IR) in the cytological model of chromomeres is still an open question. Most probable sites of palindromic sequences are the address locus area and the initiation sites of replication (proposed initiation sites in the chromomere are marked with an asterisk).

mainly from the labeling experiments both with [^{3}H]thymidine and with [^{3}H]uridine, indicates that the chromomeres or the chromomeres plus the neighboring interchromomeres can be considered as functional units of polytene chromosomes (see e.g. Pelling, 1966; Beermann, 1972). The genes of *Drosophila*, which have been localized by EM into the heavy bands *3C2-3* and *10A1-2* (see Sorsa *et al.*, 1973; Zhimulev *et al.*, 1981a,b), are evidently located in the band chromatin. It is also evident that all the recognizable gene loci in the doublets *3C2-3* and *10A1-2* are located close to the ends of chromomeric loops, which are in connection with the interbands. Though the role of interbands cannot be exactly characterized, there is some evidence that the interband DNA may be involved in the regulation of chromosomal activity (Nordheim *et al.*, 1981).

According to the general model of chromosome structure and gene activation in eukaryotes proposed by Paul (1972), the address and promoter sites are located upstream from the regulatory area in the chromatid DNA. If it is supposed that the Z-DNA sequences in the interbands represent the regulatory regions and both of the bands connected to the interband contain genes, the address site with promoters should be in the middle of the interband. The address sites and promoter areas of parallel chromatids may appear as "minibands" in the polytenized interphase chromosomes.

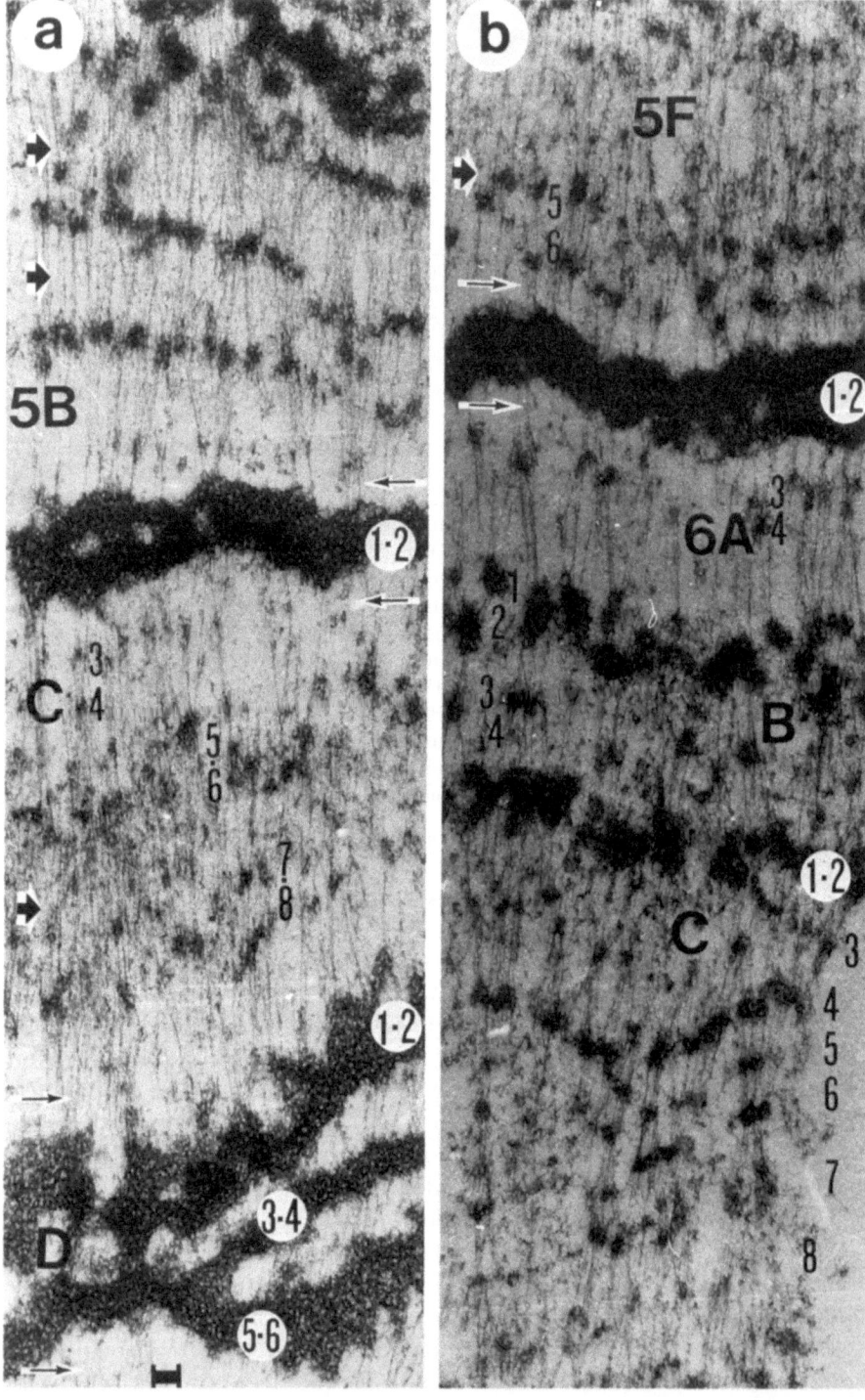

On the basis of an earlier light microscopic localization of *lethal (1)L68* in the proximal border of the doublet *10A1-2*, Lefevre (1974a) proposed that "geneless" DNA in the middle of this double band may act as an area of synaptic attraction between the homologs, but the actual recombination breaks are promoted at the edges of the bands. However, a later cytogenetic analysis of region *9F10-11* through *10A1-2* (Zhimulev *et al.*, 1981a,b) with the EM shows that *lethal (1)L68* is located much closer to the distal edge of band *10A1-2*, between *vermilion* and *lethal (1)BP4*. This new location of *L68* implies that recombination breaks take place mainly in "geneless" area, but some accumulation of rearrangement breaks in the edges of doublets does exist, according to the results of Zhimulev *et al.* (1981b) (see Figure 6A).

The shortness of interchromomeric fibrils in the interbands *3C1-3C2* and in *9F13-10A1* renders untenable the suggestion that the *white* locus (which contains 14-16 kb of DNA) and the gene complex localized in the distal end of *10A1-2* (which has still larger DNA content) can be located in those interbands. The results of the cytogenetic analyses of band doublets *3C2-3* and *10A1-2* suggest the structural and functional model for the chromomere illustrated in Figure 7.

Particularly interesting are the recent findings and experimental results concerning the *white* locus mutants with extra DNA. The total DNA content, ca. 14-16 kb (see Spradling and Rubin, 1981), estimated for the *white* locus is able to form about 70-80 nucleosomes. Potentially, it is capable of coding a polypeptide of about 5×10^3 amino acids. Though the translation product coded by the *white* gene is not known, one can presume that the total length of *white* locus DNA probably includes several intron segments which are not translated. These introns may normally act as the sites of intragenic recombination between the actual coding sequences. Proposing that the size of introns may be increased by the integration of transposable elements like *copia* sequences, we come close to the interpretation suggested by Bingham and Judd (1981) for the location of the *copia* element in the *white^a* region of the X chromosome. An interesting point is that the intraallelic recombination data (see Judd, 1976) suggest a greater DNA content for the *white* locus than the values obtained from the cloned *white* DNA (see Spradling and Rubin, 1981). If the recombination areas comprise only the DNA of introns between the coding sequences, it implies that a higher frequency of breaks is accumulated in the noncoding sequences, at least in the *white* locus of *D. melanogaster*.

◀──

Figure 8. Electron micrographs showing the variation in the number of interband fibrils in adjacent areas of the polytene chromosome X of *D. melanogaster* (see Sorsa and Saura, 1980b; Sorsa, 1982a). The number of interband fibrils, which is assumed to represent the number of parallel chromatids, is lowest in the close neighborhood of certain heavy bands like *5C1-2*, *5D1-2*, *5D3-4*, *5D5-6*, and *6A1-2*. Some of the clearest areas of assumed underpolytenization are marked with thin arrows. The number of interband fibrils is highest in the regions of narrow bands composed of small chromomeres (thick arrows). The observed variation in the number of parallel chromatids is assumed to be caused by the unequal polytenization of salivary gland chromosome. See Figure 9. (Bar = 0.1 μm.)

Another interesting result is that the rather small *copia* sequence (ca. 25 nucleosomes), which is located in the w^a area, may be responsible for the transposition of a much larger "jumping gene," *TE white-roughest*, which contains about 1000 nucleosomes, if the doublet *3C2-3* is included in this element (Gehring and Paro, 1980).

6. Local Variation in the Degree of Polyteny

Since the demonstration of underreplication of chromocentric heterochromatin in the polytene chromosomes of *D. melanogaster* (Rudkin, 1969), it became evident that the reduction of the number of parallel chromatids must appear in the border area of the chromocenter and arms. Laird (1973) proposed a branching model to illustrate this structural change in polytene chromosomes. The densitometric analyses carried out by Laird (1980) from the EM negatives of whole-mounted polytene chromosomes indicated that local differences may appear in the degree of polyteny at the chromosome arms. Lower polyteny was already expected in the late-replicating intercalary heterochromatin. A considerable variation in the number of interband fibrils entering different bands can be recognized in the thin sections of salivary gland chromosomes. The clearly lowered number of axial fibers which appear regularly in the interbands next to certain heavy bands has been interpreted to indicate a retarded polytenization in those bands (Sorsa, 1982a). Direct EM observations on the structure of individual stretched bands indicate that the number of fibers is higher in the middle of chromomeres and lower in the interbands. This suggests that the replication process is initiated in the bands and the replication forks are moving toward the interbands (Sorsa, 1974a). In the polytene chromosome regions composed of faint bands of similar size, the number of interband fibrils is usually highest and equal in the long stretches of chromosomes in the same thin section (see Figures 8 and 9).

Putting together the information concerning polytenization, it seems likely that the number of initiations of replication is probably about equal in the subsequent replicative units of polytene chromosomes. However, in the faint bands, probably because of shorter units and a less complicated structure of the chromomeres, the replications are initiated earlier, progress faster, and the cycles are completed. Conversely, in the heavy bands, the replications are initiated later, progress more slowly, and the replication cycles, especially in the last stages of polytenization, are incomplete in the tighter and more complicated structures of the chromatin. The proposal is consistent with results from labeling experiments on the early and late-replicating regions of polytenizing chromosomes (Kalisch and Hägele, 1976). The relatively greater proportion of interbands and small, separate groups of chromomeres obviously contribute to the rapid replication in the faint-band regions. The same structural diffuseness probably facilitates the transcription and puffing of bands. The molecules needed in chromosomal activities enter more easily through the sparsely packed interbands

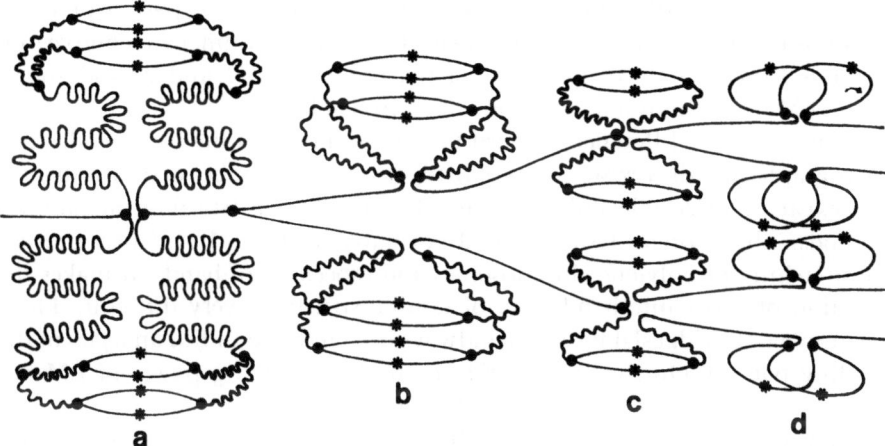

Figure 9. A model for elucidating the replication in an individual chromatid of a polytenizing chromosome. Because of the tight and complicated structure of chromatin in the huge chromomeres of heavy bands (a), the whole chromomeric loop has gone through only one complete cycle of replication. The polytenization process is more advanced in the narrower bands formed by smaller chromomeres (b, c). The number of parallel interband fibrils indicates that the replication cycles are most complete in the region of faint bands (d). The partially completed cycles of replication further increase the compact structure of heavy bands, because the sister chromomeres cannot be separated (a), while in the narrow bands the chromomeres form smaller and more separate groups (c). See Figures 1, 2a, and 8.

and tiny bands than through the dense chromatin of heavy bands (see Sorsa, 1982b). The complexity of chromomere and band structure is increased by incompleted cycles of replication which produces partially duplicated chromomeric loops (see Figure 9) particularly in heavy bands.

7. Summary

During the last 30 years, since EM was first used in chromosome studies, several different methods have been developed and applied to polytene chromosomes. EM of thin-sectioned squashed preparations has been most effectively used for the cytogenetic analyses of the salivary gland chromosomes of *D. melanogaster*. The thin-section method has made it possible to recheck at the EM level the light microscopic reference maps compiled about 40 years ago by Bridges and Bridges on the basis of camera lucida drawings. The number of bands resolved by EM in the polytene chromosomes and the correlation of electron micrographs with the revised light microscopic maps of Bridges depend to a great deal on the fixations and the methods used for preparing the polytene chromosomes for the EM. A good correlation between the electron micrographs and the revised reference maps of Bridges has been obtained by using an immediate fixation of salivary glands with FAR or AM solutions and by staining the squashed chromosomes in the UAM solution before embedding and thin-sectioning. The results of extensive EM analyses of the banding pattern of the 2L chromosome are presented as an EM map (Figure 10). A numerical comparison of bands depicted in the revised light microscopic map of 2L and of bands found in the EM of chromosome 2L is presented in Table 1.

The essential difference between the maps drawn on the basis of electron micrographs and the revised maps of Bridges is in the number of double bands. The existence of doublets of certain bands is partly a question of preparation methods. The stronger fixatives used normally in the EM evidently prevent the stretching of polytene chromosomes and especially of bands. It makes the detection of short interbands and of double structures very difficult. Thus, the number of Bridges' doublets has always been reduced in EM analyses of the banding pattern. On the other hand, the higher resolving power of EM has demonstrated a number of new, mostly faint bands.

The results of EM studies on the fine structure of chromatin in interbands

Table 1. Numerical Comparisons of the Bands Shown by Bridges (1942) in the Revised Map of 2L and of the Bands Depicted in the Present Division Maps of 2L Drawn from the Electron Micrographs of Thin-Sectioned Salivary Gland Chromosomes of *D. melanogaster*[a]

Revised reference map of Bridges (1942)				EM map of 2L chromosome				
Division number	Double bands	Single bands	Total number	Double bands	Single bands	New bands	Total number	Occasionally detected
21	4	23	31	3	25	16	47	4
22	7	20	34	4	24	10	42	7
23	9	20	38	4	27	7	42	2
24	8	22	38	4	25	—	33	6
25	14	18	46	8	27	9	52	2
26	11	24	46	5	31	8	49	3
27	8	21	37	1	30	2	34	2
28	10	25	45	—	39	—	39	1
29	10	15	35	7	19	3	36	5
30	10	24	44	4	32	2	42	7
31	8	22	38	2	31	3	38	12
32	10	8	28	4	16	—	24	5
33	13	22	48	7	22	—	36	7
34	12	25	49	9	20	—	38	13
35	12	20	44	9	21	1	40	7
36	16	21	53	12	24	3	51	6
37	9	26	44	3	32	3	41	5
38	10	25	45	7	29	—	43	8
39	7	16	30	4	17	1	26	4
40	12	7	31	9?	5?	1	24?	—
2L totals	200	404	804	106	496	69	777?	+ 106 = 883?

[a]The band numbers of the proximal subdivisions of 40 are uncertain because the region is unclear in the thin-sectioned material, and the EM map of this area is therefore tentative.

Drosophila Polytene Chromosomes

EXPLANATION OF SIGNS IN THE EM MAP:

➤   DIVISION BORDER
➤   SUBDIVISION BORDER
1·2 **:**   A DOUBLE BAND ALSO IN THE EM
1·2 **⁞**   AN OCCASIONAL BAND IN BETWEEN THE DOUBLET
1·2 **⁞**   DOUBLENESS QUESTIONABLE IN THE EM
3·4 **:**   AN UNEQUAL DOUBLET
$\genfrac{}{}{0pt}{}{3}{4}$ **:**   RATHER A PAIR OF SINGLE BANDS
➜   A NEW BAND IN THE EM
→   OCCASIONALLY DETECTED NEW BAND IN THE EM
5→   OCCASIONALLY DETECTED BRIDGES' BAND
6**?**   NOT DETECTED BRIDGES' BAND
!**{**   CORRESPONDENCE TO BRIDGES' MAP NOT CLEAR

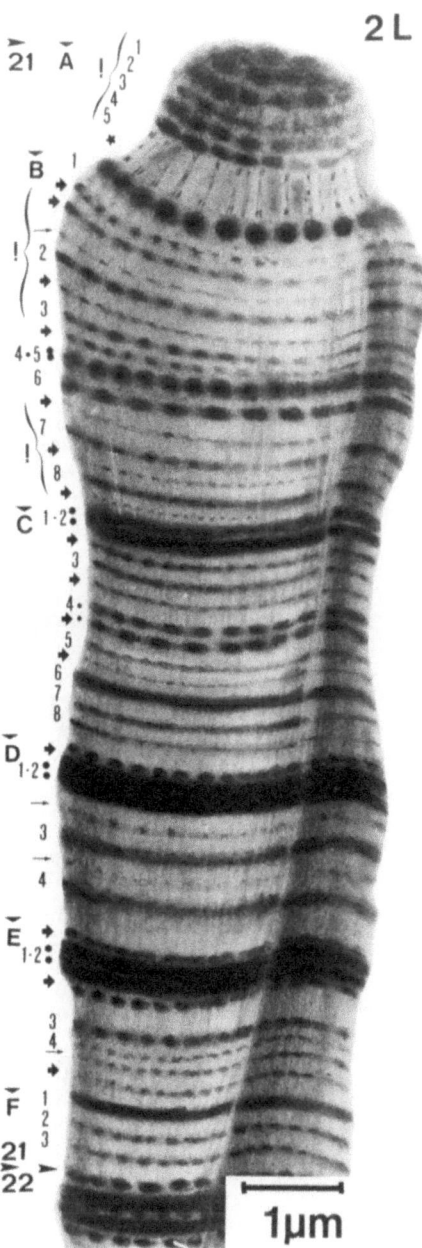

(Figure 10a)

Figure 10. An EM map for the salivary gland chromosome *2L* of *D. melanogaster*. The new maps of the Bridges' divisions *21* through *40* have been drawn on the basis of high-resolution electron micrographs, which were previously used for the EM analyses of the banding pattern in *2L* (see Saura and Sorsa, 1979a,b,c,d; Saura, 1980, and unpublished analysis of divisions *32* through *36* and *40*). The new bands and Bridges' bands which cannot always be detected in the EM are indicated with thin arrows. Extremely faint bandlike granular zones in certain interbands like *21A5–21B1* (marked with an asterisk on the map) are not counted as bands (see Saura and Sorsa, 1979a). The Bridges' doublets are interpreted as single bands if their doubleness cannot be detected (see Figure 1), and as two single bands if a clear interband exists in between the halves of doublet (see Figure 2a). A more detailed explanation of the map signs is given in the figure, and the numerical comparison of depicted bands in the EM map and in Bridges' revised map of *2L* is given in Table 1.

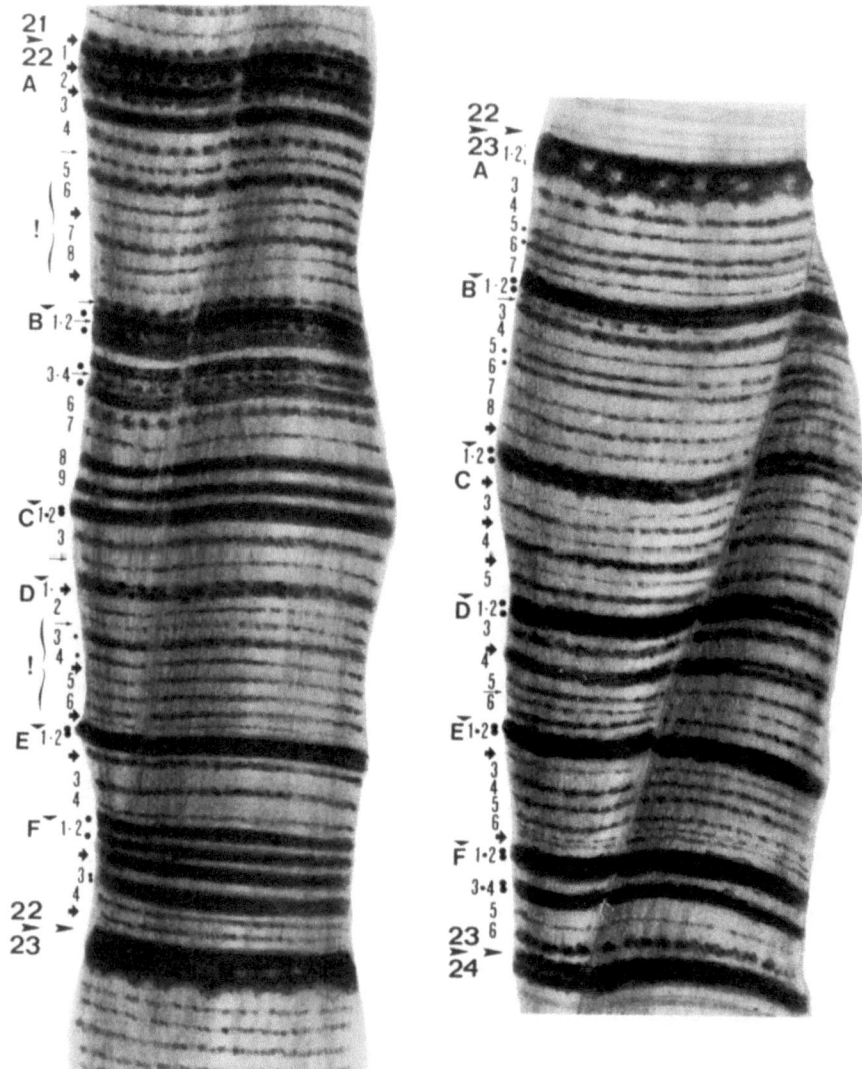

(Figure 10b)

(Figure 10c)

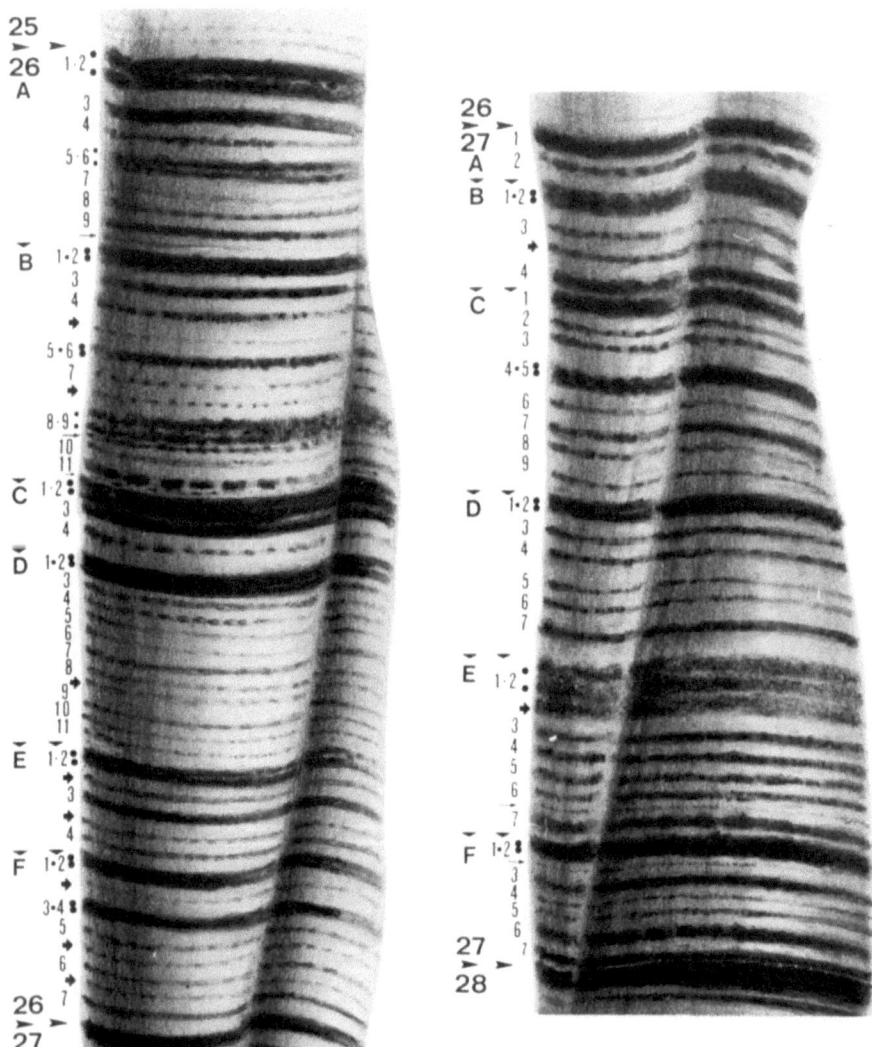

(Figure 10d)

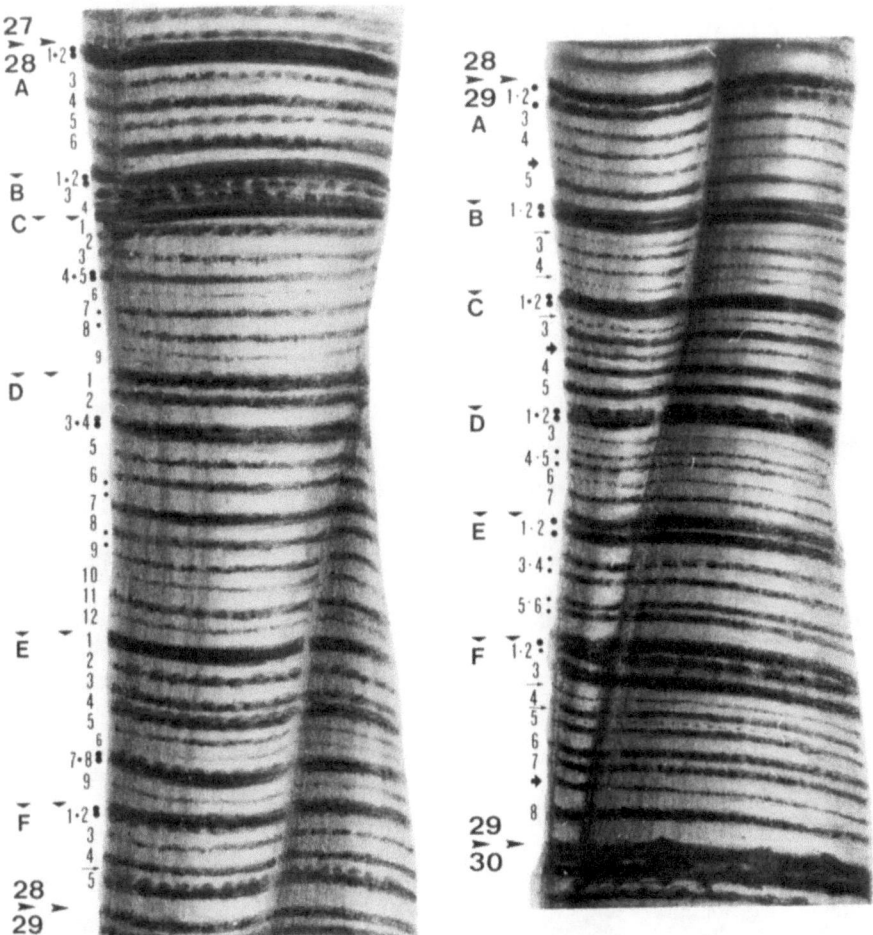

(Figure 10e)

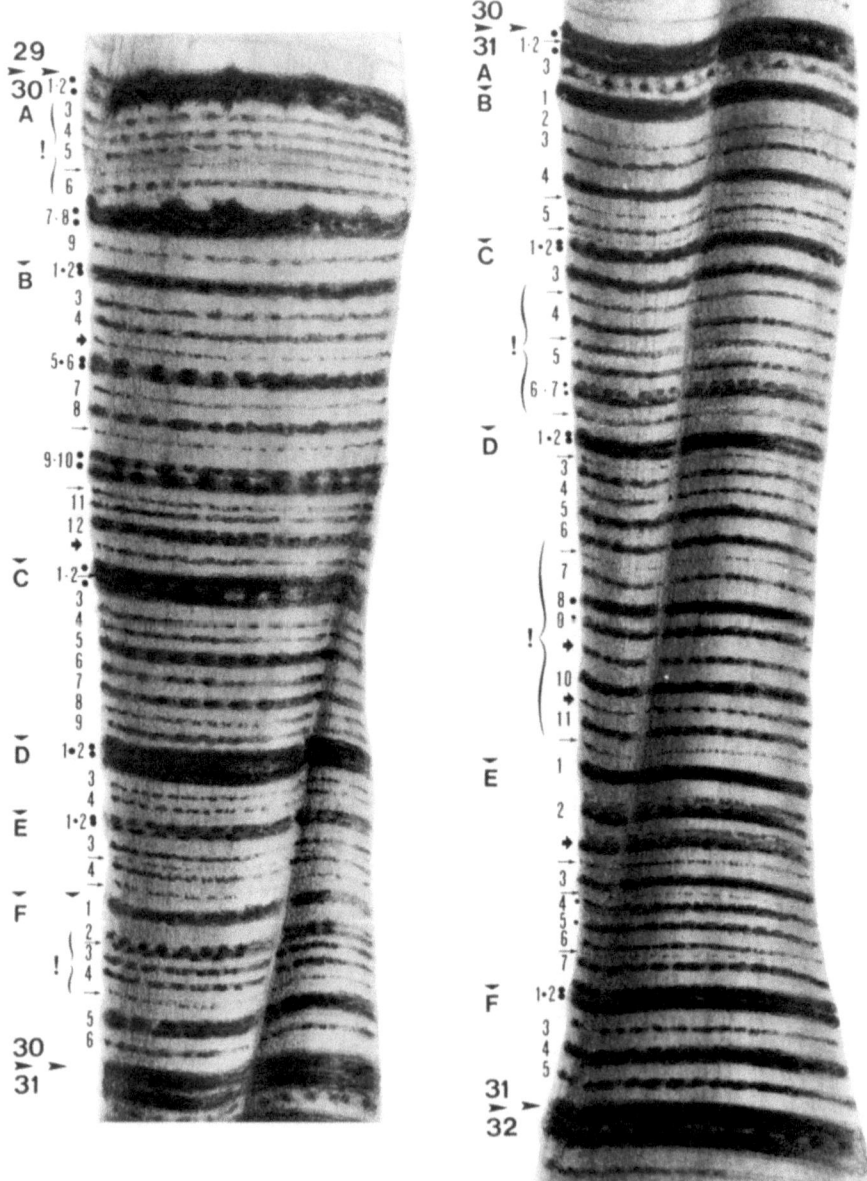

(Figure 10f)

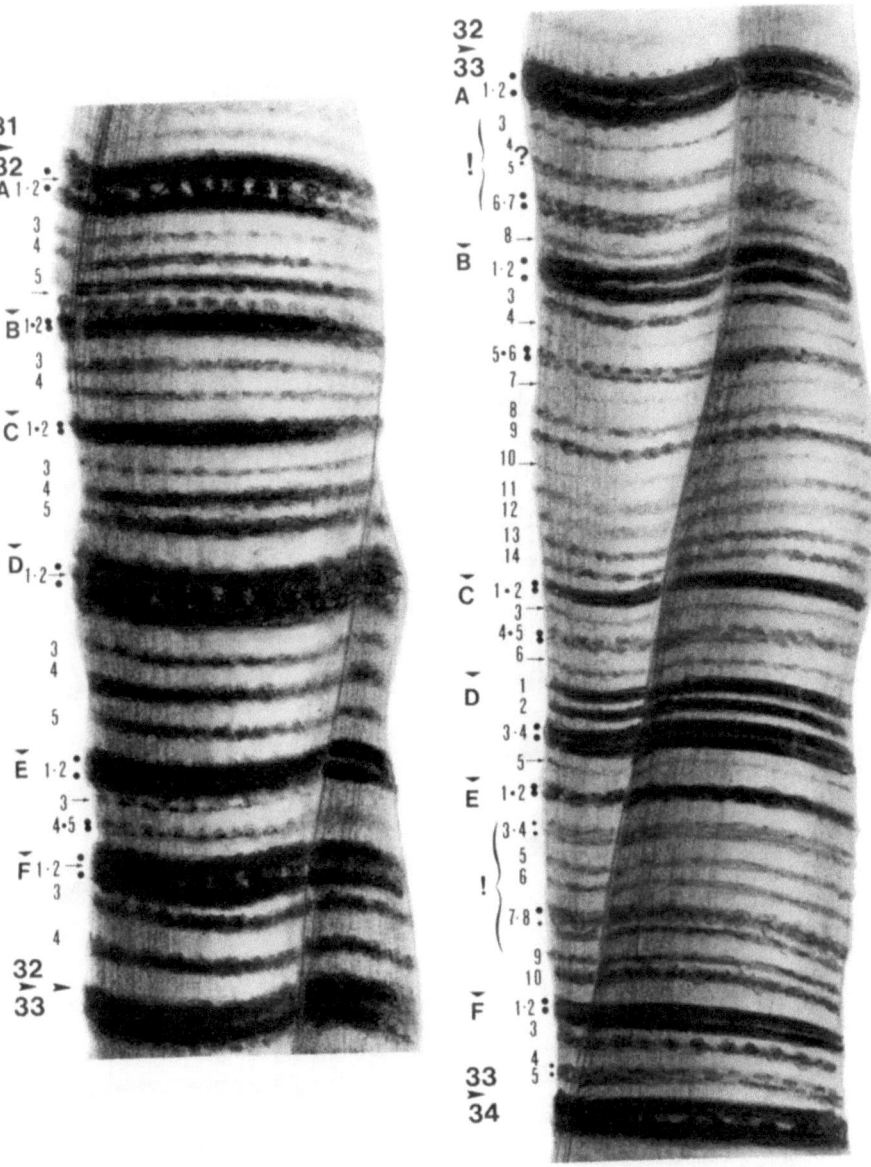

(Figure 10g)

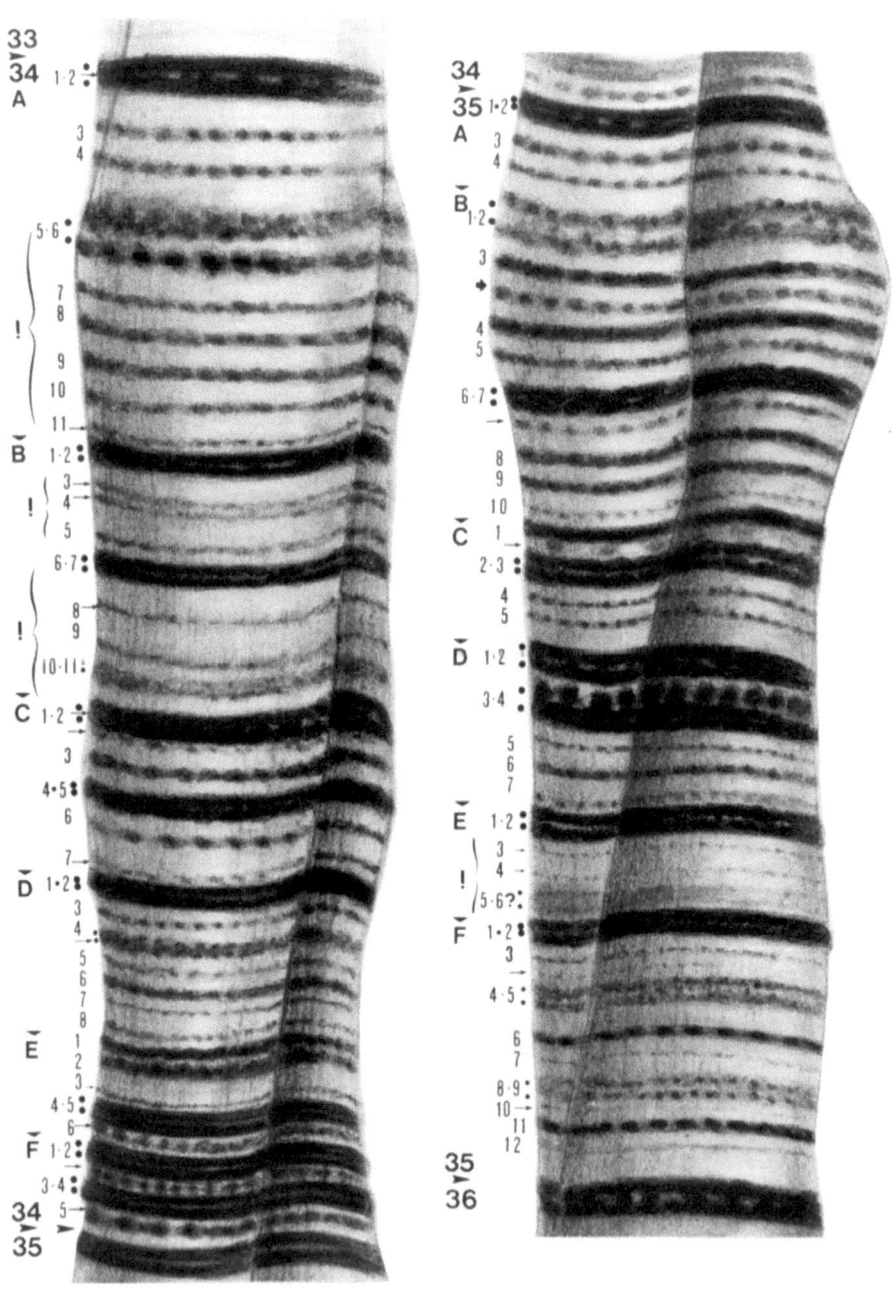

(Figure 10h)

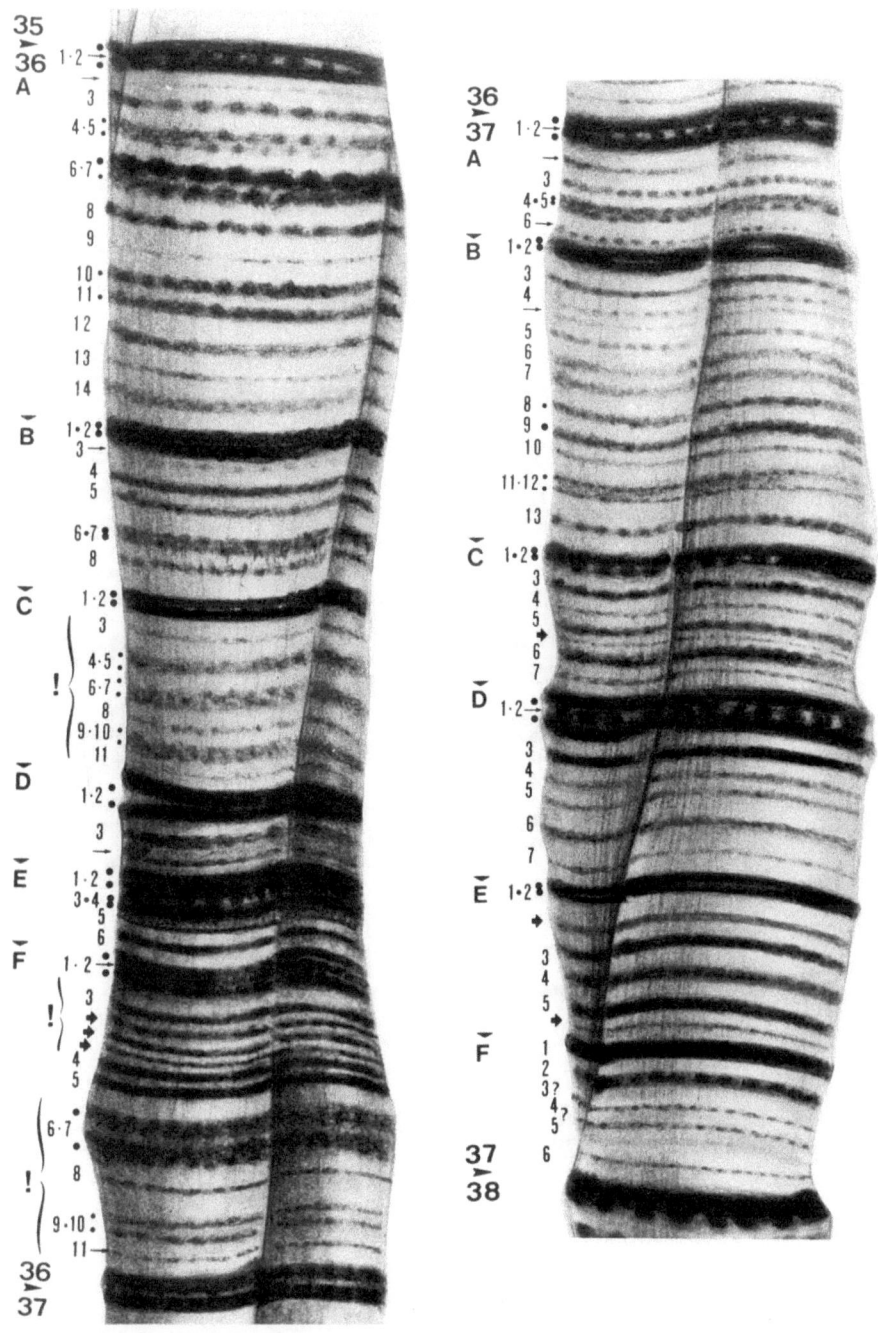

(Figure 10i)

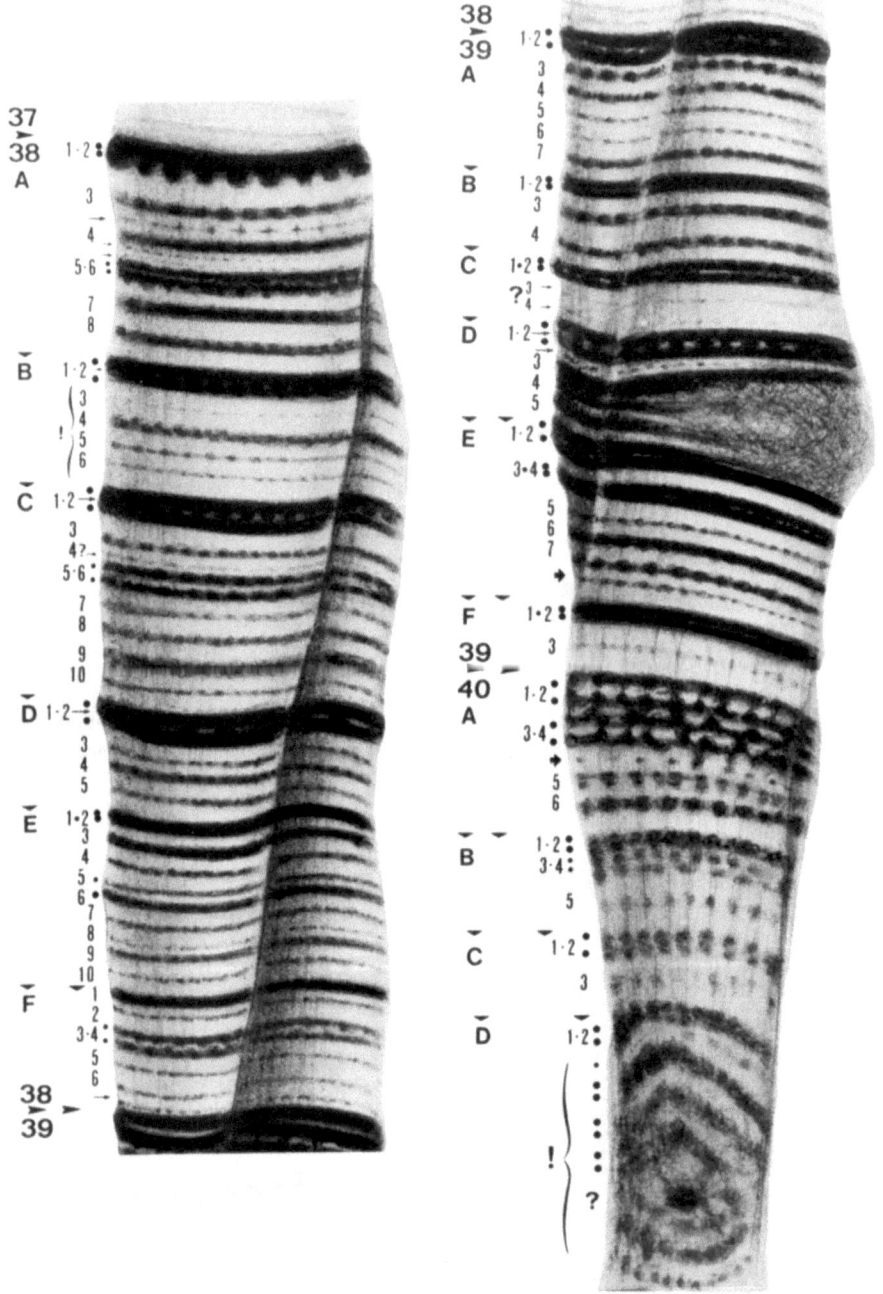

(Figure 10j)

and bands have been contradictory, depending on the methods used for the analyses. In thin sections, the interband regions consist of parallel, thin strands (diameter about 5 nm), while the band chromatin is composed of nucleosome-sized fibers. The existence of regular higher orders of coiling (even in heavy bands) of polytene chromosomes is difficult to demonstrate in thin sections.

The results of EM cytogenic localizations of genes indicate that the coding sequences are near the ends of chromomeric loops and the middle section of loops may be more active in the synaptic and recombinational events between the homologs.

ACKNOWLEDGMENTS

I wish to thank Ms. Anja O. Saura, Ph.Lic., Ms. Virpi Virrankoski-Castrodeza, M.Sc, and Ms. Riikka Santalahti for the skillful thin-sectioning of squashed salivary gland chromosomes used for the EM maps. Anja O. Saura also graciously provided me with her unpublished analyses of the rest of divisions of 2L for checking the map. The EM studies of material was carried out mainly with equipment of the Department of Electron Microscopy of the University of Helsinki. The work was supported financially by the National Research Council of Sciences of Finland.

References

Alanen, M., 1981, An integrated description of polytene chromosome structure, *Hereditas* **95**:295-321.

Alanen, M., and Sorsa, V., 1978, Identifiable regions of *Drosophila melanogaster* polytene chromosomes visualized by whole mount electron microscopy, *Hereditas* **89**:257-261.

Beermann, W., 1972, Chromomeres and genes. In *Results and Problems in Cell Differentiation*, vol. 4, edited by W. Beermann, pp. 1-33, Springer-Verlag, Berlin.

Beermann, W., and Bahr, G. F., 1954, The submicroscopic structure of Balbiani ring, *Exp. Cell Res.* **6**:195-201.

Bencze, J. L., Brash, K., and White, B. N., 1978, The location of 5S RNA genes in lampbrush polytene chromosomes from *Drosophila*, *Exp. Cell Res.* **120**:365-372.

Berendes, H. D., 1968, Electron microscopical mapping of giant chromosomes, *Drosophila Inf. Serv.* **43**:115-116.

Berendes, H. D., 1970, Polytene chromosome structure at the submicroscopic level. I. A map of region X, 1-4E of *Drosophila melanogaster*, *Chromosoma* **29**:118-130.

Bingham, P. M., and Judd, B. H., 1981, A copy of the copia transposable element is very tightly linked to the w^a allele at the *white* locus of *Drosophila melanogaster*, *Cell* **25**:705-711.

Bridges, C. B., 1935, Salivary chromosome maps with a key to the banding of the chromosomes of *Drosophila melanogaster*, *J. Hered.* **26**:60-64.

Bridges, C. B., 1938, A revised map of the salivary gland X-chromosome of *Drosophila melanogaster*, *J. Hered.* **29**:11-13.

Bridges, C. B., and Bridges, P. N., 1939, A revised map of the right limb of the second chromosome of *Drosophila melanogaster*, *J. Hered.* **30**:475-476.

Bridges, P. N., 1941a, A revised map of the left limb of the third chromosome of *Drosophila melanogaster*, *J. Hered.* **32**:64-65.

Bridges, P. N., 1941b, A revision of the salivary gland 3R-chromosome map of *Drosophila melanogaster*, *J. Hered.* **32**:299-300.
Bridges, P. N., 1942, A new map of the salivary gland 2L-chromosome of *Drosophila melanogaster*, *J. Hered.* **33**:403-408.
Burkholder, G. D., 1976, Whole mount electron microscopy of polytene chromosomes from *Drosophila melanogaster*, *Can. J. Genet. Cytol.* **18**:67-77.
Comings, D. E., and Okada, T. A., 1974, Some aspects of chromosome structure in eukaryotes, *Cold Spring Harbor Symp. Quant. Biol.* **38**:145-153.
Crick, F., 1971, General model for the chromosomes of higher organisms, *Nature (London)* **234**:25-27.
Derksen, J., and Sorsa, V., 1972, Whole mount electron miroscopy of salivary gland chromosomes after treatment with alkali-urea, *Exp. Cell Res.* **70**:246-248.
Gay, H., 1955, Serial sections of smears for electron microscopy, *Stain Technol.* **30**:239-242.
Gehring, W. J., and Paro, R., 1980, Isolation of a hybrid plasmid with homologous sequences to a transposing element of *Drosophila melanogaster*, *Cell* **19**:897-904.
Harris, W. A., Stark, W. S., and Walker, J. A., 1976, Genetic dissection of the photoreceptor system in the compound eye of *Drosophila melanogaster*, *J. Physiol. (London)* **256**:415-425.
Herskowitz, I. H., 1952, Electron microscopy of salivary gland chromosomes of *Drosophila*, *J. Hered.* **43**:226-232.
Hochman, B., 1971, Analysis of chromosome 4 in *Drosophila melanogaster*. II. Ethyl methanesulfonate induced lethals, *Genetics* **67**:235-252.
Hochman, B., 1974, Analysis of whole chromosome in *Drosophila*, *Cold Spring Harbor Symp. Quant. Biol.* **38**:581-589.
Jamrich, M., Greenleaf, A. L., and Bautz, E. K. F., 1977, Localization of RNA polymerase in polytene chromosomes of *Drosophila melanogaster*, *Proc. Natl. Acad. Sci. USA* **74**:2079-2083.
Judd, B. H., 1976, Genetic units of *Drosophila* complex loci. In *The Genetics and Biology of Drosophila*, vol. 1b, edited by M. Ashburner and E. Novitski, pp. 767-799, Academic Press, New York.
Judd, B. H., Shen, M. W., and Kaufman, T. C., 1972, The anatomy and function of a segment of the X chromosome of *Drosophila melanogaster*, *Genetics* **71**:139-156.
Kalisch, W.-E., and Hägele, K., 1976, Correspondence of banding patterns to ^{3}H-thymidine labeling patterns in polytene chromosomes, *Chromosoma* **57**:19-23.
Kalisch, W.-E., and Hägele, K., 1982, A new spread technique for polytene chromosomes and its efficiency for autoradiography, including *in situ* hybridization. In *Advances in Genetics, Development and Evolution of Drosophila*, edited by S. Lakovaara, Plenum Press, New York.
Kaufman, T. C., Shannon, M. P., Shen, M. W., and Judd, B. H., 1975, A revision of the cytology and ontogeny of several deficiencies in the 3A1-3C6 region of the X chromosome of *Drosophila melanogaster*, *Genetics* **79**:265-282.
Kavenoff, R., and Zimm, B. H., 1973, Chromosome-sized DNA molecules from *Drosophila*, *Chromosoma* **41**:1-27.
Kavenoff, R., Klotz, L. C., and Zimm, B. H., 1974, On the nature of chromosome-sized DNA molecules, *Cold Spring Harbor Symp. Quant. Biol.* **38**:1-8.
Kerkis, A. J., Zhimulev, I. F., and Belyaeva, E. S., 1977, EM autoradiographic study of ^{3}H-uridine incorporation into *D. melanogaster* salivary gland chromosomes, *Drosophila Inf. Serv.* **52**:14-17.
King, R. C., 1965, *Genetics*, 2nd ed., Oxford University Press, London.
King, R. C., 1970a, The meiotic behavior of the *Drosophila* oocyte, *Int. Rev. Cytol.* **28**:125-167.
King, R. C., 1970b, *Ovarian Development in Drosophila melanogaster*, Academic Press, New York.
King, R. C., 1975, *Drosophila melanogaster:* An introduction. In *Handbook of Genetics*, edited by R. C. King, pp. 625-652, Plenum Press, New York.
Laird, C. D., 1973, DNA of *Drosophila* chromosomes, *Annu. Rev. Genet.* **7**:177-204.
Laird, C. D., 1980, Structural paradox of polytene chromosomes, *Cell* **22**:869-874.
Laird, C. D., Ashburner, M., and Wilkinson, L., 1980, Relationship between relative dry mass and average band width in regions of polytene chromosomes of *Drosophila*, *Chromosoma* **76**:175-189.

Laird, C. D., Wilkinson, L., Johnson, D., and Sandström, C., 1981, Proposed structural principles of polytene chromosomes. In *Chromosomes Today*, vol. 7, edited by M. Bennett, M. Bobrow, and G. Hewitt, pp. 74-83, George, Allen & Unwin, London.

Lefevre, G., Jr., 1974a, The relationship between genes and polytene chromosome bands, *Annu. Rev. Genet.* **8**:51-62.

Lefevre, G., Jr., 1974b, The one band-one gene hypothesis: Evidence from a cytogenetic analysis of mutant and nonmutant rearrangement breakpoints in Drosophila melanogaster, *Cold Spring Harbor Symp. Quant. Biol.* **38**:591-599.

Lefevre, G., Jr., 1976, A photographic representation and interpretation of the polytene chromosomes of Drosophila melanogaster salivary glands. In *The Genetics and Biology of Drosophila*, vol. 1a, edited by M. Ashburner and E. Novitski, pp. 31-66, Academic Press, New York.

Lezzi, M., 1965, Die Wirkung von DNase auf isolierte polytän-chromosomen, *Exp. Cell Res.* **39**:289-292.

Lindsley, D. L., and Grell, E. H., 1968, *Genetic Variations of Drosophila melanogaster*, Carnegie Institution of Washington, Publication 627, Carnegie Institution of Washington, Washington, D.C.

Lossinsky, A. S., and Lefever, H. M., 1978, Ultrastructural banding observations in region *1A-10F* of the salivary gland X-chromosome of Drosophila melanogaster, *Drosophila Inf. Serv.* **53**:126-131.

McGhee, J. D., Rau, D. C., Charney, E., and Felsenfeld, G., 1980, Orientation of the nucleosome within the higher order structure of chromatin, *Cell* **22**:87-96.

Mott, M. R., Burnett, E. J., and Hill, R. J., 1980, Ultrastructure of polytene chromosomes of Drosophila isolated by microdissection, *J. Cell Sci.* **45**:15-30.

Nordheim, A., Pardue, M. L., Lafer, E. M., Möller, A., Stollar, B. D., and Rich, A., 1981, Antibodies to left-handed Z-DNA bind to interband regions of Drosophila polytene chromosomes, *Nature (London)* **294**:417-422.

Painter, T. S., 1933, A new method for the study of chromosome rearrangements and the plotting of chromosome maps, *Science* **78**:585-586.

Painter, T. S., 1934, Salivary chromosomes and the attacks of the gene, *J. Hered.* **25**:465-476.

Palay, S. L., and Claude, A., 1949, An electron microscope study of salivary gland chromosomes by the replica method, *J. Exp. Med.* **89**:431-438.

Paul, J., 1972, General theory of chromosome structure and gene activation in eukaryotes, *Nature (London)* **238**:444-446.

Pelling, C., 1966, A replicative and synthetic unit—The modern concept of the chromomere, *Proc. R. Soc. London Ser. B* **164**:279-289.

Rae, P. M. M., 1966, Whole mount electron microscopy of Drosophila salivary chromosomes, *Nature (London)* **212**:139-142.

Ris, H., 1961, The invitation lecture: Ultrastructure and molecular organization of genetic systems, *Can. J. Genet. Cytol.* **3**:95-120.

Ris, H., 1976, Levels of chromosome organization. In *Proc. Sixth Eur. Congr. Electron Microsc.* Jerusalem, vol. II, pp. 21-25.

Rudkin, G. T., 1969, Non-replicating DNA in Drosophila, *Genetics* **61**:227-238.

Sass, H., 1980, Hierarchy of fibrillar organization levels in the polytene interphase chromosomes of Chironomus, *J. Cell Sci.* **45**:269-293.

Sass, H., 1982, RNA polymerase B in polytene chromosomes: Immunofluorescent and autoradiographic analysis during stimulated and repressed RNA synthesis, *Cell* **28**:269-278.

Saura, A. O., 1980, Electron microscopic analysis of the banding pattern in the salivary gland chromosomes of Drosophila melanogaster: Divisions 23 through 26 of 2L, *Hereditas* **93**:295-309.

Saura, A. O., 1983, Electron microscopic analysis of the banding pattern in the salivary gland chromosomes of Drosophila melanogaster. Divisions 23 through 26 of 2L, *Hereditas* **99**: 89-114.

Saura, A. O., and Sorsa, V., 1979a, Electron microscopic analysis of the banding pattern in the salivary gland chromosomes of Drosophila melanogaster: Divisions 21 and 22 of 2L, *Hereditas* **90**:39-49.

Saura, A. O., and Sorsa, V., 1979b, Electron microscopic analysis of the banding pattern in the salivary gland chromosomes of *Drosophila melanogaster:* Divisions 30 and 31 of 2L, *Hereditas* **90:**257-267.

Saura, A. O., and Sorsa, V., 1979c, Electron microscopic analysis of the banding pattern in the salivary gland chromosomes of *Drosophila melanogaster:* Divisions 37, 38 and 39 of 2L, *Hereditas* **91:**5-18.

Saura, A. O., and Sorsa, V., 1979d, Electron microscopic analysis of the banding pattern in the salivary gland chromosomes of *Drosophila melanogaster:* Divisions 27, 28 and 29 of 2L, *Hereditas* **91:**219-230.

Semeshin, V. F., Zhimulev, I. F., and Belyaeva, E. S., 1979, Electron microscope autoradiographic study on transcriptional activity of *Drosophila melanogaster* polytene chromosomes, *Chromosoma* **73:**163-177.

Skaer, R. J., 1977, Interband transcription in *Drosophila, J. Cell Sci.* **26:**251-266.

Slizynski, B. M., 1944, A revised map of salivary gland chromosome 4, *J. Hered.* **35:**322-325.

Sorsa, M., 1969a, Ultrastructure of puffs in the proximal part of chromosome 3R in *Drosophila melanogaster, Ann. Acad. Sci. Fenn. Ser. A4* **150:**1-21.

Sorsa, M., 1969b, Ultrastructure of the polytene chromosome in *Drosophila melanogaster*, with special reference to electron microscopic mapping of chromosome 3R, *Ann. Acad. Sci. Fenn. Ser. A4* **151:**1-18.

Sorsa, M., and Sorsa, V., 1967a, The squash technique in the electron microscopic studies on the structure of polytene chromosomes, *J. Ultrastruct. Res.* **20:**302.

Sorsa, M., and Sorsa, V., 1967b, Electron microscopic observations on interband fibrils in *Drosophila* salivary chromosomes, *Chromosoma* **22:**32-41.

Sorsa, M., and Sorsa, V., 1968, Electron microscopic studies on band regions in *Drosophila* salivary chromosomes, *Ann. Acad. Sci. Fenn. Ser. A4* **127:**1-8.

Sorsa, V., 1973, Whole mount electron microscopy of small chromomeres in the salivary gland chromosomes of *Drosophila melanogaster, Hereditas* **73:**143-146.

Sorsa, V., 1974a, Organization of replicative units in salivary gland chromosome bands, *Hereditas* **78:**298-302.

Sorsa, V., 1974b, Organization of chromomeres, *Cold Spring Harbor Symp. Quant. Biol.* **38:**601-608.

Sorsa, V., 1975, A hypothesis for the origin and evolution of chromomere DNA, *Hereditas* **81:**77-84.

Sorsa, V., 1976, Beaded organization of chromatin in the salivary gland chromosome bands of *Drosophila melanogaster, Hereditas* **84:**213-220.

Sorsa, V., 1979, Electron microscopic localization and ultrastructure of certain gene loci in salivary gland chromosomes of *Drosophila melanogaster.* In *Specific Eukaryotic Genes*, edited by J. Engberg, H. Klenow, and V. Leick, pp. 55-71, Munksgaard, Copenhagen.

Sorsa, V., 1982a, An attempt to estimate DNA content and distribution in the *zeste-white* region of the X chromosome of *Drosophila melanogaster, Biol. Zentralbl.* **101:**81-96.

Sorsa, V., 1982b, Structural analysis of polytene chromosome bands and interbands. In *Advances in Genetics, Development and Evolution of Drosophila*, edited by S. Lakovaara, pp. 11-22, Plenum Press, New York.

Sorsa, V., 1982c, Electron microscopic map for the salivary gland chromosome X of *Drosophila melanogaster* divisions 1-5. In *Advances in Genetics, Development and Evolution of Drosophila*, edited by S. Lakovaara, pp. 23-32, Plenum Press, New York.

Sorsa, V., 1982d, Volume of chromatin fibers in interbands and bands of polytene chromosomes, *Hereditas* **97:**103-113.

Sorsa, V., and Saura, A. O., 1980a, Electron microscopic analysis of the banding pattern in the salivary gland chromosomes of *Drosophila melanogaster:* Divisions 1 and 2 of X, *Hereditas* **92:**73-83.

Sorsa, V., and Saura, A. O., 1980b, Electron microscopic analysis of the banding pattern in the salivary gland chromosomes of *Drosophila melanogaster:* Divisions 3, 4 and 5 of X, *Hereditas* **92:**341-351.

Sorsa, V., and Sorsa, M., 1979, Ultrastructure of induced transitions in the chromatin organization of polytene chromosomes, *Chromosoma* **31:**346-355.

Sorsa, V., and Virrankoski-Castrodeza, V., 1972, H^3-thymidine radioautography of salivary gland chromosomes treated with alkali-urea, *Hereditas* **71:**139-144.

Sorsa, V., and Virrankoski-Castrodeza, V., 1976, Whole mount electron microscopy of alkali-urea treated polytene chromosomes spread by centrifugation, *Hereditas* **82**:131-135.

Sorsa, V., Pusa, K., Virrankoski, V., and Sorsa, M, 1970, Electron microscopy of an induced "lampbrush stage" of the polytene chromosome, *Exp. Cell Res.* **60**:466-469.

Sorsa, V., Green, M. M., and Beermann, W., 1973, Cytogenic fine structure and chromosomal localization of white gene in *Drosophila melanogaster, Nature New Biol.* **245**:43-37.

Sorsa, V., Saura, A. O., and Heino, T., 1983, Electron microscopic analysis of the banding pattern in the salivary gland chromosomes of *Drosophila melanogaster:* Divisions 6 through 10 of X, *Hereditas* **98**:181-200.

Speiser, C., 1974, Eine Hypothese über die funktionelle Organization der Chromosomen der höherer Organismen, *Theor. Appl. Genet.* **44**:97-99.

Spradling, A. C., and Rubin, G. M., 1981, *Drosophila* genome organization: Conserved and dynamic aspects, *Annu. Rev. Genet.* **15**:219-264.

Steffensen, D. M., 1963, Evidence for the apparent absence of DNA in the interbands of *Drosophila* salivary chromosomes, *Genetics* **48**:1289-1301.

Swift, H., 1962, Nucleic acids and cell morphology in dipteran salivary glands. In *The Molecular Control of Cellular Activity*, edited by J. M. Allen, pp. 73-125, McGraw-Hill, New York.

Wolstenholme, D. R., 1966, Direct evidence for the presence of DNA in interbands of *Drosophila* salivary gland chromosomes, *Genetics* **53**:357-360.

Yasuzumi, G., Odate, Z., and Ota, Y., 1951, The fine structure of salivary chromosomes, *Cytologia* **16**:233-242.

Zhimulev, I. F., and Belyaeva, E. S., 1975, Proposals to the problem of structural and functional organization of polytene chromosomes, *Theor. Appl. Genet.* **45**:335-340.

Zhimulev, I. F., Semeshin, V. F., and Belyaeva, E. S., 1981a, Fine cytogenetical analysis of the band *10A1-2* and the adjoining regions in the *Drosophila melanogaster* X chromosome. I. Cytology of the region and mapping of chromosome rearrangements, *Chromosoma* **82**:9-23.

Zhimulev, I. F., Pokholkova, G. V., Bgatov, A. V., Semeshin, V. F., and Belyaeva, E. S., 1981b, Fine cytogenetical analysis of the band *10A1-2* and the adjoining regions in the *Drosophila melanogaster* X chromosome. II. Genetical analysis, *Chromosoma* **82**:25-40.

II

The Ultrastructure of the Development, Differentiation, and Functioning of Specialized Tissues and Organs

4

The Structure of Insect Muscles

DAVID SPENCER SMITH

1. Introduction

As the editors have pointed out in their prefatory remarks to Volume 1 of *Insect Ultrastructure*, the pace of fine structural work in general, and in the present context of insect cell biology in particular, has accelerated to such an extent that structures and systems now well recognized and sometimes in part biochemically characterized, were in some instances unknown a decade or two ago.

But this advance remains an acceleration, rather than a *de novo* development, since insects have, arguably, played a part second only to the human body in the development of concepts of animal structure and function through the three or four centuries of man's effective delving into the workings of his body and those of his fellow animals. The history of studies on insect tissue plan and function may be divided, for convenience, into four overlapping and approximate periods. First, the celebrated works of the early anatomists concerned either entirely or *inter alia* with insects, Van Leeuwenhoek, Hooke, Malpighi, Swammerdam, Lyonet, Straus-Durckheim, Dufour, and others, opened the insect body and established that the organ systems of these animals are describable in the terms used by the vertebrate anatomist, albeit embellished or modified by "special" features imposed by the segmental plan, exoskeleton, tracheal system and so on. Lyonet (1762) noted, without undue astonishment, that the *Cossus* larva includes three times the number of anatomically distinct muscles—not including the multiple muscles of the internal organs and the head—than the human body. Straus-Durckheim (1828) followed this work in a series of dissections of adult *Melolontha* displaying the flight muscles of the beetle, with tracheal supply, in figures that have never been surpassed (see Smith, 1980).

DAVID SPENCER SMITH • Hope Entomological Collections and Department of Zoology, Oxford University, Oxford OX1 3PS, England.

The second general period, spanning the middle decades of the last century, is relatively undocumented, but saw a dramatic burgeoning of animal design, as approached by the light microscope. The perceptive studies of Bowman, Leydig, Aubert, Kölliker, Brücke, Von Siebold, and others, during this period, contributed to contemporary and, more importantly, later concepts of muscle structure—including the structure of insect muscles (Smith, 1961a, 1972; Elder, 1975). The general background of this largely forgotten period has recently been ably recollected by Bracegirdle (1978), who has documented the wealth of instrumentation available to the "professional" and "amateur" of the middle decades of the last century. It is not generally appreciated that more than 50 section-cutting devices (microtomes) were developed between 1840 and the early 1880s, when the field of comparative histology was opened by the introduction of microtomes capable of cutting serial sections. Bracegirdle also documents, in a most interesting text, the mid-19th century heyday of the professional slide preparer, noting, for example, that an 1872 catalog of Norman offered almost 1000 slide preparations of entomological interest—40% of the entire list—and he revives, amongst the most skilled slide preparers of the period, the names of Topping, Wheeler, Enock, and others. Surviving examples of their work show that while many of their products were primarily designed to satisfy the Victorian "esthetic sense," others were of considerable scientific merit and planned to display, as adequately as possible within contemporary limitations, aspects of insect gross anatomy and histology. Examples of preparations from this period are illustrated in Figure 1.

The third arbitrary period suggested here spans the decades between the advent of histology as an organized and practical approach to tissue design and comparable developments in tissue preparation after 1950, that permitted useful examination of sections of adequately fixed and embedded sections in the electron microscope. Day (1948) compiled a "preliminary" bibliography of 4000 references to papers on insect histology, including 100 on muscle structure and arrangement, and omitting all pre-1881 references cited in the compendious work of Berlese (1909). At the end of this period, Tiegs (1955), writing on extensive histological studies of insect muscles, suggested that:

> It is important, in assessing the histology of insect muscle, that we do not extend to the latter . . . data obtained from an intensive study of vertebrate muscle; for if current views on the derivation of Arthropoda from segmented worms are well founded, then striated muscle within this group must have arisen independently of that of vertebrates. Yet the resemblances between the two are certainly very remarkable, and emphasize the need to determine the functional significance of all structural detail.

This comparative approach, established by earlier histological studies, plays an increasingly important part in our understanding of insect muscle structure, function, and evolution and augments our understanding of muscle activity in other animals, and of contractile mechanisms in general.

Studies on a wide variety of cell types, notably in the last decade, have established that actin/myosin contractility, while most conspicuous and accessible in muscle, occurs widely. It has evolved in parallel with the structurally and biochemically distinct tubulin/dynein device of the flagellar axoneme and other

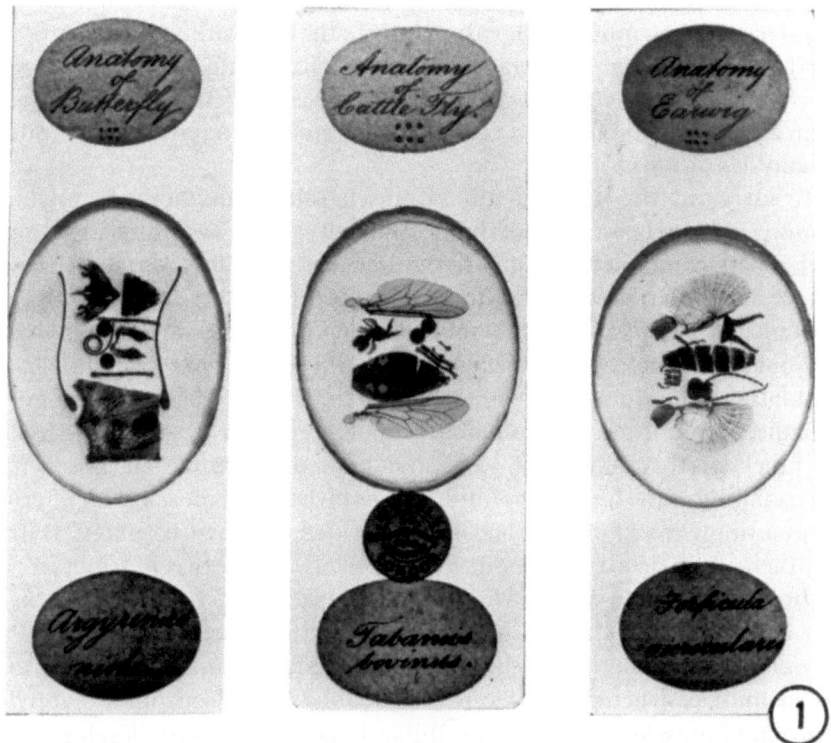

Figure 1. Examples of early microscopic slides, illustrating external features (wings, antennae, mouth parts, corneal lens, etc.) of selected insects, by the commercial preparer E. Wheeler (ca. 1865–1880).

microtubular arrays such as the ciliate axostyle (McIntosh *et al.*, 1973). Other proteins are involved in the special contractility of the vorticellid spasmonene (Amos, 1972) and the "costa" of *Trichomonas* (Amos *et al.*, 1979). The versatility of evolution of movement devices involving protein polymers familiar in muscle function has been documented in the myosin-free actin polymerization of actin in the acrosome reaction of echinoderm sperm (Tilney *et al.*, 1978). Further, actin by itself plays a nonmotile role in some cells: as in the stereocilia of auditory hair cells (Tilney *et al.*, 1980) and in the least motile situation, in the surface spines of *Schistosoma* (Cohen *et al.*, 1982).

Against this versatility, we must consider the arrangement of actin and myosin and associated proteins in muscle, noting that their deposition is very variable and, in contrast to many "primitive" contractile systems of similar biochemical basis, they do not use transient polymerization and depolymerization as an integral part of the contractile mechanism. Muscle cells are typically cylindrical and designed to ensure that the contractile force is exerted precisely along the long axis of the cell performing mechanical work via skeletal attachments or directly on a deformable organ. Cardiac muscle cells may act in a highly coordinated fashion as in the vertebrate heart, or may be less elaborately integrated as in the hydraulically simpler insect dorsal vessel. Within these general parameters, muscle cells vary greatly in the disposition of the component myofilaments, in their aggregation or partitioning into myofibrils, in their mitochondrial complement, and in the relationship between the contractile system and the membranes and compartments controlling the cycle of contraction and relaxation.

In surveying the fine structure of muscle cells throughout the phyla, we encounter an amalgam of underlying similarity and a wealth of variation in detail in all components so far recognized. In part, this variation may be interpretable in terms of the varying mechanical tasks these cells perform, but there remains a very substantial residue of variation, at present without obvious functional correlates: why, for example, should the transverse tubule (T) and sarcoplasmic reticulum (SR) membrane junctions in vertebrate skeletal muscle be triadic, and in insects and other arthropods generally dyadic? It seems clear that much of this residue is so placed as a measure of our fragmentary understanding of the functional implications of details of muscle structure: even within a single insect order this has been tackled only by Cullen (1974) in the Hemiptera and detailed assessment of correlation between structure and function of *all* muscle systems within a single species has yet to be attempted: a latter-day Lyonet, collaborating with an excellent electrophysiologist and neuropharmacologist, is needed! Some of the residual "unaccountable" variation in muscle structure, however, presumably stems from the polyphyletic evolution of muscle—a success in cellular design which, with development of appropriate skeletal devices, played a crucial role in the evolution of animal diversity.

While this chapter is primarily concerned with the structure and function of insect muscle cells—at least 75% of the muscle cells in the animal kingdom—a comparative view is essential: results of research on insect preparations may augment studies on vertebrate cells, and conversely, all results of work on other muscles are potentially relevant to the insect body. "Muscle" is perhaps best regarded as a system rather than a cell type, and all results from insects and other animals may be interrelated, contributing to our understanding of contractility throughout the phyla.

2. Contractile System

2.1. Myofilament Organization and Disposition

One of the very few generalizations about any cell type in the animal kingdom that has stood the test of time and extensive comparative studies is that in all arthropod muscles, the contractile proteins are sufficiently ordered to be termed *striated*, even in mechanical situations met, in vertebrates, by *smooth* muscles in which the myofilaments lack this degree or order. However, in common with other arthropod muscle cells, and with the addition of some special features, insect muscles vary considerably both in terms of myosin (thick) filament (and hence sarcomere) length and in the geometry and numerical ratios between these myofilaments and the accompanying actin filament orbital. As Pringle (1980) points out, it seems that increasing structural order and decreasing actin:myosin ratio is correlated with higher speed of contraction in insect muscles, but the underlying molecular features of this correlation are by no means clear.

The myofibrils or contractile system of intersegmental, visceral, and limb muscles of insects exhibit a more or less regular 12-membered actin orbital (references in Smith, 1972) in common with most arthropod leg muscles, though in some slow crustacean fibers (Hoyle and McNeill, 1968; Pringle, 1980) the actin complement is reportedly greater and the thin filaments are arranged in orbitals of >12. This degree of disorder is enhanced in vertebrate smooth fibers and also encountered in other slowly contracting invertebrate muscles, whether striated or unstriated (Smith et al., 1973). Even where a regular 12-actin array appears to be present (Figure 2), this is seldom as precise as in insect muscles with a reduced actin orbital (Figure 3; also see below): either these muscles exhibit an intrinsically variable (nonintegral) actin:myosin ratio or include an array that is precise, but very susceptible to preparative damage and disarray. The latter seems the more probable, since regions within a single cell showing the most perfect preservation of the myosin hexagonal pattern show a correspondingly close approach to actin regularity.

In contrast to these fibers, the contractile system of most if not all insect flight muscles displays a regular and readily preserved 6-actin orbital (Figures 3, 13) with the 3:1 actin:myosin ratio first described by Huxley and Hanson (1957) and differing from the 2:1 pattern of vertebrate skeletal muscle (Huxley, 1957) in the placing of each actin midway between adjacent myosin filaments rather than in the vertebrate trigonal position. This standardization of myofilament arrangement in insect flight muscles is not correlated with contraction frequency or excitation–contraction characteristics (Smith, 1968; Cullen, 1974; Elder, 1975). Reports of intermediate configurations of actin orbitals of up to 9 (Auber, 1967) in slow-flying Lepidoptera deserve further study before the possibility of imperfect lattice preservation can be ruled out. The same caveat applies to other instances of nonintegral actin:myosin ratios in insect muscle reviewed by Pringle (1972).

The events of fibrillogenesis, and notably the early establishment of the actin:myosin pattern occurring in the mature cell, have been followed in *Calliphora erythrocephala* by Auber (1969) and others (see Pringle, 1972) but as yet little is known from insect material, of the morphogenetic determinants of such features of the cell as the definitive sarcomere length, the degree of transverse alignment of the sarcomere divisions, the presence or absence of midsarcomere transverse M filaments apparently contributing to sarcomere register (Pringle, 1972), and of the developmental siting of the transverse tubular membrane invaginations at a precise level alongside the adjacent sarcomeres. Details relating to some of the above points may be inferred from preliminary studies on other muscles, notably on the behavior *in vitro* of actin and myosin molecules. Hayashi et al. (1977) showed that *in vitro* polymerization of rabbit

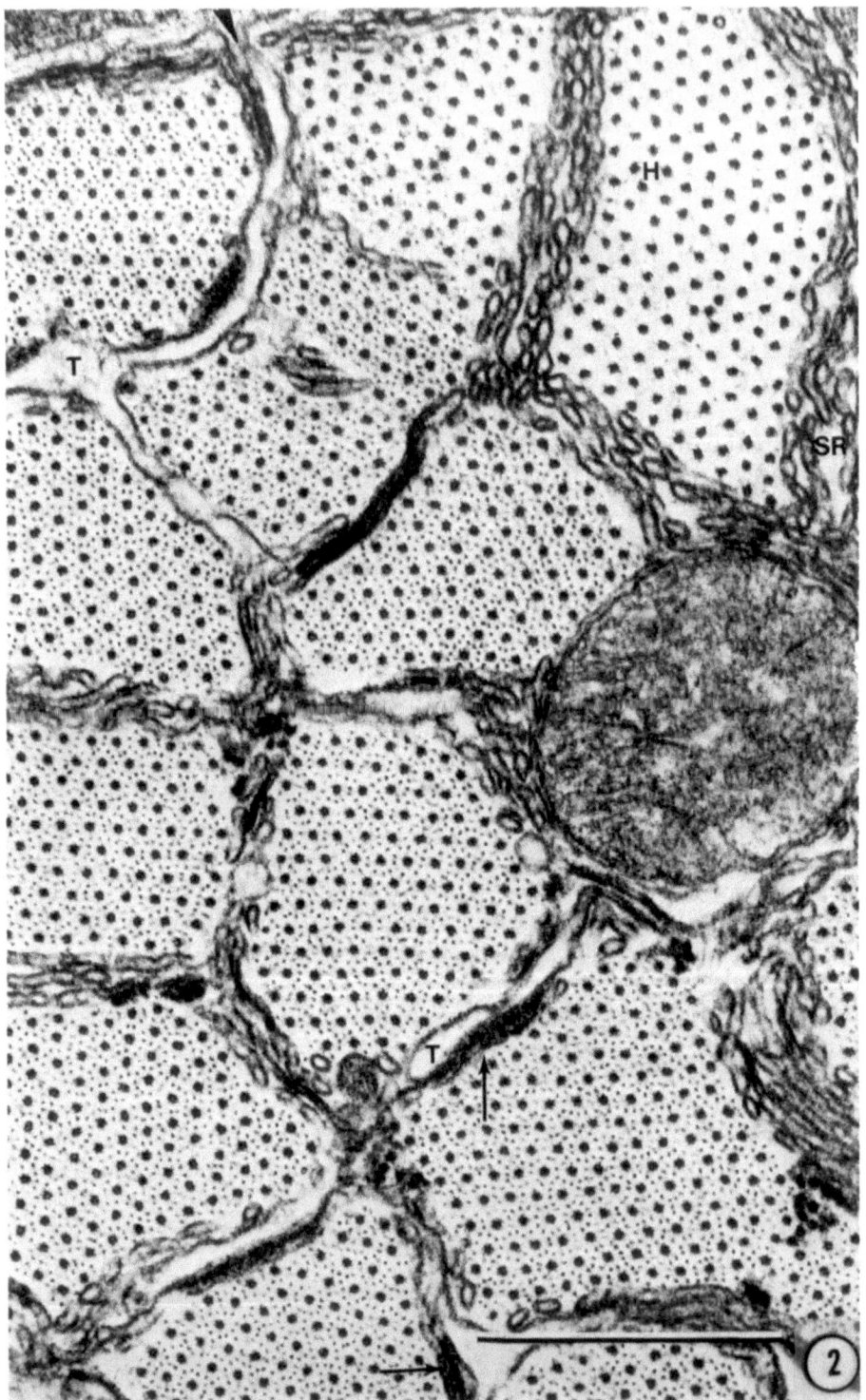

G actin on a preparation of reconstituted myosin filaments derived from rabbit skeletal muscle occurs in a regular fashion, to produce an approximation to the 6-membered actin orbital and with the actin bipolarity characteristic of the *in vivo* myofilament arrangement in the sarcomere. These findings suggest that assembly of myosin molecules into the thick filament may play a directing role, via the resulting cross-bridge pattern, in the placing of actin filaments into the array, and hence the effective and characteristic actin:myosin ratio. It is evident that fibrillogenesis is more complex in the developing muscle: Auber (1969) noted an increase in sarcomere length after establishment of the basic band pattern, and clearly the *in vitro* system of thick-filament assembly is a less than complete replica of assembly in the sarcoplasm. However, the validity of the *in vitro* studies has recently been strengthened by parallel work (Hayashi *et al.*, 1980, 1981, 1983) involving hybrid assemblies of rabbit actin and myosin filaments reconstituted from tonic crustacean muscle, which exhibits a 12-actin orbital as in many insect muscles other than those of flight. When G actin is polymerized in association with *Homarus* (lobster) thick filaments, an actin configuration approximating that in the intact muscle again results, suggesting that the myosin "species" of a particular muscle directs the actin orbital by as yet unidentified features of the myosin molecules of a specific muscle. It seems likely that a similar part is played by myosin "species" in insect muscle (and indeed in muscle in general) in determining the filament array, and it may be anticipated that additional details of muscle morphogenesis will stem from extensions of this approach, in which other protein components of the native system (e.g., paramyosin and M protein), and proteins associated with the Z bands of differentiated sarcomeres are incorporated into the *in vitro* system to provide a more complete simulacrum of the cytoplasmic environment of the developing myofibril.

A substantial amount of our knowledge of the molecular architecture of the myofilaments of insect muscle derives from studies on "fibrillar" flight muscle fibers of the giant waterbug *Lethocerus** in the Oxford Laboratory of the late J. W. S. Pringle summarized in a symposium volume edited by Tregear (1977), covering aspects of filament and sarcomere structure, cross-bridge conformation, and the mechanical and biochemical kinetics of this tissue.

Squire (1977) has discussed structural models of myosin molecule packing within the insect (*Lethocerus* flight muscle) thick filament consistent with the length and diameter of these filaments in this muscle, which differ from their

---

*Pringle (1977) noted that availability of suitable water bug flight muscle often presented a problem in these studies, compounded by muscle degeneration in old adults. While most work was carried out on *Lethocerus cordofanus* obtained from Uganda, flight muscle of the neotropical *L. maximus* and related species and genera was sometimes used. Attempts to rear giant water bugs in the laboratory have been described by Cullen (1977).

Figure 2. Transverse section of a small thoracic muscle of the whitefly *Trialeurodes* exhibiting close-packed myofibrils. The plane of section passes in places through the transverse tubule system (T) and dyad junctions (arrows). Elsewhere (e.g., at the H-zone level), extensive SR cisternae separate the myofibrils. The origin of a transverse tubule is indicated (large arrow). (Bar = 0.5 μm.)

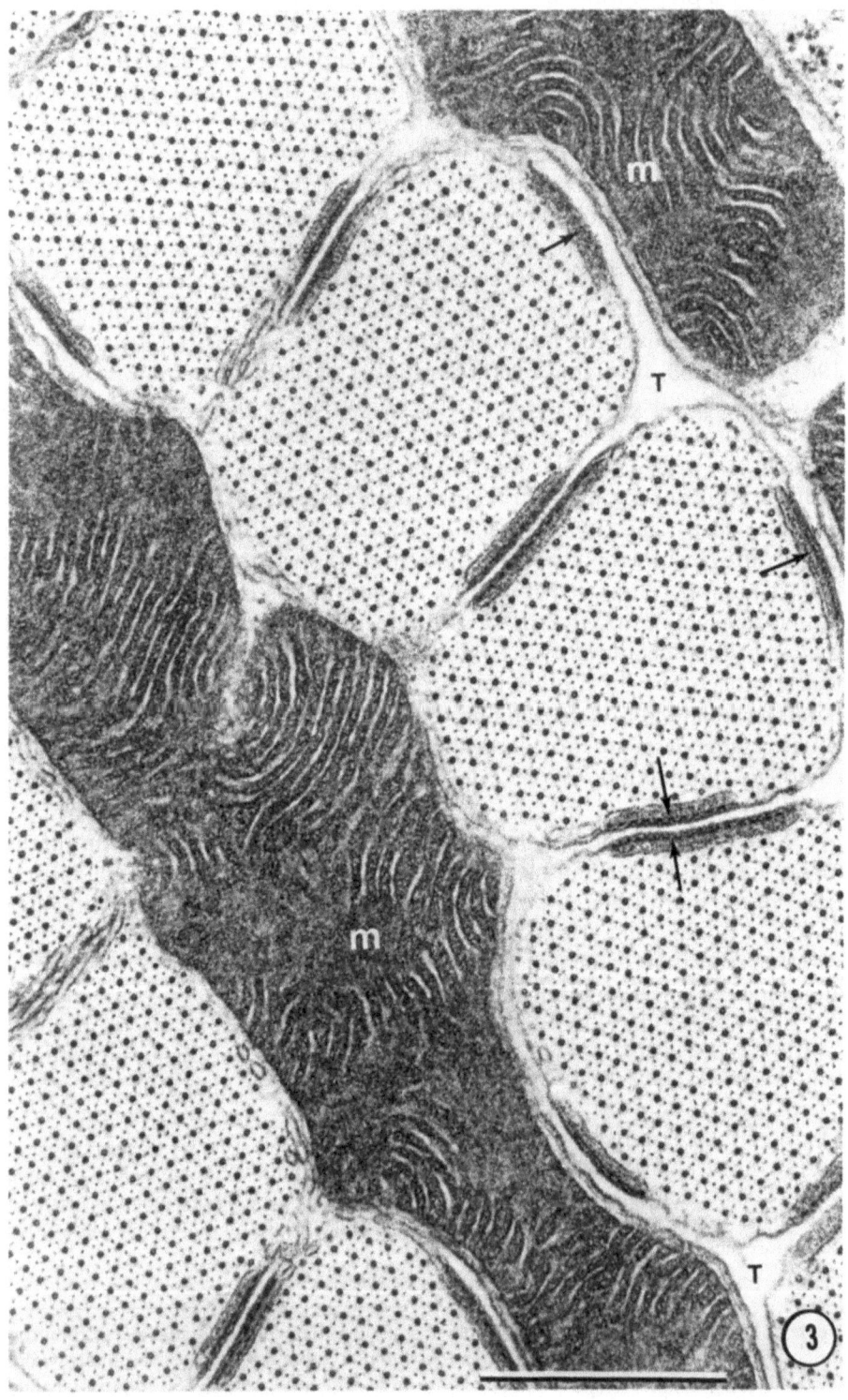

vertebrate skeletal muscle counterparts in being longer (ca. 2.2 vs. 1.57 μm) and cylindrical rather than triangular in cross-section with a greater backbone diameter (18–20 vs. 12–14 nm) and differing further from vertebrate thick filaments in frequently displaying an electron-lucent "core." With the premise that myosin heads are equivalently placed on the myosin filament surface and, further, that all myosin molecules are packed in an equivalent manner, Squire concludes that the appearance of thick filaments in thin sections may reflect either continuous-layer cylindrical packing of myosin molecules to yield a six-stranded filament or, alternatively, that myosin rods are arranged in subfilaments which are then arranged helically to constitute the cylindrical filament shell. In each instance, the model can accommodate the required bipolarity of any myosin assemblies in muscle, together with the observed cross-bridge repeat sequence.

That force generation in muscle may be effected by movement of the cross-bridges of the myosin filament, while attached to the adjacent actin filament, was first given fine structural support by Reedy *et al.* (1965) in a comparison between the cross-bridge array in relaxed and rigor preparations of *Lethocerus* flight muscle. Under rigor conditions, the cross-bridges conformed to polarized series of regular "chevrons" pointing toward the center of the sarcomere in each A band, contrasting with lack of symmetrical cross-bridge array in relaxed muscle. While these preparations did not solve the question of the details of putative cross-bridge movement attending filament "sliding," Reedy has exploited the rigor muscle preparation, by correlated fine structural analysis and X-ray diffraction studies, to construct a model of the surface lattice of the myosin filament of this muscle, and of its proposed interaction with the actin array (see Reedy and Garrett, 1977, and Squire, 1981, for detailed descriptions and discussions).

2.2. Z-Band Structure and Supercontraction

The organization of the electron-opaque Z bands/discs defining the sarcomere of striated muscles has received a good deal of attention, particularly in insect flight fibers, and it is evident that this region of the contractile system varies both in fine structure and in function. Again, the most detailed analysis has been made on flight muscle fibers of *Lethocerus*. Ashhurst (1977) traced the filament lattice progression, in transverse and slightly oblique thin sections, from the A band of one sarcomere, through the narrow I band of this muscle into the Z-band region and into the I and A lattices of the adjoining sarcomere, and discussed her results in the context of previously published models of Z-band structure. *Lethocerus*, the dung beetle *Scarabaeus femoralis*, *Calliphora*

←

Figure 3. Transverse section showing the relationship between membranes and myofibrils in the synchronous timbal muscle of a cicada (*Tibicen* sp.). The plane of section passes extensively through the T system (T) which encircles the close-packed polygonal fibrils. The tubule membranes are flanked by SR cisternae either singly or in pairs: the former (dyad) configuration occurs where the myofibrillar surface adjoins a mitochondrion (m) (single arrows) and the latter where the T tubule passes between a pair of myofibrils (paired arrows). From Smith (1972). (Bar = 0.5 μm.)

erythrocephala, and the honeybee *Apis mellifera* appear to share a common structural pattern in which I-filament lattices of adjacent sarcomeres achieve continuity via interdigitation in the Z band where actin filaments from the two half-sarcomeres are grouped into hexagons, defined by intervening opaque "Z material" and appearing as "holes" in transverse profile. The rotation of the filament lattices across the Z band has the further consequence that myosin filaments of one half-sarcomere are trigonally placed with respect to the thick filaments across the Z band. This model was supported by Sainsbury and Hulmes (1977) in optical reconstruction of electron micrographs, showing that each "hexagon" in the Z band comprises three pairs of I filaments and that the filaments of each pair are derived from adjacent half-sarcomeres.

Sainsbury and Bullard (1980) showed that actin is an important protein component of Z discs, further substantiating the proposed continuity of the thin filaments of the A band with those extending through the Z region. These authors also noted tropomyosin and α-actinin as major constituents of the Z band, the former possibly contributing to the Z-band lattice and the latter performing a cross-linking function. Further work on the protein composition of this region of the sarcomere is mentioned in the next section.

Ashhurst (1977) discussed the available evidence for and against the presence of "connecting filaments" linking the tapered ends of the thick filaments with the Z band, invoked as providing the resting elastic properties of these flight muscles (Pringle, 1974). She concluded that conventionally prepared thin sections provide no clear evidence of such structural links. Immunochemical information on this region of the sarcomere is discussed in Section 2.3.

It should be noted that the above model of Z-band structure, with minor variants ascribed by Ashhurst to differences in preparative techniques, relates only to the highly specialized "fibrillar" asynchronous flight muscle, though a similar Z-band structure has been reported for synchronous flight muscle of the lepidopteran *Antherea pernyi* (Bienz-Eisler, 1968). Of actin filament association in Z bands of slowly contracting insect muscles virtually nothing is known, and careful study of this region of insect visceral or cardiac muscles would be very useful. One modification of the Z region that has been well documented, however, involves the provision for "supercontraction."

In a conventional (vertebrate) skeletal muscle, operating under physiological conditions, the extent of linear contraction is limited by the proportion of the relaxed sarcomere occupied by the I bands, in contrast to smooth muscle, in which such transverse limiting partitions are absent and which can contract beyond 50% of resting length. It is remarkable that some animals, and notably arthropods, have evolved striated muscle fibers but have incorporated devices enabling them to circumvent the system—a most curious instance of cellular evolution! Supercontraction was first described by Hoyle *et al.* (1965) in a long-sarcomere striated barnacle muscle, capable of contracting by 80% of resting length, an ability accommodated by perforation of the Z discs through which both sets of myofilaments may penetrate into adjoining half-sarcomeres, enabling the cell to produce tension at lengths far shorter than permitted by occlusion of the I bands of the conventional sarcomere (Hoyle et al., 1973). A comparable device was described by Osborne (1967) in *C. erythrocephala* intersegmental muscle and shown by Hardie (1976) to be mechanically equivalent to the properties of unstriated fibers in other animals.

In some instances, supercontractility is reportedly achieved not by perforation of the Z disc but by fragmentation of discontinuous blocks of Z material (Candia-Carnevali, 1978; Candia-Carnevali et al., 1980), and a similar mechanism has been invoked by Jorgensen and Rice (1983) to account for the extreme extensibility and contractility of muscles controlling extension of the locust ovipositor.

While the device of a fragmented Z disc may occur in some instances, filament interpenetration via a perforated or reticular Z region is at present better documented. Hardie and Hawes (1982) have pointed out that discontinuity of Z-band material in conventional thin sections may be apparent rather than real and due to "sampling" of a reticular structure, and conversely that apparent perforations may be an artifact of sectioning a continuous but undulating Z band. These authors clearly established the reticular structure of Z discs of body wall muscle in *C. erythrocephala* larvae and larvae of a tipulid fly (Figures 4-7), by comparing the high-voltage EM aspect of "thick" sections and the thinner conventional profiles. Elder (1975) has reviewed accounts of supercontracting muscles in insects, and suggests that this ability is widespread in insect visceral, cardiac, and many larval skeletal muscles, and Osborne (1967) has considered the possible functional analogy between such fibers and muscles of other invertebrates in which blocks of Z material are present, but in which the contractile system lacks the transverse alignment of a striated cell. In a remarkable instance of convergent cellular design, Rice (1973) showed that the striated muscle controlling retraction of the chameleon tongue is able to contract reversibly to 16% of resting length—an ability accommodated by perforation of the Z disc, as in supercontracting insect fibers.

Despite structural evidence of the widespread adoption of supercontractility in muscles of the "visceral" system (Rice, 1970; Elder, 1975), some functions of the internal organs make this inappropriate, and the muscles involved differ accordingly. Rice (1970) described perforated/discontinuous Z discs in "intrinsic" visceral fibers of adult tsetse flies, for example in muscles investing the midgut and esophagus, and in the latter noted multidirectional placing of myofibrils within a single cell—apparently the basis of reports of branching and anastomosing fibers in the histological literature. However, Rice found that the "extrinsic visceral" muscles dilating the cibarial feeding pump are structurally typical of synchronous phasic fibers, with transverse sarcomere register and regular dyad disposition (see Section 4.2). Since these muscles are inserted distally in the head capsule, it seems a largely semantic point whether they are regarded as "visceral" components or skeletal fibers, coopted into the visceral system. Alary muscles of the moth *Hyalophora cecropia* (Sanger and McCann, 1968) similarly possess one exoskeletal insertion (qualifying them as extrinsic visceral cells) but they are probably supercontracting, judged on structural

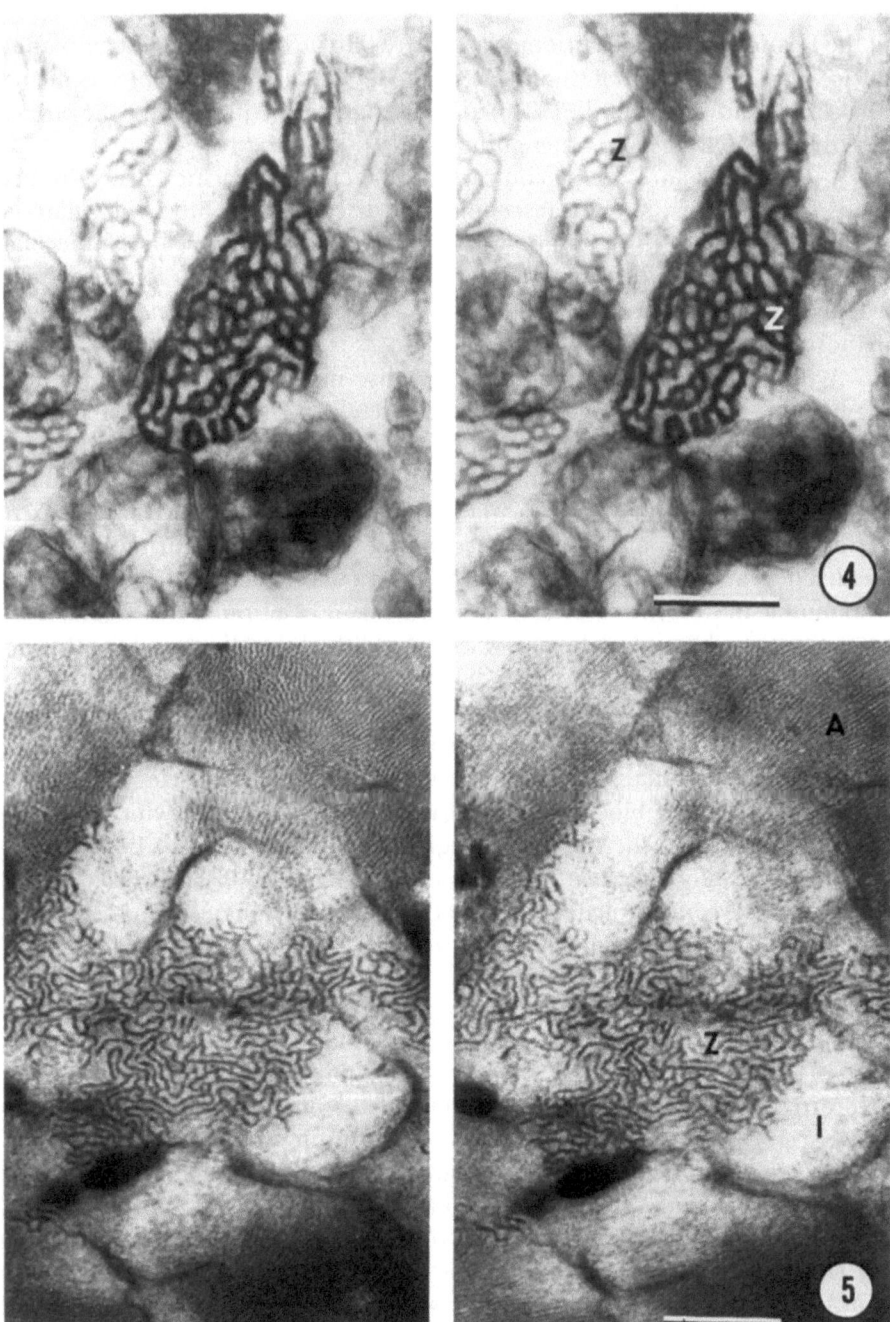

Figure 4. Stereo-pair of high-voltage (1 MV) electron micrographs of 1-μm transverse section of body wall muscle of *Calliphora*. The tissue was glycerinated and extracted with KCl: cytoplasmic and membranous components are disrupted, but myofibrillar structure is retained, notably the reticular organization of the Z bands (see also Figures 5-7). From Hardie and Hawes (1982). (Bar = 1 μm.)

Figure 5. Stereo-pair of high-voltage micrographs of transversely sectioned body wall (intersegmental) muscle of a larval tipulid, prepared as in Figure 4. Areas of A and I band are included, together with a Z-band region comprising a reticulum that extends through the 1-μm thickness of the section. From Hardie and Hawes (1982). (Bar = 1 μm.)

Structure of Insect Muscles 123

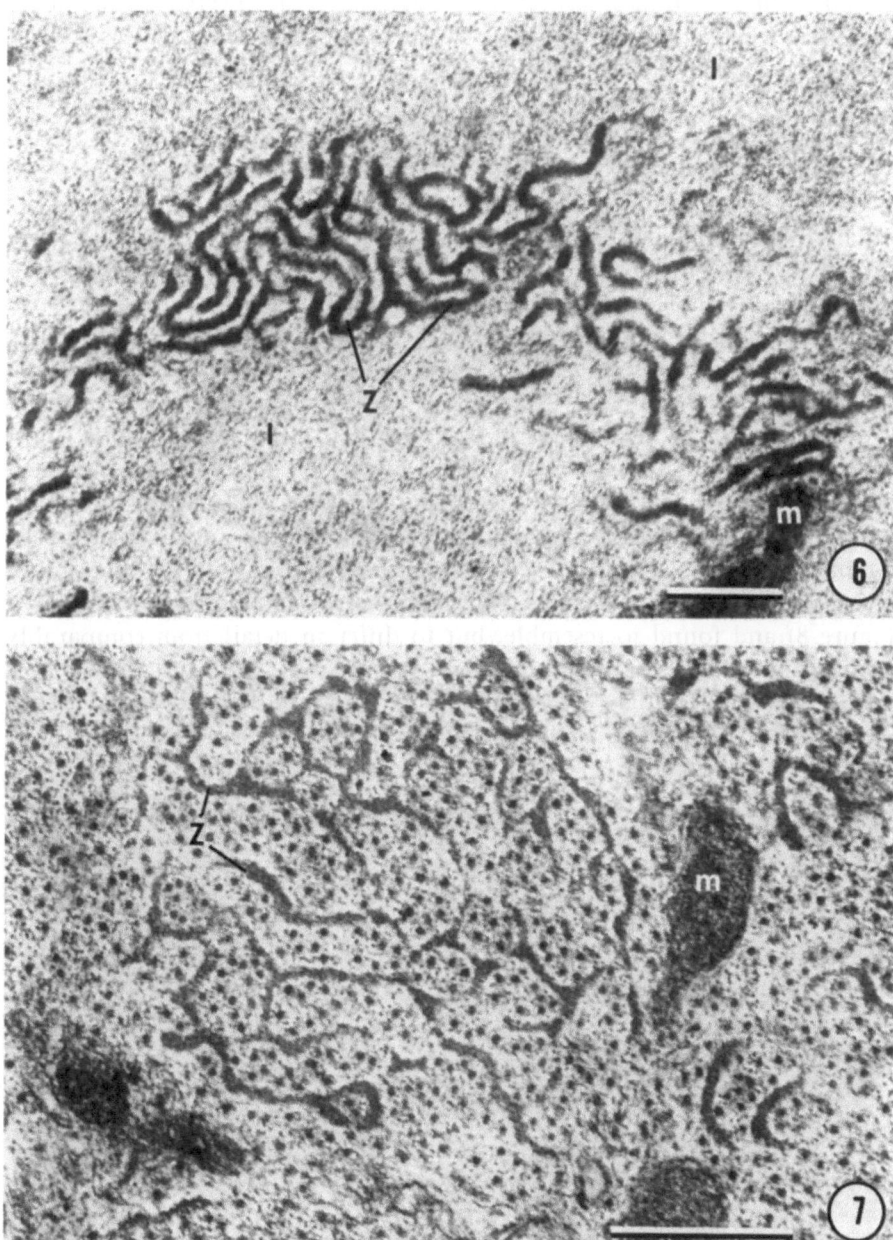

Figure 6. Conventional electron micrograph of transversely sectioned tipulid body wall muscle. In this preparation the myofibrils are relaxed and the Z band is represented by a series of discrete elongate ribbons, shown by comparison with Figure 5 to be profiles of a three-dimensional network. Mitochondria (m) are sparsely distributed in this muscle. From Hardie and Hawes (1982). (Bar = 0.5 μm.)

Figure 7. Similar to the last figure, but illustrating tipulid muscle in the "supercontracted" state. In most areas, thick and thin filaments are present between the Z-band ribbons. From Hardie and Hawes (1982). (Bar = 0.5 μm.)

grounds. These examples may serve as a reminder that in cell structure, as in other areas where taxonomy is used, a system of classification is only as good as the parameters on which it is based.

2.3. Paramyosin and Connecting Protein

During the isolation procedure for tropomyosin from insect flight muscle, Bullard *et al.* (1973) separated and characterized the fibrous protein paramyosin, first isolated from molluscan muscle and subsequently found in fibers of other invertebrates (references in Bullard *et al.*, 1973). Again, this work was carried out on asynchronous flight muscle, of which adequate amounts of starting material were available. It was estimated that the proportion of paramyosin in the thick filament varies substantially, from about 9.5% in the rose chafer *Pachnoda ephippiata* and 6.3% in *Lethocerus*, down to 1% in *Sarcophaga bullata* flight muscle (Bullard and Reedy, cited in Bullard *et al.*, 1973). It was suggested that the small amount of this protein in fly flight muscle may be related to the high wing beat frequency compared with that of other insects assessed, and it was proposed that paramyosin may constitute a core protein by analogy with the situation in molluscan muscles. Paracrystals of purified insect paramyosin were prepared (Figure 8) and found to resemble, but to differ in detail from comparably produced paracrystals of molluscan paramyosin.

Subsequently, Bullard *et al.* (1977a,b) used immunochemical techniques to localize the sites of paramyosin exposed along the thick filament of insect flight

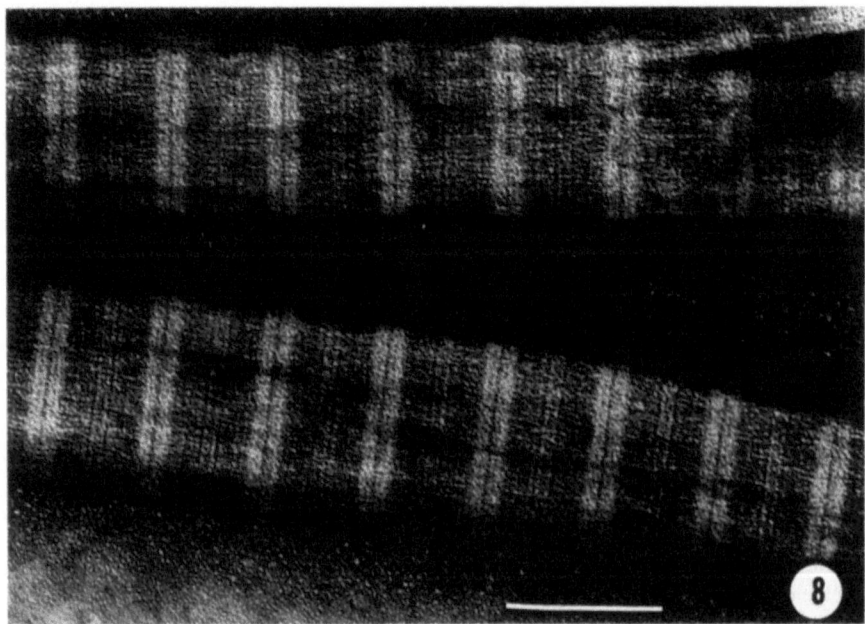

Figure 8. Negatively stained paramyosin paracrystal produced after purification of the protein from asynchronous flight muscle of *Lethocerus*. From Bullard *et al.* (1973). (Bar = 1 µm.)

muscle. Both fluorescent and EM identification of antibody to dung beetle (*Heliocopris japetus*) paramyosin was employed in sarcomeres of *Apis* and *Lethocerus* flight muscle. Binding occurred in the center of the sarcomere and at the ends of the myosin filaments in the latter (Figure 9), but in *Apis* the antibody attached in the midsarcomere region only in sarcomeres in which the M band was disrupted and it was suggested that paramyosin is exposed only in the central region of the thick filament. This observation did not preclude the possibility (Squire, 1971; Bullard *et al.*, 1973) that paramyosin is a core protein, but in a later paper (Bullard *et al.*, 1977b) it was noted that while the percentage of this protein in *Lethocerus* would fill the "hollow" core, this would not be possible in flies, with a much lower paramyosin complement, and it was concluded that paramyosin is probably concentrated in the central region of the thick filament, and the functional significance of this protein, other than the general inverse correlation between its amount and wing beat frequency, mentioned above, has yet to be determined.

The initial antiserum to paramyosin (Bullard *et al.*, 1977a) contained antibodies to another protein as a fortuitous contaminant. Since this bound at the ends of the thick filaments (Figure 9), it possibly revealed a nonmyosin connecting protein linking the thick filaments, through the I band, with the Z region. This arrangement may provide the structural basis of the high resting elasticity of these flight muscles and hence its ability to perform oscillatory contraction via stretch-activation (Pringle, 1974, 1978).

3. Muscle Insertions

The mechanical efficiency of arthropod skeletal and visceral muscle fibers, as in their counterparts in the vertebrate body, is ensured by attachment devices that are functionally analogous though differing in fine structure, especially in the case of exoskeletal insertions. Insect visceral fibers investing various organs and tissues including the midgut, Malpighian tubules, salivary and other glands, and portions of the reproductive system (Smith, 1968) are typically small in diameter, containing a contractile system showing little or no subdivision into myofibrils, and, as Elder (1975) points out, the Z-band structure of these cells suggests that they are generally capable of supercontractility. The restriction of the contractile system may be marked, as in the very small fibers associated with the midgut of a strepsipteran parasitoid illustrated in Figure 14. As in vertebrate muscles of the visceral system, these cells are more or less closely attached to the basal surface of the epithelia by sheets of extracellular material, which also hold in place the terminal tracheal branches. These connective sheets probably represent the basal laminae of the cells involved, and the collagen component of vertebrate connective tissue is generally absent. Collagen fibrils are present elsewhere in the insect body (Smith, 1968; Ashhurst, 1968), notably in the extracellular neural lamella of the central nerve cord, and collagen is involved in the spider venom gland (Smith *et al.*, 1969) both in the elaborate capsule investing the gland and in its attachment to the visceral musculature.

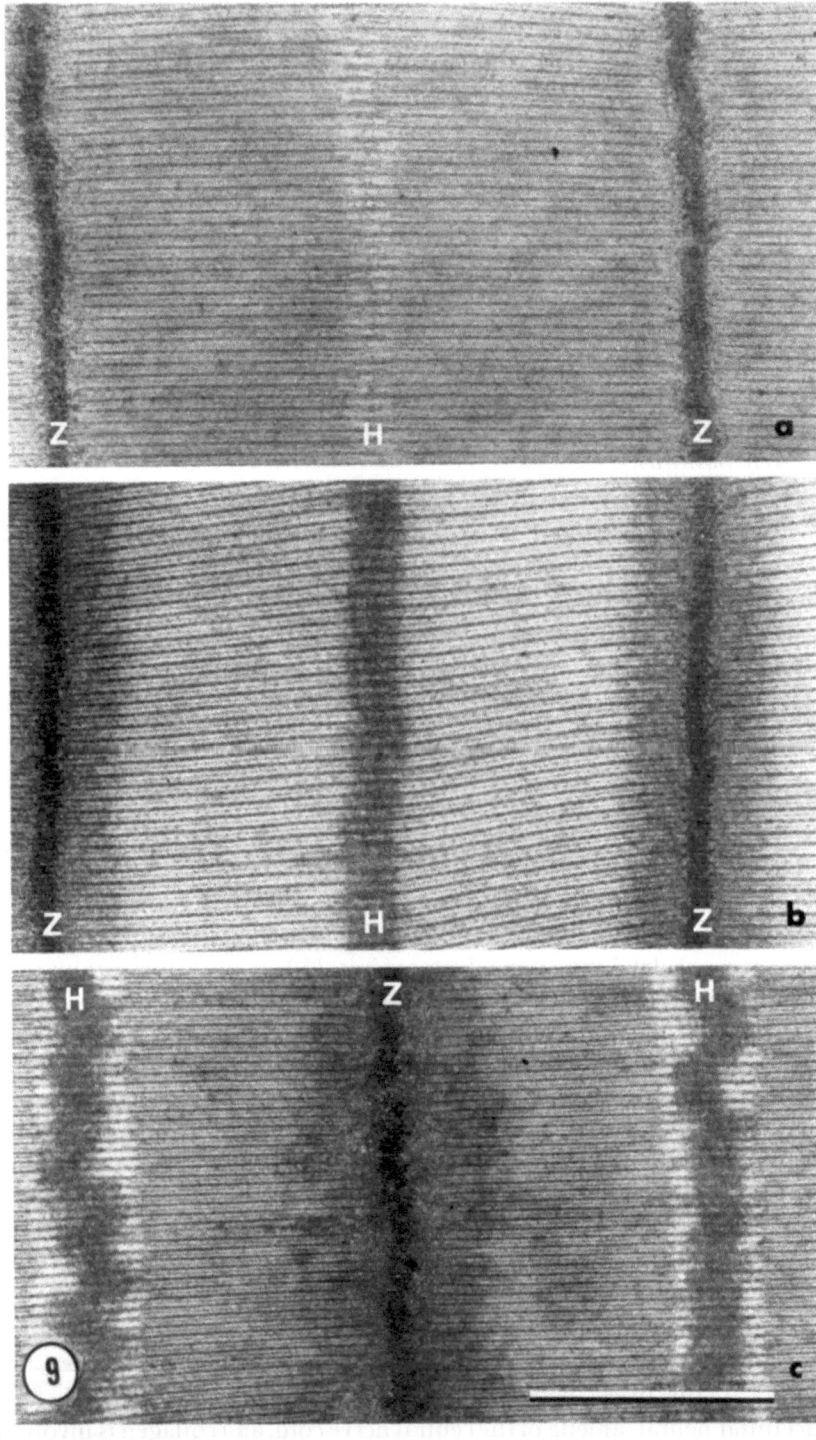

In the myotendon junction of vertebrate skeletal muscle fibers (Smith, 1972), the end of the fiber is dissected into longitudinal processes which interdigitate with the collagen-rich extracellular connective matrix; the actin filaments of the terminal half-sarcomere inserting in the junctional plasma membrane of the fiber. Skeletal muscles of arthropods are inserted either directly onto cuticular sclerites of the body surface or onto inwardly directed apodemal extensions of the cuticle—frequently mechanically analogous to tendons of the vertebrate body. The special case of myocuticular insertion in asynchronous flight muscle will be mentioned below, but it is clear that, in general, attachment of arthropod muscles to the exoskeleton is effected by means of specialized epidermal cells in which the cytoplasm is largely devoted to the elaboration of microtubules oriented in parallel with the long axis of the adjacent muscle fiber (Lai-Fook, 1967; Caveney, 1969; Smith et al., 1969) (and discussed by Pringle, 1972, and Elder, 1975). The basal surface of the epidermal cell involved bears processes that interdigitate with matching projections of the muscle cell, and the apposed plasma membranes are linked by desmosomes. While the epidermal microtubules have not been assessed chemically or by response to vinblastine, their size and morphology suggest that they represent cytoskeletal tubulin polymers. Electron micrographs show that the microtubules are more or less evenly spaced, but they do not constitute regular arrays, and the cross-links present in some instances where microtubules perform a static mechanical role (Burnside, 1975) are not conspicuous. Further, the details of attachment of these structures to the basal and apical epidermal cell membrane have yet to be investigated.

In asynchronous ("fibrillar") flight muscle, the discrete cylindrical myofibrils are inserted individually on processes of the epidermal cell containing similar microtubule bundles—the "tonofilaments" of light microscopic accounts and of the first fine structural description of these insertions in flight muscle of the hoverfly *Erystalis tenax* and *C. erythrocephala* (Auber, 1963). As in other myocuticular insertions, the intermembrane gap underlying the insertion of each myofibril is modified as a continuous desmosomal sheet incorporating a medial lamella and linking structures oriented perpendicular to the adjacent membranes. The longitudinal aspect of inserting myofibrils in the whitefly *Trialeurodes vaporariorum* is illustrated in Figure 10 (and this muscle is further discussed in Section 4.3). The terminal Z band of each myofibril is absent and in its place is interposed a band of intermediate electron opacity containing the "junctional filaments" (Auber, 1963) inserting on the muscle plasma membrane.

←

Figure 9. (a) Longitudinal section of sarcomere of belostomatid (*Lethocerus*) flight muscle: a control preparation incubated in rabbit serum prior to injection of paramyosin. No serum binding is evident. (b) and (c) represent material labeled with unfractionated antiserum to dung beetle paramyosin: (b) an unstretched *Lethocerus* myofibril treated with antibody in rigor solution, in which antibody is bound at the ends and center (H zone) of the sarcomere. (c) includes portions of two sarcomeres of a fibril treated with antibody in relaxing solution after stretching by 10% over resting length: antibody binds only in the center of the H zone and also fills the region where myosin filaments have been pulled away from the Z bands. The latter binding site is believed to represent the connecting protein. From Bullard et al. (1977a). (Bar = 1 μm.)

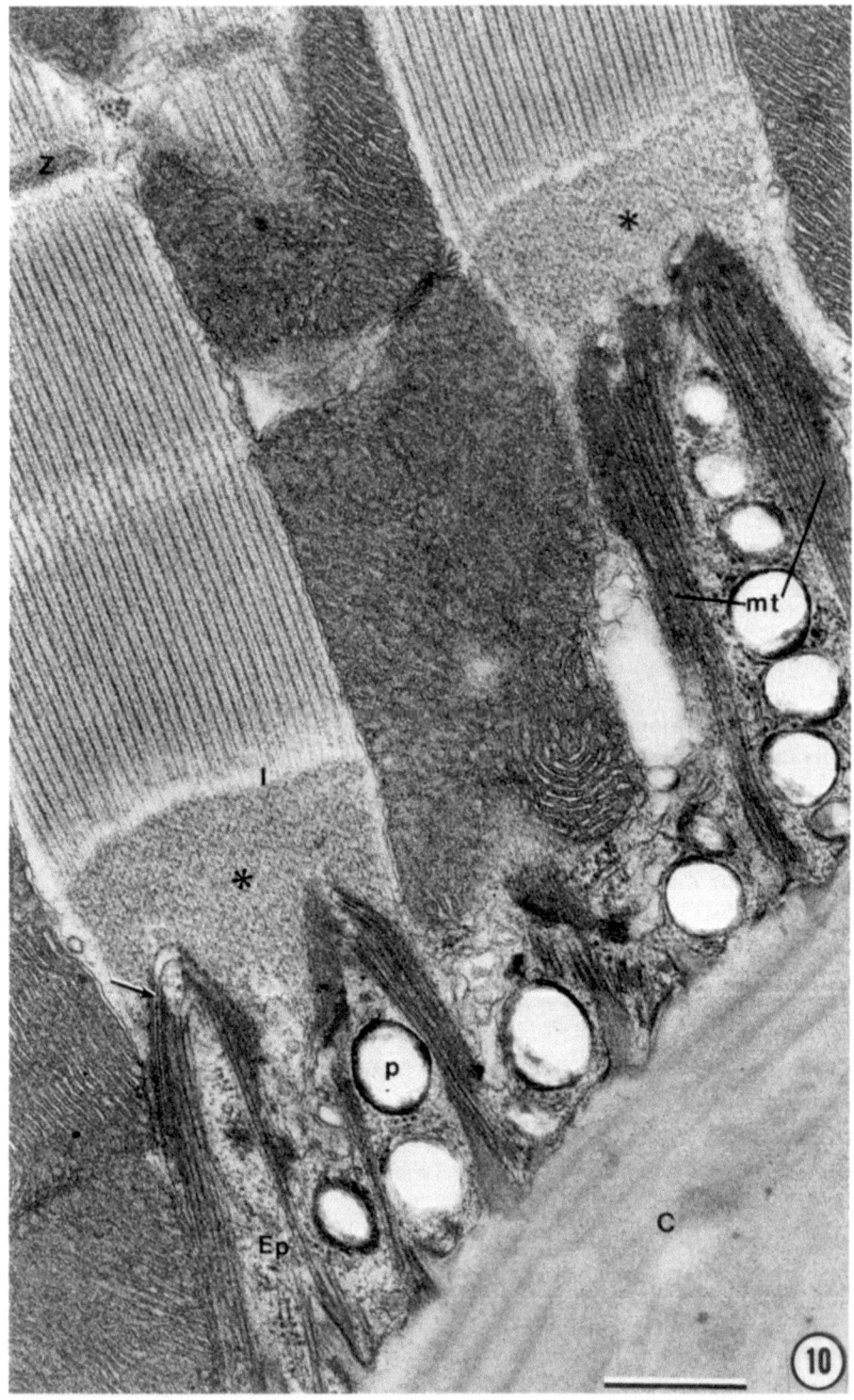

Preliminary observations in *Trialeurodes* suggest that this terminal band includes lengthened actin filaments from the terminal half-sarcomere possibly accompanied by thin extensions of the terminal thick filaments with oriented linking structures contributing to the opacity of this specialized band. Further clarification of the detailed organization of this region may be expected from freeze-fracture replication of the muscle cell plasma membrane to which the terminal filaments are attached, and from thin-sectioning augmented by immunochemical techniques such as those that have shown the presence of α-actinin (Goll *et al.*, 1977), connecting protein (Bullard *et al.*, 1977a), and actin (Sainsbury and Bullard, 1980) within the nonjunctional Z bands of flight muscle myofibrils.

4. Membrane Systems

4.1. General Features

Our knowledge of the functional morphology of the plasma membrane of insect muscle cells, in the absence of adequately extensive comparative data from freeze-fracture replication, rests primarily on documentation of its association with the contractile system via the invaginated transverse tubular (T) system (Figure 11) which, with the adjoining membrane-limited compartments of the sarcoplasmic reticulum (SR), couple surface membrane excitation with the initiation of contraction. The general role of these membranes has been reviewed elsewhere (Smith, 1966b, 1972; Elder, 1975; Pringle, 1981), and the present account is primarily concerned with comparative studies that have thrown further light on the evolution of functional diversity in insect muscles, and on cell biological aspects of the distribution and interrelations of membranes playing a role in the regulation of muscle function.

As in vertebrate visceral fibers, the smallest visceral muscles of insects (ca. 1–2 μm in diameter) as illustrated in Figure 14 are limited by a simple cylindrical plasma membrane, conforming to Hill's early model (1949) predating fine structural information, in which excitation was envisaged as involving ionic diffusion from the cell surface. Unlike their vertebrate counterparts, however, some insect visceral muscles exhibit a limited T/SR system with little or no spatial rapport with the staggered sarcomere banding of the cell (Smith *et al.*, 1966; Elder, 1975). Increased fiber diameter, as in the muscles investing the spermatheca of *Periplaneta americana* (Gupta and Smith, 1969), is accompanied by an increase in the extent and regularity of the membrane invaginations.

←

Figure 10. Longitudinally sectioned asynchronous (= "fibrillar") flight muscle of the whitefly *Trialeurodes vaporariorum* at the region of myocuticular insertion. In this muscle, each cylindrical myofibril inserts separately on conical protuberances of the epidermal cell (Ep) containing aligned bundles of microtubules (mt) and pigment granules (p), flanking the cuticle (C). The penultimate Z band of the myofibril (Z) is of normal appearance, but in place of the terminal Z band, a broader and less electron-opaque zone (∗) is interposed between the I band and the muscle plasma membrane. Desmosomal junctions (arrow) are present between the apposed muscle and epidermal cell surfaces. (Bar = 0.5 μm.)

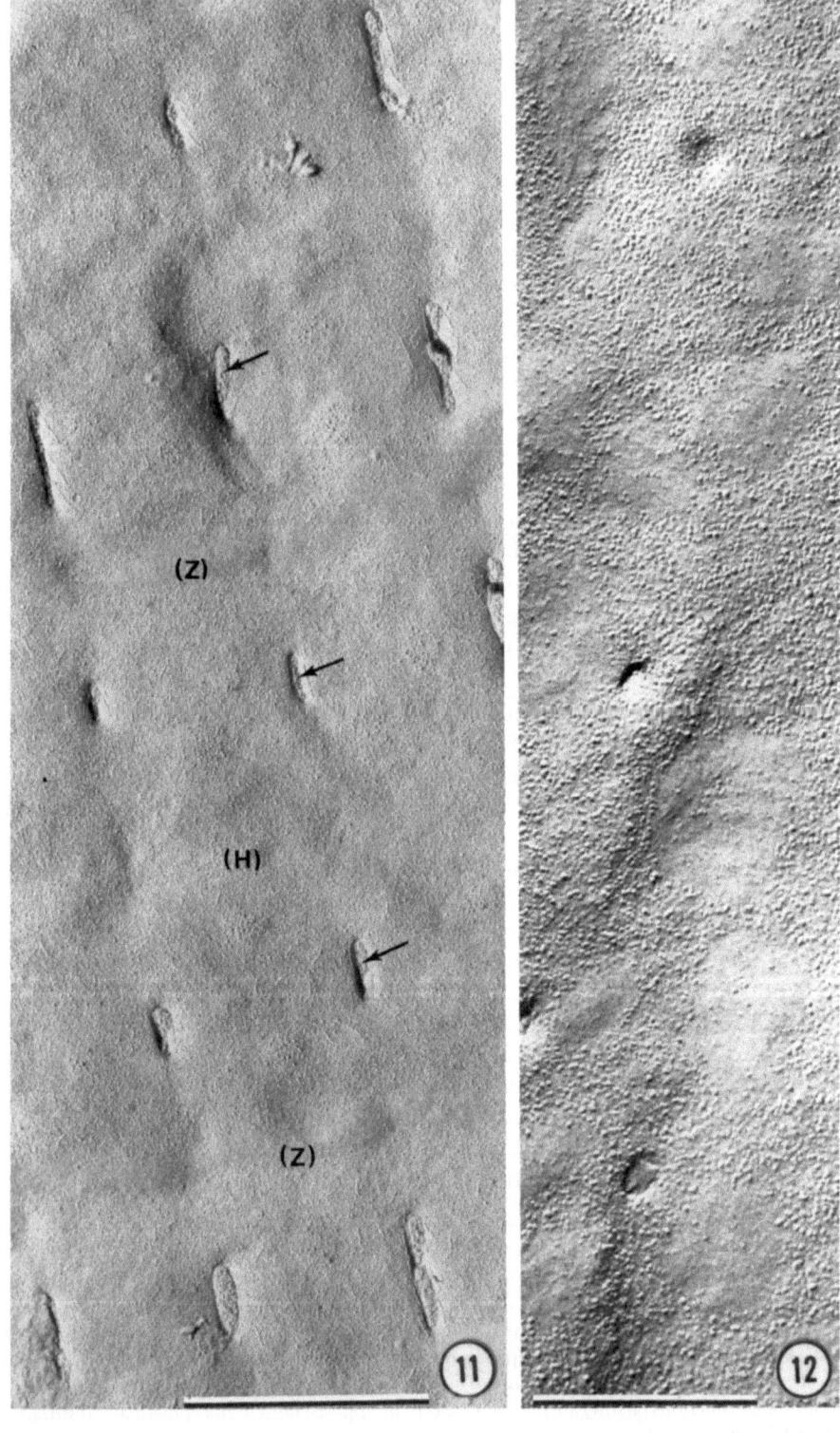

In insect muscles with precise transverse alignment of the sarcomere divisions, the positioning of the T-tubular invaginations is generally as regular as in vertebrate skeletal fibers, though the sarcomere levels selected and the geometry of the association with SR cisternae differ in the two groups of animals. However, the similarities far outweigh the differences—perhaps evolutionary quirks in cells of animals very distantly related—and clearly suggest functional equivalence wherever these membranes are present. The evolution of asynchrony in some insect muscles has been accompanied by modification in the distribution of the SR compartment that has clarified its funtion in striated muscle cells in general, and for convenience, the arrangement of membranes in synchronous and asynchronous insect fibers will be considered in sequence.

4.2. Synchronous Muscles

In vertebrate skeletal muscles, and in insect muscles other than asynchronous fibers considered in Section 4.3, the motor impulse train initiating the sequence of surface membrane depolarization is followed synchronously by a complete (phasic) contraction–relaxation cycle or by some stepwise enhancement contraction. In all these cells, the T-tubular invaginations enter the cell radially at a preferred sarcomere level, thus determining the position of the dyadic* or triadic junctions with the SR cisternae. Evidence of continuity between the T system (the "intermediary vesicle" of the triad in an early acount of the membranes of vertebrate muscle; Porter and Palade, 1957) and the surface plasma membrane was first obtained in a synchronous insect muscle (Smith, 1962) and more conclusively shown in flight muscle of an aeshnid dragonfly (Smith, 1966a). Direct evidence of confluence between the T-system compartment and the extracellular space, using ferritin as a marker, was obtained by Huxley (1964) in frog skeletal muscle. Structural continuity was demonstrated in fish skeletal muscle by Franzini-Armstrong and Porter (1964) and in frog muscle by Franzini-Armstrong (1970).

---

*Porter and Palade (1957) introduced the term *triad* to describe the group of three structures later recognized as the central T tubule and an adjoining pair of SR terminal cisternae, in skeletal muscles of vertebrates. The corresponding term *dyad* was proposed by Smith (1961b) for the two-membered T-SR configuration commonly seen in insect muscles, and defined as a "group of two." The variant *diad* is sometimes used, incorrectly.

←───

Figure 11. Freeze-fracture replica of the P face of the plasma membrane of the synchronous flight muscle in the dragonfly *Celithemis eponina* showing the origin of the aligned transverse tubule invaginations (arrows) illustrated in section in Figure 17. These pass into the fiber in a radial direction, between the Z and H levels of the sarcomeres—of which the positions underlying the plasma membrane are indicated. From Smith and Aldrich (1971). (Bar = 1 µm.)

Figure 12. A comparable replica of the plasma membrane of asynchronous flight muscle of the honeybee *Apis mellifera*. No regular array of T-tubule apertures is present, though occasional pits possibly leading to membrane invaginations occur. In this muscle, most of the T tubules arise from plasma membrane cylinders entering the cell around the invasive tracheoles. The position of the underlying sarcomeres cannot be determined in this muscle (cf. Figure 9): intramembranous particles are generally distributed over this fracture face except for irregularly placed zones where they are sparse. (Bar = 0.5 µm.)

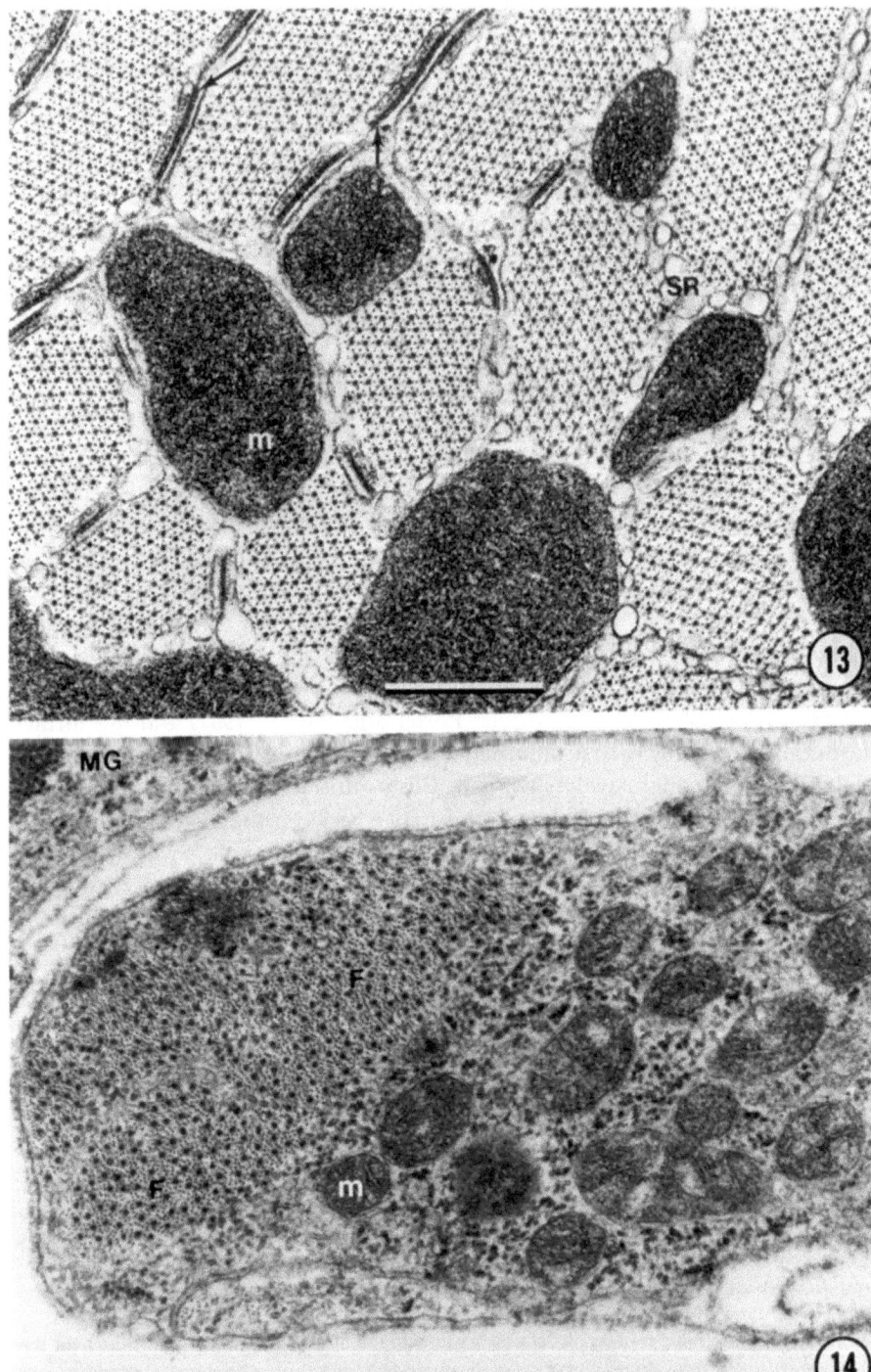

These and other studies (references in Smith, 1966b; Elder, 1975) established that while the preferred level of entry of T tubules in vertebrate skeletal fibers is either opposite the Z band or, in a configuration resulting in two membrane invaginations per sarcomere, near the A–I junction, in synchronous insect skeletal muscles the T tubules invaginate from the plasma membrane very uniformly flanking the A band, approximately midway between the Z level and the center of the sarcomere. This is true not only of flight and other fibers in which sarcomere register is good, but also of some muscles (e.g., intersegmental cells; Smith, 1968) in which band alignment is staggered. It is clear that the morphogenetic "guide" determining the position of the T tubules is rigorous for a specified muscle, in vertebrates and insects alike, representing an unsolved problem of myogenesis. As mentioned further in the next section, this problem is perhaps even more striking in asynchronous insect muscles in which the T tubules often arise from membrane cylinders entering the cell in association with invasive tracheoles yet similarly traverse the sarcoplasm at a determined sarcomere level (Figures 15, 16).

In vertebrate skeletal fibers and short-sarcomere insect fibers alike, the siting of transverse tubules brings an excitable continuation of the plasma membrane to within about 1 μm of all parts of the contractile system. No functional correlate of the variation in T-system level is obvious. The same is true of the differing topography of the SR cisternae between muscles of insects and vertebrates; however, the details of membrane apposition at the dyad/triad junction, across which T-tubule excitation initiates release of calcium from the SR, appear to be equivalent regardless of phyletic position (Smith, 1980).

Two advantages enjoyed by the insect physiologist concerned with general questions of relationships between cell structure and function are, first, the frequent accessibility of an insect tissue vs. its vertebrate counterpart, and, second, the wealth of comparative histological and physiological information from a wide range of insects may facilitate selection of an example of a cell type particularly well suited to structural study. Pringle and his colleagues (references in Tregear, 1977) selected the flight muscle of belostomatid waterbugs as optimal for work on myofibril function. For structural analysis of muscle membrane systems, the geometry of synchronous flight muscle of dragonflies, elaborated on the "tubular" plan (Tiegs, 1955) with lamellar myofibrils and mitochondria arranged in regular radial fashion and with correspondingly flat-surfaced SR

⟵―――――――――――――――――――――――――――――――――――

Figure 13. Transversely sectioned flight muscle of an embiopteran. The myofibrils are close-packed and interspersed with numerous mitochondria (m). As in other synchronous fibers, dyads (arrows) lie opposite the A band (between the Z and H levels); at the right of the field, these lie out of the plane of section, and SR cisternae separate the fibrils (as in Figure 11). Though not included in this field, internalized tracheoles are present in this muscle. Micrograph produced in collaboration with C. B. Cottrell. (Bar = 0.5 μm.)

Figure 14. Transversely sectioned visceral muscle fiber of the larval strepsipteran parasitoid *Elenchus tenuicornis*. These very small muscle cells lack T-tubule invaginations and SR cisternae and, as in this field, the contractile material (F) may be restricted to a portion of the sarcoplasm. Mitochondria (m) adjoins the group of myofilaments. Micrograph prepared in collaboration with J. Kathirithamby. (Bar = 0.5 μm.)

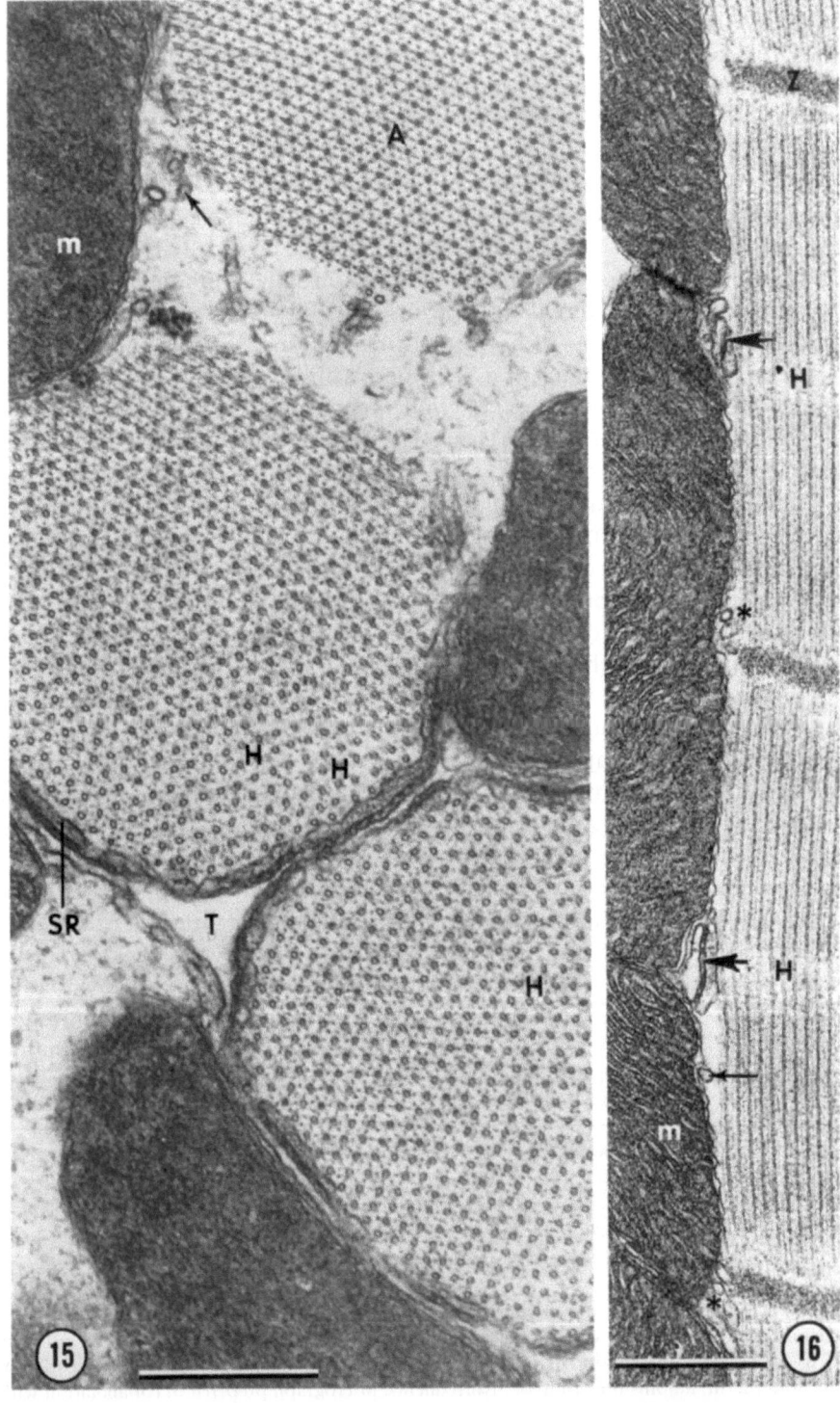

cisternal sheets and T tubules, are particularly advantageous (Smith, 1966a, 1972, 1980; Smith and Aldrich, 1971). Fine structural aspects of the plasma membrane, T system, and SR in this muscle are illustated in Figures 11 and 17–21.

While surface caveolar pits render tracing of the continuity between the plasma membrane and the T tubules difficult, in many vertebrate muscles (Franzini-Armstrong, 1970), the precise ordering of these invaginations is readily seen in freeze-fracture replicas of the dragonfly *Celithemis eponina* flight muscle (Figure 11; Smith and Aldrich, 1971). Furthermore, on the fracture face shown in this figure, the position of the underlying Z bands is reflected by corresponding transverse zones from which the arrays of intramembranous particles, occurring at other levels, are largely absent. The flattened tubular invaginations pass radially into the cell (Figure 17) and form extensive dyads with apposed surfaces of the SR cisternal sheets that cover the faces of each myofibril (Figure 18). Insect muscle dyads are similar to the triads of vertebrate muscles, differing only in the spatial relations of the membranes involved, and while most accounts relate to the triad (Peachey, 1965; Franzini-Armstrong, 1970, 1975; Kelly and Kuda, 1979; Somlyo, 1979), the straight radial course and flattened cross-section of the T tubules and facing dyadic SR in insect muscle such as that illustrated here greatly facilitate reconstruction of the junction.

The array of processes stemming from the SR surface of the triad and extending toward the T-tubule surface was shown by Franzini-Armstrong (1975) to be orthogonal in *en face* view, but the geometry of the triad, comprising a small cylindrical transverse tubule and paired SR cisternae, severely restricts the useful area in thin sections and replicas of the triad region. In dragonfly flight muscle, extensive portions of the dyads are readily obtained: the regular array of processes (ca. 20 nm apart) in a thin section is shown in Figure 18. Each process apparently comprises a "base piece" inserted on the SR membrane surface, from which projections traverse the junctional gap to the surface of the T tubule. The disposition of these structures is further illustrated in *en face* section in Figures 19 and 20. The former illustrates the disposition of blocks of processes along the periodically dilated, flattened T tubules, and at higher magnification (Figure 20) the orthogonal packing and individually quadrangular format of each process is seen (Smith, 1980). The functional significance of the shape of the T tubules in this and other insect muscles, further illustrated in a freeze-fracture replica in Figure 21, is not known, nor the functional correlate (if any) of the greater (ca.

Figure 15. Transversely sectioned "fibrillar" flight muscle of the whitefly *Trialeurodes*. Note the cylindrical and discrete myofibrils and large mitochondria (m). In this field, two fibrils are sectioned at the H level (the center of the sarcomere) and the upper fibril through the A band. The former are largely encircled by transverse tubules (T) forming dyads with SR cisternae, while only occasional SR vesicles adjoin the fibril at the A level (arrow). Details of this muscle in longitudinal section are shown in Figure 16. (Bar = 0.5 μm.)

Figure 16. Material as in Figure 15, but illustrating the longitudinal aspect. Dyads (large arrows) are located at the H level of the sarcomere; small membrane profiles occur opposite the Z bands (∗) and, very sparsely, elsewhere along the sarcomere. From Smith (1983). (Bar = 0.5 μm.)

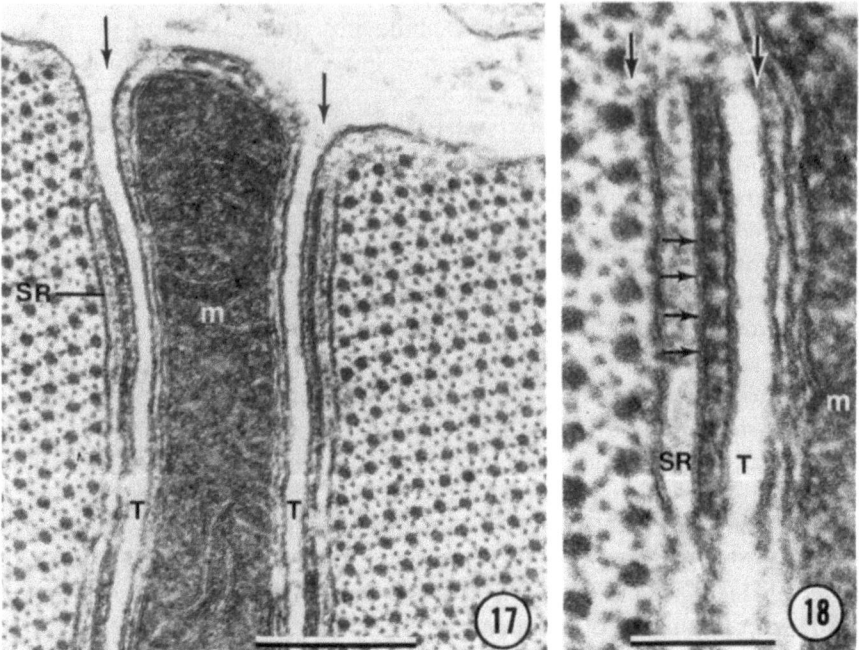

Figure 17. Transverse section of a field at the periphery of a flight muscle fiber of the dragonfly *Celithemis*. In these cells, the myofibrils and alternating mitochondria (m) are radially arranged and are accompanied, at the mid Z-H level, by open invaginations of the T-tubule system (arrows) forming dyad junctions with adjoining cisternal sheets (SR). Further details of the dyad junctions in this muscle are illustrated in Figures 18-20. See also Smith (1966a, 1968). (Bar = 0.25 μm.)

Figure 18. A section similar to the last, at higher magnification, illustrating one aspect of dyad organization. A regular array of processes stems from the SR membrane surface (arrows) and bridges the gap to the surface of the adjoining T-tubule membrane. In this orientation, each process appears to include a "base piece" and projections represented as apparent "pairs" in this orientation, probably resulting from optical alignment of quartets of projections from each quadrangular process. (Glutaraldehyde-tannic acid fixation; for explanation of large arrows, see Figure 19.) From Smith (1980). (Bar = 100 nm.)

2%) amount of the fiber volume occupied by T tubules in dragonfly flight muscle vs. the 0.3% estimated for frog sartorius (Peachey, 1965). Possibly, there is no correlation to be derived—a parameter that is obvious and readily measured by the electron microscopist is not necessarily of functional significance when cells of widely divergent evolutionary origin are considered. However, the correspondence in the volume of the fiber (ca. 5%) occupied by SR cisternae in the frog sartorius and dragonfly flight muscles reflects similar provision for an extensive calcium sequestering and releasing compartment in fast-acting muscles in which the activity cycle is synchronized with neural excitation.

In some insect muscles, the SR occupies a considerably greater portion of the cell volume: ca. 18% in the principal jumping muscle in the flea *Xenopsylla cheopis* (Cullen, 1975) and slightly more in a phasic jumping muscle of the locust (Cochrane *et al.*, 1972) and in the singing muscle of a katydid (*Neocono-*

cephalus robustus) that is able to contract synchronously at a frequency of 150-200 Hz (Josephson and Halverson, 1971; Elder, 1971). Extreme hypertrophy of the SR has been described in a crustacean muscle operating synchronously at ca. 100 Hz (Rosenbluth, 1969; Mendelson, 1969). While there is evidently some correlation between contraction frequency and SR development in synchronous fibers, this becomes conspicuous only at the upper end of the scale. Effective high frequency synchronous muscle operation involves provision for rapid curtailment of contraction, and the presence of unusually extensive SR compartments may be regarded as a device permitting rapid relaxation, as much as rapid contraction.

The membrane pattern described above for a muscle with lamellar myofibrils is in principle identical in synchronous muscles with close-packed myofibrils except that the T tubules follow the polygonal contours of the fibrils (as in the cicada timbal muscle illustrated in Figure 3) forming a reticulum of plasma membrane invaginations between the Z and H levels of the aligned sarcomeres. This micrograph serves to illustrate a further point. While the two-membered dyad is the commonest T-SR configuration in arthropod muscles, instances of association between one T tubule and a pair of SR surfaces have been noted both in synchronous and in asynchronous fibers. Pringle (1981) notes such "triads" in flight muscle of the sawfly *Tenthredo* sp., and in the cicada muscle shown in Figure 3, SR cisternae may be either single or paired apparently reflecting the distribution of mitochondria; only the T tubule facing a myofibril is involved in a junction. In dragonfly flight muscle (Smith, 1966a) for example, T tubules adjoin mitochondria and junctions are strictly dual. It seems preferable to restrict the term *triad* to the geometrically distinct configuration of membranes in vertebrate muscles in which a central T tubule is associated with a pair of terminal SR cisternae (Peachey, 1965).

Any general statement on any aspect of insect fine structure is unlikely to prove free from exceptions. In the present context, while the myofibrils are almost always, regardless of their shape, ensheathed in fenestrated SR cisternal sheets, in synchronous skeletal muscle, incursions of membranes into the midsarcomere region of the myofibrils have been described in flight muscles of Lepidoptera (Reger, 1967; Reger and Cooper, 1967; Elder, 1975). While these membranes may represent extensions of the extrafibrillar SR cisternae, this question deserves further study, and the significance of this feature remains unclear.

4.3. Asynchronous Muscles

One generalization that appears unlikely to require qualification is that some insect muscles, alone in the animal kingdom, have evolved the capacity to dissociate the frequency of motor neural excitation from that of contraction (Pringle, 1949). This is a device permitting though not necessitating a more rapid activity cycle than in synchronous muscles operating via calcium shuttling between the myofibrils and the SR, mediated by the T-tubule system. Asynchro-

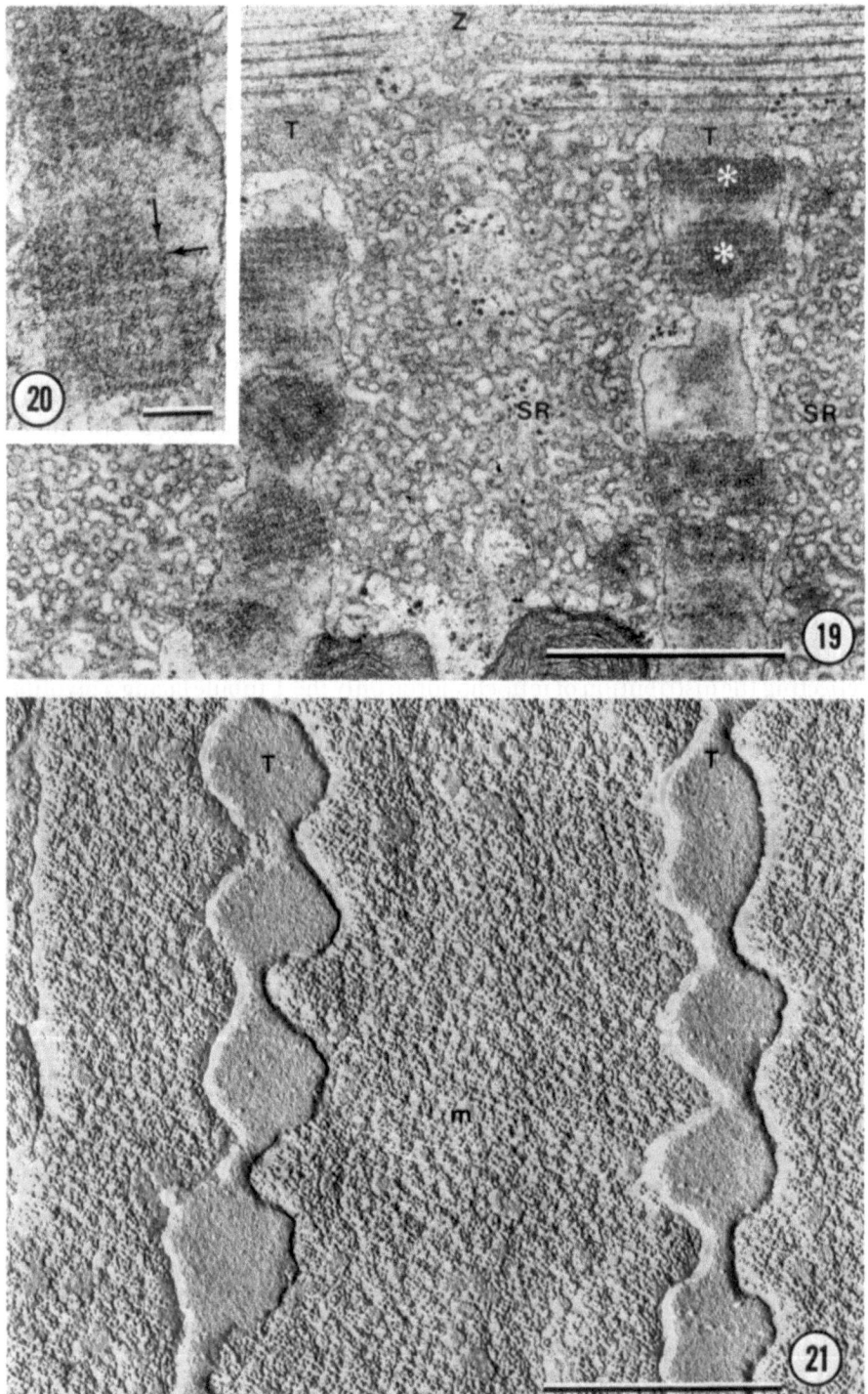

nous muscle is developed notably in the power-producing systems of some Hemiptera, most Hymenoptera, and in Coleoptera and Diptera, and also in the sound-producing timbal muscles of certain cicadas. The contractile frequency attained is determined, not by motor excitation, but by an intrafibrillar mechanism producing delayed tension after stretch, coupled with a resonant exoskeletal system that directs oscillatory activity (Pringle, 1957, 1972, 1978, 1981). As discussed by Pringle (1978), the phenomenon of stretch-activation is not restricted to these insect muscles. It is present, though unimportant, in vertebrate skeletal fibers, but prominent in cardiac cells. However, its coupling via cuticular insertions resulting in rhythmic mechanical activity has been exploited only in the evolution of some insects.

Asynchronous muscle differs conspicuously and consistently from synchronous muscle in insects not only in possessing discrete cylindrical myofibrils (for historical reviews, see Tiegs, 1955; Smith, 1961a, 1972; Elder, 1975) but in the distribution and extent of the membrane systems involved in excitation-contraction coupling, and notably in the volume occupied by the SR cisternae. Edwards and Ruska (1955) noted the reduction of the "sarcoplasmic reticulum" in belostomatid flight muscle. Subsequently, it was shown in *Tenebrio molitor* flight muscle (Smith, 1961b) that membrane invaginations corresponding to the T system are extensive and associated in dyad junctions. However, these involve only small isolated SR vesicles—a pattern later documented, with minor variations, in all asynchronous muscle cells and functionally interpretable in muscle requiring membrane excitation and an increase in the sarcoplasmic calcium ion level (Pringle, 1972) for the initiation and maintenance of contraction, but in which calcium cycling via the SR membranes evidently plays no part in determining the oscillatory contraction frequency.

Accounts and reviews of the structure of asynchronous flight muscles include those of Smith (1965a,b, 1966b, 1968, 1980), Ashhurst (1967), Smith and Sacktor (1970), Cullen (1974), Elder (1975) and Pringle (1981). While there is some variation in the extent of the SR cisternae, being somewhat more

⬅
―――――――――――――――――――――――――――――――――――

Figure 19. Further illustration of dyad organization in *Celithemis* flight muscle. This longitudinal section is almost exactly parallel with the flat surface of a myofibril, a grazing section of which is included (Z) and with the surface of the adjoining SR cisternal sheet. The section includes *en face* profiles of two transverse tubules (T) placed on either side of the Z level, together with blocks of foot-processes of the dyad junctional regions (∗) shown at higher magnification in Figure 20. *Note:* The plane of section in this figure and approximate section thickness are indicated by the pair of large arrows in Figure 18. From Smith (1980). (Bar = 0.5 μm.)

Figure 20. A section including an *en face* dyad junction, as in Figure 19, at higher magnification. The processes extending between SR- and T-membrane surfaces are orthogonally arranged (axes shown by arrows): each process appears to be square in contour. Series of processes are aligned in a section normal to this field in Figure 18. From Smith (1980). (Bar = 100 nm.)

Figure 21. Freeze-fracture replica of two transverse tubules (E face) in *Celithemis* flight muscle (T). As seen in thin sections (Figure 19), as in the replica, each tubule is periodically dilated. The tubules traverse the fiber in indentations of the flattened mitochondria, and this field includes the P-fracture face of the outer mitochondrial membrane (m). From Smith and Aldrich (1971). (Bar = 0.5 μm.)

prominent in belostomatid flight muscle than in other asynchronous fibers (Ashhurst, 1967), the limitation of this compartment is clearly diagnostic of the tissue. The T-tubule positioning is likewise somewhat variable: occasionally (Smith, 1961b; Cullen, 1974), dyads are present irregularly, in the Thysanoptera (Cullen, 1974) they are placed at the Z level, but in most instances dyads are placed at or near the center of the sarcomere (opposite the M region) rather than at the mid Z-H level characteristic of synchronous insect skeletal fibers, though the latter position has been reported in *Drosophila melanogaster* (Shafiq, 1964). Once again, no functional reason for the variable T-tubule location is obvious. The short sarcomere length of asynchronous fibers places the midsarcomere dyad at a maximal distance from the contractile material comparable with that achieved by the paired dyad disposition of synchronous muscles (see Section 4.2). This variation presents a similar (and unanswered) morphogenetic question concerning the determination of preferred T-tubule disposition in a situation involving no *structural* link with the myofilament array of the sarcomere.

In a most valuable paper, Cullen (1974) surveyed the literature documenting the distribution of asynchronous flight muscle within the insect orders, examined representatives of some orders for the first time, and, in particular, described the results of a detailed survey of the hemipteran families. He suggested an evolutionary plan for the development of flight muscle mechanisms in this order, in which both synchronous and asynchronous mechanisms occur. In summary, Cullen concluded that asynchrony has been evolved in power-producing flight muscles of Coleoptera, Diptera, most Hymenoptera, Hemiptera (Heteroptera and some Homoptera), Psocoptera, and Thysanoptera. Cullen noted that representatives of all other orders documented possess synchronous flight muscles—with the Plecoptera, Isoptera, Embioptera, and Strepsiptera awaiting fine structural study. One gap has been filled in the present chapter: flight muscle of an embiopteran (Figure 13) is clearly synchronous, with extensive SR cisternae, and of the remaining three orders, only the minute winged males of the Strepsiptera seem likely candidates to round off the list of insects that have evolved flight muscle asynchrony.* Cullen noted both types of flight muscle in the Psocoptera; in the Hymenoptera, some primitive Symphyta possibly retain synchronous power-producing flight muscle (Daly, 1963; Pringle, 1981) while most are asynchronous, but the situation in the Hemiptera is clearly the most complex. While all heteropteran families possess flight muscle with the fine structural features indicative of asynchrony, the Homoptera are divided, with a preponderance of synchrony, but with three families (Jassidae, Psyllidae, and Aphididae) in the asynchronous list.

The attribution of synchrony to the flight mechanism of whiteflies (Aleurodidae) by Cullen prompted Wootton and Newman (1979) to suggest that the wing beat frequency of these minute insects they observed (up to 181 Hz) reflected the highest known contraction rate of any nonfibrillar power-

*Since this chapter was prepared, it has been found that the dorsal longitudinal muscle of the metathoracic (wing-bearing) segment of the adult male of the strepsipteran *Elenchus tenuicollis* conforms structurally to the asynchronous pattern in other orders. However, these flight muscles are unusual (i) in their small fiber diameter, (ii) correlated lack of tracheolar invasion and (iii) in their variable myofibrillar size and contour (Smith and Kathirithamby, unpublished).

producing flight muscle, almost doubling the previously accepted limit of ca. 100 Hz for such fibers. As previously mentioned, thoracic muscles co-opted for sound production have been recorded as synchronous to ca. 200 Hz (Josephson and Halverson, 1971). Reinvestigation of the principal flight muscles of whiteflies (Smith, 1983) has shown that they conform precisely, in fine structure, to asynchronous fibers in other orders. Figures 15 and 16 illustrate dorsal longitudinal muscle of *T. vaporariorum*, respectively in transverse and longitudinal section and show reduction of the SR and its retention in dyadic association with T tubules at the midsarcomere level. This correction places one more homopteran family in the 'asynchronous' list, removes the anomaly of reported synchrony in flight muscle of the smallest Hemiptera (where high beat frequency coupled with asynchrony would be expected; Weis-Fogh, 1977) and reinstates ca. 100 Hz as the probable upper limit of myoneural synchrony in the power-producing muscle fibers of insect flight.

Cullen (1974) and Pringle (1981) considered the evolutionary inferences that may be drawn from the distribution of the two types of flight muscle within the Hemiptera, and Cullen concluded that asynchrony may have evolved three times within the flight system of Homoptera which, with the acquisition of this tissue in Heteroptera and the occurrence of both types of muscle in sound-producing cicada muscles (see below), suggests five parallel but independent evolutionary events (now perhaps six, to include the whiteflies) within a single order. He points out that the transition from synchrony to asynchrony may be a less radical step than might appear: that the ability to perform oscillatory work is inherent in the contractile system via stretch-activation if so directed by the resonant frequency of the system—in this instance the pterothorax. In a discussion on the evolution of asynchrony, Pringle (1981) notes the economy of SR membrane reduction in making available more of the fiber volume to accommodate the large cylindrical myofibrils, characteristic of the tissue. Evidently, asynchrony has evolved at least a dozen times.

In general, the aerodynamic problems of flight in very small insects are met by high wing beat frequency (Sotavalta, 1947; Greenawalt, 1962) and by special wing postures (Weis-Fogh, 1977). Asynchrony makes possible contraction frequencies of up to 1000 Hz (Sotavalta, 1947). Once evolved within an evolutionary line, however, asynchronous muscle imposes no limitation on wing beat frequency: the flight muscles of *Lethocerus* spp. and related waterbugs, some of the bulkiest insects, operate at modest contraction frequencies of 38 Hz and below (Barber and Pringle, 1966). At present, clearwing (sesiid) and sphingid moths remain at the top of the list of insects thought to be powered by synchronous flight muscles, with wing beat frequencies approaching 100 Hz (Sotavalta, 1947), and, among the Lepidoptera, the very small nepticulids are perhaps the most obvious candidates in a further search for the development of asynchrony in insect flight muscle.

Cicada Timbal Muscle

Other than in the flight mechanism, asynchronous muscle has been evolved only in the sound-producing (timbal) muscle of some cicadas. Pringle (1954) described myoneural asynchrony in a Sri Lankan species *Platypleura capitata* and later noted (Pringle, 1957) that sound production in other species is effected by synchronous fibers. Muscle of a cicada (*Tibicen*) employing the latter mechanism is illustrated in Figure 3: the fibrils are close-packed with conspicuous SR cisternae forming dyads with T tubules within the A-band level.

Recently, Josephson and Young (1981) have compared the fine structure of *Platypleura* timbal muscle with that of an Australian cicada (*Cyclochila australasiae*) which produces its song by a synchronous mechanism, and the physiological distinction between the two is precisely mirrored in their muscle fine structure. In *Cyclochila*, dyads are present as in *Tibicen* and similarly accompanied by an extensive SR whereas in *Platypleura* dyads are placed at the midsarcomere level and associated with isolated SR vesicles—a configuration common to many asynchronous flight muscles. The occurrence of asynchrony in cicada timbal muscle was included in a survey of the evolution of this physiological device in general by Pringle (1981). In a further investigation of sound-production in several Australian cicadas, Young and Josephson (1983) found that while most synchronous timbal muscles operate at ca. 100 Hz, the remarkably high frequency of 224 Hz is achieved synchronously in *Psaltoda claripennis*.

5. Tracheation

While no detailed comparative survey of insect muscle tracheation has been made, some general features are clear. First, the smaller muscle, including visceral and intersegmental fibers, some tonic limb fibers, and the flight muscles of Odonata are served by a variable plexus of tracheoles passing close to the cell surface (references in Smith, 1968; Elder, 1975) but not indenting the plasma membrane. Second, all other flight muscles, whether synchronous or asynchronous, and many other skeletal fibers bring the terminal tracheoles close to the mitochondria by their invagination from the cell surface and ramification within the sarcoplasm. This tracheolar invasion was proposed from light microscopic studies by Tiegs (1955) and confirmed in early fine structural work by Edwards *et al.* (1958), Smith (1961b), and others. The functionally critical aspect of this supply is that all tracheal processes invaginate with a concentric cylinder of the muscle plasma membrane and are never intracellular. Bücher (1965) has described the development of this internalized tracheolar supply accompanying metamorphosis in flight muscles of *Locusta migratoria* adults.

The association of tracheolar invagination of insect muscles with the T-tubule system, and hence the distribution of dyads within the cell, is an important aspect of the device that has received relatively little attention. Smith (1961b, 1965a) described the origin of the T tubules from circumtracheolar muscle plasma membrane in *Tenebrio* and *Megoura viciae* (aphid) flight muscle, and this has received incidental note in later studies (Elder, 1975). The quantitative use of tracheolar incursions vs. the peripheral plasma membrane has yet to be established: while T tubules may sometimes originate in these cells by direct invagination from the surface (as in fibers lacking internal tracheoles; Figure 11) and as proposed by Ashhurst (1967) in flight muscle of *Lethocerus*, it seems likely that the circumtracheolar pathway has been extensively employed. Since the surface plasma membrane is not complicated by surface caveolae, the obvious approach to answering the above question is by comparison of replicas of this membrane in a range of muscles invaded, to varying extents, by the tracheolar system. Wigglesworth and Lee (1982) illustrate a small area of the surface membrane in flight muscle of the butterfly *Pieris brassicae*, noting T-tubule invaginations; these, however, appear to be less regular than in dragonfly flight muscle (Figure 11), probably reflecting the extensive contribution made by invasive tracheoles. A preliminary study of *Apis* flight muscle surface membrane showed substantial areas devoid of regular T-tubule invaginations (Figure 12) suggesting that circumtracheolar origin is more important. That the annular space between the concentric membranes of muscle and tracheolar plasma membrane is confluent with the ambient hemolymph has been shown by Smith and Sacktor (1970) by ingress of ferritin marker injected into the living insect.

In a recent study, Wigglesworth and Lee (1982) further examined the invasion by tracheoles of the more metabolically active insect muscles, using as an infiltration marker a mixture of myrcene and light petroleum. This offers advantages over the earlier use of cobalt sulfide as a tracer (Wigglesworth, 1950); notably, the osmiophilic properties of myrcene facilitate mapping of the finest tracheolar branches in thin-sectioned survey fields. In asynchronous flight muscle of Diptera, Hymenoptera, and Coleoptera, these authors traced the finest endings of the system to smooth-surfaced tubes about 50 nm in diameter passing more or less regularly between the large mitochondria. As described in early accounts (Smith, 1961b), the tracheoles parallel the course of the surrounding muscle membrane, but more extensively than previously recognized. Regular placing of terminal tracheoles in *Pieris* flight muscle was also described. The possibility of tracheolar invasion of damselfly flight muscle requires further substantiation, since tracheoles are restricted to the surface of the small-diameter fibers in dragonfly muscles (Smith, 1966a). Wigglesworth and Lee propose that tracheoles are more closely juxtaposed to mitochondria than has been supposed, minimizing the gap between the two surfaces bridged by diffusion in ambient fluid (Weis-Fogh, 1964). A full understanding of the three-dimensional topography of muscle membrane incursion, when complicated by association with the tracheal system, requires a morphometric analysis of one or more selected muscles. The approach described above offers promise as a tracer method that can yield quantitative results.

While the fine structure of the tracheae and tracheolar branches has been extensively documented, less is known of the organization of the air sac dilatations of the tracheal system often present in association with active muscles, notably of the flight system, augmenting the movement of respiratory gases. The general morphology of these has been investigated from the time of Straus-Durckheim (1828) onwards. Figure 22 illustrates a freeze-fracture replica of the cuticular intima of a tracheal dilatation associated with flight muscle of *Apis* showing replacement of the familiar helical disposition of taenidia by a series of looping taenidial branches, linked by a "microtaenidial reticulum"—a peripheral observation in an account of muscle fine structure, but serving as a

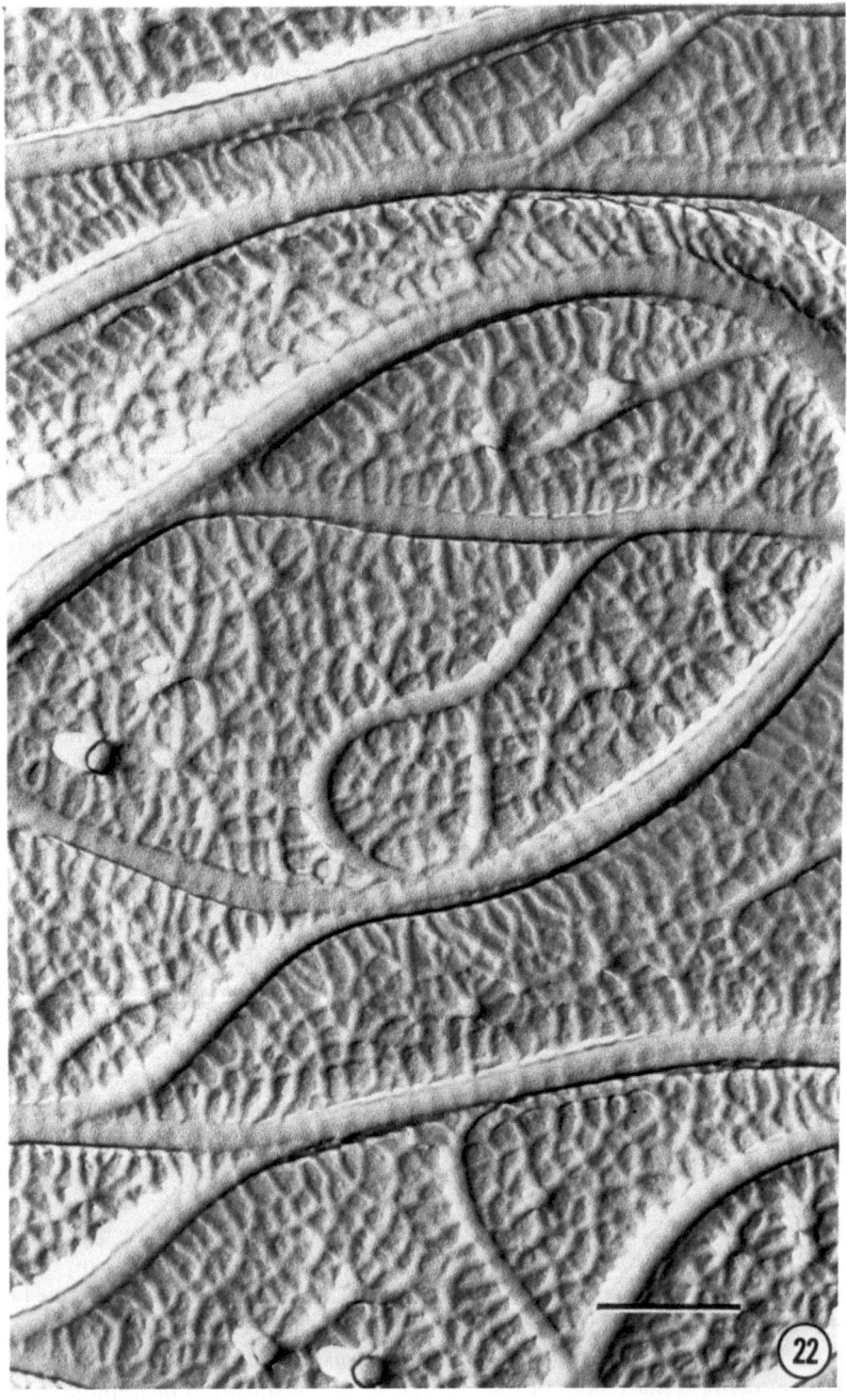

reminder that function of insect muscles may involve special modifications ensuring adequate respiratory supply.

6. Summary

Comparison between the fine structure and function of muscles of insects and other animals shows an amalgam of features of general occurrence in contractile systems and special modifications and variations in detail, some clearly interpretable and others less obviously so. While the generalization that all insect muscles are "striated" in the classical histological sense holds good, the structural range of fibers in insects is at least as great as in the vertebrate body, which employs cells both with and without sarcomeric disposition of the myofilaments. The myofibrillar array of insect muscles varies both in filament alignment and sarcomere length and in thick: thin filament numerical ratio. While the morphogenetic determination and functional significance of this variation have yet to be fully established, speed of contraction appears to be one correlate. The morphogenetic involvement of the thick filaments in the arrays of insect fibers may be inferred from *in vitro* myosin–actin assembly studies on proteins from crustacean and mammalian muscle. Work on selected insect flight muscles has provided models of thick-filament structure and of their association with actin of general importance and has revealed details of localization of some ancillary myofibrillar proteins. The variable structure of Z bands has been elucidated in one muscle type, and Z-band modification permitting a range of mechanical responses comparable with those of vertebrate smooth muscle has been described. Myofibrillar insertion on the integument, in the absence of collagenous tendons, promises to provide information on the general question of actin–plasma membrane attachment mechanisms. The distribution of membrane systems of insect muscle cells involved in excitation–contraction coupling is comparable with that of other large-diameter synchronous fibers, and the membrane association of the dyad (corresponding to the vertebrate triad) junction is ideally studied in selected geometrically appropriate insect flight muscles. Modification of the common plan of SR and T-system disposition, concomitant with the evolution of asynchrony in some insect muscles, is not only functionally interpretable in terms of dissociation between myoneural excitation and the contraction cycle, but substantiates the interpretation of the role of these membranes in synchronous cells. Where tracheolar incursion into insect fibers occurs, the circumtracheolar muscle cell membrane cylinders

Figure 22. Freeze-fracture replica of the cuticular intima of a large tracheal air sac associated with flight muscle of *Apis*. The field includes looping patterns of taenidia linked by an intervening reticular array of "microtaenidia". The surface replicated here cannot be identified with certainty, and may represent either the outer aspect of the unfractured epicuticle or a cleavage surface within the epicuticle. Micrograph prepared in collaboration with C. Noirot and C. Noirot-Timothée. (Bar = 1 µm.)

provide an alternative to peripheral origin of T tubules. Quantitative assessment of this feature, and of the terminal diffusion pathway in insect muscle respiration are potentially amenable to morphometric analysis. Information on the functional significance of variation in plasma membrane structure in physiologically varied insect muscles, and between these and vertebrate fibers, may be expected from further comparative freeze-fracture studies.

From the earliest histological work, use of insect material has played a substantial role in the recognition of principal structural features of muscle cells. Continuing studies are of importance not only in the context of insect physiology, but they also contribute to our general understanding of correlations between structure and function in a "cell type" with a uniform generalized role through the phyla, but vying with the neuron in adaptive versatility and associated fine structural diversification.

References

Amos. W. B., 1972, Structure and coiling of the stalk of the peritrich ciliates *Vorticella* and *Carchesium*, *J. Cell Sci.* **10**:95-122.

Amos, W. B., Grimstone, A. W., Rothschild, J. L., and Allen, R. D., 1979, Structure, protein composition and birefringence of the costa: A mobile flagellar root fibre in the flagellate *Trichomonas*, *J. Cell Sci.* **35**:139-164.

Ashhurst, D. E., 1967, The fibrillar flight muscles of giant water bugs: An electron microscope study, *J. Cell Sci.* **2**:435-444.

Ashhurst, D. E., 1968, The connective tissues of insects, *Annu. Rev. Entomol.* **13**:45-74.

Ashhurst, D. E., 1977, The Z-line: Its structure and evidence for the presence of connecting filaments. In *Insect Flight Muscle*, edited by R. T. Tregear, pp. 55-69, North-Holland, Amsterdam.

Auber, J., 1963, Ultrastructure de la jonction myo-épdidermique chez les Diptères, *J. Microsc. (Paris)* **2**:325-336.

Auber, J., 1967, Particularités ultrastructurales des myofibrilles des muscles du vol chez les Lepidopterès, *C.R. Acad. Sci.* **264**:621-624.

Auber, J., 1969, La myofibrillogenèse du muscle strié. 1. Insectes, *J. Microsc. (Paris)* **8**:197-232.

Barber, S. B., and Pringle, J. W. S., 1966, Functional aspects of flight in belostomatid bugs (Heteroptera), *Proc. R. Soc. London Ser. B* **164**:21-39.

Berlese, E., 1909, *Gli Insetti*, Vol. 1, Societa Editrice Libraria, Milan.

Bienz-Eisler, G., 1968, Elektronenmikroskopische Untersuchungen über der imaginale Struktur der dorsolongitudinalen Flugsmuskeln von *Antherea pernyi* Guer., *Acta Anat.* **70**:416-433.

Bracegirdle, B., 1978, *A History of Microtechnique*, Cornell University Press, Ithaca, N.Y.

Bücher, T., 1965, Formation of the specific structural and enzymic pattern of the insect flight muscle, *Biochem. Soc. Symp.* **25**:15-28.

Bullard, B., Luke, B., and Winkelman, L., 1973, The paramyosin of insect flight muscle, *J. Mol. Biol.* **75**:359-367.

Bullard, B., Hammond, K. S., and Luke, B. M., 1977a, The site of paramyosin in insect flight muscle and the presence of an unidentified protein between myosin filaments and the Z-line, *J. Mol. Biol.* **115**:417-440.

Bullard, B., Bell, J. L., and Luke, B. M., 1977b, Immunological investigation of proteins associated with thick filaments of insect flight muscle. In *Insect Flight Muscle*, edited by R. T. Tregear, pp. 41-52, North-Holland, Amsterdam.

Burnside, B., 1975, The form and arrangement of microtubules: An historical, primarily morphological, review, *Ann. N.Y. Acad. Sci.* **253**:14-26.

Candia-Carnevali, M. D., 1978, Z-line and supercontraction in the hydraulic muscular systems of insect larvae, *J. Exp. Zool.* **203**:15-30.

Candia-Carnevali, M. D., Eguileor, M., and Valvassori, R., 1980, Z-line morphology of functionally diverse insect skeletal muscles, *J. Submicrosc. Cytol.* **12**:427-446.

Caveney, S., 1969, Muscle attachment related to cuticle architecture in Apterygota, *J. Cell Sci.* **4**:531-559.

Cochrane, D. G., Elder, H. Y., and Usherwood, P. N. R., 1972, Physiology and ultrastructure of phasic and tonic skeletal muscle fibres in the locust, *Schistocerca gregaria, J. Cell Sci.* **10**:419-441.

Cohen, C., Reinhardt, B., Castellani, L., Norton, P., and Stirewalt, M., 1982, Schistosome surface spines are 'crystals' of actin, *J. Cell Biol.* **95**:987-988.

Cullen, M. J., 1974, The distribution of asynchronous muscle in insects with particular reference to the Hemiptera: An electron microscope study, *J. Entomol. Ser. A* **49**:17-41.

Cullen, M. J., 1975, The jumping mechanism of *Xenopsylla cheopis*. II. The fine structure of the jumping muscle, *Philos. Trans. R. Soc. London Ser. B* **271**:491-497.

Cullen, M. J., 1977, The breeding of giant water bugs in the laboratory. In *Insect Flight Muscle*, edited by R. T. Tregear, pp. 357-366, North-Holland, Amsterdam.

Daly, H. V., 1963, Close-packed and fibrillar muscles in the Hymenoptera, *Ann. Entomol. Soc. Am.* **56**:295-306.

Day, M. F., 1948, References for an outline of insect histology, CSIRO Division of Economic Entomology, pp. 1-223, Canberra.

Edwards, G. A., and Ruska, H., 1955, The function and metabolism of certain insect muscles in relation to their structure, *Q. J. Microsc. Sci.* **96**:151-159.

Edwards, G. A., Ruska, H., and De Harven, E., 1958, The fine structure of insect tracheoblasts, tracheae and tracheoles, *Q. J. Microsc. Sci.* **96**:151-159.

Elder, H. Y., 1971, High frequency muscles used in sound production by a katydid. II. Ultrastructure of the singing muscles. *Biol. Bull.* **141**:434-448.

Elder, H. Y., 1975, Muscle structure. In *The Structure and Function of Insect Muscles*, edited by P. N. R. Usherwood, pp. 1-74, Academic Press, New York.

Franzini-Armstrong, C., 1970, Studies on the triad. I. Structure of the junction in frog twitch fibers, *J. Cell Biol.* **47**:488-499.

Franzini-Armstrong, C., 1975, Membrane particles and transmission at the triad, *Fed. Proc.* **34**:1382-1389.

Franzini-Armstrong, C., and Porter, K. R., 1964, Sarcolemmal invaginations constituting the T-system in fish muscle fibers, *J. Cell Biol.* **22**:675-696.

Goll, D. E., Stromer, M. H., Robson, R. M., Luke, B. M., and Hammond, K. S., 1977, Extraction, purification and localization of α-actinin from asynchronous insect flight muscle. In *Insect Flight Muscle*, edited by R. T. Tregear, pp. 15-40, North-Holland, Amsterdam.

Greenawalt, C. H., 1962, Dimensional relationships for flying animals, *Smithson. Misc. Collect.* **144**:1-46.

Gupta, B. L., and Smith, D. S., 1969, Fine structural organization of the spermatheca in the cockroach, *Periplaneta americana, Tissue Cell* **1**:295-324.

Hardie, J., 1976, The tension/length relationship of an insect (*Calliphora erythrocephala*) supercontracting muscle, *Experientia* **32**:714-716.

Hardie, J., and Hawes, C., 1982, The three-dimensional structure of the Z disc in insect supercontracting muscles, *Tissue Cell* **14**:309-317.

Hayashi, T., Silver, R. B., Ip, W., Cayer, M. L., and Smith, D. S., 1977, Actin-myosin interaction: Self-assembly into a bipolar contractile unit, *J. Mol. Biol.* **111**:159-171.

Hayashi, T., Hinssen, H., and Smith, D. S., 1980, Comparative actin-myosin self-assembly studies, *Eur. J. Cell Biol.* **22**:323 (abstract).

Hayashi, T., Hinssen, H., Cayer, M. L., and Smith, D. S., 1981, Organization of native and *in vitro*-reassembled myosin filaments from lobster tonic muscle, *Tissue Cell* **13**:35-44.

Hayashi, T., Wozniak, P., Cayer, M. L., and Smith, D. S., 1983, Actin-myosin interaction: The role of myosin in determining the actin pattern in self-assembled 'hybrid' contractile units. *Tissue & Cell* **15**:955-963.

Hill, A. V., 1949, The abrupt transition from rest to activity in muscle, *Proc. R. Soc. London Ser. B* **136**:339-419.

Hoyle, G., and McNeill, P. A., 1968, Correlated physiological and ultrastructural studies on specialised muscles: Ultrastructure of white and pink fibres of the levator of the eyestalk of *Podophthalamus vigil* (Weber), *J. Exp. Zool.* **167**:487-522.

Hoyle, G., McAlear, J. H., and Selverston, A., 1965, Mechanism of supercontraction in a striated muscle, *J. Cell Biol.* **26**:621-640.

Hoyle, G., McNeill, P. A., and Selverston, A., 1973, Ultrastructure of barnacle giant muscle fibres, *J. Cell Biol.* **56**:74-91.

Huang, B., and Pitelka, D. R., 1973, The contractile process in the ciliate, *Stentor coeruleus*, *J. Cell Biol.* **57**:704-728.

Huxley, H. E., 1957, The double array of filaments in cross-striated muscle, *J. Biophys. Biochem. Cytol.* **7**:255-318.

Huxley, H. E., 1964, Evidence for continuity between the central elements of the triads and extracellular space in frog sartorius muscle, *Nature (London)* **202**:1067-1071.

Huxley, H. E., and Hanson, J., 1957, Preliminary observations on the structure of insect flight muscle. In *Electron Microscopy*, Proc. Stockholm Conf. 1956, pp. 202-203, Almqvist & Wiksells, Stockholm.

Jorgensen, W. K., and Rice, M. J., 1983, Superextension and supercontraction in locust ovipositor muscles. *J. Insect Physiol.* **29**:437-448.

Josephson, R. K., and Halverson, R. C., 1971, High frequency muscles used in sound production by a katydid. I. Organization of the motor system, *Biol. Bull.* **141**:411-433.

Josephson, R. K., and Young, D., 1981, Synchronous and asynchronous muscles in cicadas, *J. Exp. Biol.* **91**:219-237.

Kelly, D. E., and Kuda, A. M., 1979, Subunits of the triadic junction in fast skeletal muscle as revealed by freeze-fracture, *J. Ultrastruct. Res.* **68**:220-233.

Lai-Fook, J., 1967, The structure of developing muscle insertions in insects, *J. Morphol.* **123**:503-527.

Lyonet, P., 1762, *Traite Anatomique de la Chenille qui Ronge le Bois de Saule*, Cosse, Pinet & Rey, The Hague.

McIntosh, J. R., Ogata, E. S., and Landis, S. C., 1973, The axostyle of *Saccinobacculus*. 1. Structure of the organism and its microtubule bundle, *J. Cell Biol.* **56**:304-323.

Mendelson, M., 1969, Electrical and mechanical characteristics of a very fast lobster muscle, *J. Cell Biol.* **42**:548-561.

Osborne, M. P., 1967, Supercontraction in the muscles of the blowfly larva: An ultrastructural study, *J. Insect Physiol.* **13**:1471-1482.

Peachey, L. D., 1965, The sarcoplasmic reticulum and transverse tubules of the frog's sartorius, *J. Cell Biol.* **25**:209-232.

Porter, K. R., and Palade, G. E., 1957, Studies on the endoplasmic reticulum. III. Its form and distribution in striated muscle cells, *J. Biophys. Biochem. Cytol.* **3**:269-300.

Pringle, J. W. S., 1949, The excitation and contraction of the flight muscles of insects, *J. Physiol. (London)* **108**:226-232.

Pringle, J. W. S., 1954, A physiological analysis of cicada song, *J. Exp. Biol.* **31**:525-560.

Pringle, J. W. S., 1957, *Insect Flight*, Cambridge University Press, Cambridge.

Pringle, J. W. S., 1972, Arthropod muscle. In *The Structure and Function of Muscle*, edited by G. H. Bourne, vol. 1, pp. 491-541, Academic Press, New York.

Pringle, J. W. S., 1974, The resting elasticity of insect flight muscle, *Symp. Biol. Hung.* **17**:67-78.

Pringle, J. W. S., 1977, The availability of insect fibrillar muscle. In *Insect Flight Muscle*, edited by R. T. Tregear, pp. 337-355, North-Holland, Amsterdam.

Pringle, J. W. S., 1978, Stretch activation of muscle: Function and mechanism, *Proc. R. Soc. London Ser. B* **201**:107-130

Pringle, J. W. S., 1980, A review of arthropod muscle. In *Development and Specialization of Skeletal Muscle*, edited by D. F. Goldspink, pp. 91-105, Cambridge University Press, London.

Pringle, J. W. S., 1981, The evolution of fibrillar muscle in insects, *J. Exp. Biol.* **94**:1-14.

Reedy, M. K., and Garrett, W. E., 1977, Electron microscope studies of *Lethocerus* flight muscle in rigor. In *Insect Flight Muscle*, edited by R. T. Tregear, pp. 115-136, North-Holland, Amsterdam.

Reedy, M. K., Holmes, K. C., and Tregear, R. T., 1965, Induced changes in orientation of the cross-bridges in glycerinated insect flight muscle, *Nature (London)* **207**:1276-1280.

Reger, J. F., 1967, The organization of the sarcoplasmic reticulum in direct flight muscle of the lepidopteran *Achalarus lyciades*, *J. Ultrastruct. Res.* **18**:595-599.

Reger, J. F., and Cooper, D. P., 1967, A comparative study on the fine structure of the basalar muscle of the wing and the tibial extensor muscle of the leg of the lepidopteran *Achalarus lyciades*, *J. Cell Biol.* **33**:531-542.

Rice, M. J., 1970, Supercontracting and non-supercontracting visceral muscles in the tsetse fly, *Glossina austeni*, *J. Insect Physiol.* **16**:1109-1122.

Rice, M. J., 1973, Supercontracting striated muscle in a vertebrate, *Nature (London)* **243**:238-240.

Rosenbluth, J., 1969, Sarcoplasmic reticulum of an unusually fast-acting crustacean muscle, *J. Cell Biol.* **42**:534-547.

Sainsbury, G. M., and Bullard, B., 1980, New proline-rich proteins in isolated Z-discs, *Biochem. J.* **191**:333-339.

Sainsbury, G. M., and Hulmes, D., 1977, Notes on the structure of the Z-discs of insect flight muscle. In *Insect Flight Muscle*, edited by R. T. Tregear, pp. 75-78, North-Holland, Amsterdam.

Sanger, J. W., and McCann, F. V., 1968, Ultrastructure of moth alary muscles and their attachment to the heart wall, *J. Insect Physiol.* **14**:1539-1544.

Shafiq, S. A., 1964, An electron microscopical study of the innervation and sarcoplasmic reticulum of the fibrillar flight muscle in *Drosophila melanogaster*, *Q. J. Microsc. Sci.* **105**:1-6.

Smith, D. S., 1961a, Reticular organizations within the striated muscle cell, *J. Biophys. Biochem. Cytol.* **10**(Suppl.):61-87.

Smith, D. S., 1961b, The structure of fibrillar flight muscle: A study made with special reference to the membrane systems of the fibre, *J. Biophys. Biochem. Cytol.* **10**(Suppl.):123-158.

Smith, D. S., 1962, The sarcoplasmic reticulum of insect muscles. In *Fifth Int. Congr. Electron Microsc.* Philadelphia, TT-3.

Smith, D. S., 1965a, The organization of flight muscle in an aphid, *Megoura viciae* (Homoptera): With a discussion on the fine structure of synchronous and asynchronous striated muscle fibers, *J. Cell Biol.* **27**:379-393.

Smith, D. S., 1965b, Insect flight muscles, *Sci. Am.* **212**:77-88.

Smith, D. S., 1966a, The organization of flight muscle fibers in the Odonata, *J. Cell Biol.* **28**:109-126.

Smith, D. S., 1966b, The organization and function of the sarcoplasmic reticulum and T-system of muscle cells, *Prog. Biophys. Mol. Biol.* **16**:107-142.

Smith, D. S., 1968, *Insect Cells: Their Structure and Function*, Oliver & Boyd, Edinburgh.

Smith, D. S., 1972, *Muscle: A Monograph*, Academic Press, New York.

Smith, D. S., 1980, The past and future of insect muscles. In *Insect Biology in the Future*, edited by M. Locke and D. S. Smith, pp. 797-818, Academic Press, New York.

Smith, D. S., 1983, 100Hz remains the upper limit of synchronous muscle contraction: An anomaly resolved, *Nature (London)*.

Smith, D. S., and Aldrich, H. C., 1971, Membrane systems of freeze-etched striated muscle, *Tissue Cell* **3**:261-281.

Smith, D. S., and Kathirithamby, J., 1984, Asynchronous flight muscle in Strepsiptera, *Tissue Cell*, **16**:(in preparation).

Smith, D. S., and Sacktor, B., 1970, Disposition of membranes and the entry of hemolymph-borne ferritin in flight muscle fibers of the fly *Phormia regina*, *Tissue Cell* **2**:355-374.

Smith, D. S., Gupta, B. L., and Smith, U., 1966, The organization and myofilament array of insect visceral fibers, *J. Cell Sci.* **1**:48-57.

Smith, D. S., Jarlfors, U., and Russell, F. E., 1969, The fine structure of muscle attachments in a spider (*Latrodectus mactans* Fabr.), *Tissue Cell* **1**:673-687.

Smith, D. S., del Castillo, J., and Anderson, M., 1973, Fine structure and innervation of an annelid muscle with the longest recorded sarcomere, *Tissue Cell* **5**:281-302.

Somlyo, A. V., 1979, Bridging structures spanning the junctional gap at the triad of skeletal muscle, *J. Cell Biol.* **80**:743-750.

Sotavalta, O., 1947, The flight-tone (wing stroke frequency) of insects, *Acta Entomol. Fenn.* **4**:1-117.

Squire, J., 1971, General model for the structure of all myosin-containing filaments, *Nature (London)* **233**:457–462.
Squire, J. M., 1977, The structure of insect thick filaments. In *Insect Flight Muscle*, edited by R. T. Tregear, pp. 91–112, North-Holland, Amsterdam.
Squire, J. M., 1981, Comparative ultrastructures of diverse muscle types. In *The Structural Basis of Muscular Contraction*, edited by J. Squire, pp. 381–414, Plenum Press, New York.
Straus-Durckheim, H., 1828, *Considérations Générales sur l'Anatomie Comparée des Animaux Articulés*, Levrault, Paris.
Tiegs, O. W., 1955, The flight muscle of some insects. Their anatomy and histology: with some observations on the structure of striated muscle in general, *Philos. Trans. R. Soc. London Ser. B* **238**:221–347.
Tilney, L. G., Kiehart, D. P., Sardet, C., and Tilney, M., 1978, Polymerization of actin. IV. Role of Ca^{++} and H^+ in the assembly of actin and in membrane fusion in the acrosomal reaction of echinoderm sperm, *J. Cell Biol.* **77**:536–550.
Tilney, L. G., De Rosier, D. J., and Mulroy, M. J., 1980, The organization of actin filaments in the stereocilia of cochlear hair cells, *J. Cell Biol.* **86**:244–259.
Tregear, R. T. (ed.), 1977, *Insect Flight Muscle*, North-Holland, Amsterdam.
Weis-Fogh, T., 1964, Diffusion in insect wing muscle, the most active tissue known, *J. Exp. Biol.* **41**:229–256.
Weig-Fogh, T., 1977, Dimensional analysis of hovering flight. In *Scale Effects in Animal Locomotion*, edited by T. J. Pedley, pp. 405–420, Academic Press, New York.
Wigglesworth, V. B., 1950, A new method for injecting the tracheae and tracheoles of insects, *Q. J. Microsc. Sci.* **91**:217–224.
Wigglesworth, V. B., and Lee, W. M., 1982, The supply of oxygen to the flight muscle of insects: A theory of tracheole physiology, *Tissue Cell* **14**:501–518.
Wootton, R. J., and Newman, D. J. S., 1979, Whitefly have the highest contraction frequencies yet recorded in nonfibrillar muscles, *Nature (London)* **280**:402–403.
Young, D., and Josephson, R. K., 1983, Mechanisms of sound production and muscle contraction kinetics in cicadas, *J. Comp. Physiol.* **152**:183–195.

5

The Structure and Development of the Vacuolar System in the Fat Body of Insects

MICHAEL LOCKE

1. Introduction

1.1. The Fat Body

The fat body is the center of intermediary metabolism and nutrient storage in insects. It consists of sheets or ribbons of cells slung in the hemocoel between the gut and the integument by elastic extensions of the basal lamina and by tracheae. The sheets are in two layers, one close to the integument and the other nearer the gut. Each layer is rarely more than two cells thick and is surrounded by a sac of basal lamina to which the cells attach by hemidesmosomes. The cells are united by desmosomes and gap junctions and are often separated from one another by lymph spaces. The fat body is totally dependent on the bathing hemolymph for its informational signals and for all its raw materials. Conversely, the composition of the hemolymph is to a very large degree a reflection of fat body syntheses and secretions. The loose texture and elasticity of the fat body encourage to the full its interchanges with the hemolymph.

In many insects, the fat body contains only a single cell type, the trophocyte, adapted primarily for protein synthesis and secretion, lipid and glycogen metabolism, and intermittently for the storage and breakdown of protein and urate. There may be some regional differences in structure and function but these are more quantitative than qualitative. In other insects, separate cell

---

MICHAEL LOCKE • Cell Science Laboratories, Department of Zoology, University of Western Ontario, London, Ontario N6A 5B7, Canada.

types are differentiated for urate metabolism or the maintenance of symbionts in addition to trophocytes. Oenocytes may also find a home between fat body cells. These general features of fat body biology are discussed in Dean *et al.* (1984). This chapter will be concerned with fat body ultrastructure.

1.2. The Ultrastructure of Fat Body Cells

The importance of the fat body in insect physiology and the relative ease with which it can be isolated in quantity for biochemical procedures or maintained in tissue culture for developmental studies have resulted in much experimental work (Wyatt, 1980). This has not been matched by corresponding progress in ultrastructural studies, largely because fat body cells at some stages are little more than storage containers for lipid, glycogen, and protein granules, which make fixation difficult. Ultrastructural studies on most cell components have been neglected. There has not yet been any work on the cytoskeleton and cell adhesions in relation to the changes in cell and tissue shape that accompany molting and metamorphosis. Nor has there been any work on the ultrastructure of the nuclei and the problems of nuclear replication followed by polyploidy rather than cell division. The structural differentiation of mitochondria and their cycle of replication has been described (Larsen, 1976), but this has yet to be correlated with biochemical events. As in most cells, the main structural correlates to intermediary metabolism are in the elaboration of the vacuolar system, and most fat body ultrastructural studies are on this topic. The variety of synthetic capabilities of the fat body is matched by an equal variety of compartments or vacuoles derived from the vacuolar system.

1.3. The Fat Body Vacuolar System

During its life history, the insect fat body displays most of the principles of membrane topology on which an understanding of eukaryote cell biology is based. The vacuolar system is completely divided into two hierarchies of compartments by the lock gates of the transition vesicles. The endoplasmic reticulum (ER) may attach to the plasma membrane by confronting cisternae, but the only fusion between the ER-nuclear envelope and the rest of the vacuolar system is through transition vesicles and the Golgi complex. There is a flow of membranes from their sites of synthesis in the RER, through the Golgi complex to form the plasma membrane. At particular regions along the route, notably in the Golgi complex, the membranes are modified. There are return routes for membrane recycling and for membrane turnover that may meet with the Golgi complex or its derivatives (the inner saccule in recycling and primary lysosomes in turnover). The flow of membranes necessarily allows the concurrent movement of material within the compartments enclosed by the membranes, depending on the binding properties of the luminal surfaces and the nature of the various environments within the compartments (Figure 1). Mem-

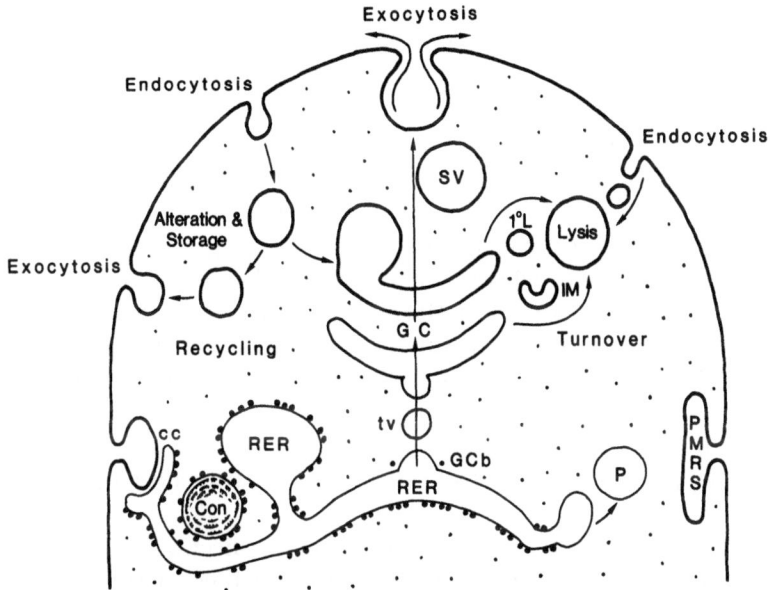

Figure 1. A scheme for the interrelations between vacuoles and the vacuolar system in a generalized fat body cell. The thin pretransition membranes connect with the thick posttransition region membranes only through the lock gates of the microvesicles between them. These transition vesicles arise by budding through the rings of Golgi complex beads on the smooth surface of RER that marks the forming face of a Golgi complex. Apart from this connection, the two classes of membrane interact only through confronting cisternae and not by fusion. Within the pretransition compartment, regions may differentiate to make confronting cisternae or may separate into RER sacs and peroxisomes. RER may also give rise to concretions and to SER, but these have not been observed in fat body cells. The thick posttransition membranes give rise to the plasma membrane and a variety of vacuoles. The plasma membrane may be confronted by ER, and it may form semipermanent infolds into a plasma membrane reticular system. Eleven kinds of vacuole can be distinguished according to their function and the origin of their contents and membranes. In this diagram the origin and fate of these vacuoles have been simplified into three groups. There are those for alteration and storage that involve recycling of the plasma membrane (tyrosine storage vacuoles, Section 2.1, possibly urate storage vacuoles, 2.2, and probably vacuoles containing symbionts, 2.6). There are secretory vesicles passing in one direction from the Golgi complex to the plasma membrane (2.5). And there are vacuoles that at some stage become lytic chambers with an input from primary lysosomes (urate granules, 2.3, protein storage granules, 2.4, some stages of vacuoles containing symbionts, 2.6, and a variety of phagic vacuoles, 2.8 to 2.11, differing in the material that they digest). Table 1 gives a key to all abbreviations.

branes become thicker as they pass through the Golgi complex, allowing pre- and posttransition membranes to be distinguished from one another (Pelttari and Helminen, 1979; McDermid and Locke, 1983). Once past the Golgi complex, pre- and posttransition membranes remain separate and interact only through confronting cisternae. The differentiation of enlarged compartments to form vacuoles that are at least temporarily separated from the rest of the system, occurs in both regions. Vacuoles are defined here as pre- or posttransition compartments having more than transient spatially separate existences that allow structural and functional differentiation of membrane and contents.

The elements from which this simplified picture of the movement and modification of membranes and their contents are derived, can nearly all be found in insect fat body cells. The only important route known from studies on eukaryotic cells, particularly vertebrates, that has yet to be demonstrated experimentally in the fat body, is the recycling of plasma membrane through the Golgi complex (Farquhar and Palade, 1981). Large fluid-filled vacuoles are common in plant cells where they are bounded by thin membranes derived from ER (Matile and Moor, 1968; Marty, 1978). Such large structures are rare in animal cells, and the giant tyrosine storage vacuoles of the fat body are the exception. Most animal vacuoles are relatively inconspicuous and nearly always derived from thick posttransition vesicle membranes. Nevertheless, these smaller vacuoles exist in considerable variety. The fat body has at least 11 specializations of particular posttransition region compartments and two or three in the pretransition region.

The object of this essay is to make a balance sheet of the membranous structures present in fat body cells and to suggest how they relate to the development of the rest of the vacuolar system. It is about cell ultrastructure. Changes in relation to the development of the whole organism are dealt with elsewhere (Locke, 1980, 1981, 1983; Locke and Collins, 1980; Dean *et al.*, 1980, 1983). The interpretation depends very much on research on comparable structures in the cells of other organisms, but this essay is concerned with insects and detailed references will not be made to vertebrate work. The synchrony of development in *Calpodes ethlius* (and other insect) tissues makes them particularly suitable for studying events that may be common in other cells, but rarely observed because of their transience. The synchrony of development can be further increased by brief periods of starvation followed by feeding to initiate the intermolt/molt sequence (Kunkel, 1966). For example, mitochondrial division and tyrosine vacuole formation can be synchronized in this way (Wigglesworth, 1967; Locke, 1980; McDermid and Locke, 1983). The account below is illustrated by electron micrographs of the fat body from the larva of the hesperiid butterfly, *C. ethlius*, and where noted of the cockroach, *Periplaneta americana*. The conclusions, which are mainly from observations on *Calpodes*, probably have general relevance for fat body cells.

1.4. Kinds of Fat Body Vacuole

Vacuoles have presumably evolved to meet the need for several different and independently controllable kinds of environment in compartments separated from the cell sap. Control is exercised by the addition of material, usually in microvesicles, and through the nature of the bounding membrane that determines surface enzyme activity and transmembrane transport. Among the posttransition region vacuoles we can distinguish two kinds of vacuole function (Figure 2). There are vacuoles for digestion (autophagic vacuoles, multivesicular bodies, lamellate bodies, protein granules, phagocytic vacuoles, and

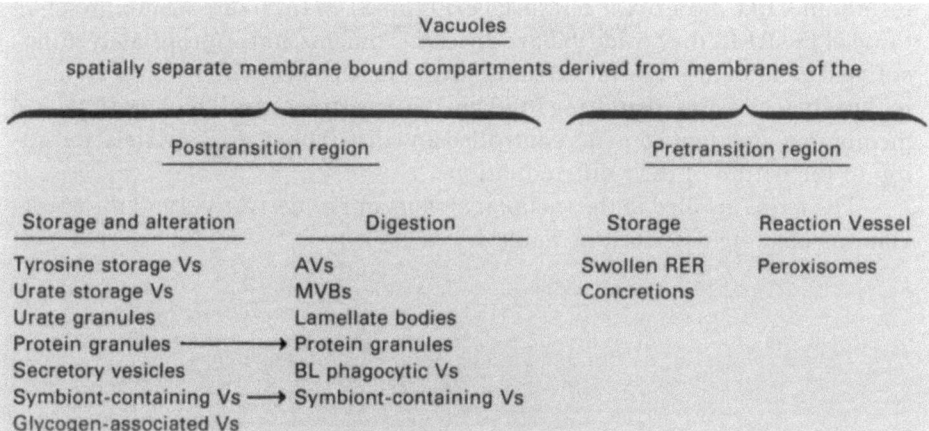

Figure 2. A grouping of the specialized compartments or vacuoles in fat body cells according to their function and relation to pre- and posttransition membranes. Concretions derived from the RER are not known from the fat body, but they do occur in Malpighian tubules and midgut cells.

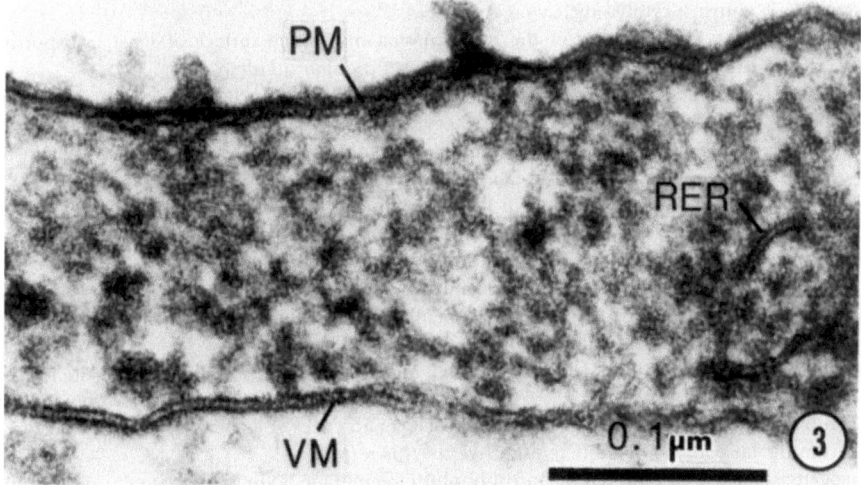

Figure 3. The thickness of representative pre- and posttransition membranes. A profile of part of a mid-fourth-stage fat body cell showing the relatively thick posttransition membranes around the vacuole and at the cell surface, compared with the thin pretransition membrane of the RER.

vacuoles digesting symbionts), and there are vacuoles for less drastic alteration of the contents, often in relation to storage (tyrosine storage and urate storage vacuoles, urate granules, protein granules, secretory vesicles, vacuoles maintaining symbionts, and vacuoles associated with glycogen). In the pretransition region, the vacuoles may be primarily reaction vessels for molecules that they contain only transiently (peroxisomes) or rather specialized storage containers (Malpighian tubule or midgut concretions and distensions of the RER). Although these vacuoles have common developmental origins, either from thick membranes like those of the cell surface (Figure 3), or from thin membranes like those of the RER, they come to have different contents and appropriately differentiated membranes. The often simultaneous occurrence of a variety of vacuoles in the same cell poses interesting fundamental questions of cell biology. How are membranes and their contents controlled and distributed appropriately according to the nature of their differentiation?

The terms applied to the vacuolar system and its derivatives and the abbreviations used are described in Table 1.

Table 1. The Terminology Applied to the Vacuolar System and Other Components Commonly Found in Fat Body Cells

Autophagic vacuole (AV): A secondary lysosome derived from isolation bodies and primary lysosomes. Synonyms: cytolysome, cytosegresome.

Basal infolds (Bi): An infolding of the basal plasma membrane surface of most transporting epithelia such as the gut, salivary glands, and Malpighian tubules.

Basal lamina (Bl): The heterogeneous layer of polysaccharides reinforced with collagen and elastic fibers that separates all tissues except hemocytes from the hemolymph.

Basal lamina phagocytic vacuole (Blv): A digestive vacuole formed by invagination of the plasma membrane together with the basal lamina to which it is attached by hemidesmosomes.

Concretions (Con): Spherical dilations of RER containing concentric shells mainly composed of calcium phosphate. They occur in Malpighian tubules and midgut, but not in the fat body.

Condensing vacuole (CV): A compartment derived from the forming face of the Golgi complex that becomes a secretory vesicle by consolidation.

Confronting cisternae (cc): ER apposed to the membranes of other structures such as peroxisomes, mitochondria, or localized regions of the plasma membrane.

Endoplasmic reticulum (ER): Membrane-bound compartments continuous with the nuclear envelope that connect with Golgi complexes through transition vesicles. ER varies in form (tubular or lamellate) and in the presence (RER) or absence (SER) of ribosomes. SER is absent from the fat body, which differs from liver in this respect.

Glycogen-associated vacuole (Glv): An osmiophilic membrane-bound compartment often found in or near masses of glycogen.

Golgi complex (GC): Golgi complexes encompass the region from the rings of Golgi complex beads on the RER where transition vesicles arise, through three or four saccules to the secretory vesicles and primary lysosomes at the secretory face. There are usually two outer saccules (OS), and two inner saccules (IS).

Golgi complex beads (GCb): Ten-nanometer particles arranged in rings around the transition vesicles on the smooth surface of RER where Golgi complexes arise.

Hemidesmosome (HD): An adhesion between the basal lamina through the plasma membrane to the cytoskeleton.

Heterophagic vacuole (HV): A membrane-bound vacuole containing extracellular material scheduled for lysis. Derived from endocytosed material and primary lysosomes, i.e., a secondary lysosome.

Isolation body (IB): The fragment of cytoplasm isolated by the paired isolation membranes. It becomes an autophagic vacuole after fusing with a primary lysosome. Several isolation bodies often fuse together.

Table 1. (Continued)

Isolation membranes (IM): The paired membranes of the envelope which invests organelles scheduled for lysis.

Lamellar body (LB): A membrane-bound vacuole containing myelinlike debris derived from the plasma membrane, a secondary lysosome concerned in membrane degradation.

Lymph (Ly): The fluid in intercellular spaces below the basal lamina through which it has filtered from the hemolymph.

Multivesicular body (MVB): A structure bound by a single unit membrane containing material carried to it by pinocytosis vesicles and primary lysosomes. Multivesicular bodies have inner vesicles arising by invagination at their surface and are concerned in membrane turnover and protein digestion.

Nuclear envelope (NE): The envelope of vacuolar system membranes around the nucleus that has a cisterna continuous with the ER.

Peroxisome (P): A particle with a single unit membrane enclosing dense contents and a core, containing catalase and oxidases.

Pinocytotic vesicle (Pv): A microvesicle arising at the plasma membrane surface that is involved in conveying membrane, adsorbed protein, and luminal contents into the cell.

Plasma membrane reticular system (PMRS): The system of plasma membrane infolds forming lymph spaces in insect cells concerned with lipid metabolism.

Posttransition region: The compartments derived from membranes that have thickened in their passage through the Golgi complex after their exit from the ER in transition vesicles.

Pretransition region: The thin-membrane compartments extending from the nuclear envelope through the ER and ending with the smooth surface where transition vesicles arise to form the Golgi complex.

Primary lysosome (1° L): Microvesicles from the Golgi complex carrying lytic enzymes or proenzymes.

Protein granule (PG): Membrane-bound vacuole storing protein that has been pinocytosed from the hemolymph. The protein is often crystalline. Protein granules are often composite structures, having fused with autophagic vacuoles.

Provacuoles (Pv): Plasma membrane-derived compartments that fuse together to form tyrosine or other storage vacuoles.

Residual body (RB): A membrane-bound inclusion characterized by undigested residues (membrane fragments or whorls, myelin figures, ferritinlike particles, etc.). The term is used here for the late stages of any phagic vacuole whose origin can no longer be determined from the morphology of the contents.

Secondary lysosome (2° L): A membrane-bound vacuole resulting from the fusion of a primary lysosome with a vacuole carrying material to be digested.

Secretory vesicle (SV): A membrane-bound vesicle of protein derived from a condensing vacuole of the GC that is destined for secretion outside the cell by exocytosis. Synonyms: secretory granules, zymogen granules.

Symbiont-containing vacuole (Syv): A membrane-bound compartment within which microorganisms exist in symbiosis.

Transition vesicle (tv): Microvesicles between the RER and the outer saccule of the Golgi complex. They bud off through the centers of rings of Golgi complex beads.

Tyrosine vacuole (Tv): A large vacuole storing tyrosine in larval fat body cells.

Urate granule (Ug): A membrane-bound structure sometimes with fibrous and vesicular contents storing urate and uric acid at metamorphosis. Of uncertain origin.

Urate vacuole (Uv): A large vacuole storing urate in cockroach urate cells. Structurally similar to urate granules in Lepidoptera but larger.

Vacuole (V): Usually post-, but sometimes pretransition region compartment differentiated for the storage and/or processing of its contents.

Other abbreviations: Cell wall of symbiont (CW), glycogen (Gl), lipid (L), mycetocyte (My), plasma membrane (PM), symbiont (S), symbiont membrane (SM), urocyte (U), vacuole membrane (VM).

2. Posttransition Compartments

One aspect of the maturation of membranes as they pass through the Golgi complex is their increase in thickness. The thickness of membrane profiles seen in electron micrographs varies with the fixation and staining procedure but after similar treatments, posttransition membranes are usually in the 8- to 10-nm range and pretransition membranes are nearer 6 to 8 nm. Figure 3 shows profiles of the plasma membrane, a tyrosine storage vacuole membrane, and a membrane of the RER in a fat body cell. The ER, as a representative of a pretransition membrane, is clearly about 20% thinner than the other two. There are, no doubt, other differences between membranes from the two classes, but in general, the vacuoles described in this section are characterized by having thick membranes compared to the pretransition structures described in Section 3.

2.1. Tyrosine Storage Vacuoles and Provacuoles

Large fluid-filled vacuoles were first described in the fat body of live *Chironomus* (species undetermined) larvae by Voinov (1927). Wigglesworth (1942) found vacuoles in newly ecdysed *Aedes aegypti* larvae and in *Rhodnius prolixus* that had been fed after prolonged starvation (Wigglesworth, 1967). The vacuoles in *Leptinotarsa decemlineata* (Labour, 1970, 1974; De Loof, 1972; Dortland and Hogen Esch, 1979), *Calliphora erythrocephala* larvae (de Priester

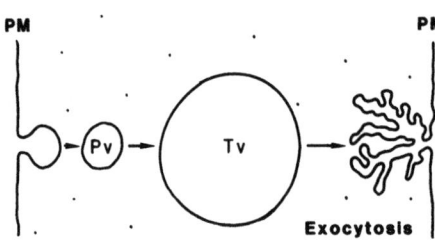

Figure 4. The origin and fate of tyrosine storage vacuoles. Shortly after the third to fourth ecdysis, plasma membrane infolds bud into the cell to form provacuoles. During the intermolt, these fuse and grow to become giant tyrosine storage vacuoles. At molting, the vacuoles contribute their contents to the hemolymph to supply the phenolics needed for cuticle formation, and their membranes return to the plasma membrane.

Figure 5. Tyrosine storage vacuoles in a mid-fourth-instar larva. By this time, each cell has only a single giant vacuole that dwarfs all other cell components. In live cells, the vacuoles are spherical, but they are very fragile and susceptible to shrinkage during fixation. They burst easily and their lumens may become contaminated with cell debris. Light micrograph of an epoxy section.

Figure 6. Vacuoles in a late-fifth-stage larva. These vacuoles arise very quickly prior to pupation rather than by steady growth during the intermolt. They are easily overlooked by confusion with the spaces left by unfixed or shrunken lipid droplets. Light micrograph of an epoxy section.

Figure 7. A tyrosine storage vacuole. In carefully fixed preparations, the vacuole membranes are intact and the lumen is featureless. More commonly, the membrane breaks and mobile cytoplasmic components, such as glycogen granules, leak in.

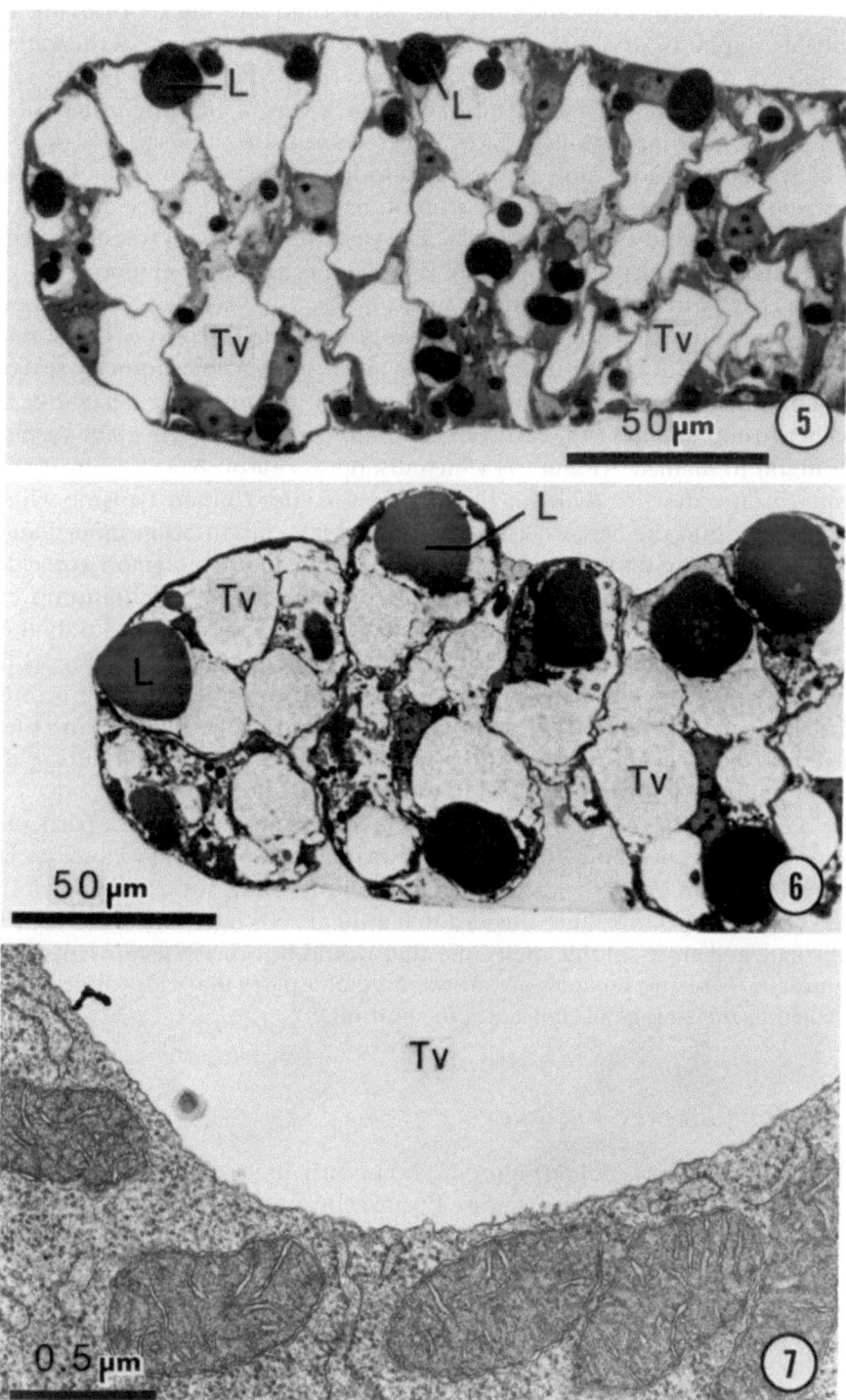

and van der Molen, 1979) and *Calpodes* (McDermid and Locke, 1983) are all probably similar in structure and origin (Figure 4) although some of the watery vacuoles in *Rhodnius* fat body may be derived from enlarged autophagic vacuoles that have lost their contents except for a rim of lipofuscin deposited around the inner membrane surface (Wigglesworth, 1982). In larval *Calpodes* fat body, giant vacuoles store tyrosine and follow a cycle of growth and release of phenolics to the hemolymph in relation to the needs of cuticle tanning at molting (McDermid and Locke, 1983). The vacuoles are largest toward the end of the intermolt just prior to molting (Figure 5) or pupation (Figure 6).

At ecdysis to the fourth instar in *Calpodes*, the vacuoles occupy about 40% of the volume of each cell. They disappear within the next 6-12 hr and a new population of fourth-instar vacuoles arises. Provacuoles form by surface infolding of plasma membrane and fuse to form vacuoles (Figures 8-10). By further fusion and growth, each cell comes to contain a single giant vacuole 20-40 μm in diameter (Figure 7) which disappears abruptly in a 12-hr period centered on ecdysis to the fifth instar. The vacuoles contain tyrosine which they release into the hemocoel to give a maximal concentration immediately after ecdysis. The disappearance of tyrosine from the hemolymph coincides with a need for phenolic precursors by the epidermis for cuticular tanning.

Vacuoles are not found in the fifth stadium until the molt and only then for a brief period and more markedly in some cells than in others (Figure 6). These prepupal vacuoles are presumed to be tyrosine storage vacuoles because of their structural similarity to those in the fourth stage but tyrosine titers have not yet been measured at this time. Vacuoles are not found during the pupal stadium or at the pupal-adult molt (Larsen, 1976).

These vacuoles are of interest in cell biology as an example of a particular kind of plasma membrane recycling. Plasma membrane is endocytosed to become the membrane of the vacuole and this returns to the surface at exocytosis. The vacuoles also introduce the idea that animal cells have evolved structures to isolate and store soluble molecules that would be otherwise difficult to accumulate. Tyrosine vacuoles are an example of a particular class of organelle devoted to the storage of molecules in solution.

2.2. Urate Storage Vacuoles

Urate storage vacuoles (Figure 11) occur only in urocytes such as those of cockroach fat body. Urocytes (Figure 12) are cells specializing in the accumulation and release of urate (Cochran, 1975; Cochran *et al.*, 1979; Mullins, 1979). The urocytes are packed between trophocytes, often near a third kind of fat body cell that is specialized to house symbionts, the mycetocyte. Unlike the trophocytes that have abundant lipid, glycogen, protein granules, and the machinery for protein synthesis and secretion, the urocytes are mainly vacuoles with all other cell components much reduced. The vacuoles have a cortex of fibrous material and a characteristic core (Figure 13) that may be connected to sacs of fibrous material similar to that of the cortex. Cochran and colleagues

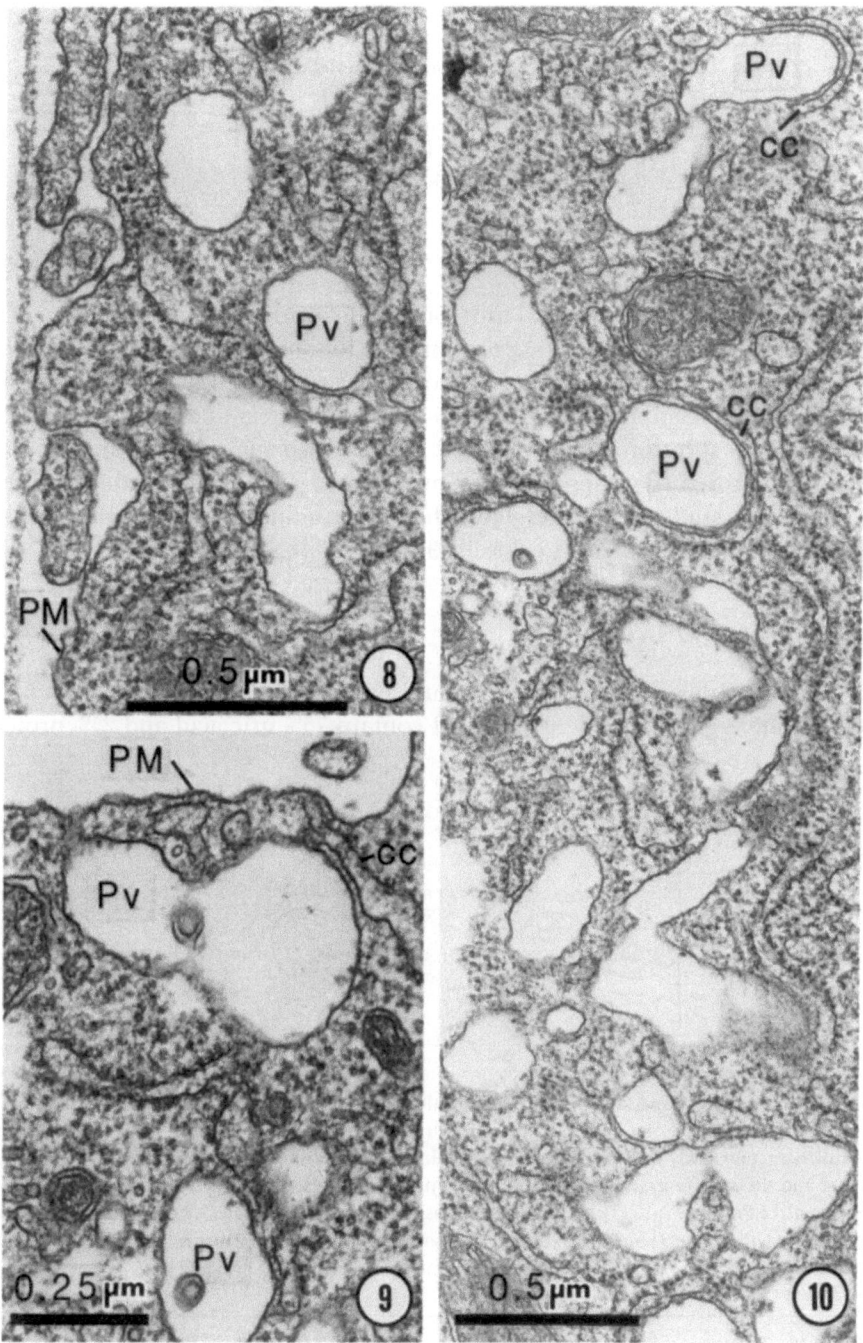

Figures 8-10. The origin of provacuoles from infolds of the plasma membrane. The development of early fourth-stage larvae was synchronized by brief starvation followed by feeding. Plasma membrane infolds, provacuoles, and vacuoles occur in sequence between 10 and 24 hr after ecdysis. Both plasma membrane infolds and provacuoles may initially have confronting cisternae of ER.

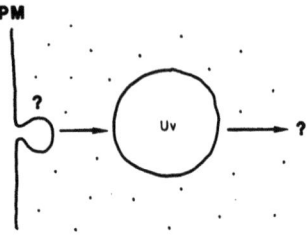

Figure 11. The origin and fate of the urate storage vacuoles found in *Periplaneta* urocytes. The thick membranes of these vacuoles suggest that they may be derived from the plasma membrane as in tyrosine storage vacuoles. The luminal tents have an ordered structure making it unlikely that they release material by exocytosis.

(1979) call this a urate structural unit, but nothing is known of its significance. The urate, which is not preserved in electron micrographs, occupies the clearer area between the core and the cortex. There is no obvious vesiculation of the vacuole membrane, and urate presumably enters and leaves across the membrane in solution. The structured interior also suggests that the vacuole is not exocytosed like those for tyrosine storage. The origin of the vacuoles has not been studied, but they have thick membranes similar to the plasma membrane (McDermid and Locke, 1983; Dean et al., 1983).

2.3. Urate Granules

At particular stages of development, Lepidoptera store urate in all fat body cells in urate granules (Figure 14) that contain 75% uric acid and 25% protein

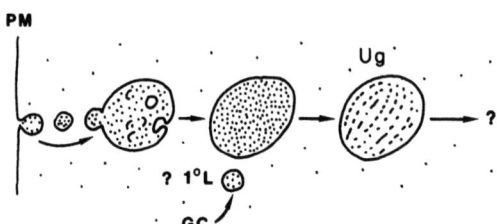

Figure 14. The origin and fate of the urate granules found in the fat body of lepidopterans at metamorphosis. In a period of a few hours prior to pupation, the fat body sequesters protein in a kind of multivesicular body that has a different fate from the intermolt multivesicular bodies formed earlier and the protein granules that are formed just afterwards. These vacuoles have a moderately dense matrix that very quickly becomes fibrous as they accumulate urate. They contain acid phosphatase and therefore presumably receive a lysosomal component. During the development of the adult in the pupa, they lose their contents and disappear by the time of imaginal eclosion.

Figure 12. The three cell types in cockroach fat body. Trophocytes form the bulk of the fat body and have a structure appropriate for the synthesis and storage of lipid, glycogen, and protein. They are like the cells found in most undifferentiated fat bodies. Within the mass of trophocytes are occasional urocytes, distinguished by their lack of dense contents, since they are mainly composed of urate vacuoles. The mycetocytes are almost completely filled with vacuoles containing symbionts. Light micrograph of an epoxy section.

Figure 13. A urate storage vacuole in a cockroach urocyte. The vacuole has a fibrous cortex and a characteristic core. The urate, which does not survive preparation for electron microscopy, occupies the clear spaces.

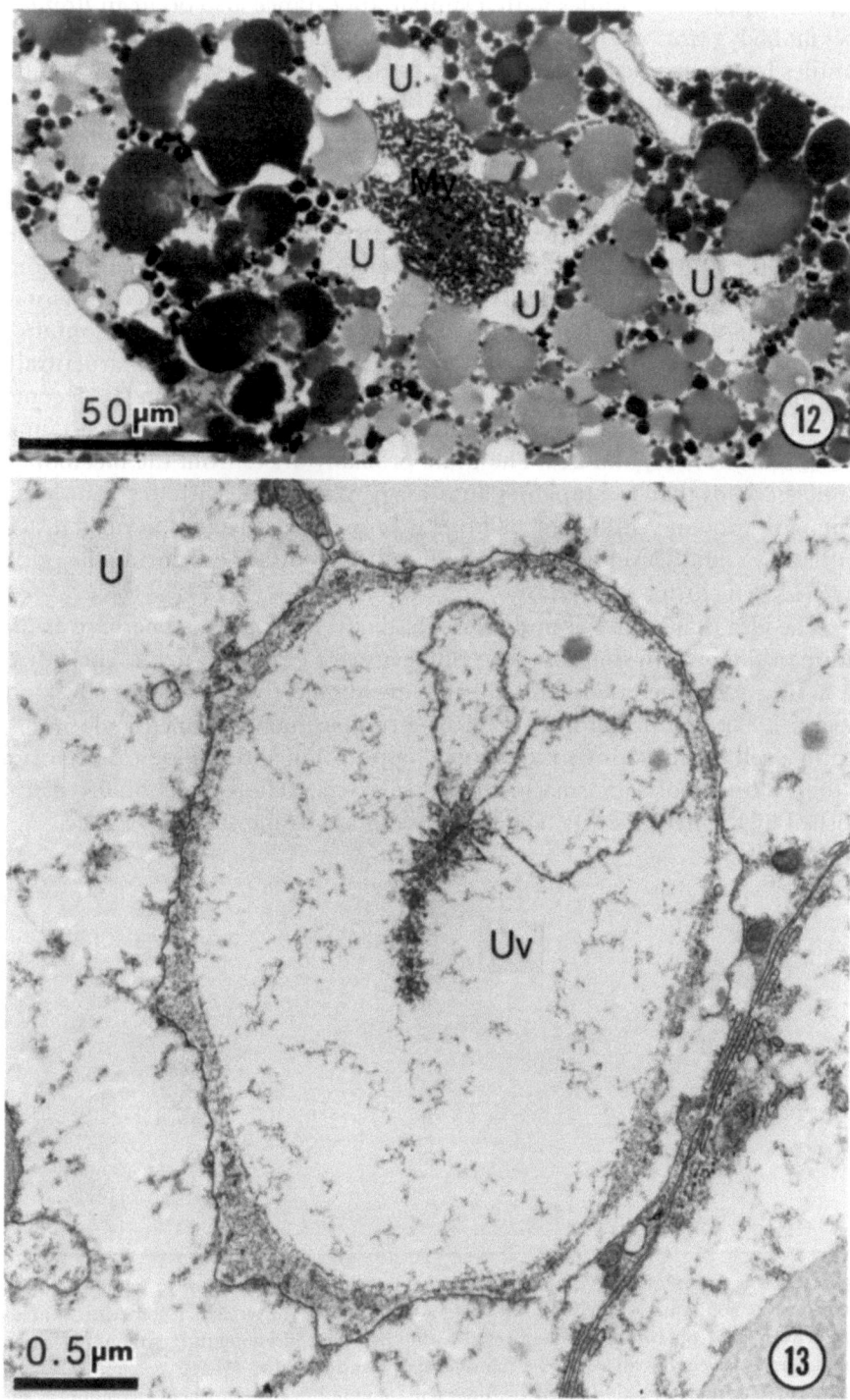

(Tojo et al., 1978). Granules with a similar appearance also occur in *Bombyx mori* fat body (Mori et al., 1970; Waku and Sumimoto, 1969). In *Calpodes*, the granules have a characteristic fibrous matrix that develops at a time when the uric acid accumulates (Dean et al., 1983; Locke, 1984) which is just after peroxisomes containing urate oxidase disappear (Locke and McMahon, 1971). The origin of the urate storage granules is of interest (Figure 15). Initially, they probably arise mainly from protein sequestered into a special kind of multivesicular body containing different proteins from those in the protein granules (Tojo et al., 1978). Since urates may be transported on protein-bound complexes (Cochran, 1975), the multivesicular body urate granule precursors may contain hemolymph proteins with that property. They may also contain a lysosomal component. Very soon after the pro-urate granules are structurally recognizable as a distinct class of multivesicular body, their matrix becomes fibrous (Figure 16). Most of the protein then disappears as the urate accumulates (Figures 17, 18, and 22). The urate probably arises from the metabolism of nucleic acids that accompanies autophagy of the RER with the completion of massive protein synthesis. The urate is probably used again in nucleic acid synthesis for adult development when the granules disappear during the pupal stage (Larsen, 1976).

The idea that a kind of multivesicular body, distinguished perhaps by the nature of the protein that has been pinocytosed to make it, my be specialized for a function other than protein and membrane turnover, is a novel one (Locke, 1984). The principle of creating a compartment of functional proteins within a cell by pinocytosis is of general application and is a logical extension of the evolution of a plasma membrane-derived vacuole functionally distinguished by the nature of its membrane, as in a tyrosine storage vacuole.

Figure 15. The origin of urate granules. At the end of the intermolt, prior to pupation, the fat body comes to contain newly sequestered protein in a kind of multivesicular body with a moderately dense matrix containing few microvesicles as shown here. Surviving intermolt multivesicular bodies tend to have more microvesicles than matrix (see also Figures 43, 44).

Figure 16. The characteristic fibrous texture of urate granules early in their formation as they begin to accumulate urate at a stage shortly after that shown in Figure 15.

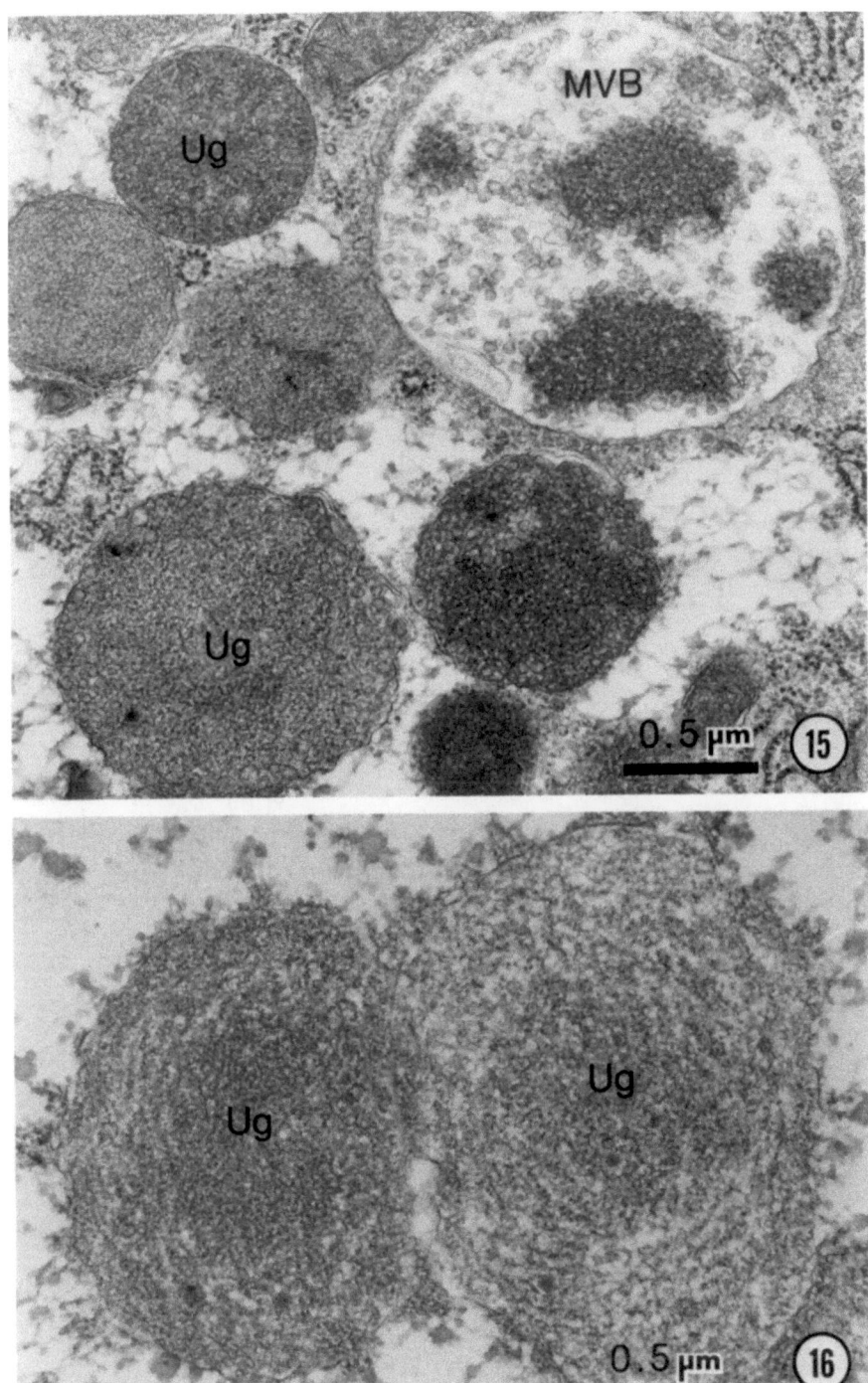

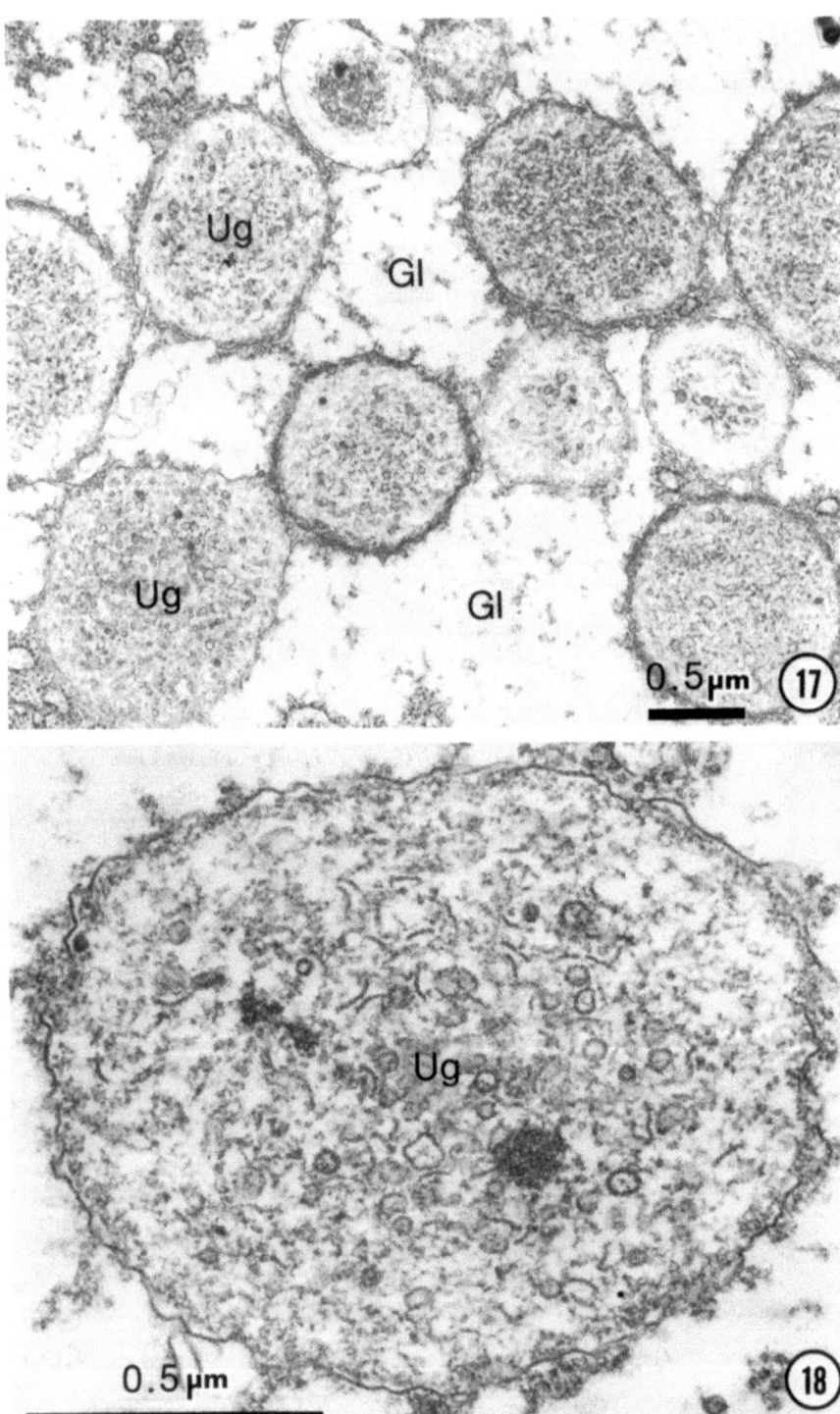

2.4. Protein Storage Granules

Almost all cells in an insect take up hemolymph protein, particularly at metamorphosis (Locke and Collins, 1967), but the process is most clearly displayed in the fat body (Figure 19) (Locke and Collins, 1968; Collins, 1969, 1974, 1975; Collins and Downe, 1970; Locke *et al.*, 1982; Lauverjat, 1977; Dean *et al.*, 1984; Butterworth *et al.*, 1979; Waku and Sumimoto, 1969). The initial pinocytosis, usually from coated vesicles, may take place from any surface—the plasma membrane below the basal lamina, the plasma membrane reticular system, the lateral surfaces, or the interior lymph spaces. The uptake of foreign proteins has been demonstrated by electron microscopy, but under natural circumstances, the fat body preferentially takes up particular hemolymph proteins for storage. One might expect that all storage proteins would be retrieved from the hemolymph at metamorphosis but that some lipophorin would be left circulating. This occurs in *Calpodes* where the two storage proteins are sequestered together with some lipophorin (Webster, 1982).

Protein granules and multivesicular bodies receive three categories of material: membrane, membrane-bound molecules, and, in the core, a sampling of the molecules present in the extracellular space. Surface-to-volume ratios dictate that very small vesicles contain more membrane than lumen. Structures formed from them must therefore always be autophagic, turning over more membrane than contents. Small vesicles derived from plasma membrane that has bound specific proteins have little luminal volume left to carry unbound molecules. Vacuoles fed by such small vesicles may therefore have specific contents without a mechanism for excluding molecules. The selective uptake of proteins that has been described for the fat body (Chippendale and Kilby, 1969; Loughton and West, 1965; Chippendale, 1970; Collins and Downe, 1970; Locke *et al.*, 1982; Webster, 1982; Tojo *et al.*, 1978, 1980, 1981; Kramer *et al.*, 1980) is probably the result of specific binding to the plasma membrane surface above the coated pits that precede pinocytosis.

Pinocytosis varies during development in amount, probably in specificity, and certainly in the fate of the coated vesicles. In *Calpodes*, pinocytosis occurs throughout the fifth stadium. During the intermolt, the pinocytotic vesicles form multivesicular bodies and the proteins are turned over. At molting, there is a switch from lysis to storage and protein is sequestered in large membrane-bound granules (Figure 20) (Locke and Collins, 1968; Dean *et al.*, 1984; Locke *et al.*, 1982). The fat body becomes the main protein store of an insect when the hemolymph storage proteins are transferred to it.

A side effect of the pinocytosis of massive amounts of protein is the possibility for plasma membrane turnover. Assuming that there is membrane

---

Figures 17 and 18. The structure of urate granules after they have begun to accumulate urate. The fibrous matrix is soon replaced by the remains of microvesicles and the urate that is not preserved in these electron micrographs. Urate granules are typically in masses of glycogen rather than RER and have crenate edges rather than the smooth surfaces of multivesicular bodies and immature protein granules.

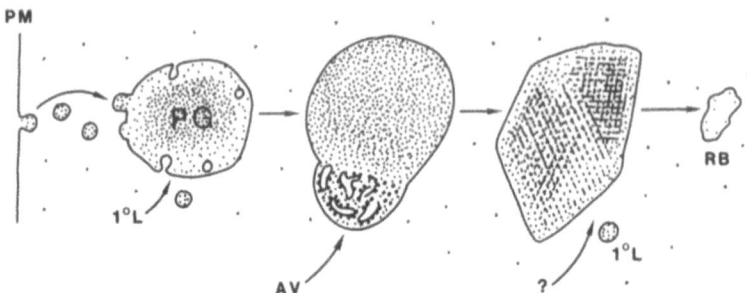

Figure 19. The formation and fate of protein storage granules. At metamorphosis in endopterygotes, there is a major shift in the disposition of storage proteins from the hemolymph into the fat body. In exopterygotes, immature stages as well as adults store protein in their fat body. Protein pinocytosed from all surfaces collects in vacuoles that are secondary lysosomes. Within them the extra membrane from the transporting microvesicles is digested and the protein condenses. Digestion may continue within them when they fuse with autophagic vacuoles, but the bulk of the protein remains intact and is stored in crystals that determine the vacuole shape. A second round of digestion takes place for the crystalline protein to be utilized during development.

turnover rather than recycling, order of magnitude calculations show that the fat body plasma membrane would be replaced every 10 min during the uptake of protein. Even with some recycling and a margin of error in the calculations, it seems likely that fat body membranes are replaced rather rapidly at metamorphosis.

Protein granules, like multivesicular bodies, receive a complement of lytic enzymes from primary lysosomes (they are secondary lysosomes, indeed they often fuse with RER autophagic vacuoles, Figure 21) but the result is condensation and crystallization of the protein contents (Figures 22, 23). These crystalline protein granules are stored and used later in the pupa and in the adult (Larsen, 1976). Crystalline protein storage granules have been reported in many other insects (*Locusta migratoria*, Lauverjat, 1977; *P. americana*, Dean et al., 1984). Although the protein granules in *Calpodes* pupae are mainly composed of crystals of a single type, the fat body continues to contain two kinds of storage protein and lipophorin which are all made from different polypeptides (Webster, 1982). The crystals may therefore be mixed polypeptides as seems to be the case in *B. mori* (Tojo et al., 1980).

In the context of this essay, these protein storage granules are vacuoles bounded by thick membranes. They are reaction vessels that have two phases of activity. At first, they maintain the conditions for autophagic digestion and

Figure 20. A very early stage protein granule at the beginning of pupation. Protein granules have a dense matrix from very early in their formation and microvesicles are limited to the periphery. Many of the pinocytosis vesicles that carry sequestered protein to them are completely filled.

Figure 21. Protein granules may fuse with autophagic vacuoles and consolidation of the pinocytosed protein goes on side by side with digestion of RER. The elaboration of the plasma membrane into a reticular system is not an adaptation to allow an increased surface area for the pinocytosis of protein. The plasma membrane reticular system is reduced by the beginning of granule formation and absent when granules are still growing as in this preparation. Pinocytosis occurs from any plasma membrane surface.

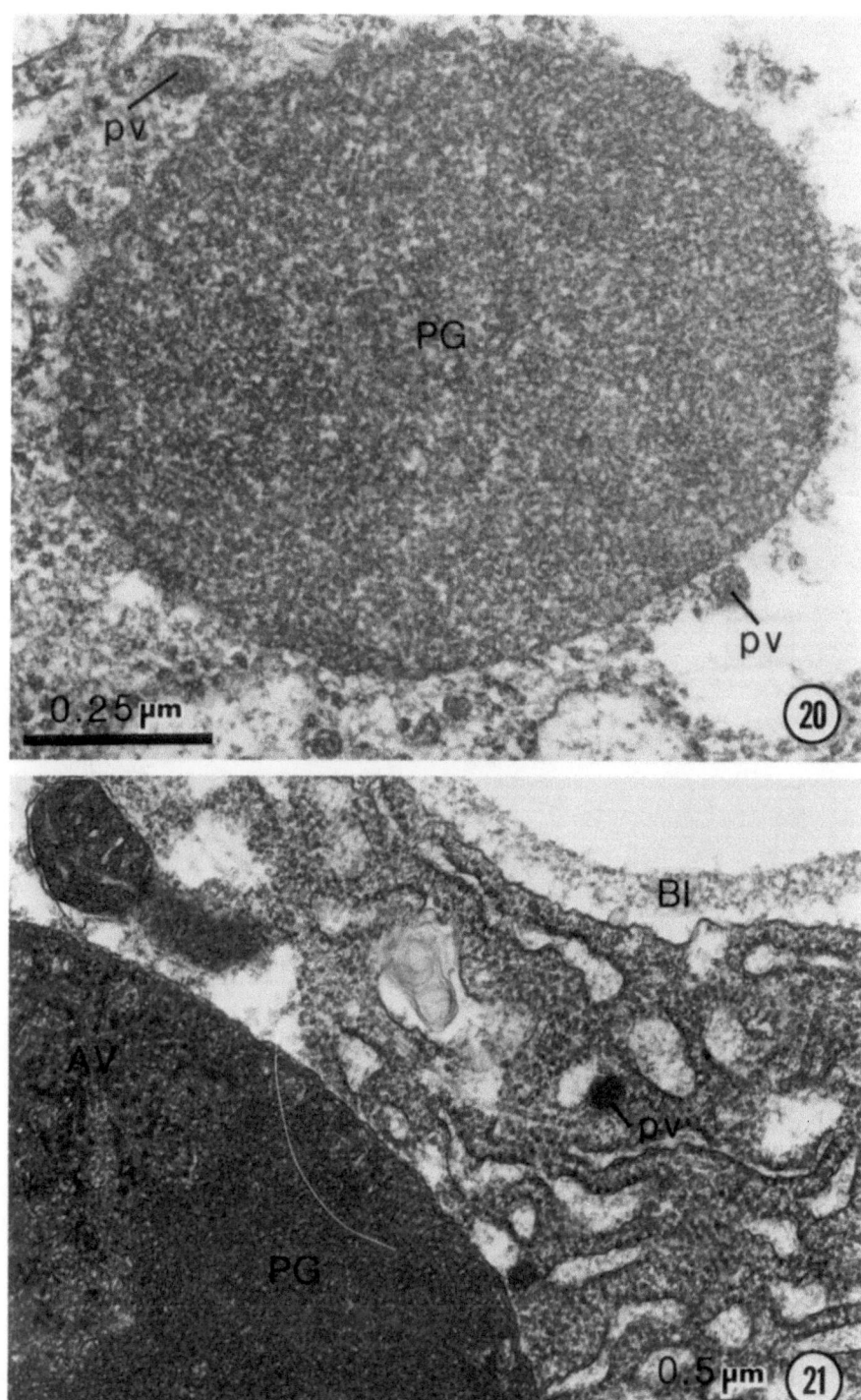

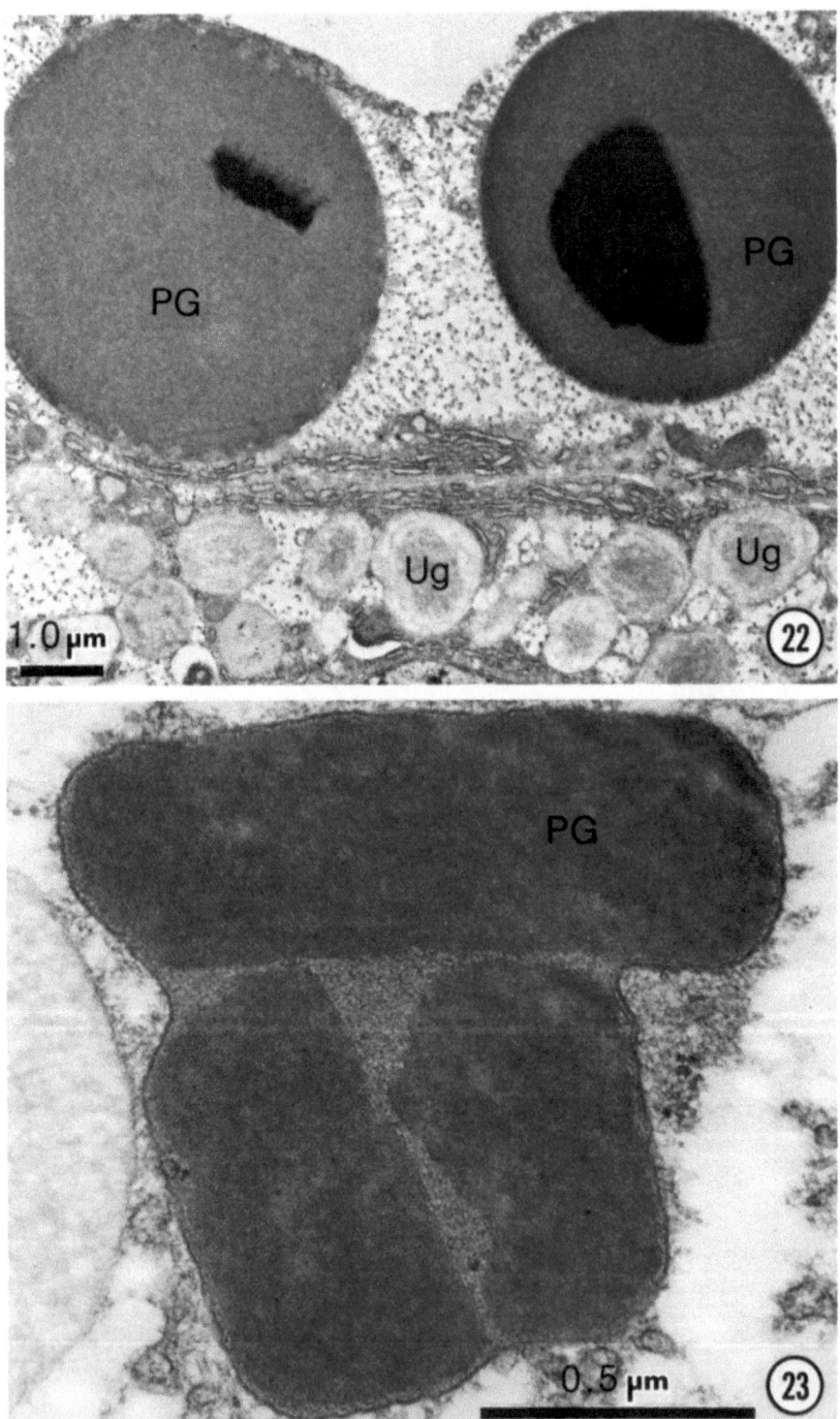

the condensation and crystallization of proteins for storage. Later, they create conditions to hydrolyze the crystals to amino acids used for adult development. The two phases presumably require two separate rounds of primary lysosome addition, each with the appropriate enzymes.

2.5. Secretory Vesicles

Secretory vesicles are among the first compartments to differentiate from membranes that have newly traversed the Golgi complex (Figure 24). The Golgi complexes of insect fat body are structurally simpler than those found in vertebrate or plant cells. In the fat body of mid-fifth-instar *Calpodes* larvae, the Golgi complex (Figure 25) begins with the transition vesicles that bud from the smooth face of the RER through the rings of Golgi complex beads (Locke and Huie, 1976a; Locke, 1980). There are then one or two outer saccules and two or three inner saccules, and depending on the stage of development, condensing vacuoles, secretory vesicles, primary lysosomes, and isolation envelope precursor vesicles (McClintock and Locke, 1982). As in the Golgi complexes of other organisms (Pelttari and Helminen, 1979), the differentiation of membranes as they pass through the Golgi complex involves an increase in their thickness. The membrane of the RER is only about 85% of the thickness of membranes leaving the Golgi complex to become secretory vesicles and

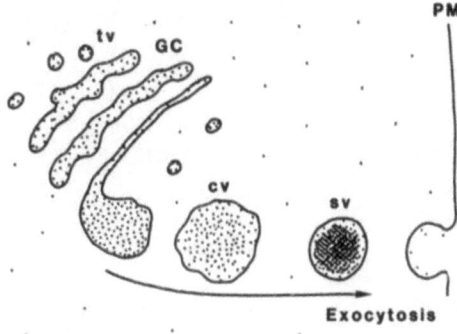

Figure 24. Secretory vesicles and the Golgi complex. Fat body Golgi complexes have a relatively small number of compartments compared to those in most secretory cells. A rather broad outer saccule arises from the transition vesicles. There may then be another broad saccule, or a thinner one followed by one or two thin saccules, making three or four in all. Condensing vacuoles arise from the edges of the thin innermost saccule. These become smaller and more spherical, often with crystallization of the contents as they mature into secretory vesicles. The secretory vesicles are not stored but are exocytosed within 2 hr.

←―――

Figure 22. A few hours before ecdysis to the pupa, the protein has begun to crystallize out in the center of the spherical granules that may be many micrometers in diameter. By this stage, the accumulation of urate in the urate granules is complete.

Figure 23. In the fat body of both nymphal and adult cockroaches, the protein granules contain crystals that distort the shape of the vacuole.

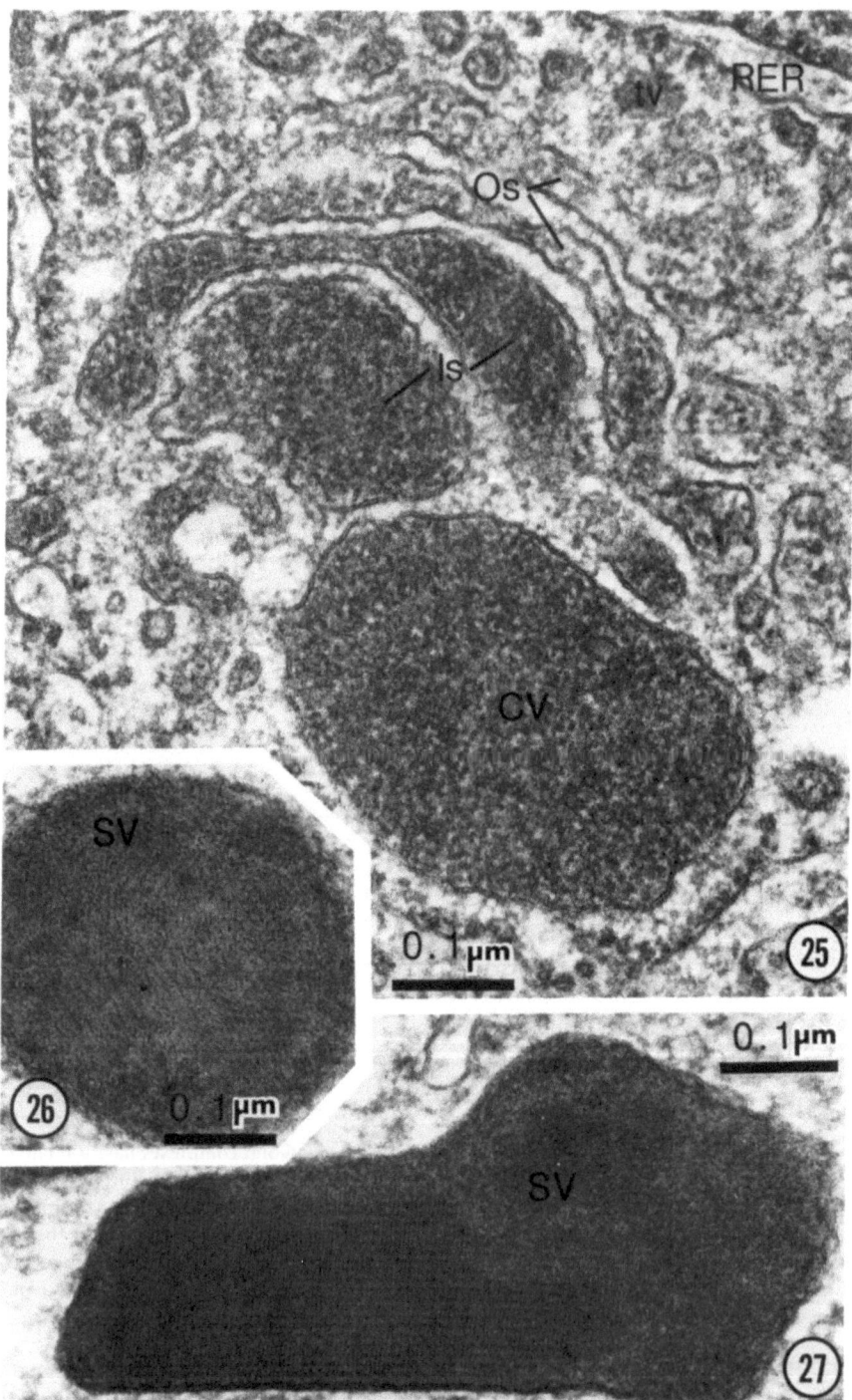

plasma membrane (McDermid and Locke, 1983). Material from the Golgi complex compartments fills the condensing vacuoles which at first contain protein with a loosely fitting membrane. As the condensing vacuoles become smaller to form secretory vesicles, the membrane becomes more nearly spherical (Figure 26). In the latter part of the intermolt, a crystalline core may eventually occupy the whole secretory vesicle and distort its shape (Figure 27). These secretory vesicles contain at least three proteins—lipophorin and two storage proteins—each constructed from different polypeptides (Webster, 1982). The crystals are therefore probably mixed, as in protein storage granules. Secretory vesicles are thus vacuoles maintaining an environment in which proteins can change in a way that allows condensation and crystallization. The need for the crystallization of the contents of secretory vesicles is not clear since they are exocytosed and not stored. A natural pulse of [^{14}C]leucine in the hemolymph is incorporated by the fat body into proteins which reach equilibrium in the hemolymph within 2 hr (Webster, 1982). Although the fat body may need hormonal stimulation for protein synthesis, it is not like vertebrate pancreas which requires hormonal stimulation to cause the release of zymogen granules.

2.6. Vacuoles Containing Symbionts

Many insects live in symbiotic relationships with microorganisms lodged within some of their cells which may form specialized mycetocytes (Houk and Griffiths, 1980). Fat body cells differentiated as mycetocytes occur most commonly in orthopteroids and hemipteroids (Dean *et al.*, 1983). The problem of coexistence is probably similar at the cellular level whatever cell types are involved. The host cell must keep its symbiont in a controlled environment to which the symbiont has access. For this purpose, fat body cells have evolved vacuoles that are presumably derived from the plasma membrane (Figure 28). In *Periplaneta*, the cells occur toward the center of fat body lobes (Figure 12) (Bodenstein, 1953). In *Blatella germanica*, they develop even in the absence of symbionts (Brooks and Richards, 1955) although such mycetocytes probably lack normal vacuoles. Superficially, electron micrographs of *P. americana* mycetocytes seem to show the symbionts surrounded by envelopes of two unit membranes, like organelles in the process of isolation in the first stage of au-

---

Figures 25-27. Secretory vesicles and the Golgi complex. The Golgi complex begins with the transition vesicles that arise by budding from a smooth face of the RER. The budding takes place through rings of Golgi complex beads (not shown in this electron micrograph) that can be thought of as organizers for Golgi complexes. There are then three or four saccules. The outer saccules and the transition vesicles differ from nearby compartments in being osmiophilic in the hot osmium tetroxide reaction (Locke and Sykes, 1975). The inner saccules are shown here in oblique profile but typically the innermost one is thin with a straight profile suggesting rigidity. The inner saccules contain acid phosphatase (McClintock and Locke, 1982). Condensing vacuoles and microvesicles that are presumed to be primary lysosomes are often seen in continuity with the edge of the innermost saccule. Secretory vesicles with more rounded profiles and crystalline cores occur next to the condensing vacuoles. The crystallization may sometimes distort the secretory vesicle shape.

tophagy (Figures 29, 30). However, these symbionts are gram-negative bacteria which have a membranelike envelope (Locke, 1982) surrounding their cell wall within which is their plasma membrane. The host cell contributes a single thick membrane to make the vacuole housing the bacterium (Figure 31). Since mycetocytes have to be infected by their bacteria, the vacuole membrane is presumably derived from host cell plasma membrane as it envelopes its guest in a mechanism similar to basal lamina phagocytosis. Once within the vacuole, communication between host and symbiont must take place through the fluid space of the vacuole. In *Acyrthosiphon pisum*, it has been claimed that microvesicles may be involved (Griffiths and Beck, 1973, 1975, 1977). These authors provide some of the only experimental evidence that may show the gain derived by the host from the relationship. Electron microscope autoradiography, coupled with the incorporation of [^{3}H]mevalonate, suggested that the microorganisms synthesize cholesterol. The majority of silver grains were at first associated with the symbiote membranes, but the frequency of grains over surrounding mycetocyte tissue increased with increasing incubation time. These symbiotes may export cholesterol to the host cell. In all mycetocytes, the host may digest its symbionts from time to time, presumably by treating the vacuoles as digestive compartments by fusing primary lysosomes with them.

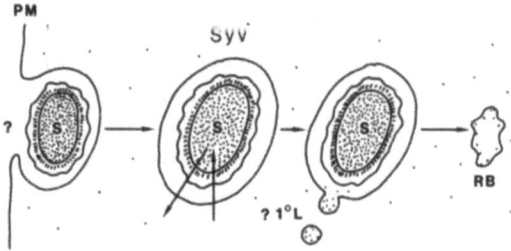

Figure 28. Vacuoles containing symbionts in the mycetocytes of *Periplaneta*. The symbionts occur in vacuoles. Since the mycetocytes have to become infected during their development, the vacuole membrane is most probably derived from the plasma membrane. Vacuoles at some stages contain degenerating symbionts and these have presumably become digestive vacuoles by the addition of primary lysosomes. Gram-negative symbionts have a very plasma-membrane-like outer covering that can easily be confused with a host cell membrane.

———————————————————————————————————————→

Figures 29-31. The vacuoles containing symbionts in the mycetocytes of *Periplaneta*.
Figure 29. The symbionts in the mycetocytes (see Figure 12) are all within vacuoles (Syv) that take up most of the cell.
Figure 30. The vacuole membrane surrounds the outer envelope of the gram-negative bacterium. Superficially, the bacterium seems to be within an envelope like an isolation body, but gram-negative bacteria are characterized by having a membranelike outer covering (OE) around the cell wall.
Figure 31. Enlargement of the vacuole region showing the symbiont, the symbiont membrane, the symbiont cell wall, the outer covering or envelope of the symbiont, the vacuole, and the vacuole membrane of the host cell.

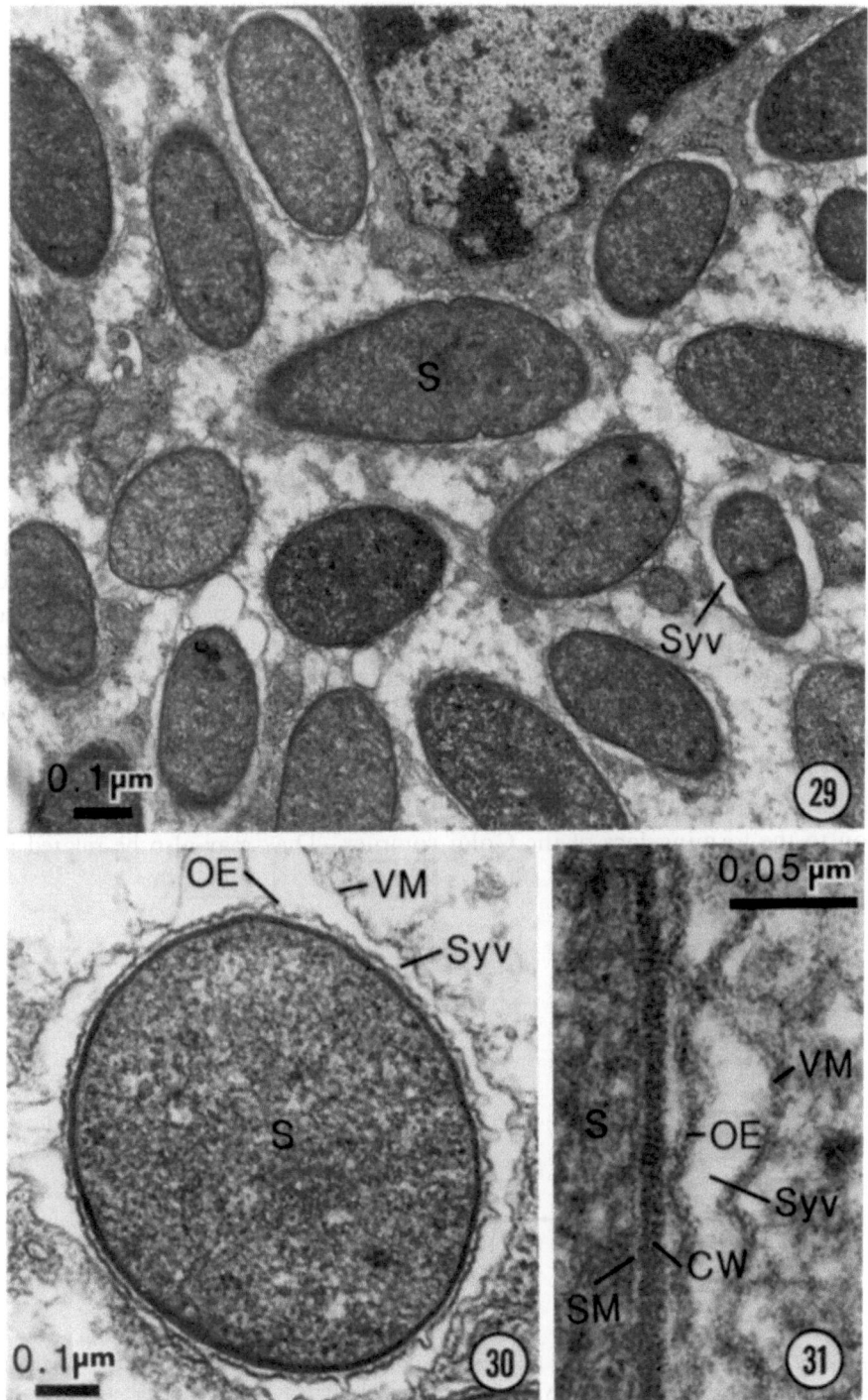

Most vacuoles have separate phases of filling, processing, and emptying as in protein storage granules for example. The host relationship with its vacuoles containing symbionts is more subtle. They are compartments that must be both fed and milked at the same time. The complexity of this interaction may explain the evolution of mycetocytes that are specialized for this one purpose.

2.7. Vacuoles Associated with Glycogen

The fat body and many tissues accumulate glycogen. In *Calpodes*, its accumulation accelerates in the fifth stadium at the time of commitment to pupation. As much as 30% of a cell may be glycogen by the beginning of molting. At its maximal extent, perhaps coincident with its conversion to soluble molecules that can be easily moved, there are vacuoles in the heart of these masses of glycogen (Figures 32, 33). The vacuoles also occur in glycogen of other tissues such as the epidermis (Locke, 1984). If they prove to be a constant feature, then they may have to be incorporated into hypotheses for glycogen redeployment. It is usually assumed that glycogen is converted to glucose as it may be required (Steele, 1982), but the presence of vacuoles suggests a mechanism to allow sugar storage to meet sudden large demands. If the location of these vacuoles near glycogen is a functional one, it might be explained if they are akin to tyrosine storage vacuoles containing phenolics conjugated as glucosides, since glucosides have been found in the fat body and hemolymph (Brunet, 1980).

The origin of the vacuoles has not been seen. Their characteristic osmiophilia (Figure 34) is similar to that of isolation envelopes and suggests a relation to the forming face of the Golgi complex. They may also be related to tyrosine storage vacuoles.

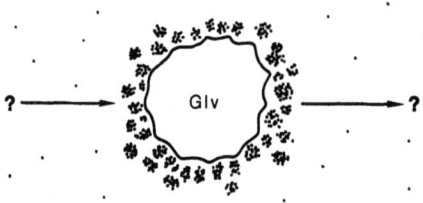

Figure 32. Vacuoles associated with glycogen. Within the masses of glycogen built up by the fat body, there are often rather irregularly shaped membrane-bound compartments. They seem to be too constant a feature to be merely an artifact of fixation of myelinlike lipid. Nothing is known of their contents, origin, or fate.

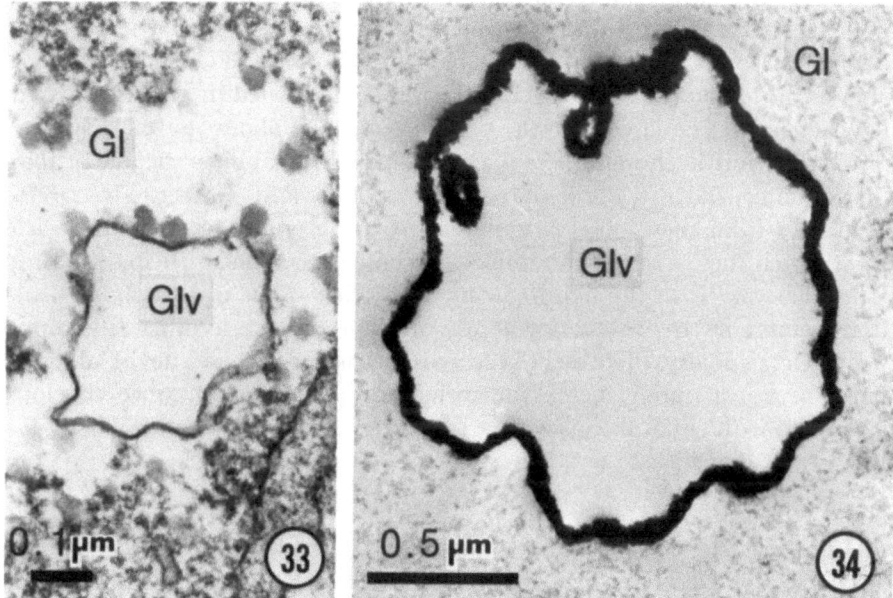

Figures 33 and 34. The vacuoles associated with glycogen.
Figure 33. The vacuoles often have an irregular profile, presumably due to shrinkage and fixation damage. They are often much larger but are easily broken.
Figure 34. Vacuoles associated with glycogen have osmiophilic membranes after treatment with hot osmium tetroxide or osmium-zinc iodide, like the outer saccules and transition vesicles of the Golgi complex.

2.8. Autophagic Vacuoles

The problem facing a cell that is intent on destroying part of itself is how to transfer organelles that are scheduled for destruction from the cytoplasm into digestive vacuoles without letting the digestive environment out into the cell. The answer, from studies on metamorphosing fat body of *Calpodes*, is that components to be digested are first isolated from the cytoplasm (Locke and Collins, 1965, 1968; Locke, 1980, 1981), and only then do primary lysosomes fuse with the new, topologically external compartments, to digest their contents (Locke and Sykes, 1975). Autophagy is a two-step process, first isolation, then digestion (Locke, 1980, 1981, 1984), and autophagic vacuole membranes are derived from two different posttransition regions (Figure 35).

The first indication that an organelle is to be destroyed is the presence of a tiny vesicle closely apposed to its surface (Figures 36, 37, 39, and 40). As more vesicles fuse with it, the isolating envelope seems to creep over the surface until investment is complete (Figures 37, 38). The close apposition between the envelope and its prey, suggests that there may be a special kind of adhesion between their surfaces. The specificity of this adhesion may explain the specificity of destruction. The contents of autophagic vacuoles are not a random sample of the cytoplasmic constituents of the cell. Although most cell components are on the menu, they are eaten separately and in a particular order: first peroxisomes (Figures 37, 38), then mitochondria (Figures 39, 40), and after a gap of about 12 hr the RER. The RER is isolated in a particularly interesting way. The envelope cuts off a fragment of about the same size as a mitochondrion, as though there is a mechanism for making a vacuole of about that size independently of the shape and area of the RER surface to be covered. It follows from these observations that isolating envelopes and the vesicles from which they arise are autophagic vacuole precursors with specific properties that vary with the requirements of the first phase of autophagy. Isolation membranes are unique in the way that they adhere to other cell components. Occasionally, there seem to be particles between these sites of adhesion, but it is not a purely membrane-membrane phenomenon, since envelopes also creep with equal intimacy over ribosomes on the RER.

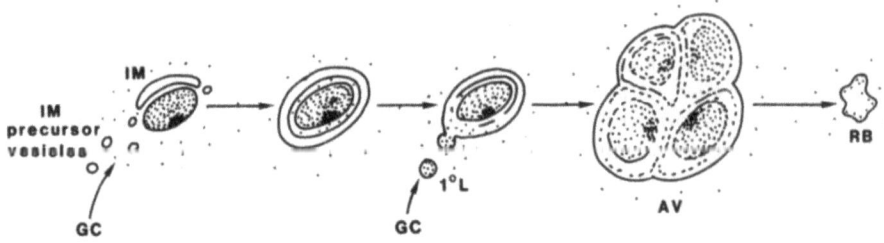

Figure 35. The origin and fate of autophagic vacuoles. Autophagy is a two-step process. Organelles scheduled for lysis are first made external to the cell by enclosing them in an envelope. Only then are hydrolytic enzymes added in the second step by fusion with primary lysosomes. The isolating step is organelle specific. Not all cell components are invested equally at all stages. Several isolated cell fragments, or isolation bodies, may fuse together to make autophagic vacuoles before or after the addition of primary lysosomes. Autophagic vacuoles also fuse with heterophagic vacuoles such as the protein granules.

Figures 36-41. Autophagic vacuoles in prepupal metamorphosis.

Figure 36. The new envelopes that arise in the first step of autophagy often have microvesicles next to them. They are presumed to grow by fusion with these microvesicles. Propinquity and cytochemistry both suggest a relation between the envelopes and the forming face of the Golgi complex.

Figure 37. The sequence of destruction is organelle specific. In the early stages of metamorphosis to the pupa, the isolating envelopes attach only to peroxisomes. Although the outer surfaces of these envelopes may be near mitochondria, only their inner surfaces have developed a very close relationship with the peroxisomes.

Figure 38. The end result of the first step in autophagy is an organelle, in this case a peroxisome, made completely external to the cytoplasm. In the next step, primary lysosomes add lytic enzymes to eliminate the whole population of peroxisomes.

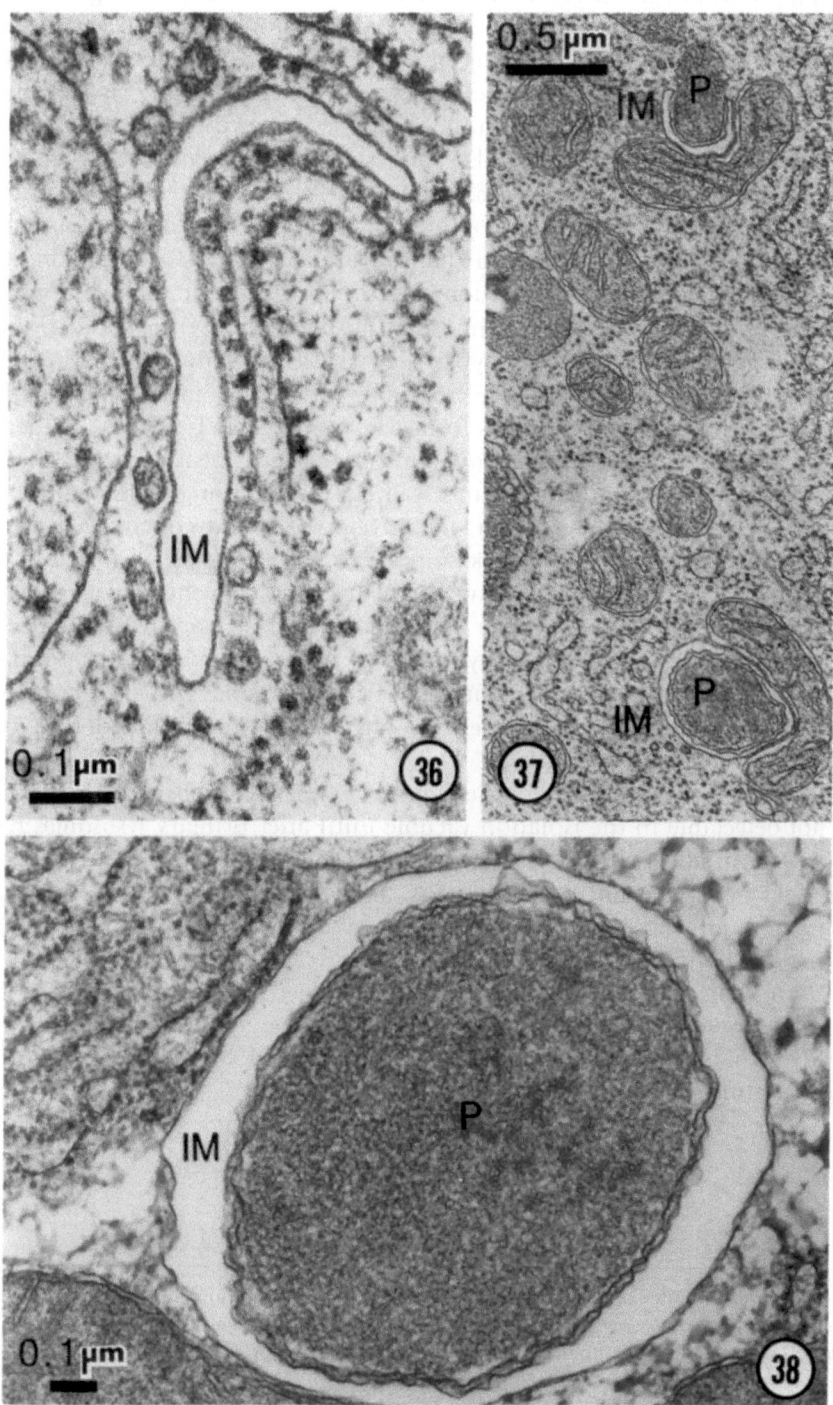

The morphology suggests that the membrane of the isolating envelope arises from the forming face of the Golgi complex or nearby ER (Locke and Collins, 1965, 1980). The isolating envelopes also share some cytochemical features with the transition vesicles and nearby ER and Golgi complex outer saccules. They all stain with osmium after the hot osmium reaction (Locke and Sykes, 1975). Lead staining also suggests a relation with the Golgi complex (McClintock and Locke, 1982). The cisternal contents of isolation envelopes and parts of the Golgi complex both stain with lead in mouse and *Calpodes* tissue. Both osmiophilia and lead staining are lost after fusion with primary lysosomes and the transition to become autophagic vacuoles.

The chief characteristics of the isolating phase are: (1) its specificity for certain organelles (in some cell types such as the fat body), (2) the intimate relation between the envelope and the object that it encloses, (3) the rather uniform size of the compartments formed independently from their contents, and (4) the emptiness of the envelope lumen after most treatments, except hot osmium which reacts intensely, and lead staining (prior to the addition of primary lysosomes). In particular, the envelopes do not contain acid phosphatase, the marker enzyme for lysosomes.

Once the cell components have been externalized in their own compartments in this way, they may either fuse directly with primary lysosomes to become autophagic vacuoles or they may first fuse with one another to make larger autophagic vacuoles (Figure 41). In this second step of autophagy, primary lysosomes carry digestive enzymes from the maturing face of the Golgi complex (Locke and Sykes, 1975).

The chief characteristics of the digestive phase are: (1) hydrolytic enzymes are not detectable in the isolation vacuole until after primary lysosomes have fused with it, after which lead staining and osmiophilia disappear, (2) digestion does not begin until this happens, and (3) isolation vacuoles may fuse with one another before and after fusion with primary lysosomes but not with any other membrane.

Autophagy is a metamorphic event stimulated by β-ecdysone in the absence of juvenile hormone. It normally takes place as the hemolymph titer of ecdysteroids rises to its prepupal peak (Dean *et al.*, 1980) and can be induced by β-ecdysone *in vitro* (Dean, 1978).

Figures 39 and 40. After peroxisome autophagy, most, but not all, mitochondria are destroyed. The isolating envelopes now develop a very close relationship with mitochondria.

Figure 41. An early stage autophagic vacuole. Several isolated organelles may fuse together to form a composite structure that later becomes an autophagic vacuole by the addition of primary lysosomes.

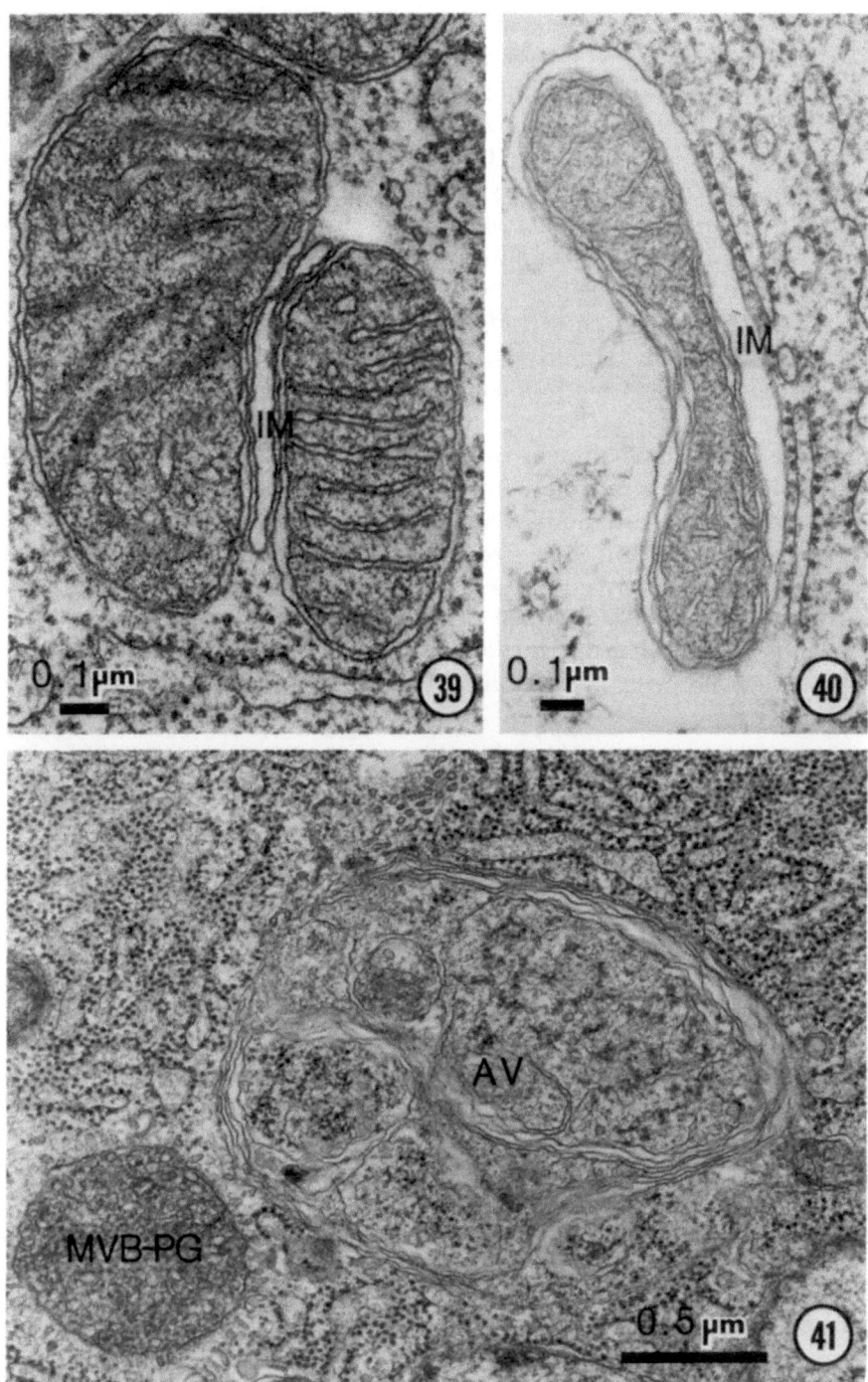

2.9. Multivesicular Bodies

Multivesicular bodies are vacuoles containing material from two sources: pinocytotic vesicles from the plasma membrane surface and primary lysosomes from the Golgi complex (Figure 42). Multivesicular bodies are present in most cells most of the time (Locke and Collins, 1967), continually receiving plasma membrane with specific proteins bound to surface receptors together with molecules trapped in the center of the pinocytotic vesicle. Depending on the stage of fat body development, multivesicular bodies may be mainly involved in membrane turnover or in the digestion of extracellular protein or both. Early in the fifth stadium of *Calpodes* when hemolymph protein titers are low, there are very few multivesicular bodies and these have little contents. They become larger with denser contents prior to metamorphosis (Figures 43, 44), before switching to become storage granules (Figure 20). At pupation when the hemolymph protein titers are again depleted, the multivesicular bodies return to their intermolt appearance (Locke and Collins, 1968). Hemolymph proteins are also sampled and fat body plasma membrane turned over at the pupal/adult molt. The frequency of multivesicular bodies is highest at about the time of pupal/adult ecdysis (Larsen, 1976).

The specific uptake of storage proteins and lipophorin into protein granules has been investigated in the fat body but little is known of the specificity of pinocytosis to make multivesicular bodies. The extent to which they may

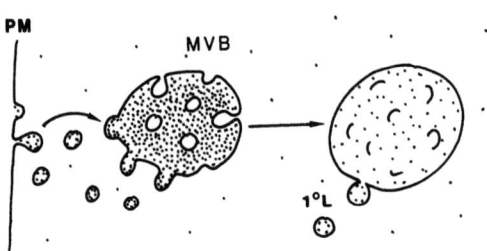

Figure 42. The origin and fate of multivesicular bodies. Multivesicular bodies are present in fat body cells most of the time. They are made from the fusion of pinocytotic vesicles coming from the cell surface carrying plasma membrane with a sampling of the extracellular environment. Once formed, they begin digestion with the receipt of primary lysosomes. The relative role of plasma membrane turnover and the digestion of extracellular protein varies with stage of development. Prior to pupation, there is a switch to the formation of protein granules which are structurally giant multivesicular bodies adapted for protein storage (Figure 19). At other times, most multivesicular bodies are probably in an equilibrium between the reception and digestion of material rather than each having a separate life cycle of formation and degradation.

Figures 43 and 44. Multivesicular bodies vary in composition with stage of development. Early in the stadium, they are mainly membranous and presumably concerned with membrane turnover as in the structure at the top of Figure 43. At pupation, as in the other larger multivesicular bodies shown, more protein is pinocytosed before the complete switch to protein granule formation is made.

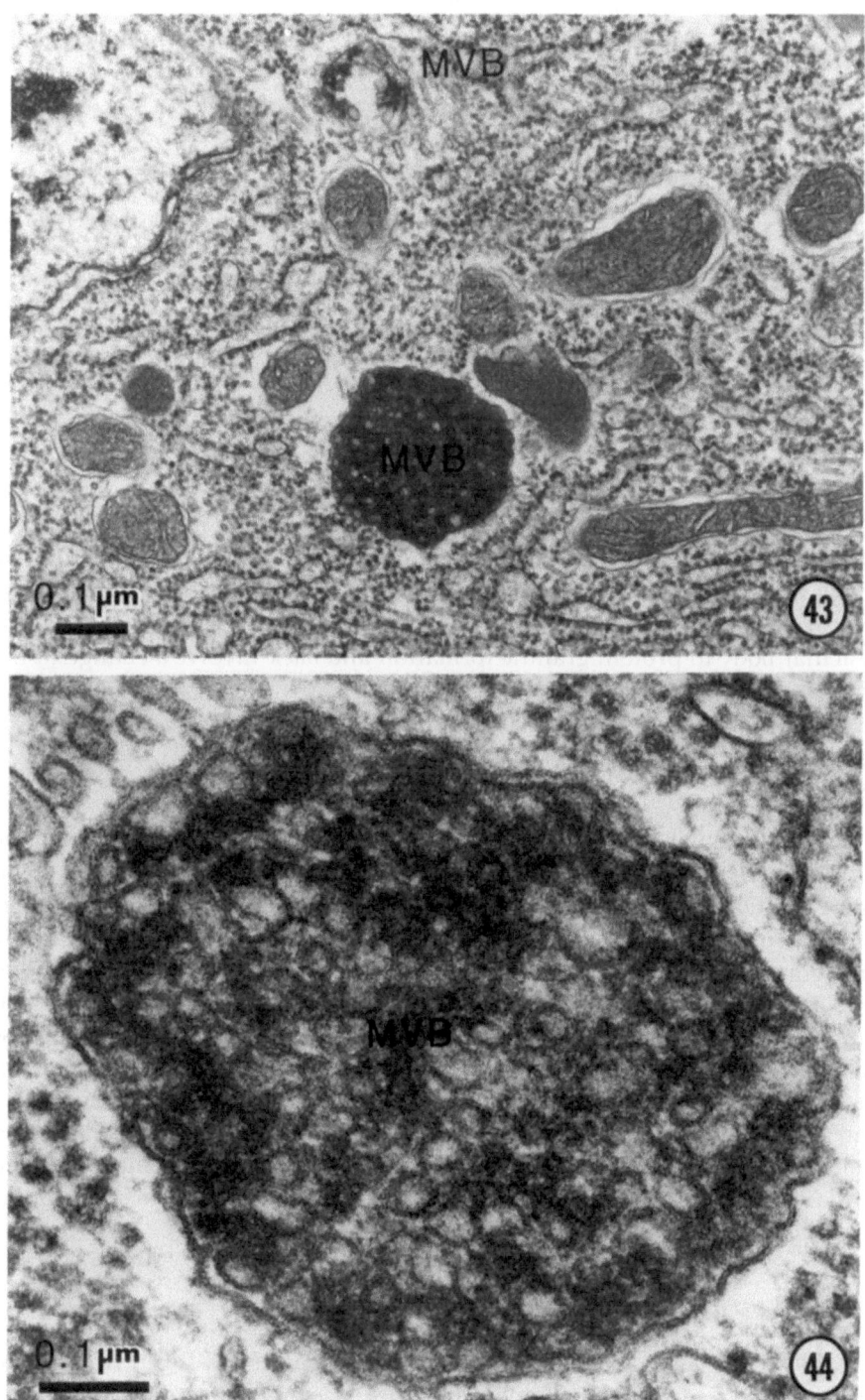

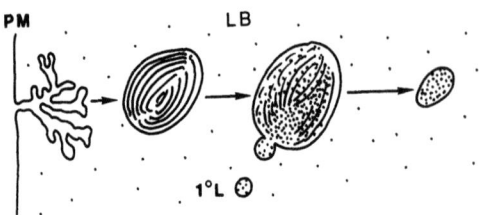

Figure 45. The origin and fate of lamellar bodies. At some stages of fat body development, there are myelinlike structures that can sometimes be traced to show continuity with the plasma membrane. They are associated with vacuoles containing concentric whorls of membrane in various stages of degradation. Later stages of these lamellar bodies contain acid phosphatase and have presumably fused with primary lysosomes. Since the lamellar bodies occur when extra membrane has been added suddenly to the cell surface, as in the exocytosis of tyrosine storage vacuoles, they are presumed to be a special kind of vacuole for the digestion of membranes.

be concerned in sampling the 40–50 protein species present in *Calpodes* hemolymph remains to be determined. We also need to know if there are qualitative changes in the primary lysosomes. Do multivesicular bodies receive the same primary lysosomes as autophagic vacuoles and does the switch from multivesicular body to protein granule formation and back involve a change in the kind of primary lysosome?

2.10. Lamellar Bodies

Multivesicular bodies are not the only organelle concerned in membrane turnover. Lamellar bodies form in the fat body when large amounts of membrane are released suddenly as the giant tyrosine storage vacuoles pass their contents to the hemolymph (McDermid and Locke, 1983). Excess vacuole membrane folds inwards from the surface and collapses to form lamellar bodies (Figures 45, 46) that develop membrane showing various degrees of degradation as in some autophagic vacuoles. Late stages contain acid phosphatase and have presumably received lytic enzymes from primary lysosomes (McClintock, 1982). Lamellar bodies occur particularly at the beginning of the fourth and fifth stadia when tyrosine storage vacuoles disappear but also at other times and in other cells. They may be of general importance for massive membrane turnover as in the loss of surface infolds and the plasma membrane reticular system.

2.11. Basal Lamina Phagocytic Vacuoles

The basal lamina is attached by hemidesmosomes to all cell surfaces exposed to the hemolymph except hemocytes. At some surfaces (e.g., nerves), the basal lamina is in layers suggesting that it grows by addition from stadium to stadium. In other tissues (fat body, epidermis), the basal lamina is not layered, and there is no separate visible record of the laminae laid down in

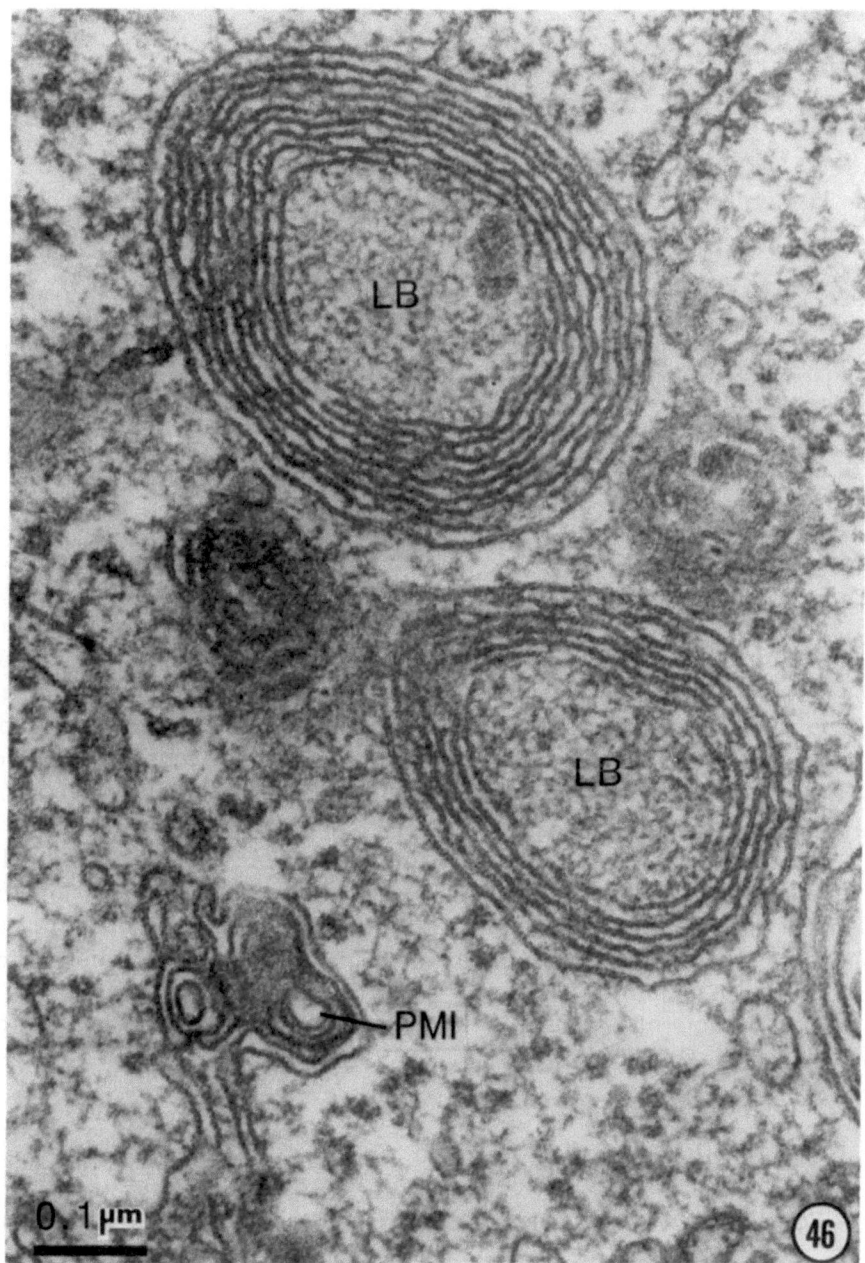

Figure 46. Lamellar bodies in the fat body of a very early fourth-instar larva just after the loss of tyrosine storage vacuoles. Exocytosis of the tyrosine storage vacuoles leaves an excess of surface membrane that collapses upon itself locally to form infolds (PMI) that are presumed to give rise to the lamellar bodies.

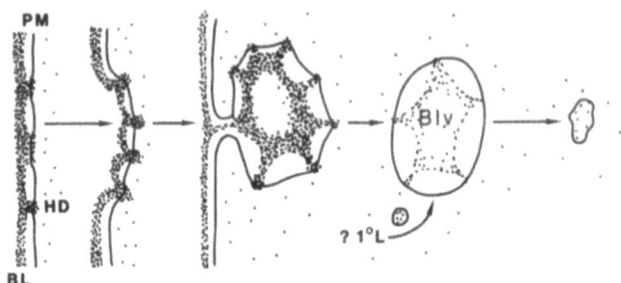

Figure 47. Basal lamina phagocytic vacuoles. At the beginning of the fourth and fifth stadia and presumably at other times, the fat body phagocytoses its basal lamina. Hemidesmosomes attaching the lamina to the plasma membrane concentrate in areas that fold inwards to become phagocytic vacuoles. Since the contents disperse and the vacuoles ultimately disappear, they presumably receive primary lysosomes. There is phagocytosis of the lamina strands between cells as well as the lamina between the outer cell surface and the hemolymph.

earlier stages. The reason for this is that the old basal lamina is phagocytosed before secretion of a new one (Figure 47). All stages of phagocytosis of the basal lamina into vacuoles can be made out in the fat body at the third to fourth molt (Figures 48–52). The basal lamina is held to the plasma membrane by hemidesmosomes which concentrate in patches that are to become vacuoles, dragging the lamina into the vacuole lumen with them. Both the basal lamina forming strands between cells and the surface covering are endocytosed. Later vacuoles show various stages of degradation presumably as a result of receiving primary lysosomes.

Phagocytosis is perhaps a fundamental property of most eukaryote cells. Fat body cells in Lepidoptera and Coleoptera may be able to phagocytose the debris of histolyzed tissues (Wigglesworth, 1972). It is therefore not surprising that the fat body can phagocytose its basal lamina from time to time. The loss of basal lamina in this way may occur generally in other tissues, particularly at metamorphosis when new shapes expose different surfaces to the hemolymph.

3. Pretransition Compartments

Posttransition vacuoles nearly all have a dual origin. For the most part, they are compartments receiving material from outside the cell with a con-

---

Figures 48–52. Stages in the phagocytosis and degradation of basal lamina by the fat body at the beginning of the fourth stadium.

Figure 48. Soon after ecdysis, the lamina between cells ceases to extend in strands and rounds up in masses of about the size for phagocytosis.

Figure 49. At this time, the hemidesmosomes are found with lamina attached in pockets below the plasma membrane surface.

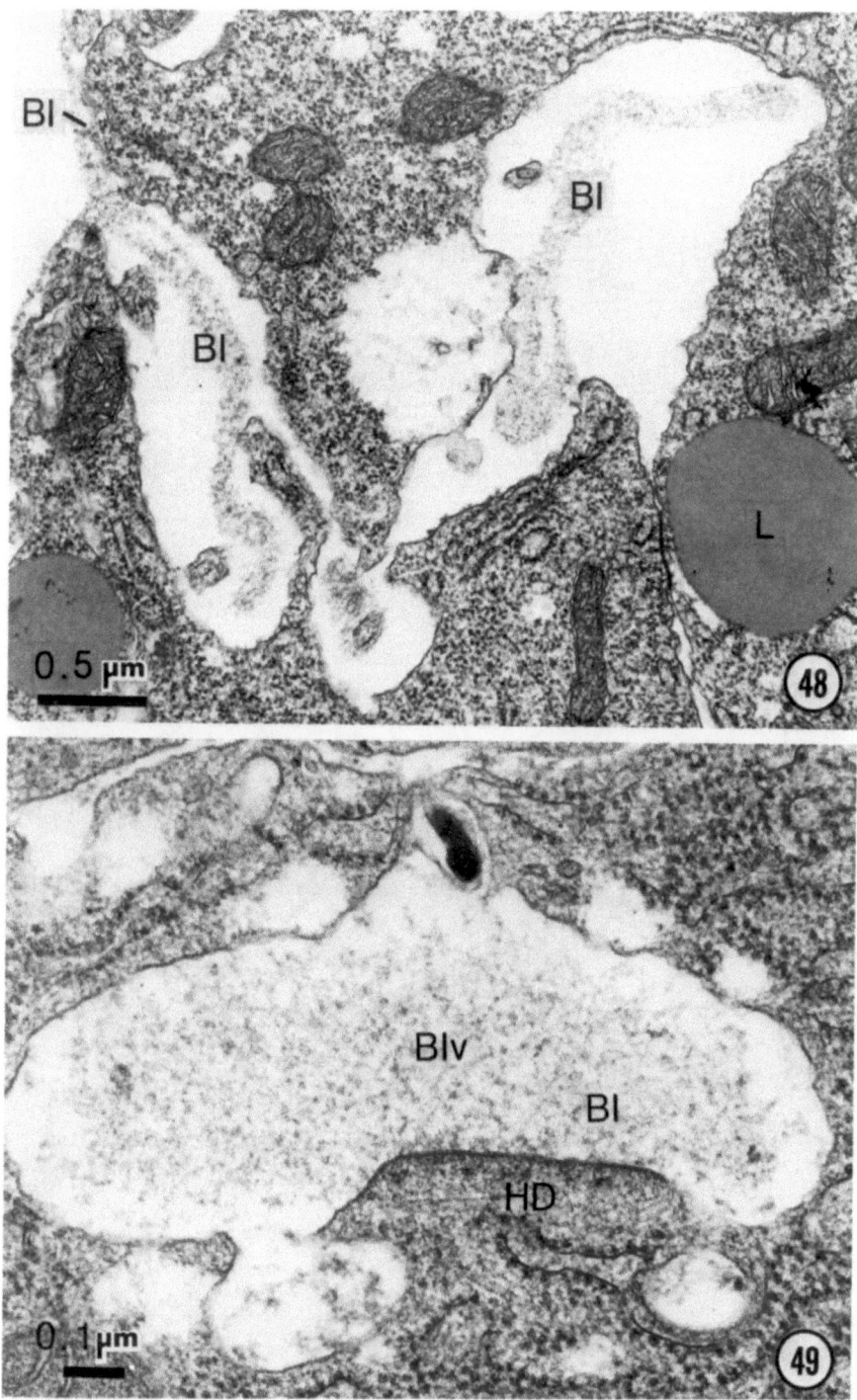

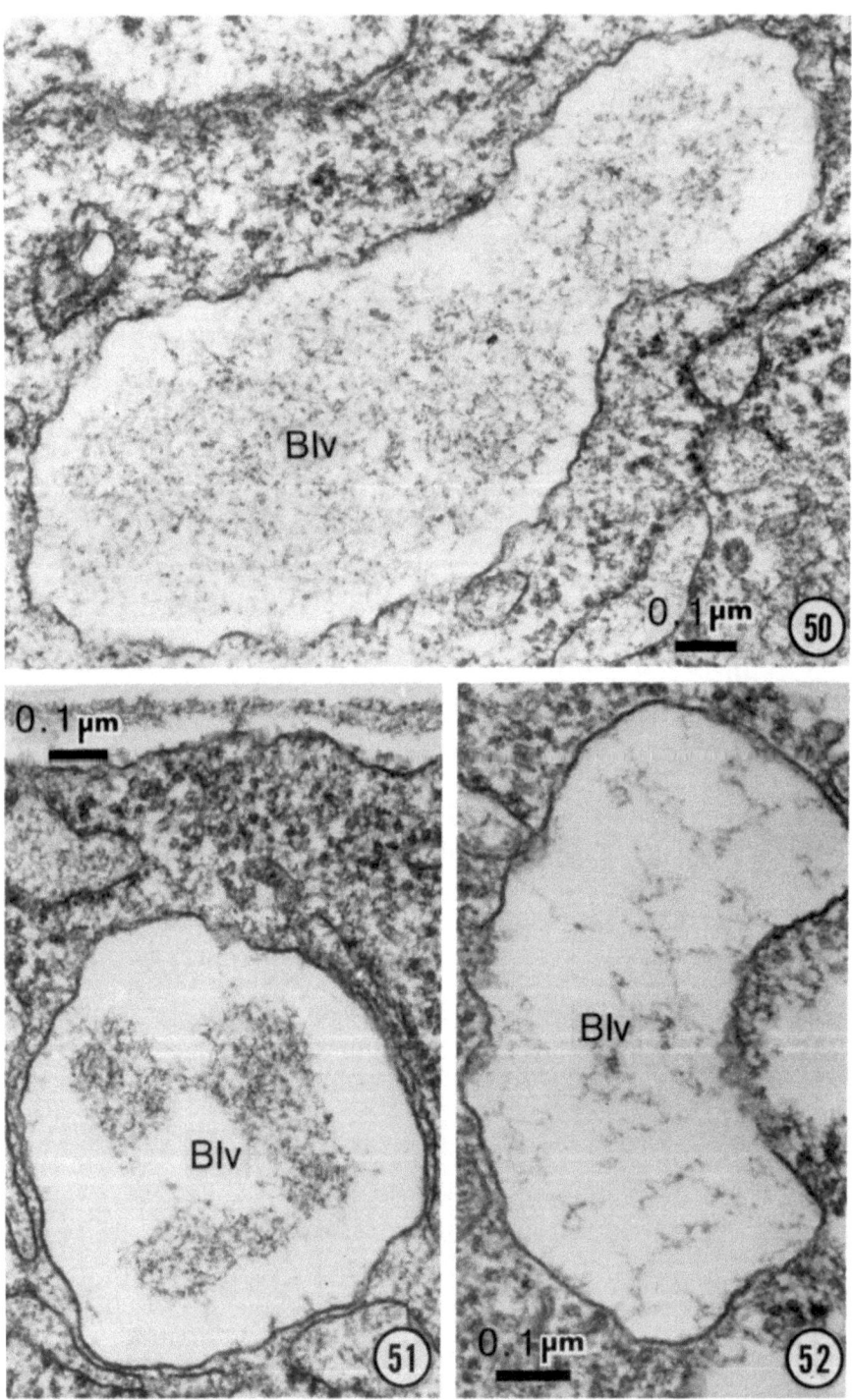

tribution from the region of the Golgi complex. Pretransition vacuoles are simpler. They are only specialized compartments of the ER.

3.1. Distensions of the RER

The RER of *Calpodes* fat body has a characteristic response to both natural and administered rising titers of ecdysteroid. When hemolymph protein synthesis is maximal as the ecdysteroid titer rises to its prepupal peak (Dean *et al.*, 1980), a few cisternae are greatly distended (Figure 53) (Dean *et al.*, 1984; Locke, 1984). The ER, particularly in the cisternae juxtaposed to Golgi complexes, distends similarly earlier in the stadium coincident with the ecdysteroid pulse

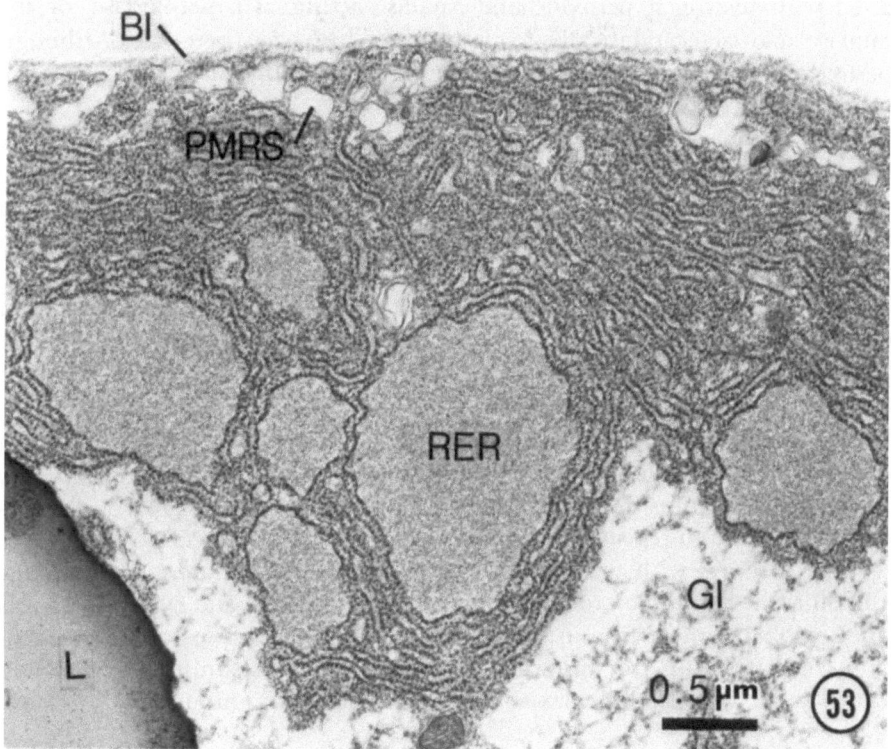

Figure 53. The distension of the RER that occurs at the end of the intermolt during the prepupal rise of hemolymph ecdysteroid when protein synthesis is maximal. A similar distension can be induced by injecting ecdysteroids. A somewhat smaller swelling occurs in the ER adjacent to Golgi complexes when protein synthesis for secretion is elevated at the beginning of the intermolt.

←───────────────────────────────────

Figures 50-52. The edges of these phagocytotic vacuoles are at first crenate as though pulled by the remaining strands of lamina attached to them. In later stages, the contents progressively disappear.

marking the beginning of intermolt protein synthesis and export through the Golgi complex. ER distension may be a temporary imbalance between the rate of entry of newly synthesized protein and its exit in secretion. Such a simple modification of the ER scarcely falls within the definition of a vacuole, but it does suggest how a compartment functioning to store newly synthesized protein could evolve.

3.2. Peroxisomes

Peroxisome is the name given by biochemists to the structure called a *microbody* by electron microscopists and is now the more widely used term. Peroxisomes are vacuoles containing a variable number of oxidases, usually with catalase, that work by oxidizing soluble substrates which have diffused into their interior. The first step may oxidize a molecule that is further oxidized with hydrogen peroxide and catalase acting as a peroxidase, or the catalase may act catalytically to destroy the hydrogen peroxide. Although peroxisomes occur in other cell types (Locke, 1984), they are characteristic of most fat body cells.

Peroxisomes were first described in insects from fifth-stage *Calpodes* fat body (Figures 37, 38) where they contain catalase and urate oxidase (Locke and McMahon, 1971). They arise as diverticula of the RER that mature while in association with RER-confronting cisternae. They are destroyed by isolation and autophagy separately and independently from mitochondrial autophagy. Peroxisomes undergo a repeating cycle of formation and destruction from molt to molt. They form at the beginning of the fourth stadium and are lost after ecdysis to the fifth. The same process is presumed to take place in earlier stadia but this has not been studied. Peroxisomes form again at the beginning of the fifth stage and are lost prior to pupation. During formation of the adult at the end of the pupal stage, they form once more (Larsen, 1976). Peroxisomes are thus organelles that develop only for use in a particular stadium and undergo repeating cycles of formation and destruction.

Peroxisomes with urate oxidase function to convert urate to allantoate and other more soluble products that can be transported and excreted. This is presumably their function in lepidopteran fat body. Their loss prior to pupation correlates with the appearance of urate storage granules. Peroxisomes with urate oxidase do not coexist with urate storage in granules.

3.3. Concretions in the RER

Concretions in the RER have not been described from the fat body, but they occur in midgut and Malpighian tubule cells, illustrating how vacuoles specialized for ion storage can evolve. In *Calpodes* Malpighian tubules, spherical concretions within limiting membranes are made primarily of calcium phosphate (Ryerse, 1979). Since the RER is the site of origin for membrane proteins conferring specificity of function, it is perhaps surprising that there are not more examples of specialized compartments like peroxisomes and these concretions that are derived directly from the RER.

4. Discussion

The fat body has a diversity of functions that may be unequaled in the Metazoa (Wyatt, 1980), and the compartmentation of its cells is correspondingly complex. A first step relating its physiology to cell biology requires a knowledge of the differentiation of the vacuolar system. An objective of this essay has been to describe the different membrane-bound compartments present in the fat body and their development in relation to the rest of the vacuolar system. The description is far from complete and some of the relationships are still uncertain, but the requirement is clear. Developmental sequences of structural change in the vacuolar system must be disentangled from one another and matched with physiological processes.

A second object has been to show how studies on the fat body can and have illuminated cell biology problems of general interest. The sequential development of the fat body makes it possible to observe transient events like the investment of organelles for autophagy in a way that is difficult in vertebrate cells that are in a dynamic equilibrium of growth and destruction. The fat body is one of only a few cell types that have an organelle specific sequence of autophagy. Isolation envelopes had been described in the fat body (Locke and Collins, 1965) when the topological problem was being hypothesized by some workers on vertebrate cells (De Duve and Wattiaux, 1966), although others had realized the importance of the sequestration of cytoplasmic components (Hruban et al., 1963; Swift and Hruban, 1964). The fat body is also the most favored cell type for studies on the Golgi complex beads (Brodie, 1981; Brodie et al., 1982; Locke and Huie, 1976a, 1977) although beads do occur in vertebrate cells (Locke and Huie, 1976b; Brodie, 1982).

Although 11 different kinds of vacuole can be made from the posttransition membranes of fat body cells, they do not all occur in the same cell or at the same time. There are clearly limitations in the number of concurrent functions that even a fat body cell can have. The limitations tell us something about the control of vacuole activity. Nevertheless, it is a remarkable tribute to what can be accomplished by compartmentation of the vacuolar system that so few insects have fat bodies with differentiated cell types as in cockroaches.

All insect tissues endocytose extracellular material from the hemolymph to some extent, but the fat body is distinguished by the magnitude of the process. It becomes a major storage organ of the hemolymph protein that it has previously synthesized. It may be asked why fat body cells should synthesize proteins for export to the hemolymph only to take them back up again for storage. One kind of answer may be that while cells do have mechanisms for selectively pinocytosing proteins, as in the formation of the storage granules, they do not have mechanisms for sorting proteins at the level of the Golgi complex into different kinds of secretory vesicles, some to contain hemolymph proteins like lipophorin destined for export, others containing storage proteins for retention. The rather cumbersome sequence of secretion and selective recovery may be the price that is paid for having several functionally different proteins secreted by the same cell.

The fat body is distinguished by the functional variety of its vacuoles. Even those derived from the plasma membrane surface such as multivesicular bodies, protein granules, basal lamina phagocytic vacuoles, and tyrosine storage vacuoles, have very different fates. Several kinds of differentiation must contribute to this functional variety. We can group them under four main headings. First, there are the permeability and transporting properties of the vacuole membranes. The membranes of a vacuole concentrating tyrosine need to be very different from those around a phagic vacuole, for example. Second, there are the properties of the external surface, for example, those that distinguish between protein molecules for pinocytosis and basal lamina for phagocytosis. Third, there is the relation with primary lysosomes. The different kinds of contents to be digested suggest a need for different kinds of primary lysosome. The characteristic timing of occurrence of some vacuoles suggests that there is precise control of the release of primary lysosomes for digestion. Last, there is the integration with the cytoskeleton that results in appropriate kinds of vacuole movement, fusion, exocytosis, and endocytosis. A consideration of vacuoles and the vacuolar system shows us a new world within a cell as complicated in its interactions and as difficult to understand as that inside a large organism.

The fat body has led us to the description of various kinds of vacuole that pose a general problem of differentiation: whether, how, and for what purpose membranes and the spaces that they enclose become different. We may hope that the fat body may lead us on further to the answers.

5. Summary and Conclusions

The fat body consists of strips of cells suspended in the hemolymph upon which they are solely dependent for the variety of metabolic processes that they perform. The ease with which they can be isolated and maintained *in vitro* makes them a promising cell type for correlating ultrastructure with physiology and development. As in other cells, the vacuolar system of a fat body cell is separated into two hierarchies of compartments by transition vesicles. Transition vesicles arise from the ER by budding through the rings of Golgi complex beads that can be thought of as Golgi complex organizers. As membranes traverse the Golgi complex, they become thicker, causing an observable structural difference between pre- and posttransition membranes. It is in the differentiation of the vacuolar system to match its diverse metabolic functions that the fat body has a special interest for cell biologists. At various times and in different kinds of fat body cell, the posttransition membranes differentiate into 11 kinds of vacuole. The pretransition membranes have several specializations and at least two separate compartments.

First, in the posttransition hierarchy, there are secretory vesicles that carry plasma membrane to the surface and take secretory material out of the cell in exocytosis.

There is a class of vacuole that utilizes the property of the plasma membrane for binding and endocytosing extracellular material (protein storage granules, multivesicular bodies, lamellar bodies, and basal lamina phagocytic vacuoles). These have as variables the kind of material that is to be broken down, and the timing and nature of the digestion controlled by fusion with primary lysosomes. In all of them the vacuole contents are released to the cell through the membrane. Autophagic vacuoles are a special example within this category, in which a cell component that has been made topologically external forms a compartment that can be used for digestion. Symbiont vacuoles are another special case within this category, distinguished by the long period during which the symbionts are endocytosed but not digested.

There are vacuoles with insoluble contents specialized for the storage and utilization of urate (urate vacuoles and granules). They occur either in cell types specialized for their use or more briefly in trophocytes at particular stages of development. The vacuole contents with the property of accumulating urate may have come initially from outside the cell. Urate movement takes place across the membranes and loss is not by exocytosis.

There is a class of vacuole (tyrosine storage vacuoles and perhaps those associated with glycogen) that concentrates and stores molecules such as tyrosine in solution, probably as more soluble derivatives. Their membranes must be impermeable to the stored molecule but adapted for its transport and perhaps its conjugation with other molecules. The vacuole contents are released to the outside of the cell by exocytosis with a return of membrane to the plasma membrane in a particular kind of recycling that does not seem to involve the Golgi complex.

Pretransition membranes forming the ER give rise to peroxisomes, Golgi complexes, distensions of various forms, and confronting cisternae that interact particularly with posttransition membranes.

This description outlines the nature of the differentiation of the fat body vacuolar system. The problems are now to correlate the ultrastructure with biochemistry and physiology.

Acknowledgments

I am very grateful to Mr. Philip Huie and Ms. Lora Wilkie for technical assistance, and to Mrs. Jane Sexsmith for word processing. This work was supported by Grant A6607 from the Natural Sciences and Engineering Research Council of Canada.

References

Bodenstein, D., 1953, Studies on the humoral mechanisms in growth and metamorphosis of the cockroach, *Periplaneta americana*. III. Humoral effects on metabolism, *J. Exp. Zool.* **124**:105-115.
Brodie, D. A., 1981, Bead rings at the endoplasmic reticulum-Golgi complex boundary: Morphological changes accompanying inhibition of intracellular transport of secretory proteins in arthropod fat body tissue, *J. Cell Biol.* **90**:92-100.
Brodie, D. A., 1982, Golgi complex beads in vertebrates and their relationships with clathrin coats, *Tissue Cell* **14**:253-262.
Brodie, D. A., Locke, M., and Ottensmeyer, F. P., 1982, High resolution microanalysis for phosphorus in Golgi complex beads of insect fat body tissue by electron spectroscopic imaging, *Tissue Cell* **14**:1-11.
Brooks, M., and Richards, A. G., 1955, Intracellular symbiosis in cockroaches. II. Mitotic division of mycetocytes, *Science* **122**:242.
Brunet, P. C. J., 1980, The metabolism of the aromatic amino acids concerned in the cross-linking of insect cuticle, *Insect Biochem.* **10**:467-500.
Butterworth, F. M., Tysell, B., and Waclawski, I., 1979, The effect of 20-hydroxyecdysone and protein on granule formation in the *in vitro* cultured fat body of *Drosophila*, *J. Insect Physiol.* **25**:855-860.
Chippendale, G. M., 1970, Metamorphic changes in fat body proteins of the Southwestern corn borer, *Diatraea grandiosella*, *J. Insect Physiol.* **16**:1057-1068.
Chippendale, G. M., and Kilby, B. A., 1969, Relationship between the proteins of the hemolymph and fat body during development of *Pieris brassicae*, *J. Insect Physiol.* **15**:905-926.
Cochran, D. G., 1975, Excretion in insects. In *Insect Biochemistry and Function*, edited by D. J. Candy and B. A. Kirby, pp. 177-282, Chapman & Hall, London.
Cochran, D. G., Mullins, D. E., and Mullins, K. J., 1979, Cytological changes in the fat body of the American cockroach, *Periplaneta americana*, in relation to dietary nitrogen levels, *Ann. Entomol. Soc. Am.* **72**:197-205.
Collins, J. V., 1969, The hormonal control of fat body development in *Calpodes ethlius* Stoll (Lepidoptera, Hesperiidae), *J. Insect Physiol.* **15**:341-352.
Collins, J. V., 1974, Hormonal control of protein sequestration in the fat body of *Calpodes ethlius* Stoll, *Can. J. Zool.* **52**:639-642.
Collins, J. V., 1975, Secretion and uptake of ^{14}C proteins by fat body of *Calpodes ethlius* Stoll (Lepidoptera, Hesperiidae), *Differentiation* **3**:143-148.
Collins, J. V., and Downe, A. E. R., 1970, Selective accumulation of hemolymph proteins by the fat body of *Galleria mellonella*, *J. Insect Physiol.* **16**:1697-1708.
Dean, R. L., 1978, The induction of autophagy in isolated insect fat body of β-ecdysone, *J. Insect Physiol.* **24**:439-447.
Dean, R. L., Bollenbacher, W. E., Locke, M., Smith, S. L., and Gilbert, L. I., 1980, Hemolymph ecdysteroid levels and cellular events in the intermolt/molt sequence of *Calpodes ethlius*, *J. Insect Physiol.* **26**:267-280.
Dean, R. L., Collins, J. V., and Locke, M. 1984, Structure of fat body. In *Comprehensive Insect Physiology, Biochemistry and Pharmacology*, edited by G. A. Kerkut and L. I. Gilbert, Vol. 3, Pergamon Press, Elmsford, N.Y.

De Duve, C., and Wattiaux, R., 1966, Functions of lysosomes, *Annu. Rev. Physiol.* **28**:435-492.
De Loof, A., 1972, Diapause phenomena in non-diapausing last instar larvae. pupae and pharate adults of the Colorado beetle, *J. Insect Physiol.* **18**:1039-1047.
de Priester, W., and van der Molen, L. G., 1979, Premetamorphic changes in the ultrastructure of *Calliphora* fat cells, *Cell Tissue Res.* **198**:79-93.
Dortland, J. F., and Hogen Esch, T., 1979, A fine structural survey of the development of the adult fat body of *Leptinotarsa decemlineata*, *Cell Tissue Res.* **201**:423-430.
Farquhar, M. G., and Palade, G. E., 1981, The Golgi apparatus (complex)—(1954-1981)—from artifact to center stage, *J. Cell. Biol.* **91**:77s-102s.
Griffiths, G. W., and Beck, S. D., 1973, Intracellular symbiotes of the pea aphid, *Acyrthosiphon pisum*, *J. Insect Physiol.* **19**:75-84.
Griffiths, G. W., and Beck, S. D., 1975, Ultrastructure of pea aphid mycetocytes: Evidence for symbiote secretion, *Cell Tissue Res.* **159**:351-367.
Griffiths, G. W., and Beck, S. D., 1977, *In vivo* sterol biosynthesis by pea aphid symbiotes as determined by digitonin and electron microscope autoradiography, *Cell Tissue Res.* **176**:179-190.
Houk, E. J., and Griffiths, G. W., 1980, Intracellular symbiotes of the Homoptera, *Annu. Rev. Entomol.* **25**:161-187.
Hruban, Z., Spargo, B., Swift, H., Wissler, R. W., and Kleinfeld, R. G., 1963, Focal cytoplasmic degradation, *Am. J. Pathol.* **42**:657-683.
Kramer, S. J., Mundall, E. C., and Law, J. H., 1980, Purification and properties of manducin, an amino acid storage protein of the hemolymph of larval and pupal *Manduca sexta*, *Insect Biochem.* **10**:279-288.
Kunkel, J., 1966, Development and the availability of food in the German cockroach, *Blattella germanica* (L.), *J. Insect Physiol.* **12**:227-235.
Labour, G., 1970, Étude histologique et histochimique du corps adipeux au cours de développement post-embryonnaire du Doryphore, *Arch. Anat. Microsc.* **59**:235-252.
Labour, G., 1974, Etude ultrastructurale de l'évolution du tissu adipeux au cours du developpement larvaire et nymphal chez le Doryphore, *Ann. Soc. Entomol. Fr. (N.S.)* **10**:943-958.
Larsen, W. J., 1976, Cell remodeling in the fat body of an insect, *Tissue Cell* **8**:73-92.
Lauverjat, S., 1977, L'évolution post-imaginale de tissu adipeux femelle de *Locusta migratoria* et son controle endocrine, *Gen. Comp. Endocrinol.* **33**:13-34.
Locke, J., McDermid, H., Brac, T., and Atkinson, B. G., 1982, Developmental changes in the synthesis of hemolymph polypeptides and their sequestration by the prepupal fat body in *Calpodes ethlius* Stoll, Lepidoptera, Hesperiidae, *Insect Biochem.* **12**:431-440.
Locke, M., 1980, The cell biology of fat body development. In *Insect Biology in the Future*, edited by M. Locke and D. S. Smith, pp. 227-252, Academic Press, New York.
Locke, M., 1981, Cell structure during insect metamorphosis. In *Metamorphosis: A Problem in Developmental Biology*, edited by E. Frieden and L. I. Gilbert, pp. 75-103, Plenum Press, New York.
Locke, M., 1982, Envelopes at cell surfaces—A confused area of research of general importance. In *Parasites: Their World and Ours*, edited by D. F. Mettrick and S. S. Desser, pp. 73-88, Elsevier, Amsterdam.
Locke, M., 1984, A structural analysis of post-embryonic development. In *Comprehensive Insect Physiology and Biochemistry*, edited by G. A. Kerkut and L. I. Gilbert, Vol. 2, Pergamon Press, Elmsford, N.Y.
Locke, M., and Collins, J. V., 1965, The structure and formation of protein granules in the fat body of an insect, *J. Cell Biol.* **26**:857-885.
Locke, M., and Collins, J. V., 1967, Protein uptake in multivesicular bodies in the molt/intermolt cycle of an insect, *Science* **155**:467-469.
Locke, M., and Collins, J. V., 1968, Protein uptake into multivesicular bodies and storage granules in the fat body of an insect, *J. Cell Biol.* **36**:453-483.
Locke, M., and Collins, J. V., 1980, Organelle turnover in insect metamorphosis. In *Pathologic*

Aspects of Cell Membranes, edited by B. F. Trump and A. Arstila, pp. 223-248, Academic Press, New York.

Locke, M., and Huie, P., 1976a, The beads in the Golgi complex/endoplasmic reticulum region, *J. Cell Biol.* **70**:384-394.

Locke, M., and Huie, P., 1976b, Vertebrate Golgi complexes have beads in a similar position to those found in arthropods, *Tissue Cell* **8**:739-743.

Locke, M., and Huie, P., 1977, Bismuth staining of the Golgi complex is a characteristic arthropod feature lacking in *Peripatus, Nature (London)* **270**:341-343.

Locke, M., and McMahon, J. T., 1971, The origin and fate of microbodies in the fat body of an insect, *J. Cell Biol.* **48**:61-78.

Locke, M., and Sykes, A. K., 1975, The role of the Golgi complex in the isolation and digestion of organelles, *Tissue Cell* **7**:143-158.

Loughton, B. G., and West, A. S., 1965, The development and distribution of hemolymph proteins in Lepidoptera, *J. Insect Physiol.* **11**:919-932.

McClintock, J., 1982, Lead staining and phosphatase localizations in the vacuolar system. M.Sc. thesis, Department of Zoology, University of Western Ontario, London, Ontario, Canada.

McClintock, J., and Locke, M., 1982, Lead staining in the Golgi complex, *Tissue Cell* **14**:541-554.

McDermid, H., and Locke, M., 1983, Tyrosine storage vacuoles in insect fat body, *Tissue Cell* **15**:137-158.

Marty, F., 1978, Cytochemical studies on GERL, provacuoles, and vacuoles in root meristematic cells of *Euphorbia, Proc. Natl. Acad. Sci. USA* **75**:852-856.

Matile, P., and Moor, H., 1968, Vacuolation: Origin and development of the lysosomal apparatus in root-tip cells, *Planta* **80**:159-175.

Mori, T., Akai, H., and Kobayashi, M., 1970, Ultrastructural changes of the fat body in the silkworm during postembryonic development, *J. Sericult. Sci. Jpn.* **39**:51-61.

Mullins, D. E., 1979, Isolation and partial characterization of uric acid spherules obtained from cockroach tissues (Dictyoptera), *Comp. Biochem. Physiol. A.* **62**:699-705.

Pelttari, A., and Helminen, H. J., 1979, The relative thickness of intracellular membranes in epithelial cells of the ventral lobe of the rat prostate, *Histochem. J.* **11**:613-624.

Ryerse, J. S., 1979, Developmental changes in Malpighian tubule cell structure, *Tissue Cell* **11**:533-551.

Steele, J. E., 1982, Glycogen phosphorylase in insects, *Insect Biochem.* **12**:131-147.

Swift, H., and Hruban, Z., 1964, Focal degradation as a biological process, *Fed. Proc.* **23**:1026-1037.

Tojo, S., Betchaku, T., Ziccardi, V. J., and Wyatt, G. R., 1978, Fat body protein granules and storage proteins in the silkmoth, *Hyalophora cecropia, J. Cell Biol.* **78**:823-838.

Tojo, S., Nagata, M., and Kobayashi, M., 1980, Storage proteins in the silkworm, *Bombyx mori, Insect Biochem.* **10**:289-304.

Tojo, S., Kiguchi, K., and Kimura, S., 1981, Hormonal control of storage protein synthesis and uptake by the fat body in the silkworm, *Bombyx mori, J. Insect Physiol.* **27**:491-497.

Voinov, V., 1927, Sur l'éxistence d'un tissu mesenchymateux vacuolaire dans les larves de *Chironomus, C.R. Soc. Biol.* **96**:1015-1017.

Waku, Y., and Sumimoto, K., 1969, Light and electron microscopical study of the fat body cells in the metamorphosing silkworm, *Bull. Fac. Text. Fibers Kyoto Univ. Ind. Arts Text. Fibers* **5**:256-287.

Webster, D., 1982, The major hemolymph proteins of an insect, M.Sc. thesis, Department of Zoology, University of Western Ontario, London, Ontario, Canada.

Wigglesworth, V. B., 1942, The storage of protein, fat, glycogen and uric acid in the fat body and other tissues of mosquito larvae, *J. Exp. Biol.* **19**:56-77.

Wigglesworth, V. B., 1967, Cytological changes in the fat body of *Rhodnius* during starvation, feeding and oxygen want, *J. Cell Sci.* **2**:243-256.

Wigglesworth, V. B., 1972, *The Principles of Insect Physiology*, 7th ed., Chapman & Hall, London.

Wigglesworth, V. B., 1982, Fine structural changes in the fat body cells of *Rhodnius* (Hemiptera) during extreme starvation and recovery, *J. Cell Sci.* **53**:337-346.

Wyatt, G. R., 1980, The fat body as a protein factory. In *Insect Biology in the Future*, edited by M. Locke and D. S. Smith, pp. 201-225, Academic Press, New York.

6

The Ultrastructure of the Digestive and Excretory Organs

ROGER MARTOJA AND
CHRISTIANE BALLAN-DUFRANÇAIS

1. Introduction

Within the digestive tract, the ingested food is broken down into simpler molecules by the action of enzymes and thus rendered absorbable through the gut wall. The midgut is composed of a long straight tube which often contains blind pouches (caeca). It is the main site of production of digestive enzymes and of absorption. In many insect species, it secretes a peritrophic membrane which is permeable to the digestive enzymes and to the products of digestion. In most insects, the structure of the midgut seems to be rather uniform, and thus the same cells are both secretory and absorptive. However, ultrastructural and cytochemical studies have shown a variety of differences which are related sometimes to cycles of cellular activities. In other cases, the midgut exhibits a clear division into functional segments or a mixture of different kinds of specialized cells may be present. Physiological experiments combined with ultrastructural and cytochemical investigations have sometimes clearly established the functions of these segments, or cells. It was studies of this type that showed the important role of the midgut epithelium in ionic regulation and mineral accumulation in insects belonging to several different orders. The transfer of fluid and ions through the wall of the midgut sometimes leads to curious arrangements that bring into contact a midgut segment with the Malpighian tubules, and concurrently certain specialized gut cells lose their digestive function in order to undertake an absorptive role.

ROGER MARTOJA and CHRISTIANE BALLAN-DUFRANÇAIS • Laboratory of Fundamental and Applied Histophysiology, Pierre and Marie Curie University, 75005 Paris, France.

Peculiar segments of the hindgut may play an important role in the digestive processes in the species where digestion of some materials, such as cellulose, is made possible by the activities of microorganisms. In such cases, the hindgut is enlarged to form a pouch (the fermentation chamber) containing protozoans or bacteria. Ultrastructural studies agree with other experimental data which demonstrate that the products of cellulose digestion are absorbed through the wall of the fermentation chamber.

In most insects, saliva produced by labial glands plays a part in the breakdown of food materials. It also serves to moisten the food and to lubricate and clean the mouth parts. Investigations analyzing the chemistry of salivary products have revealed both enzymes and various proteins and mucous secretions. Physiological experiments have demonstrated that some insects produce a dilute saliva. The osmolarity of this is very low compared with hemolymph, and the formation of this dilute saliva results from ionic transport. Studies emphasizing correlations between structure and physiology have been performed primarily on glands from insects belonging to the Dictyoptera, Orthoptera, Lepidoptera, and higher Diptera, and interpretations concerning the function of the different cell types are satisfactory for only the last two groups.

Excretory organs maintain the constancy of composition of the internal environment through the regulation of salt and water balance and the breakdown of worn-out proteins. Excretory products are very diversified (purinic compounds, mineral salts, and organic complexes), but the identification of such components in a tissue is not enough to conclude in favor of an excretory function. Thus, purinic and mineral compounds stored in organs such as the epidermis or in oocytes may possess ethological functions or serve as nutriments. Ultrastructural studies are essential, and they have often demonstrated an excretory function even when excretory products remain unknown. The discharge of excretory products into the environment is carried out by the "Malpighian tubules-proctodeum" complex, excretion being the net result of Malpighian tubule secretion after modification by the proctodeum. Ultrastructural and physiological data emphasize resemblances with the kidney of mammals, though the Malpighian tubules-proctodeum complex often excretes solid compounds in the form of natural casts. Likewise, nephrocytes are similar to the mononuclear macrophage system. But in contrast to vertebrates, many insects possess an important storage kidney which accumulates wastes without discharging them outside. Finally, some insects make do with the primitive excretory organs they inherited from other arthropods, whereas others have improved the excretory functions at the expense of glandular systems.

Only studies demonstrating or strongly suggesting correlations between fine structure and function are considered in this review. In spite of their scientific importance, data concerning the Collembola, which are now separated from the true insects, are not presented. Those dealing with the development of the organs are omitted, and so are studies concerning the salivary glands and the Malpighian tubules where the secretion has no digestive or excretory function. Salivary glands are often devoted to the production of silk or glue, and those of dipteran larvae provide an excellent tool for researches on gene activity, since the production of specific secretory products can be correlated with changes in the polytene chromosomes of the organ (see Daneholt, 1982).

2. The Midgut: General Organization

2.1. Structure of the Wall

Situated between the stomodeum and the proctodeum, the midgut consists of an epithelium which rests on a layer of connective tissue and muscle. In many species, the food is separated from the apical surface of the epithelium by an extracellular sheath termed the *peritrophic membrane*, which is secreted by the epithelial cells.

The midgut epithelium is made up of three types of cells: (1) several kinds of differentiated cells that are involved in the digestive processes and accessory mechanisms, (2) undifferentiated replacement cells which are often collected together in nests, and (3) enigmatical cells that are generally believed to have an endocrine function. In some insects, no regenerative cells are present; in others, an additional layer of epithelial cells is formed during metamorphosis. The apical part of the differentiated cells possesses a well-developed striated border composed of microvilli. The cells are linked by several types of junctions: long, smooth, septate junctions (Figure 1), small desmosomes, and gap junctions. Lane's review (1982) should be consulted for discussion of the fine

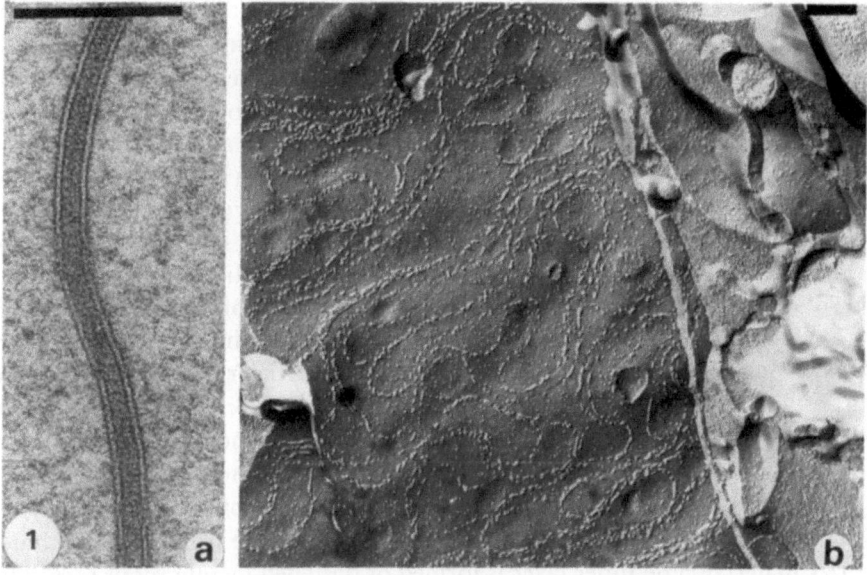

Figure 1. Smooth septate junctions between midgut cells. (a) Thin section from *Kalotermes flavicollis.* (b) Freeze-fracture replica from *Blaberus cranifer.* Courtesy of Professor C. Noirot. (Bars = 150 nm.)

structure and the physiological role of such junctions. In a few cases, autocellular junctions are present between the plasma membrane of the microvilli.

The connective tissue layer is bordered inside by the basement membrane and outside by a continuous limiting layer which resembles a basement membrane and segregates the connective tissue sheath from the hemocele (François, 1978). The connective sheath is traversed by longitudinal and circular muscle fibers, tracheae, and nerves, and it contains numerous collagen fibrils and fibroblasts and occasional hemocytes. The presence of elastic fibers has been observed in *Periplaneta americana* (François, 1978; see also Ashhurst, 1982).

2.2. Striated Border

The striated border is represented by a great number of cylindrical processes or microvilli that project into the gut lumen. Generally, the microvilli are about 0.1-0.2 μm in diameter, reach a length of several micrometers, and are arranged in a regular hexagonal array. The cytoplasm within the villi is composed of fine fibrous material which shows a preferred orientation along the axis of the microvilli and gives the region a filamentous aspect (Figure 2a). The external surface is covered with a fuzzy material referred to as a cell coat that is rich in glycoproteins (Noirot and Noirot-Timothée, 1972).

Many insects show interesting variations in the basic theme of microvillar fine structure. The branching of the microvilli has been observed in *Sarcophaga bullata* (Nopanitaya and Misch, 1974) and is considered a further means for increasing surface area. Mitochondria occur in the villi of the goblet cells. In the Aphidida (Forbes, 1964) and in certain Diptera (the lipophilic cells of *Lucilia cuprina*, Waterhouse and Wright, 1960 and the cuprophilitic cells of *Drosophila melanogaster*, Filshie et al., 1971) the striated border sometimes shows an unusual structure, being formed of lamellae instead of microvilli.

More unusual are the extracellular structures that replace the typical cell coat. In many aphids, the area surrounding each microvillus is made up of numerous extracellular "microtubules" that lie close to and parallel with the microvilli. They attain the same length as the microvilli (1.5-2 μm) (O'Loughlin and Chambers, 1972), and their function is unknown. In contrast to true microtubules, they are stable in cold solvents. In certain species of Thysanoptera, forklike projections cover the microvilli and are believed to be a peculiar type of cell coat (Kitajima, 1975). In other insects, additional membranes cover the apical surface of the epithelial cells and project into the gut lumen (Figures 2b,c) (Homoptera: Marshall and Cheung, 1970; Reger, 1971; Heteroptera; Burgos and Gutierrez, 1976; Gutierrez and Burgos, 1978; Lane and Harrison, 1979; Odonata: Andriès and Torpier, 1982). The precise morphology, of course, depends on the species. In *Fulgora candelaria* (Homoptera), this "plexiform surface coat" has two layers. The outer layer consists of saccules and membranes, and each saccule contains a membranous vesicle from which radiates a fuzzy material. The inner layer is formed of flattened membranous sacs. The surface coat has been shown by cytochemical tests to contain acid mucopolysaccharides. In *Triatoma infestans* (Heteroptera), the surface coat has a tubular, plexiform structure. A trilaminar membrane covers the apical surface of the epithelial cells and projects tubular expansions into the lumen. The inter-

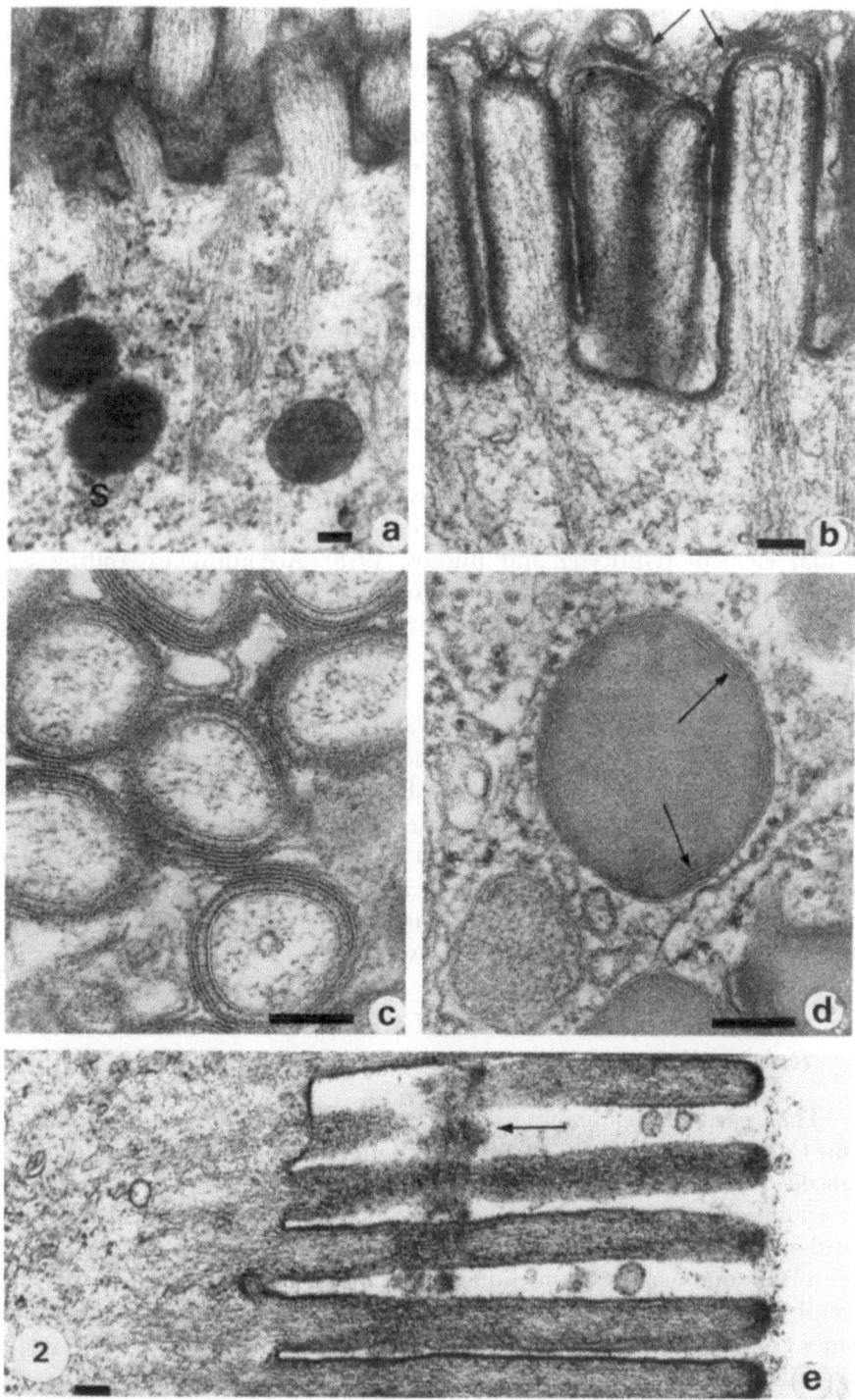

Figure 2. The striated border of the midgut. (a) Usual appearance in *Formica polyctena*. S, secretory granules. (b, c) Extracellular organelles in *Nepa cinerea*. Note the laminar aspect of these structures (arrows in b), especially in transverse section (c). (d) A secretory granule with internal membranes from *N. cinerea* (arrows). (e) Extrusion of the peritrophic membrane (arrow) between the microvilli in *Petrobius maritimus*. Courtesy of Drs. A. Y. Jeantet and J. C. Andriès and Professor A. M. Fain-Maurel. (Bars = 0.1 μm.)

villar space is occupied by a meshwork of septa continuous with the trilaminar membrane. The extracellular coating is thought to contain glycophospholipids and acid phosphatases. In *Nepa cinerea* (Heteroptera), the multilaminar sheets are separated from the plasma membrane by a regular space of 9 nm and are believed to be lipidic membranes from which membrane proteins have been excluded. The membranes contain both phospholipids and glycolipids. Likewise, in *Rhodnius prolixus* (Heteroptera), the cell membrane is surrounded by a second, outer membrane, and the two are separated from one another by a highly regular space about 10 nm wide. This very precise distance is maintained by a cluster of inclined columns, so that the arrangement forms a sort of continuous junction. The microvilli are often closely associated with one another, and their adjacent outer membranes form gaplike junctions with one another. In addition, spokelike projections and a system of myelinlike sheets radiate from the outer membrane.

The origin of these membranous systems remains uncertain. They might be a delamination product of the epithelial cells or an unusual form of extracellular element (Gutierrez and Burgos, 1978). According to Andriès and Torpier (1982), they may originate from secretory granules in which additional membranes are present, or from lipidic molecules released by exocytosis that aggregate into myelinlike sheets (Figure 2d).

The function of such coatings can hardly be imagined. Although such membranes show chemical differences and perhaps may be produced in a variety of ways, they still may be included within a loose definition of the peritrophic membrane (Marshall and Cheung, 1970). Taking the elaborated junctions into consideration, the double-membrane system could contribute to increase the strength of the microvilli in blood-sucking insects (like *Rhodnius*) during the blood meals which create a great pressure on the cell layer. In sap-sucking species, these structures might facilitate the rapid entry of the fluid meal into the gut (Lane and Harrison, 1979).

2.3. Basement Membrane (Basal Lamina) and Related Structures

The basement membranes examined so far fall into several categories. In some Coleoptera (Holter, 1970), the membrane is simply composed of granules embedded in a uniform substance. In Diptera (Nopanitaya and Misch, 1974), the granules are located in an inner sublayer, whereas the outer layer is composed of amorphous fibrous material. In contrast, in the larvae of Lepidoptera, the inner component is filamentous, and the outer one is finely granular. In several species of Coleoptera and Orthoptera, thin unbanded fibrils are randomly arranged in a homogeneous matrix, whereas distinctly banded fibrils, recognizable as collagen, are seen in Dictyoptera (Ashhurst, 1982).

In some insects, the epithelium lies on a specialized layer of connective tissue which differs from a true basement membrane, but is often referred to as a basal lamina of unusual structure. In aquatic Heteroptera (Nepidae, Gouranton, 1970), the inner part shows a discontinuous sheet formed by a system of hexagonal plates connected by filaments. These plates represent a peculiar organization of collagen embedded in a polysaccharide matrix. In the larvae of the beetle (*Oryctes nasicornis*, Bayon and François, 1976), the lamina consists of three layers, each of which contains two structural elements arranged to form a regular hexagonal network or grid surrounded by a homogeneous matrix. The structures forming the grid are believed to be composed of proteins, whereas the surrounding matrix contains mucoproteins and mucopolysaccharides. A somewhat similar structure has been described in a mosquito, *Aedes dorsalis* (Houk et al., 1980). The basal lamina is divided into two distinct halves which are made up of a series of interconnected pillars. Each pillar is composed of six to eight superimposed cuboidal units, which are made up of globular protein or lipoprotein complexes arranged in a ring around a less-electron-dense center. The most frequently observed arrangement is gridlike, and this has been described in many blood-sucking species of Diptera and Siphonaptera (Terzakis, 1967; Reinhardt et al., 1972; Reinhardt and Hecker, 1973; Houk, 1977). The basal lamina consists of several layers which are interconnected by particles. These are approximately 15-17 nm in diameter in *Culex tarsalis* (Houk, 1977).

Comparative investigations have been performed on relaxed and distended midguts of blood-feeding dipterans. The results suggest that these peculiar basal membranes are structural specializations that facilitate the elasticity exhibited by midguts that are subjected to considerable changes in volume. But the mechanical properties seem to differ according to the structural organization of the membrane. In gridlike ones, a role of the basal lamina in response to distension was considered, because of the report by Reinhardt and Hecker (1973) of a symmetry distortion in *Aedes aegypti*. But in the lamina of *A. dorsalis*, this symmetry distortion was not observed. According to Houk et al. (1980), the division of the basal lamina in two halves coupled with an excess of basal lamina immediately adjacent to the epithelial cells allows the lamina to move short distances upon itself in response to increases of the gut volume.

However, it is obviously impossible to ascribe such functions for the basement membranes of insects that do not periodically engorge themselves with food.

3. Types of Epithelial Cells in the Midgut

3.1. Columnar Cells

The most common type is the columnar cell. It is present in all insects, though it may be called by a variety of names (i.e., *cuboidal cell, lipophilic cell*, etc.). Its fine structure is highly variable and depends on the species, the position in the alimentary cycle, and particularly on its multiple functions during the life cycle, i.e., secretion of enzymes, mucopolysaccharides, and peritrophic membrane; absorption; and storage of inorganic and organic products.

Whatever the species, the nucleus is located in the center or in the midbasal part of the cell. The cytoplasm is richly supplied with cisternae of RER and with Golgi material, especially in the vicinity of the nucleus.

The nucleus generally exhibits the classical structure. However, in a few cases, peculiarities have been observed: a fibrous zone surrounding the nuclear envelope in *Cixius nervosus* (Homoptera; Gouranton, 1968a); intranuclear bodies rich in both DNA and RNA located in the anterior midgut of *Bacillus rossius* (Phasmidae) and considered to arise by gene amplification (Scali and Montanelli, 1975); and in several species, intranuclear protein granules (see below, p. 215).

Under the border of microvilli, the cytoplasm often shows two regions. The apical zone is devoid of organelles, contains fibrous material arising from the core of fine filaments located in each microvillus (Figures 2a, b), and a network of cisternae of SER. This apical zone lies above a region rich in mitochondria. In larvae of *Aeshna cyanea* (Andriès, 1976c), the structure of the apical zone is very complex, and its appearance depends on the feeding cycle. This region contains stacks of unfenestered, parallel cisternae which often surround the rootlets of the microvilli. These stacked membranes are devoid of ribosomes and are sometimes connected with profiles of RER. This observation suggests that these membranes constitute a specialization of the RER. The morphologies of the stacks differ in fed and starved insects. After feeding, the most apical stacks are in contact with vesicles of the SER, and small lipid droplets exist in their vicinity. The stacks located more deeply may be involved in intracellular transport and in the synthesis of triglycerides.

Below the microvilli, pinocytotic vesicles are often observed, especially during the periods of feeding, sometimes together with multivesicular bodies (Lepidoptera, Smith *et al.*, 1969). In Diptera, the coated vesicles originate from pinocytotic activity. They fuse with primary lysosomes, and the resultant profile is that of a multivesicular body in which hydrolase activities can be detected (de Priester, 1972).

RER consists of short, single cisternae and vesicles, or most often of cisternae arranged in parallel arrays. During the feeding periods of hematophagous dipterans, cisternae are disposed in tight whorls (Figure 11b).

Golgi bodies produce membrane-bound secretion granules (Figure 2a) which certainly constitute a heterogeneous population. Some are zymogen granules which release exoenzymes into the lumen of the gut, others contain mucopolysaccharides, and others include the components of the peritrophic membrane.

Lysosomal figures may be abundant. They consist of dense bodies and autophagic and heterophagic vacuoles; but their abundance depends on the position in the cycle. Most generally, autophagic vacuoles appear in cells which are fated to die (Fain-Maurel *et al.*, 1973). However, in Diptera, they may serve to eliminate superfluous secretory products (de Priester, 1972).

In the basal area of the cell, mitochondria are abundant and are associated with basal infoldings. In many pterygote insects, the basal plasma membrane forms extensive, irregularly shaped and branched infoldings which can reach almost to the microvillar border. They constitute a labyrinth of irregular channels, rich in mitochondria, which is involved in transport processes (Figure 3). However, the labyrinth does not exist in Thysanura. Its expansion shows im-

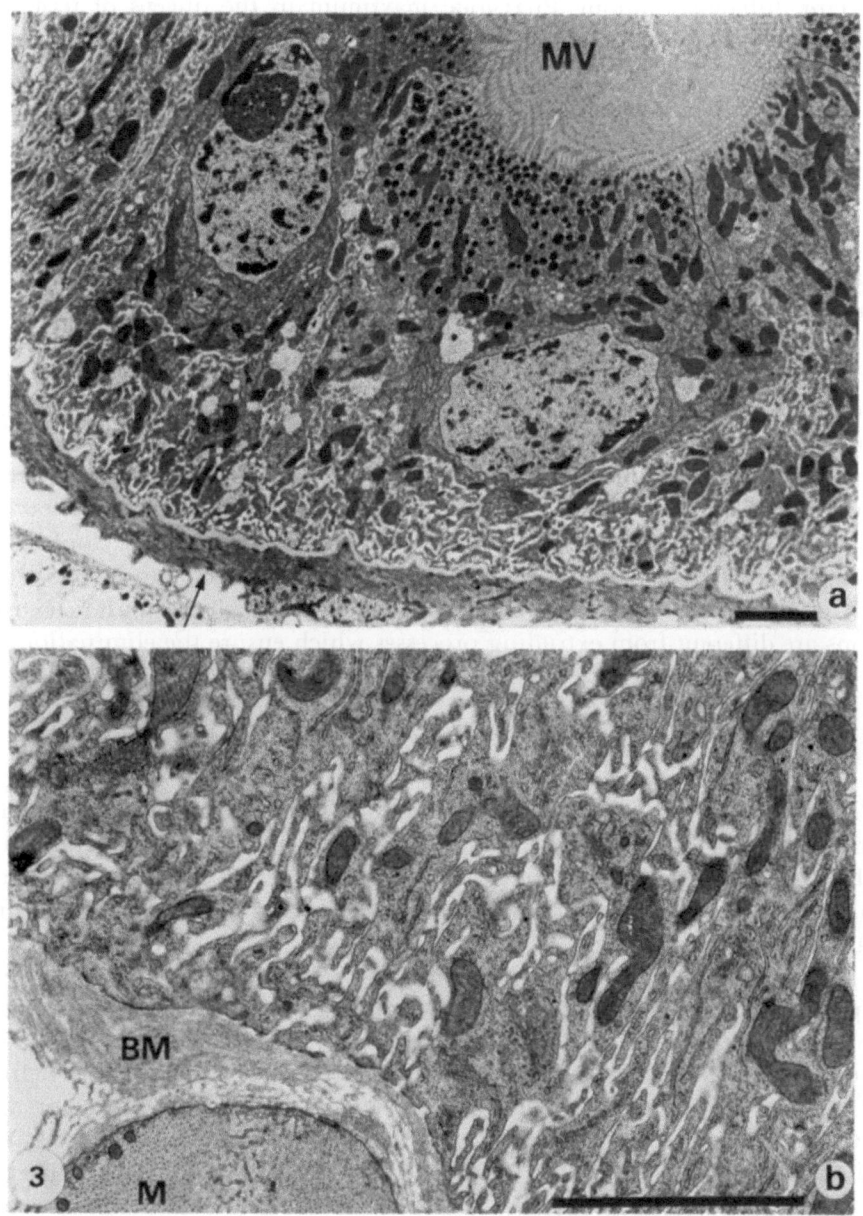

Figure 3. Sections of columnar cells. (a) Low magnification in *Anopheles gambiae*. The border of microvilli (MV) is well developed, large populations of dense secretory granules occur in the apical cytoplasm, infoldings of the basal plasma membrane generate a labyrinth, and the basement membrane (arrow) is thick. (b) The labyrinth in *Calliphora erythrocephala*. BM, basement membrane. M, muscle fiber. Courtesy of Professor H. Hecker and Dr. W. de Priester. (Bars = 3 μm.)

portant differences among Pterygota: maximum in the queens of termites (Noirot-Timothée and Noirot, 1965), minimum in the workers of ants (Jeantet, 1971).

An important question is the mode of release of the secretory products into the lumen of the gut. Exocytosis has been observed in some species (Thysanura, Fain-Maurel et al., 1973; Odonata, Andriès, 1976b; Diptera, Lehane, 1976a,b) and considered as a mechanism for the release of exoenzymes and components of the peritrophic membrane. The suggestion has also been made that secretion is apocrine, involving the splitting off or extrusion of a part of the cytoplasm of the columnar epithelial cell. However, it is also possible that cytoplasmic extrusions are due to fixation artifacts or that they occur in response to a fasting period. While it is true that unsuitable fixation procedures may produce artifacts, the different fixatives do not modify the appearance of extrusions, and they occur in normal conditions as well as after a prolonged fasting period. The extrusions are correlated with a specific physiological state, since they appear at a precise period of the cellular cycle. In cells containing spherocrystals (see below), apical extrusions release a cytoplasmic area filled with concretions. Thus, apocrine secretion seems to be the only way of allowing the cell to extrude bulky inclusions. Such cytoplasmic extrusions are different from extruding processes which ensure the elimination of degenerating cells.

3.2. Goblet Cells

So-called goblet cells, which differ from the goblet cells of vertebrates, are present in the midguts of some insects (i.e., Ephemeroptera and Lepidoptera). The characteristic feature of these cells is the invagination of the apical border to form a deep cavity, the goblet chamber, which confines the nucleus to the basal region.

The phenomenon of membrane infolding reaches a maximum in this cell type. In the goblet cells of Lepidoptera, four areas of extensive plications may be observed (Anderson and Harvey, 1966; Smith et al., 1969; Akai, 1970). At the base of the cell, the plasma membrane is evaginated to form podocytelike extensions. Along the lateral boundary, evaginations are more prevalent and create a laminar pattern between the basal two-thirds of the goblet cell and adjacent columnar cells. The extracellular space in the region of the lateral evaginations is very close to the goblet chamber and hence to the midgut lumen via this chamber. Lining the goblet chamber, but best developed on its basal and lateral regions, are numerous cytoplasmic projections which resemble large microvilli. Each projection contains a mitochondrion. Toward the apical end of the chamber, projections become smaller, contain no mitochondria, and are like typical microvilli. In contrast to microvilli of the columnar cells, the inner leaflet of the membrane of the cytoplasmic projections possesses a coat of fine spikelike units (5-6 nm long), oriented perpendicular to the membrane. Finally, in the apical region of the cell, the cytoplasm is drawn out into a few large villuslike projections which converge toward the middle of the goblet chamber and form a canal by which the chamber becomes confluent with the lumen of the gut. As in normal microvilli, cytoplasmic filaments run along the length of some of the projections (but not in those which are present within the chamber).

A peculiar junction structure is present in the projections located at the apical tip of the goblet chamber in regions where projections abut each other. These junctions occur between two areas of the same plasma membrane rather than between the membranes of two different cells (Flower and Filshie, 1976). The few mitochondria present in the cytoplasm of the main cell body are randomly oriented, and RER is much less abundant than in the columnar cells.

The ultrastructural features of the goblet cells reflect a role in ion transport. The cytoplasmic projections appear to act as a valve that isolates the cavity from the gut lumen. The cell is thought to be involved in the active transport of potassium ions out of the hemolymph and of calcium ions from adjacent columnar cells into the gut lumen (Waku and Sumimoto, 1974).

3.3. Peculiar Secretory Cell Types: The Goblet Cells of Thysanura and the Bottle-Shaped Cells of Odonata

Fain-Maurel *et al.* (1973) discovered in the midgut of *Petrobius maritimus* peculiar goblet cells which were scattered among columnar cells that contained concretions. These cells (Figure 4) resemble both the goblet cells of other insects, because they possess a sort of goblet chamber, and the true goblet cells of vertebrates, because they are secretory cells. But, in contrast to other insects, the chamber is lined by microvilli that lack mitochondria, and in contrast to vertebrates, the cells do not secrete mucous products.

The cytoplasm shows ultrastructural organelles characteristic of secretory cells. RER and Golgi bodies are abundant and synthesize granules which are transferred to the lumen of the gut in a merocrine fashion. Histochemical tests demonstrate that the granules contain proteins, and Fain-Maurel *et al.* suggest that these granules represent stored enzymes. So, we think that these cells are peculiar columnar cells with an apical invagination of the membrane and in which an important storage of secretory product occurs. These goblet cells sometimes contain lipid droplets and mineral spherocrystals like adjacent columnar cells.

Though they lack an invagination of the apical membrane, the bottle-shaped cells of Odonata resemble the goblet ones of Thysanura in terms of the storage of secretory material (Andriès, 1976a). Their secretory vesicles are more or less electron-dense and are believed to contain, at least in part, neutral mucopolysaccharides. Large numbers of microtubules are located between the area of formation of the secretory granules and their region of storage. As in exocrine cells of vertebrates, these microtubules are certainly involved in the intracellular translocation of secretory material.

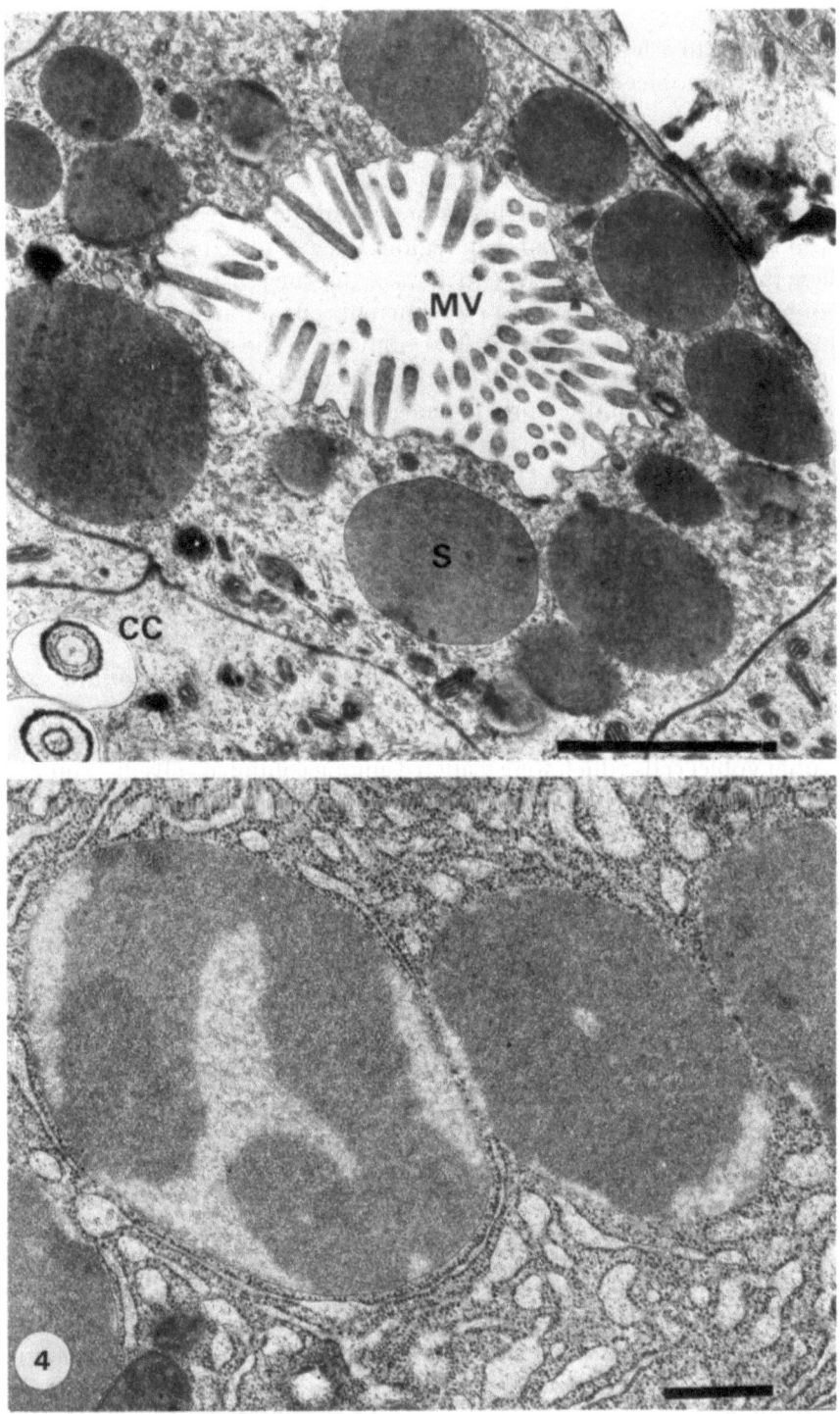

Figure 4. The goblet cell of *Petrobius maritimus*. (top) A transverse section of the apical part of the cell, showing the usual microvilli (MV) and secretory granules (S). Note the presence of spherocrystals in the adjacent columnar cell (CC). (bottom) A more enlarged view of secretory granules and the abundant RER. Courtesy of Professor P. Cassier. (Bars = 2 μm.)

3.4. Endocrine Cells

Gut cells that are thought to have endocrine function have been described in a wide variety of insects: Apterygota (Cassier et al., 1972), Dictyoptera (Cassier and Fain-Maurel, 1977; Nishiitsutsuji-Uwo and Endo, 1981), Orthoptera, Coleoptera, and Heteroptera (Cassier and Fain-Maurel, 1977; Bayon, 1981), Odonata (Andriès, 1976a), Lepidoptera (Endo and Nishiitsutsuji-Uwo, 1981), and Diptera (de Priester, 1971; Platzer-Schultz and Reiss, 1970). They are sometimes called clear cells, granular cells, or secretory cells, and they are characterized by a light cytoplasm, the absence of a basal labyrinth, and the presence of granules (Figure 5a). The apices of these cells may or may not be in contact with the lumen of the gut. Considering the shape and the location of the cells, the shape and the electron-opacity of the granules, several types have been observed which coexist in the same species. For instance, six types of granules and four cell types were identified in the cockroach (Nishiitsutsuji-Uwo and Endo, 1981). By using immunohistochemical technique, Iwanaga et al. (1981) demonstrated immunoreactivities to antibodies raised against polypeptide hormones such as pancreatic polypeptides, somatostatin, and enteroglucagon. The pancreatic polypeptide immunoreactive cells are most numerous and the reactive material is located inside the granules (Endo et al., 1982). Exocytotic release of the granules has been observed along the basal part of the cells in Thysanura (Cassier et al., 1972) and in Dictyoptera (Endo and Nishiitsutsuji-Uwo, 1982) (Figures 5e, f). Iwanaga et al. (1981) suggest that these cells in the insect gut are functionally similar to the gut endocrine cells of vertebrates. Both resemble neurosecretory cells in synthesizing and releasing a variety of polypeptide hormones.

3.5. Storage Products of the Columnar Cells

The columnar cells often stores glycogen, globules of fat and mineral salts or metalloproteins. Accumulation depends on the species, the functional subdivisions of the midgut, and cell cycle.

Organic components. Fat inclusions are often associated with glycogen-rich areas. Fat droplets may be so numerous that the columnar cells resemble the fat body cells. In some species (*Lucilia cuprina*), the term *lipophilic cell* is used. The role of these storage reserves is not understood. In Culicidae, the amount of lipids and glycogen is reduced when digestion is completed, and Hecker (1977) has suggested that these storage products are mobilized and transported to the hemolymph. In the fly *Rhynchosciara americana*, starvation produces the disappearance of the lipid spheres (Ferreira et al., 1981). Likewise, in the stablefly, *Stomoxys calcitrans*, the cells of the "lipid zone" of the midgut absorb fatty acids after a blood meal. These acids enter the RER and are transported to the Golgi bodies where they are converted to triglycerides and phospholipids (Lehane, 1977). Depending on the degree to which the transportation systems are loaded, these molecules may be transferred to "lipoid spheres" for temporary storage, and then the lipids of the lipoids spheres are mobilized by ele-

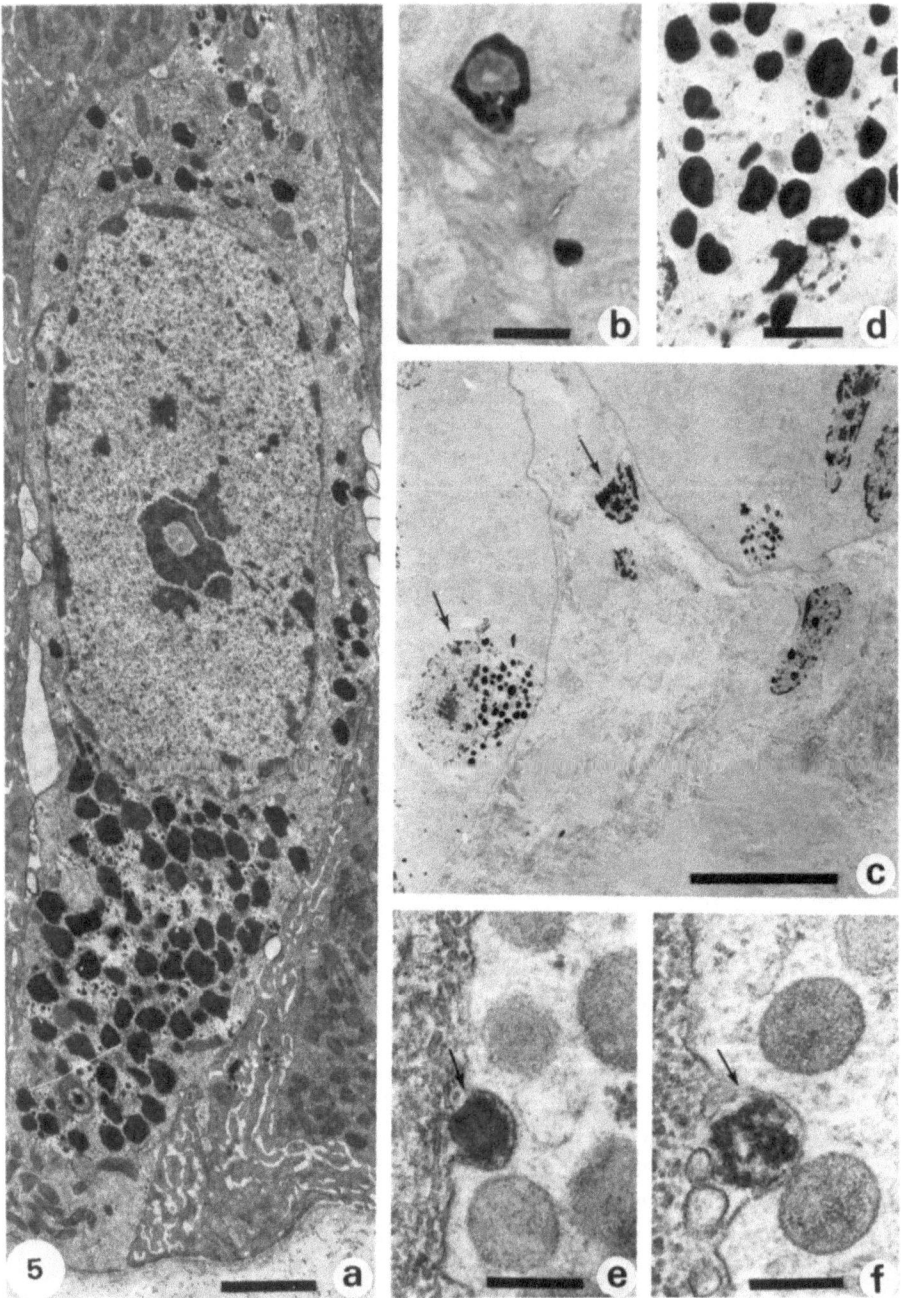

Figure 5. The endocrine cells in the gut of *Periplaneta americana.* (a) Low magnification showing the ultrastructural aspect of a type II-b cell. (b) Pancreatic polypeptide-immunoreactivity in a semithin section of a type II-b cell. (c) Electron micrograph of a thin section adjacent to (b). (d) Higher magnification of (c) showing the secretory granules. (e, f) Two stages of exocytosis of granules: immediately after opening of fused membranes (e, type I-b granule) and during degradation of an extruded granule (f, type I-a granule). Courtesy of Dr. J. Nishiitsutsuji-Uwo. [Bars = 2 μm (a), 1 μm (b, c, d), and 0.2 μm (e, f).]

ments of the RER and transferred to the basal region of the cells and finally to the hemocele. However, in the thysanuran *P. maritimus*, according to Fain-Maurel et al. (1973) (Figure 13), the storage of fat increases gradually until the time when the cell degenerates and releases its inclusions in the lumen of the gut.

Mineralized concretions. Very often the columnar cells (or even the regenerative cells) of the midgut contain intracytoplasmic mineral concretions. In most orders (Thysanura, Planipenna, Homoptera, Trichoptera, Lepidoptera, Hymenoptera, and some Diptera), mineral salts constitute spherocrystals with numerous concentric layers (Figures 6a, 7). Granules of ferritin are found in homopterans. Spherocrystals originate in cisternae of the RER and are formed by the binding of mineral salts on polyanionic stroma. A precipitation, resulting in the formation of more or less thick concentric strata, takes place while the RER enlarges and looses its ribosomes (Gouranton, 1968b; Jeantet, 1971). However, in Thysanura and Lepidoptera, spherocrystals seem to originate from Golgi vesicles (Fain-Maurel et al., 1973; Waku and Sumimoto, 1974) (Figure 7). The spherocrystals are composed of a complicated collection of metals (Ca, Mg, K, Mn, Fe, Zn, and Sr) in the form of phosphates, carbonates,

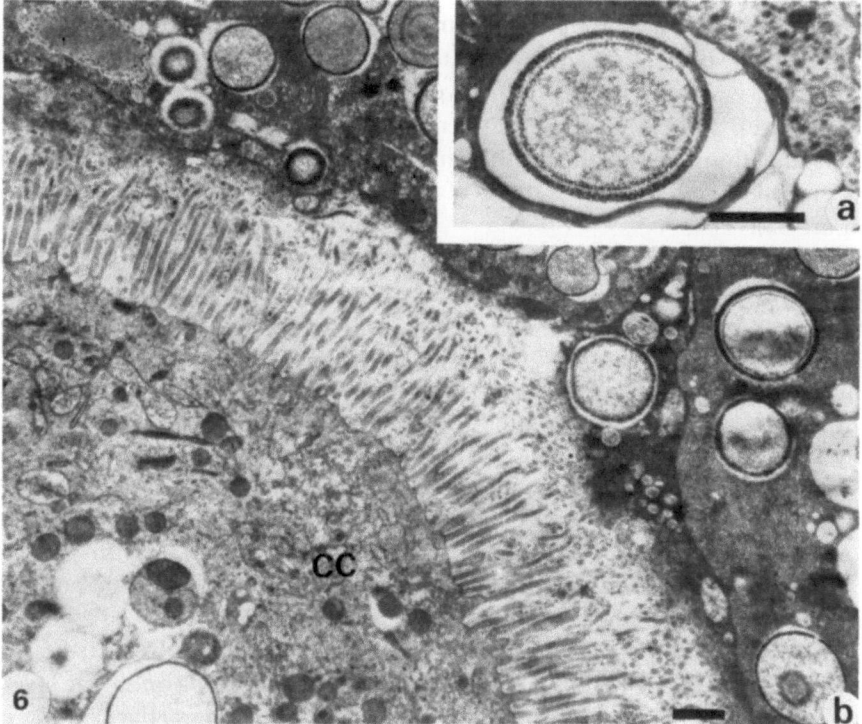

Figure 6. Spherocrystals in the columnar cells of a larval worker of *Formica polyctena*. (a) Spherocrystal in an ergastoplasmic cisterna. (b) Apical part of the cell (CC) and lumen of the midgut containing numerous crystals ejected together with a part of cytoplasm. Courtesy of Dr. A. Y. Jeantet. (Bars = 1 μm.)

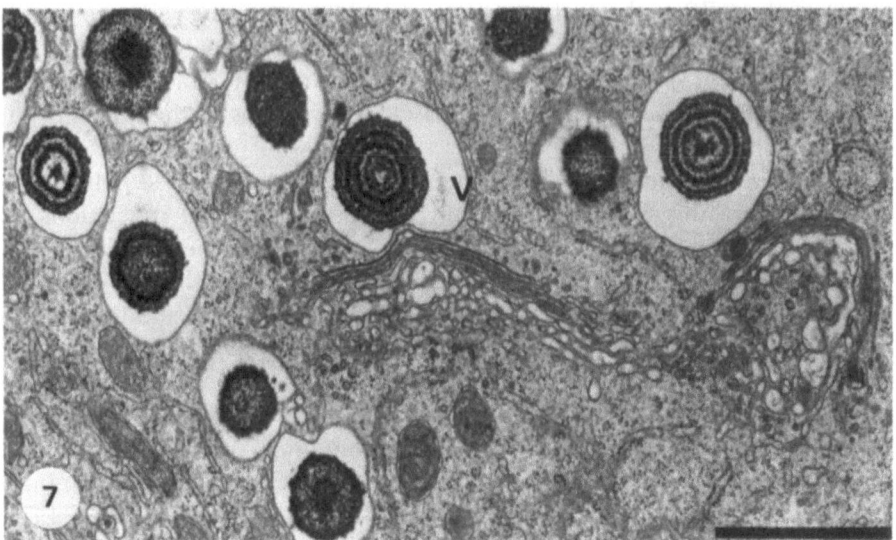

Figure 7. Spherocrystals arising from Golgi vesicles (V) in a columnar cell of *Petrobius maritimus*. Note that the spherocrystals are made up of concentric strata. Courtesy of Professor P. Cassier. (Bar = 1 μm.)

or even chlorides, and the chemical complexity depends on the species. Spherocrystals are devoid of uric acid, and thus the name *urospherite*, which is sometimes used, is wrong. Spherite turnover is high in many insects (Homoptera for instance), and therefore new spherites are constantly being formed while others are eliminated in the gut lumen. In *Formica polyctena*, spherocrystals are ejected together with a part of the apical cytoplasm (Jeantet, 1971) (Figure 6b).

The biological accumulation of minerals in the spherocrystals is an important mechanism regulating the composition of the internal environment. Experimental data using ^{45}Ca show that, in the silkworm, the columnar cells absorb calcium ions from the hemolymph primarily via the goblet cells (Waku and Sumimoto, 1974). However, the observed transport of calcium from hemolymph to lumen does not exclude the possibility that movement of calcium and other ions may occur in the reverse direction. Besides, most insects possess columnar cells with spherocrystals and lack adjacent goblet cells. When unusual cations like Cd or Pb used as tracers are ingested in the food, they accumulate in the spherocrystals. So, the midgut forms a partial intestinal barrier picking up cations coming from the food and preventing them from entering the hemolymph (Jeantet *et al.*, 1977).

In the dense bodies of the housefly, metals, such as Ca, Fe, and Cu, are less numerous. The concretionary material is not arranged in concentric layers. It is initially deposited within Golgi vesicles, and also in lamellar bodies and residual bodies. Concretions accumulate with age, and metals are more abundant when the food contains an excess of them (Sohal *et al.*, 1977).

Granules of ferritin are free in the hyaloplasm and in the nuclei of the columnar cells of some homopterans. These inclusions consist of aggregates of individual ferritin molecules, 5–6 nm in size (Gouranton, 1967; Gouranton and Folliot, 1968) (Figures 10b, c).

When ants are contaminated with toxic metals, many lysosomes are formed in the apical part of the columnar cells of the gut. They, together with spherocrystals, accumulate these metals and so ensure detoxification (Jeantet, 1981).

The cuprophilic cells of the midgut of larval flies. The midguts of fly larvae provide striking examples of morphological and functional differentation, so that several segments are recognizable in which epithelial cells have received peculiar names. In the middle midgut of *D. melanogaster*, in which a low pH is maintained within the lumen, Filshie *et al.* (1971) distinguish a mosaic of "interstitial" cells and cup-shaped, copper-accumulating cells. These cuprophilic cells accumulate histochemically detectable copper. They are inserted between the columnar interstitial cells. The infoldings of the basal plasma membrane are less developed in cuprophilic cells than in interstitial cells, and microvilli are remarkable in that they possess a large electron-dense layer (12 nm in thickness) adjacent to the anterior face of the plasma membrane. In some mutants, microvilli coexist with lamellae. It appears probable that copper is sequestered into lysosomes whose distribution in the cytoplasm is uniform (Figure 8).

The function of the cuprophilic cell is quite obscure. There are neither associations of mitochondria with microvilli nor stacks of infolded lateral plasma membranes which are known to have an active function in the movement of ions. Moreover, in the midgut of larvae of *Lucilia cuprina* and others flies where, as in *Drosophila*, a change from alkaline to acid pH occurs in the lumen, there are no cells corresponding to the cuprophilic cells. So, the release into the midgut lumen of hydrogen ions is not necessarily related to the occurrence of cuprophilic cells. Cuprophilic cells may pick up cations coming from the food, because the number of lysosomes in these cells is increased in larvae raised on a diet rich in copper.

Protein crystals within the nuclei of columnar cells. Intranuclear inclusions were first reported at the end of the last century, for several species of insects, and recent studies have clearly established their fine structure, their chemical composition, their rate of formation, and their site of synthesis. The intranuclear crystals of midgut cells are not restricted to a specific group of insects. They have been observed in species belonging to different orders, and two species of the same genus may be very different. In the case of *Gyrinus marinus* and *G. natator* for instance, the crystals are present only in *marinus*. The classical data (Gouranton, 1969; Gouranton and Thomas, 1972; Thomas and Gouranton, 1972; Gouranton and Thomas, 1974) have been obtained on the midgut cells of *T. molitor* and *G. marinus* (Coleoptera) and *Sympetrum depressiusculum* (Odonata) in which crystals are present both in larvae and in adults, though in *Sympetrum* they seem to be less frequent in adults. Crystals

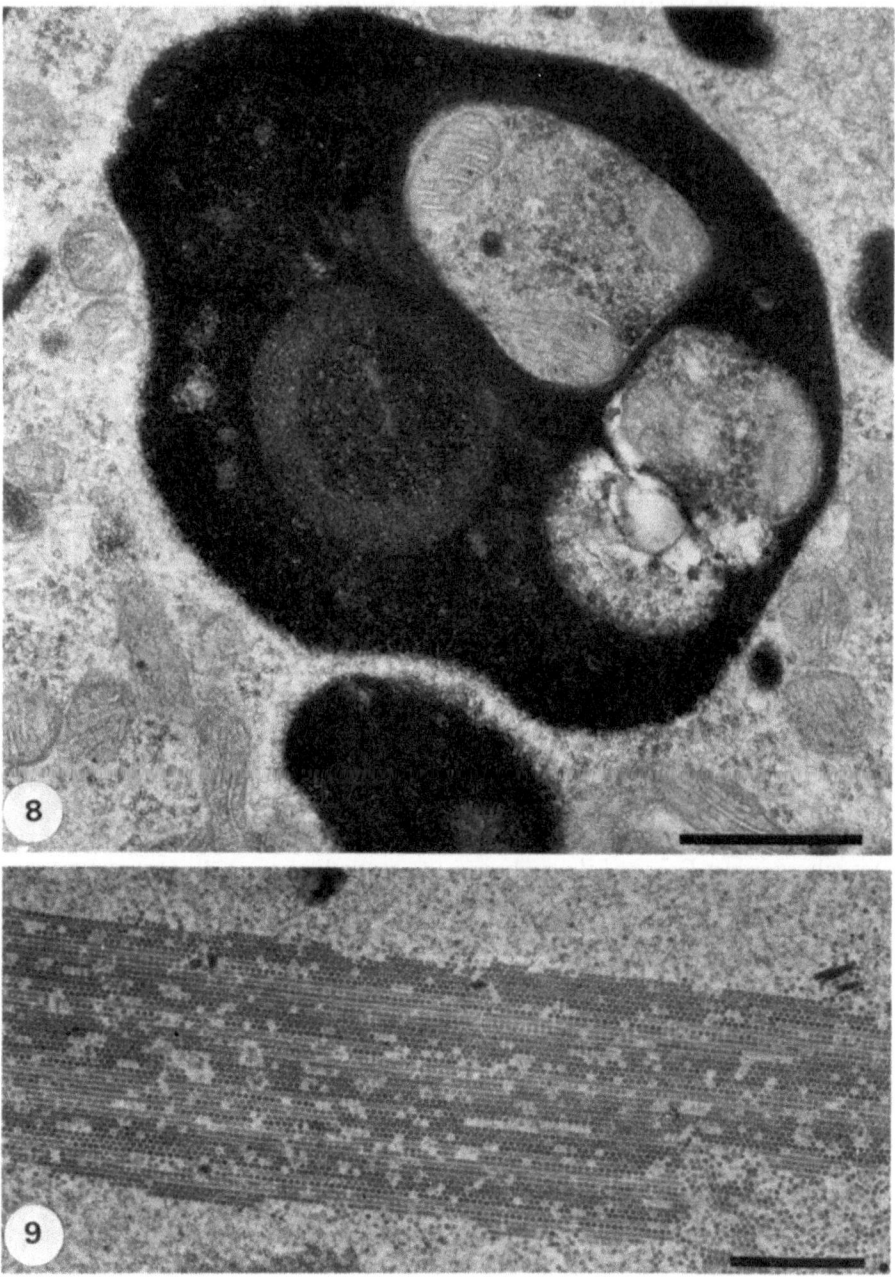

Figure 8. Lysosome within a cuprophilic cell of a larva of *Drosophila melanogaster* (Canberra strain) raised on a diet with 200 µg/g added copper. The cytolysome contains cytoplasmic components (ribosomes and a mitochondrion) and a phospholipidic whorl. This organelle is believed to sequester the copper. Courtesy of Dr. B. K. Filshie. (Bar = 0.4 µm.)

Figure 9. Crystal of viruslike particles in the nucleus of a columnar cell of a larva of *Drosophila melanogaster* (Noumea strain) raised on a diet with 300 µg/g added copper. Courtesy of Dr. B. K. Filshie. (Bar = 1 µm.)

are exclusively located in the nuclei of the cells of *Tenebrio*, whereas similar ones are often found in the cytoplasm of the cells of the other species. Whatever their localization, the crystals are always devoid of a surrounding membrane. They are absent in the regenerative cells. They appear in the differentiating cells, increase in size and are present without changes until the time when the cells degenerate. In *Tenebrio*, they break up before being ejected into the gut lumen together with the cell (Devauchelle, 1970).

In *Tenebrio* (Figure 10a) and *Sympetrum*, the crystals are generally single with a species-specific morphology. But in *Gyrinus*, they are twin crystals which exhibit an irregular shape. In all cases, they reveal an ordered periodicity. In *Tenebrio*, the pattern appears as a series of parallel bands spaced 8 nm apart; in *Sympetrum* and *Gyrinus*, the distance between the parallel bands is quite similar (8–9 nm), but generally the crystal sectioned shows crossed bands, and the angles between these latter vary according to the plane of section through the crystal.

Cytochemical and enzyme digestion tests show that crystals are composed of protein. In *Tenebrio*, this protein appears to be very reactive to cytochemical tests for cysteine and tryptophan (Gouranton, 1969). When prepared from isolated crystals, the protein is acidic and has a molecular weight of only 25,000 daltons (Thomas *et al.*, 1977). In the adult of *Tenebrio*, the crystals are formed in less than 4 days (Thomas and Gouranton, 1973a). During cell differentiation, they increase in size, while the nucleoli become smaller. Thus, the site of synthesis of the crystals is a subject of interest, because it is generally agreed that most of the nuclear proteins are synthesized in the cytoplasm and then migrate to the nucleus. Here it might be postulated that the crystals develop from nucleolar materials. However, in *Tenebrio* and *Gyrinus*, autoradiographic studies show a decrease in cytoplasmic labeling and a concomitant increase in the labeling of intranuclear crystals after the injection of tritiated lysine. This transfer of radioactivity suggests that the crystal proteins are synthesized in the cytoplasm and that they migrate to the nucleus where they crystallize (Thomas and Gouranton, 1973b; Gouranton and Thomas, 1976). The nucleolus labels vary little, and so the decrease in size of the nucleolus is not related to development of the crystal.

The significance of such crystals remains unclear. The ability of viruses to induce nuclear inclusions is well known, but virus particles are not observed in these nuclei (Devauchelle, 1970). It cannot be suggested that environmental conditions are responsible for the occurrence of these inclusions, since there are instances where two species belonging to the same genus live in the same habitat and yet intranuclear crystals occur only in insects of one species. The crystal formation could represent the storage, without further use, of an excess of proteins synthesized by the midgut cells.

Intranuclear protein spherules. Most nuclei of the midgut cells of the phasmid *Carausius morosus* contain one or several clusters of spherical electron-dense bodies. The spherules are never surrounded by a membrane. In contrast with intranuclear crystals, their chemical composition is very complex. At least 16 different proteins have been detected, and some originate

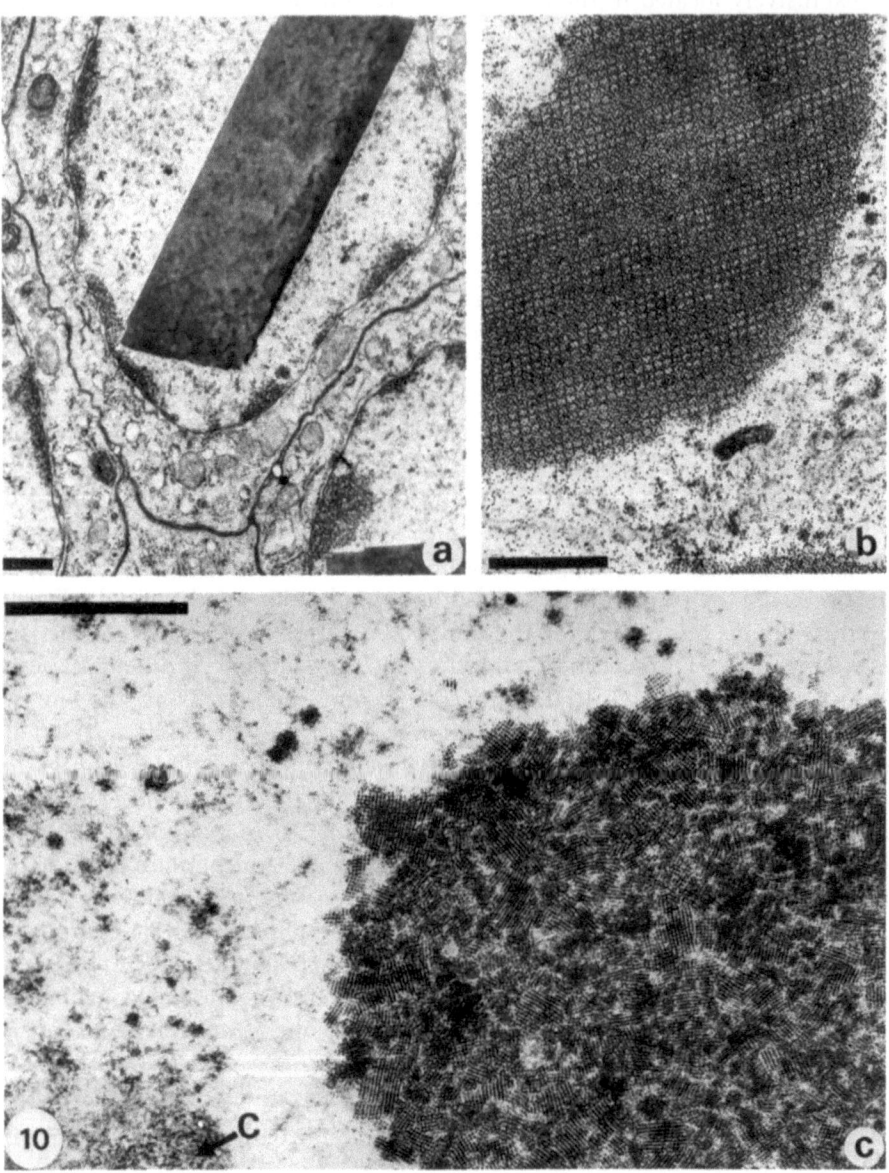

Figure 10. (a) Intranuclear protein crystal in a columnar cell from *Tenebrio molitor*. (b, c) Aggregates of ferritin granules in the nuclei of the homopterans *Campylenchia latipes* (b) and *Philaenus spumarius* (c). C, chromatin. Courtesy of Professor J. Gouranton. (Bars = 0.5 μm.)

from cytoplasmic proteins that migrate to the nucleus (Thomas and Gouranton, 1980).

Viruslike inclusions. In *D. melanogaster*, the nuclei of cuprophilic cells and interstitial cells contain viruslike particles which tend to aggregate into crystalline inclusions (Figure 9). These inclusions are present in larvae raised on normal diets, and the frequency of their occurrence and the size of the crystals increase with the increase in the amount of copper in the diet (Filshie et al., 1971).

3.6. Secretion of the Peritrophic Membrane

In many insects, a peritrophic membrane exists between the food bolus and the midgut epithelium. This structure secreted by the midgut cells is believed to protect the midgut cells from mechanical damage caused by food particles. However, the membrane is present in some fluid-feeding species. In holometabolous species, structural differences between the membranes of larvae and adults have been observed (Richards and Richards, 1969).

Several layers are always resolved. In *D. melanogaster*, three layers can be recognized. Bordering the lumen is an electron-dense, granular layer (8-12 nm in thickness), separated by a less-dense layer (10 nm) from a granular layer (100-150 nm) (Filshie et al., 1971). At high magnification, some of the layers (or even all the layers) appear to contain chitin-protein units which are aggregated into layers of microfibers. The fibers commonly have an orthogonal or hexagonal arrangement, but they may also be randomly oriented.

The peritrophic membrane is formed in two main ways. In most cases, it is secreted along the length of the midgut. In *Thysanurans*, the cells discharge secretory granules whose contents gather in the spaces between the microvilli (Fain-Maurel et al., 1973) (Figure 2e). In flies, the membrane is secreted by a belt of cells located at the junction of the stomodeum and the midgut and often named the *proventriculus*. In *Chironomus thummi*, these cells are characterized by an abundant RER and synthesize secretory granules. By a process of classic apocrine secretion, granules are discharged. The peritrophic membrane is formed by the secretion granules and also by the transformed and pinched-off upper part of the cell. The membrane consists of two layers. The one facing the lumen of the gut exhibits a honeycomb texture and is probably formed by the secretion. The other layer may arise from the transformed apical cytoplasm of the cell (Platzer-Schultz and Welsch, 1970). In *Calliphora erythrocephala*, three layers are present in the peritrophic membrane. Each component is distinct in fine structure and origin, and it must be noted that in contrast to *Chironomus*, no degeneration of apical cytoplasm occurs during the formation of the membrane (Smith, 1968). The most anterior cells of the midgut epithelium elaborate pale secretory granules. The outer ring of cells manufactures a more homogeneous, electron-dense secretion, which is probably the second constituent of the membrane that forms in the extracellular cleft between the two rings. Behind these specialized cells, the midgut cells secrete the third component of the membrane. This is a very thin sheet which is composed of a dense 20-nm layer to which are attached wisps of fibrous material. The most complex peritrophic membrane yet described is that of the stablefly *S. calcitrans* (Lehane, 1976b). It is formed of five layers, and each of them is secreted by precise cells of the proventriculus, and each exhibits peculiar chemical characteristics. The first layer, which lies nearest the food, contains "pores" and is probably lipidic; the second one contains chitin-protein units in a non-fibrous form; layers 3 and 4 are made of neutral mucopolysaccharides; and layer 5 contains sulfated mucopolysaccharides.

It is not clear why and how the chitin-protein units aggregate to form geometric array. The suggestion has been often accepted that the arrangement of microfibers is organized by the tips of the microvilli. Data concerning the origin of the peritrophic membrane of *A. aegypti* (Richards and Richards, 1969) do not agree with this opinion. The cells of the proventriculus secrete a material that is or contains a chitin-protein component. This material is of unknown physical state, but it appears at first to be granular. Then, as it moves posteriorly, a postsecretion aggregation produces layers of microfibers within the secretion. Thus, the fiber-inducing and fiber-orienting mechanisms do not seem to involve a direct control by the microvilli, and they remain unknown. Richards and Richards suggest a phase change in a liquid crystal system.

4. Changes in Fine Structure

4.1. Functional Subdivisions of the Gut and Cellular Cycles

In insects where the histological structure of the epithelium appears more or less uniform, ultrastructural examination of the columnar cells shows differences which represent a true functional subdivision of the gut or indicate that the columnar epithelia are a heterogeneous population of cells.

In female mosquitoes (Hecker, 1977), the midgut consists of a narrow anterior part and a wide posterior part (the stomach) into which the blood is taken for digestion. In both parts, the epithelium is composed of columnar cells, but ultrastructural differences can be observed between them. In the anterior part, the cells have more microvilli, a higher density of membranes of the SER, and a tendency for more channels in the basal labyrinth. The cells of the posterior part are characterized by more RER and mitochondria. Gap junctions, septate junctions, and hemidesmosomes are present in the cells from both parts, whereas desmosomes occur only in the cells of the stomach. Hecker proposes that absorption of sugar occurs in the anterior part, whereas the posterior segment synthesizes the peritrophic membrane components, releases digestive enzymes, and absorbs the remaining nutrients.

The midgut of *Rhynchosciara americana* larvae (Ferreira *et al.*, 1981) consists of a cylindrical ventriculus from which protrude two gastric caeca. The epithelial cells of the caeca and of the posterior part of the ventriculus are typical columnar cells. Those of the ventriculus contain lipid inclusions, as is the rule in many Diptera. In contrast, the anterior ventricular cells possess very small microvilli and remarkably developed infolding of the basal plasma membrane with associated mitochondria. The ER and Golgi bodies are poorly developed. These cells exhibit the ultrastructural characteristics of absorptive cells, and according to Ferreira *et al.*, they function to transport endogenous fluids into the lumen of the gut. Thus, their functions resemble those of goblet cells.

In the larvae of the beetle *O. nasicornis* (Bayon, 1981), the midgut shows three rings of caeca and a ventral groove. The cells of the first ring have the typical appearance of columnar cells. But in the second and the third ring of caeca and in the groove, the cells exhibit small amounts of ER and few Golgi bodies, while the border of microvilli is very well developed and mitochondria are very numerous in the apical part of the cytoplasm. Basal infoldings of the plasma membrane form an archlike arrangement. Considering these ultrastructural aspects and microanalytical data, Bayon suggests that these cells could be involved in the transport of ions, especially potassium, and so their function is similar to that of goblet cells.

In *Locusta migratoria*, four types of cells can be recognized among the population of columnar cells of the straight part of the gut and of the ring of caeca (Papillon *et al.*, 1974). "Clear cells" (type I) have a poorly developed RER; in "dense cells" (type II), RER is more abundant; "very dense cells" (type III) show expanded RER vesicles; and "acidophilic cells" (type IV) are characterized by the extrusion of secretory granules. Though these four cell types correspond to four periods of the cellular cycle, the percentage of each type is not the same in the different parts of the midgut. In *P. maritimus*, Fain-Maurel *et al.* (1973) recognize three stages of the columnar cells which contain spherocrystals. In the first one, Golgi bodies are active and synthesize mucopolysaccharide granules. During the second stage, RER increases, microbodies (peroxisomes) appear, and spherocrystals are formed. Finally, during the third period, a lipid storage occurs, Golgi bodies produce vesicles which contain acid phosphatases and curious, rippled lamellae. As in *Locusta*, all these types are present in different frequencies in the different parts of the midgut. So, in both species, ultrastructural examination allows one to distinguish several segments in the digestive tract, although the structure appears rather uniform upon superficial examination.

4.2. Effects of Unsuitable Diets and Starvation

When the wasp *Nasonia vitripennis* is fed on a diet of sterile sugar water, there are extensive alterations in cell organelles (mitochondria, RER) and lipid inclusions. Many lysosomes and residual bodies are produced (Davies, 1977).

As a general rule, starvation causes a disappearance of the stored products, but in addition, organelles and physiological processes may be affected. In locusts fasting for 52 hr, the type IV cells which release secretions are more numerous than in insects fed *ad libitum*. The secretion of the peritrophic membrane is not stopped, so that the lumen of the gut contains a network of membranes and secretory products. When fasting is longer (72 hr), many cells degenerate, even in the regenerative nests, and cell parts including pycnotic nuclei are cast into the lumen (Papillon *et al.*, 1974).

4.3. Blood-Sucking Insects

Adult hematophagous Diptera, because of their feeding habit, periodically engorge with large quantities of blood. A sexual dimorphism of the midgut cells, uncommon in insects, has been described in *A. aegypti* (Hecker *et al.*, 1971). In the posterior part of the midgut (the P part, or stomach), the dimorphism in fine structure is an expression of the structural and mechanical adaptations to digestion of blood by the females. A more complex RER, desmosomes between epithelial columnar cells, and a gridlike structured basal lamina are found only in female cells. In addition, the more complex functions of the female stomach are reflected by the higher concentrations of other organelles and membrane systems, such as mitochondria and basal labyrinth (Rudin and Hecker, 1976).

After a blood meal, the midgut becomes distended and cuboidal symmetry of the basement membrane is disrupted. In the columnar cells which become distended laterally, organelles show drastic changes. When bloodfed females of *Culex tarsalis* were observed 1 week after ingesting the meal, the electron-opacity of the hyaloplasm was increased. Mitochondria were much enlarged, and lamellar lysosomal figures and residual bodies were abundant (Figure 11a). Nuclear size was reduced (whereas it enlarged in *Aedes*) and the nuclei were more basally located (Houk, 1977). But the most important changes demonstrated in many species concerned the RER. Rough-surfaced vesicles disappeared, and reticular whorls were formed (Figure 11b). According to

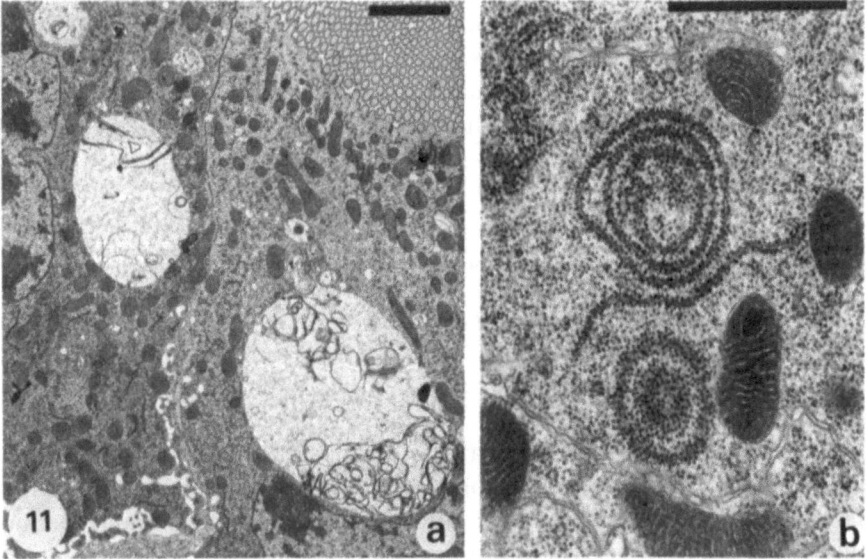

Figure 11. Columnar cells in the anterior midgut of *Culex tarsalis* after a blood meal. Note the presence of two large lamellar lysosomal figures (a) and the whorls of RER (b). Courtesy of Dr. E. J. Houk. (Bar = 1 μm.)

Staübli et al. (1966), the whorls and apical vesicles may contain proteases, but this hypothesis has not been substantiated.

The thickness of the basal lamina is reduced by an expansion of the gridlike structure of the layers. This would indicate a mechanical function for the basal lamina, and the capacity to expand suggests that it may be a limiting factor in midgut engorgement. The stretching is reversible. A mechanical role has been suggested for the desmosomes in cell adhesion, especially important during the distension of the epithelium due to blood intake. But the absence of desmosomes in the gut of *Anopheles stephensi* shows that their mechanical role in the gut epithelium of hematophagous insects should not be overestimated.

5. Cell Degeneration and Renewal

In the digestive tracts of most insects, cell division is continuous, worn-out cells are cast into the lumen, and they are replaced by the differentiation of underlying regenerative cells. With the exception of dipterans, such epithelia are replaced during each molt by such regenerative cells, and such renewal also occurs during metamorphosis. Ultrastructural aspects of degeneration have been studied at various periods during the gut cycle, whereas cell differentiation has been best observed during molts and metamorphosis.

5.1. Cell Degeneration

Degeneration processes are very similar for cytoplasmic components in all cases that have been studied: degeneration of isolated cells during the intermolt (*P. maritimus*, Fain-Maurel et al., 1973; *L. migratoria*, Papillon et al., 1974), and degeneration of the epithelium during metamorphosis (*F. polyctena*, Jeantet, 1971; *A. cyanea*, Andriès, 1975). Degeneration is characterized by a vesiculation of the RER, a decrease in the number of ribosomes, a disappearance of Golgi bodies, and an increase in the number and volume of autophagic vacuoles (Figure 12) and sometimes of residual bodies. The background hyaloplasm shows an increase in electron density, and so cells which are destined to die appear as "dark cells" (Figure 13). In *Aeshna*, the appearance of mitochondria is modified. They become very polymorphic, their cristae are enlarged, and sometimes the matrix contains aggregates of filaments or tubules. Papillon et al. observed in *Locusta* that the nucleus became pycnotic at the last stage of degeneration (Figure 13), whereas Andriès reported that a karyolytic process occurred which reduced the basophilia of the nucleoplasm.

5.2. Cell Differentiation

In Pterygota, the most extensive study has been performed on the differentiation of the epithelium of the imaginal digestive tract of *A. cyanea* (Andriès, 1975). The replacement cells have a very restricted cytoplasm with unattached

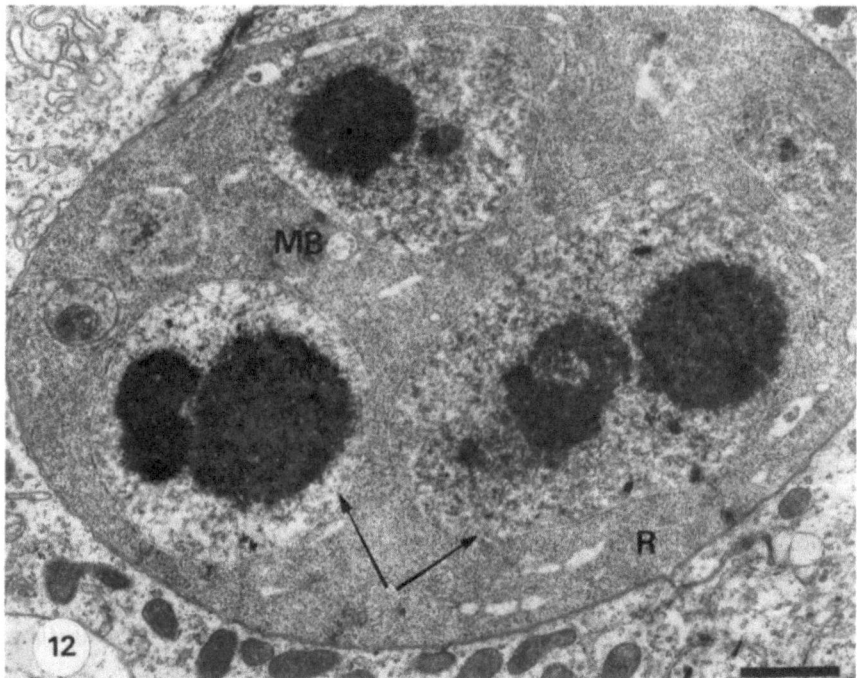

Figure 12. Degenerating columnar cell. an autophagic vacuole in an old larva of *Formica polyctena*, with numerous ribosomes (R), multivesicular bodies (MB), and large amounts of degenerative RER (arrows). Courtesy of Dr. A. Y. Jeantet. (Bar = 1 μm.)

ribosomes, polysomes, a few mitochondria, and Golgi bodies. At the beginning of the process of differentiation, many granules (40 nm in size) are transferred from the nucleoplasm. These are probably ribonucleoprotein particles. Then, while the cell elongates, the nucleus produces many buds which pinch off from it and become large vacuoles. These vacuoles are devoid of masses of chromatin, but they retain some of their nuclear characteristics. The outer membrane is covered with ribosomes and generates small vesicles which constitute new Golgi bodies. In the cytoplasm, the mitochondria divide, and Golgi bodies are produced by the RER, the nucleus, and the buds.

The development of the microvilli is quite fascinating (Andriès, 1972). It takes place in the intercellular space between the differentiating cells and the degenerative cells. The intercellular space appears dilated and filled with an electron-dense glycoprotein material (Figure 15) arising from the content of dense vesicles secreted by the differentiating cells. Then, the membrane which separates the intercellular space from the lumen of the gut breaks down, and the microvilli expand. By this time they are covered with a cell coat.

In other insects (*L. migratoria*, Papillon *et al.*, 1974; *F. polyctena*, Jeantet, 1971), the cytological aspects seem quite similar, especially concerning the development of the microvilli (Figure 14). However, nuclear budding has not been observed.

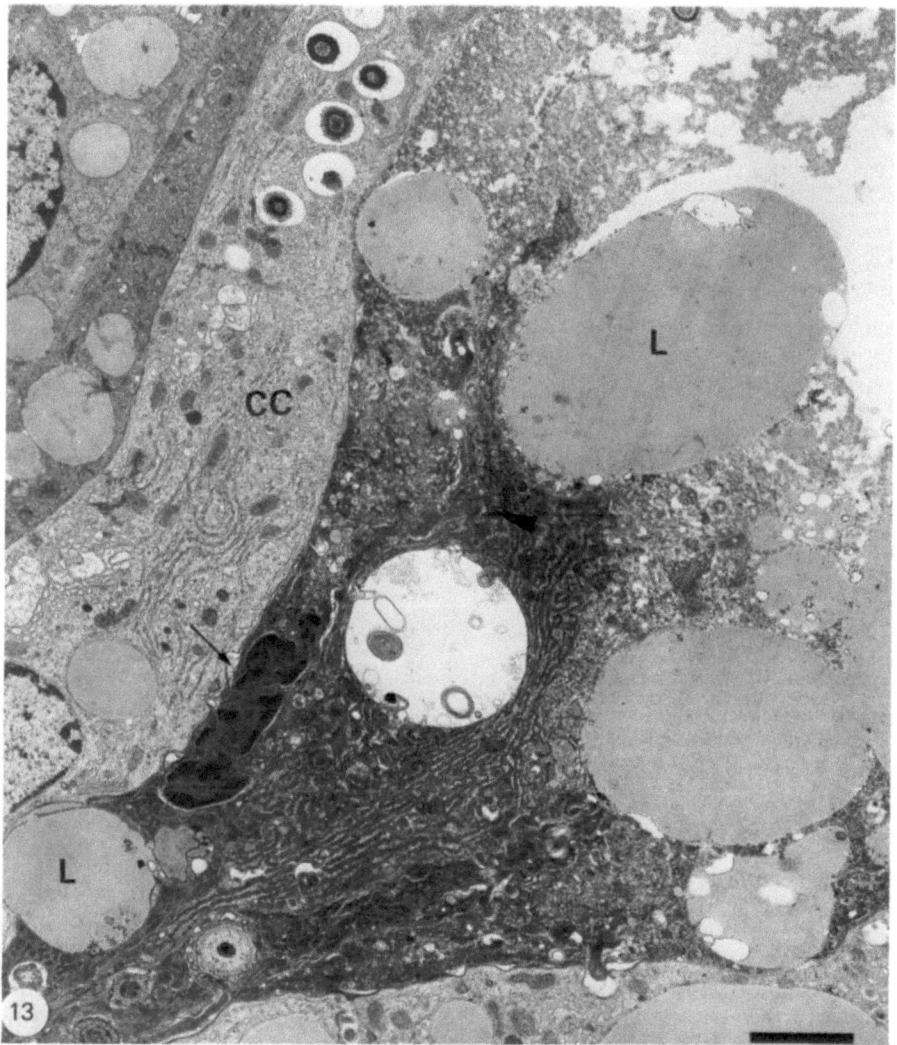

Figure 13. Degeneration of a columnar cell in *Petrobius maritimus*. In contrast to the normal columnar cell (CC), the degenerative cell has a dark cytoplasm, a pycnotic nucleus (arrow), and a vacuole with lamellar figures. Lipid spheres (L) are more numerous in the degenerating cell. Courtesy of Professor P. Cassier. (Bar = 2 μm.)

In Apterygota, the nuclear budding does not seem to exist, and microvilli are not formed in a large extracellular space (Fain-Maurel *et al.*, 1973). During the molt, small microvilli appear on the flat apical part of the replacement cells located under the degenerative epithelium.

5.3. Differentiation of Microvilli in Embryonic Cells

During the differentiation of the epithelium of the embryonic gut, microvilli are formed in some insects as previously described (Phasmida, Kadiri,

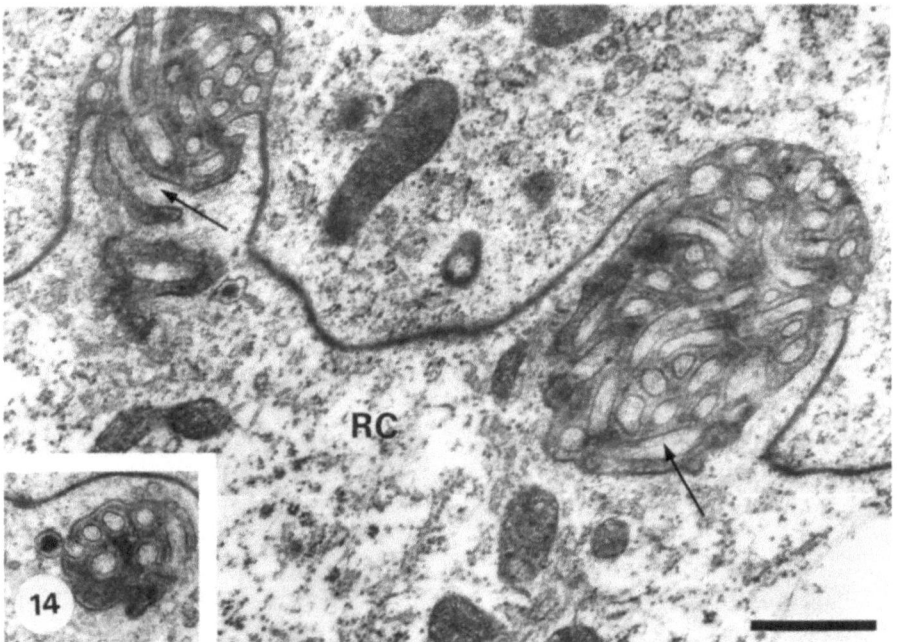

Figure 14. Development of the microvilli in a regenerative cell (RC) of *Formica polyctena*. The extracellular space around the microvilli is filled with an electron-opaque material, which arises from secretory granules (inset). The newly formed microvilli contain the usual fibrous material (arrows). See also Figure 15 for the dense extracellular material. Courtesy of Dr. A. Y. Jeantet. (Bar = 0.5 μm.)

1979; Orthoptera, Martoja, 1981). In Diptera, microvilli arise in a different way. During the embryonic development of the midgut cells, dense vesicles occur prior to microvilli formation, and these vesicles provide membrane and coating material for the modling of the microvilli. First, some of the vesicles fuse with the apical cell membrane. This results in an increase of the cell surface, part of which is coated with filamentous material derived from other dense vesicles. Short microvilli appear, which then stiffen and elongate. Tubular elements, located in the apical cytoplasm, could accomplish the erection of the microvilli perpendicular to the cell surface (van der Starre *et al.*, 1972).

5.4. Peculiar Aspects of Metamorphosis in Some Holometabola

In some Holometabola, there is a true pupal epithelium in the gut between those of the larva and the adult (Figure 15). The pupal epithelium (also called the *reticulated tissue*) and the imaginal epithelium result from the differentiation of regenerative cells. The reticulated tissue appears before the imaginal epithelium and forms a compact layer between the larval and imaginal epithelia. Both the larval epithelium and the reticulated tissue undergo histolysis, whereas the imaginal epithelium differentiates. So, there is a period during

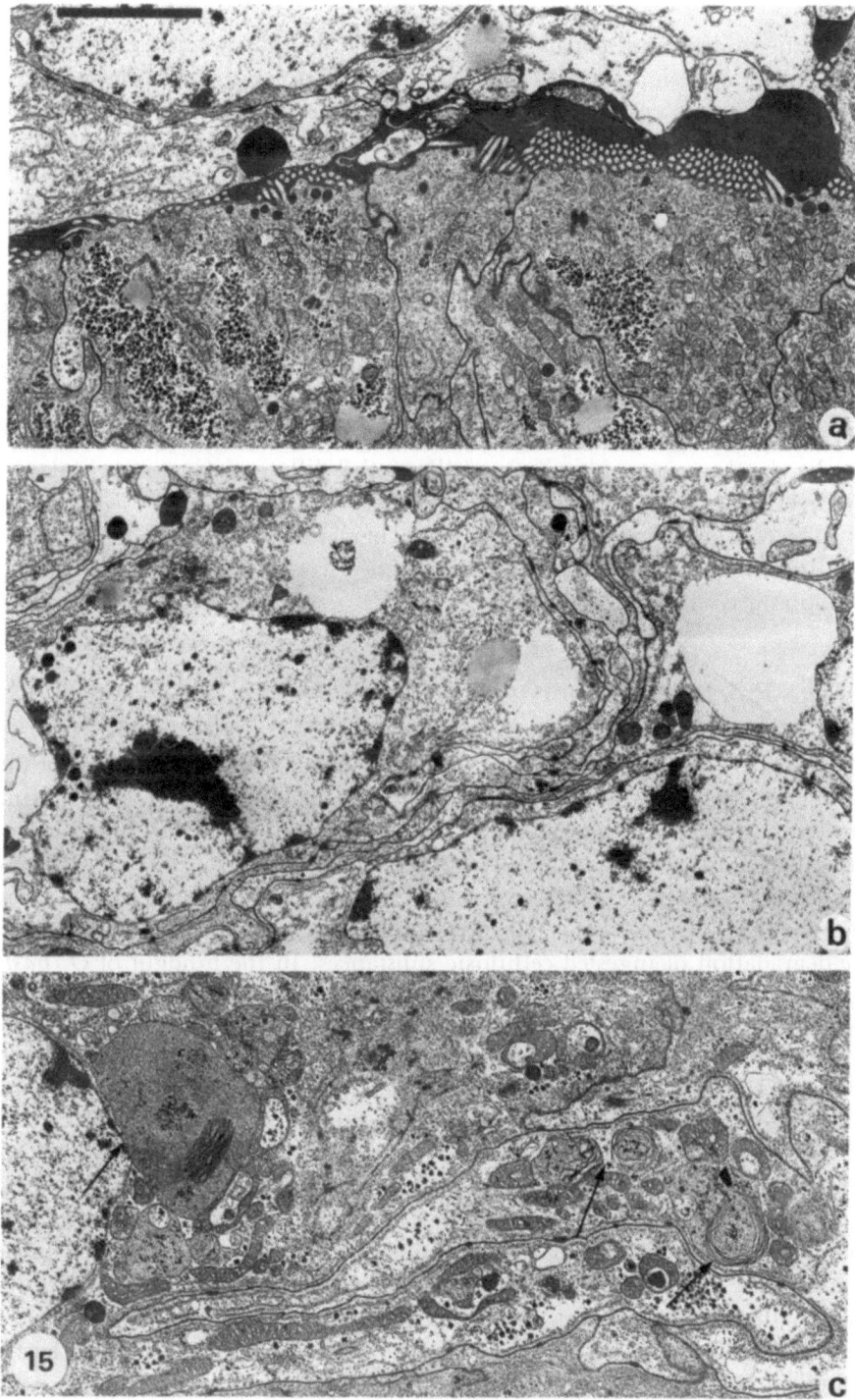

Figure 15. Metamorphosis of the midgut epithelium of *Aeshna cyaneae*. The wall consists of three epithelia: new epithelium (a), pupal tissue (b), and old epithelium (c). The extracellular material is abundant around the differentiating microvilli of the new epithelium. In the reticulated tissue, the hyaloplasm is pale and vacuoles are numerous. Many myelin figures (arrows) can be seen in the old epithelium. Courtesy of Dr. J. C. Andriès. (Bar = 3 μm.)

which the midgut wall consists of three epithelia, and several kinds of junctions ensure adhesion between the cells of the same epithelium and between those of different epithelia (Andriès, 1979). The cellular contacts between larval or imaginal cells involve smooth septate junctions, interrupted by small desmosomes, and eventually gap junctions. Cell-to-cell contacts in the reticulated tissue consist only in desmosomes with poorly defined filaments. Small desmosomes bind the cells of the reticulated tissue to those of the larval epithelium. Contact between the reticulated tissue and the imaginal epithelium are made by desmosomes and by projections of the reticulated tissue into the imaginal cells. In these regions, the intracellular spaces are wide (20–30 nm) and are spanned by electron-dense bridges. Andriès suggests that the anchoring projections delay the casting of the reticulated tissue into the lumen and that they are essential to the formation of the intercellular space which is required for the development of the microvilli of the differentiating imaginal cell (Figure 15a).

The cells of the reticulated tissues are characterized by numerous vacuoles and lipid droplets. During degeneration, the electron density of the ground hyaloplasm is not increased (Figure 15b).

6. The Proctodeal Fermentation Chambers

In many insects feeding on cellulose-rich diets, a peculiar segment of the hindgut shelters a large population of protozoans and/or bacteria. The fine structure of such segments was first studied in termites (Noirot and Noirot-Timothée, 1967), then in lamellicorn larvae (Bayon, 1971, 1981), and finally in cockroaches (Bignell, 1980). In *P. americana*, the ileum (termed *colon* by Bignell, 1980) is a saclike organ in which the microorganisms participate in the degradation of ingested cellulosic substances. In the hindgut of *O. nasicornis*, there is a dilated pouch which teems with many kinds of bacteria and behaves as a fermentation chamber in which bacteria break down cellulose. In both cases, it has been established that the wall is permeable to the breakdown products *in vitro*, and ultrastructural examination has shown many features characteristic of an adsorbing epithelium.

In *Periplaneta*, only one kind of cell is present, and the epithelium shows similarities with the rectal pads. Both types of cells have extensive folding of the apical plasma membrane, associated with mitochondria and an internal coating of 12-nm-diameter particles. Many apical infoldings are dilated at the tip, and these dilatations make a significant contribution to extracellular space within the epithelium. At their lateral margin, adjacent cells are linked by septate junctions. On the basal side of each junction is a narrow extracellular space running to the basal zone and thus forming a channel separating the cells. At the basal end of the channels, mitochondria and membranes sometimes form a structure which resembles a plasmalemma–mitochondria complex. The surface densities of apical plasma membranes are not significantly different in the posterior midgut and in the ileum. Similarly, the volume densities of mitochondria are equal in the ileal epithelium and rectal pads, but it must be noted that in the pad cells the mitochondria are almost exclusively confined to the cytoplasm bordering the plasma membranes. Overlying the epithelium is a lamellae cuticle lacking a sclerotized layer. The epicuticle supports epicuticular filaments, but in contrast with rectal pads, it is thinned at regular interval to form domelike protrusions into the lumen instead of the typical depressions in the surface.

In *Oryctes*, the epithelium of the pouch consists of two cellular types. Cells are linked at their apical borders, but they are separated from each other at their basal side by deep invaginations of the connective layer. Cubic cells are very similar to the ileal cells of *Periplaneta*. However, the epicuticle shows true depressions similar to the rectal ones (Figures 16a, d, e). Flat cells have a low mitochondria content, and their apical plasma membranes are devoid of any particulate coating. In definite areas, their cuticle exhibits various differentiations: spines or protrusions. Some areas bear branched spines which have a sensory function (Figures 16b, c).

So, an ultrastructural examination suggests that the ileal cells of *Periplaneta* and the cubic cells of *Oryctes* function in absorption. According to Bignell (1980), the ileal cells play a minor role in ionic regulation, but function in active transport. In both species, the cells presumably absorb organic solutes, as well components entering the hindgut with the primary urine, either as organic materials that are not absorbed in the midgut or as the products of microbial fermentations. In *Oryctes*, the invaginations of the connective layer increase the elasticity of the pouch. Some spines or cuticular protrusions keep the gut material and the bacterial flora near the wall, while others would help with the transit. Sensory spines could control the admission of food to the rectum.

An earlier study on the fermentation chamber of *Cephalotermes rectangularis* (Noirot and Noirot-Timothée, 1967) clearly demonstrates that the epithelial cells resemble the rectal pad cells and have an absorptive function.

7. Filter Chambers and Related Organs

7.1. Midgut Filter Chambers

Homopterans which feed on watery juices possess an enlargement of the midgut, named a *filter chamber*. Ultrastructural studies have been performed on the filter chambers of bugs belonging to the Cicadidae, Cercopidae, and Jassidae (Gouranton, 1968c; Munk, 1968; Foldi, 1973; Marshall and Cheung, 1974; Lindsay and Marshall, 1980). The results reveal similarities in ultrastructure which point to water and ion transport as the function of the cells.

The filter chamber comprises the saclike anterior segment of the midgut, around which are looped a posterior segment of the midgut (the internal midgut) and the terminal ends of the Malpighian tubules (the internal Malpighian

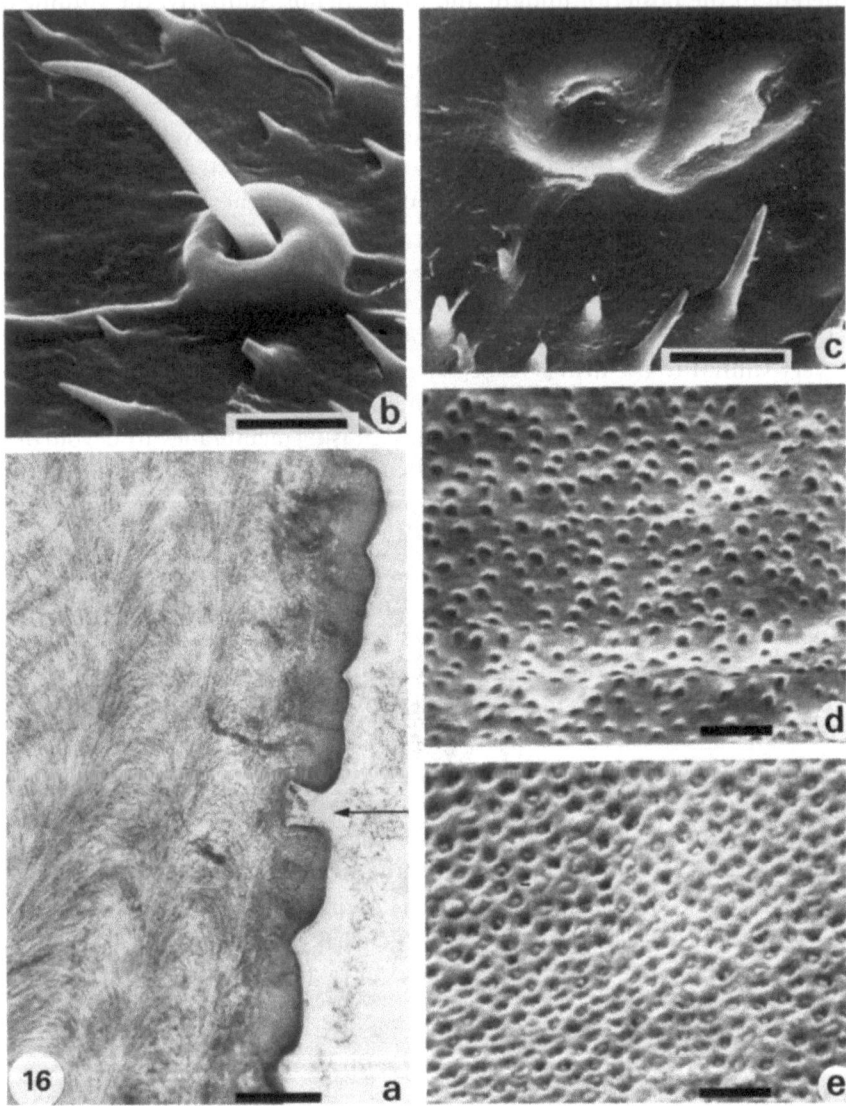

Figure 16. The proctodeal fermentation chamber of *Oryctes nasicornis*. (a) TEM; (b-e) SEM. (a) Epicuticular depression (arrow). (b) Articulated hair. (c) Campaniform sensilla. (d) and (e) allow a comparison of the density of epicuticular depressions in the fermentation chamber (d) and in the rectum (e). Courtesy of Dr. C. Bayon. [Bars = 0.2 μm (a), 10 μm (b, c), and 1 μm (d, e).]

tubules) (Figure 17a). The epithelium of the anterior part of the midgut consists of large cuboidal cells, flattened elongate cells and cells which are intermediate between these two types. The cuboidal cells are secretory and produce granules which contain mucoproteins. In the flattened cells, passage of water takes place from the apex to the basal part. The cells exhibit a border of microvilli, deep infoldings of the basal plasma membrane which reach the apical

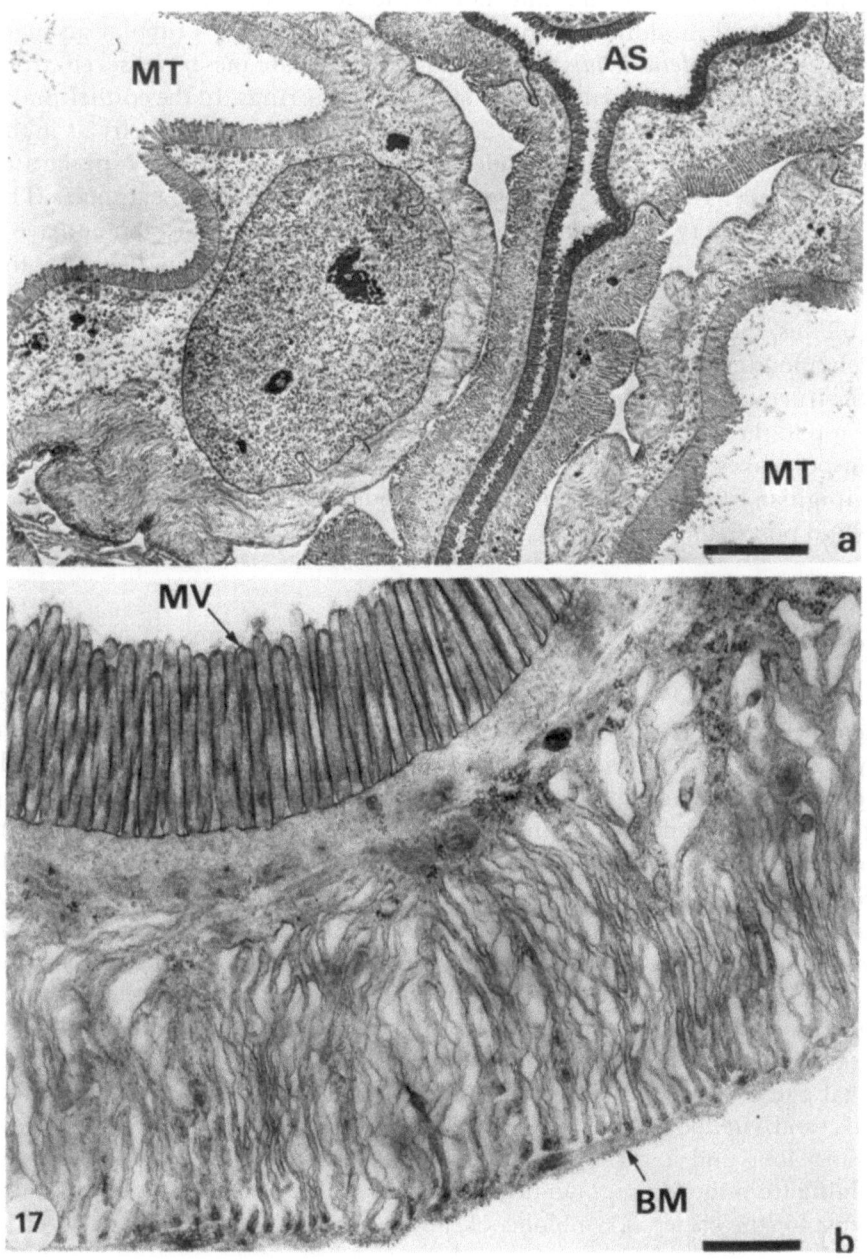

Figure 17. The filter chamber of *Cicadella viridis*. (a) Two Malpighian tubules (MT) are associated with an anterior segment (AS) of the midgut. Note that the wall of this midgut segment resembles that of the Malpighian tubules. (b) Enlarged view of the wall of the midgut, showing the deep infoldings of the plasma membrane (BM) and the border of microvilli (MV). Courtesy of Professor J. Gouranton. [Bars = 5 μm (a) and 0.5 μm (b).]

cytoplasm (Figure 17b), and mitochondria located under the border of microvilli. The basal infoldings of the plasma membrane show a tubular arrangement, and in *Cicadella viridis* the outer part of the membranes possesses electron-dense areas, 6 nm in thickness, which form periodic rings. In the epithelium of the internal midgut, water is absorbed from the basal part of the cells, as in the internal Malpighian tubules. As one would expect, there are many similarities between the cells of this part of the gut and those of the Malpighian tubules. The microvilli are very high and the infoldings of the plasma membrane have a lamellar shape. However, in *Cicadella* cytochemical tests show that only the microvilli of the gut cells are covered with a coating rich in polysaccharides.

Thus, the midgut cells which are inserted in the filter complex are extensively modified. The flat cells are adapted for water and ion transport, and ultrastructural differences appear between cells where transport occurs from the top to the bottom (in the anterior part of the midgut) and those where the transport takes place in the opposite direction (internal midgut and internal Malpighian tubules). The cuboidal cells produce a secretion which is believed to bind potassium, the major cation of xylem fluid on which the insects feed.

7.2. The Composite Segment of Termite Intestines

In the digestive system of many species of Termitidae, a composite segment occurs which lies at the boundary of the midgut and the proctodeum and includes a strip of midgut cells and one of proctodeal cells. The middle segments of the Malpighian tubules constitute a ball which clings to the strip of midgut cells, and a thin connective layer encloses this ball and the composite segment. Therefore, the complex appears to be isolated from the hemocele (Noirot *et al.*, 1967). Whereas the other parts of the midgut contain typical columnar cells, those of the midgut cell strip show an uncommon specialization. The border of microvilli is short, the RER is scarce, and no secretion granules can be seen. The basal part of the cells contains very numerous mitochondria, which are often arranged in plasmalemma-mitochondria complexes, and coated vesicles are also abundant. Thus, these cells resemble both rectal pad cells and pinocytotically active cells such as hemocytes, pericardial cells, and vitellogenic oocytes. Noirot *et al.*, suggest that these midgut cells absorb ions and a variety of organic substances, including purine wastes, coming from the Malpighian tubules. These compounds would serve as nutriments for the bacterial symbionts of the intestine.

8. The Salivary Glands

The salivary glands vary a great deal in their morphologies. In higher Diptera, simple elongated tubes join to form a common duct. In many other orders, the glands appear as grapelike clusters composed of acini, which make up the bulk of the glands and connect to the salivary ducts by a complex network of intercalary ducts. The wall consists of an epithelium which is lined by a cuticle layer (in the ducts and sometimes in the other parts of the glands) and rests on a basal membrane. Very little attention has been paid to the organization of this extracellular material. It has been shown that tracheoles are embedded in the basal membrane, but the occurrence of nerves seems to differ according to the species studied (Oschman and Berridge, 1970).

8.1. The Tubular Glands of Adult Flies

In adult *C. erythrocephala*, the salivary glands are a pair of thin tubes that extend into the abdomen on either side of the gut and join in the anterior region of the thorax to form a common duct. Both have a single layer of epithelial cells with an extracellular border similar to cuticulin. Each tube consists of two parts: a long distal secretory segment and a short proximal reabsorptive segment. The cells of each segment exhibit characteristic ultrastructural morphologies (Oschman and Berridge, 1970). The common duct that conveys the saliva to the mouth parts will not be considered here.

The cells of the secretory part are characterized by the presence of deep involutions of the apical plasma membrane that form a branching system of channels, named *secretory canaliculi*. At the apical surface, the canaliculi open into the lumen and often extend to within less than 1 μm of the basal plasma membrane of the cell. The canaliculi are lined with closely packed leaflets, 60-80 nm thick, in which the inner surface of the plasma membrane is coated with particles 15 nm in diameter. By the side of canaliculi, the apical surface of the cell is also folded into leaflets similar to those lining the canaliculi. Infoldings also occur in the basal plasma surface, but they are not as striking as those of the cellular apex. There are regions where mitochondria are closely associated with the infolded membranes, and a dense material occludes the extracellular channels. Mitochondria are very numerous and are found throughout the cytoplasm, although those of the basal region are associated with plasma membranes. They are of an unusual type (with cristae containing circular perforations or fenestrae about 45 nm in diameter) and are quite similar to the mitochondria of flight muscles. In addition to an abundant RER, free ribosomes, and Golgi bodies, the cells contain secretory granules with a pale content. However, the secretory activity in the epithelium seems to be asynchronous, since some cells lack granules, while others contain many.

The cells of the reabsorptive region are flattened. The apical plasma membrane is folded and coated with particles, but canaliculi are not present. The infolded plasma membrane is asymmetric. Mitochondria have conventional cristae. The cells lack secretory granules.

Cell junctions consist of pleated septate junctions in the reabsorptive region. The secretory cells are linked by three types of junctions: the one closest to the lumen is a desmosome, interior to this is a pleated septate junction, and finally plaques of gap junctions occur.

Considering the correlations known between structure and physiology, the cells of the secretory segment exhibit two main functions. They elaborate granules which are believed to contain amylase, and they are specialized for ion transport (Na and especially K). Thus, the secretory region produces a potassium-rich primary saliva. The cells of the reabsorptive segment are specialized for transport alone; they reabsorb potassium from the primary saliva, and this results in the formation of a dilute saliva.

8.2. The Tubular Salivary Glands of Adult Lepidoptera

The salivary glands of the moth *Manduca sexta* produce a dilute saliva containing invertase. In the adults, each one of the paired glands consists of a single layer of epithelial cells coated with a cuticular lining. The two glands unite to form a common duct which appears to have no function other than conducting saliva. As in the higher Diptera, ultrastructural investigations show a satisfactory correlation between fine structure and function (Leslie and Robertson, 1973).

Each gland can be divided into five morphologically distinct segments. The distal segment is a protein-secreting region. Its cells are characterized by an extensive RER, numerous Golgi bodies, and large secretion-filled vesicles containing a pale product. In contrast to *C. erythrocephala*, the secretory cells lack the ultrastructural characteristics of transport cells. The neighboring cells are linked by four types of junctions (desmosomes, tight junctions, gap junctions, and pleated septate junctions). The next segment is the "fluid-secreting region." The outstanding feature of its cells is the extensive deep mitochondria-lined infolds of the apical plasma membrane which extend almost to the basal membrane. In contrast, basal infolds are short. The third and fourth segments (the thin duct and the bulbous duct) constitute the reabsorptive region. Here the cells exhibit leaflets or fingerlike projections of the apical plasma membranes.

Thus, whereas the secretory cells of *Calliphora* perform two main functions (protein and fluid secretion), a separation of these functions occurs in the tubular gland of the moth. The secretory cells produce the enzyme invertase while the cells of the second segment secrete fluid. Therefore, both segments are concerned with the production of an enzyme-containing primary saliva, isoosmotic with the hemolymph. The saliva is modified in the remaining regions of the gland, by ionic reabsorption, to yield a dilute saliva.

8.3. The Tubular Salivary Glands of Mosquitos

The salivary glands of adult *A. aegypti* (Janzen and Wright, 1971) consist of several tubes, or lobes, that are made, like those of higher dipterans, of several segments. In the lateral lobes, an intermediate portion, consisting of transport cells, is located between two glandular segments. In the median lobes, a distal glandular portion is followed by a segment containing transport cells. The nature of the secretion is unknown, and it is likely that the nonglandular portions of the tube function to hydrate the secretory material.

8.4. The Acinar Glands of Dictyoptera and Orthoptera

Various aspects of the fine structure of the acinar salivary glands have been described. For Dictyoptera, there are studies by Kessel and Beams (1963) on *P. americana*, by Bland and House (1971) on *Nauphoeta cinerea*, by Dailey and Crang (1977) on *Blaberus discoidalis*, and by Dailey and Crang (1978a,b) on *Gromphadorhina portentosa*. In the Orthoptera, there are studies by Kendall (1969) on *Schistocerca gregaria*, by Lauverjat (1972, 1973) on *L. migratoria*, and by Anstee (1975) on *Homorochoryphus nitidulus*.

The salivary glands contain three basic cell types: central cells (zymogen-secreting cells), parietal cells (peripheral cells), and duct cells. Central and peripheral cells are located exclusively in the acini. They generally constitute a simple layer, though in *Homorochoryphus*, the parietal cells are peripherally situated and surround the central cells. The duct cells form a branched duct system with the smaller branches terminating within the granular acini. Thus, a few duct cells are included in the acini and are sometimes considered as acinar cells. Interstitial cells, observed in *Locusta*, have not been studied.

8.4.1. Central Cells

The central cells give the salivary acinus its characteristic spongy appearance in light microscopy, a feature due to the presence in the cytoplasm of large amounts of secretion (Figure 18a). Whatever the species, the central cell possesses a border of microvilli, a moderately infolded basal plasma membrane, and few mitochondria. The most prominent organelle during early secretory activity in the RER, often in both a lamellar and a vesicular form. Some ultrastructural aspects, and particularly that of the secretory products, show quite important species-specific differences. Golgi bodies are very numerous in *Locusta*, but not in *Gromphadorhina*, and they have not been observed in *Schistocerca* or *Blaberus*. So even in insects of the same order (locusts), the secretory vesicles are believed to arise directly from the ER (*Schistocerca*) or from Golgi material (*Locusta*). In *Gromphadorhina*, a variety of lamellar bodies are abundant, and spherical organelles, possibly representing microbodies, are observed in a few cells. In *Locusta*, a correlation between the ultrastructural organization of the secretory granule and its chemical composition has been established. Each granule consists of an electron-opaque part which is rich in protein and of a paler part which is fibrous, granular, or microtubular in nature and rich in acidic mucopolysaccharides (Figure 18b). The release of secretory vesicles by exocytosis has been observed in *Locusta*, the two parts of the granules separating immediately after the extrusion (Figure 19c).

8.4.2. Parietal Cells

The parietal cells, inserted between the central cells, contain an intracellular ductule which is lined with numerous microvilli (Figures 19a, b). It

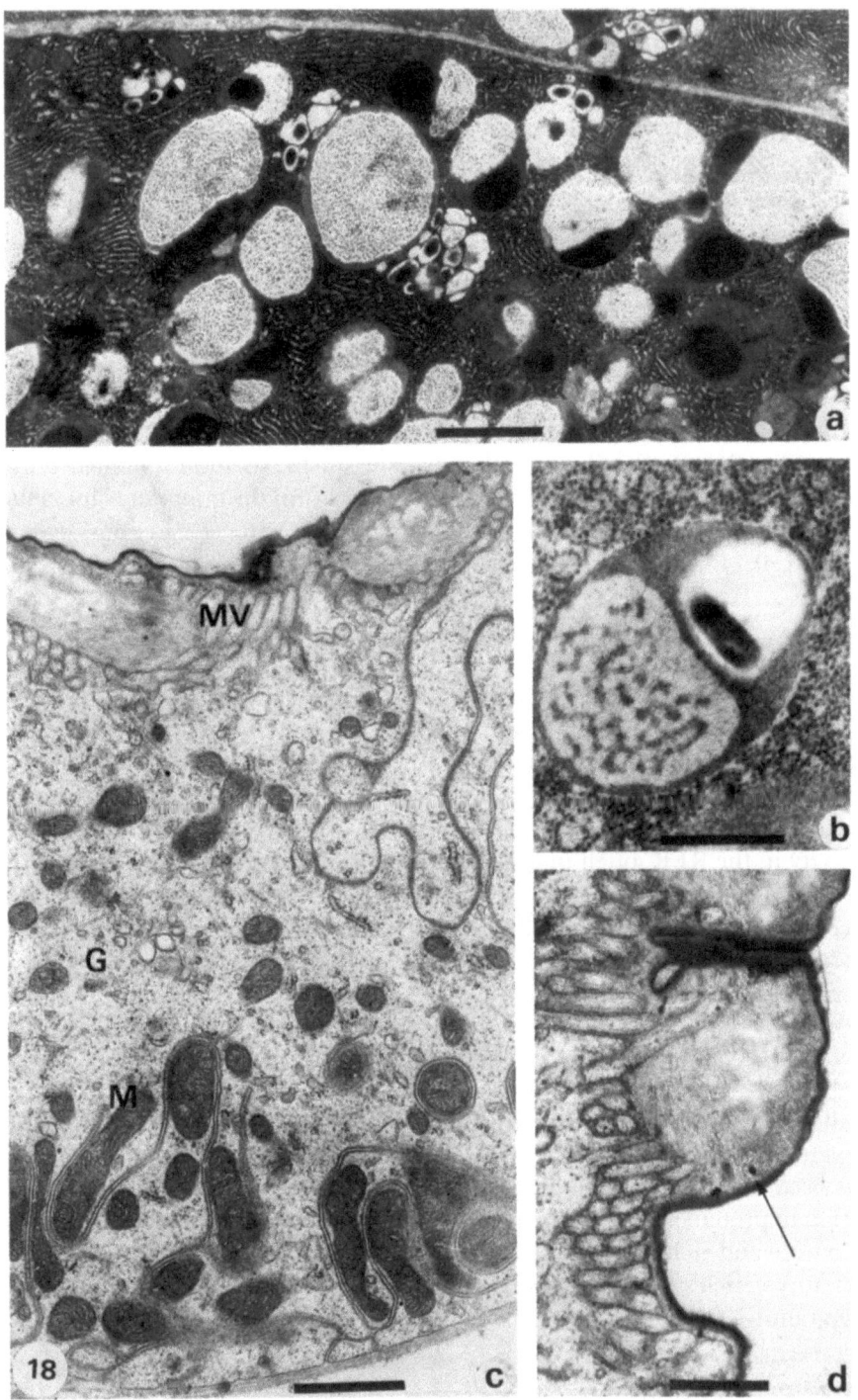

Figure 18. The salivary gland of *Locusta migratoria*. (a) A part of a central cell filled with RER and secretory granules. (b) Secretory granule of a central cell showing the electron-opaque protein part and the fibrous mucopolysaccharide part. (c) A cell from the duct. G, Golgi material; M, mitochondrion; MV, microvilli. (d) Cuticular lining of a duct cell with a depression and epicuticular filaments (arrow). Courtesy of Dr. S. Lauverjat. [Bars = 4 μm (a), 2 μm (b, c), and 1 μm (d).]

leaves the cell at its top and becomes continuous with the intercellular canaliculus which opens into the extracellular duct. The ductule can be expanded in the cytoplasm, forming a small goblet chamber, or branched into smaller ductules. In *Schistocerca*, each cell contains several ductules which are not confluent with the ducts, but open individually in the extracellular space (see Lauverjat, 1973).

The border of microvilli is often developed to such a degree that it obscures the ductule lumen. The basal portion of the plasma membrane is considerably infolded, so that invaginations may extend long distances into the cytoplasm. Numerous mitochondria are present throughout the cytoplasm, and concentrations of microtubules can often be seen in the immediate vicinity of the intracellular ductule. Along the basal plasma membrane of *Locusta*, coated vesicles arise by micropinocytosis and gather into microvesicular bodies.

The cells of Dictyoptera are devoid of secretory vesicles; RER and Golgi bodies are rare. In contrast, orthopteran cells possess a well-developed RER, and in *Locusta* cisternae produce voluminous vacuoles which aggregate under the apical plasma membrane (Figures 19a, b).

8.4.3. Duct Cells

Unlike the central and parietal cells, the duct cells possess a cuticular lining. It has been clearly established that the ducts of *Nauphoeta* contain secretory cells and nonsecretory cells, whereas those of orthopterans are formed by nonsecretory cells alone.

The secretory duct cells of *Nauphoeta* only occur in close proximity to the acinus, some of them being even included in the acini. The basal plasma membrane shows no pronounced infoldings although the apical membrane does invaginate under the cuticular layer. Numerous mitochondria tend to be associated with the apical rather than with the basal plasma membrane. Although the ER in mature cells is not well developed, it contains large droplets of dense granular material. These reach diameters of 5 μm, and a mucous component containing sialoglycans and a tryptophan-rich component have been identified. This secretory material appears to dissolve immediately upon its release.

The nonsecretory cells occupy the major portion (*Nauphoeta*) or the totality (Orthoptera) (Figure 18c) of the length of the ducts. The apical plasma membrane is highly invaginated underneath the cuticle, and according to Bland and House (1971) its infoldings form a weblike structure containing vacuolar spaces. In *Nauphoeta*, the cytoplasmic sides of the membrane about these invaginations are coated with numerous, fairly regularly spaced, electron-dense particles, similar to those of many transport cells of insects. A small number of pinocytotic vesicles form at the base of the microvilli of cockroaches. In *Locusta*, the cuticular lining shows deep depressions and epicuticular filaments (Figure 19d). There is also a great deal of infolding of the basal plasma membrane, and numerous mitochondria are found between them. The RER not prominent and Golgi bodies are rare. Aggregates of true microtubules are often seen directly below microvilli.

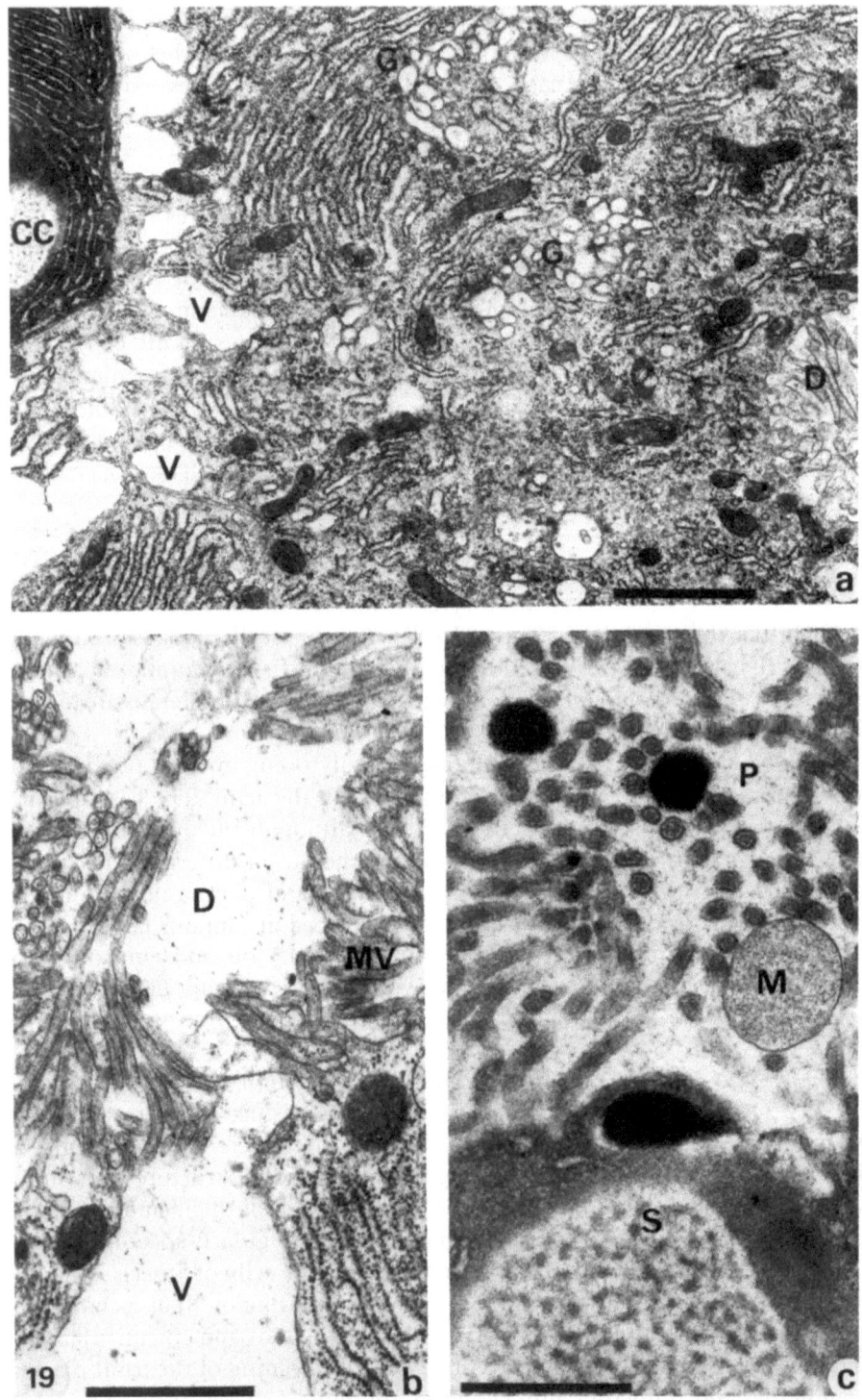

In the acini, pleated septate junctions alone have been observed. They occur in the vicinity of the lumen, between central cells and their neighbors (the central cells, parietal cells, and eventually secretory duct cells), and, in *Nauphoeta*, between parietal cells and secretory duct cells. In the epithelium of the extraacinar ducts, there coexist under the border of microvilli a desmosome and underneath it a pleated septate junction. No gap junctions have been observed.

Acidic mucopolysaccharides are synthesized both in the central cells and in the secretory duct cells, but the role of the mucus remains unclear. With regard to the nonsecretory duct cells, it is obvious that they are transport cells that modify, like the reabsorptive cells of *Drosophila*, the ionic composition of the primary saliva.

The origin of enzymatic secretion and therefore the role of central and parietal cells are a matter of debate. Considering that the secretion material of the central cells of *Periplaneta* is protein in nature, Kessel and Beams (1963) suggested that proteases are secreted by the central cells, while parietal cells produce amylase. According to Lauverjat (1972), the protein part in the granules of the central cells of *Locusta* is merely the carrier of the saliva enzymes, and Bland and House (1971) propose that the central cells of *Nauphoeta* are the source of amylase and possibly the other enzymes.

The parietal cells are concerned with the transport of ions and water from the hemolymph to the lumen of the gland (Bland and House, 1971). Anstee (1975) maintains that they produce a potassium-rich fluid into which the central cells, and perhaps the parietal cells themselves, pass their secretions. Kendall (1969) suggests that in *Schistocerca*, the parietal cells elaborate a primary secretion which is transferred to the neighboring central cells.

Since the origin of the enzymes is not yet established, the term *zymogenic cell* must be avoided. It is likely that the parietal cells are generally transport cells, and that in addition they produce secretory vesicles in some species. The content of these resembles the amylase-containing granules of *Drosophila*.

8.5. The Sac-Shaped and Tubular Salivary Glands of Lower Diptera

The paired salivary glands of *Chironomus thummi* larvae are flattened sacs containing about 30 large cells and a few thin cells, arranged in a single layer around a central lumen. A duct carries the secretion to the mouth. Though the sac consists of several cell types, none of them appears to be absorptive. The "large polytenic cells," the "special lobe cells," and the "thin

---

Figure 19. Salivary gland of *Locusta migratoria*. (a) Parietal cell, showing a part of the ductule (D), a well-developed RER, Golgi bodies (G), and numerous vacuoles (V). CC, a part of a central cell. (b) Contact between the ductule and a vacuole. MV, microvilli. (c) Extrusion of granules from a central cell. The secretory granule (S) located in the bottom of the micrograph is about to be ejected. In the lumen, the two parts of each granule have separated. M, mucopolysaccharide part; P, protein part. Courtesy of Dr. S. Lauverjat. [Bars = 5 μm (a), 2 μm (b), and 1 μm (c).]

layer cells" display distinct specializations related to the production of the secretion, and each type provides characteristic compounds of the saliva. The cells of the duct have a smooth plasma membrane, and coated vesicles are often seen in the apical plasma membrane (Kloetzel and Laufer, 1969).

Likewise, secretory cells alone have been observed and described in the tubular glands of *Sciara coprophila* (Phillips and Swift, 1965). They secrete three morphologically distinct types of granules: the lucent granules may contain enzymes, while the ellipsoid and spherical ones presumably provide the slime coating of the larva and the mucoprotein of the pupal cocoon.

Thus, the cytological studies reveal a wide variety of secretory products in the saliva of the larvae of lower Diptera, but their role is uncertain, and the different processes involved in the formation of the final saliva remain to be discovered.

9. The Malpighian Tubules

The Malpighian tubules may be similar in structure throughout their length or divided into morphologically distinct segments. Two types of tubules may coexist in the same species. In others, a close association occurs between the tubules and the midgut (see Section 7) or the rectum (cryptonephridial arrangement).

9.1. The General Organization of the Malpighian Tubules

The tubules consist of an epithelium which rests on a connective tissue sheath containing tracheoles. Circular muscular fibers are frequent, although they may be rare or absent (Crowder and Shankland, 1972; Eichelberg and Wessing, 1975). The basement membrane is composed of granular and fibrillar layers. The epithelial cells are joined by septate, gap, and sometimes scalariform junctions. Whatever the species, a cellular type that we term the *principal Malpighian cell* exhibits similar structural features associated with fluid secretion as well as storage processes. Other kinds of cells are present together with the principal ones. Many papers are concerned with the fine structure of the epithelial cells. The most important of these are listed in Table 1.

9.2. Types of Cells

9.2.1. The Principal Malpighian Cell

This cell (Figure 20) exhibits a structural polarity. The basal region shows numerous infoldings of the plasma membrane which are linked to the basement membrane by hemidesmosomes. These infoldings generate a labyrinth

Table 1. Occurrence of the Different Types of Cells and of Concretions in the Malpighian Tubules[a]

	Principal cells			Other types of cells		
	Localization of the cells studied	Concretions	Luminal concretions	Name	Localization	References
Dictyoptera						
Periplaneta americana	Middle seg.	+ (m)	—	Stellate c.	Middle seg.	Wall et al. (1975)
				Proximal c.	Proximal seg.	
Blatella germanica	Distal seg.	+ (m)	—	Mucous c.	Distal seg.	Ballan-Dufrançais (1974, 1975)
				Proximal c.	Proximal seg.	
Cheleutoptera						
Carausius morosus	Inferior tub.	—		Mucous c.	Inferior tub.	Taylor (1971a,b)
Orthoptera						
Locusta migratoria	Main seg.	+ (m)		Mucous c.	Main seg.	Ballan-Dufrançais and Martoja (1971); Bell and Anstee (1977); Charnley (1982)
Gryllus bimaculatus	Main seg.	+ (m,u)	+ (u)	Mucous c.	Main seg.	Berkaloff (1960); Ballan-Dufrançais and Martoja (1971)
Gryllotalpa gryllotalpa	Yellow tub.	+ (m)	+ (m)	Reabsorb. c.?	Seg. 2	Lhonoré (1973)
	White tub.	—	+ (u)	Mucous c.	Seg. 3	
Heteroptera						
Rhodnius prolixus	Upper seg.	+	—	Reabsorb. c.	Lower seg.	Wigglesworth and Salpeter (1962)
Cenocorixa bifida	Middle (3) seg.	+	+ (u)	Undeterm. c.		Jarial and Scudder (1970)
Hymenoptera						
Formica polyctena	Proximal seg.	+ (m)	+ (u)			Jeantet (1981)
Lepidoptera						
Pieris brassicae (adult)	Distal seg.	—	+ (m)	Reabsorb. c.?	Proximal seg.	Lhonoré (1976)
			+ (u)			

(*Continued*)

Table 1. (Continued)

	Principal cells			Other types of cells		References
	Localization of the cells studied	Concretions	Luminal concretions	Name	Localization	
Calpodes ethlius	Yellow seg.	+	+			Ryerse (1979)
Bombyx mori						Teigler and Arnott (1972); Waku (1974)
Larva	Coelomic seg.	−	+ (m)			
Pupa	Coelomic seg.	+ (m)	−			
Young adult	Coelomic seg.	−				
Diptera						
Calliphora erythrocephala	Posterior tub.	+ (m,u)		Stellate c.	Posterior tub.	Berridge and Oschman (1969); Martoja and Seureau (1972)
Musca domestica	Main seg.	+ (m)	+ (m)	Stellate c.	Main seg.	Sohal (1974); Sohal et al. (1976)
				Secretory c.	Main seg.	
Drosophila melanogaster	Distal seg.	−	+ (m)	Reabsorb. c.?	Proximal seg.	Eichelberg and Wessing (1975)
Arachnocampa luminosa	Seg. 2	+ (m,u?)		Reabsorb. c.?	Seg. 1	Green (1979)

[a]Abbreviations: c., cells; m, mineral components; reabsorb. c., reabsorbing cells; seg., segment; tub., tubule; u, uric acid or urate; undeterm., undetermined.

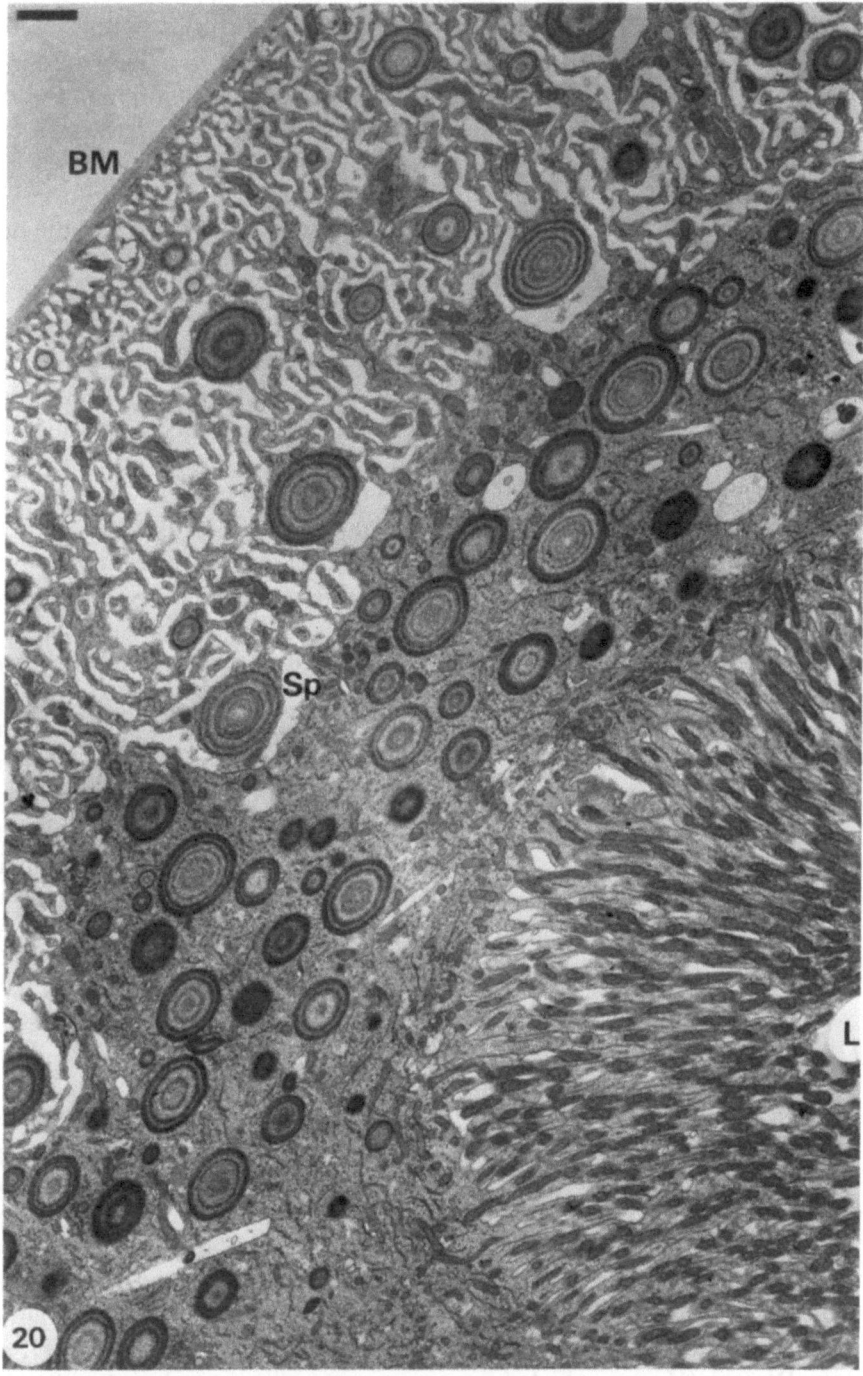

Figure 20. The principal cell in the Malpighian tubules of *Calpodes ethlius* (larva, yellow segment). Note the extension of the basal labyrinth, the abundance of spherocrystals (Sp), the tightly packed microvilli. BM, basement membrane; L, lumen. Courtesy of Dr. J. S. Ryerse. (Bar = 1 μm.)

which extends from one-quarter (*P. americana*) to three-quarters (*C. erythrocephala*) the height of the cell. Mitochondria are numerous, while microtubules, ribosomes, and ER are also present. Pinocytotic vesicles may occur along the membrane. The apical region possesses long (1-10 μm) and tightly packed microvilli with numerous (*C. morosus*) or scarce (*Blatella germanica*) mitochondria inside them. Extensions of the smooth-surfaced, tubular ER (canaliculi) are present in microvilli that lack mitochondria. Pinocytotic vesicles are frequent, but no cell coat can be observed. The midpart of the cell contains the nucleus, RER, few Golgi bodies, mitochondria, lysosomes. Mineral spherocrystals are frequently abundant (Figures 20, 21a, b, c, 28a). Like the midgut concretions, they are generally formed in the RER, sometimes in the nuclear envelope (*P. americana, F. polyctena*), or in the lysosomes (*M. domestica*) (Figure 21e). They are always made of calcium and magnesium phosphates. Other elements (Na, K, Mn, Fe, Zn, Cu) or compounds (such as uric acid) occur, according to the species and the stage. Concretions may store unusual elements added to the diet (Jeantet *et al.*, 1977). Moreover, the principal cells sometimes sequester organic components such as granules of ommochrome containing calcium phosphates (Orthoptera, Diptera), or flavins. Flavins have been described in Dictyoptera: in *Blatella*, they are in the form of needles located in mitochondria or lysosomes (Figures 21f, g). Such needles can be recognized in micrographs of the Malpighian tubules cells of *Periplaneta, Calpodes ethlius*, and *Calliphora*.

Minor variations in fine structure are obvious when considering different species. They mostly concern the development of microvilli, the number of mitochondria, and the occurrence of concretions. In contrast, important developmental changes influence the tubules of holometabolous species during the pupal stage (*B. mori, C. ethlius*). In *Calpodes*, when fluid secretion is switched off, the cells show a disappearance of the microvilli and a degeneration of mitochondria, basal infoldings are reduced, and the basement membrane doubles in thickness (Ryerse, 1979).

9.2.2. Other Cell Types

In Dictyoptera, Lepidoptera, Diptera, and probably Homoptera (see micrographs by Smith and Littau, 1960), stellate (= flat) cells are scattered among the principal ones. Stellate cells have a well-developed basal labyrinth and are devoid of apical mitochondria, intramicrovillar canaliculi, and concretions. A proximal segment, made of numerous cells which resemble the stellate ones, may also be present. In Dictyoptera and Orthoptera, the tubules contain mucous cells. Their appearance is characteristic of secretory cells, with large vesicles containing a granular material. Rather similar is the "type 3" cell of certain Diptera (*Musca*) which forms a proteinaceous secretion. Finally, in Homoptera, peculiar cells manufacture brochosomes which are bodies that contain lipids and proteins (Smith and Littau, 1960; Gouranton and Maillet, 1967).

9.3. Compounds Deposited in the Lumen of the Tubules

The cells of the Malpighian tubules release solid products such as mucous secretions, brochosomes, and perhaps spherocrystals into the lumen, sometimes during the whole life cycle, sometimes just at ecdysis. But in some cases, the lumina contain crystals even when the cells are devoid of concretions. For example, in *Bombyx*, the whole lumen is filled with calcium oxalate which is formed during the intermolt and is excreted at each molt. In *Pieris brassicae* and *D. melanogaster*, the distal segment contains mineral concretions. In the white tubules of *Gryllotalpa gryllotalpa* and in those of *Rhodnius prolixus* and *Formica* (Figure 21d), concretions made up of purinic and mineral components are present in the lumen of the proximal segment. Whatever the species, uric acid is the main component of the purinic crystals and is present as an Na or K salt.

With regard to correlations between structure and function, the principal cells are transport cells which are believed to be involved in the formation of the primary urine. Changes in ionic composition of the fluid are linked to a gradation of physiological properties in the tubules, and the proximal segment in some species functions in reabsorption. For stellate cells and proximal cells which resemble them, such a function has been suggested without being demonstrated. Moreover, the principal cells of many species concentrate mineral components. The degree to which minerals are stored is in inverse ratio to the complexity of the cytoplasmic membranous differentiations. This suggests that the precipitation of minerals is important when there is little fluid transport. Perhaps minerals can be transported directly from the hemolymph, but there is as yet no experimental support for this opinion, and they may be reabsorbed from the tubule fluid. Likewise, the extrusion of the spherocrystals remains to be proved. Considering the concretions which are born in the lumen, it is obvious that precipitation is initiated by a reabsorption of water and also by lowering the pH of the solvent (Stobbart and Shaw, 1964). Whatever the location and origin the concretions may have, it must be emphasized that the storage of excretory products is very important in many species: in *Blatella*, the stored wastes constitute up to 30% of the dry weight of the tubules (Ballan-Dufrançais, 1977). Mucous secretions serve as nucleation sites for the growth of luminal crystals (*Gryllotalpa*) while in other cases (*L. migratoria*) they may function as a lubricant. As for peculiar storage products (ommochromes, flavins, lipoproteins), their physiological meaning remains unknown.

10. The Proctodeum

The proctodeum generally consists of three segments (the ileum, colon, and rectum). In the rectum, reabsorptive processes have been assigned to the principal cells which are sometimes arranged in pads. The organization of the proctodeum may differ between larvae and adults of the same species. Concerning the cellular differentiations, the ileum may be similar to the rectum,

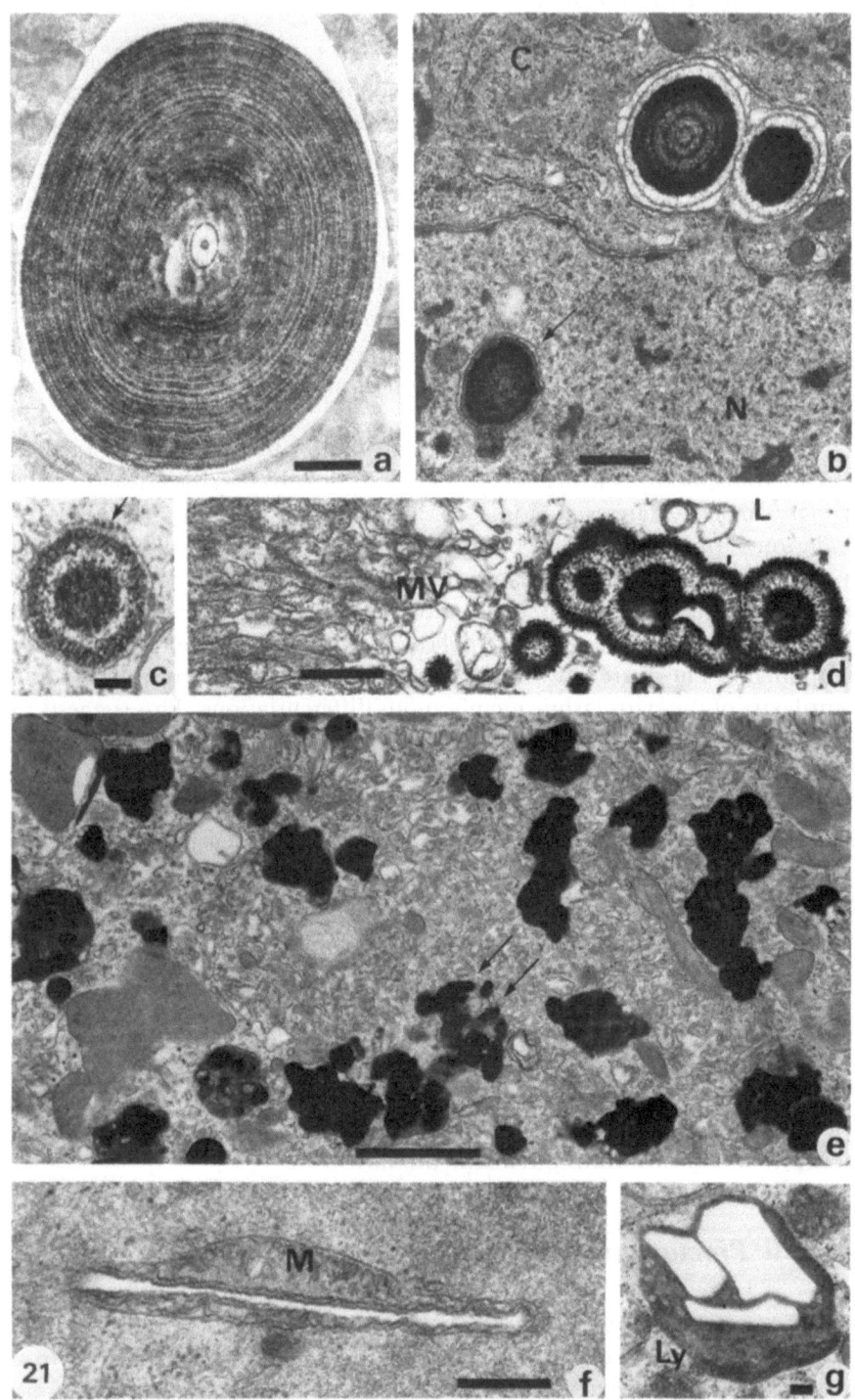

and even may be the only segment which exhibits the ultrastructural features of reabsorption. In a few cases, similar differentiations are observed in the colon and in the anal papillae (Table 2).

10.1 General Organization of the Epithelium

The rectal epithelium shows a great structural diversity. In the simplest system, the reabsorbing epithelium is made up of principal cells which are separated from the simple cells of the organ by a ring of narrow cells (Thysanura, Coleoptera). In other cases, the principal cells are very tall and protrude into the lumen, forming rectal pads. The ordinary epithelium is then separated from the pads by short junctional cells (Diptera) or by long and thin sheath cells (Dictyoptera, Isoptera) which lie beneath the principal ones. In this instance, basal cells are located under the principal ones, and they may be scattered or may form a continuous layer which isolates the principal cells from the hemolymph (see Caveney and Berdan, 1982, their Figure 17). Finally, in the most complex system, observed in one species only (*Apis mellifica*), the basal cells are separated from the reabsorbing epithelium by a closed papillary lumen.

The pads are often separated from the musculo-connective tunic of the rectum by a subepithelial sinus. Concretions containing mineral and purinic wastes may be present in the lumen of the proctodeum.

In coleopteran and lepidopteran larvae, the rectum and the distal ends of the Malpighian tubules are isolated from the hemolymph by a common sheath to form a cryptonephridial complex. This system does not exhibit important ultrastructural peculiarities (Grimstone *et al.*, 1968). Noirot *et al.* (1979) have proposed this system to represent the final stage in the evolution of sheath cells which extend to form a continuous layer around the Malpighian tubules.

10.2. Types of Epithelial Cells

10.2.1. The Principal Cells

These cells have often apical and lateral differentiations, and basal ones occur less frequently. The mitochondria are very numerous. Spherocrystals are present in the ileal cells of Dictyoptera and Orthoptera.

←

Figure 21. Storage structures in the Malpighian tubules. (a) An intracytoplasmic spherocrystal in *Blatella germanica*, with numerous mineral strata. (b) The principal cell in *Formica polyctena* showing two spherocrystals in the cytoplasm (C) and another one (arrow) in the nucleus (N). Courtesy of Dr. A. Y. Jeantet. (c) The formation of a spherocrystal in the RER (arrow). (d) Intraluminal spherocrystals in *F. polyctena*. MV, microvilli; L, lumen. Courtesy of Dr. A. Y. Jeantet. (e) A principal cell in *Musca domestica*, in which the concretions are formed in the lysosomes (arrows). Courtesy of Dr. R. S. Sohal. (f, g) Crystals of flavin which appear as ghosts in a mitochondrion (M) and a lysosome (Ly) in a cell of *B. germanica*. [Bars = 0.5 µm (a, b, d), 0.1 µm (c), 1 µm (e), and 0.2 µm (f, g).]

Table 2. Occurrence and Features of the Different Types of Cells in the Proctodeum[a,b]

			Principal cells				Types of junctional cells	References
	Organ	Epicuticular depressions	Apical region	Lateral regions	Basal region	Basal cells		
Thysanura								
Petrobius maritimus and *Lepismodes inquilinus*	Rectum	+	No Part.	SJ, no IS	BI	—	Narrow c.	Gabe et al. (1973); Noirot and Noirot-Timothée (1971a)
Lepismodes inquilinus	Ileum	—	No Part.	No SJ, no IS	BI	—	Narrow c.	Noirot and Noirot-Timothée (1971a)
	Anal sac	—	Part. + M	No SJ, no IS	No BI	—		Noirot and Noirot-Timothée (1971b)
Dictyoptera								
Periplaneta americana	Rectal pads	+	Part. + M	SJ + M	BI + SJ	+ (cont.)	Sheath c.	Oschman and Wall (1969); Noirot and Noirot-Timothée (1976)
Supella supellectilium and *Blatella germanica*	Rectal pads	+	Part. + M	SJ + M	BI + SJ	+ (discont.)	Sheath c.	Wall and Oschman (1973); Noirot and Noirot-Timothée (1976)
Blatella germanica	Ileum	—	Part. + M	SJ + M	BI	—	Narrow c.	Ballan-Dufrançais (1972)
Orthoptera								
Aiolopus strepens	Ileum	—	No Part.	No SJ, no IS	BI	—	Junctional c.	Bacetti (1960)
	Rectum	+	Part. + M	SJ + M	BI + SJ	—		Bacetti (1962)
Jamaicana flava	Rectal pads		Part. + M	SJ + M	No BI		Narrow c.	Peacock and Anstee (1977)
Isoptera								
13 species	Rectal pads	+	Part. + M	SJ + M	BI + SJ	+ (cont.)	Sheath c.	Noirot and Noirot-Timothée (1977)

Homoptera						
Gaeana maculata, etc.	Ileum		No Part.	No SJ, no IS	BI	—	Narrow c.	Marshall and Cheung (1973)
Heteroptera						
Cenocorixa bifida	Ileum		Part. + M	No SJ, no IS	BI	—	Narrow c.	Jarial and Scudder (1970)
Coleoptera						
Troglodromus bucheti	Ileum	+	No Part.	No SJ, no IS	BI	—	Narrow c.	Strambi and Zylberberg (1972)
	Colon (proximal)	+	No Part.	No SJ, no IS	BI	—	Narrow c.	
	Rectum	+	Part. + M	SJ + M	No BI	—	Narrow c.	
Hymenoptera						
Apis mellifica	Rectal pads	—	No Part.	SJ + IS	No BI	+ (2 layers)	Sheath c.	Kummel and Zerbst-Boroffka (1974)
Diptera						
Aedes aegypti and *A. campestris* larva	Rectum and anal papillae		Part., no M	No SJ, no IS	BI	—		Copeland (1964); Meredith and Phillips (1973a,b)
Aedes aegypti adult	Rectal pads		No Part.	SJ + IS	No BI	—		Hopkins (1967)
Calliphora erythrocephala	Rectal pads	—	Part., no M	SJ + IS	No BI	—	Junctional c.	Berridge and Gupta (1967)
Drosophila melanogaster	Rectal pads		Part., no M	SJ + IS	No BI	—	Junctional c.	Wessing and Eichelberg (1978)

[a] Abbreviations: BI, basal infoldings; c., cells; cont., continuous; discont., discontinuous; IS, intercellular spaces; M, mitochondria associated with membranes or junctions; Part., particles coating the membranes; SJ, scalariform junction.
[b] See also Noirot and Bayon (1969), Noirot and Noirot-Timothée (1970), Noirot et al. (1979), and Noirot-Timothée et al. (1979).

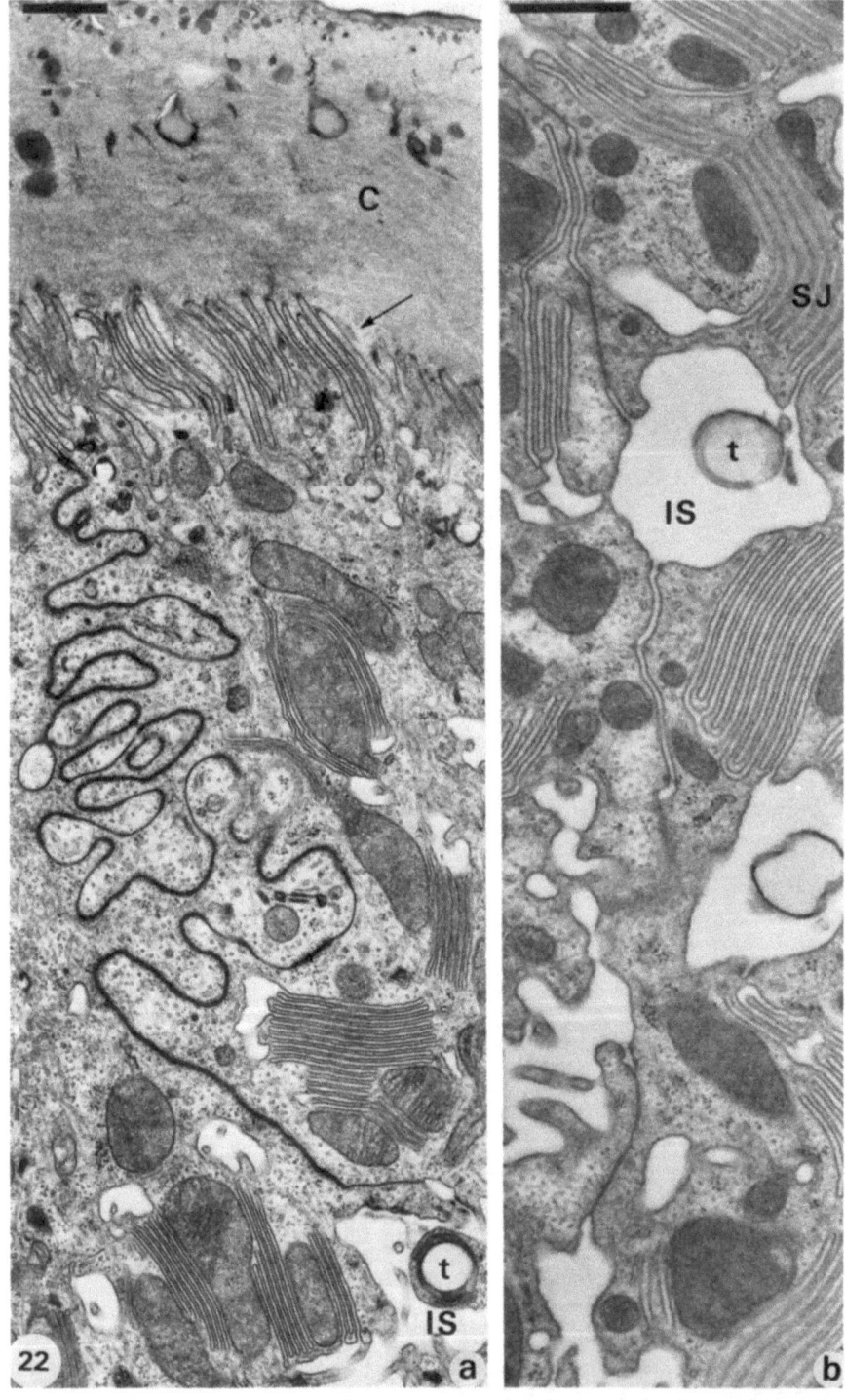

Apical differentiations. Under a cuticle which often contains epicuticular depressions (Figure 16e), the plasma membrane is enlarged with plications, invaginations, microvilli, or leaflets and usually has an internal coating of 12- to 15-nm-diameter particles similar to that of the Malpighian cells of *C. erythrocephala* (Figures 22a, 23, and 24a). Mitochondria are often associated with the membrane. Considering the species and the proctodeal segment, three types of apical morphologies have been observed: (1) plasma membranes without particles, (2) plasma membranes with particles and no mitochondria, and (3) plasma membranes with particles and mitochondria.

Lateral differentiations. The cells are linked by an exceptional variety of junctions: *zonula adhaerens*, septate junctions, gap junctions, and "mitochondria-scalariform junction complexes" (Figure 24b), structures which appear to exist only in Arthropoda (see Lane, 1982). Lateral intercellular spaces are more or less developed, sometimes with tracheoles in them (Diptera) (Figure 22b). Four types of lateral morphologies are observed among various species: (1) no scalariform junctions and no intercellular spaces, (2) short scalariform junctions with a few associated mitochondria and no intercellular spaces, (3) long scalariform junctions with few mitochondria and large spaces, and (4) long scalariform junctions with numerous mitochondria and intercellular spaces of a variety of sizes.

Basal differentiations. Infoldings of the plasma membrane exist in the ileum and the colon, whereas the rectal pads of numerous species lack these structures. When they are present, they are associated with scalariform junctions.

Because variations concern individually the apical, lateral, and basal regions, a great morphological complexity characterizes the principal cells. We classify the most frequent variations into four types:

Type A. Particles are attached to the apical membrane. Apical mitochondria are infrequent; lateral ones are not associated with scalariform junctions. Voluminous intercellular spaces are connected with a basal sinus. A well-developed tracheolar system is present. This type exists in the rectal pads and anal papillae of Diptera, although the adults of *A. aegypti* deviate from this scheme.

Type B. Numerous apical mitochondria are present; lateral ones are associated with scalariform junctions. Intercellular spaces are poorly developed; they contain tracheoles and are separated from the basal sinus by basal cells. This types is commonly seen in the Dictyoptera, Orthoptera, and Isoptera.

Type C. The apical and lateral aspects are similar to those of type B. There are no basal cells, and basal infoldings are well developed. This type has been described in the ileum of Dictyoptera and Orthoptera.

Figure 22. The principal cell in the rectal pads of *Calliphora erythrocephala*. (a) Under the cuticle (C), the apical leaflets lack mitochondria (arrow). Note the tortuous profiles of lateral plasma membranes. (b) Enlarged view showing the large intercellular spaces (IS) with trachea (t) and the scalariform junctions (SJ). Courtesy of Professor M. J. Berridge. (Bars = 0.5 μm.)

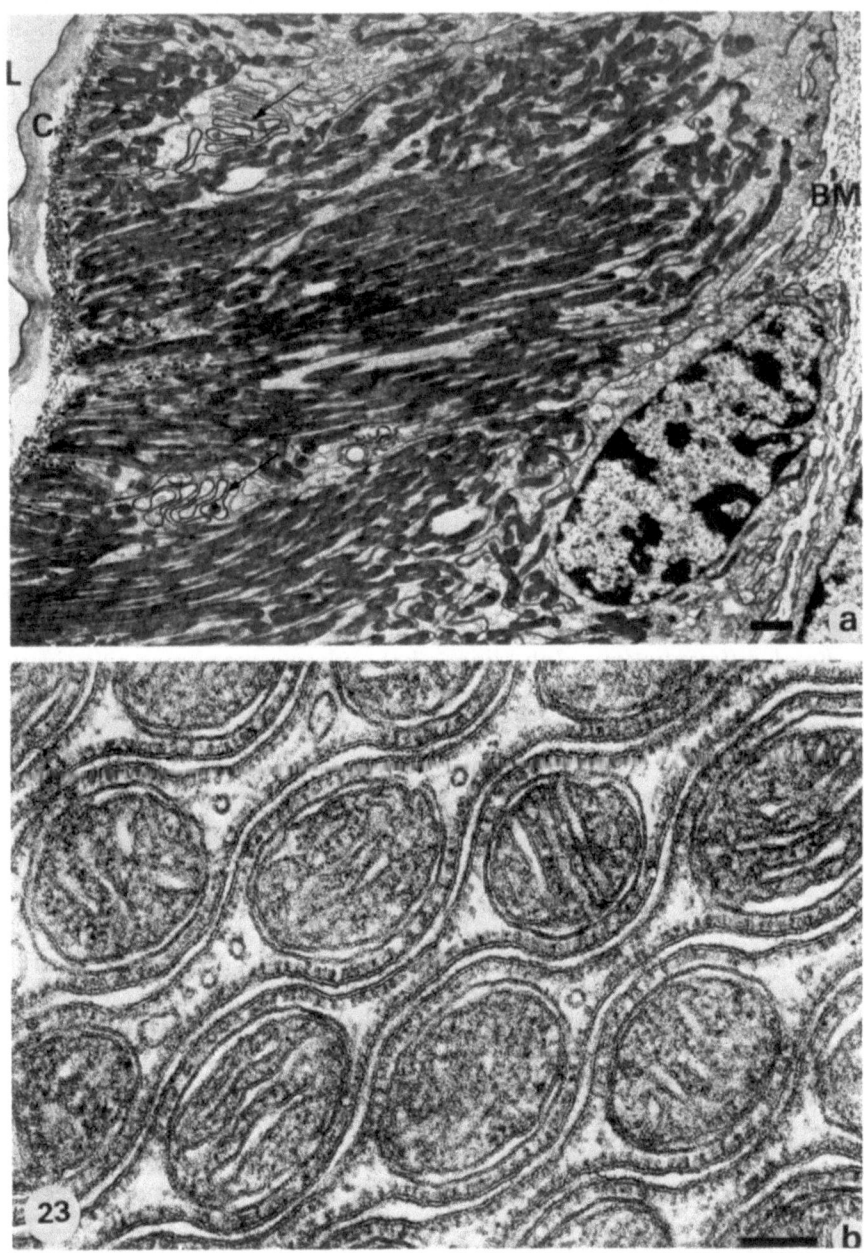

Figure 23. The principal cell in the anal sac of *Lepismodes inquilinus*. (a) Low magnification. Note the great development of the apical leaflets and mitochondria which invade the entire cytoplasm. L, lumen; C, cuticle; BM, basement membrane. Arrows indicate the lateral membranes. (b) Enlarged view of the apical region showing the mitochondria associated with membranes covered with particles. Courtesy of Professor C. Noirot. [Bars = 1 μm (a) and 0.1 μm (b).]

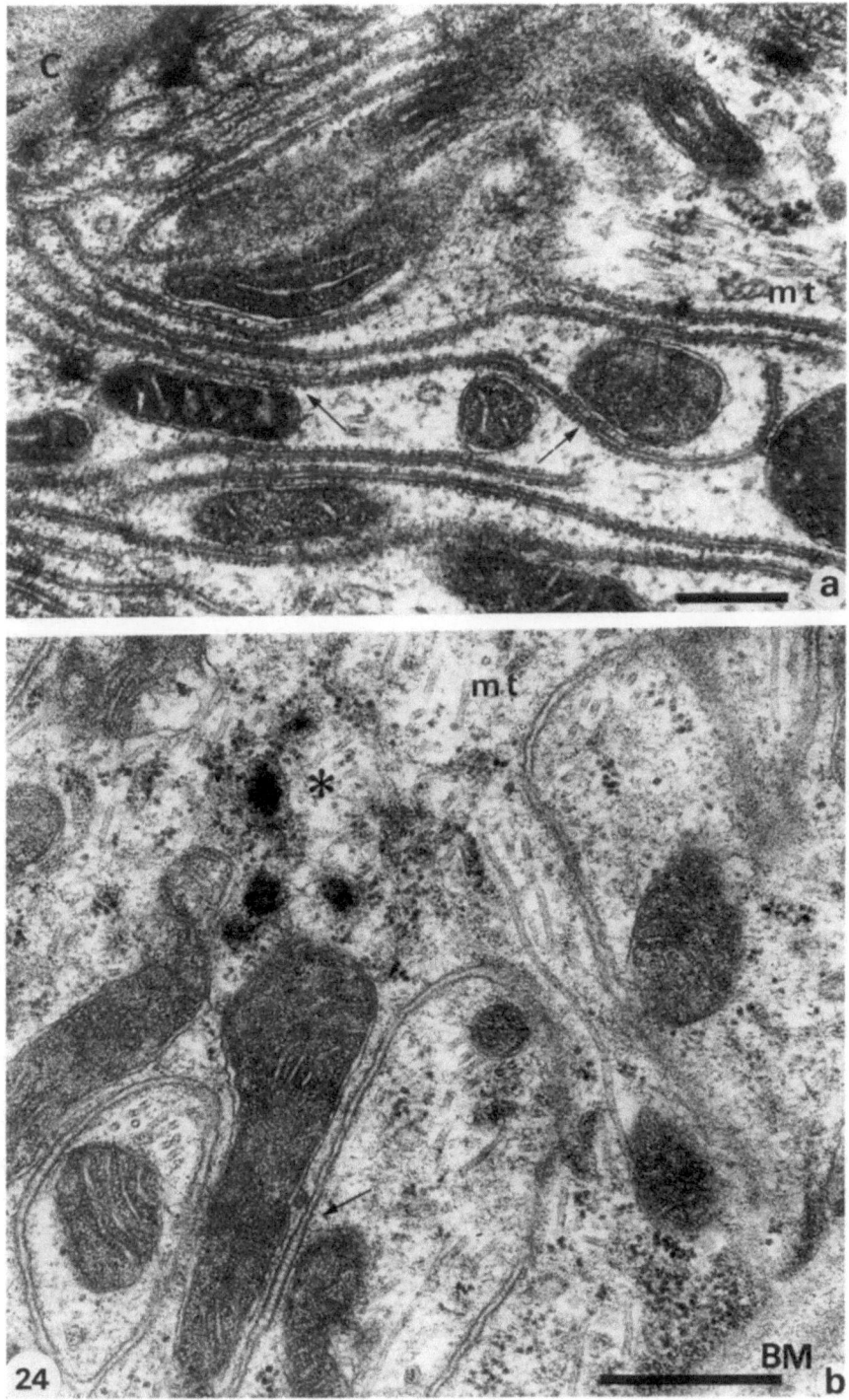

Figure 24. The principal cell in the ileum of *Blatella germanica*. (a) The apical region. Note the presence of particles on the apical membrane and associated mitochondria (arrows). C, cuticle; mt. microtubules. (b) The basal region. Note the presence of mitochondria-scalariform junction complexes on the lateral membranes (arrow). Young spherocrystals are forming in the RER (asterisk). [Bars = 0.25 μm (a) and 0.5 μm (b).]

Type D. Here, apical differentiations invade the whole cell. This very peculiar arrangement characterizes the anal sac of the bristle tail, *Lepismodes inquilinus*.

10.2.2. Junctional or Sheath Cells: Basal Cells

Junctional cells contain numerous microtubules and attach firmly to cuticular areas which lack endocuticle and constitute a sclerotized framework around the pads. The cells are linked to each other by a long septate junction and by atypical junctions (see Noirot-Timothée *et al.*, 1979). When they are as long as the principal cells, their extreme thinness reduces them to cytoplasmic leaflets, and they are termed *sheath cells* (Noirot *et al.*, 1979).

The basal cells generally contain pinocytotic vesicles along their basal plasma membrane. They are linked to the principal cells by gap and scalariform junctions. In Dictyoptera and Isoptera, they are joined to the sheath cells by septate junctions, an arrangement which isolates completely the principal cells from the hemolymph.

Considering the function of the different types of cells, it appears that the principal ones are transport cells, but the roles played by the peculiar morphological differentiations are still speculative. The cells may accomplish the same functions according to different schemes which may act together or be exclusive. In the anal sac of *Lepismodes*, which differs from all proctodeal segments when considering its physiological role (the absorption of atmospheric water), the cells lack lateral and basal differentiations, whereas the apical leaflets invade the whole cytoplasm. In contrast, the lateral differentiations are very well developed in the rectal pads of Diptera. Thanks to the subepithelial sinus, the pad constitutes a system in which local osmosis and ion recycling contribute to the reabsorption of fluid, whereas the junctional cells have a supportive role (Gupta *et al.*, 1980). In the pads of cockroaches, the characteristics of the apical differentiations point to a role in active and selective transport. The mitochondria–scalariform junction complexes would be adapted for the establishment of osmotic gradients. A functional relation between the subepithelial sinus and the intercellular spaces was proposed by Oschman and Wall (1969) in an attempt to explain the reabsorption of water against a concentration gradient. However, Noirot and Noirot-Timothée (1976) attach a greater importance to the continuous septate junctions between the tracheal and the basal cells. They consider that the basal cells transfer fluid to the intercellular spaces, and conversely recycle solutes secreted by the mitochondria–scalariform junction complex by absorption from the subepithelial sinus and transfer to the principal cell.

When other proctodeal segments ensure reabsorption, their cells are similar to the principal cells of the pads. Experimental data showing that mercury injected in the hemolymph is stored in the ileal cells of dictyopterans (Figure 25) suggest a basal filtration which may coexist with apical reabsorp-

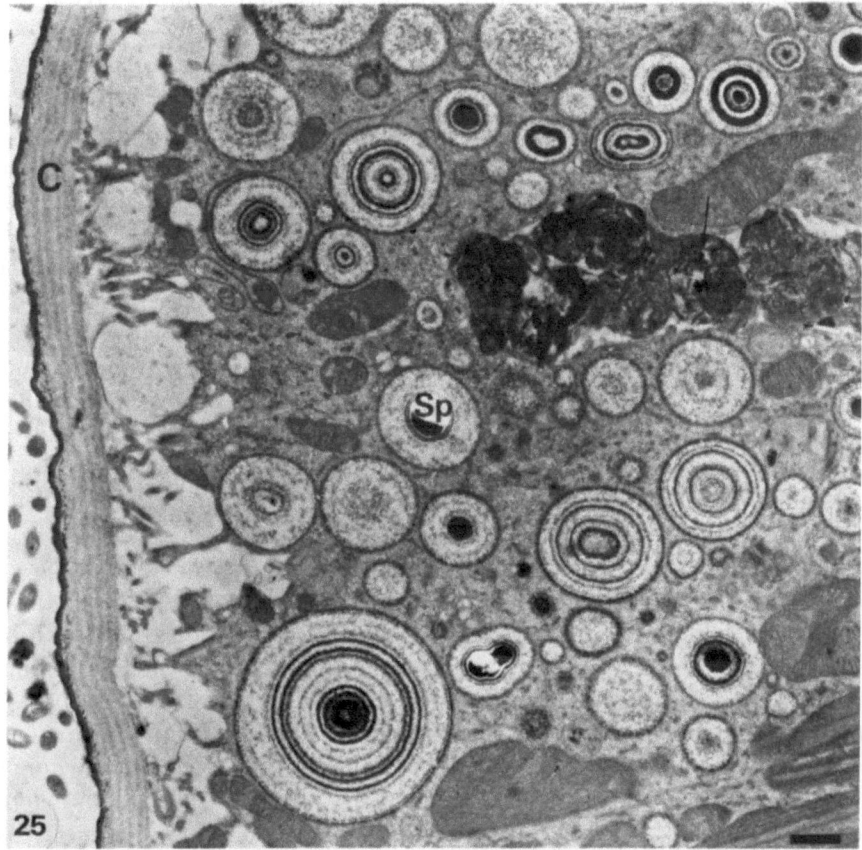

Figure 25. The principal cell in the ileum of *Blatella germanica* contaminated with mercury. Note the abundance of spherocrystals (Sp) and the occurrence of a lysosome with dense inclusions (arrow) composed of Hg, Cu, Zn, and S. C, cuticle. (Bar = 0.5 μm.)

tion (Jeantet *et al.*, 1980). When purinic and mineral concretions are present in the lumen (Figure 28d), they precipitate in response to the reabsorption and acidification of the fluid.

11. The Pericardial Cells

Clusters of phagocytic cells (nephrocytes) lie suspended in the hemolymph. An important group, termed *pericardial* or *diaphragm* cells, is located on the lateral wall of the heart, and has been studied with the electron microscope in Thysanura (Porter *et al.*, 1967; Gabe *et al.*, 1973), Dictyoptera (Edwards and Challice, 1960; Smith, 1968), Orthoptera (Kessel, 1961, 1962; Hoffmann, 1966), Homoptera (Bowers, 1964), Lepidoptera (Tanaka and Ikeda, 1982), and Diptera (Mills and King, 1965; Crossley, 1972). Analogous cells (wreath or garland cells) appear to be identical to the pericardial cells (Aggarwal and King, 1966; Crossley, 1972).

The pericardial cells are spherical and surrounded by a basement membrane. The organelles are arranged in concentric zones. The cortex is composed of a peculiar labyrinth which consists of a series of folds and fingerlike processes. In these processes, the apposed edges of the infolded membrane are separated by a gap which is bridged by a diaphragm of cementing material (see Figure 26a for a similar structure). Along the channels of the labyrinth, invaginations of the membrane occur as coated vesicles. Numerous tubular elements weave between the channels which are connected either with the ER (Dictyoptera) or with large vacuoles (Thysanura, Diptera). A narrow zone with numerous mitochondria exists under the cortex. Then, passing from this zone toward the center of the cell containing the nucleus, RER, and Golgi bodies, a large region of vacuoles is encountered. In most species, several kinds of vacuoles, often termed *vesicles*, coexist in this region: in *C. erythrocephala*, α, β, and γ vacuoles differ from one another in size and content (homogeneity; electron-opacity; occurrence of acid phosphatases and of crystals). The electron-lucent vacuoles are confined to the periphery of the vacuolar region and are connected with the tubular elements. These tubules do not seem to arise from the plasma membrane or from the coated vesicles. The vacuoles which exhibit a high electron-opacity are often termed *granules*, and those of Orthoptera contain a pigment (ommochrome) which gives the cell a brown color. In *B. mori*, and certainly in most lepidopteran species, a cycle of formation and consumption of granules is closely parallel to the molting cycle and metamorphosis. In contrast, the molt does not affect the aspect of the cells in heterometabolous species; but in *Melanoplus differentialis*, starvation induces a disappearance of the granules.

As for the function, it is now well established that the pericardial cells play both phagocytic and synthetic roles, which are both involved in regulation of hemolymph composition. The cells ensure the breakdown of ingested hemolymph proteins in the vacuolar apparatus. The bloodborne molecules enter the labyrinth channels, are uptaken by coated vesicles, sequestered, and digested in vacuoles. In this process, the function of the tubular elements remains uncertain. The phagocytic process has certainly different meanings among the different species. In Orthoptera, there is no evidence for protein turnover. In *L. migratoria*, for example, the amounts of essential metals (Cu, Zn) originating from digested proteins and those of toxic metals (Ag, Cd) linked to blood proteins in contaminated locusts, increase with age, and these metals are bound to ommochromes. In that species, the cells play a role in detoxification, serve as a storage kidney and the granules do not disappear (Martoja and Truchet, 1983). In contrast, the cells of lepidopterans sequester proteins which are then consumed during molt and metamorphosis. This turnover makes the pericardial cells rather analogous to adipose cells which in the same manner store and release blood proteins (Locke and Collins, 1965). The synthetic activities concern first the synthesis of the intracellular lytic enzymes and pigments. In addition, the cells release secretory products into the hemolymph. Many of them remain to be identified, but it has been demonstrated that the cells of *Calliphora* produce the enzyme lysozyme. The γ vacuoles, which are devoid of acid phosphatase, could be the site of secretion of this protein.

12. The Other Excretory Organs

The *labial kidneys* of thysanurans are regarded as excretory organs homologous with the cephalic nephridia of myriapods and crustaceans, though nothing is known of the normal products of their excretion. This opinion is supported by the fine structure of the organ and by identifying ammonia as the excretory product in the cephalic kidneys of myriapods. The labial kidneys consist of an upper region, or saccule, communicating with a long coiled tube or labyrinth, which is connected to an unpaired excretory duct. The saccule is made up of podocytes (Figure 26a) with a fine structure rather similar to that of the pericardial cells (Fain-Maurel and Cassier, 1971, 1972). Fingerlike processes are bridged by a diaphragm (Figure 26b), and the cells contain many pinocytotic vesicles, various vacuoles, and lysosomes. The cells of the labyrinth bear basal infoldings with numerous mitochondria and distal microvilli. The plasma membranes are linked by scalariform junctions. Considering the apical parts of the cells, three regions can be distinguished, which suggest a gradation of physiological properties, like that of Malpighian tubules. As for the fine structure of the excretory duct, it appears to be similar to that of the rectum. Fain-Maurel and Cassier suggest that molecules are absorbed by pinocytosis and degraded in the saccule, whereas solutes could be reabsorbed in the labyrinth and in the excretory duct.

A storage excretion is often carried out by the *fat body*. Uric acid is formed either within the fat cells or in urate cells which are distinct from the ordinary cells of the tissue, and it appears in the form of spherocrystals (Figures 27, 28b). These cells have been considered in many physiological studies, and concretions have been analyzed with new techniques (see Ballan-Dufrançais *et al.*, 1979). Unfortunately, satisfactory descriptions of their fine structure are still missing, because of technical difficulties in fixation and sectioning the samples. The concretions, composed of urates and mineral salts, may nearly fill up the whole cells. As a rule, the concretions are enclosed by the RER.

The *utricules*, male accessory glands of *Blatella germanica*, are both secretory and excretory organs. They are made up of a single epithelium with cells of one type which perform both functions. The plasma membrane of the basal region has many infoldings with numerous mitochondria, whereas the apical region has microvilli which lack mitochondria but, like those of the Malpighian tubules, contain a canaliculus. The middle part of the cells contains the nucleus, the RER, Golgi bodies, and spherocrystals which form in the cisternae of the RER. The cells release glycoproteins and spherocrystals, whereas other concretions are formed in the lumen (Figure 28c) (Ballan-

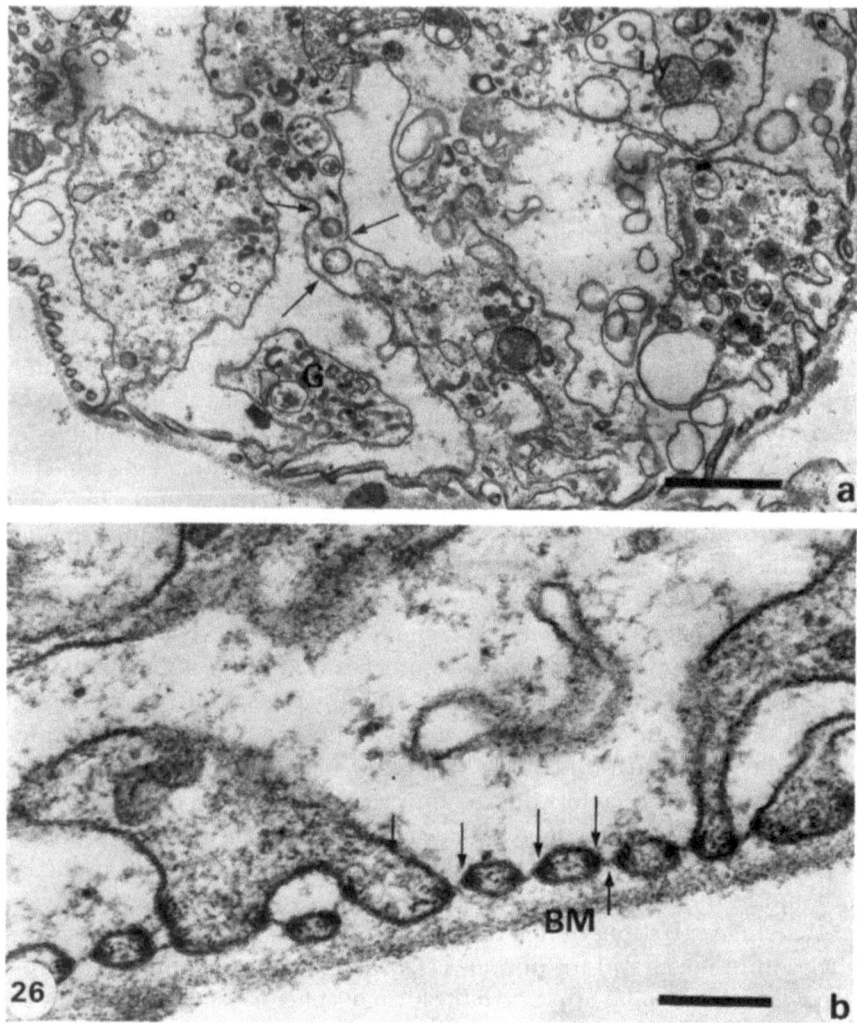

Figure 26. The labial kidney of *Petrobius maritimus*. (a) The podocytes with pinocytotic vesicles (arrows), Golgi bodies (G), and lysosomes (Ly). (b) The saccular sieve formed by the basement membrane (BM) and by a thin, single membrane (short arrow) anchored on annular specializations of foot processes (long arrows). Courtesy of Professor M. A. Fain-Maurel. [Bars = 1 μm (a) and 0.2 μm (b).]

Dufrançais, 1970). Concretions are made of uric acid, K and Na urates, xanthine and undetermined Ca and Mg salts (Ballan-Dufrançais *et al.*, 1979).

13. Concluding Remarks

Ultrastructural data have demonstrated the existence of a cellular cycle for gut epithelial cells which previously were believed to constitute a rather

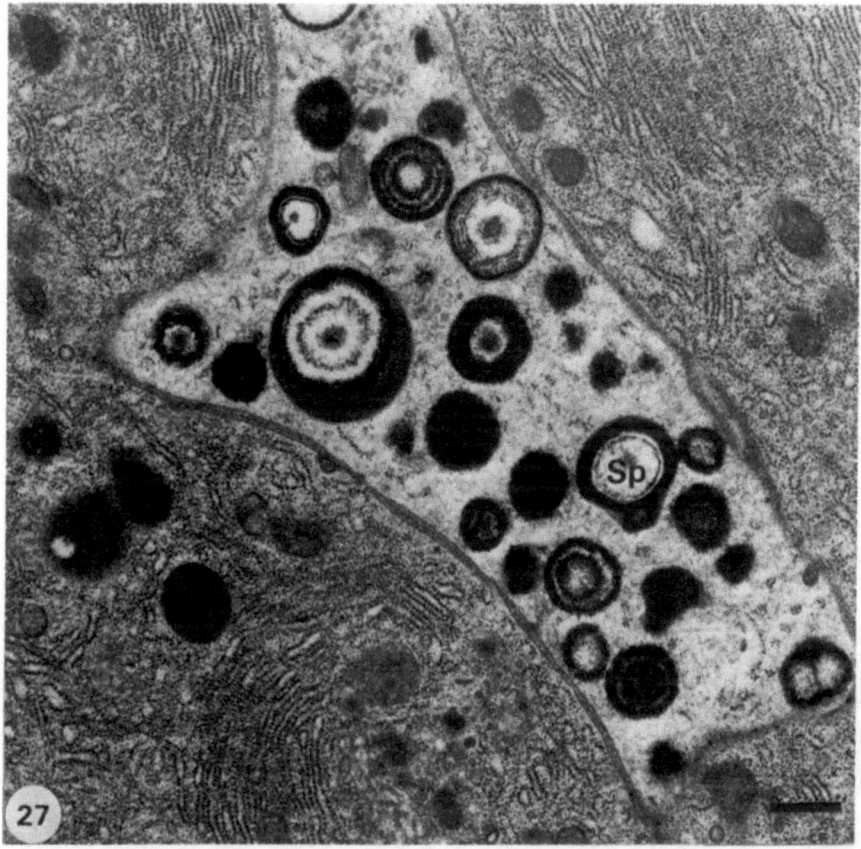

Figure 27. A urate cell in the fat body of *Locusta migratoria* with numerous spherocrystals (Sp). Courtesy of Dr. S. Lauverjat. (Bar = 0.5 μm.)

uniform cellular layer and have provided important information concerning intercellular junctions, cell degeneration, and cell differentiation. Studies which combine the electron microscopic techniques with physiological, chemical, and cytochemical methods have shown many cases where function is unequivocally correlated with structure.

However, the ultrastructural researches have often raised as many questions as they have answered. Because of the diversities in the fixation and staining procedures and of the magnifications used for the observations, the distinctions in function that fine structural differences must signify remain quite speculative. Comparisons between studies performed on the same structures show that the interpretations may sometimes be equivocal. Moreover, little attention has been paid to important organs such as the cryptonephridial system or the different rings of midgut caeca which in some orders (i.e., Coleoptera) could be involved in peculiar physiological processes. Descriptions often concern cells for which physiological data are missing (i.e., endocrine cells and acinar cells of the salivary glands) or have been performed without considering

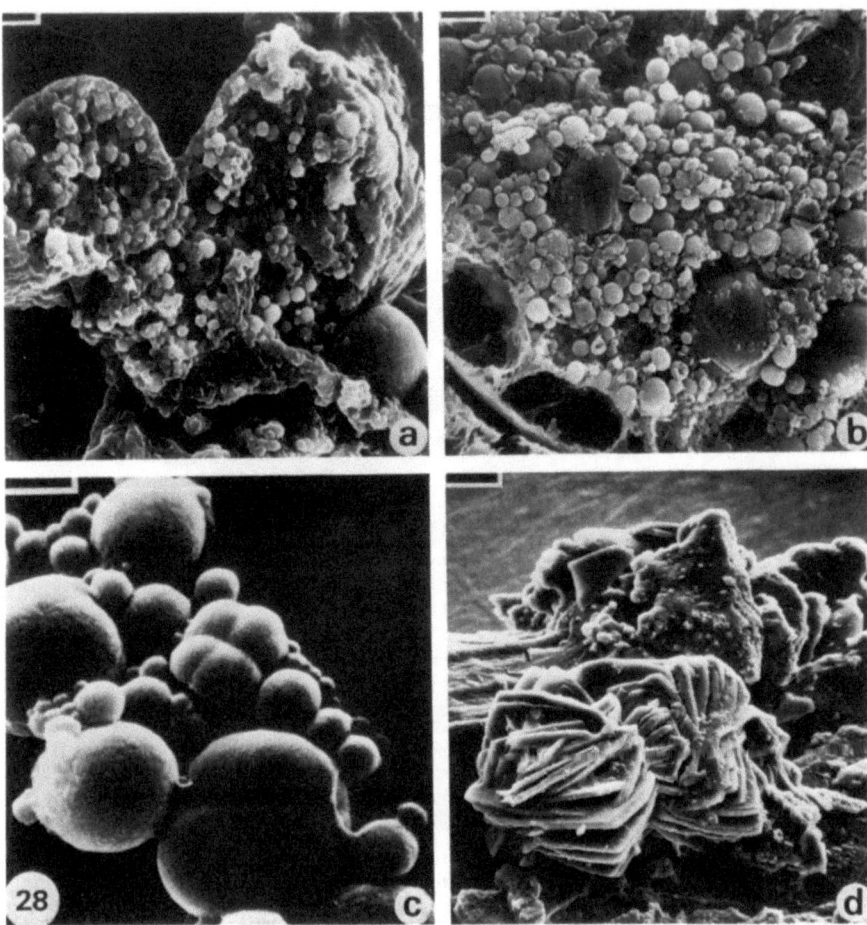

Figure 28. SEM. (a) A Malpighian tubule of *Blatella germanica*. The tubule is broken so that the numerous intracytoplasmic spherocrystals are clearly visible. (b) A section in a fat cell of *Blatella* showing the numerous spherocrystals. (c) Spherocrystals in the lumen of the utricules of *Blatella*. (d) Crystallized wastes in the proctodeal lumen of *Locusta migratoria*. [Bars = 1 µm (a, c, d) and 2 µm (b).]

the diversified aspects of a problem (i.e., Malpighian tubules). It is to be hoped that a "homogenization" of the cytological techniques and the development of new physiological methods will elucidate the remaining problems connected with the digestive and excretory functions of these organs. Tissues of the insect digestive and excretory systems are ideal subjects for studies on the functioning of intercellular junctions, on the physical changes that accompany secretion, intracellular storage, and gene activation, and on the differentiation and eventual death of cells. It is safe to predict that exciting advances in our understanding of these and other aspects of cell biology will result from future research in this field.

ACKNOWLEDGMENTS

We are grateful to Dr. King for correction of the manuscript language and to colleagues who kindly provided the illustrations for this chapter. Because it was necessary to make a selection from a very large literature, we have undoubtedly omitted some important work, and we apologize to the authors for such shortcomings.

References

Aggarwal, S. K., and King, R. C., 1966, The ultrastructure of the wreath cells of *Drosophila melanogaster* larvae, *Protoplasma* **63**:343-352.

Akai, H., 1970, An electron microscopic study of the alimentary canal of the silkworm, *Bombyx mori* L. I. The ultrastructure of the midgut epithelium, *Bull. Sericult. Exp. Stn.* **24**:303-344 (Japanese with English summary).

Anderson, E., and Harvey, W. R., 1966, Active transport by the cecropia midgut: Fine structure of the midgut epithelium, *J. Cell Biol.* **31**:107-134.

Andriès, J. C., 1972, Genèse intraépithéliale des microvillosités de l'épithélium mésentérique de la larve d'*Aeshna cyanea*, *J. Microsc. (Paris)* **15**:181-204.

Andriès, J. C., 1975, Différenciation et mort cellulaire au cours de la métamorphose mésentérique de la larve d'*Aeshna cyanea*, *J. Microsc. Biol. Cell.* **24**: 327-350.

Andriès, J. C., 1976a, Présence de deux types cellulaires endocrines et d'un type exocrine au sein du mésentéron de la larve d'*Aeshna cyanea* Muller (Odonata: Aeshnidae). *Int. J. Insect Morphol. Embryol.* **5**:393-407.

Andriès, J. C., 1976b, Variations ultrastructurales au sein des cellules épithéliales mésentériques d'*Aeshna cyanea* (Insecte, Odonate) en fonction de la prise de nourriture, *Cytobiologie* **13**: 451-468.

Andriès, J. C., 1976c, Specialization of the endoplasmic reticulum in the apex of the midgut cells of *Aeshna cyanea* (Insecta, Odonata), *Cell Tissue Res.* **198**:97-101.

Andriès, J. C., 1979, Junctional structures in the metamorphosing midgut of *Aeshna cyanea* (Insecta, Odonata), *Cell Tissue Res.* **202**:9-15.

Andriès, J. C., and Torpier, G., 1982, An extracellular brush border coat of lipid membranes in the midgut of *Nepa cinerea* (Insecta, Heteroptera): Ultrastructure and genesis, *Biol. Cell.* **46**: 195-202.

Anstee, J. H., 1975, An electron microscopical study of the salivary glands of the tettigoniid, *Homorochoryphus nitidulus*, *J. Insect Physiol.* **21**:1073-80.

Ashhurst, D. E., 1982, The structure and development of insect connective tissue. In *Insect Ultrastructure*, edited by R. C. King and H. Akai, vol. 1, pp. 313-350, Plenum Press, New York.

Bacetti, B., 1960, Ricerche sull' ultrastruttura dell' intestino degli insetti. I. L'ileo di un ortottero adulto, *Redia* **45**:263-278.

Bacetti, B., 1962, Ricerche sull' ultrastruttura dell' intestino degli insetti. IV. Le papille rettali in un ortottero adulto, *Redia* **57**:105-118.

Ballan-Dufrançais, C., 1970, Données cytophysiologiques sur un organe excréteur particulier d'un Insecte: *Blatella germanica* (L.) (Dictyoptère), *Z. Zellforsch. Mikrosk. Anat.* **109**:336-355.

Ballan-Dufrançais, C., 1972, Ultrastructure de l'iléon de *Blatella germanica* L. (Dictyoptère): Localisation, genèse et composition des concrétions minérales intracytoplasmiques, *Z. Zellforsch. Mikrosk. Anat.* **133**:163-179.

Ballan-Dufrançais, C., 1974, Accumulations minérales et puriques chez trois espèces d'insectes dictyoptères, *Cellule* **70**:317-330.

Ballan-Dufrançais, C., 1975, Bioaccumulation minérale, purique et flavinique chez les Insectes: Méthodes d'étude, importance physiologique, Thèse de Doctorat d'Etat, Université Paris VI, Biologie animale.

Ballan-Dufrançais, C., 1977, An important way of ionic regulation in insects: Mineral and purinic bioaccumulation, *Bull. Soc. Zool. Fr.* **102**:318-319.

Ballan-Dufrançais, C., and Martoja, R., 1971, Analyse chimique d'inclusions minérales par spectrographie des rayons X et par cytochimie: Application à quelques organes d'insectes orthoptères, *J. Microsc. (Paris)* **11**:219-248.

Ballan-Dufrançais, C., Truchet, M., and Dhamelincourt, P., 1979, Interest of Raman laser microprobe (Mole) for the identification of purinic concretions in histological sections, *Biol. Cell.* **36**:51-58.

Bayon, C., 1971, La cuticule proctodeale de la larve d'*Oryctes nasicornis* L.: Etude au microscope electronique a balayage, *J. Microsc. (Paris)* **11**:353-370.

Bayon, C., 1981, Ultrastructure de l'épithélium intestinal et flore pariétale chez la larva xylophage d'*Oryctes nasicornis* L. (Coleoptera: Scarabaeidae), *Int. J. Insect Morphol. Embryol.* **10**:359-371.

Bayon, C., and François, J., 1976, Ultrastructure de la lame basale du mésentéron chez la larve d'*Oryctes nasicornis* L. (Coleoptera: Scarabaeidae), *Int. J. Insect Morphol. Embryol.* **5**:205-217.

Bell, D. M., and Anstee, J. H., 1977, A study of the Malpighian tubules of *Locusta migratoria* by scanning and transmission electron microscopy, *Micron* **8**:123-134.

Berkaloff, A., 1960, Contribution à l'étude des tubes de Malpighi et de l'excrétion chez les Insectes, *Ann. Sci. Nat. Zool.* **12**:869-947.

Berridge, M. J., and Gupta, B. L., 1967, Fine-structural changes in relation to ion and water transport in the rectal papillae of the blowfly *Calliphora*, *J. Cell Sci.* **2**:89-112.

Berridge, M. J., and Oschman, J. L., 1969, A structural basis for fluid secretion by the Malpighian tubules, *Tissue Cell* **1**:247-272.

Bignell, D. E., 1980, An ultrastructural study and stereological analysis of the colon wall in the cockroach *Periplaneta americana*, *Tissue Cell* **12**:153-164.

Bland, K. P., and House, C. R., 1971, Function of the salivary glands of the cockroach *Nauphoeta cinerea*, *J. Insect Physiol.* **17**:2069-2084.

Bowers, B., 1964, Coated vesicles in the pericardial cells of the aphid (*Myzus persicae* Sulz.), *Protoplasma* **59**:351-367.

Burgos, M. H., and Gutierrez, L. S., 1976, The intestine of *Triatoma infestans*. I. Cytology of the midgut, *J. Ultrastruct. Res.* **57**:1-9.

Cassier, P., and Fain-Maurel, M. A., 1977, Sur la présence d'un système endocrine diffus dans le mésentéron de quelques Insectes, *Arch. Zool. Exp. Gen.* **118**:197-209.

Cassier, P., Alibert, J., and Fain-Maurel, M., 1972, Sur la présence de cellules de type endocrine dans l'intestin moyen de *Petrobius maritimus* Leach (Insecte Aptérygote, Thysanoure), *C.R. Acad. Sci. Ser. D* **275**:2691-2693.

Caveney, S., and Berdan, R., 1982, Selectivity in junctional coupling between cells of insect tissues. In *Insect Ultrastructure*, edited by R. C. King and H. Akai, vol. 1, pp. 434-465, Plenum Press, New York.

Charnley, A. K., 1982, The ultrastructure of the type-2 cell in the Malpighian tubules of *Locusta migratoria*, *Micron* **13**:45-47.

Copeland, E., 1964, A mitochondrial pump in the cells of the anal papillae of mosquito larvae, *J. Cell Biol.* **23**:253-264.

Crossley, A. C., 1972, The ultrastructure and function of pericardial cells and other nephrocytes in an insect: *Calliphora erythrocephala*, *Tissue Cell* **4**:529-560.

Crowder, L. A., and Shankland, D. L., 1972, Structure of the Malpighian tubule muscle of the American cockroach *Periplaneta americana*, *Ann. Entomol. Soc. Am.* **65**:614-619.

Dailey, P. J., and Crang, R. E., 1977, Ultrastructure of salivary glands of the cockroach *Blaberus discoidalis* Serville (Blattaria: Blaberidae), *Int. J. Insect Morphol. Embryol.* **6**:61-66.

Dailey, P. J., and Crang, R. E., 1978a, The fine structure of the salivary glands in the cockroach *Gromphadorhina portentosa*: Secretion, *J. Morphol.* **156**:157-172.

Dailey, P. J., and Crang, R. E., 1978b, The fine structure of the salivary glands in the cockroach *Gromphadorhina portentosa*: Duct transport system, *J. Morphol.* **157**:329-346.

Daneholt, B., 1982, Structural and functional analysis of Balbiani ring genes in the salivary glands of *Chironomus tentans*. In *Insect Ultrastructure*, edited by R. C. King and H. Akai, vol. 1, pp. 382-401, Plenum Press, New York.

Davies, I., 1977, The effect of diet on the ultrastructure of the midgut cells of *Nasonia vitripennis* (Walk) (Insecta, Hymenoptera) at various ages, *Cell Tissue Res.* **184**:529-538.

de Priester, W., 1971, Ultrastructure of the midgut epithelial cells in the fly *Calliphora erythrocephala*, *J. Ultrastruct. Res.* **36**:783-805.

de Priester, W., 1972, Lysosomes in the midgut of *Calliphora erythrocephala* Meigen, *Z. Zellforsch. Mikrosk. Anat.* **129**:430-446.

Devauchelle, G., 1970, Inclusions cristallines et particules d'allure virale dans les noyaux des cellules de l'intestin moyen du Coléoptère *Tenebrio molitor*, *J. Ultrastruct. Res.* **33**:263-277.

Edwards, G. A., and Challice, C. E., 1960, The ultrastructure of the heart of the cockroach *Blatella germanica*, *Ann. Entomol. Soc. Am.* **53**:369-383.

Eichelberg, D., and Wessing, A., 1975, Morphology of the Malpighian tubules of insects, *Fortschr. Zool.* **23**:124-147.

Endo, Y., and Nishiitsutsuji-Uwo, J., 1981, Gut endocrine cells in insects: The ultrastructure of the gut endocrine cells of the lepidopterous species, *Biomed, Res.* **2**:270-280.

Endo, Y., and Nishiitsutsuji-Uwo, J., 1982, Exocytic release of secretory granules from endocrine cells in the midgut of insects, *Cell Tissue Res.* **222**:515-522.

Endo, Y., Nishiitsutsuji-Uwo, J., Iwanaga, T., and Fujita, T., 1982, Ultrastructural and immunohistochemical identification of pancreatic polypeptide-immunoreactive endocrine cells in the cockroach midgut, *Biomed, Res.* **3**:454-456.

Fain-Maurel, M. A., and Cassier, P., 1971, Différenciations cytoplasmiques en relation avec la fonction excrétice dans les reins céphaliques de *Petrobius maritimus* Leach (Insecte, Aptérygote), *J. Microsc. (Paris)* **10**:163-178.

Fain-Maurel, M. A., and Cassier, P., 1972, Un nouveau type de jonctions: Les jonctions scalariformes. Etude ultrastructurale et cytochimique, *J. Ultrastruct. Res.* **39**:222-238.

Fain-Maurel, M. A., Cassier, P., and Alibert, J., 1973, Étude infrastructurale et cytochimique de l'intestin moyen de *Petrobius maritimus* Leach en rapport avec ses fonctions excrétrices et digestives, *Tissue Cell* **5**:603-631.

Ferreira, C., Ribeiro, A. F., and Terra, W. R., 1981, Fine structure of the larval midgut of the fly *Rhynchosciara* and its physiological implications, *J. Insect Physiol.* **27**:559-570.

Filshie, B. K., Poulson, D. F., and Waterhouse, D. F., 1971, Ultrastructure of the copper-accumulating region of *Drosophila* larval midgut, *Tissue Cell* **3**:77-102.

Flower, N. E., and Filshie, B. K., 1976, Goblet cell membrane differentiation in the midgut of a lepidopteran larva, *J. Cell Sci.* **20**:357-375.

Foldi, I., 1973, Etude de la chambre filtrante de *Planococcus citri* (Insecta, Homoptera): Histochimie et ultrastructure, *Z. Zellforsch. Mikrosk, Anat.* **143**:549-568.

Forbes, A. R., 1964, The morphology, histology and fine structure of the gut of the green peach aphid, *Myzus persicae* (Sulzer) (Homoptera: Aphididae), *Mem. Entomol. Soc. Can.* **36**:1-74.

François, J., 1978, The ultrastructure and histochemistry of the mesenteric connective tissue of the cockroach *Periplaneta americana* L. (Insecta, Dictyoptera), *Cell Tissue Res.* **189**:91-107.

Gabe, M., Cassier, P., and Fain-Maurel, M. A., 1973, Données morphologiques sur les organes excréteurs abdominaux de *Petrobius maritimus* Leach (Insecte, Aptérygote), *Arch. Anat. Microsc. Morphol. Exp.* **62**:101-143.

Gouranton, J., 1967, Accumulation de ferritine dans les noyaux et le cytoplasme de certaines cellules du mésentéron chez des Homoptères Cercopides âgés, *C.R. Acad. Sci.* **264**:2657-2660.

Gouranton, J., 1968a, Présence d'une zône cytoplasmique différenciée autour des noyaux dans les cellules de l'intestin moyen de *Cixius nervosus* L. (Homoptera: Fulguroïdea), *C.R. Acad. Sci.* **266**:818-819.

Gouranton, J., 1968b, Composition, structure et mode de formation des concrétions minérales dans l'intestin moyen des Homoptères Cercopides, *J. Cell Biol.* **37**:316-328.

Gouranton, J., 1968c, Ultrastructures en rapport avec un transit d'eau: Etude de la "chambre filtrante" de *Cicadella viridis* L. (Homoptera, Jassidae), *J. Microsc. (Paris)* **7**:559-574.

Gouranton, J., 1969, Observations cytochimiques et ultrastructurales sur les cristaux intranucléaires de l'intestin moyen de la larve de *Tenebrio molitor* L., *C.R. Acad. Sci.* **268**:2948-2951.

Gouranton, J., 1970, Etude d'une lame basale presentant une structure d'un type nouveau, *J. Microsc. (Paris)* **9**:1029-1040.

Gouranton, J., and Folliot, R., 1968, Présence de cristaux de ferritine de grande taille dans les cellules de l'intestin moyen de *Campylenchia latipes* Say (Homoptera, Membracidae), *Rev. Can. Biol.* **27**:77-81.

Gouranton, J., and Maillet, P. L., 1967, Origine et structure des brochosomes, *J. Microsc. (Paris)* **6**:53-64.

Gouranton, J., and Thomas, D., 1972, Présence de cristaux protéiques intranucléaires et intracytoplasmiques dans l'intestin moyen de *Sympetrum depressiusculum* Sel. (Odonate), *C.R. Acad. Sci. Ser. D* **274**:1843-1845.

Gouranton, J., and Thomas, D., 1974, Cytochemical, ultrastructural and autoradiographic study of the intranuclear crystals in the midgut cells of *Gyrinus marinus* Gyll, *J. Ultrastruct. Res.* **48**:227-241.

Gouranton, J., and Thomas, D., 1976, Observations on intranuclear crystal and nucleolar size at different stages of cell differentiation in the midgut epithelium of several insects, *J. Cell Sci.* **22**:87-98.

Green, L. F. B., 1979, Regional specialization in the Malpighian tubules of the New Zealand glowworm *Arachnocampa luminosa* (Diptera: Mycetophilidae): The structure and function of the type I and II cells, *Tissue Cell* **11**:673-702.

Grimstone, A. V., Mullinger, A. M., and Ramsay, J. A., 1968, Further studies on the rectal complex of the mealworm *Tenebrio molitor* L. (Coleoptera, Tenebrionidae), *Philos. Trans. R. Soc. London* **253**:343-382.

Gupta, B. L., Wall, B. J., Oschman, J. L., and Hall, T. A., 1980, Direct microprobe evidence of local concentration gradients and recycling of electrolytes during fluid absorption in the rectal papillae of *Calliphora*, *J. Exp. Biol.* **88**:21-49.

Gutierrez, L. S., and Burgos, M. H., 1978, The intestine of *Triatoma infestans*. II. The surface coat of the midgut, *J. Ultrastruct. Res.* **63**:244-251.

Hecker, H., 1977, Structure and function of midgut epithelial cells in Culicidae mosquitoes (Insecta, Diptera), *Cell Tissue Res.* **184**:321-342.

Hecker, H., Freyvogel, T. A., Briegel, H., and Steiger, R., 1971, The ultrastructure of midgut epithelium in *Aedes aegypti* (L.) (Insecta, Diptera) males, *Acta Trop.* **28**:275-290.

Hoffmann, J., 1966, Etude ultrastructurale de l'absorption de saccharate de fer par les cellules péricardiales de *Locusta migratoria* (Orthoptère), *C.R. Acad. Sci.* **262**:1469-1472.

Holter, P., 1970, Regular grid-like substructures in the midgut epithelial basement membrane of some Coleoptera, *Z. Zellforsch. Mikrosk. Anat.* **110**:373-385.

Hopkins, C. R., 1967, The fine-structural changes observed in the rectal papillae of the mosquito *Aedes aegypti* L. and their relation to the epithelial transport of water and inorganic ions, *J. R. Microsc. Soc.* **86**:235-252.

Houk, E. J., 1977, Midgut ultrastructure of *Culex tarsalis* (Diptera: Culicidae) before and after a blood meal, *Tissue Cell* **9**:103-118.

Houk, E. J., Chiles, R. E., and Hardy, J. L., 1980, Unique midgut basal lamina in the mosquito, *Aedes dorsalis* (Meigen) (Insecta: Diptera), *Int. J. Insect Morphol. Embryol.* **9**:161-164.

Iwanaga, T., Fujita, T., Nishiitsutsuji-Uwo, J., and Endo, Y., 1981, Immunohistochemical demonstration of PP-, somatostatin-, enteroglucagon- and VIP-like immunoreactivities in the cockroach midgut, *Biomed. Res.* **2**:202-207.

Janzen, H. G., and Wright, K. A., 1971, The salivary glands of *Aedes aegypti* (L.): An electron microscope study, *Can. J. Zool.* **49**:1343-1345.

Jarial, H. S., and Scudder, G. E., 1970, The morphology and ultrastructure of the Malpighian tubules and hindgut in *Cenocorixa bifida* (Hemiptera, Corixidae), *Z. Morphol. Tiere* **68**:269-299.

Jeantet, A. Y., 1971, Recherches histophysiologiques sur le développement post-embryonnaire et le cycle annuel de *Formica* (Hyménoptère). II. Particularités histochimiques et ultrastructurales de l'intestin moyen de *Formica polyctena* Foerst, *Z. Zellforsch. Mikrosk. Anat.* **116**:405-424.

Jeantet, A. Y., 1981, Principaux aspects de l'accumulation de composés minéraux et de réserves organiques par un insecte social, *Formica polyctena* F. Implications physiologiques et écotoxicologiques, Thèse de Doctorat d'Etat, Université Paris 6, Biologie animale.

Jeantet, A. Y., Ballan-Dufrançais, C., and Martoja, R., 1977, Insects resistance to mineral pollution: Importance of the spherocrystal in ionic regulation, *Rev. Ecol. Biol. Sol* **14**:563-582.

Jeantet, A. Y., Ballan-Dufrançais, C., and Ruste, J., 1980, Quantitative electron probe microanalysis on insects exposed to mercury. II. Involvement of the lysosomal system in detoxification processes, *Biol. Cell.* **39**:325-334.

Kadiri, Z., 1979, Cytodifferencitation embryonnaire et postnatale de l'intestin moyen chez le phasme *Clitumnus extradentatus* Br. Analyse ultrastructurale et experimentale, Thèse de Doctorat de 3e cycle, Université de Bordeaux, Biologie animale.

Kendall, M. D., 1969, The fine structure of the salivary glands of the desert locust, *Shistocerca gregaria* F., *Z. Zellforsch. Mikrosk. Anat.* **98**:399-420.

Kessel, R. G., 1961, Electron microscope observations on the submicroscopic vesicular component of the subesophageal body and pericardial cells of the grasshopper *Melanoplus differentialis* Thomas, *Exp. Cell Res.* **22**:108-119.

Kessel, R. G., 1962, Light and electron microscope studies on the pericardial cells of nymphal and adult grasshoppers, *Melanoplus differentialis* Thomas, *J. Morphol.* **110**:79-103.

Kessel, R. G., and Beams, H. W., 1963, Electron microscope observations on the salivary gland of the cockroach, *Periplaneta americana*, *Z. Zellforsch. Mikrosk. Anat.* **59**:857-877.

Kitajima, W. E., 1975, A peculiar type of glycocalyx on the microvilli of the midgut epithelial cells of the thrips *Frankliniella* sp. (Thysanoptera, Thripidae), *Cytobiology* **2**:299-303.

Kloetzel, J. A., and Laufer, H., 1969, A fine structural analysis of larval salivary gland function in *Chironomus thummi*, *J. Ultrastruct. Res.* **29**:15-36.

Kummel, G., and Zerbst-Boroffka, I., 1974, Electron microscopic and physiological studies on the rectal pads in *Apis mellifica*, *Cytobiology* **9**:432-459.

Lane, N. J., 1982, Insect intercellular junctions: Their structure and development. In *Insect Ultrastructure*, edited by R. C. King and H. Akai, vol. 1, pp. 402-433, Plenum Press, New York.

Lane, N. J., and Harrison, J. B., 1979, An unusual cell surface modification: A double plasma membrane, *J. Cell Sic.* **39**:355-372.

Lauverjat, S., 1972, Rôle des cellules zymogènes dans les secretions salivaires de *Locusta migratoria* (Orthoptères, Acridoidea), *Tissue Cell* **4**:301-310.

Lauverjat, S., 1973, Ultrastructure des glandes salivaires de *Locusta migratoria*, *Arch. Zool. Exp. Gen.* **114**:129-147.

Lehane, M. J., 1976a, Digestive enzyme secretion in *Stomoxys calcitrans* (Diptera, Muscidae), *Cell Tissue Res.* **170**:275-287.

Lehane, M. J., 1976b, Formation and histochemical structure of the peritrophic membrane in the stablefly *Stomoxys calcitrans*, *J. Insect Physiol.* **22**:1551-1557.

Lehane, M. J., 1977, Transcellular absorption of lipids in the midgut of the stablefly *Stomoxys calcitrans*, *J. Insect Physiol.* **23**:945-954.

Leslie, R. A., and Robertson, H. A., 1973, The structure of the salivary gland of the moth (*Manduca sexta*), *Z. Zellforsch. Mikrosk. Anat.* **146**:553-564.

Lhonoré, J., 1973, Application conjointe des méthodes morphologiques, cytochimiques et d'analyse par spectrographie des rayons X à l'étude de l'appareil excréteur de *Gryllotalpa gryllotalpa* Latr. (Orthoptere, Gryllotalpidae), *Arch. Zool. Exp. Gen.* **114**:439-474.

Lhonoré, J., 1976, Données morphologiques et histochimiques sur les tubes de Malpighi des imagos de *Pieris brassicae* L. (Lépidoptère), *Ann. Sci. Nat. Zool.* **12**:275-293.

Lindsay, K. L., and Marshall, A. T., 1980, Ultrastructure of the filter chamber complex in the alimentary canal of *Eurymela distincta* Signoret (Homoptera, Eurymelidae), *Int. J. Insect Morphol. Embryol.* **9**:179-198.

Locke, M., and Collins, J. V., 1965, The structure and function of protein granules in the fat body of an insect, *J. Cell Biol.* **26**:857-884.

Marshall, A. T., and Cheung, W. W. K., 1970, Ultrastructure and cytochemistry of an extensive plexiform surface coat on the midgut cells of a fulgorid insect, *J. Ultrastruct. Res.* **33**:161-172.

Marshall, A. T., and Cheung, W. W. K., 1973, Studies on water and ion transport in homopteran insects: Ultrastructure and cytochemistry of the cicadoid and cercopoid hindgut, *Tissue Cell* **5**:671-678.

Marshall, A. T., and Cheung, W. W. K., 1974, Studies on water and ion transport in homopteran insects: Ultrastructure and cytochemistry of the cicadoid and cercopoid Malpighian tubules and filter chambers, *Tissue Cell* **6**:153-171.

Martoja, R., 1981, Contribution des vitellophages à l'édification du mésentéron du premier stade larvaire de *Locusta migratoria* (Orthoptère): Signification physiologique, *Arch. Zool. Exp. Gen.* **122**:39-46.

Martoja, R., and Seureau, C., 1972, Répartition des accumulations de métaux et de déchets puriques chez *Calliphora erythrocephala* (Diptère, Brachyptère), *C.R. Acad. Sci. Sér. D* **274**:1534-1537.

Martoja, R. and Truchet, M., 1983, Role d'un ommochrome dans l'excretion de metaux essentiels (Cu,Zn) et dans la detoxication a l'egard de contaminants metalliques (Ag,Cd) chez un Insecte (*Locusta migratoria*, Orthoptère). *C.R. Acad. Sci. Sér. III* **297**:219-224.

Meredith, J., and Phillips, J. E., 1973a, Rectal ultrastructure in salt- and fresh-water mosquito larvae in relation to physiological state, *Z. Zellforsch. Mikrosk. Anat.* **138**:1-22.

Meredith, J., and Phillips, J. E., 1973b, Ultrastructure of the anal papillae of a salt water mosquito larva, *Aedes campestris, J. Insect Physiol.* **19**:1157-1172.

Mills, R. P., and King, R. C., 1965, The pericardial cells of *Drosophila melanogaster, Q. J. Microsc. Sci.* **106**:261-268.

Munk, R., 1968, Ueber den Feinbau der Kleinzikade *Euscelidius variegatus* KBM (Jassidae), *Z. Zellforsch. Mikrosk. Anat.* **85**:210-224.

Nishiitsutsuji-Uwo, J., and Endo, Y., 1981, Gut endocrine cells in insects: The ultrastructure of the endocrine cells in the cockroach midgut, *Biomed. Res.* **2**:30-44.

Noirot, C., and Bayon, C., 1969, La cuticule proctodeale des Insectes: Mise en évidence de "dépressions épicuticulaires" par le microscope électronique à balayage, *C.R. Acad. Sci.* **269**:996-999.

Noirot, C., and Noirot-Timothée, C., 1967, L'épithélium absorbant de la panse d'un Termite supérieur: Ultrastructure et rapport avec la symbiose bactérienne, *Ann. Soc. Entomol. Fr.* **3**:577-592.

Noirot, C., and Noirot-Timothée, C., 1970, Revêtement particulaire de la membrane plasmique et absorption dans le rectum des insectes, *Proc. 7th Int. Electron Microsc.* Grenoble, pp. 37-38.

Noirot, C., and Noirot-Timothée, C., 1971a, Ultrastructure du proctodeum chez le thysanoure *Lepismodes inquilinus* Newman (*Thermobia domestica* Packard). I. La region anterieure (ileon et rectum), *J. Ultrastruct. Res.* **37**:119-137.

Noirot, C., and Noirot-Timothée, C., 1971b, Ultrastructure du proctodeum chez le thysanoure *Lepismodes inquilinus* Newman (*Thermobia domestica* Packard). II. Le sac anal. *J. Ultrastruct. Res.* **37**:335-350.

Noirot, C., and Noirot-Timothée, C., 1972, Structure fine de la bordure en brosse de l'intestin moyen chez les Insectes, *J. Microsc. (Paris)* **13**:85-96.

Noirot, C., and Noirot-Timothée, C., 1976, Fine structure of the rectum in cockroaches (Dictyoptera): General organization and intercellular junctions, *Tissue Cell* **8**:345-368.

Noirot, C., and Noirot-Timothée, C., 1977, Fine structure of the rectum in termites (Isoptera): A comparative study, *Tissue Cell* **9**:693-710.

Noirot, C., Noirot-Timothée, C., and Kovoor, J., 1967, Revêtement particulaire de la membrane plasmique en rapport avec l'excrétion dans une région specialisée de l'intestin moyen des Termites supérieurs, *C.R. Acad. Sci.* **264**:722-725.

Noirot, C., Smith, D., Cayer, M., and Noirot-Timothee, C., 1979, The organization and isolation of insect rectal sheath cells: A freeze-fracture study, *Tissue Cell* **17**:325-336.

Noirot-Timothée, C., and Noirot, C., 1965, L'intestin moyen chez la reine des Termites superieurs: Etude au microscope électronique, *Ann. Sci. Nat. Zool.* **7**:185-208.

Noirot-Timothée, C., Noirot, C., Smith, D. S., and Cayer, M. L., 1979, Jonctions et contacts intercellulaires chez les Insectes. II. Jonctions scalariformes et complexes formés avec les mitochondries: Etude par coupes fines et cryofractures, *Biol. Cell.* **34**:127-136.

Nopanitaya, W., and Misch, D. W., 1974, Developmental cytology of the midgut in the flesh fly *Sarcophaga bullata* (Parker), *Tissue Cell* **6**:487-502.

O'Loughlin, G. T., and Chambers, T. C., 1972, Extracellular microtubules in the aphid gut, *J. Cell Biol.* **53**:575-578.

Oschman, J. L., and Berridge, M. J., 1970, Structural and functional aspects of salivary fluid secreation in *Calliphora, Tissue Cell* **2**:281-310.

Oschman, J. L., and Wall, B. J., 1969, The structure of the rectal pads of *Periplaneta americana* L. with regard to fluid transport, *J. Morphol.* **127**:475-510.

Papillon, M., Fain-Maurel, M. A., and Cassier, P. 1974, Contribution à l'étude morphologique de l'intestin moyen de *Locusta migratoria migratorioides* (R. et F.). In *Recherches Biologiques Contemporaines* (L. Arvy, ed.) pp. 119-138, Vagner, Nancy.

Peacock, A. J., and Anstee, J. H., 1977, Anatomical and ultrastructural study of the rectum of *Jamaicana flava* (Caudell), *Micron* **8**:9-18.

Phillips, D., and Swift, H., 1965, Cytoplasmic fine structure of *Sciara* salivary glands, *J. Cell Biol.* **27**:395-409.

Platzer-Schultz, I., and Reiss, F., 1970, Zur Histologie der Bildungszone der peritrophischen Membian einiger Chironomidenlarven (Diptera), *Arch. Hydrobiol.* **67**:396-411.

Platzer-Schultz, I., and Welsch, U., 1970, Apokrine Sekretion der peritrophischen Membran von *Chironomus thummi piger* Str. (Diptera), *Z. Zellforsch. Mikrosk. Anat.* **104**:530-540.

Porter, K., Kenyon, K., and Badenhausen, S., 1967, Specializations of the unit membrane, *Protoplasma* **63**:262-274.

Reger, J. F., 1971, Fine structure of the surface coat of midgut epithelial cells in the homopteran, *Phylloscelis atra* (Fulgorid), *J. Submicrosc. Cytol.* **3**:353-358.

Reinhardt, C. A., and Hecker, H., 1973, Structure and function of the basal lamina and of the cell junction in the midgut epithelium (stomach) of female *Aedes aegypti* L. (Insecta, Diptera), *Acta Trop.* **30**:193-212.

Reinhardt, C. A., Schulz, Y., Hecker, H., and Freyvogel, T. A., 1972, Zur Ultrastruktur des Mitteldarmepithels bei Flöhen (Insecta, Siphonaptera), *Rev. Suisse Zool.* **79**:1130-1137.

Richards, A., and Richards, P., 1969, Development of microfibers in the peritrophic membrane of a mosquito larva, *Proc. 27th Annu. Meet. Electron Microsc. Soc. Am.* pp. 256-257.

Rudin, W., and Hecker, H., 1976, Morphometric comparison of the midgut epithelial cells in male and female *Aedes aegypti* L. (Insecta, Diptera), *Tissue Cell* **8**:459-470.

Ryerse, J. S., 1979, Developmental changes in Malpighian tubule cell structure, *Tissue Cell* **11**:533-552.

Scali, V., and Montanelli, E., 1975, DNA-RNA bodies in midgut cells of the stick insect, *Bacillus rossius*, *J. Exp. Zool.* **193**:361-367.

Smith, D. S., 1968, *Insect Cells: Their Structure and Function*. Oliver & Boyd, Edinburgh.

Smith, D. S., and Littau, N. C., 1960, Cellular specialization in the excretory epithelia of an insect, *Macrosteles fascifrons*, *J. Biophys. Biochem. Cytol.* **8**:103-133.

Smith, D. S., Compher, K., Janners, M., Lipton, C., and Wittle, L. W., 1969, Cellular organization and ferritin uptake in the midgut epithelium of a moth *Ephestia kühniella*, *J. Morphol.* **127**:31-72.

Sohal, R. S., 1974, Fine structure of the Malpighian tubules in the housefly, *Musca domestica*, *Tissue Cell* **6**:719-728.

Sohal, R. S., Peters, P. D., and Hall, T. A., 1976, Fine structure and X-ray microanalysis of mineralized concretions in the Malpighian tubules of the housefly, *Musca domestica*, *Tissue Cell* **8**:447-458.

Sohal, R. S., Peters, P. D., and Hall, T. A., 1977, Origin, ultrastructure, composition and age-dependence of mineralized dense bodies (concretions) in the midgut epithelium of the adult housefly, *Musca domestica*, *Tissue Cell* **9**:87-102.

Staübli, W., Freyvogel, T. A., and Suter, J., 1966, Structural modification of the endoplasmic reticulum of midgut epithelium cells of mosquitoes in relation to blood intake, *J. Microsc. (Paris)* **5**:189-204.

Stobbart, R. H., and Shaw, J., 1964, Salt and water balance: Excretion. In *The Physiology of Insecta*, edited by M. Rockstein, vol. 3, pp. 189-258, Academic Press, New York.

Strambi, C., and Zylberberg, L., 1972, Histologie et ultrastructure du proctodeum des Coléoptères Catopides (imagos), *Ann. Sci. Nat. Zool.* **14**:241-284.

Tanaka, K., and Ikeda, M., 1982, Electron microscopic studies of the pericardial cells of *Bombyx mori*. In *Ultrastructure and Functioning of Insect Cells*, edited by H. Akai, R. C. King, and S. Morohoshi, pp. 131-134, Society for Insect Cells, Japan.

Taylor, H. H., 1971a, Water and solute transport by the Malpighian tubules of the stick insect, *Carausius morosus:* The normal ultrastructure of the type-1 cell, *Z. Zellforsch. Mikrosk. Anat.* **118**:333-368.

Taylor, H. H., 1971b, The fine structure of the type-2 cell in the Malpighian tubules of the stick insect *Carausius morosus*, *Z. Zellforsch. Mikrosk. Anat.* **122**:411-424.

Teigler, D. J., and Arnott, H. J., 1972, Crystal development in the Malpighian tubules of *Bombyx mori* (L.), *Tissue Cell* **4**:173-185.

Terzakis, J. A., 1967, Substructure in an epithelial basal lamina (basement membrane), *J. Cell Biol.* **35**:273-278.

Thomas, D., and Gouranton, J., 1972, Isolement des cristaux intranucléaires de l'intestin moyen du ver de farine *Tenebrio molitor* L. et observation au microscope à balayage, *J. Microsc. (Paris)* **14**:125-128.

Thomas, D., and Gouranton, J., 1973a, Durée de formation des cristaux protéiques intranucléaires de l'intestin moyen de *Tenebrio molitor*, *J. Insect Physiol.* **19**:515-522.

Thomas, D., and Gouranton, J., 1973b, Etude au moyen de précurseurs marqués de la synthese de cristaux protéiques intranucléaires chez *Tenebrio molitor* L., *J. Microsc. (Paris)* **16**:287-298.

Thomas, D., and Gouranton, J., 1980, Globular intranuclear inclusions in the midgut cells of *Carausius morosus:* Ultrastructure, composition and kinetics of growth, *J. Ultrastruct. Res.* **70**:137-152.

Thomas, D., Gouranton, J., and Wroblewski, H., 1977, Etude de la proteine des cristaux intranucléaires de l'intestin moyen de *Tenebrio molitor* L., *Biol. Cell.* **28**:195-206.

van. der Starre, van der Molen, L. G., and de Priester, W., 1972, Brush-border formation in the midgut of an insect, *Calliphora erythrocephala* Meigen, *Z. Zellforsch. Mikrosk. Anat.* **125**: 295-305.

Waku, Y., 1974, Ultrastructure of Malpighian tubule cells on the silkworm, *Bombyx mori* L., with special regard to metamorphosis, *Zool. Mag.* **83**:152-162 (Japanese with English summary).

Waku, Y., and Sumimoto, K., 1974, Metamorphosis of midgut epithelial cells in the silkworm (*Bombyx mori* L.) with special regard to the calcium salt deposits in the cytoplasm. II. Electron microscope, *Tissue Cell* **6**:127-136.

Wall, B. J., and Oschman, J. L., 1973, Structure and function of rectal pads in *Blatella* and *Blaberus* with respect to the mechanism of water uptake, *J. Morphol.* **140**:105-118.

Wall, B. J., Oschman, J. L., and Schmidt, B. A., 1975, Morphology and function of Malpighian tubules and associated structures in the cockroach *Periplaneta americana*, *J. Morphol.* **146**: 265-306.

Waterhouse, D. F., and Wright, M., 1960, The fine structure of the mosaic midgut epithelium of blowfly larvae, *J. Insect Physiol.* **5**:230-239.

Wessing, G., and Eichelberg, G., 1978, Malpighian tubules, rectal papillae and excretion. In *The Genetics and Biology of Drosophila*, edited by M. Ashburner and T. R. F. Wright, vol. 2c, pp. 1-42, Academic Press, New York.

Wigglesworth, V. B., and Salpeter, M. M., 1962, Histology of the Malpighian tubules in *Rhodnius prolixus* Stal (Hemiptera), *J. Insect Physiol.* **8**:299-307.

The Ultrastructure of Interacting Endocrine and Target Cells

BONNIE JOY SEDLAK

1. Introduction

A study of cell structure alone is inherently interesting, but it is far less informative than a study which attempts to elucidate the functions reflected in cellular morphology. In developmental biology, the focus is on cellular changes occurring with time. How does the ultrastructure of a cell change as it matures and what is the functional significance of these changes? Since the development of insects is controlled by hormonal cues, a study of the developing endocrine glands and their target tissues interpreted in the context of their own endocrine environment presents a unique opportunity to determine a structural basis for endocrine activity and to define interactions between these tissues. One of the best studied insects from this point of view is *Manduca sexta*. In this tobacco hornworm, the titers of various hormones have been carefully determined and the ultrastructures of certain endocrine glands and target tissues have been recorded. This review will concentrate on the work carried out with this insect, but will refer to work with other insects where appropriate.

M. sexta is a holometabolous lepidopteran which is easily reared on an artificial diet (Bell and Joachim, 1976). *Manduca* develop quite synchronously when maintained on a specific photoperiod. Reared on a long-day light period (16L:8D), the insect will not enter diapause and will be competent to develop only at specific periods during each day (i.e., at specific "gates") (Truman, 1972). Thus, these animals can be precisely staged. The time it takes a larva to develop

---

BONNIE JOY SEDLAK • Developmental Biology Center, University of California, Irvine, California 92717, USA.

from the fourth to the fifth instar is 3 days, and it takes 9 days to develop from the fifth larval instar to the pupa. The critical timing of animals is important if correlations are to be made between changes in hormone titers and ultrastructural morphologies of endocrine-producing and target cells.

2. The Insect Endocrine System

A simplistic description of the endocrine system which controls insect development usually begins with the role of neurosecretory cells (NSC) of the brain. These nerve cells, which are capable of generating electrical impulses (Raabe, 1982), are specialized to also perform endocrine functions. One of the hormones synthesized in the brain is the prothoracicotropic hormone (PTTH). It is one factor which controls hormonal activity of the prothoracic glands (PTG); and these glands responds to PTTH stimulus by secreting ecdysone. This prohormone is converted to its active form ecdysterone (20-hydroxyecdysone; 20HE) in the peripheral tissues such as the fat body. It is 20HE that gives all responding tissues the signal to molt. Juvenile hormone (JH) is secreted by the corpora allata (CA), and it determines whether the insect will molt to a larva, a pupa, or an adult. Epidermal cells are the targets which respond to 20HE and JH by secreting the cuticle characteristic for each stage. The actual titer of JH influences the qualitative properties of each molt in holometabolous insects. The larval-to-larval molts occur in response to 20HE and a relatively high level of JH, a lower level of JH plus 20HE will cue the pupal type of metamorphosis, and in the presence of 20HE without JH, the insect will respond by transforming into an adult. In addition to its role as a hormone controlling growth and development, JH is also a gonadotropic hormone. In many adult insects, it is needed for the uptake of vitellogenin into developing oocytes (Bownes, 1982).

This simplistic view of the functions of endocrine glands has been complicated by recent studies that show there are interactions between the hormones and the glands which secrete them; interactions which influence certain developmental programs. Early in the study of insect endocrinology, Bodenstein (1955) emphasized the role of humoral balance in influencing metamorphosis, e.g., CA transplanted into successively older last-instar host larvae of *Galleria* lose their characteristic effectiveness in producing a larval morphology, presumably because the high titer of PTG hormone found late in the instar inactivates the CA. Such hormone interactions have continued to interest endocrinologists and there is evidence for a stimulation by 20HE of both the PTG and the CA and a stimulation of both glands by JH. Specifically, the molting hormone stimulates RNA synthesis, initially in the PTG and a few hours later in the CA of the moths *Philosamia cynthia* and *Hyalophora cecropia*. Similarly, JH stimulates first the CA and then the PTG (Siew and Gilbert, 1971). That 20HE may stimulate PTTH secretion is also suggested from a study of the cabbage armyworm *Mamestra brassicae* where the PTTH secretion was assessed by noting the degree to which NSC stained with aldehyde fuchsin; a reduction in the ability of cells to stain was interpeted to indicate a release of NSC product.

When 20HE was injected into larvae that had their PTG removed by cauterization, synthesis and release of NSC material was stimulated. Agui and Hiruma (1977) interpreted their results to indicate that an indirect stimulation of the PTG by 20HE occurred via the NSC of the brain.

Hiruma et al. (1978) showed that the response of PTG to JH analogs (JHA) is time dependent. If JHA are applied to neck-ligated larvae of *M. brassicae* during the fourth or fifth day of the last instar, the animals pupate presumably because of stimulation of the PTG, but when JHA are applied during the second or third day, pupation is not induced. Thus, PTG activation by JHA only occurs during the latter part of the ultimate larval instar. In contrast, PTG activation by the PTTH was not temporally restricted and occurred throughout the last larval instar.

20HE may suppress the action of JH since Friedel et al. (1980) provided evidence that ecdysteroids play an allatostatic role during normal reproductive cycles in the cockroach *Diploptera punctata*. When 20HE was injected into female adults, there was a decline in the biosynthesis of JH as reflected in the small size of vitellogenic oocytes and in the reduced content of oocyte vitellin. The *in vitro* experiments suggested that the suppression of JH synthesis was not the result of a direct action of 20HE on the CA, since the glands incubated with 20HE showed no decrease in rate of JH synthesis. The authors suggest that the negative influence of ecdysteroids on JH synthesis occurs at the level of the NSC.

In several lepidopterans, the kinds of effect that JH has on PTG secretory activity differ depending on the dose of hormone and/or the developmental stage when hormone is applied to the insect. In general, JH or its analogs can either accelerate or delay the larval-to-pupal metamorphosis. Cymborowski and Stolarz (1979) applied JHA to the moth *Spodoptera littoralis* at the time when the larvae attained their maximum body weight and found that metamorphosis was delayed. JHA application after this time accelerated metamorphosis. Similarly, Safranek et al. (1980) reported that in *M. sexta*, the effect of JH on metamorphosis (and, by inference, the stimulation of the PTG by JH) depends on the period when JH and JHA are injected into developing larvae. If JH or JHA are injected into neck-ligated larvae before the onset of the wandering behavior, metamorphosis is delayed, whereas if the hormones are injected after wandering has begun, further development is accelerated. Since it is known that ecdysteroids are at a low level prior to wandering but reach a maximal peak afterwards, the JH effects are thought to be ecdysone dependent. The authors conclude that the brain (specifically the NSC which secrete PTTH) is not the target of JH, since the ability of JH to accelerate or delay a molt is the same in both brainless and intact larvae.

Sehnal et al. (1981) also reported that JH can either accelerate or delay the larval-to-pupal molt in the wax moth *Galleria mellonella*. In this insect, if JHA are topically applied to larvae early in the last larval instar, the molt is accelerated, whereas a moderate dose of hormone administered later (at mid instar) will delay the molt. JH may normally suppress PTG activity in *Galleria* since in decapitated larvae the PTG will become spontaneously activated to secrete ecdysteroids, but the amount of molting hormone secreted is three times more than that which is found in intact larvae, i.e., in animals where the JH source has not been removed. The authors conclude that molting hormone secretion depends on both JH and PTTH since JH can accelerate or delay the release of PTTH *in situ*, but if PTG have already been stimulated by PTTH, JH can directly inhibit ecdysone secretion. Thus, it is the balance between JH and PTTH which is essential in PTG regulation.

The elucidation of such complex hormonal interactions points to the need to understand the ultrastructural changes that take place in the cells of each gland, not only with respect to the synthesis of their specific hormones, but also in response to their hormonal environment. The primary reason for focusing on *M. sexta* in this review of the structural basis of endocrine function is that the ecdysteroid titers (Bollenbacher *et al.*, 1981), JH titers (Bergot *et al.*, 1983), and cycles of PTTH activity (Agui *et al.*, 1980) have been defined for this moth, and studies of the ultrastructure of CA and PTG have been completed (Sedlak *et al.*, 1983).

3. The Ultrastructure of the Endocrine Glands of Manduca

The ultrastructure of the two glands that synthesize and secrete the major hormones which stimulate growth and differentiation will be described separately. PTG are composed of single cells that are linked to each other in rows forming a Y-shaped organ. The CA are paired glands that are connected to the brain via axonal tracts which first traverse a neurohemal organ called the *corpus cardiacum*.

Although the ultrastructure of both glands have been described for many insects (for reviews see Herman, 1967; Cassier, 1979; Sedlak, 1983), the relationships between ultrastructural changes and hormone production can only be determined where either good bioassays for endocrine activity or the actual titers of the hormones have been determined. *M. sexta* is the only insect where both the ultrastructure and the hormone titers have been described (Figure 1). The ecdysteroid titers determined by Bollenbacher *et al.* (1981) were expressed in micrograms per milliliter, whereas the JH titers (JH I, JH II, JH III) determined by Bergot *et al.* (1983) were presented in nanograms per milliliter. In Figure 1 the JH values from both studies were averaged, and both JH and ecdysteroid titers are represented as relative titers drawn with maximal peaks of each titer set at 100%. The graph shows the fluctuations of each hormone with time. A single cell of each gland is drawn in a temporal sequence to represent the major changes in ultrastructure which occur during development.

4. The Corpora Allata

The CA, which synthesize and secrete JH, are also the release site for PTTH in *Manduca* (Agui *et al.*, 1980; Sedlak, 1981). The JH titer fluctuates at precise developmental stages during both the fourth and the fifth larval instar, then it falls to a minimum level late in the ultimate larval instar, and remains low

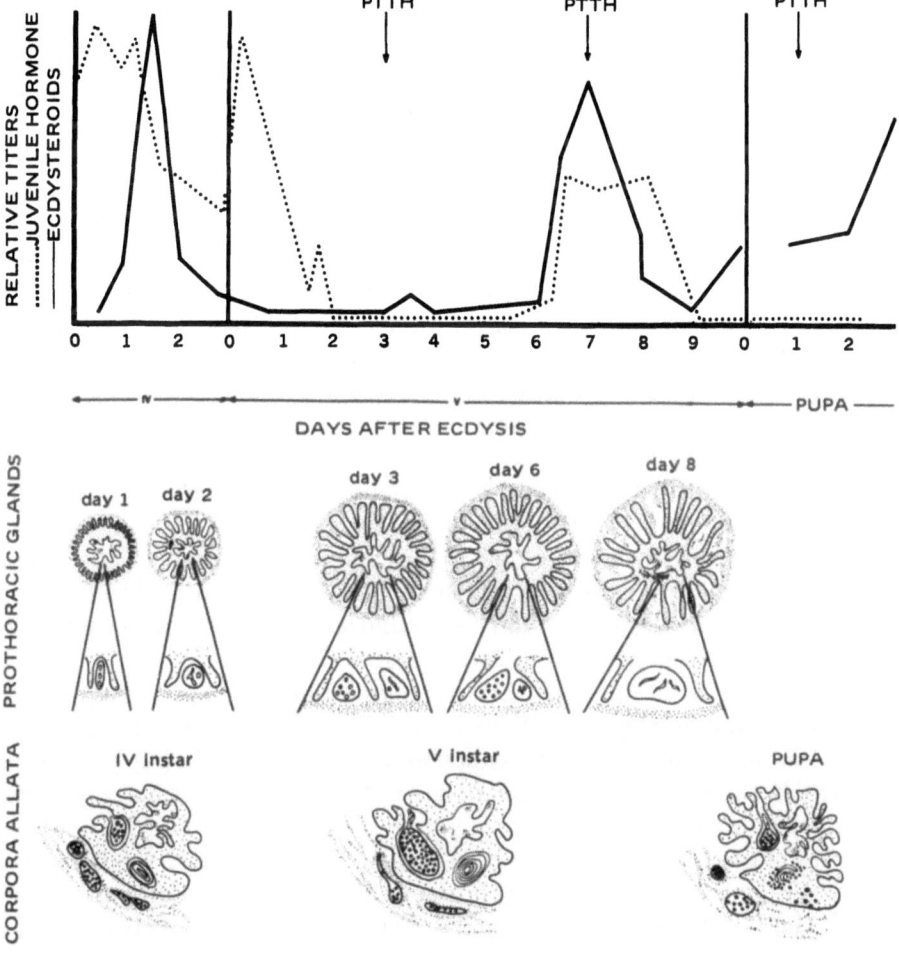

Figure 1. A diagram summarizing the relationship between hormone levels and the cytology of the endocrine cells of *Manduca sexta*. The relative titers of both juvenile hormone and ecdysteroids are indicated for the fourth (IV) and fifth (V) larval instars plus the first few days of pupal life. The times of peak prothoracicotropic hormone (PTTH) activity are indicated by arrows. Individual cells of the prothoracic gland and the corpora allata are drawn along the time axis to indicate the major morphological changes which are found coincident with fluctuations in hormone titers. The major structural changes in prothoracic gland (PTG) cells occur early and late in each larval instar. The invaginations of each PTG cell plasma membrane increase to form large intercellular spaces (ICS) on day 2 of the fourth larval instar and day 8 of the fifth larval instar. A portion of each PTG cell is enlarged to emphasize both the changes in ICS dimensions and the alterations in multivesicular sacs (MVS) which lie within these ICS. Both intact MVS and MVS depleted of their internal vesicles are found throughout both larval instars, but on day 8 most MVS lack inner vesicles. The nucleus in these cells retains a highly irregular outline during the time studied. The major alterations in corpora allata cells occur between larval and pupal development. The gland consists of cells which interdigitate with their neighbors, plus a stromal layer of extracellular material. Neurosecretory cell axons (NSC) are found both within the gland proper and the stromal regions. The NSC reach a maximum diameter in the middle of the fifth larval instar (by days 5–6). Concentric whorls of SER are present in the fourth and fifth larval instars but are not found in pupal glands. The pupal glands show a high degree of plasma membrane infolding and contain more Golgi complexes and dense bodies than are found in the larval cells. The nuclei retain an irregular outline throughout development (Sedlak *et al.*, 1983).

during early pupal development (Bergot *et al.*, 1983). Structural changes occur within the CA coincident with peaks and valleys of hormone titer which suggest a structural basis for the fluctuations in endocrine activity (Figure 1). There are periods of synthesis, secretion, and breakdown of JH, and each of these physiologically separate events is associated with a specific cellular morphology.

There is great variation seen in cell structure when comparisons are made between the CA of different species. Therefore, it is difficult to generalize and define the ultrastructural organelles responsible for hormone synthesis and secretion in all CA. However, in spite of this variability, there is consistent evidence that the SER is the site of JH production. It also seems clear that the glands are stimulated by NSC axons which have their cell bodies in the brain. The CA of *Manduca* are comprised of gland cells and their surrounding stroma. Within the extracellular stroma are tracheoles and NSC axons, and there are also NSC axons located within the gland proper. The CA cells of larvae, which presumably synthesize and secrete JH, have numerous mitochondria and free ribosomes, but relatively little RER and few Golgi complexes. SER is found in larval CA, but in pupae where the JH titer is low, gland cells have far fewer organelles than do the cells of larval glands, and the SER is notably missing in pupal CA. The only organelles which are found in increased numbers in the cells of pupal CA are the Golgi complexes and dense bodies.

In *Manduca*, the CA cells vary in volume depending on the age of the animal; e.g., the larval gland cells are large with a cytoplasm rich in organelles, in contrast to pupal glands. The reduction of cell cytoplasm in pupal glands is reflected by the increased interdigitation of cell membranes accommodating the decreased volume (Figure 1).

In almost every insect that has been studied, the glands which are thought to be actively synthesizing or secreting hormone have a large cytoplasm-to-nucleus ratio. The overall volume of the gland is, in fact, large in such "active" glands. Examples where CA are considered to be active on the basis of gland size are the locust *Schistocera gregaria* (Odhiambo, 1966a), the roaches *Leucophaea maderae* (Scharrer, 1964a) and *Periplaneta americana* (Khan *et al.*, 1978), the bug *Eurygaster integriceps* (Panov and Bassurmanova, 1970), the aphid *Aphis craccivora* (Elliott, 1976), the beetles *Galeruca tanaceti* (Siew, 1965), *Choleva augustata* (Deleurance and Charpin, 1978), *Leptinotarsa decemlineata* (Schooneveld, 1979), and *Hypera postica* (Tombes and Smith, 1970), the lepidopterans *Monema flavescens* (Takeda, 1977) and *Pectinophora gossypiella* (Raina and Borg, 1980), and the flies *Drosophila melanogaster* (Aggarwal and King, 1969) and *Calliphora erythrocephala* (Thomsen and Thomsen, 1970).

In some insects other than *Manduca*, the volume changes have also been correlated with hormone synthesis, e.g., in the cockroach *Nauphoeta cinerea* the CA volume increases at the same time that the JH titer rises, as assayed by the *Galleria* wax test (Lanzrein *et al.*, 1978). Schooneveld (1979) also reports that the increase in CA volume in active (preovulatory) females of the Colorado potato beetle, *L. decemlineata*, occurred coincidentally with high JH titers and high rates of JH synthesis. The volume of the CA does not increase when JH titers are high in every insect studied. For example, in adult female *S. gregaria*, there is a cyclic production of JH, but the volume of the CA does not change; whereas the volume of the CA in the male locust increases with age (Odhiambo, 1966a). Tobe and Pratt (1975) also report that there is no correlation between CA size and JH synthesis in S. gregaria. In their study, JH biosynthesis was monitored by in vitro radioimmunoassay (RIA). In the mealworm *Tenebrio molitor* (Mordue, 1965) and *Locusta migratoria migratoriodes* (Johnson and Hill, 1973), changes in volume were found, but these were not correlated with times of JH synthesis and release.

4.1. Synthesis of Juvenile Hormone

The most conspicuous organelle present during synthesis of JH is the SER, e.g. in the adult CA of the blatteran insect *Leucophaea maderae* (Scharrer, 1978), the coleopteran *Hypera postica* (Tombes and Smith, 1970), and the collembolan *Folsomia candida* (Palevody and Grimal, 1976) the SER is present in reproductively active females. In the adult insect where JH acts as a gonadotrophic hormone required for the uptake of vitellogenin into developing oocytes, the formation of a mature yolky oocyte can be used as an assay for the presence of JH. Scharrer's study showed that the amount of SER increased in castrated females. In fact, the glands from these operated animals resembled CA from normally active female glands in terms of their SER content. Removal of the ovaries is presumed to release the CA from a negative feedback regulation, and thus the SER in the glands of castrates is available to synthesize JH.

Another line of evidence suggesting that the SER is the site of JH biosynthesis comes from studies with precocenes. The precocenes are natural plant products which have an anti-JH effect on both larvae and adult insects. If milkweed bug (*Oncopeltus fasciatus*) larvae are exposed to precocenes, the animals will molt to the adult precociously as if the CA had been surgically removed. Precocenes block oocyte development at vitellogenic stages as if the treated females had insufficient JH (Bowers et al., 1976). These effects can be reversed with the application of JH or JHA, which suggests that the precocenes alter the ability of the CA to produce JH. However, precocenes also affect the NSC of the brain. This effect is seen in the decreased staining of NSC cells. Thus, it is possible that the effect of these drugs is at the level of the NSC regulation of the CA (Unnithan et al., 1977).

Prolonged administration of precocenes will cause the CA to atrophy (Unnithan et al., 1980); but application for a short time alters the cellular morphology in a more specific manner. Feyereisen et al. (1981) have demonstrated that in *Diploptera punctata*, the SER aggregates and eventually degenerates in precocene-treated animals. These authors suggest that individual CA cells can respond differently depending on the dose and time of drug application. In the CA of *O. fasciatus*, there are two distinct cell populations, termed *light cells* and *dark cells*. The dark cells appear more electron dense and are rich in ribosomes and RER, whereas the light cells are filled with vesicles to give a more electron-lucent appearance. Dorn (1973) concluded that the light cell population represented the "active" cells. The assumption that the cells were secreting JH was based on the finding that these light cells had larger nuclei and more cytoplasm in comparison with the dark cells. Note that the JH titer is not known for this insect. In contrast, Liechty and Sedlak (1978) concluded that the light cells are "inactive" but the dark cells "active" in terms of JH synthesis or secretion. They observed that in normal glands, there are more light cells present on the day of eclosion when the gland is not synthesizing JH (as determined by ovarian bioassay). Five days later, when yolky oocytes are present (indicating that JH has been secreted), the dark cell population predominates. Furthermore, in precocene-treated animals, light cells are the dominant type. Feyereisen et al. (1981) proposed that the light cell vesicles described in the study of Liechty and Sedlak are actually vesiculated SER, which is consistent with the idea that the SER degenerated in precocene-treated glands.

Tobe and Saleuddin (1977) incubated the CA of S. gregaria with ^{3}H-labeled JH precursor ([^{3}H]farnesenic acid), and time was allowed for the conversion to the hormone before the CA were processed for autoradiography. The conversion to the final JH product was not actually demonstrated in this study, and although the grain distribution showed localization over the SER, this was not the only cellular site labeled, i.e., silver grains were also localized over mitochondria, Golgi complexes, and even nuclei. These studies provide weak support for the hypothesis that the SER is the site of JH synthesis.

Nonetheless, it is likely that the SER is a site of JH biosynthesis based on the facts that (1) SER is present in reproductively active females where JH has been synthesized and released; (2) SER degeneration is found in insects treated with the precocenes which seem to decrease JH production and/or release by CA; and (3) SER is one of the sites labeled during the autoradiographic localization of a JH precursor. SER has been reported in the CA of a number of species: the collembolan *Folsomia candida* (Palévody and Grimal, 1976), the roach *L. maderae* (Scharrer, 1964a, 1978), the locust *S. gregaria* (Odhiambo, 1966b), the beetles, *H. postica* (Tombes and Smith, 1970) and *C. augustata* (Deleurance and Charpin, 1978), several lepidopterans, including *Hyphantria cunea* (Melnikova and Panov, 1975), *M. flavescens* (Takeda, 1977), *Diatrea grandoisella* (Yin and Chippendale, 1979), *P. gossypiella* (Raina and Borg, 1980), and *M. sexta* (Sedlak et al., 1983), and the flies *D. melanogaster* (King et al., 1966; Aggarwal and King, 1969) and *C. erythrocephala* (Thomsen and Thomsen, 1970).

In several insects, i.e., *D. melanogaster* (King et al., 1966), *C. augustata* (Deleurance and Charpin, 1978), and *D. grandiosella* (Yin and Chippendale, 1979), the SER is arranged as whorls of concentric cisternae (Figure 2). SER whorls are also found in *M. sexta*. They are present during the fifth larval instar up until the time when the JH peak is seen on day 7 but absent from pupal glands when the JH titer is minimal. Thus, in *Manduca*, the organelle is present at all times required for JH synthesis and is probably a site of hormone production.

A well-developed RER has also been observed in CA from a number of species: *F. candida* (Palévody and Grimal, 1976), *L. maderae* (Scharrer, 1964a), *C. augustata* (Charpin, 1975), *Eurygaster integriceps* (Panov and Bassurmanova, 1970), *Rhodnius prolixus* (Baehr et al., 1973), *O. fasciatus* (Dorn, 1973), *H. cecropia* (Waku and Gilbert, 1964). The role of the RER in JH production must be indirect, since JH is a sesquiterpenoid, not a protein, and RER has long been established as the site of synthesis of protein for export. Among the polypeptides produced in the CA could be carrier proteins for the hormone but the product could not be the hormone itself. It is obvious that the mere presence of either SER or RER at a time of presumed glandular activity does not indicate endocrine synthesis.

4.2. Secretion of Juvenile Hormone

Stay and Tobe (1977) have shown that JH production is continuous in the roach *D. punctata*. If this is also true for *Manduca*, what accounts for the rise and fall in larval JH titer? Since the decrease in SER occurs at the same time as a rise in the ecdysteroid titer in *Manduca*, the molting hormone may suppress JH production. Friedel *et al.* (1980) have suggested that 20HE inhibits JH synthesis in *D. punctata*. Since no inhibition of JH synthesis was found when CA were incubated with 20HE *in vitro*, the authors suggest that 20HE affects the brain NSC cells to cause a decrease in JH.

How then is the elevation in JH titer accounted for in *Manduca*? If synthesis and secretion of hormone are indeed separate events, there may be structural evidence for a stimulus to secretion. The CA are regulated by NSC and non-NSC sources in many insects. The NSC components are usually stimulatory, and the non-NSC are usually inhibitory (deKort and Granger, 1981). In *M. sexta*, there are NSC axons terminating within the CA (Figure 3). NSC axons have also been found in the CA of the following insects: the locusts *L. migratoria migratoriodes* (Goltzene and Porte, 1978) and *S. gregaria* (Odhiambo, 1966b), the bug *O. fasciatus* (Dorn, 1973; Unnithan *et al.*, 1977), the aphid *A. craccivora* (Elliott, 1976), the weevil *H. postica* (Tombes and Smith, 1970), and the fruitfly *D. melanogaster* (Aggarwal and King, 1969). These axons contain NSC granules of variable sizes and variable staining intensities (i.e., electron densities). Although axonal types have been classified according to such morphological characteristics, these criteria are not a reliable indication that the axons differ in their hormonal content. Nordmann (1977) has demonstrated that the staining intensity of a population of granules can vary depending on the pH of the fixative used. At pH 7.0, the granules appear pleomorphic, in contrast to the uniformly staining electron-dense granules seen in cells fixed at either pH 5.0 or 6.0. Thus, the granule type is not a reliable indicator of the physiological status of a given axon.

It would be useful in assessing secretory activity in the CA if there were differences noted in the release of material from the gland at specific times. The two most commonly observed morphological markers indicating release are omega figures, associated with exocytosis of granules, and synaptoids.

Exocytosis is not frequently seen, unless glands are stimulated. In the corpora cardiaca (CC; the glands which store NSC material from the brain), there are many more NSC axons than are found in the CA, and thus the CC have been used to study the mode of granule release from axons. Stimulation of exocytosis by electrical shock in *C. erythrocephala* (Norman, 1969) and *S. gregaria* (Krogh and Normann, 1977), or by serotonin administration in the cockroach *L. maderae* (Scharrer and Wurzelmann, 1974) increased the number of omega figures seen in neurosecretory neurons within the CC. The authors of these

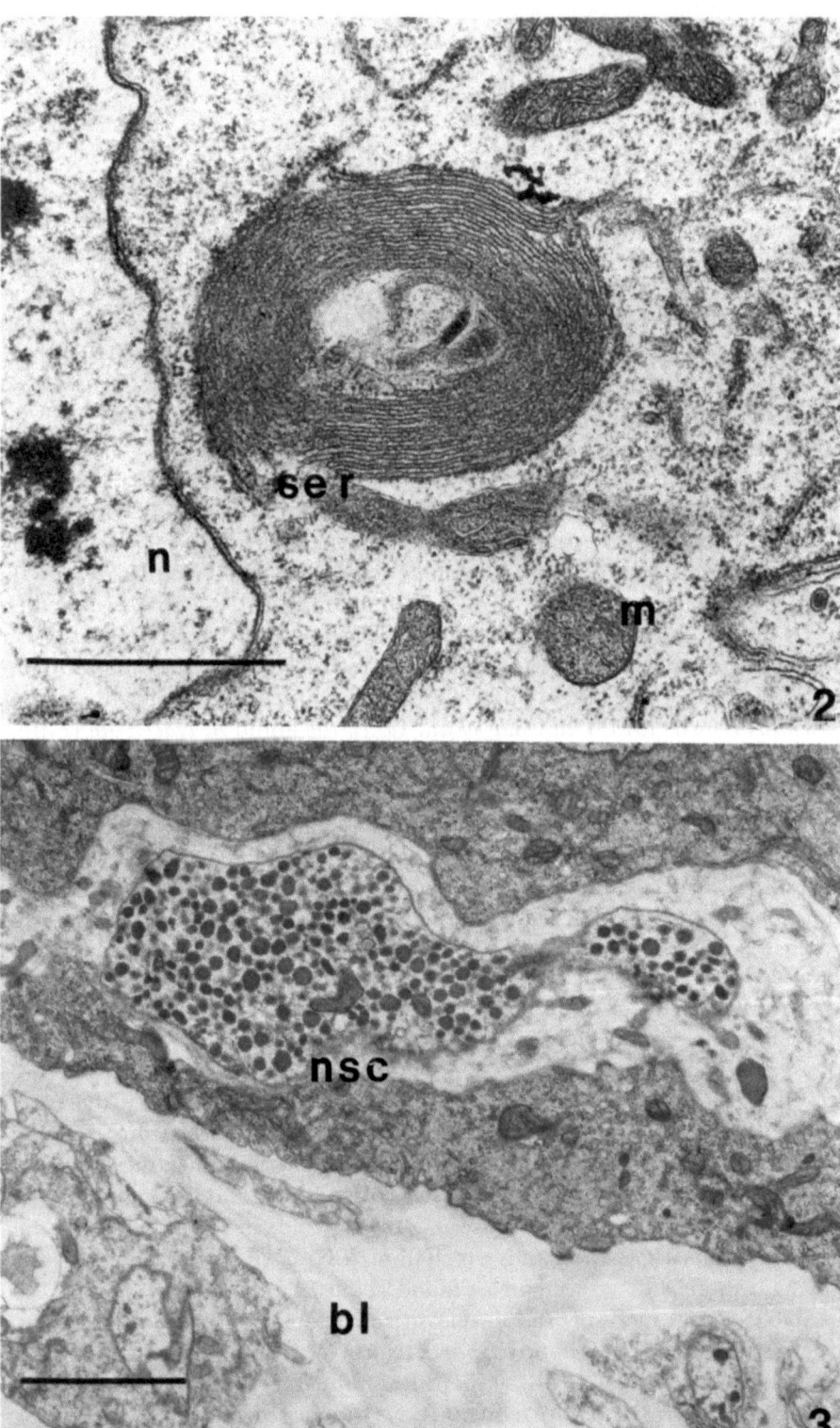

studies concluded that exocytosis is a rapid event, not readily seen in most preparations. It is reasonable to assume that exocytosis in the CA is an equally rare phenomenon. In *Manduca* CA, omega figures are not seen whereas synaptoids are plentiful at the junctions betweeen the NSC axon and the gland (Figure 4). Norman (1969) maintains that synaptoids represent the release sites of the hormone, but whether these vesicles result from fragmentation of the remaining omega figure following granule exocytosis is not clear. There was no fluctuation in either synaptoid number or morphology in *Manduca* CA during the developmental sequence studied (Sedlak, 1981).

However, the diameters of *Manduca* NSC axons change during development. They are small (0.4-1.9 μm) at the beginning of the fifth larval instar, much larger (1.4-4.4 μm) by day 4-5, and small (1.2 μm) again in the early pupal stages (Sedlak, 1981). This size variation corresponds to the rise and fall in the JH titer. The titer is low early in larval life, rises on day 6 (just following the time when large axons are noted), and then falls to a minimal level at pupation. Thus, the structural evidence indicates that there may be a stimulus of the CA by NSC axons which can account for the rise in JH titer. A change in NSC diameter was also seen in *A. craccivora* CA (Elliott, 1976). The axons enlarged during the larval stages when the glands presumably were producing JH, but decreased in the adult stages. Localized axonal bulges were found in both the CC and the CA of *D. melanogaster* (King *et al.*, 1966). Such axonal swellings in insects resemble the Herring bodies (local accumulations of neurosecretory material) of the vertebrate neurohypophysis (Bloom and Fawcett, 1975).

4.3. Axons of PTTH Release

There is a second population of NSC axons in the *Manduca* CA which is regionally distinct from the axons discussed in Section 4.2. This population resides within the relatively thick stromal sheath surrounding the CA gland proper (Figure 5). These axons may be the release sites for PTTH, the hormone which stimulates the PTG to secrete ecdysone (Sedlak, 1981).

The tropic hormone is synthesized by two lateral NSCs located within the brain (Agui *et al.*, 1979) and is sent down axonal tracts and released from the CA (Agui *et al.*, 1980). This surprising conclusion resulted from studies in which first brain extracts, then extracts from either the CC or the CA were used to stimulate the *in vitro* synthesis of ecdysone by the PTG. The ecdysone released from the PTG was detected by RIA, and measuring ecdysone levels gave an indirect assay for PTTH activity. The results showed that the PTTH activity was highest in the CA of 2-day-old pupae (Agui *et al.*, 1980).

Figure 2. Whorls of concentric cisternae of smooth endoplasmic reticulum (ser) are found with a corpus allatum cell of *Manduca* on the second day of the fifth larval instar. n, nucleus; m, mitochondrion. (Bar = 1 μm.)

Figure 3. A relatively large neurosecretory cell axon (nsc) located within the corpus allatum proper on day 5 of the fifth larval instar of *Manduca*. A stromal layer composed of basal lamina (bl) and nonglandular cells surrounds the gland proper. (Bar = 2 μm.)

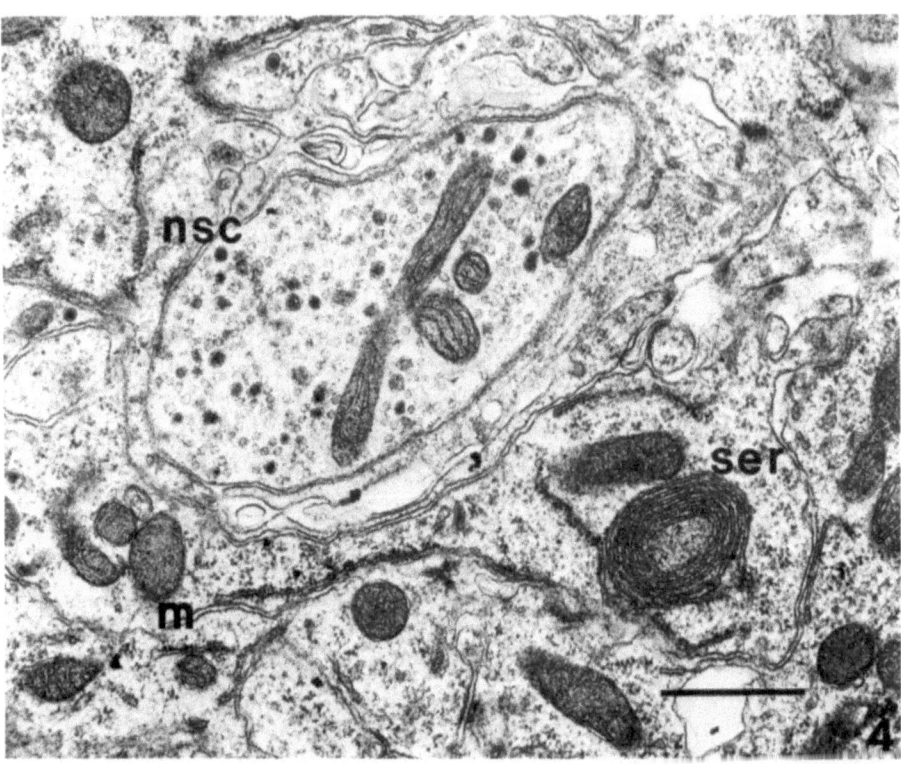

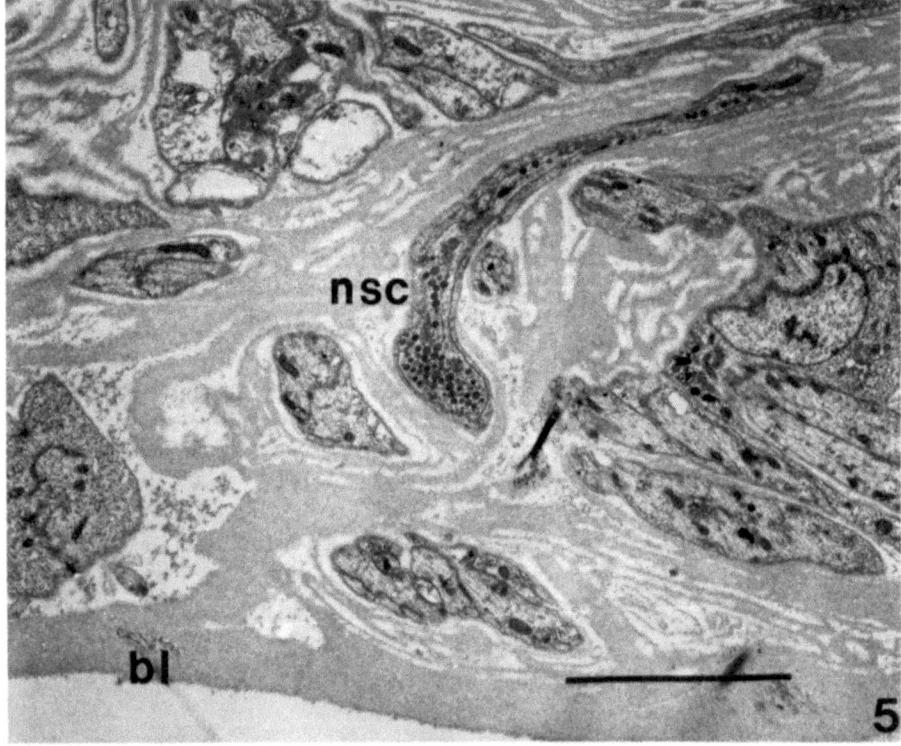

Based on the ultrastructure of CA axons, the stromal population of axons may be the PTTH release sites in the CA. The evidence for this is that diameter changes are only seen in the glandular NSC axons, and these changes correspond to fluctuations in JH titers. That is, the largest axon fibers were seen just preceding the major JH peak late in the fifth larval instar. In contrast, the stromal population of axons does not change during development, and their position in the extracellular sheath provides direct access for the NSC product to the hemolymph. No striking changes were found in the numbers of synaptoids present in the peripheral axons at the different developmental stages studied and exocytosis figures were not seen in these peripheral axons (Sedlak, 1981).

4.4. Degradation of Juvenile Hormone

Even though the biochemical degradation of JH by hemolymph esterases has been well documented and is probably a major mechanism to lower the hormone levels in *Manduca* (Vince and Gilbert, 1977), there is some structural indication that the hormone is broken down within the gland itself. *In situ* degradation of the mammotrophic hormone via lysosomes has been proposed for the vertebrate pituitary (Smith and Farquhar, 1966).

Golgi complexes and dense bodies are much more numerous in pupal than in larval gland cells, and this difference may reflect turnover of JH. The dense bodies could function as lysosomes which arise at the Golgi complexes. What is the role of these lysosomes in pupal cells? Since the JH titer has fallen to its minimal level at this time, these organelles may be responsible for eliminating the excess hormone, hormone precursors, and the enzymes necessary for hormone biosynthesis that are present in the pupal glands (Sedlak *et al.*, 1983).

5. Prothoracic Glands

The PTG of *M. sexta* secrete ecdysone (King *et al.*, 1974). In other insects, this hormone may be secreted by other tissues since ecdysteroids are present in animals after the PTG have degenerated. Degeneration of PTG occurs in adult *P. americana* and *L. maderae* (Scharrer, 1964b, 1966), *R. prolixus* (Herman, 1967), and *H. cecropia* (Herman and Gilbert, 1966). Degeneration of pupal PTG is found in *T. molitor* (Srivastava, 1958; Glitho *et al.*, 1979), *G. mellonella* (Blazsek *et al.*, 1975), *D. melanogaster*, and *C. erythrocephala* (King, 1970, p. 27).

The vertebrate sex glands secrete steroids and are rich in SER (Christensen, 1965; Bjersing, 1967). Based on analogies with these vertebrate glands, some

---

Figure 4. A neurosecretory cell axon (nsc) containing only a few electron-dense granules and numerous electron-lucent synaptoids from the corpus allatum of a day 2 fifth-instar larva of *M. sexta*. m, mitochondrion; ser, smooth endoplasmic reticulum. (Bar = 1 μm.)

Figure 5. A neurosecretory cell axon (nsc) located within the thick stromal sheath area surrounding a day 0 pupal corpus allatum of *Manduca sexta*. bl, basal lamina. (Bar = 10 μm.)

authors have proposed that the insect oenocytes (which are rich in SER) secrete ecdysone (Locke, 1969a; Romer, 1974). The vertebrate ovary and testis are not, however, the only glands which synthesize and secrete a steroid product. The adrenal cortex is also the site of steroid hormone production but this gland is not particularly rich in SER, and Giacomelli *et al.* (1965) suggest that the mitochondria associated with vacuoles are sites of steroid synthesis (see also discussion in Section 5.1). Therefore, it is not clear which vertebrate gland is analogous with the PTG. Furthermore, the vertebrates synthesize cholesterol and then convert this to their steroid hormones, whereas the insects merely convert dietary cholesterol to the ecdysteroids. Therefore, it is difficult to draw accurate analogies from the vertebrate to the invertebrate systems.

PTGs from different species show some similarities in the ultrastructural characteristics which are associated with either the synthesis or the secretion of ecdysone. For example, a peripheral channel system undergoes striking alterations during development, microvesicles are present within the intercellular spaces, and the morphology of mitochondria undergoes cyclical changes in many insects.

In *Manduca*, major peaks in the titers of hemolymph ecdysteroids occur late in both the fourth and the fifth instar. Early in the fifth larval instar, there is an additional minor peak of hormone (Bollenbacher *et al.*, 1981; Figure 1). Studies of the ecdysteroid levels in *T. molitor* (Delbecque *et al.*, 1978), *Philosamia cynthia* (Calvez *et al.*, 1976), *Pieris brassicae* (Lafont *et al.*, 1977), *G. mellonella* (Bollenbacher *et al.*, 1978), *Calpodes ethlius* (Dean *et al.*, 1980), and *D. melanogaster* (Richards, 1981) also show early minor and late major ecdysteroid peaks in the ultimate larval instars. In some of these insects, e.g., *Tenebrio* (Romer, 1971), *Galleria* (Gersch *et al.*, 1975; Blazsek *et al.*, 1975), and *Drosophila* (Aggarwal and King, 1969), the ultrastructure of the PTG has also been studied.

5.1. Synthesis of Ecdysone

There are relatively few cytoplasmic organelles in the PTG of *Manduca*, and there is a notable lack of SER. SER is found in the PTG of some insects which secrete the molting hormone, e.g., the cricket *Gryllus bimaculatus* (Romer, 1974), the moths *Antheraea pernyi* and *B. mori* (Beaulaton 1968), and the fly *D. melanogaster* (Aggarwal and King, 1969). Aggarwal and King (1969) noted in *D. melanogaster* that the amount of RER increased after each molt but SER was rare, whereas prior to the next molt the amount of RER decreased and SER increased. Although the authors suggest that there is a conversion of RER to SER, it is equally likely that the difference in proportion between these organelles represents a simple conversion in biosynthesis in these glands. After the molt, RER may be exporting protein to perhaps accommodate the increase in gland size found in the larger larvae and, since the ecdysteroid titers increase prior to the molt (Richards, 1981), SER may be involved in the synthesis of the steroid hormone. In support of the hypothesis that SER is involved in ecdysone synthesis in *Drosophila*, Aggarwal and King (1969) observed that SER increases during the prepupal period, i.e., about the time that ecdysteroid titers increase. In contrast, there is very little SER found in PTG cells of *lethal (2) giant larvae*, a mutant of *Drosophila* which does not molt, presumably because the PTG are subnormal in size (Aggarwal and King, 1969).

In *Manduca*, it has been demonstrated that ecdysone is secreted from PTG *in vitro* even though these glands lack SER. Thus, SER is not necessary for ecdysone production in this species, and it is not wise to assume that the presence or absence of an organelle defines the functional capabilities of a gland. It is only with studies that compare the structural characteristics of a gland with known hormone titers in the same insect that structure can be accurately, but still cautiously, assigned a function.

There is a suggestion from work with vertebrates that SER is not the sole site of steroid synthesis. The vertebrate adrenal cortex synthesizes and secretes steroids, and Giacomelli *et al.* (1965) suggested that in the rat adrenal glands, mitochondria associated with vacuoles may be involved in hormone synthesis along with a smooth membrane system, which could be either SER and/or Golgi complexes. In a similar manner, free ribosomes, mitochondria, and Golgi complexes are cellular components which could provide the enzymes used in ecdysone biosynthesis in *Manduca* PTG. There is no significant change in the number of ribosomes or Golgi complexes present during development, but mitochondrial morphologies differ somewhat between the fourth and the fifth larval instar. In fourth-instar larvae, the mitochondrial matrix appears more electron opaque and amorphous than the matrix of mitochondria from fifth-instar gland cells. What physiological significance is reflected in these changes is not clear.

In both *A. peryni* and *B. mori*, changes in mitochondrial morphology were studied in the context of fluctuations in hormone levels. Molting hormone titers were only determined for *Bombyx* using the *Calliphora* bioassay (Beaulaton, 1968) but in both lepidopterans, the mitochondira change from organelles with rodlike profiles to much enlarged macromitochondria which lack inner cristae. Since these larger organelles were associated with SER and were found at the time the molting hormone titer increased significantly, Beaulaton concluded that both mitochondria and SER were sites of ecdysone biosynthesis.

5.2. Secretion of Ecdysone

Ecdysone is a steroid which can presumably diffuse out of cells simply by dissolving within the lipid bilayer of the plasma membrane. There is no evidence of hormone storage within the gland.

What then is the structural evidence for secretion? The most common characteristic of PTG is the intercellular spaces formed by invaginations of the plasma membrane. These intercellular channels vary in depth and width throughout the larval-to-pupal transformation in many insects (Herman, 1967). In *Manduca*, the channels are narrow but gradually increase in size late in both the fourth and the fifth larval instar (Figure 6). On day 8 of the fifth instar, enormous extracellular chambers are formed (Figure 7). This final increase in surface area follows closely upon the appearance of the major ecdysteroid peak and suggests that the membrane changes reflect hormone secretion (Figure 1).

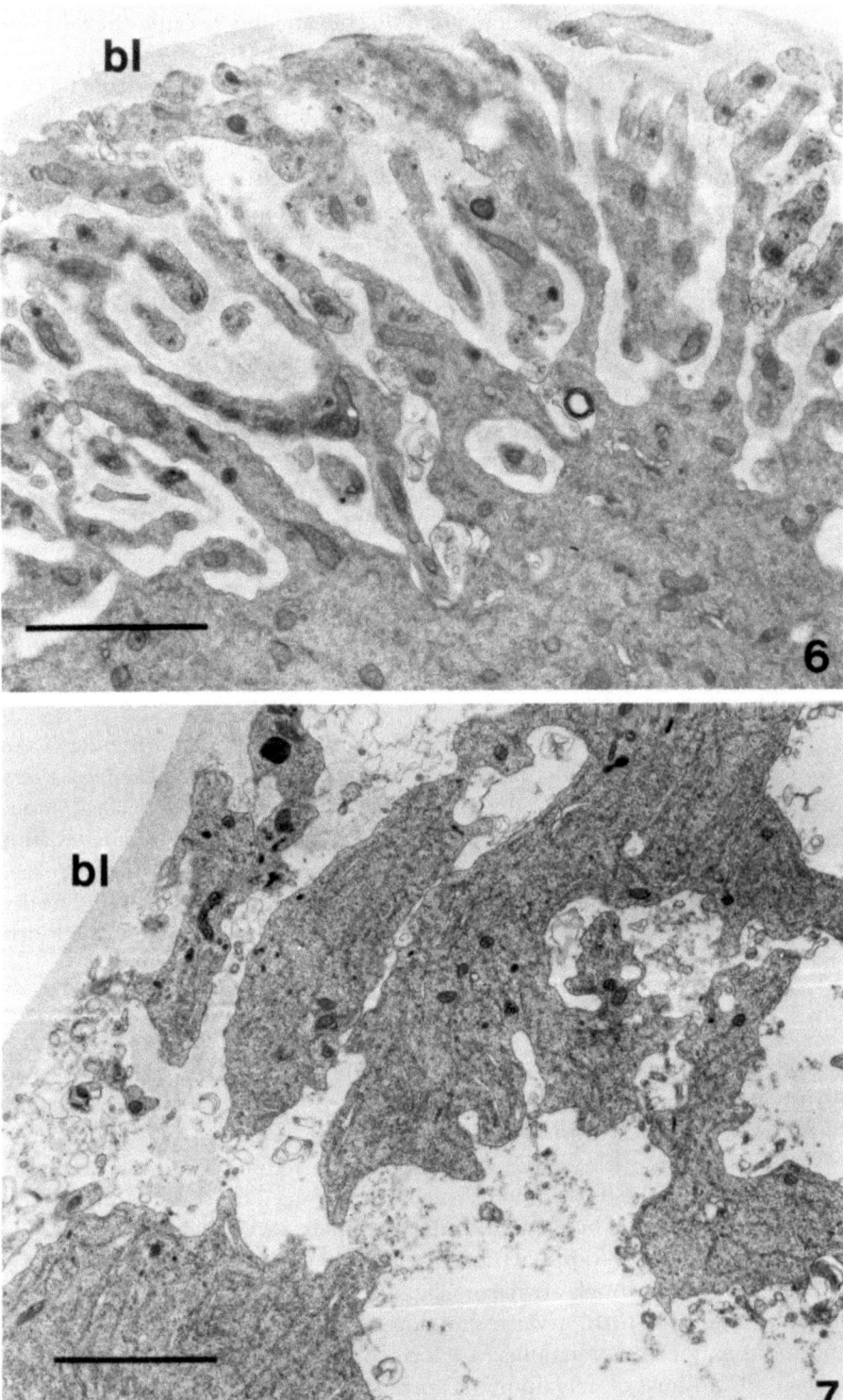

Similarly, channels increase with time and reach their maximal value following the major ecdysteroid peaks in the following species: *T. molitor* (Romer, 1971), *G. mellonella* (Gersch et al., 1975; Blazsek et al., 1975), and *D. melanogaster* (Aggarwal and King, 1969). Thus, alterations in the surface area of the PTG are generally associated with ecdysone secretion.

Is there any cell component associated with the controlled release of hormone? Microvesicles have been observed in many species and may be the morphological representatives of exocytosis of a cell product, such as the hormone itself or a precursor of the hormone.

Blazsek and Mala (1978) reported populations of microvesicles within the extracellular spaces of *G. mellonella* PTG. In order to test the hypothesis that these membranes were involved with hormone synthesis, the digitonin method was used to cytochemically identify the presence of sterols. Tissue was incubated with digitonin which binds specifically to the free -OH groups of sterols such as cholesterol and ecdysone. The digitonin-positive crystals are electron dense, and such crystals were deposited within extracellular spaces near microvesicles. Blazsek and Mala interpreted their data to show that first there was uptake of cholesterol into gland cells, and later ecdysone was exported. The assumption that there was a changeover from cholesterol uptake to hormone export was based on the fact that the hormone must be synthesized from exogenous cholesterol and the observation that there was a major release of hormone from glands as detected by RIA in this and other species. This inferential evidence of hormone export within microvesicles gains support from observations of the PTG in *M. sexta*, where the timing of stimulation of the PTG is known.

5.3. Stimulation of Prothoracic Glands by PTTH

It was mentioned earlier that the vertebrate adrenal cortex rather than the testis or ovary may be analogous to the insect PTG. Rhodin (1971), studying the adrenal cortex of the rat, assumed that corticosteroids are dissolved within the lipid inclusions which are eliminated from cells via exocytosis. Supporting evidence for the idea that these hormones are liberated during the extrusion of lipids came from an experiment where the drug dexamethasone was used to block hormone secretion. Corticosteroid release is stimulated by ACTH. When dexamethasone was used to inhibit secretion of ACTH, the exocytosis of the lipid droplets also ceased. This result supports Rhodin's suggestion that steroid was exported within lipid droplets. In *Manduca*, the hormone may be stored transiently in microvesicles which are exported in response to PTTH stimulation.

As mentioned in Section 4.3, *Manduca* PTG are stimulated by a hormone (PTTH) produced in one of two lateral NSC of the brain (Agui et al., 1979) and

Figure 6. The plasma membrane invaginates to form a system of peripheral intercellular channels in prothoracic glands of *Manduca* on day 3 of the fifth larval instar. The basal lamina (bl) which forms a layer surrounding each cell is also present within the intercellular spaces. (Bar = 2 μm.)

Figure 7. The intercellular spaces of *Manduca* prothoracic glands are greatly enlarged on day 8 of the fifth larval instar. bl, basal lamina. (Bar = 2 μm.)

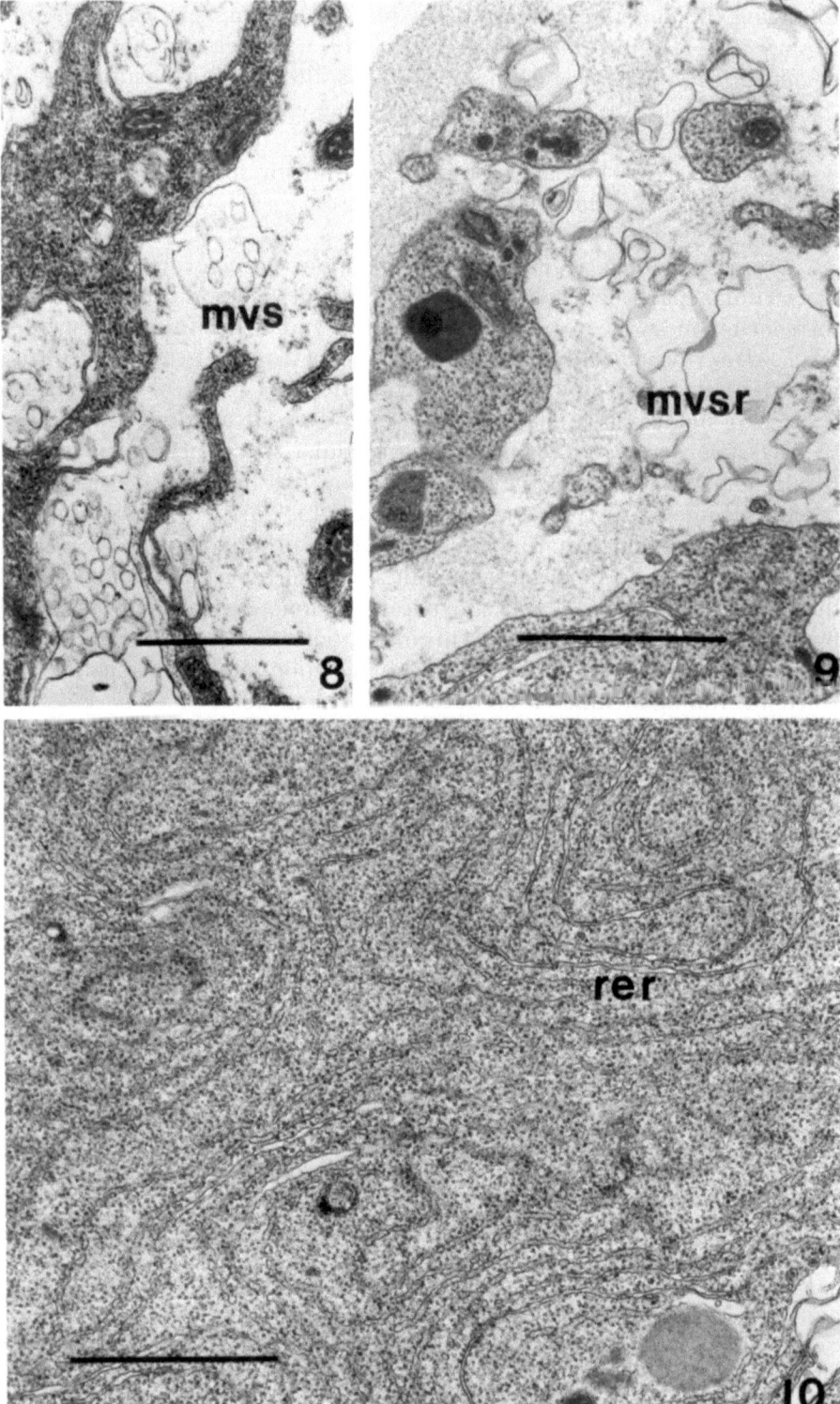

released from axons in the CA at specific times in development (Agui *et al.*, 1980). The cycles of PTTH secretion occurred in the last larval instar just prior to both the minor and the major peaks of ecdysteroids (Agui *et al.*, 1980).

The microvesicles found in *M. sexta* PTG are contained in sacs (Figure 8). These multivesicular sacs (MVS) are located within the intercellular spaces along with other sacs apparently devoid of their internal vesicles, i.e., multivesicular sac remnants (MVSR; Figure 9). The ratio of MVS to MVSR changes with time. Although both MVS and MVSR are present during most of larval development, MVSR are the dominant components following the major ecdysteroid peak. Since the change to MVSR coincides with the second cycle of PTTH activity which stimulates PTG secretion, the MVS may release either ecdysone or a precursor of the hormone (Sedlak *et al.*, 1983). A similar observation was made in *G. mellonella*, where the timing of the appearance of microvesicles and their subsequent absence from the intercellular spaces of the PTG also corresponded to fluctuations in the concentrations of ecdysteroids in the blood (J. Mala, personal communication).

5.4. Stimulation of Prothoracic Glands by 20HE

In *Manduca*, the RER becomes arranged in concentric cisternae following the major ecdysteroid peak (Figure 10). The epidermal cells (Sedlak and Gilbert, 1979) and fat body (Sedlak, unpublished observations) show similar arrangements of their RER during this period. The timing of the increase in the amount of RER suggests that these organelles are stimulated to produce exportable protein in response to 20HE. The proteins produced may be translocated to the cell surface of the pupal gland. These pupal PTG cells lack the deep plasma membrane invaginations typically seen in the actively secreting cells of the larval gland cells.

6. Summary of Gland Cell Structure

The CA and the PTG are capable of continuous synthesis, but the cyclical fluctuations in secretion may be controlled by the brain. The synthetic capabilities of both glands remain constant throughout larval life. For example, the free ribosomes, mitochondria, and Golgi complexes which do not change in either their morphology or number could provide the enzymes needed for ecdysone synthesis of larval PTG; in larval CA, the SER and mitochondria may be the sites of biosynthesis for JH production.

Figure 8. Multivesicular sacs (mvs) in a *Manduca* prothoracic gland cell on day 6 of the fifth larval instar. (Bar = 1 µm.)

Figure 9. On day 8 of the fifth larval instar in *Manduca*, multivesicular sac remnants (mvsr) predominate the intercellular spaces of the prothoracic gland. (Bar = 1 µm.)

Figure 10. Layers of rough endoplasmic reticulum (rer) cisternae from a *Manduca* prothoracic gland cell on day 8 of the fifth larval instar. (Bar = 1 µm.)

The fluctuations in secretion which result in peaks in hormone levels are associated with specific changes in cell morphology in each gland. Secretion may be directly visualized in PTG as an emptying of the MVS located within the enlarged intercellular spaces. The PTG are under brain control, and the structural evidence supporting this statement is seen in the change from MVS to MVSR which occurs concurrently with the second cycle of PTTH activity. No comparable exocytosis-like mechanism has been observed during the development of the larval CA that reflects an elevated secretion of JH. There is, however, a notable increase in the diameter of NSC axons located within the gland proper which occurs prior to the late JH surge. The increased carrying capacity of larger axons may indicate a stimulation of hormone secretion by the brain. Thus, in both glands, secretion of hormone may be stimulated indirectly by NSC located within the brain.

The proposed interactions between molting hormone and JH provide information about structure–function relationships. For example, 20HE may indirectly inhibit JH (Friedel et al., 1980). Structural evidence of such an interaction in Manduca CA is seen as a decrease in the size of the SER whorls and the decrease in NSC axonal diameter which occurs concurrent with the rise in ecdysteroid levels. Evidence of a stimulation by the molting hormone of its own gland is seen in the increase in RER which occurs when the ecdysteroid titer rises.

7. Target Tissue: Epidermal Cells and Cuticle Secretion

The epidermis is a target tissue which responds to both molting hormone and JH, and it is one of the most thoroughly studied tissues from the structural point of view. The responses to each hormone are clearly and permanently recorded by the cuticle produced by epidermal cells.

The epidermis is a single layer of epithelial tissue which secretes the insect exoskeleton. This protective cuticle is a complex layer of materials which are manufactured and secreted in a specific sequence, and the epithelial cell morphology changes significantly during the formation of each layer. For a review of cuticle structure, see Filshie (1982).

The first layer of epicuticle to be secreted, the cuticulin, is a trilaminar structure. It is 10–17 nm thick in C. ethlius (Locke, 1966). Cuticulin first appears as small disconnected deposits at the apex of the plasma membrane, and as more material accumulates, it fuses into a continuous cuticulin layer. Two other components of the epicuticle are the cement layer, produced by dermal glands, and the wax layer, made by epidermal cells and secreted via pore canals. In many cases, neither of these is seen in the electron microscope because they are extracted during the preparative phases. In C. ethlius, the inner epicuticle forms from vesicles arising at the Golgi complexes and is a homogeneous electron-dense layer adjacent to the cuticulin (Locke, 1966, 1969b). Thus, the epicuticle can consist of cuticulin, the cement layer, the wax layer, and inner cuticle.

Following deposition of these epicuticle layers, the endocuticle is produced.

Some endocuticle is secreted prior to each molt, but the bulk of endocuticle is laid down after ecdysis. The chitin-protein microfibers of endocuticle are laid down cyclically by epidermal cells at the plasma membrane, and lamellated and nonlamellated bands of endocuticle are alternatively deposited in some insects according to a photoperiodic rhythm (Noble-Nesbitt, 1967). The lamellae often appears to be arranged in a parabolic conformation. This arrangement is actually an optical artifact arising from cutting the cuticle at an angle slightly oblique to the perpendicular. In fact, each layer is laid down with a slight rotation with respect to the preceding layer (Bouligand, 1965). The layers of endocuticle which were laid down first are eventually stabilized; this quinone-tanned cuticle layer is called the *exocuticle* and is shed at each molt (Locke, 1974).

In addition to these cuticle layers, the epidermal cells also synthesize and secrete the molting fluid, which digests the majority of old endocuticle at each molt. These old layers are reabsorbed by the epidermal cells after being broken down to their simple components and then reutilized by the cell in making new cuticle. At the cell level, the molting fluid is identified as ecdysial droplets, which contain electron-dense digestive enzymes arising at the Golgi complex (Locke and Krishnan, 1973).

7.1. Hormonal Influences on the Epidermis

Locke, in his pioneering studies on cuticle deposition in *C. ethlius*, has shown that the epidermal plasma membrane changes in a very specific manner when it is secreting each cuticle layer. Regularly oriented microvilli with plasma membrane plaques are found at the cell apex when either cuticulin or endocuticle is being formed. In contrast, both ecdysial droplets and epicuticle arise at the Golgi vesicles, and these products are discharged at the cell apex between microvilli (Locke, 1976). However, the alternation between cuticle secretion at the plasma membrane plaques and secretion from the Golgi complexes is not universal. In *T. molitor* (Delbecque *et al.*, 1978) and *M. sexta* (Sedlak and Gilbert, 1979), these structures are present concurrently.

In many insects, the microvilli of epidermal cells vary in number and length during cuticle deposition. For example, in *T. molitor* the numerous microvilli are of a uniform height, oriented parallel to one another during cuticle secretion, but the height of these structures decreases in some areas during the deposition of inner epicuticle (Delbecque *et al.*, 1978). Similar variations have been reported for several other species, e.g., the number of epidermal microvilli increases after puparium formation in *D. melanogaster* leg disc cells (Poodry and Schneiderman, 1970). In the sheep blowfly *Lucilia cuprina* (Filshie, 1970), the termite *Kalotermes flavicollis* (Noirot and Noirot-Timothée, 1971), and the moth *H. cecropia* (Greenstein, 1972), the microvillar border undergoes structural variation when cuticle is being deposited.

Each cellular change gains significance when the data are interpreted in an endocrine context. Again, *M. sexta* will be used as the primary example to describe the epidermal cell responses to endogenous ecdysteroids and JH (Figure

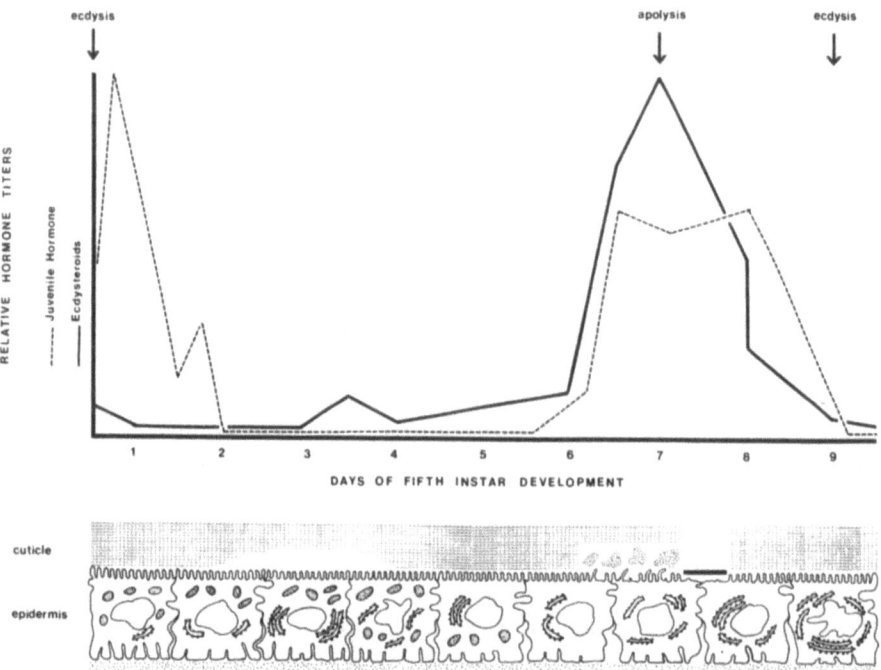

Figure 11. A diagram summarizing the relationship between hormone levels and the cytology of the target epidermal cells of *Manduca sexta*. Both juvenile hormone and ecdysteroids are expressed as relative titers and are shown for the fifth larval instar. Changes in individual epidermal cells plus their overlying cuticel are shown for each day of the instar. Larval endocuticle firmly adheres to the epidermis until days 2-4, when there is a partial separation between cells and the cuticle. On day 6 the amorphous ecdysial droplets are secreted, and a thin dense line of pupal cuticulin is deposited by day 7. The apical plasma membrane remains oriented in characteristic microvillar folds when secreting endocuticle, but this orientation is altered when both ecdysial droplets are secreted and the cuticulin is laid down. Some cisternae of RER are present throughout development but on day 3 and days 8-9 the amount of RER increases. The nuclei have an irregular outline on days 3 and 9. Dense bodies are found in cells from days 1 to 5. Initially distributed in the apical region on days 1-3, they are found throughout the cell on day 4, basally on day 5 and are absent thereafter. The basal lamina is relatively thick except on days 4, 8, and 9. From Sedlak and Gilbert (1979).

11). Figure 11 shows the fluctuations in relative ecdysteroid and JH titers for *M. sexta* presented as in Figure 1. Epidermal cells and their overlying cuticle are drawn along the temporal axis during the fifth larval instar. The major changes in cell and cuticle morphology are shown for each day.

Before estimates of the endogenous titers of these hormones were available, however, the role that the growth hormones play in controlling cellular changes in the epidermis was shown in a study where either 20HE or JH was injected into developing *H. cecropia* pupae (Sedlak and Gilbert, 1976a,b). The responses indicated that there were specific effects elicited by each hormone; i.e., 20HE affected cuticle deposition and accelerated the normal developmental program, while JH retarded cellular differentiation.

7.2. Effects of Ecdysteroids on the Epidermis

The cue for laying down cuticle both *in vivo* and *in vitro* comes from 20HE (Agui *et al.*, 1969). Oberlander *et al.* (1973) suggest from their studies with ecdysone and 20HE on the lepidopterans *Cadra cautella*, *Paramyelois transitella*, and *Plodia interpunctella* that 20HE may be the major molting hormone, whereas ecdysone initiates metamorphosis. In contrast, in *D. melanogaster* imaginal leg discs, cuticle layers were secreted *in vitro* in response to both ecdysone and 20HE (Mandaron, 1976). Ryerse and Locke (1978) studied the effect of 20HE on tracheal cuticle deposition in *C. ethlius* and found that although the hormone stimulated cuticle deposition *in vitro*, 20HE did not support cell growth. Cuticle secretion is seen coincident with an ecdysteroid peak in both *C. ethlius* (Dean *et al.*, 1980) and *R. prolixus* (Steel *et al.*, 1982).

When 20HE was injected into diapausing *H. cecropia*, cuticle secretion was precociously induced with a dose-dependent response (Sedlak and Gilbert, 1976b). During pupal-to-adult metamorphosis in this insect, adult cuticulin is normally deposited on day 10 of the 21-day period of metamorphosis, and endocuticle appears within the next 24 hr (Sedlak and Gilbert, 1976a). When a physiological dose of 20HE (5 µg/g) was injected into diapausing pupae, adult cuticulin appeared 5 days prematurely. After injection of a low dose (2 µg/g) of JH (C_{18}; mixed isomers), the cuticulin layers were deposited 2 days earlier than in normal animals. Since JH can stimulate the PTG (Gilbert and Schneiderman, 1959), the precocious cuticle deposition by JH-injected animals was believed to be a response to the molting hormone and not due directly to JH. Another response of the epidermis to 20HE was the precocious stimulation of adult differentiation as shown by the premature formation of adult scales. In the epidermis of *H. cecropia*, autophagic vacuoles appeared prematurely after 20HE injection. Since these organelles were normally seen just prior to the appearance of scales, lysosomes are probably involved in the process of cellular remodeling which results in epidermal differentiation. The autophagic vacuoles were found on day 1 in pupae injected with 5 µg/g 20HE in comparison with their appearance on day 7 in normal animals. Thus, 20HE elicited precocious differentiation as well as premature cuticle deposition.

This study with *H. cecropia* did not provide information on the effects that endogenous levels of hormone had on cell structure, and therefore when estimates of the ecdysteroid titers became available for larval *M. sexta* (Bollenbacher *et al.*, 1975), the epidermal target tissue of this insect was investigated (Sedlak and Gilbert, 1979). In *Manduca*, the events of apolysis and ecdysis coincided with the major peak of hemolymph ecdysteroids (Sedlak and Gilbert, 1979). The release of molting fluid needed to digest old larval endocuticle occurred at the time of a rise in ecdysteroids, and pupal endocuticle secretion followed within 24 hr. The ecdysial droplets appeared at the cell border adjacent to old endocuticle layers by day 6 (Figure 12). The secretion of pupal cuticulin was not observed. Cuticulin is formed within a few hours in *Calpodes* (Locke, 1966). We assumed that in *Manduca*, cuticulin secretion is also a rapid event, and that since samples were taken at 6-hr intervals, cuticulin deposition was missed.

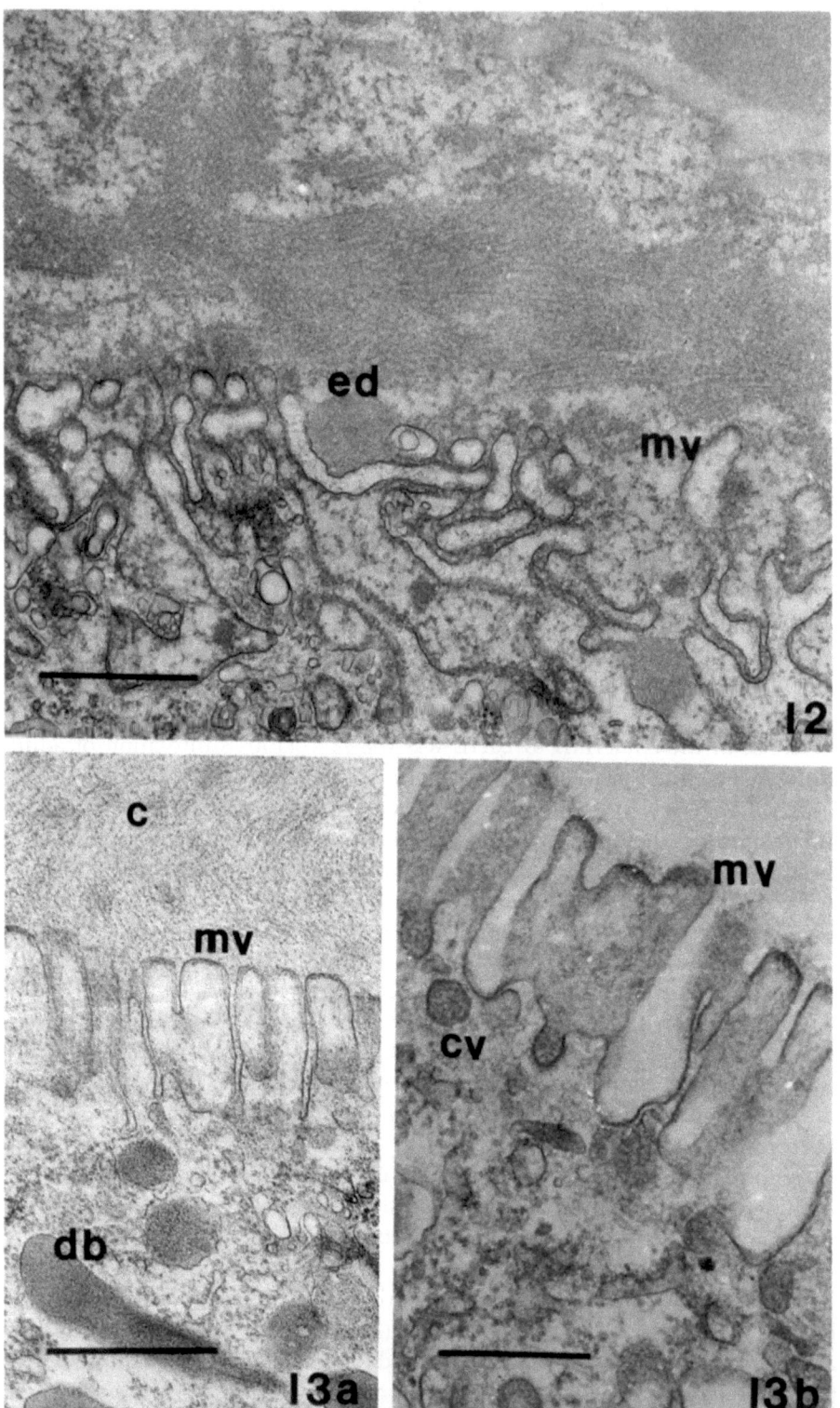

Apolysis does not always coincide with the major surge of ecdysteroids. In *T. molitor*, cuticle retraction during the larval-to-pupal ecdysis occurs long before the molting hormone peak is seen (Delbecque *et al.*, 1978). Similarly, in *Gryllus bimaculatus*, cuticulin secretion, but not apolysis, coincides with the molting hormone peak (Gnatzy and Romer, 1981). In contrast, during the pupal-to-adult transition in *Tenebrio*, apolysis is immediately followed by cuticle secretion, and during this final molt, both events coincide with peak ecdysteroid secretion (Delbecque *et al.*, 1978).

There is no doubt that cuticle secretion and apolysis occur in response to the molting hormone in *Manduca*. What is perhaps more interesting is the epidermal cell response to the minor ecdysteroid peak which occurs 3-4 days after the last larval ecdysis. Riddiford (1976) reported that the epidermal cell layer separated from overlying cuticle on days 2-4 after ecdysis, and then the epidermis and cuticle reattached. She referred to the phenomenon as pseudoapolysis. An ultrastructural analysis (Sedlak and Gilbert, 1979) has shown that organized microvilli with plasma membrane plaques are found whether or not endocuticle is present at the cell apex (Figure 13). In the regions where the cuticle had detached, fibrous material still adheres to the plasma membrane, and numerous coated vesicles are present at the base of microvilli. The morphology suggests that endocuticle was being deposited in these regions but was not adherent. The small peak of ecdysteroids which occurs concurrently with the pseudoapolysis presumably signaled the detachment of cuticle. After the ecdysteroid titer fell again on day 4, the cuticle reattached and remained close to the epidermal cells until the actual apolysis occurred 2-3 days later. Thus, apolysis seems to be a general response to the molting hormone even when small quantities of hormone are available, and a small increase elicits a pseudoapolysis and the large peak signals the actual apolysis.

Riddiford (1976) has called the first ecdysteroid peak a reprogramming peak. She concludes that the larval epidermis in *M. sexta* becomes committed to a pupal developmental program when there is a basal level of 20HE in the absence of JH and that the actual pupal syntheses are triggered by a larger dose of 20HE. Both RNA and protein synthesis are essential for the reprogramming of epidermal tissue (Riddiford, 1981). The minor ecdysteroid peak is also correlated with a second round of DNA synthesis that results in a polyploid epidermis (Wielgus *et al.*, 1979). This DNA synthesis is followed, 6 days after ecdysis, by a wave of cell divisions in the epidermis (Wielgus and Gilbert, 1978).

Another cellular response in *Manduca* which is apparently induced by

---

Figure 12. On day 7 of the fifth larval instar in *Manduca*, the ecdysial droplets (ed), seen as amorphous dense deposits, fill the space between the microvilli (mv) of the epidermal plasma membrane and the larval endocuticle. (Bar = 1 μm.)

Figure 13. Two areas of the epidermis on day 3 of the fifth larval instar in *Manduca*. In one area, microvilli (mv) with their plasma membrane plaques adhere to fibrous endocuticule (c); whereas in an adjacent area, cuticle is absent and only wisps of fibrous cuticle are found. There are numerous coated vesicles (cv) located at the base of microvilli in the area where there is no endocuticle. db, dense bodies. [Bars = 1 μm (a) and 0.5 μm (b).]

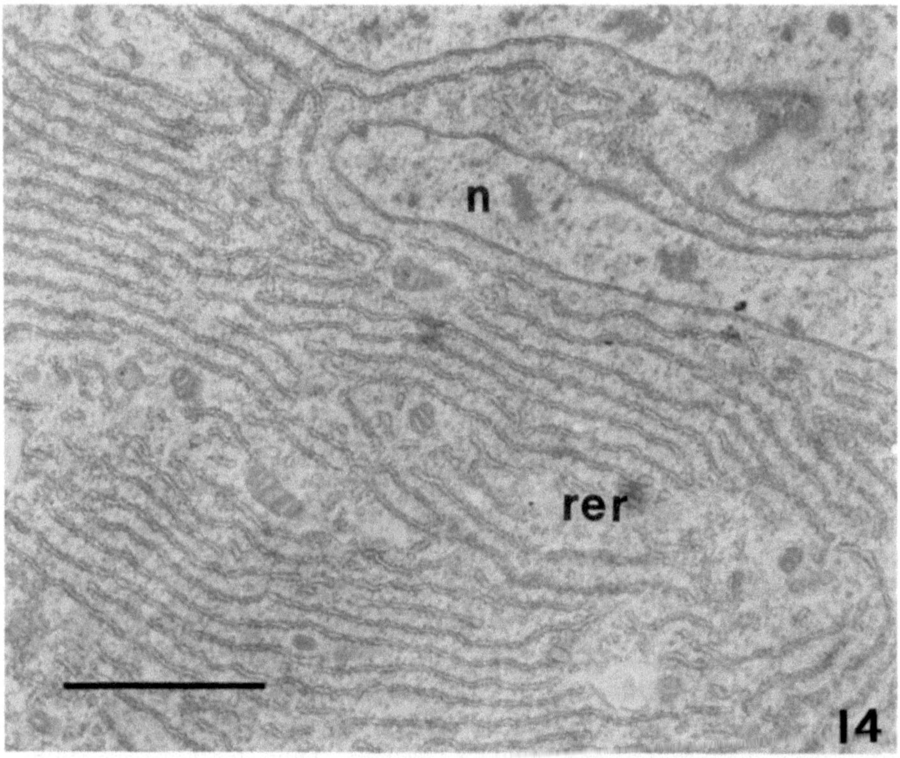

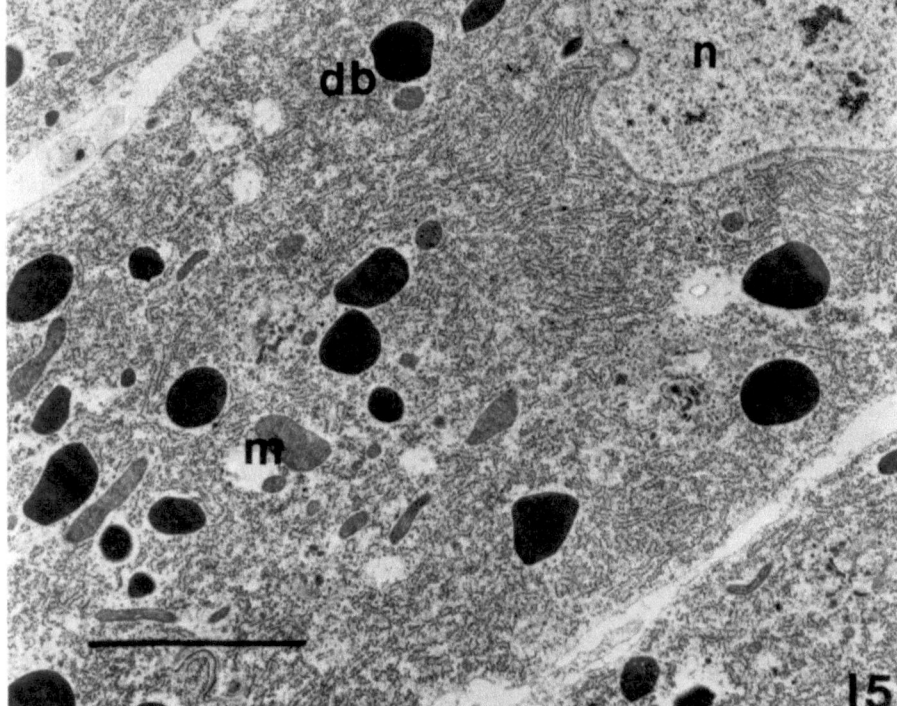

ecdysteroids is the increase in quantity and organization of the RER. The RER at most stages of development is found as single cisternae randomly distributed throughout a cell. However, 9 days after ecdysis, the amount of RER increases, especially in the nuclear area, and it is arranged as concentric cisternae (Figure 14). Since a similar but less dramatic increase in RER occurs concurrently with the minor ecdysteroid peak 3 days after ecdysis, the RER probably produces exportable endocuticle protein in response to the steroid hormone (Sedlak and Gilbert, 1979). Both *Calpodes* (Locke, 1974) and *Rhodnius* (Wigglesworth, 1964) show a similar increase in endocuticle deposition in response to the molting hormone.

Each ecdysteroid peak detected in *C. ethlius* is accompanied by specific events in the epidermal cells. The first peak (the intermolt peak) is associated with organelle formation. The second surge (the commitment peak) signals DNA synthesis and mitosis, and is responsible for reprogramming the tissue from larval to pupal syntheses. This latter peak may also trigger the secretion of lamellated cuticle. The late major ecdysteroid peak (the prepupal peak) coincides with the decrease in synthetic activity, with the secretion of ecdysial droplets, loss of plasma membrane plaques, and the initiation of pupal cuticle synthesis (Dean *et al.*, 1980).

7.3. Effects of Juvenile Hormone on the Epidermis

JH is thought to maintain larval characteristics and the absence of JH allows progress to the next stage. For example, when *H. cecropia* pupae are injected with JH, autophagic vacuoles are not seen, cell restructuring does not occur, and adult scales are not formed by epidermal cells (Sedlak and Gilbert, 1976b). Thus, JH maintains a more juvenile, undifferentiated morphology of pupal cells, even on the day of metamorphosis to the intermediate adult. These intermediate adults (formed under the influence of the excess JH) were partially adult and partially pupae. For example, in the adominal area, which was the area of epidermis studied, the insect was scaleless, indicating a pupal morphology (Sedlak and Gilbert, 1976b).

In *Manduca*, a larval characteristic which seems to be maintained by JH is the epidermal pigmentation. *Manduca* larvae have a blue pigmentation throughout all larval instars up until the wandering phase of the fifth (last) larval instar. At wandering, when the animal prepares to pupate, there is a color change in the larvae; the blue color disappears and the larvae become yellow. The cellular components which represent blue pigment granules are the dense bodies found within each epidermal cell (Figure 15). These dense bodies are located only in the cell apex early in the instar when the JH titer is high, but after the JH titer falls to a minimal value, the dense bodies migrate to the basal side of each cell and are

◄ ───

Figure 14. On day 8 of the fifth larval instar in *Manduca*, numerous stacks of rough endoplasmic reticulum (rer) fill epidermal cells. n, nuclear projection. (Bar = 1 μm.)

Figure 15. Dense bodies (db) found in an epidermal cell during the early part of the fifth larval instar. These organelles are probably pigment granules. m, mitochondrion; n, nucleus. (Bar = 5 μm.)

subsequently eliminated from the epidermis at the wandering phase (Sedlak and Gilbert, 1979).

8. Summary of Target Tissue Responses to Hormones

The epidermis, as a major target for both 20HE and JH, responds to each hormone with specific cellular changes. In *H. cecropia*, exogenously applied 20HE accelerates normal development, including both cuticle deposition and the cell restructuring which transforms the pupa to the adult. Excess JH, on the other hand, retards or prevents differentiation by inhibiting the formation of the autophagic vacuoles necessary for restructuring.

In *M. sexta*, where estimates of the endogenous levels of hormone are available, it is clear that apolysis occurs and cuticle is deposited in response to the ecdysteroids and that JH maintains such larval characteristics as pigmentation. A pseudoapolysis occurs in response to the minor reprogramming peak of ecdysteroids.

9. Conclusions

In endocrine glands, the processes of hormones synthesis and secretion are separate and structurally definable events, and the responses to either ecdysteroids or JH can be visually defined in the responding epidermis.

Conclusions about the organelles responsible for functional endocrine activity have been based on the temporal relationships between fluctuations in endogenous titers of the growth hormones and the coincident presence of specific organelles. A rather complete summary of such indirect evidence exists for *M. sexta* PTG, CA, and at least one target tissue, the epidermis. These data are consistent with those gathered for other insects. Conclusions concerning whether or not (1) the MVS contain ecdysone or a precursor which is secreted in response to PTTH, (2) the SER is a site of JH synthesis, and (3) the JH is secreted in response to stimulation from NSC require further studies. More direct proofs, such as the immunocytochemical localization of the growth hormones at precise organelle sites, could provide definitive answers to the question of where the growth hormones are synthesized and released, and could determine the sites where they have an effect in target tissues.

ACKNOWLEDGMENTS

The author was supported by Grant AG 01979 from the National Institutes of Health while writing this chapter.

References

Aggarwal, S. K., and King, R. C., 1969, A comparative study on ring glands from wild type and *l(2)gl* mutant *Drosophila melanogaster*, *J. Morphol.* **129**:171-200.

Agui, N., and Hiruma, K., 1977, Ecdysone as a feedback regulator for the neurosecretory brain cells in *Mamestra brassicae*, *J. Insect Physiol.* **23**:1393-1396.

Agui, N., Yagi, S., and Fukaya, N., 1969, Induction of molting of cultivated integuments taken from a diapausing rice stem borer larva in presence of ecdysterone (Lipidoptera, Pyralidae), *Appl. Entomol. Zool.* **4**:156-157.

Agui, N., Granger, N. A., Gilbert, L. I., and Bollenbacher, W. E., 1979, Cellular localization of the insect prothoracicotropic hormone: *In vitro* assay of a single neurosecretory cell, *Proc. Natl. Acad. Sci. USA* **76**:5694-5698.

Agui, N., Bollenbacher, W. E., Granger, N. A., and Gilbert, L. I., 1980, Corpus allatum is release site for insect prothoracicotropic hormone, *Nature (London)* **285**:669-670.

Baehr, J. C., Cassier, P., and Fain-Maurel, M. A., 1973, Contribution expérimentale et infrastructurale à l'etude de la dynamique de corpus allatum de *Rhodnius prolixus* Stal: Influence de la nutrition, de l'activité ovarienne, de la pars intercerebralis et de ses connections, *Arch. Zool. Exp. Gen.* **114**:611-626.

Beaulaton, J., 1968, Modifications ultrastructurales des cellules sécrétrices de la glande prothoracique de vers à soie au cours des deux derniers ages larvaires. I. Le chondriome et ses relations avec le réticulum agranulaire, *J. Cell Biol.* **39**:501-525.

Bell, R. A., and Joachim, F. G., 1976, Techniques for rearing laboratory colonies of tobacco hornworms and pink bollworms, *Ann. Entomol. Soc. Am.* **69**:365-373.

Bergot, B. J., Hall, M. S., Furrer, A. A., and Schooley, D. A., 1984, Juvenile hormone titers in *Manduca sexta* throughout the life cycle, *J. Insect Physiol.* Submitted.

Bjersing, L., 1967, On the ultrastructure of granulosa lutein cells in porcine corpus luteum, *Z. Zellforsch. Mikrosk. Anatl.* **82**:187-211.

Blazsek, I., and Mala, J., 1978, Steroid transport through the surface of the prothoracic gland cells in *Galleria mellonella* L., *Cell Tissue Res.* **187**:507-513.

Blazsek, I., Balaz, A., Novak, J. A., and Mala, J., 1975, Ultrastructural study of prothoracic glands of *Galleria mellonella* L. in the penultimate, last larval, and pupal stages, *Cell Tissue Res.* **158**:269-280.

Bloom, W., and Fawcett, D. W., 1975, *A Textbook of Histology*, pp. 518-519, Saunders, Philadelphia.

Bodenstein, D., 1955, Humoral agents in insect metamorphosis. In *Aspects of Synthesis and Order in Growth*, edited by D. Rudnick, Princeton University Press, Princeton, N.J.

Bollenbacher, W. E., Vedeckis, W., Gilbert, L. I., and O'Connor, J. D., 1975, Ecdysone titers and prothoracic gland activity during the larval-pupal development of *Manduca sexta*, *Dev. Biol.* **44**:46-53.

Bollenbacher, W. E., Zvenko, H., Kumaran, A. K., and Gilbert, L. I., 1978, Changes in ecdysone content during post-embryonic development of the wax moth, *Galleria mellonella*: The role of the ovary, *Gen. Comp. Endocrinol.* **34**:169-179.

Bollenbacher, W. E., Smith, S. L., Goodman, W., and Gilbert, L. I., 1981, Ecdysteroid titer during larval-pupal-adult development of the tobacco hornworm, *Manduca sexta*, *Gen. Comp. Endocrinol.* **44**:302-306.

Bouligand, M. Y., 1965, Sur une architecture torsadée répandue dans de nombreuses cuticles d'Arthropodes, *C.R. Acad. Sci.* **261**:3665-3668.

Bowers, W. S., Ohta, T., Cleere, J. S., and Marsella, P. A., 1976, Discovery of insect antijuvenile hormones in plants, *Science* **193**:542-547.

Bownes, M., 1982, Hormonal and genetic regulation of vitellogenesis in *Drosophila*, *Q. Rev. Biol.* **57**:247-274.

Calvez, B., Hirn, M., and DeReggi, M., 1976, Ecdysone changes in the haemolymph of two silkworms (*Bombyx mori* and *Philosamia cynthia*) during larval and pupal development, *FEBS Lett.* **71**:57-61.

Cassier, P., 1979, The corpora allata of insects, *Int. Rev. Cytol.* **57**:1-73.

Charpin, P., 1975, Evolution ultrastructurale des corps allatum au cours du dernier stade larvaire chez *Choleva augustata* Fab. (Coléoptères Catopidae de la sous-famille des Catopuae), *C.R. Acad. Sci. Ser. D* **280**:1997-1999.

Christensen, A. K., 1965, The fine structure of testicular interstitial cells in guinea pigs, *J. Cell Biol.* **26**:911-934.

Cymborowski, B., and Stolarz, G., 1979, The role of juvenile hormone during larval-pupal transformation of *Spodoptera littoralis:* Switchover in the sensitivity of the prothoracic gland to juvenile hormone, *J. Insect Physiol.* **25**:939-942.

Dean, R. L., Bollenbacher, W. E., Locke, M., Smith, S. L., and Gilbert, L. I., 1980, Haemolymph ecdysteroid levels and cellular events in the intermoult/moult sequence of *Calpodes ethlius, J. Insect Physiol.* **26**:267-280.

deKort, C. A. D., and Granger, N. A., 1981, Regulation of the juvenile hormone titer, *Annu. Rev. Entomol.* **26**:1-28.

Delbecque, J. P., Hirn, M., Delachambre, J., and DeReggi, M., 1978, Cuticular cycle and molting hormone levels during the metamorphosis of *Tenebrio molitor* (Insecta: Coleoptera), *Dev. Biol.* **64**:11-30.

Deleurance, S., and Charpin, S., 1978, Ultrastructural dynamics of the corpus allatum of *Choleva augustata* Fab. (Coleoptera, Catopidae), *Cell Tissue Res.* **191**:151-160.

Dorn, A., 1973, Electron microscopic study on the larval and adult corpus allatum of *Oncopeltus fasciatus* Dallas, *Z. Zellforsch. Mikrosk. Anat.* **145**:447-458.

Elliott, H. J., 1976, Structural analysis of the corpus allatum of an aphid *Aphis craccivora, J. Insect Physiol.* **22**:1275-1280.

Feyereisen, R., Johnson, G., Koener, J., Stay, B., and Tobe, S. S., 1981, Precocenes as pro-allatocidins in adult female *Diploptera punctata:* A functional and ultrastructural study, *J. Insect Physiol.* **27**:855-866.

Filshie, B. K., 1970, The fine structure and deposition of the larval cuticle of the sheep blowfly (*Lucilia cuprina*), *Tissue Cell* **2**:479-498.

Filshie, B. K., 1982, Fine structure of the cuticle of insects and other arthropods. In *Insect Ultrastructure,* edited by R. C. King and H. Akai, vol. 1, pp. 281-312, Plenum Press, New York.

Friedel, T., Feyereisen, R., Meindall, E. C., and Tobe, S., 1980, The allatostatic effect of 20-hydroxyecdysone on the adult viviparous cockroach, *Diploptera punctata, J. Insect Physiol.* **26**:665-670.

Gersch, M., Birkenbeil, H., and Ude, J., 1975, Ultrastructure of the prothoracic gland cells of the last instar of *Galleria mellonella* in relation to the state of development, *Cell Tissue Res.* **160**:389-397.

Giacomelli, F., Wiener, J., and Spiro, D., 1965, Cytological alterations related to stimulation of the zona glomerulosa of the adrenal gland, *J. Cell Biol.* **26**:499-522.

Gilbert, L. I., and Schneiderman, H. A., 1959, Prothroacic gland stimulation by juvenile hormone extracts of insects, *Nature (London)* **184**:171-173.

Glitho, I., Delbecque, J. P., and Delachambre, J., 1979, Prothoracic gland involution related to molting hormone levels during the metamorphosis of *Tenebrio molitor, J. Insect Physiol.* **25**:187-192.

Gnatzy, W., and Romer, F., 1981, Morphogenesis of mechanoreceptor and epidermal cells of crickets *Gryllus bimaculatus* during the last instar and its relation to molting hormone level, *Cell Tissue Res.* **213**:369-392.

Goltzene, F., and Porte, A., 1978, Endocrine control by neurosecretory cells of the pars intercerebralis and the corpora allata during the earlier phases of vitellogenesis in *Locusta-migratoria-migratorioides* (Orthoptera), *Gen. Comp. Endocrinol.* **35**:35-45.

Greenstein, M. E., 1972, The ultrastructure of developing wings in the giant silkmoth, *Hyalophora cecropia.* I. Generalized epidermal cells, *J. Morphol.* **136**:1-22.

Herman, W. S., 1967, The ecdysial glands of arthropods, *Int. Rev. Cytol.* **22**:269-374.

Herman, W. S., and Gilbert, L. I., 1966, The neuroendocrine system of *Hyalophora cecropia* (L.) (Lepidoptera: Saturniidae). I. The anatomy and histology of the ecdysial glands, *Gen. Comp. Endocrinol.* **7**:275-291.

Hiruma, K., Shimada, H. and Yagi, S., 1978, Activation of the prothoracic gland by juvenile hormone and prothoracicotropic hormone in *Mamestra brassicae, J. Insect Physiol.* **24**:215-220.

Johnson, R. A., and Hill, S., 1973, The activity of the corpora allata in the fourth and fifth larval instars of the migratory locust, *J. Insect Physiol.* **19**:1921-1932.

Khan, T. R., Singh, S. B., Singh, R. K., and Singh, T. K., 1978, Neurosecretory control of corpora allata activity in cockroach *Periplaneta americana, Experientia* **34**:49-51

King, D. S., Bollenbacher, W. E., Borst, D. W., Vedeckis, W. V., O'Connor, J. D., Ittycheriah, P. I., and Gilbert, L. I., 1974, The secretion of ecdysone by the prothoracid glands of *Manduca sexta in vitro*, *Proc. Natl. Acad. Sci. USA* **71**:793-796.

King, R. C., 1970, *Ovarian Development in Drosophila melanogaster*, Academic Press, New York.

King, R. C., Aggarwal, S. K., and Bodenstein, D., 1966, The comparative submicroscopic cytology of the corpus allatum-corpus cardiacum complex of *Drosophila melanogaster*, *J. Exp. Zool.* **161**:151-176.

Krogh, I. M., and Normann, I. C., 1977, The corpus cardiacum neurosecretory cells of *Schistocerca gregaria*: Electron microscopy of resting and secreting cells, *Acta Zool.* **58**:69-78.

Lafont, R., Mauchamp, B., Blais, C., and Pennetier, J. L., 1977, Ecdysone and imaginal disc development during the last larval instar of *Pieris brassicae*, *J. Insect Physiol.* **23**:277-283.

Lanzrein, B., Gentinetta, V., Fehr, R., and Lüscher, M., 1978, Correlation between hemolymph juvenile hormone titer, corpus allatum volume and corpus allatum *in vivo* and *in vitro* activity during occyte maturation in a cockroach (*Nauphoeta cinerea*), *Gen. Comp. Endocrinol.* **36**:339-345.

Liechty, L., and Sedlak, B. J., 1978, The ultrastructure of precocene-induced effects on the corpora allata of the adult milkweed bug, *Oncopeltus fasciatus*, *Gen. Comp. Endocrinol.* **36**:433-436.

Locke, M., 1966, The structure and formation of the cuticulin layer in the epicuticle of an insect, *Calpodes ethlius* (Lepidoptera, Herperiidae), *J. Morphol.* **118**:461-494.

Locke, M., 1969a, The ultrastructure of the oenocytes in the molt/intermolt cycle of an insect, *Tissue Cell* **1**:103-154.

Locke, M., 1969b, The structure of an epidermal cell during the development of the protein epicuticle and the uptake of molting fluid in an insect, *J. Morphol.* **127**:7-40.

Locke, M., 1974, The structure and formation of the integument in insects. In *The Physiology of Insecta*, edited by M. Rockstein, Vol. 6, 123-213, Academic Press, New York

Locke, M., 1976, The role of plasma membrane plaques and Golgi complex vesicles in cuticle deposition during the molt-intermolt cycle. In *The Insect Integument*, edited by H. R. Hepburn, pp. 237-258, Elsevier/North-Holland, Amsterdam.

Locke, M., and Krishnan, N., 1973, The formation of the ecdysial droplets and the ecdysial membrane in an insect, *Tissue Cell* **5**:441-450.

Mandaron, P., 1976, Ultrastructure des disques de patte de drosophile cultives *in vitro:* Evagination, sécrétion de la cuticle nymphale et apolysis, *Wilhelm Roux Arch. Dev. Biol.* **179**:185-196.

Melnikova, F. J., and Panov, A. A., 1975, Ultrastructure of the larval corpus allatum of *Hyphantria cunea*, *Cell Tissue Res.* **162**:395-410.

Mordue, W., 1965, Studies on oocyte production and associated histological changes in the neuro-endocrine sytem in *Tenebrio molitor* L., *J. Insect Physiol.* **11**:493-503.

Noble-Nesbitt, J., 1967, Aspects of the structure, formation, and function of some insect cuticles. In *Insects and Physiology*, edited by J. W. L. Beament and J. E. Treherne, pp. 3-16, Oliver & Boyd, Edinburgh.

Noirot, C., and Noirot-Timothée, C., 1971, La cuticle proctodéale des insectes. II. Formation durant la mue, *Z. Zellforsch. Mikrosk. Anat.* **113**:361-387.

Nordmann, J. J., 1977, Ultrastructural appearance of neurosecretory granules in the sinus gland of the crab after different fixation procedures, *Cell Tissue Res.* **185**:557-563.

Norman, T. C., 1969, Experimentally induced exocytosis of neurosecretory granules, *Exp. Cell Res.* **55**:285-287.

Oberlander, H., Leach, C. E., and Tomblin, C., 1973, Cuticle deposition in imaginal discs of three species of Lepidoptera: Effect of ecdysones *in vitro*, *J. Insect Physiol.* **19**:993-998.

Odhiambo, T. R., 1966a, Morphometric changes and the hormonal activity of the corpus allatum in the adult male of the desert locust, *J. Insect Physiol.* **12**:655-665.

Odhiambo, T. R., 1966b, The fine structure of the corpus allatum of the sexually inactive male of the desert locust, *J. Insect Physiol.* **12**:819-828.

Palévody, C., and Grimal, A., 1976, Variations cytologiques des corps allates au cours du cycle reproducteur du collembole *Folsomia candida*, *J. Insect Physiol.* **22**:63-72.

Panov, A. A., and Bassurmanova, O. K., 1970, Fine structure of the glands in inactive and active corpus allatum of the bug, *Eurygaster integriceps*, *J. Insect Physiol.* **16**:1265-1281.

Poodry, C. A., and Schneiderman, H. A., 1970, The ultrastructure of the developing leg of *Drosophila melanogaster*, *Wilhelm Roux Arch. Dev. Biol.* **166:**1-44.

Raabe, M., 1982, *Insect Neurohormones*, Plenum Press, New York.

Raina, A. K., and Borg, T. K., 1980, Corpora cardiaca allata complex of the larvae of the pink bollworm *Pectinophora gossypiella:* An ultrastructural study in relation to diapause, *Acta. Zool.* **61:**65-78.

Rhodin, J. A. G., 1971, Ultrastructure of the adrenal cortex of the rat under normal and experimental conditions, *J. Ultrastruct. Res.* **24:**23-71.

Richards, G., 1981, The radioimmune assay of ecdysteroid titres in *Drosophila melanogaster*, *Mol. Cell Endocrinol.* **21:**181-197.

Riddiford, L. M., 1976, Hormonal control of insect epidermal cell commitment *in vitro*, *Nature (London)* **259:**115-117.

Riddiford, L. M., 1981, Hormonal control of epidermal cell development, *Am. Zool.* **21:**751-768.

Romer, F., 1971, Die Prothorakaldrusen der Larvae von *Tenebrio molitor* L. (Tenebrionidae, Coleoptera) und ihre Veranderungen warhend eines Häutungszyklus, *Z. Zellforsch. Mikrosk. Anat.* **122:**425-455.

Romer, F., 1974, Ultrastructural changes of the oenocytes of *Gryllus bimaculatus* (Saltatoria, Insecta) during the molting cycle, *Cell Tissue Res.* **114:**27-46.

Ryerse, J. S., and Locke, M., 1978, Ecdysterone-mediated cuticle deposition and the control of growth in insect trachea, *J. Insect Physiol.* **24:**541-550.

Safranek, L., Cymborowski, B., and Williams, C., 1980, Effects of juvenile hormone on ecdysone-dependent development in the tobacco hornworm, *Manduca sexta, Biol. Bull.* **301:**248-256.

Scharrer, B., 1964a, Histophysiological studies on the corpus allatum of *Leucophaea maderae*. IV. Ultrastructure during normal activity cycle, *Z. Zellforsch. Mikrosk. Anat.* **62:**125-148.

Scharrer, B., 1964b, The fine structure of blattarian prothoracic glands, *Z. Zellforsch. Mikrosk. Anat.* **64:**301-326.

Scharrer, B., 1966, Ultrastructural study of the regressing prothoracic glands of blattarian insects, *Z. Zellforsch. Mikrosk. Anat.* **69:**1-21.

Scharrer, B., 1978, Histophysiological studies on the corpus allatum of *Leuophaea maderae*. VI. Ultrastructural characteristics in gonadectomized females, *Cell Tissue Res.* **194:**533-545.

Scharrer, B., and Wurzelmann, S., 1974, Observations on synaptoid vesicles in insect neurons, *Zool. Jahrb. Physiol.* **78:**387-396.

Schooneveld, H., 1979, Precocene induced necrosis and hemocyte mediated breakdown of corpora allata in nymphs of the locust *Locusta migratoria*, *Cell Tissue Res.* **203:**25-34.

Sedlak, B. J., 1981, An ultrastructural study of neurosecretory fibers within the corpora allata of *Manduca sexta*, *Gen. Comp. Endocrinol.* **44:**207-218.

Sedlak, B. J., 1984, The structure of endocrine glands. In *Comprehensive Insect Physiology, Biochemistry, and Pharmacology*, edited by G. A. Kerkut and L. I. Gilbert, vol. 7, Pergamon Press, Elmsford, N.Y.

Sedlak, B. J., and Gilbert, L. I., 1976a, Epidermal cell development during the pupal-adult metamorphosis of *Hyalophora cecropia*, *Tissue Cell* **8:**637-648.

Sedlak, B. J., and Gilbert, L. I., 1976b, Effects of ecdysone and juvenile hormone on epidermal cell development in *Hyalophora cecropia*, *Tissue Cell* **8:**649-658.

Sedlak, B. J., and Gilbert, L. I., 1979, Correlations between epidermal cell structure and endogenous hormone titers during the fifth larval instar of the tobacco hornworm, *Manduca sexta*, *Tissue Cell* **11:**643-653.

Sedlak, B. J., Marchione, L., Devokin, B., and Davino, R., 1983, Correlations between endocrine gland ultrastructure and hormone titers in the fifth larval instar of *Manduca sexta*, *Gen. Comp. Endocrinol.* **52:**291-310.

Sehnal, F., Maroy, P., and Mala, J., 1981, Regulation and significance of ecdysteroid titre fluctuations in lepidopterous larvae and pupae, *J. Insect Physiol.* **27:**535-544.

Siew, Y. C., 1965, The endocrine control of adult reproductive diapause in the crysomelid beetle, *Galeruca tanaceti* L., *J. Insect Physiol.* **11:**463-479.

Siew, Y. C., and Gilbert, L. I., 1971, Effects of moulting hormone and juvenile hormone on insect endocrine gland activity, *J. Insect Physiol.* **17:**2095-2104.

Smith, R. E., and Farquhar, M. G., 1966, Lysosome functions in the regulation of the secretory process in cells of the anterior pituitary gland, *J. Cell Biol.* **31**:319-347.

Srivastava, U. S., 1958, Prothoracic glands in *Tenebrio molitor* L. (Coleoptera, Tenebrionidae), *Nature (London)* **181**:1668.

Stay, B., and Tobe, S. S., 1977, Control of juvenile hormone biosynthesis during the reproductive cycle of a viviparous cockroach. I. Activation and inhibition of corpora allata, *Gen. Comp. Endocrinol.* **33**:531-540.

Steel, C. G. H., Bollenbacher, W. E., Smith, S. L., and Gilbert, L. I., 1982, Haemolymph ecdysteroid titers during larval adult development in *Rhodnius prolixus:* Correlations with moulting hormone action and brain neurosecretory cell activity, *J. Insect Physiol.* **28**:519-525.

Takeda, N., 1977, Histo-physiological studies on the corpus allatum during prepupal dispause in *Monema flavescens, J. Morphol.* **153**:245-262.

Thomsen, E., and Thomsen, M., 1970, Fine structure of the corpus allatum of the female blowfly, *Calliphora erythrocephala, Z. Zellforsch. Mikrosk. Anat.* **110**:40-60.

Tobe, S. S., and Pratt, G., 1975, The synthetic activity and glandular volume of the corpus allatum during ovarian maturation in the desert locust *Schistocerca gregaria, Life Sci.* **17**:417-422.

Tobe, S. S., and Saleuddin, A. S. M., 1977, Ultrastructural localization of juvenile hormone biosynthesis by insect corpora allata, *Cell Tissue Res.* **183**:25-32.

Tombes, A. S., and Smith, D. S., 1970, Ultrastructural studies on the corpora cardiaca-allata complex of the adult alfalfa weevil, *Hypera postica, J. Morphol.* **132**:137-148.

Truman, J. W., 1972, Physiology of insect rhythms. I. Circadian organization of the endocrine events underlying the moulting cycle of larval tobacco hornworms. *J. Exp. Biol.* **57**:805-820.

Unnithan, G. C., Bern, H. A., and Nair, K. K., 1977, Ultrastructural analysis of the neuroendocrine apparatus of *Oncopeltus fasciatus* (Heteroptera), *Acta Zool.* **52**:117-143.

Unnithan, G. C., Nair, K. K., and Syed, A., 1980, Precocene-induced metamorphosis in the desert locust, *Schistocerca gregaria, Experientia* **36**:135-156.

Vince, R. K., and Gilbert, L. I., 1977, Juvenile hormone esterase activity in precisely timed last instar larval and pharate pupae of *Manduca sexta, Insect Biochem.* **7**:115-120.

Waku, Y., and Gilbert, L. I., 1964. The corpora allata of the silkmoth, *Hyalophora cecropia:* An ultrastructural study, *J. Morphol.* **115**:69-96.

Wielgus, J. J., and Gilbert, L. I., 1978, Epidermal cell development and control of cuticle deposition during the last instar of *Manduca sexta, J. Insect Physiol.* **24**:629-637.

Wielgus, J. J., Bollenbacher, W. E., and Gilbert, L. I., 1979, Correlations between epidermal DNA synthesis and haemolymph ecdysteroid titer during the last larval instar of the tobacco hornworm, *Manduca sexta, J. Insect Physiol.* **25**:9-16.

Wigglesworth, V. B., 1964, The hormonal regulation of growth and reproduction in insects, *Adv. Insect Physiol.* **2**:247-336.

Yin, C. M., and Chippendale, G. M., 1979, Ultrastructural characteristics of insect corpora allata in relation to larval diapause, *Cell Tissue Res.* **197**:453-462.

8

The Fine Structure of Insect Glands Secreting Waxy Substances

YOSHIO WAKU AND IMRÉ FOLDI

1. The Waxy Secretions of Insects

1.1. Chemical Properties

Chemically, waxes are defined as esters of long-chain alcohols and fatty acids. Insects are the most remarkable wax producers among terrestrial animals. Marine crustaceans like copepods utilize the waxes they synthesize as metabolic fuels necessary for a pelagic life. However, instead of being metabolized, insect waxes are inert. They accumulate on the body surface where they function to protect the insect against desiccation and mechanical damage. With respect to insects, the term *wax* cannot be used in the strict sense of the chemist's definition. For example, the most familiar insect wax, beeswax, is composed to 70% esters of monocarboxylic acids and about 30% of lipoidal substances other than wax (Richards, 1978). The situation is even more complex since the combwax, which is produced by the abdominal wax gland cells, and cuticular wax, which is produced by the general epidermal cells, are different in their quantitative composition. Blomquist *et al.* (1980) have shown that esters are the predominant molecules in combwax, while hydrocarbons are the most abundant molecules in cuticular wax. The "greaselike waxy substance" on the cuticle of *Periplaneta americana* is composed of about 75% hydrocarbons and only 5% alkyl esters (Richards, 1978). As shown by these examples, the native "wax" of the insects is

---

YOSHIO WAKU • Biology Laboratory, Kyoto Technical University, Matsugasaki, Sakyo-ku, Kyoto 606, Japan.   IMRÉ FOLDI • Department of General and Applied Entomology, National Museum of Natural History, 75005 Paris, France.

not a simple class of molecules but is actually a complex mixture of true waxes and other substances, especially lipids such as hydrocarbons and free fatty acids. The relative amounts of true wax are quite variable from one species to another.

Moreover, the term *insect wax* is often used to refer to the bulky covering substance produced by the epidermis and accumulated on the body surface where it can play a protective role. Homopterans including scale insects and aphids are the most conspicuous examples of such wax producers. Again, such "waxes" are complex mixtures of various organic substances, and relative amounts of true waxes are variable (Beadsley and Gonzalez, 1975). The unusually simple wax of the psyllid *Anomoneura mori* contains as much as 93.8% of a true wax, lacceryl laccerate (Kuwahara, 1980). The "wax" covering of the scale insect *Ceroplastes pseudoceriferus* contains 64% wax and 30% resin (Tamaki, 1970), while the shellac of *Kerria lacca* contains only 5% true wax and 76% resin (Brown, 1975). Even the lipid extract from the cast cuticle of the cricket *Anabrus* contains about 12% resin which is similar to that of shellac (Baker *et al.*, 1960). Therefore, the reader should keep in mind that the waxy secretions of insects are actually mixtures of not only a wide variety of lipids ranging from long-chain hydrocarbons (paraffins), fatty acids, alcohols, sterols to true waxes, but also nonlipoidal substances such as natural resins, free amino acids, and proteins.

Since the chemical composition of waxy secretions is extremely variable, it is not surprising that the glands producing these substances are similarly diverse in morphology. Nevertheless, they share the common characteristic that they are derivatives of epidermal cells and produce at least some true wax in their secretions. We use in this review the term *wax gland* to refer to a gland that produces a secretion containing some wax as its constituent. Even those glands that have not been proved to produce wax show fundamental morphological similarities with other glands producing wax, and therefore it is reasonable to add them to the glands producing wax substances. The number of papers dealing with the fine structure of wax glands is not very large, but recent advances in this field permit us to give a general idea on the common characteristics and taxonomical significance of these glands in the Insecta.

1.2. Site of Synthesis of Waxy Secretions

Cuticle-secreting general epidermal cells also secrete wax. However, the quantity of wax which is produced and secreted by these epidermal cells is very small and the thickness of the wax layer ranges only from 0.1 to 0.4 μm (Lockey, 1960). Accordingly, SER, which is thought to be responsible for wax or lipid production, is not very prominent in epidermal cells. Cuticular wax is sometimes so voluminous that it can be seen with the naked eye. Hibernating pupae of lepidopterans like *Antheraea pernyi* are examples. Each hibernating pupae of *Manduca sexta* produces about 1.0 mg whereas a nonhibernating pupa produces only about 0.3 mg (Bell *et al.*, 1975). In such cases, epidermal cells produce voluminous wax, but their fine structure has not been investigated from the viewpoint of wax production.

Apart from general epidermal cells, oenocytes clustering under them have been frequently suggested as the site of synthesis of waxes or their precursors which then are transported to the epidermal cells. As early as 1914, Hollande (cited in Wigglesworth, 1972) reported that oenocytes contain crystals in their cytoplasm which were thought to be wax. The developing wax gland in the scale insect *Eriosoma* always contains a basal layer of oenocytes, suggesting an intimate functional relationship between the two tissues (Rogojanu, 1935). Piek (1964) postulated that the esters in beeswax are synthesized by the fat body cells, while both hydrocarbons and fatty acids are synthesized by the oenocytes. Reimann (1952) found that both wax gland cells and oenocytes go through similar developmental process during the wax secretion in the honeybee. Locke (1969) suggested that the oenocytes in the caterpillar *Calpodes ethlius* could be the site of many of the reactions of intermediary metabolism especially those of lipids and waxes. The cytoplasm of oenocytes is characterized by tubular SER. From those findings, Wigglesworth (1972) proposed that the oenocytes be regarded as epidermal cells that have been specialized for the production of the lipoprotein components of cuticle.

Wax glands, apart from those epidermal and subepidermal cells, are groups of cells specialized for the massive secretion of wax. The description of the fine structure of wax glands is the principal subject of this review.

1.3. The Function of Insect Waxes

Thin layers of cuticular wax on the body surface play an important role in protection against desiccation. This would take place inevitably in such small animals as insects, if the cuticular wax was absent. Wax also plays critical roles in the permeability of lipid-soluble substances of vital importance, and such phenomena have been discussed in many reviews (Wigglesworth, 1972; Ebeling, 1974; Jackson and Blomquist, 1976; Richards, 1978). It is likely that the bulky wax "blooms" covering the body surface of many homopterans, lepidopterans, and insects belonging to other orders confer considerable protection against parasites, predators, and diffusion of toxic substances such as insecticides, but direct evidence is scarce. In one case, a water extract of the covering of the scale insect *Aonidiella* was found to elicit oviposition by the parasite *Aphystis* (Miller and Kosztarab, 1979).

The sticky wax filaments covering coccinellid larvae seem to entangle the legs and coat the sense organs of the ants attacking them. Larvae of some species even use their wax filaments to attract the ants they prey upon. Wax coverings of many coccinellid larvae, scale insects, mealybugs, and wooly aphids reflect UV light strongly. Therefore, the color of the covering would be seen as "insect white" by predacious insects, and this signal may have some warning or detering effect (Pope, 1979).

The female adults of the bark beetle *Melanophila* oviposit on the trees recently scorched by forest fires. These insects possess special sense organs on their mesothorax to detect infrared radiation. The organ is covered by the mass of wax filaments which are sloughed off periodically, thus keeping the surface of sense organs clean (Evans, 1975).

Perhaps the most peculiar role played by wax is seen during the change in the body color of the desert beetle, *Cryptoglossa verrucosa*. During the periods of low humidity, this species is whitish blue. It turns black as the humidity increases, and the state of the cuticular wax is the cause of the color change (Hadley, 1979).

1.4. The Economical Importance of Insect Waxes

Honeycombs, made of beeswax, are melted down and utilized as the source of candles, furniture polish, car wax, cosmetics, etc. (Atkins, 1978). The scale insect *K. lacca* is the producer of commercial shellac. Their heavy infestation on *Ficus* trees in India results in the accumulation of thick cylinders of shellac resins along the branches. The purified shellac is used in sealing wax, inks, varnishes, etc. (Brown, 1975; Atkins, 1978). The wax from the Chinese scale insect *Ericela pela* is used to make candles (Miller and Kosztarab, 1979). In Central and South America, scale insects belonging to the genera *Gascardia* and *Ceroplastes* are used to produce wax.

2. Classification of Wax Glands

2.1. Classification of Epidermal Glands

The ectodermal exocrine glands can be grouped into two general categories: (1) simple glands which are either unicellular or made up of an aggregation of uniform cells, and (2) compound glands which contain a variety of cell types (the secretory cells, and the cells of the duct, aperture, and reservoir) (Richards and Davies, 1977).

Noirot and Quennedey (1974) divided the epidermal glands of insects into three classes depending on their cuticular covering and the way their secretions were emitted. In class 1, each secretory cell is simply covered by a cuticle which is produced by the secretory cell itself. Here the secretion must cross this barrier of cuticle. In class 2 which is rather rare, the gland cells are surrounded by other epidermal cells, and the secretion must be conveyed to the outside through the epidermal cells. Cuticle is never in contact with these gland cells. In class 3, the canal cell secretes a cuticular canal which penetrates the gland cell and opens to the outside. Here the modification is the greatest, and the gland is called *complex*. This useful classification is widely accepted, and wax glands also can be grouped according to these criteria.

2.2. Simple Wax Glands

Figure 1 shows a typical wax gland of class 1. Each larva of the sawfly *Eriocampa ovata* secretes massive amounts of wax from cup-shaped wax

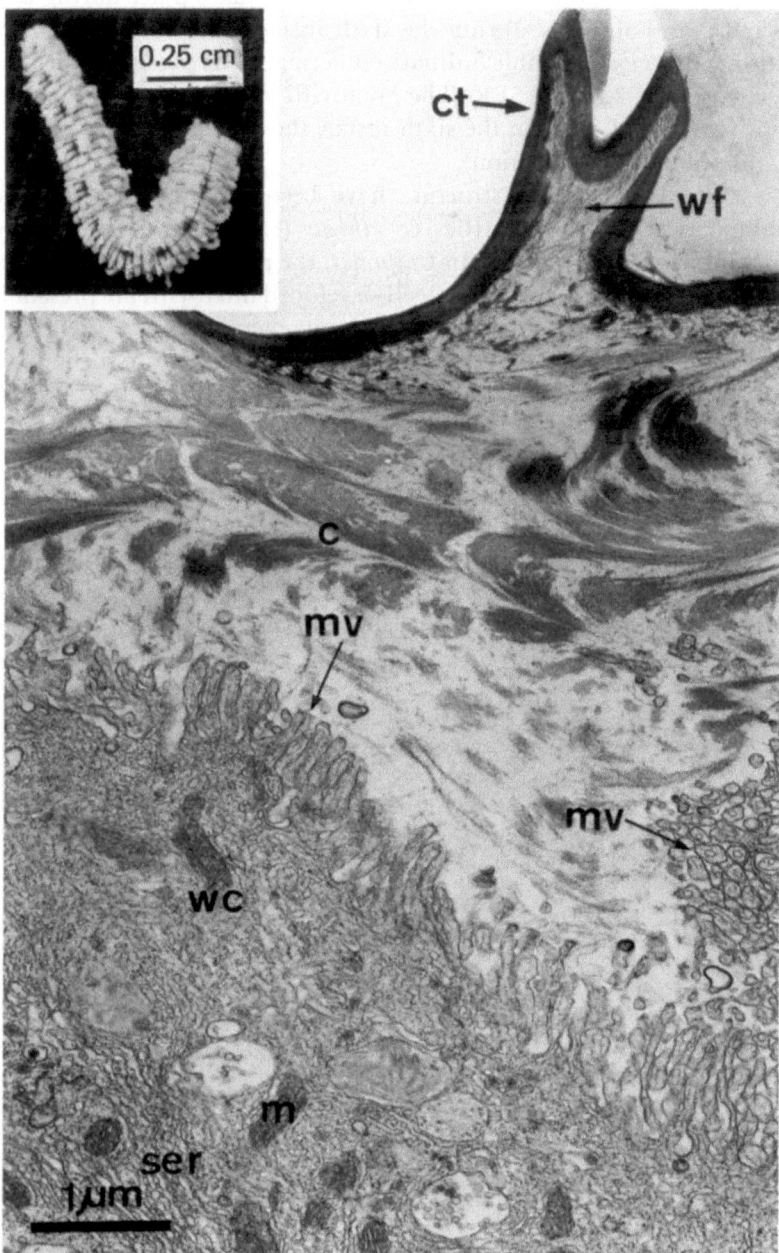

Figure 1. An electron micrograph of the apical border of a wax gland cell and its cuticle from a fifth-instar larva of the sawfly *Eriocampa ovata*. (Inset) A larva with wax blooms attached to its surface. c, cuticle; ct, cup-shaped tubercle which extrudes wax; m, mitochondria; mv, microvilli; ser, smooth-surfaced endoplasmic reticulum; wc, wax cell; wf, wax filaments. Courtesy of Dr. Jean Percy.

tubercles distributed on its body surface. Secretion takes place during the fifth instar and ceases abruptly during the sixth instar. The secreting cells in the fifth-instar epidermis resemble ordinary epidermal cells, except for the presence of an extensive network of SER. The microvilli are well developed, and each contains a tubule of SER. In the sixth instar, the ER becomes rough-surfaced (Percy, unpublished observation).

Wax glands of similar structure have been reported in other species, including larvae of the butterflies *C. ethlius* (Locke, 1974) and *Epipyrops anomala* (Marshall et al., 1974). In *Calpodes*, the plasma membrane under the cuticle is lobed, while in *Epipyrops* well-developed microvilli are present in the secretory cells. Such modifications of the plasma membrane ensure the efficient secretory activity of the waxy substances. SER is characteristic of such wax gland cells. An exception are the cells of the honeybee wax gland in which Sanford and Dietz (1976) found only RER. They suggested that the cells of the epidermal wax gland only transport the wax which is synthesized by oenocytes.

The wax gland of psyllids represents a gland of intermediate complexity. The gland has only one type of wax-secreting cell, and each is tall and slender (Figure 2) with an invagination extending its entire length. This invagination acts as the collecting and transporting duct for the flocculent wax substance and communicates with the overlying cuticle by finely branched canaliculi. The cells have well-developed SER, in contrast to the RER of the interstitial cells. The wax

Figure 2. A diagram of a wax gland from a psyllid, *Anomoneura mori*. The gland is composed of tall wax cells (wc) and flat interstitial cells (ic). The overlying wax-secreting cuticle (wsc) is finely sculptured and is divided into rim cuticle (rc) and canaliculate cuticle (cc), bm, basement membrane; wd, wax duct.

cells do not possess surface microvilli in the plasma membrane. There is a space between the cuticle and the plasma membrane of each wax cell. The plasma membrane retracts from the overlying cuticle during development of the wax gland (Waku, 1978; see Figure 20B).

Morphologically similar wax glands occur in many scale insects including those species belonging to the subterranean family Margarodidae (Foldi, 1981) and the shellac-producing species *K. lacca* (Haque, 1975a,b). In each integumentary wax gland of *Porphyrophora critmi* (Margarodidae), there are 18 identical secretory cells with branched collecting canals and extensive network of SER (Figures 3A, 4, and 5). Between the overlying cuticle and secretory cells, there is a subcuticular space in which the secretion is stored. The overlying cuticle is disk-shaped and has 25 excretory holes closed at their base. The basal cuticle is composed of characteristic endocuticle made of a network of fine tubular structures and thin epicuticle with many epicuticular pores (10 nm in diameter) (see Figure 3A-3). The endocuticle sends many polymorphic projections into the subcuticular cavity. They bear a filament brush which faces the cavity, and each filament is about 0.1 μm in diameter and 0.6 μm in length. The filaments, the tubular endocuticle, and epicuticular pores allow the transit of the secretion across the cuticle. Each excretory hole produces a hollow wax fibril of the same diameter as the hole (about 0.8 μm), and 25 fibrils extruded from 25 holes are fused together to form a thick wax thread.

Since the cuticle of these simple glands is the product of the wax-secreting cells and each cell faces the cuticle directly, there must be a special pathway to transport and emit the wax through the cuticle. Locke (1960, 1961) was the first to find this pathway. In *Calpodes* larvae (Locke, 1974), the wax forms liquid-crystalline thin filaments only 6 nm in diameter in the epicuticular "wax canals" (Figure 19A). Wax filaments in larvae of the butterfly *Epipyrops* measure 9-10 nm in diameter and are very prominent in the wax-secreting papillae of the cuticle (Marshall *et al.*, 1974). Sometimes the wax filaments can have a central canal as in the case of tergal wax produced by the fly *Dacus tryoni* (Evans, 1967a,b). On the other hand, Filshie (1970a,b) proposed the less precise term *epicuticular filaments* in place of *wax filaments*. He observed similar filaments in the cuticle of larvae of the fly *Lucilia cuprina* which lack a layer of surface wax, and since it was impossible to dissolve the filaments in *Calpodes* cuticle by organic solvents, Filshie concluded that while the filaments might provide a path for the wax precursors through the epicuticle, it was unlikely that the filaments themselves were composed of wax precursor molecules. Locke and Krishnan (1971) found that the epicuticular filaments in *Calpodes* were associated with polyphenols and suggested that the filaments could provide a channel for the transport of the wax.

The epicuticular wax filaments in the wax-secreting tubercle of *Eriocampa* cuticle are shown in Figure 1, and wax canals and filaments in the epidermal cuticle of the scale insect *Coccus hesperidum* are shown in Figures 6 and 7. However, such filaments cannot always be seen in every wax-secreting cuticle. The psyllid *A. mori* has no apparent wax filaments and canals with uniform size in its wax-secreting cuticle (Waku, 1978).

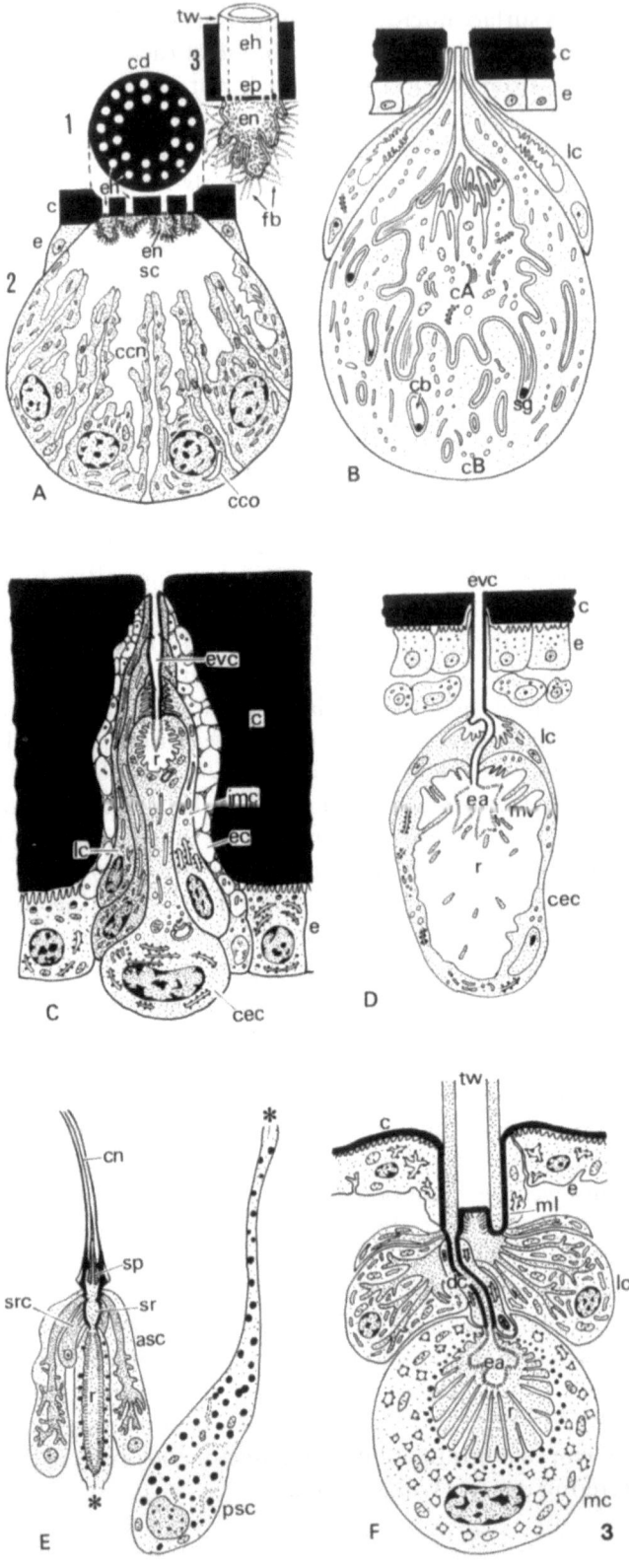

The cuticle of glands of this simple type is characterized by delicate surface sculptures, papillae and indentations, etc. (Figures 2, 19A, B). The function of these structures will be discussed in Section 2.4

2.3. Complex Wax Glands

Complex wax glands commonly occur in the scale insects (Figures 3B-F). Such glands are composed of duct and secretory cells, and their secretory cells may be of different types, with a different secretion produced by cells of each type. Generally, one large central cell is present at the basal extremity of the gland. Its cytoplasm is rich in RER, mitochondria, Golgi apparatus, and dense secretory globules (Figure 9). The secretion of this cell is stored in a large reservoir (Figure 12) which is formed by a deep, pouchlike invagination of the apical plasma membrane. Long microvilli project into the reservoir. Cytochemical tests demonstrate that the stored secretion contains proteins and glycoproteins. In one case (Waku and Manabe, 1981), this secretion contains lipid also. The central cell also secretes a cuticular end apparatus (Figures 3D and E, ea) which attaches to the secretory duct (Noirot and Quennedey, 1974). The end apparatus is porous since it is pierced by a network of canals, each 3-3.5 nm in diameter (Figure 10). The secretion is transported to the exterior via a cuticular duct which is the product of small ductule cells (Figure 3F, dc). Sometimes, as in the case of *Aonidiella aurantii* (Pesson and Foldi, 1978), the ductule cell can produce a second secretion in a second reservoir (Figure 3E, sr).

Around the ductule cell, there are several cells called *lateral cells* or *accessory cells* (Figures 3C, D, F, lc; Figure 3E, asc). These cells secrete the lipoidal moiety of the final covering substance of the insect. The cytoplasm of the lateral cell is

←

Figure 3. Diagram of the wax glands of various scale insects. (A) Integumentary type-A gland from *Porphyrophora crithmi* (Margarodidae). (1) Top view of discoid cuticle which extrudes wax. (2) Sagittal section of the gland. (3) Detail of an excretory hole. c, cuticle; ccn, central canal; cco, collecting canal; cd, discoid cuticle in which 25 excretory holes are arrayed circularly; e, epidermal cell; eh, excretory hole closed at its base; en, endocuticle with tubular substructure; ep, epicuticle with peripheral pores; fb, cuticular filament brush; tw, tubular wax fibril. Adapted from Foldi (1981). (B) Wax gland in *Ceroplastes sinensis* (Coccidae). The complicated arrangement of the secretory cells is characteristic of this species. c, cuticle; cA, secretory A cell; cB, secretory B cell; cb, cytoplasmic branch of A cell; e, epidermal cell; lc, lateral cell; sg, secretory globule. (C) Lac gland in *Coccus hesperidum* (Coccidae). This gland is composed of cells of several different types. c, cuticle, cec, central cell; e, epidermal cell; ec, envelope cell; evc, evacuating canal; imc, intermediate cell; lc, lateral cell. Adapted from Foldi (1978). (D) Tubular wax gland in *Lecanodiaspisis sardoa* (Lecanodiaspididae). The central cell has an extremely large reservoir. c, cuticle; cec, central cell; e, epidermal cell; ea, end apparatus; evc, evacuating canal; lc, lateral cell; mv, microvilli; r, reservoir. (E) Pygidial gland in *Aonidiella aurantii* (Diaspididae). A most advanced and complicated gland with a secondary reservoir and spinneret. The reader must mentally join the left and right figures together at the asterisk. asc, accessory cell; cn, canal; psc, principal secretory cell; r, reservoir; sp, spinneret; sr, secondary reservoir; src, secondary reservoir cell. Adapted from Pesson and Foldi (1978). (F) Wax gland in *Eriococcus lagerstraemiae* (Eriococcidae). The secretion is cast by cuticular mold as tubular wax. c, cuticle; dc, ductule cell; e, epidermal cell; ea, end apparatus; lc, lateral cell; mc, main cell; ml, molding cuticle; tw, tubular wax. Adapted from Waku and Manabe (1981).

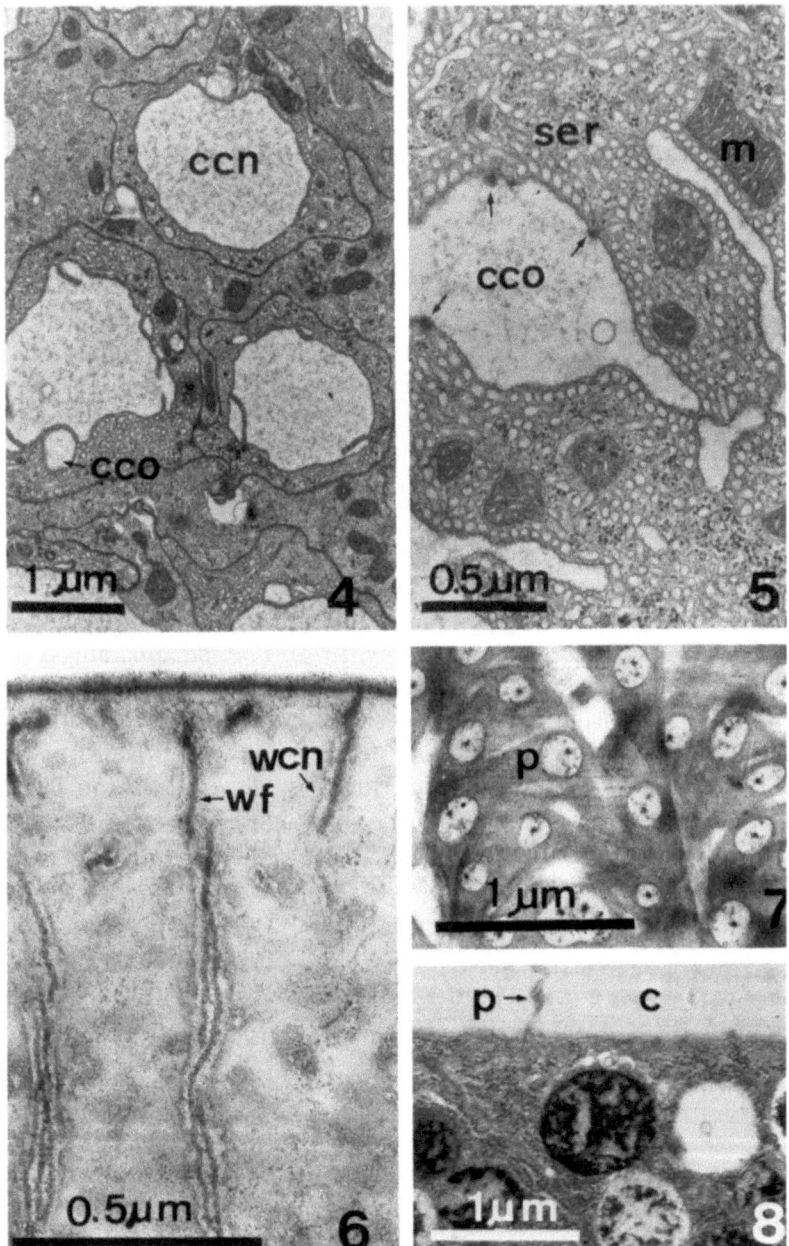

Figure 4. A transverse sectional view of a portion of integumentary wax gland in *Porphyrophora crithmi* (Margarodidae). Identical wax cells, each with a large central canal (ccn) and a branched collecting canal (cco), interdigitate with each other. Reproduced from Foldi (1981).
Figure 5. A higher-magnification view of an area similar to that shown in Figure 4. Note the prominent smooth-surfaced endoplasmic reticulum (ser) and the collecting canals (cco) which contain many discharging masses of the secretory substance (arrows). m, mitochondria. Reproduced from Foldi (1981).
Figure 6. An epicuticular wax canal (wcn) and wax filaments (wf) in the epidermal cuticle of *Coccus hesperidum*.
Figure 7. Cross-sectioned pore canals (p) in the endocuticle of *C. hesperidum*. Note the wax filaments in their lumina.
Figure 8. The cytoplasm of an epidermal cell just under the cuticle of *C. hesperidum* showing lipofuscin granules. c, cuticle; p, pore canal.

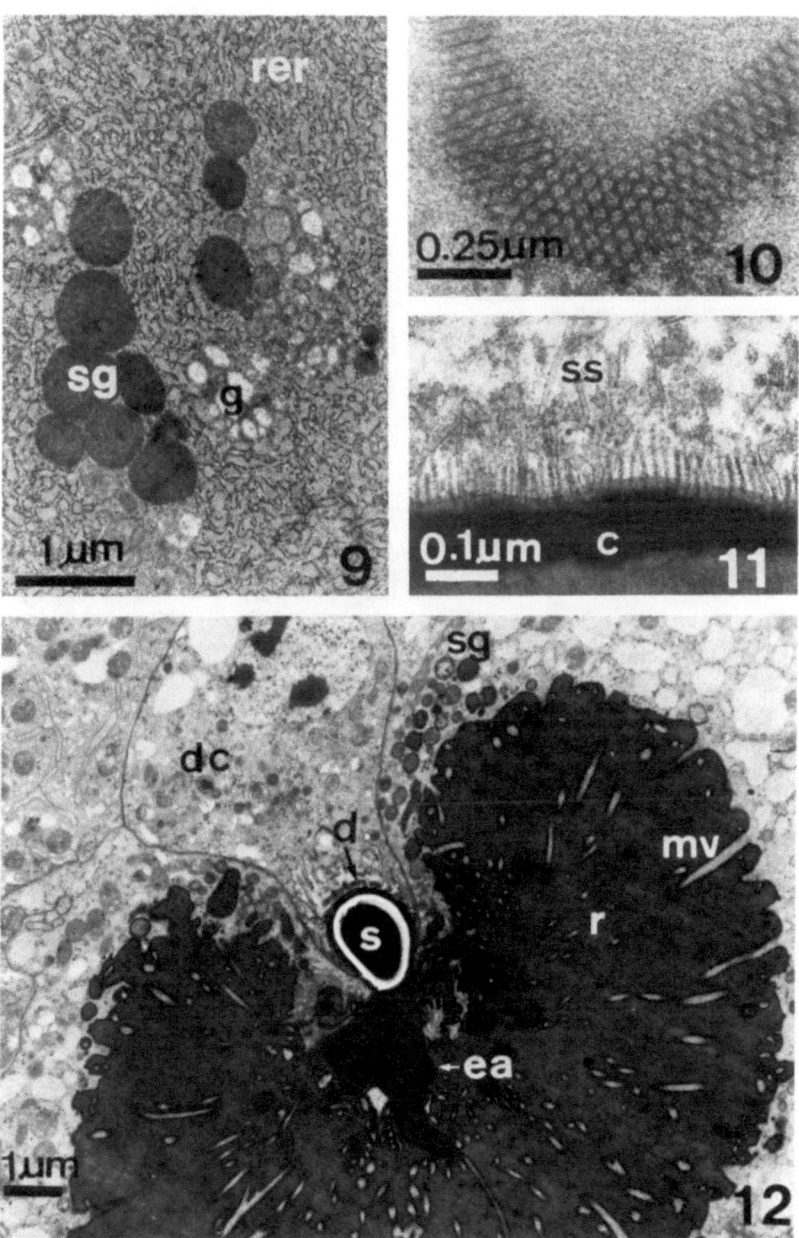

Figure 9. The cytoplasm of the principal secretory cell in *Aonidiella aurantii*. g, Golgi apparatus; rer, rough-surfaced endoplasmic reticulum; sg, secretory globule. Reproduced from Pesson and Foldi (1978).
Figure 10. The end apparatus of the pygidial gland in *A. aurantii*. Reproduced from Pesson and Foldi (1978).
Figure 11. The cuticular filament brush of the pygidial gland of *A. aurantii*. c, cuticle; ss, subcuticular space. Reproduced from Pesson and Foldi (1978).
Figure 12. The reservoir (r) and outlet ductule (d) in the wax gland of *Eriococcus lagerstraemiae*. ea, end apparatus; dc, ductule cell; mv, microvilli; s, secretion of the main cell; sg, secretory globule. Reproduced from Waku and Manabe (1981).

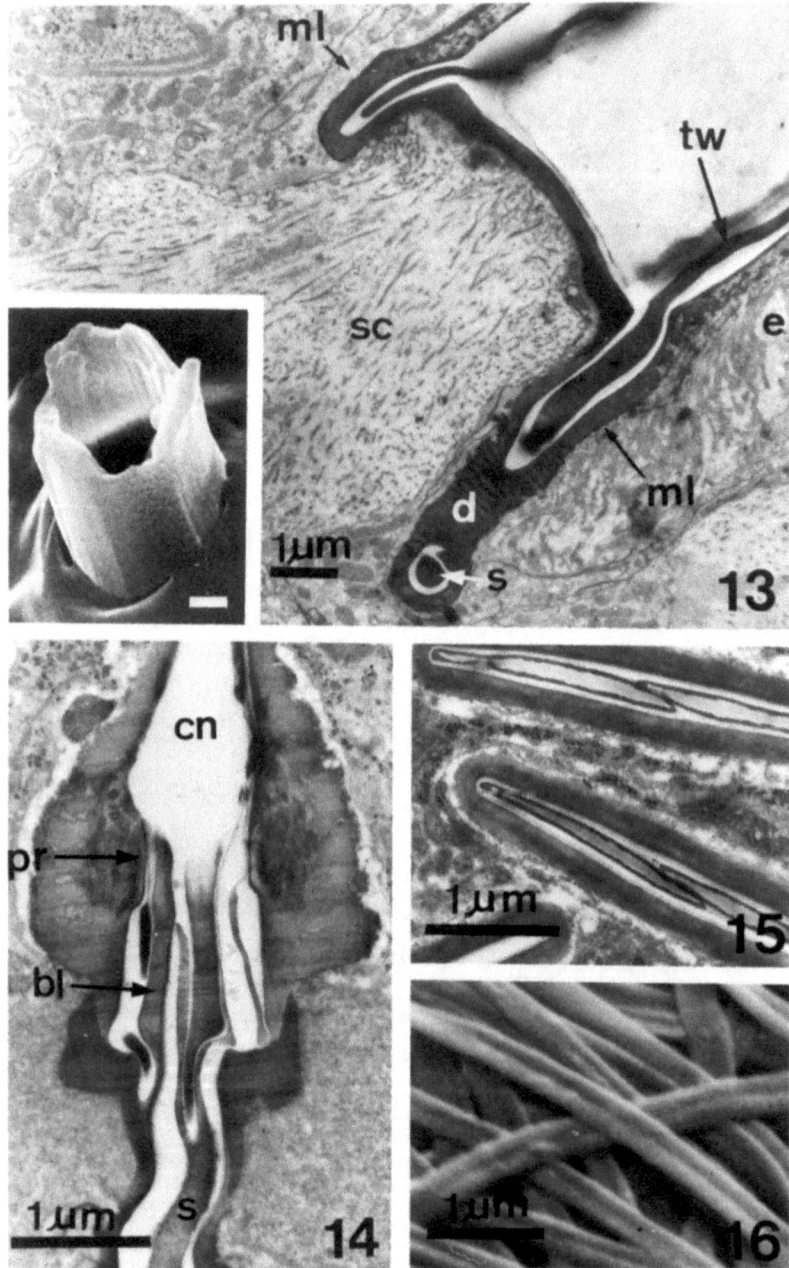

Figure 13. The molding structure in the wax gland of *Eriococcus lagerstraemiae*. The secretion of the main cell (s) is transported via the ductule (d) to the bottom of the evacuating canal where the cuticle has a troughlike invagination which functions as a mold (ml). e, epidermal cell; sc, subcuticular cavity; tw, tubular wax. (Inset) A SEM photograph of a tubular wax thread of *E. lagerstraemiae* protruding from the cuticular surface. (Bar = 1 μm.) Reproduced from Waku and Manabe (1981).

Figure 14. The molding structure in the pygidial gland in *Aonidiella aurantii*. The columnar secretion (s) is divided and molded into two strands by the complicated spinneret cuticle with two blades (bl) and presses (pr). cn, canal. Reproduced from Pesson and Foldi (1978).

osmiophilic (Tamaki, 1970) and contains an extensive network of SER (Foldi, 1978, 1981; Pesson and Foldi, 1978; Waku and Manabe, 1981). The plasma membrane is deeply infolded in a manner reminiscent of the collecting canal in the cells of the simple wax gland of *P. crithmi* (Figure 3A). These cells face the cuticular canal that carries the secretion to the body surface. There is a fluid-filled space between the folded plasma membrane and the cuticle in which is suspended a flocculent substance (Pesson and Foldi, 1978; Foldi, 1981; Waku and Manabe, 1981), and this probably represents the secretion of the lateral cells.

The cuticle facing the lateral cells bears a brushlike collection of filaments which project into the subcuticular space (Figure 11). A similar brushlike configuration of filaments is also present in the simple wax gland of the Margarodidae (Figure 3A) and thus seems to be common in scale insects. The structural similarities between the lateral cells of the complex wax glands in the advanced group of scale insects and the simple wax cells in a primitive group like Margarodidae suggest the two cell types are homologous. The number of lateral cells per wax gland varies from species to species: there are 10 in *Planococcus*, six in *Eriococcus*, four in *Lecanodiaspisis*, and only two in *Aonidiella*.

The secretions of the central and lateral cells converge apically and mix at the base of the evacuating canal to form the final product of the gland. Sometimes there is a special molding structure of the cuticle which casts the secretions into particular shapes, such as tubes (Figures 13-16). But in other cases, like the figure-8-shaped gland of *Lecanodiaspisis*, the mixing takes place on the outer surface where the canal opens and the lateral cells secrete their products through the overlying cuticle (Foldi, 1982).

Perhaps the most complex cellular organization found among the wax glands in the scale insects is that of *Ceroplastes sinensis* (Foldi, 1982). The main body of the gland is composed of two cells: A cells and B cells (Figure 3B). The lipid-secreting B cell completely surrounds the A cell, which is less dense than the B cell. The apical plasma membrane of the B cell is deeply infolded forming many cavities into which the A cell extends its cytoplasmic processes. The secretion of the B cell may traverse the plasma membranes and accumulate in the tip of the A cell process as a highly osmiophilic globule (Figures 3B, 17, and 18). This secretion mingles with the secretion of the A cell and then is released through a much ramified evacuating canal. The secretion of the lateral cells joins that of the A and B cells at the aperture (Figure 3B).

According to Tamaki (1970), the general epidermal cells in *C. pseudoceriferus* seem to play a role in wax production along with the specialized gland cells. In this case, the epidermal cells may secrete a dry wax which mingles with the liquid secretion of the gland to form a wet wax which covers the body surface. The presence of osmiophilic granules and pore canals in the epidermal cells and their cuticle supports this suggestion. Similar granules and pore canals also are present in the epidermis of a related species, *Coccus hesperidum* (Figure 8), but these granules may be lipofuscins or pigments and not precursors of wax. The

---

Figure 15. The two-stranded secretion of *A. aurantii* formed by the molding structure shown in Figure 14. Reproduced from Pesson and Foldi (1978).

Figure 16. A SEM photograph of the ribbon-shaped, two-stranded secretion of *A. aurantii*. Reproduced from Pesson and Foldi (1978).

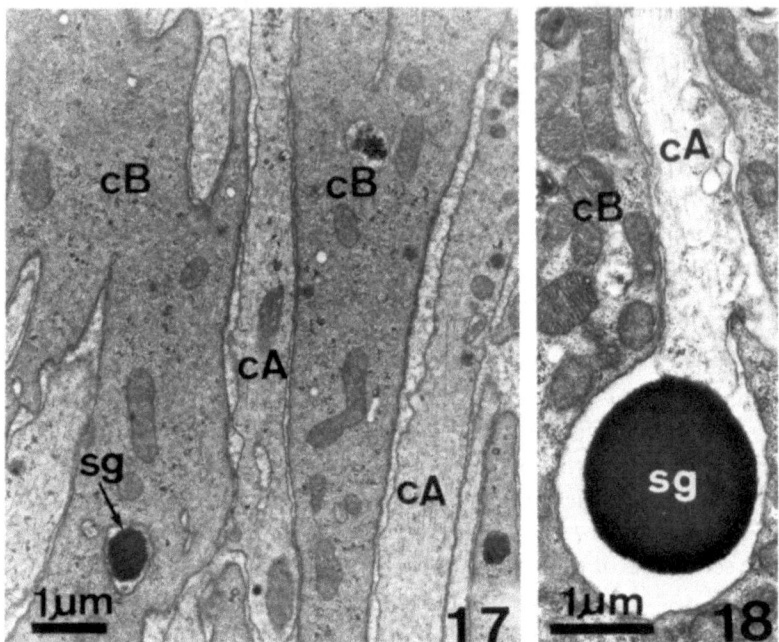

Figure 17. Interdigitated cytoplasmic processes of an A cell (cA) and a B cell (cB) in the wax gland of *Ceroplastes sinensis*.
Figure 18. A secretory globule (sg) of a B cell (cB) which has been transferred to the tip of a cytoplasmic process of an A cell (cA).

tergal gland of the fly *D. tryoni* is a case in which a bristle-bearing trichogen cell produces a waxy substance (Evans, 1967a,b).

The cuticle of scale insects contains many apertures leading to complex wax glands. For instance, those in *C. pseudoceriferus* are trilocular, quinquelocular, or multilocular (Kawai and Tamaki, 1967; Tamaki *et al.*, 1969). The various shapes of these apertures may serve as useful characters in future taxonomic studies (Miller and Kosztarab, 1979).

2.4. The Mechanism of Pattern Formation by the Products of Wax Glands

The "wax bloom," the final product of wax glands, often takes an appearance characteristic for each species (Hashimoto and Kitaoka, 1971; Locke, 1974; Marshall *et al.*, 1974; Hamon *et al.*, 1975; Pesson and Foldi, 1978; Pope, 1979; Foldi, 1981; Waku and Manabe, 1981). The structure of the cuticle-secreting wax blooms exerts a decisive effect on their morphology. The tubular wax bloom of *Calpodes* larva is produced through crater-shaped tubercles in which rows of epicuticular wax filaments are present (Locke, 1974; Figure 19A). Presumably, the tubular wax of coccinellid larvae is spun out by a mechanism similar to that in *Calpodes* (Pope, 1979). The paraffin-secreting cuticle of *Epipyrops* contains numerous lanceolate papillae each with several ridges on its

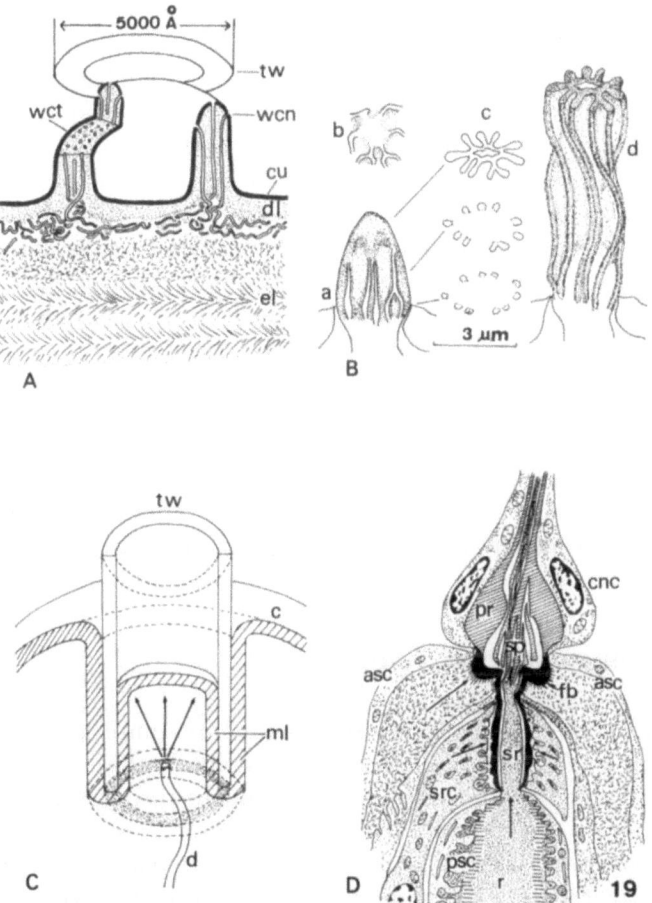

Figure 19. Diagrams of the various mechanisms employed in some wax glands for the formation of patterned secretions. (A) Crater-shaped epicuticular tubercle from a *Calpodes* larva in which rows of thin wax canals are present. The wax is extruded as a tube (500 nm in diameter) from the rim of the crater. cu, cuticulin; dl, dense layer; el, endocuticular lamellae; tw, tubular wax; wcn, wax canal; wct, wax canals shown in transverse section. Adapted from Locke (1974). (B) A paraffin-secreting papilla from an *Epipyrops anomala* larva. As in the case of *Calpodes*, there are numerous thin wax filaments in each cuticular papilla. (a) A lateral view of a papilla. (b) Another viewed obliquely from above. (c) A secreted paraffin tubule cross-sectioned at different levels, which are referred to in the papilla (a). (d) A general view of a sculptured tubule growing from a papilla. Adapted from Marshall *et al.* (1974). (C) The mechanism of molding the wax tube in *Eriococcus lagerstraemiae*. The secretion comes up through a thin outlet ductule (d) to the bottom of the evacuating canal (stippled area) and is molded there as a tube. Arrows show the directions of spreading of the secretion. c, cuticle; ml, mold; tw, tubular wax. (D) The complicated organ forming the ribbonlike secretion in *Aonidiella aurantii*. The secretions of prinipal (psc), secondary reservoir (src), and accessory secretory (asc) cells unite at the base of the evacuating canal where the press (pr) and spinneret (sp) mold it to the double-stranded ribbon. Arrows indicate the directions of the secretions in each secretory cell. cnc, canal; fb, cuticular filament brush; r, reservoir; sr, secondary reservoir. Adapted from Pesson and Foldi (1978).

surface. Within them are numerous epicuticular filaments. The paraffin tubules are secreted through such papillae and their pattern is related to that of ridges (Marshall et al., 1974; Figure 19B).

The formation mechanism of tubular wax blooms in some scale insects is quite different from the above cases. The tubes are "molded" by some special cuticular casts present close to the aperture. In the Margarodidae, there are 25 excretory holes in the disk-shaped area of the cuticle from which wax is molded and extruded. Each hole produces a thin and hollow wax tube and the 25 tubes together form a thick bundle of wax bloom (Foldi, 1981; Figure 3A). In *Eriococcus*, the cuticular mold lies at the bottom of the evacuating canal where the secretions of both main and lateral cells mix, and the secretions are molded by this structure as a comparatively thick (5 μm in diameter) but hollow tube (Waku and Manabe, 1981; Figures 13 and 19C). In *Aonidiella aurantii*, the wax tubules are double-stranded ribbons. At the distal end of the gland, there are two cells producing a complex chitinous press with blades, which work together as a spinneret to transform the secreted cylinder to a double-stranded ribbon (Pesson and Foldi, 1978; Figures 14-16, 19D).

The wax bloom which covers the woolly aphid, *Adelges piceae*, is ribbonlike, and its pattern formation results from the location of wax pores and the shape of the dorsal epidermis (Retnakaran et al., 1979).

3. Development of Wax Glands

Pollister (1937) studied the development of the complex wax gland of the scale insect *Pseudococcus maritimus*. She described mitosis of the gland cells during the nymphal molt. However, no electron microscopic studies have been carried out on the development of complex glands.

The development of the simpler wax gland of *A. mori* was studied in detail by Waku (1978; Figure 20). Here the gland is composed of two cell types, tall wax cells and flat interstitial cells. Both differentiate from a common primordial cell group of epidermal origin. The primordial wax cell can produce its own cuticle at the early molting period, but subsequently it shifts its role to wax production completely and therefore becomes incapable not only of producing cuticle but also of shedding its cuticle at the next molt. The interstitial cell helps the wax cell to accomplish this by severing the neck of the slender wax cell as it grows, and the result is the degeneration of all wax cells of the previous instar at each molt. The interstitial cell, like the usual epidermal cell, has RER, while the actively secreting wax cell contains only SER. In this respect, the finding of Percy (unpublished) that in the fifth-instar larva of *Eriocampa* the simplest wax gland cell (which produces much wax) is rich in SER while the epidermal cell of the sixth-instar larva (which does not produce wax) has only RER is of special interest. Similarly, paraffin-secreting cells in *Epipyrops* contain much SER while the same cells synthesizing new cuticle do not (Marshall et al., 1974). Probably the simple and primitive wax cell of *Eriocampa* may be able to shift its role from cuticle production to wax production at each larval molt except for the last one, thus surviving through successive molts. On the contrary, the specialized wax cell in *Anomoneura* cannot do so. This reminds us of Locke's

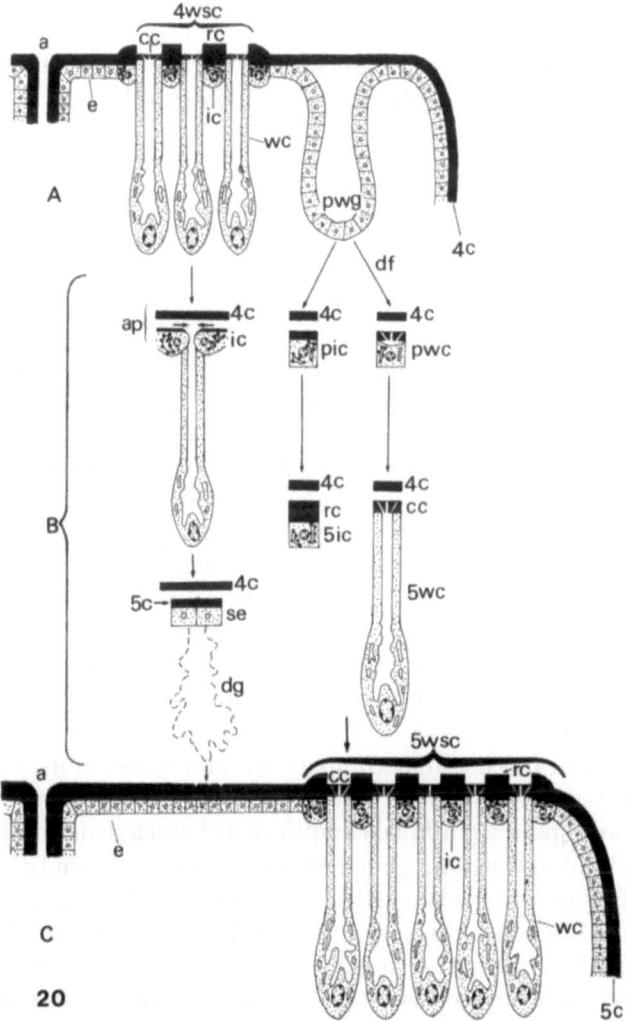

Figure 20. A diagrammatic representation of stages in the development of the perianal wax gland of *Anomoneura mori* during the final nymphal molt. (A) Fourth-instar nymph. The wax gland is composed of wax cells (wc) and interstitial cells (ic). The overlying wax-secreting cuticle of the fourth-instar nymph (4wsc) consists of thick rim cuticle (rc) and thin canaliculate cuticle (cc). Surrounding the wax gland, there is a pleated sheet of epidermal tissue (pwg) which is the primordial wax gland for the next (fifth) instar. a, anus; e, epidermal cell; 4c, cuticle of the fourth instar. (B) Final nymphal stage during the molt from the fourth to the fifth instar. The interstitial cells (ic) grow toward the center of the wax cell and sever it at its neck, resulting in the degeneration of the wax cell. The interstitial cells now become secondary epidermal cells (se) and secrete the cuticle of the fifth instar (5c) under the fourth instar cuticle (4c). On the other hand, primordial wax gland cells (pwg, A) differentiate to the primordial interstitial cells (pic) and promordial wax cells (pwc), respectively. Then the former cells become the interstitial cells of the fifth instar (5ic), secreting thick rim cuticle (rc). The latter cells secrete thin canaliculate cuticle (cc) and then become the wax cells of the fifth instar (5wc), shifting their role completely to wax secretion. ap, apolysis; df, differentiation; dg, degeneration. (C) Fifth-instar nymph. The wax gland is composed of wax cells (wc) and interstitial cells (ic) as in the fourth instar. The gland develops in the outer peripheral area of the previous fourth-instar wax gland which has degenerated, and therefore the gland grows outward. The fifth-instar nymph does not produce the primordial wax gland cells for the adult, and thus the adult lacks the wax gland. a, anus; cc, canaliculate cuticle; e, epidermal cell, rc, rim cuticle; 5c, cuticle of the fifth instar; 5wsc, wax-secreting cuticle of the fifth instar.

suggestion (1965, 1974) that the production of wax in epidermal cells is a cyclic phenomenon occurring repeatedly during molt-intermolt cycles as a response to molting hormones.

4. Taxonomic Importance of Wax Glands

Pope (1979) carried out a comparative and taxonomical study of wax blooms and wax-secreting cuticles from larvae belonging to 20 species from 10 tribes of coccinellid beetles using the scanning electron microscope. These larvae possess on their cuticle many small disks with delicate sculpture, cylindrical projections, papillae and spicules of various shapes, and shallow depressions. All these wax-secreting structures have a specific pattern for each species. The 10 tribes can be segregated, by the characteristics of these structures, into four distinct groups. The most primitive and simple structure appears in the noviine/cocciduline group which forms a strong contrast to the efficient and complex structure in the advanced scymnine/hyperaspine group. Within the same group, Pope could classify many genera according to the characteristics of the wax-secreting surfaces. He established a new tribe for genus *Scymodes* which has very peculiar wax-secreting structures. These wax-secreting structures may play an important role in the insect systematics.

Foldi (1982) carried out a comparative and gross morphological and ultrastructural study on the gland types of scale insects belonging to 20 families. Each family and species exhibited complicated characteristics in the organization and distribution of various glands, depending on the sex and developmental stage studied. With respect to the cell organization and ultrastructure, the glands can be classified into two categories: (1) simple glands with similar type cells having SER and ramified collecting invaginations, but no evacuating cuticular duct, and (2) complex glands with cells of different types having both SER and RER according to their functions, together with a distinct evacuating canal made by another cell. The former appears in all glands of Margarodidae, the stigmatic and perivulvar glands of Diaspididae, and the stigmatic and multilocular glands of numerous families. The latter appears in all families except for Margarodidae, and its most complex case is seen in the pygidial gland of *Aonidiella* (Diaspididae) which has three kinds of secretory cells and one kind of spinneret cell. There is a general evolutionary trend toward a more complex wax gland morphology and toward secretory products that become harder and more durable. These observations support the concept that wax gland morphology can be regarded as an effective criterion and should be included as important phenotypes in the classification of scale insects.

ACKNOWLEDGMENTS

We are grateful to Dr. Jean Percy, Forest Pest Management Institute, Canadian Forestry Service, Sault Ste. Marie, Ontario, Canada, for letting us use her unpublished data and photographs (Figure 1 and inset) in this review.

Permission to reproduce published illustrations was generously provided by the Japanese Society of Applied Entomology and Zoology (for Figures 12 and 13), the Longman Group (for Figures 9, 10, 11, 14, 15, and 16), and the Wistar Institute Press (for Figures 4 and 5).

References

Atkins, M. D., 1978, *Insects in Perspective*, Macmillan Co., New York.
Baker, G., Pepper, J. H., Johnson, L. H., and Hastings, E., 1960, Estimation of the composition of the cuticular wax of the Mormon cricket, *Anabrus simplex* Hald, *J. Insect Physiol.* 5:47-60.
Beadsley, J. H., Jr., and Gonzalez, R. H., 1975, The biology and ecology of armored scales, *Annu. Rev. Entomol.* 20:47-73.
Bell, R. A., Nelson, D. R., Borg, T. K., and Cardwell, L. T., 1975, Wax-secretion in non-diapausing and dispausing pupae of the tobacco hornworm, *Manduca sexta*, *J. Insect Physiol.* 21:1725-1729.
Blomquist, G. T., Chu, A. J., and Ramaley, S., 1980, Biosynthesis of wax in the honeybee, *Apis mellifera* L., *Insect Biochem.* 10:313-321.
Brown, K. S., Jr., 1975, The chemistry of aphids and scale insects, *Chem. Soc. Rev.* 4:263-288.
Ebeling, W., 1974, Permeability of insect cuticle. In *The Physiology of Insecta*, 2nd ed., edited by M. Rockstein, vol. 6, pp. 271-345, Academic Press, New York.
Evans, J. J. T., 1967a, The integument of the Queensland fruit fly, *Dacus tryoni* (Frogg.). I. The tergal glands, *Z. Zellforsch. Mikrosk. Anat.* 81:18-33.
Evans, J. J. T., 1967b, The integument of the Queensland fruit fly, *Dacus tryoni* (Frogg.). II. Development and ultrastructure of the abdominal integument and bristles, *Z. Zellforsch. Mikrosk. Anat.* 81:34-48.
Evans, W. G., 1975, Wax secretion in the infrared sensory pit of *Melanophila acuminata* (Coleoptera: Bupalidae), *Quaest. Entomol.* 11:587-589.
Filshie, B. K., 1970a, The resistance of epicuticular components of an insect to extraction with lipid solvents, *Tissue Cell* 2:181-190.
Filshie, B. K., 1970b, The fine structure and deposition of the larval cuticle of the sheep blowfly (*Lucilia cuprina*), *Tissue Cell* 2:479-488.
Foldi, I., 1978, Ultrastructure des glandes tégumentaires dorsalis, sécrétrices de la "laque" chez la femelle de *Coccus hesperidum* L. (Homoptera: Coccidae), *Int. J. Insect Morphol. Embryol.* 7:155-163.
Foldi, I., 1981, Ultrastructure of the wax-gland system in subterranean scale insects (Homoptera, Coccidae, Margarodidae), *J. Morphol.* 168:159-170.
Foldi, I., 1982, Les glandes tégumentaires des Cochenilles (Homoptera: Coccoidea). Ultrastructure comparée et signification phylogénétique, Thesis, D.Sc., Faculty of Science, University of Paris VI.
Hadley, N. F., 1979, Wax secretion and color phases of the desert tenebrioniid beetle *Cryptoglossa verrucosa* (La Conte), *Science* 203:367-369.
Hamon, A. B., Lambdin, P. L., and Kosztarab, M., 1975, Eggs and wax secretion of *Kermes kingi* (Hom., Kermesidae), *Ann. Entomol. Soc. Am.* 68:1077-1078.
Haque, M. S., 1975a, Non-resinous secretions of *Kerria lacca* (Homoptera-Coccoidea), *J. Zool.* 176:1-25.
Haque, M. S., 1975b, Cells secreting non-resinous substances in the lac insect *Kerria lacca* (Homoptera-Coccoidea), *J. Zool.* 176:27-38.
Hashimoto, A., and Kitaoka, S., 1971, Scanning electron microscopic observation of the waxy substances secreted by some scale insects, *Jpn. J. Appl. Entomol. Zool.* 15:76-86 (Japanese with English summary).
Jackson, L. L., and Blomquist, G. J., 1976, Insect waxes, In *Chemistry and Biochemistry of Natural Waxes*, edited by P. E. Kolattukudy, pp. 201-233, Elsevier, Amsterdam.
Kawai, S., and Tamaki, Y., 1967, Morphology of *Ceroplastes pseudoceriferus* Green with special reference to the wax secretion, *Appl. Entomol. Zool.* 23:61-68.

Kuwahara, Y., 1980, Lacceryl lacerate: Major wax component produced by *Anomoneura mori* Schwartz (Homoptera: Psyllidae), *Agric. Biol. Chem.* **44**:1297-1300.

Locke, M., 1960, The cuticle and wax secretion in *Calpodes ethlius* (Lep.), *Q. J. Microsc. Sci.* **101**:333-338.

Locke, M., 1961, Pore canals and related structures in insect cuticle, *J. Biophys. Biochem. Cytol.* **10**:589-618.

Locke, M., 1965, The hormonal control of wax secretion in an insect *Calpodes ethlius* Stoll (Lepidoptera, Hesperidae), *J. Insect Physiol.* **11**:641-658.

Locke, M., 1969, The ultrastructure of the oenocytes in the molt/intermolt cycle of an insect, *Tissue Cell* **1**:103-154.

Locke, M., 1974, The structure and formation of the integument in insects. In *The Physiology of Insecta*, 2nd ed., edited by M. Rockstein, vol. 6, pp. 124-215, Academic Press, New York.

Locke, M., and Krishnan, N., 1971, The distribution of phenol-oxidases and polyphenols during cuticle formation, *Tissue Cell* **3**:103-126.

Lockey, K. H., 1960, The thickness of some insect epicuticular wax layers, *J. Exp. Biol.* **37**:316-329.

Marshall, A. T., Lewis, C. T., and Parry, G., 1974, Paraffin tubules secreted by the cuticle of an insect *Epipyrops anomala* (Epipyropidae: Lepidoptera), *J. Ultrastruct. Res.* **47**:41-60.

Miller, D. R., and Kosztarab, M., 1979, Recent advances in the study of scale insects, *Annu. Rev. Entomol.* **24**:1-27.

Noirot, C., and Quennedey, A., 1974, Fine structure of the insect epidermal glands, *Annu. Rev. Entomol.* **19**:61-80.

Pesson, P., and Foldi, I., 1978, Fine structure of the tegumentary glands secreting the protective 'shield' in a sessile insect (Homoptera, Diaspididae), *Tissue Cell* **10**:389-399.

Piek, T., 1964, Synthesis of wax in the honeybee (*Apis mellifera* L.), *J. Insect Physiol.* **10**:563-572.

Pollister, P. F., 1937, The structure and development of wax glands of *Pseudococcus maritimus* (Homoptera, Coccidae), *Q. J. Microsc. Sci.* **80**:127-152.

Pope, R. D., 1979, Wax production by coccinellid larvae (Coleoptera), *Syst. Entomol.* **4**:171-196.

Reimann, K., 1952, Neue Untersuchungen über die Wachsdrüse der Honigbiene, *Zool. Jahrb. Abt. Anat.* **72**:251-272.

Retnakaran, A., Ennis, T., Jobin, L., and Granett, J., 1979, Scanning electron microscopic study of wax distribution on the balsam wooly aphid, *Adelges piceae* (Homoptera: Adelgidae), *Can. Entomol.* **111**:67-72.

Richards, A. G., 1978, The chemistry of insect cuticle. In *Biochemistry of Insects*, edited by M. Rockstein, pp. 205-232, Academic Press, New York.

Richards, O. W., and Davies, R. G., 1977, *Imms' General Textbook of Entomology*, 10th ed., vol. 1, Chapman & Hall, London.

Rogojanu, P., 1935, Untersuchungen über die Wachsdrüsen und der Wachsabsonderung bei der Gattungen Schizoneura und Orthezia, *Z. Mikrosk. Anat. Forsch.* **37**:151-171.

Sanford, M. T., and Dietz, A., 1976, The fine structure of the wax gland of the honeybee (*Apis mellifera* L.), *Apidologie* **7**:197-207.

Tamaki, Y., 1970, Studies on waxy covering of *Ceroplastes* scale insects, *Bull. Natl. Inst. Agric. Sci. Ser. C* **24**:1-111 (Japanese with English summary).

Tamaki, Y., Yushima, T., and Kawai, S., 1969, Wax secretion in a scale insect, *Ceroplastes pseudoceriferus* Green (Homoptera: Coccidae), *Appl. Entomol. Zool.* **4**:126-134.

Waku, Y., 1978, Fine structure and metamorphosis of the wax gland cells in a psyllid insect, *Anomoneura mori* Schwartz (Homoptera), *J. Morphol.* **158**:243-273.

Waku, Y., and Manabe, Y., 1981, Fine structure of the wax gland in a scale insect, *Eriococcus lagerstraemiae* Kuwana (Homoptera: Eriococcidae), *Appl. Entomol. Zool.* **16**:94-102.

Wigglesworth, V. B., 1972, *The Principles of Insect Physiology*, 7th ed., Halsted Press, New York.

9

The Ultrastructure and Functions of the Silk Gland Cells of *Bombyx mori*

HIROMU AKAI

1. Introduction

The domesticated silkmoth, *Bombyx mori* L., belongs to the order Lepidoptera, the suborder Heteroneura, and the family Bombycidae (Latreille, 1802; Tazima et al., 1977). Its closest relatives are *Bombyx (Theophila) mandarina* which is found in China, Korea, and Japan and *Theophila religiosae* which inhabits Himalayan mountainsides (Tazima, 1964). *B. mori* was first domesticated for the purpose of silk production in China more than 4000 years ago, and it has been subjected to selection and cross-breeding for hundreds of generations. During the last 70 years, as a result of research carried out by Japanese sericulturists, cocoon shell weight, the total amount of silk spun by a larva, has increased threefold in popular commercial races. Silkworms of special genotypes treated by juvenoids under laboratory conditions can produce twice as much silk as the best commercial races (Akai, 1982).

B. mori has been domesticated to the extent that it depends entirely upon being bred and reared by humans for its survival. The domesticated silkworms are less active than wild silkworms, and the adults have lost their ability to fly. Larvae produce such an overabundance of silk that it is impossible for the adults to escape through the cocoon shell themselves. In addition to its commercial value, *Bombyx* serves as a valuable laboratory animal. They are large and therefore easier to manipulate than insects the size of *Drosophila*. About 320 mutants are available for genetic and physiological studies (Chikushi, 1972).

HIROMU AKAI • The Sericultural Experiment Station, Yatabe, Ibaraki 305, Japan.

Bombyx embryos developing from fertilized eggs normally enter diapause after about 2 days of development. These diapause "eggs" can be stored for long periods, and techniques are available for terminating the diapause. Therefore, larvae can be produced at any desired time, and they can be reared throughout the year on artificial diets (Ito and Kobayashi, 1978; Ito, 1981). Larval development and silk production in *Bombyx* can be controlled by the administration of synthetic juvenile hormones (juvenoids) (Akai *et al.*, 1973), and the secretion of silk in *Bombyx* has served as a model system for studying the regulation of protein synthesis and gene regulation (Suzuki, 1977; Garel, 1982).

During the later period of the last century, histologists started to characterize the nuclei and cytoplasmic components of the silk gland cells (Helm, 1876; Blanc, 1889; Meves, 1897). Subsequent studies were made of the silk proteins found in the lumen of various portions of the glands (Maziarski, 1911; Tanaka, 1911; Nakahara, 1917; Yamanouchi, 1922; Machida, 1926; Oba, 1950, 1957; Shibukawa, 1959). Next came electron miroscopic studies of the silk gland which characterized the intracellular organelles of the component cells (Akai, 1963b, 1964, 1965; Voigt, 1965a,b). Since the relatively recent isolation of fibroin mRNA (Suzuki and Brown, 1972), research in molecular biology has advanced rapidly. Finally, the results from endocrinological and cytological studies on *B. mori* were applied to improve techniques for silk production in sericulture.

2. The General Cytology of the Silk Gland

The silk gland of the larva of *B. mori* is a typical exocrine gland designed to secrete large amounts of silk proteins. At the completion of larval maturation during the fifth instar, the silk glands are so large that they reach a length of about 25 cm and make up about 40% of the body weight (Tazima *et al.*, 1977). The gland consists of three divisions (Figure 1). The posterior silk glands secrete fibroin, which is the main silk component, and the gelatinous silk components, consisting of three types of sericin which coat the fibroin, are secreted by different

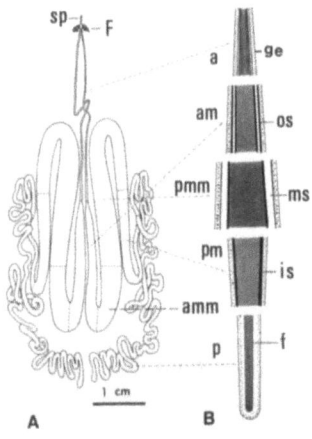

Figure 1. (A) A diagram of the silk gland of *Bombyx mori*. (B) The secretory materials found in various regions of the gland. a, anterior silk gland; am, anterior division of the middle silk gland; amm, anterior part of the middle division of the middle silk gland; F, Filippi's gland; f, fibroin; ge, gland epithelium; is, inner layer sericin; ms, middle layer sericin; os, outer layer sericin; p, posterior silk gland; pm, posterior division of the middle silk gland; pmm, posterior part of the middle division of the middle silk gland; sp, spinneret. From Akai (1976a).

regions of the middle silk gland (Machida, 1926). Anterior silk glands are merely ducts lined with a thick cuticular intima. These cells do not contribute to the secretion of silk. Tracheal branches are distributed along the posterior and middle silk glands.

The glands are derived from the ectoderm, and they begin as invaginations of the basal parts of the second maxillae (Nunome, 1937). At 6 hr after the beginning of the invagination, cell division takes place actively in both sides of the glands; the anterior region forms the anterior and middle segments of the silk glands, and the posterior region forms the posterior silk glands. The details of the embryonic development of the glands have been described by Nunome (1937). About 48 hr after the invagination, the spinneret forms, and at 60 hr, the middle silk glands become S-shaped and show distinctions between the anterior, middle, and posterior regions. At the same time, cell divisions terminate, and cross-sections at various levels of the gland show only two cells. Gland cells now start their hypertrophic growth, and liquid silk is first seen in the gland lumen at about 132 hr (Ono, 1942).

Many scientists have published estimates concerning the number of cells in various regions of the gland (Nunome, 1937; Ono, 1942, 1951; Shimizu and Horiuchi, 1952; Shigematsu and Takeshita, 1968; Suzuki, 1977; Fournier, 1979). Although the number of cells varies between larvae of different sexes and strains, there are approximately 300 cells in the anterior, 250 in the middle, and 500 cells in the posterior silk gland.

During larval development the cells in the different regions of the gland grow to different degrees. From the data given in Table 1, I have estimated that the cells of the anterior region increase in volume by a 5000-fold factor, while the cells in the middle and posterior regions increase about 86,000 and 57,000 times, respectively (Table 1).

Cross-sections of the gland show that the lumen is embraced by two huge cells. Each cell of the posterior silk gland is characterized by a thick basement membrane, a ramified nucleus containing numerous nucleoli, and a brush border facing the lumen which is packed with fibroin (Akai, 1976a). Ichimura *et al.* (1982) have developed methods for isolating intact nuclei from the posterior silk gland cells. Nuclear ramification develops gradually as the larvae grow and is most pronounced in the fifth-instar larva (Figure 2). The ramification enlarges the nuclear surface and apparently facilitates the transfer of RNAs related to silk synthesis from the nucleus to the cytoplasm.

The gland lumen contains a large column of liquid fibroin. The amino acid composition of fibroin has been studied often (Kirimura and Suzuki, 1962; Hayashiya *et al.*, 1964; Sasaki and Noda, 1973; Sprague, 1975; Shimura *et al.*, 1976; Gamo *et al.*, 1977), and it is very rich in glycine (44%), alanine (30%), serine (12%), and tyrosine (6%) (Fournier, 1979). Three types of sericin secreted from different parts of the middle silk gland (Yamanouchi, 1922; Oba, 1950, 1957; Shibukawa, 1959) are called the inner, middle, and outer layer sericins. The sericins have been less studied than fibroin (Fukuda *et al.*, 1955; Kirimura, 1962; Hayashiya *et al.*, 1964; Komatsu and Yamada, 1975; Gamo *et al.*, 1977; Komatsu, 1982). The preponderant amino acids are serine (31%), glycine (14%), and aspartic acid (14%) (Fournier, 1979).

Table 1. Dimensions (μm) of the Silk Gland Cells of Larval *Bombyx mori*[a]

Stage	Anterior silk gland			Middle silk gland			Posterior silk gland		
	Width	Length	Thickness	Width	Length	Thickness	Width	Length	Thickness
Newly hatched larva	11	36	5	3	104	10	11	61	9
6-day-old, fifth-instar larva	437	412	52	147	4800	171	1449	1578	151

[a]From Ono (1951).

The middle silk gland is divided into posterior, middle, and anterior divisions. The inner layer sericin is secreted exclusively in the posterior division, the middle layer sericin by the middle division, and the outer layer sericin both from the anterior part of the middle division and from the anterior division (see Figure 1).

The anterior portion of each silk gland functions as a duct. The paired ducts join to form a common duct that enters the spinneret (Figure 1). The duct plays no role in secretion and has no tracheae attached to it. It is lined with a thick cuticular intima (Akai, 1960, 1973). Filippi's glands (also called Lyonet's glands) connect to the region where the paired ducts fuse (Figure 1). The function of this gland will be discussed in Section 3.4.

The spinneret of *B. mori* is located in the apical part of the labium near the mouth cavity (Akai, 1976b). The morphology of the spinneret shows interesting variations among the larvae of different silk-producing lepidopterans (Nakajima, 1941). The liquid silk transferred to the spinneret passes through the thread press which consists of two concave chitinous plates. The resulting cocoon filament contains two cylinders of fibroin, each surrounded by three layers of sericins (Akai, 1976b).

Tracheae attach to the middle and posterior portions of the silk glands. The tracheal branches reach the basal surfaces of the cells, and the fine branches penetrate deeply in their cytoplasm. The tracheal end cells are observed in the basal regions of larval gland cells during the fourth molting stage and early in the fifth instar (Akai, 1976a). Nakagawa (1949, 1970) has made detailed comparative studies of the tracheal systems of lepidopteran silk glands and has shown that changes in an ancestral pattern has occurred during evolution.

3. The Ultrastructure and Functioning of the Silk Gland Cells

3.1. Posterior Silk Gland

Most ultrastructural studies of the silk glands concern the posterior portion and focus on the synthesis of fibroin and its intracellular transport and secretion (Akai, 1963b, 1964, 1965, 1970, 1971a, 1976a; Akai and Kobayashi, 1965a,b, 1966; Voigt 1965a; Matsuura *et al.*, 1968; Morimoto *et al.*, 1968; Tashiro *et al.*, 1968a,b; Iijima, 1971a,b, 1972; Matsuura and Tashiro, 1976; Sasaki and Tashiro, 1976; Sasaki, 1977; Akai and Kataoka, 1978; Sasaki and Nakagaki, 1980; Sasaki *et al.*, 1981, 1982; Tashiro *et al.*, 1982).

3.1.1. The Nucleus and Nucleic Acid Synthesis

The nuclei of the posterior silk gland cells of the mature larva are extremely convoluted (Figure 2e), and they contain numerous nucleoli. In the newly hatched larva, the nucleus is smooth-surfaced, but convolutions develop with time and become more complex (Akai, 1976a), especially during the fourth and fifth instars (Figure 2).

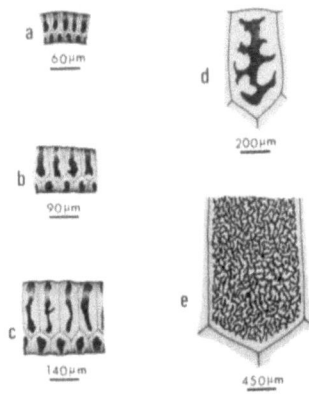

Figure 2. Developmental changes in the nuclear morphology of the cells of the posterior division of the silk glands of *Bombyx* larvae. (a) First instar; (b) second instar; (c) third instar; (d) fourth instar; (e) fifth instar. From Akai (1981).

During the fifth instar, there are changes in the sizes and shapes of the nucleoli that can be correlated with changes in RNA synthesis (Figures 3, 4, and 5) (Akai and Kobayashi, 1966; Akai, 1976a). Early in the fifth instar, the nucleolus is dense, compact, and rich in RNP granules (Akai, 1964, 1976a). Next the nucleolus expands (Figure 4b), and the concentration of RNP granules decreases gradually, and they disappear by the 9th day (Figure 4e). Studies utilizing

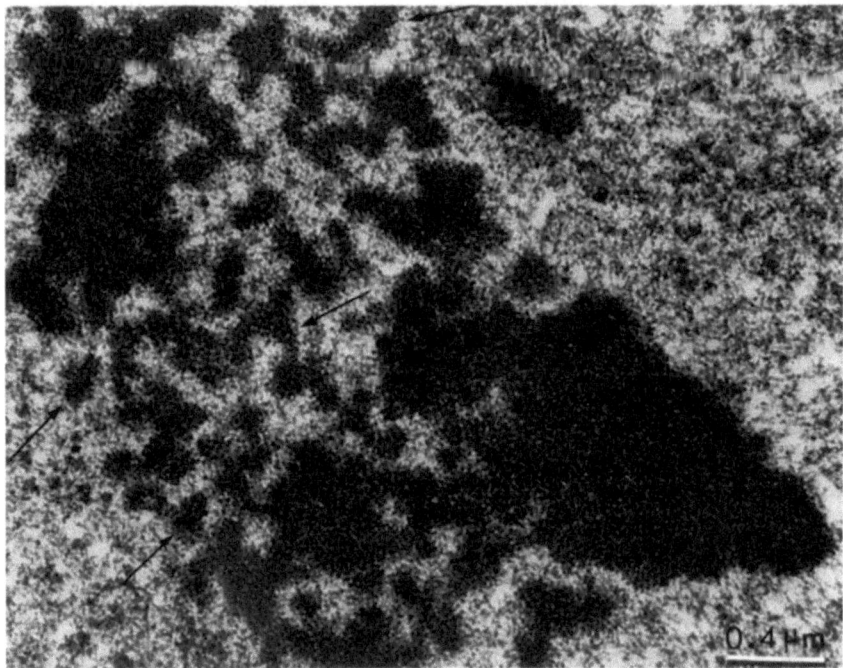

Figure 3. An electron micrograph of the nucleolus in a cell from the posterior division of the silk gland of a larva at the 4th day of the fifth instar. Numerous nucleolonemata (arrows) appear during active RNA synthesis. From Akai (1976a).

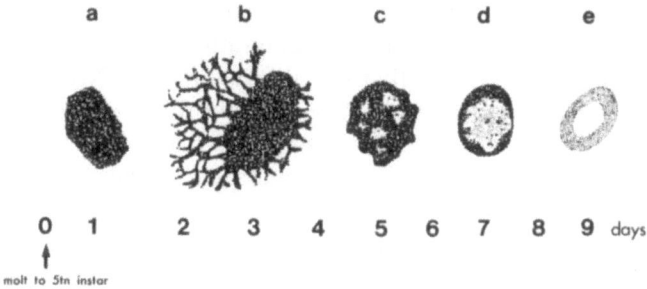

Figure 4. Changes in the nucleoli of the cells of the posterior silk gland during the fifth instar. (a) Early fifth instar (1 day after fourth molt); (b) stage of active RNA synthesis (2-3 days); (c) middle fifth instar (5 days); (d) mature to spinning stage (7 days); (e) prepupal stage (9 days). From Akai (1976a).

quantitative autoradiography have shown that the most active period of RNA synthesis occurs during days 3 and 4 of the fifth instar. The rate of synthesis decreases from the 5th to 7th day, and ceases by the 8th day (Figure 5). Iijima (1971b) reported that there were no RNP granules in the nucleoli of posterior silk gland cells of *Bombyx* larvae which were poisoned with actinomycin D.

When its maximum size is reached near the 7th day of the fifth instar, the silk

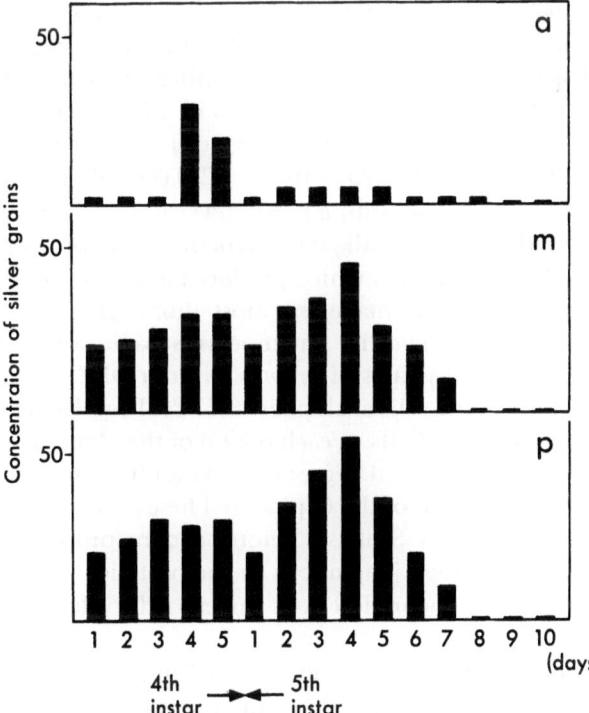

Figure 5. Incorporation of [³H]uridine into the silk gland cells of *Bombyx* larvae during the fourth and fifth instars. a, anterior silk gland cells; m, middle silk gland cells; p, posterior silk gland cells. From Akai (1976a).

gland contains large quantities of messenger and transfer RNAs used in protein synthesis (Shigematsu et al., 1978) and a huge number of ribosomes (Prudhomme and Couble, 1979). During the fifth instar, the quantity of RNA increases 15 fold between days 1 and 7 and then decreases during the final period of cocoon spinning (Fournier, 1979). After spinning, the decrease in RNA is even more rapid, since tissue histolysis related to metamorphosis takes place (Matsuura et al., 1968; Okabe et al., 1975).

Tritiated thymidine is incorporated into the diffuse chromatin and chromatin bodies before each endomitosis (Akai, 1965; Akai and Kobayashi, 1965a). During the fourth instar, the incorporation continues at a moderately high level. In both the fourth molting stage and the early fifth instar, extremely high activities are seen, and then the activity decreases gradually and finally disappears at day 6 of the fifth instar. By observing electron microscopic autoradiographs, silver grains can be located above dispersed chromatin, but not above the condensed chromatin body (Akai, 1965).

Mitosis of the silk gland cells terminates during embryonic development, and subsequently the cells grow by endomitosis (Akai, 1965; Perdrix-Gillot, 1979). Gillot and Daillie (1968) determined that there was 0.4 μg of DNA per single silk gland at the end of the second instar and 150 μg at the end of the fifth instar. The DNA content of the posterior silk gland is about 1.5 μg early in the fourth instar and 100 μg in the middle of the fifth instar (Morimoto et al., 1968; Tashiro et al., 1968b). Rasch (1974) determined by microspectrophotometric analyses of sperm that the haploid amount of DNA for B. mori is 0.52×10^{-12} g, and Gage (1974) demonstrated that the endoreduplication of the DNA of the cells of the posterior silk gland involves replication of all the genome. During the larval period, the DNA content of the average cell in the posterior silk gland increases from 1.04×10^{-12} g to 0.25×10^{-6} g, a 250,000-fold increase (Fournier, 1979; Perdrix-Gillot, 1979). According to Kurata et al. (1974), in silkworm races which produce small amounts of silk, the cells of the posterior silk gland contain half as much DNA as in the races which produce large amounts of silk.

Perdrix-Gillot (1979) has made the most thorough study of the endopolyploidization of silk gland cells. The process is cyclic, alternating between phases of synthesis (S) and phases of growth without DNA synthesis (G). She showed that in all silk gland cells at all phases of larval development, the length of the S phase was constant. Cells in each region of the gland undergo three or four S cycles per instar. The final degree of polyploidization in a given region is determined by the duration of the G phases. These regions with the shortest G phases have time for more S phases before cocoon spinning and therefore have the highest DNA values. The nuclei in the posterior silk gland undergo 18–19 endomitotic cycles, the middle cells undergo 20 cycles, and the anterior cells only 13.

The sex chromatin body (SB) is maintained in a polytene configuration during the G phase, while the rest of the genome is dispersed. The SB enters and completes the S phase later than the autosomes. Since the SB is relatively condensed during the G phase, Perdrix-Gillot suggests that it may be transcriptionally inactive. The SB can be observed in acetoorcein-stained squash preparations of silk gland nuclei. In diploids, female nuclei contain one, male nuclei none; in tetraploids, female nulei contain two, male nuclei none (Gillot, 1968; Ito, 1977). Obviously, the SB represents the W chromosome, which causes the heterogametic sex to develop into females.

Tritiated uridine is incorporated into the nucleus 5-10 min after ingestion (Akai, 1965). The grain concentration in the nucleus reaches a peak 30-60 min after the injection, and the nucleolus shows the highest grain concentration (Akai, 1976a). Silver grains appear in the cytoplasmic regions packed with endoplasmic reticulum 24 hr after ingestion.

The relative proportions of the different species of stable RNA do not change significantly during the fifth instar. Fournier (1979) found that 88% of the RNA was ribosomal, 11% was tRNA, and 1% was mRNA.

Pioneering studies of the fibrion mRNA transcribed by the cells of the posterior silk gland of *Bombyx* larvae were carried out by Suzuki and Brown (1972). Suzuki and Suzuki (1974), and Suzuki (1975). Fibroin mRNA has also been isolated from the posterior silk glands of the wild silkworms, *Antheraea yamamai* and *A. pernyi* (Tamura et al., 1982). In *Bombyx*, the fibroin mRNA accumulates in the posterior silk gland at an almost constant rate (7-10 molecules gene per min) throughout the fifth instar. During this period, rRNA synthesis decreased from 7 to 0.2 molecule/gene per minute, and synthesis of heterogeneous RNA also decreased. Thus, during the 2nd, 4th, and 5th days of the fifth instar, fibroin mRNA comprises about 1, 3, and 8% of the RNA synthesized in a 30-min period, respectively (Suzuki and Giza, 1976).

A large quantity of fibroin mRNA accumulates during the fifth instar for the massive synthesis of fibroin. However, the efficiency of fibroin gene transcription and of the accumulation of mature mRNA is about the same in the third, fourth, and fifth instars (Shimura, 1978). The size of nuclear fibroin mRNA is essentially the same as that of mature cytoplasmic mRNA (Lizardi, 1976). There is a genetic polymorphism in the size of the fibroin mRNAs isolated from a number of *B. mori* strains (Lizardi, 1979). The molecular weights range from 5.5 to 6.0×10^6. Ohmachi et al. (1982) isolated a new poly(A)-containing RNA of 5.0×10^5 molecular weight from the posterior silk gland of *Bombyx*. The size of the new RNA (corresponding to about 1500 nucleotides) is sufficiently large to code for the small (300 amino acid) subunit of fibroin. The subunit structure of fibroin is discussed further in Section 3.1.2.

Recent studies have elucidated our understanding of the structure of the fibroin gene (Suzuki *et al.*, 1972; Lizardi, 1975; Gage and Manning, 1976; Ohshima and Suzuki, 1977; Suzuki and Ohshima, 1978; Tsujimoto and Suzuki, 1979a,b; Beyer *et al.*, 1979; Gage and Manning, 1980; Suzuki, 1982). In the cells of the posterior silk gland, the fibroin gene is present in one copy per haploid genome, as well as in other tissues (Suzuki *et al.*, 1972; Gage and Manning, 1976). Techniques of recombinant DNA research have generated several clones containing all or part of the fibroin gene and its flanking sequences (Ohshima and Suzuki, 1977; Suzuki and Ohshima, 1978). The silk fibroin transcription unit is only slightly larger than necessary to encode the 16-kb mRNA molecule. Maekawa and Suzuki (1980) have demonstrated that the transcription of the fibroin gene is repeatedly turned on and off during the development of the silk gland. During periods of maximum transcription, 13 molecules of mRNA are formed per fibroin gene per minute.

Sprague et al. (1979) have shown that silkworms producing fibroin proteins of different lengths possess variant alleles of the fibroin gene, and three fibroin phenotypes in *Bombyx* have been demonstrated by comparing the electrophoretic mobilities of the large fibroin subunit from different strains (Hyodo and Shimura, 1980). A linkage analysis was performed utilizing the electrophoretic fibroin phenotypes, and the fibroin gene was shown to be linked to the *Nd* gene which is located on the 23rd chromosome (Hyodo et al., 1980).

McKnight et al. (1976) used chromatin spreading techniques for electron microscopic visualization of the nuclear contents of posterior silk gland cells of fifth-instar larvae. These electron micrographs contained structures thought to be the fibroin transcription units. These were not tandemly repeated, but rather occurred singly, which is consistent with data showing that the fibroin gene exists only once per haploid genome equivalent. These transcription units (Figure 6) are characterized by their length and the tight packing of nascent RNP fibers (Beyer et al., 1979). The length of the transcription unit corresponds to about 18 kb.

Tsuda and Suzuki (1981) reported an *in vitro* system utilizing a HeLa cell extract in which exogenously added fibroin gene was faithfully transcribed. Using this system and a series of deletion mutants of the fibroin gene, Tsujimoto et al. (1981) were able to assign a promoter sequence to the fibroin gene.

Matsuzaki (1963, 1970) was the first to investigate the silk gland tRNAs. She studied the specificity of each tRNA and showed that a correlation exists between the amino acid composition of the fibroin and the extent of acylation of each tRNA. Kawakami and Shimura (1973) were the first to isolate a pure tRNA from the posterior silk gland of fifth-instar larvae. The population of RNAs in the silk gland cells is composed of at least 53 molecular species (Garel et al., 1977; Chevallier and Garel, 1979). The nucleotide sequences for various pure tRNAs have been determined (Sprague et al., 1977; Garel and Keith, 1977; Zuniga and Steitz, 1977; Kawakami et al., 1978). The anticodon loop sequences have been determined for tRNA and tRNA (Hentzen et al., 1976).

The nucleus of the silk gland cell is covered by a typical annulate double membrane. Early cytologists believed that basophilic material was transported to the cytoplasm through the nuclear membrane (Helm, 1876; Maziarski, 1911; Nakahara, 1917). Electron microscopic studies support this concept (Akai, 1964; Tashiro et al., 1968a).

3.1.2. The Cytoplasm and Fibroin Synthesis

In early electron microscopic studies of the posterior silk gland cells, RER, free ribosomes, Golgi complexes, fibroin globules, mitochondria, microvilli, bundles of microfilaments, and microtubules were observed in the cytoplasm (Akai, 1963b, 1964, 1965). The RER is the site of fibroin synthesis. Early in the fifth instar, only a few tubules of RER are scattered in the cytoplasm, and large numbers of free ribosomes are suspended in the cytoplasm. However, free

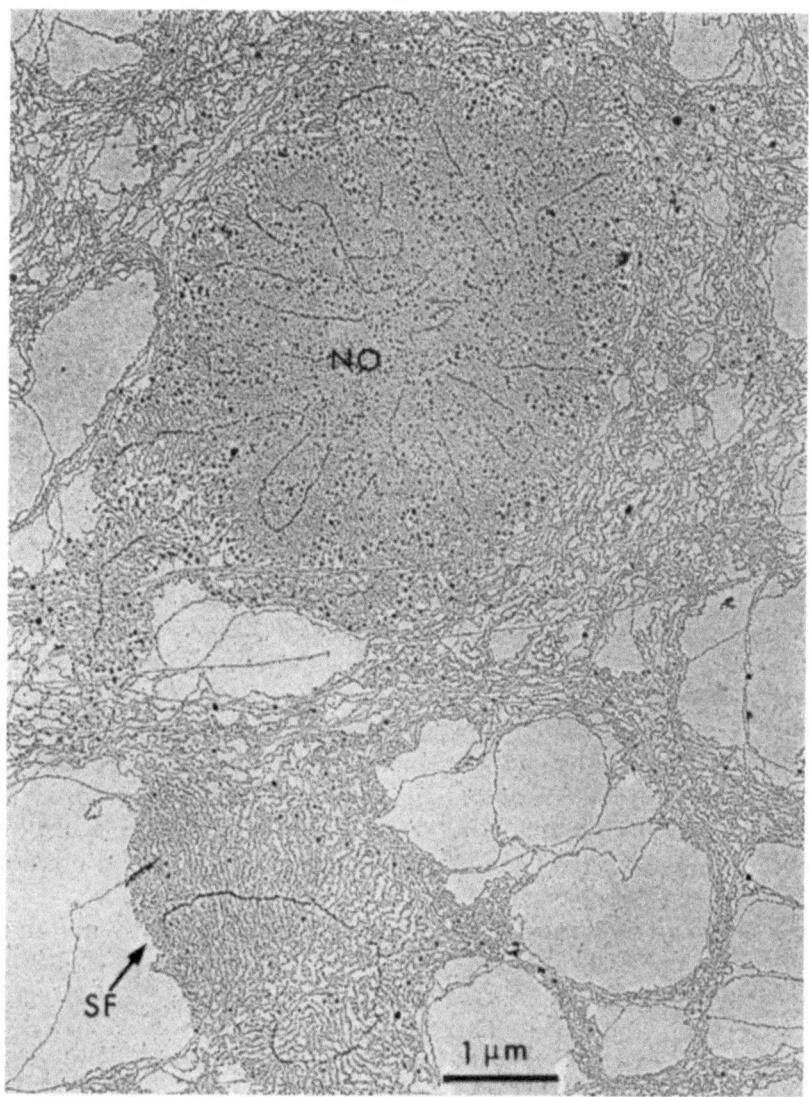

Figure 6. Electron micrograph of transcriptionally active rDNA and putative silk fibroin DNA from the silkworm *Bombyx mori*. The chromatin was prepared from posterior silk gland tissue of a fifth-instar larva. The nucleolus organizer region (NO) containing repeated ribosomal precursor RNA genes is in an intermediate stage of dispersion. The silk fibroin transcription unit (SF) measures approximately twice the length of ribosomal precursor RNA gene. Note the high density of nascent transcripts associated with the fibroin transcription unit. From Beyer *et al.* (1979).

ribosomes disappear and RER expands and occupies the whole cytoplasm from the middle of the instar to the spinning stage (Akai, 1964, 1965). During the larval molts and metamorphosis, the lamellated RER regresses (Tashiro *et al.*, 1968b; Matsuura *et al.*, 1968; Akai, 1973, 1976a). Similar morphological changes are also detected in glands cultured *in vitro* (Iijima, 1971a).

There are about 1 trillion ribosomes per cell when the silk synthesis is maximal. In such cells, most of the ribosomes are linked together by the mRNAs to form the polysomes. There are two kinds of polysomes: those that are suspended in the cytoplasm and those that are bound to the membranes of the ER. Membrane-bounded polysomes synthesize secretory proteins, whereas free polysomes synthesize proteins utilized by the cell itself. When the rate of fibroin production is at a maximum, secretion is close to 2 mg/hr per pair of glands or four amino acids incorporated per ribosome per second (Prudhomme and Couble, 1979). Fibroin is synthesized 60 times faster than serum albumin is synthesized by liver cells, for an equal weight of tissue (Suzuki, 1977).

Lizardi (1980) isolated the giant silk fibroin polyribosomes from the posterior silk gland of *B. mori*. The polyribosomes contained between 45 and 112 ribosomal particles. Treatment of giant fibroin polyribosomes with EDTA released a particle that sedimented at 125 S. This mRNP particle contained biologically active silk fibroin mRNA, as judged by cell-free translation in an mRNA-dependent, reticulocyte system.

Golgi complexes (Figure 7) consisting of several Golgi vacuoles and vesicles lacking typical Golgi membranes are scattered in an area containing RER (Akai, 1976a). The vesicles are thought to arise by budding from the RER and to be transferred to the Golgi complex. In early stages of each instar, fibers cannot be seen in the Golgi vacuoles, but they appear later and increase gradually as the larva grows (Akai, 1971a). The elementary fibroin fibers which float in Golgi vacuoles (Figure 8) are about 13 nm thick and appear to contain a helical bundle of filaments (Akai, 1971a). The elementary fibroin fibers were also detected in fibroin globules, in silk layer, and in the columnar fibroin in the lumen (Akai and Kataoka, 1978). McKnight *et al.* (1976) observed polysomes isolated from the posterior silk gland cells (Figure 9). The attached fibrils were assumed to be nascent silk fibroin polypeptides.

Green *et al.* (1975) isolated and purified the mRNA for silk fibroin from mature posterior silk glands of *Bombyx* larvae. This mRNA was used to direct polypeptide synthesis in an Ehrlich ascites cell-free extract. The mRNA increased [^{3}H]alanine incorporation three- to fourfold, and the polypeptide products were heterogeneous in size, including some as large as 100,000 daltons. The amino acid sequences of certain segments of the polypeptides isolated from the cell-free systems were similar to those from native fibroin.

Silk fibroin consists of more than one polypeptide. The molecular weight generally given is about 3.6×10^5 (Tashiro and Otsuki, 1970a,b; Sasaki and Noda, 1973). Sprague (1975) reported that fibroin consists of approximately equimolar amounts of two large (350,000 dalton) polypeptide chains. These chains are connected by disulfide bonds to smaller proteins (Tashiro and Otsuki, 1970b; Tashiro *et al.*, 1972). The molecular weights of the small components were estimated to be 2.6×10^4 (Sasaki and Noda, 1973; Gamo *et al.*, 1977). Shimura *et al.* (1976, 1982) and Ohmachi and Shimura (1981) isolated the small component proteins from cocoons of *B. mori* and assayed for amino acid composition and molecular size. Most of the small proteins possessed the same characteristics as those isolated from posterior silk gland fibroin.

Chymotrypsin digestion of fibroin produces an insoluble or crystalline

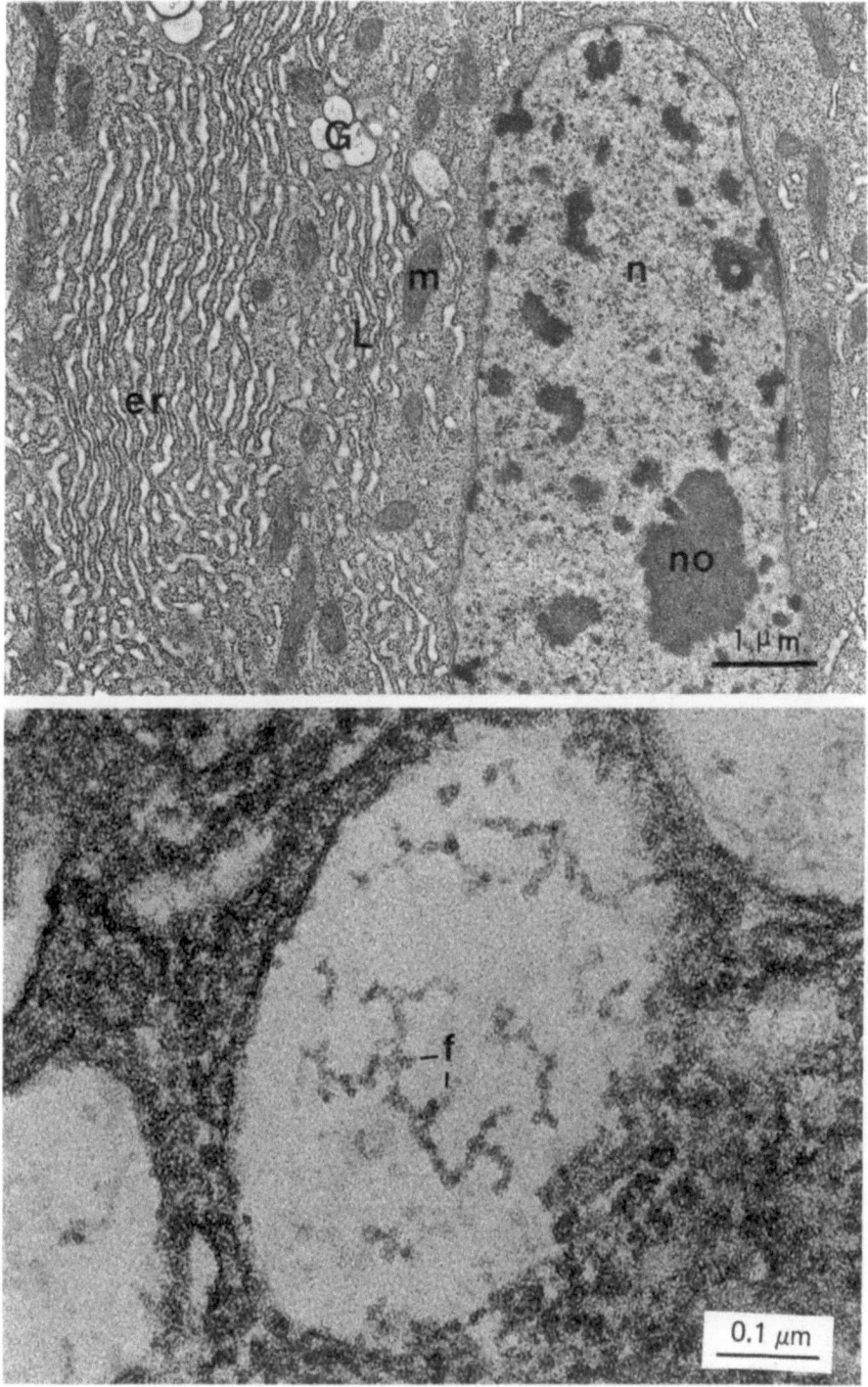

Figure 7. An electron micrograph of a typical cell from the posterior silk gland of *Bombyx mori* showing the distribution of organelles. er, rough-surfaced endoplasmic reticulum; G, Golgi complex; m, mitochondrion; n, a projection from the ramified nucleus; no, nucleolus.
Figure 8. A higher-magnification electron micrograph showing the elementary fibroin fibers (f) in a Golgi vacuole. Each fibroin fiber appears to be composed of several subunits. From Akai (1971a).

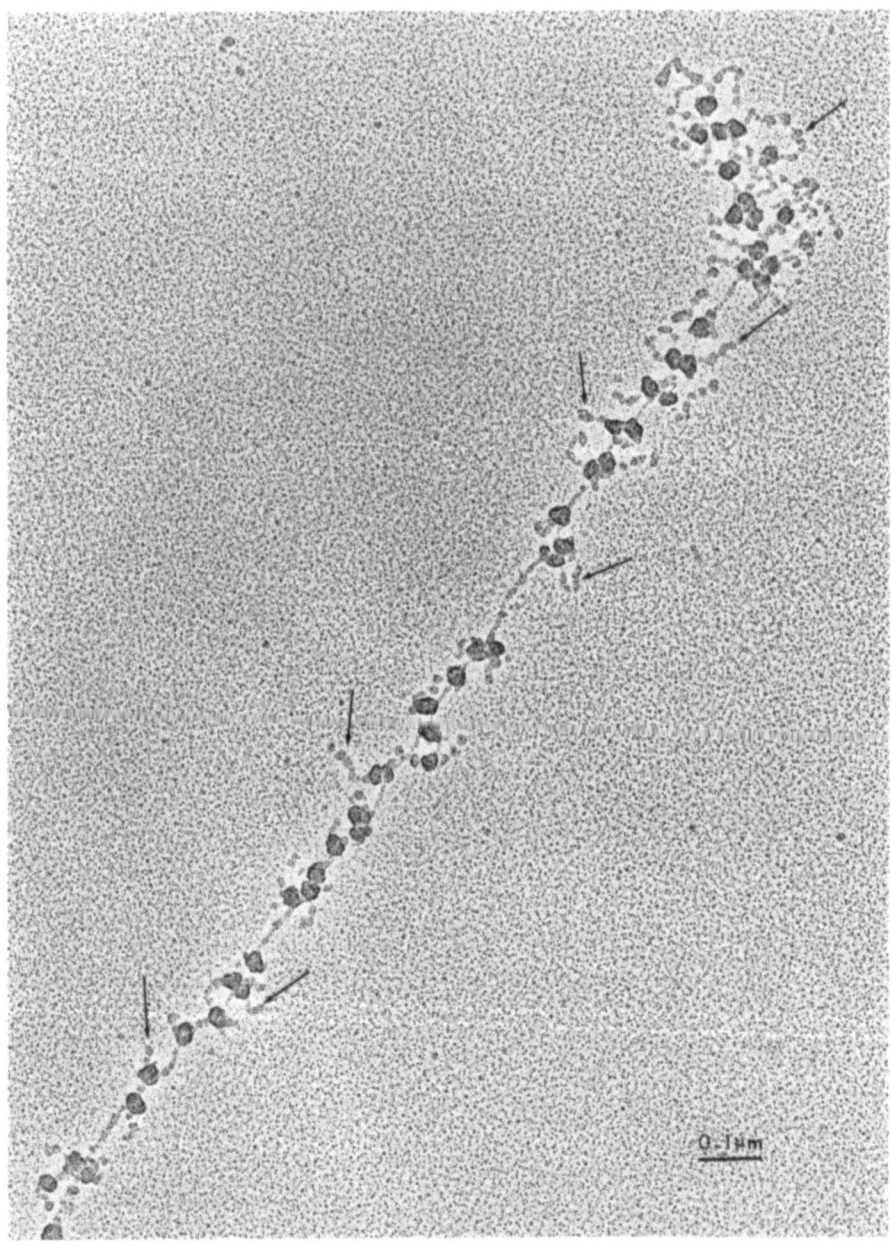

Figure 9. An electron micrograph of a polysome from posterior silk gland cells of a late fifth-instar *Bombyx* larva. The fibroin mRNA is being translated simultaneously by many ribosomes. The polypeptide chains (arrows) become longer as the ribosomes move toward the 5′ end of the RNA molecule. (Courtesy of S. L. McKnight and O. L. Miller Jr., University of Virginia.)

fraction which comprises about 60% of the protein (Lucas et al., 1957). This polypeptide fraction is composed of only four amino acids arranged in a very simple repeating sequence. The main feature of the polypeptide is that glycine residues alternate with three other amino acids almost throughout its entire sequence. The remaining 40% of fibroin, the soluble or amorphous region of the protein, contains the minor amino acids. However, even in these polypeptides, glycine residues mostly alternate with other amino acids and no adjacent glycine residues were found (Lucas et al., 1962).

Fibroin molecules must satisfy at least two criteria in order to function properly. They must (1) remain soluble in the silk gland lumen prior to spinning, and (2) be able to crystallize promptly upon being spun. It is the crystalline domain of fibroin which forms the β-pleated sheet, a crystalline structure that is characteristic of silk, while the amorphous domains, with their bulkier amino acid residues, do not participate. The solubility of fibroin may be determined by the arrangement of crystalline and amorphous sequences.

The small apparent complexity of the fibroin molecule suggests that the fibroin gene resembles a satellite DNA sequence that is expressed. The gene is fundamentally an extensive array of 18-bp crystalline repeats coding for Gly-Ala-Gly-Ala-Gly-Ser which in turn comprises longer repeats coding for the major crystalline region, chymotrypsin peptide which is about 180 to 240 bp in length. The similarity of the fibroin gene core to that of a satellite DNA suggests that the gene grew to its current size and continues to evolve via a process of unequal crossing over between similar homologs (Gage and Manning, 1980; Manning and Gage, 1980).

Quantitative analyses of the mulberry leaves silkworms ingest have shown that the leaves contain insufficient quantities of the four principal amino acids (glycine, alanine, serine, and tyrosine) necessary for fibroin synthesis (Prudhomme and Couble, 1979). However, the silkworm has evolved the necessary systems of enzymes to provide it with the amino acids it needs during silk production (Prudhomme and Couble, 1979). The silkworm synthesizes alanine, glycine, and serine in the silk gland itself, using aspartic acid, glutamic acid, asparagine, and glutamine as sources of carbon and nitrogen. The necessary tyrosine is given off as a product of phenylalanine metabolism in other tissues, and the hemolymph carries it to the silk gland.

Mature Golgi vacuoles containing high concentrations of elementary fibroin fibers leave the Golgi complex as a fibroin globule, and these accumulate in the apical cytoplasm (Figure 10) (Akai, 1976a; Akai and Kataoka, 1978). The fibers in the fibroin globules are released into the silk layer by reverse pinocytosis and are stored as massive fibroin fibers (Figures 10-14). The silk layer is separated from the lumen by perforated membranes. The fibroin masses pass through the spaces in these membranes and accumulate in the lumen as a column of liquid fibroin (Figures 11, 14) (Akai and Kataoka, 1978). These membranes are composed of fine fibrous materials and appear to be the degenerating cuticular intima (Figures 11, 14) (Akai, 1983).

During the fourth instar and at the very beginning of the fifth instar, round or oval mitochondria predominate, and they are distributed randomly in the

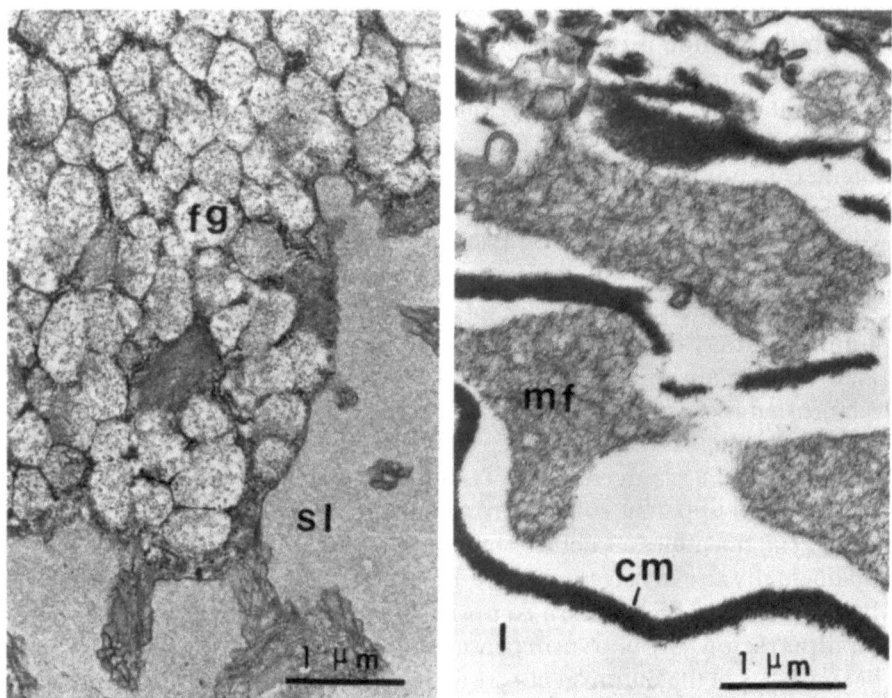

Figure 10. An apical region from a cell of the posterior silk gland showing numerous fibroin globules (fg). These contain elementary fibroin fibers. sl, silk layer.

Figure 11. Part of the silk layer containing masses of fibroin fibers (mf). cm, cuticular membrane; l, lumen.

cytoplasm. Later in the fifth instar, elongated mitochondria predominate, and they are arranged radially (Matsuura and Tashiro, 1976). During the prepupal stage, a number of cup-shaped mitochondria appear.

ATPase is concentrated in the microvilli which form the apical surface of the gland cells (Akai, 1970). Bundles of microfilaments are arranged periodically along the inner, apical surface of the cells (Figure 15) (Akai, 1976a). Large numbers of microtubules are also distributed in the cytoplasm (Sasaki and Tashiro, 1976). Sasaki *et al.* (1981) have proposed that the radial system of microtubules is responsible for the intracellular transport of fibroin globules from the Golgi complexes to the apical cytoplasm and that a circular system of microtubules and microfilaments is responsible for the exocytosis of fibroin at the luminal surface. Sasaki *et al.* (1982) suggested that release of Ca into the cytoplasm triggers the circular microtubule-microfilament system to discharge the secretory fibroin globules from the apical portion of the cell.

The times spent during the synthesis and the intracellular transport of fibroin have been studied by both light and electron microscopic autoradiography (Rabinovitch and Vugman, 1959; Akai, 1963a, 1964; Akai and Kobayashi, 1965b, 1966; Akai, 1973). Tritiated glycine is incorporated into the RER of the

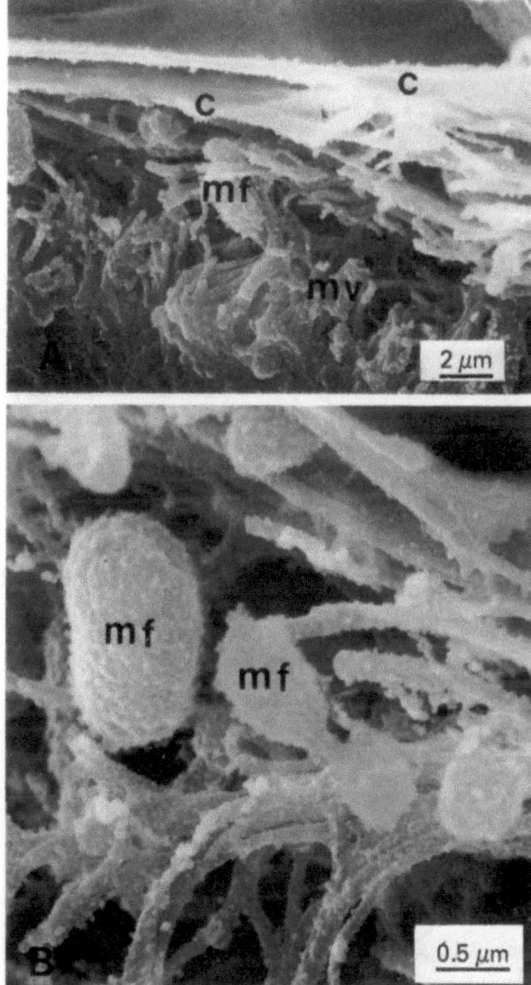

Figure 12. Scanning electron micrographs of the silk layer in the posterior silk gland. (A) A region which includes the cuticular membranes (c); masses of fibroin fibers (mf), and microvilli (mv). (B) Masses of fibroin fibers shown at higher magnification.

posterior silk gland cells 10 min after ingestion (Akai, 1965; Akai and Kobayashi, 1965b). Silver grains appear above Golgi complexes by 15 min, and labeled fibroin globules accumulate in the apical cytoplasm by 45 min. Next the fibroin masses stored in the silk layer and finally the columnar fibroin in the lumen become labeled (see Figure 14). Sasaki *et al.* (1981) have studied the intracellular transport of fibroin *in vitro*. Silver grains appeared above the RER after 10 min, above the Golgi complexes after 25 min, and above fibroin globules in apical cytoplasm after 40 min.

The incorporation of labeled amino acids into subcellular fractions has been studied (Shigematsu *et al.*, 1966a,b). Fibroin was present in large and small microsomes (Shigematsu and Takeshita, 1966). Actinomycin blocks fibroin synthesis, but mitomycin does not (Shigematsu *et al.*, 1965).

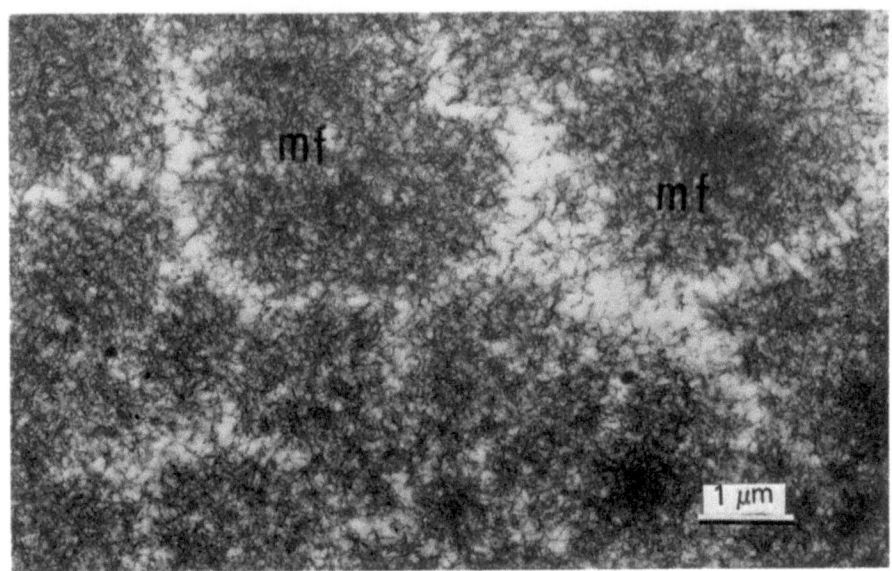

Figure 13. The masses of fibroin fibers (mf) in the column of fibroin accumulating in the lumen of the posterior silk gland. The masses range in diameter from 1 to 3 μm. From Akai and Kataoka (1978).

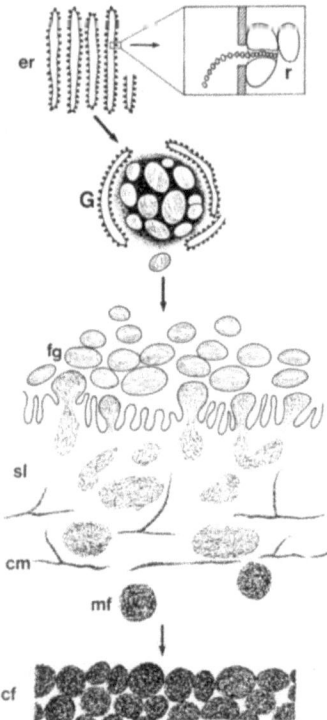

Figure 14. A schematic drawing illustrating fibroin synthesis, transport, and secretion in the posterior silk gland. cf, columnar fibroin composed of masses of fibroin fibers (mf) in the gland lumen; cm, cuticular membranes, er, rough-surfaced endoplasmic reticulum; fg, fibroin globules; G. Golgi complex; r, ribosome, sl, silk layer. From Akai (1981).

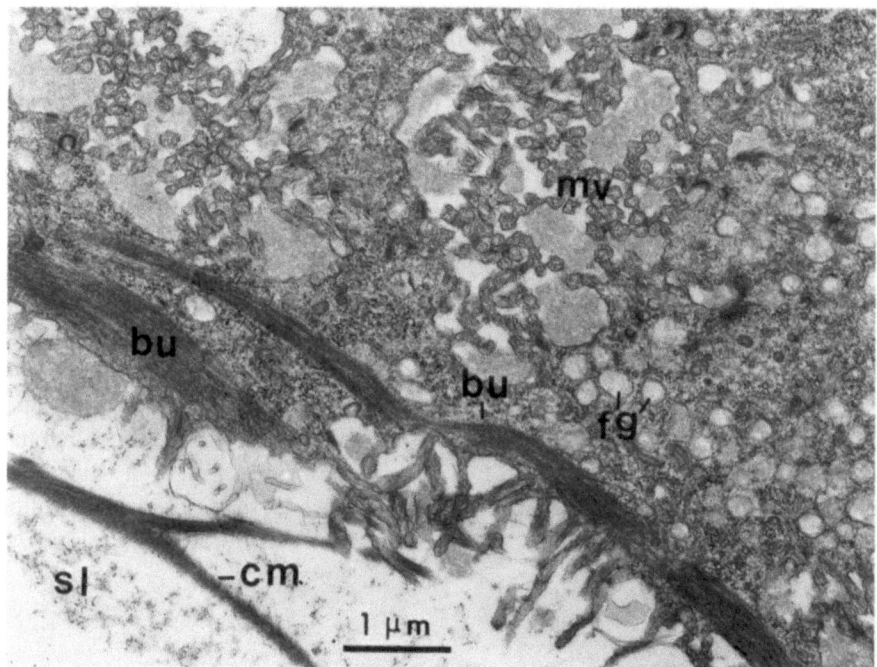

Figure 15. A longitudinal section of the bundles of microfilaments (bu) located in the apical cytoplasm of a posterior silk gland cell. cm, cuticular membrane; fg, fibroin globule; mv, microvilli; sl, silk layer.

3.1.3. The Postembryonic Development of the Gland

The gland cells from newly hatched larvae have been studied by Akai (1976a). The spherical nucleus contains several nucleoli and chromatin bodies. The cytoplasm contains moderate amounts of RER, Golgi complexes, and mitochondria. On the basal surface a basement membrane, about 0.25 μm thick, has already formed. Small quantities of the fibroin fibers are stored in the lumen.

Drastic morphological changes are first seen in the silk gland cells during the fourth molt (Akai, 1965). The nucleus moves toward the basal side of the cell, and the nucleolus and chromatin bodies aggregate at the basal side in the nucleus. Golgi complexes and RER decrease in concentration, and the fibroin globules disappear. Large vacuoles and lysosomes appear in the cytoplasm.

In the middle period of the molt, the large vacuoles gradually disappear. The nucleus returns to the central part of the cell and the nucleoli and chromatin bodies relocate. Very little RER and Golgi complexes are seen, and the cytoplasm is filled with free ribosomes. Part of the stored silk is spun out just before the molting stage and the remainder is digested. Late in the molt the cell recovers its synthetic functions and begins to secrete a small amount of silk in the lumen.

One of the interesting facts concerning the molting stage is that the stored silk in the lumen is digested, just like the endocuticle of the integument (Akai,

1965). Silk seems to be broken down to its component amino acids which are transported to the hemolymph. Similar ultrastructural changes occur in Eri-silkworm during larval molting (Akai, 1971b).

The cytolytic changes that occur in the posterior silk gland cells during metamorphosis have been studied at the light and electron microscopic levels (Ito, 1915; Akai, 1965, 1976a; Matsuura et al., 1968, 1976; Tashiro et al., 1976). During metamorphosis, the wet weight of the gland and the amounts of RNA and protein it contains decrease rapidly and markedly, while the amount of DNA decreases only slightly (Matsuura et al., 1968). At the beginning of the prepupal stage, autophagosomes containing RER appear and are gradually transformed to autolysosomes. In the middle of the prepupal stage, a number of smooth, membrane-bounded vacuoles appear in the cytoplasm. Toward the end of the prepupal stage, a partitioning of cytoplasmic areas was observed, and large autophagosomes containing cytoplasmic organelles are formed. The nucleus is also partitioned into smaller units by smooth membranes, and then autophagosomes containing condensed chromatin blocks are formed. These autophagosomes are continuously released into the hemolymph until the gland has disintegrated completely.

3.2. Middle Silk Gland

3.2.1. The Sericins

These proteins receive their name because of the abundance of serine, which make up over 30% of the total amino acids. Three types of sericin which differ in their staining properties are secreted from different parts of the middle silk gland of B. mori (Yamanouchi, 1922; Oba, 1950; Shibukawa, 1959). They are called the inner, the middle, and the outer sericins (see Figure 1B). Shibukawa (1959) reported that the inner layer of sericin gave positive reactions for lipids, polysaccharides, and arginine-rich proteins, while the middle layer gave strong reactions to tests for tyrosine-rich proteins. However, the outer layer sericin showed strong reactions for lipids and polysaccharides, but weak ones for proteins. Three types of sericin have also been detected by X-ray analysis (Shimizu, 1941).

When sericin is obtained from cocoon filaments, it can be fractionated into four classes which differ in their dissolution velocities (Komatsu, 1975). The amino acid composition is different between sericins I, II, III, and IV. The proportion of the amino acids having polar side chains in the sericin classes becomes progressively greater as the dissolution velocities increase. Komatsu also reported that the posterior division of the middle silk gland secreted an insoluble material that contained about 30% of the cocoon wax. Komatsu and Yamada (1975) compared the cocoon sericins of wild and domesticated silkmoths and found that the wild strains contained more acidic and basic amino acids and less hydroxylated amino acids than the domesticated strain.

According to Sprague (1975), sericin is composed of at least the three large polypeptides ranging in molecular weight from 1.3 to 2.2×10^5. On addition of SDS, the sericins are broken down to subunits of between 3.0 and 4.0×10^4

(Komatsu, 1982). Gamo *et al.* (1977) studied the sericins extracted by disulfide cleavage from the lumen of silk glands of mature *Bombyx* larvae at acid pH containing 4 M urea. The sericin could be separated by gel electrophoresis into five polypeptides of molecular weights from 8×10^4 to 3×10^5. The posterior division of the middle silk gland secreted one polypeptide (s-4), the middle division two (s-1 and s-3), and the anterior division two (s-2 and s-5). All five sericin polypeptides secreted are rich in serine, glycine, and aspartic acid, and two polypeptides (s-1 and s-2) are glycoproteins. Recently, Gamo (1982) reported that the s-2 and s-3 polypeptides are similar in molecular weight (226,500 and 218,800, respectively), and that a variant s-2 polypeptide occurs in the *Nd* mutant which has a molecular weight of 164,000. The *Src-2* gene which codes for the sericin-2 polypeptide resides on chromosome 11.

Okamoto *et al.* (1982) isolated two sericin mRNAs (11.0 and 9.6 kb long) from the middle silk gland, and determined the site of initiation of sericin gene transcription at the nucleotide level.

3.2.2. The Posterior Division

All cells of this division produce only the inner layer sericin. In mutants that spin pink cocoons, a red carotene pigment diffuses into the silk formed in the posterior division of the middle silk gland (Harizuka, 1953). This division can be clearly distinguished from the middle division because of the red pigment in the cells. Kawai (1976, 1978) showed that the molting and juvenile hormones secreted at each instar control the uptake of the carotenoid.

The ultrastructural characteristics of these silk gland cells have been described (Akai, 1973, 1976a, 1980). During the fifth instar, the nucleus is very convoluted and contains numerous nucleoli. The cytoplasm contains highly concentrated RER, Golgi complexes, and sericin globules. The RER is present as long tubules during the early to middle stages, and late in the instar they change to short tubules. During spinning, sericin globules form in the Golgi complexes, and the globules increase in size by accumulation of dense material derived from the Golgi membranes (Figure 16). The sericin globules are 200-300 nm in diameter and are packed with the fibers, 11-13 nm in diameter. The globules move to the apical surface and release these materials into the silk layer by exocytosis.

The mitochondria are larger than those found in other regions of the silk gland (Figure 16). Bundles of microfilaments are positioned regularly along the apical cell surface, and these presumably function similarly to the microfilaments of the cells of the posterior division. Perforated cuticular membranes separate the silk layer from the lumen. Stored sericin in the silk layer passes through the spaces in the cuticular membranes and accumulates on the columnar fibroin mass being transferred from the posterior silk gland.

3.2.3. The Posterior Part of the Middle Division

Middle layer sericin is secreted in this part of the silk gland. The general ultrastructure resembles that of the posterior division, except for the secretory

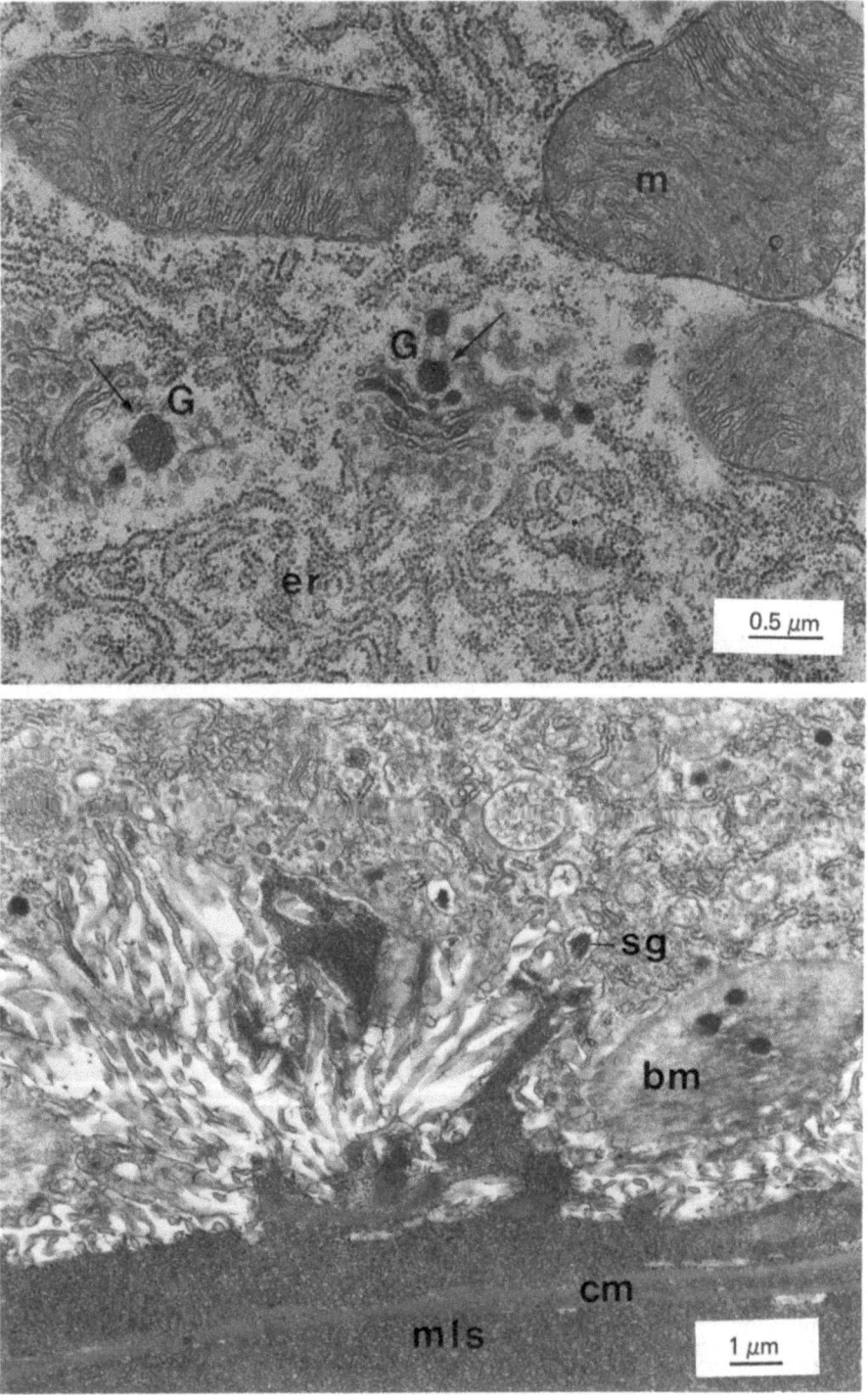

Figure 16. Part of the cytoplasm of a cell from the posterior division in the middle silk gland showing Golgi complexes (G) and large mitochondria (m). Sericin globules are formed in the Golgi complexes (arrows). er, rough ER.

Figure 17. An apical cytoplasmic region from a silk gland cell secreting middle layer sericin (mls). bm, bundle of microfilaments; cm, cuticular membrane; sg, sericin globules. From Akai (1976a).

materials (Figure 17) (Akai, 1973, 1976a). The sericin globules are granular and less dense than those of the posterior division, and they are formed by condensation of the materials within the Golgi membranes. Fibrous molecules floating in the sericin globules are released and stored temporarily in the silk layer (Figure 17). The lumen contains the inner layer sericin from the posterior division of the middle silk gland and fibroin from the posterior silk gland. The middle layer sericin passes through the cuticular membranes and accumulates on the surface of the inner layer sericin. Both sericins are clearly distinguishable, since the middle layer sericin is less dense and more granular than the inner layer sericin.

3.2.4. The Anterior Part of the Middle Division

Shibukawa (1959) concluded that this region secretes both middle and outer layer sericin. However, it is not clear whether these two types of sericin are secreted from the same cell simultaneously. Akai (1973, 1976a) studied these cells at the spinning stage. At the ultrastructural level, they resemble the cells from the other parts of the middle division. However, the Golgi vacuoles contain numerous fine fibers (15-20 nm in diameter) and these are also seen in the sericin globules that accumulate in apical cytoplasm (Figure 18).

3.2.5. The Anterior Division

Cells from this division have also been observed at the spinning stage (Akai, 1973, 1976a). Outer layer sericin is secreted in this division. The gland cells are very large and are characterized by the presence of lipid bodies. The swollen cisternae of the RER contain fine granular materials. Golgi vacuoles contain fine fibers 10-12 nm in diameter. Mature Golgi vacuoles leave the Golgi complex as globules containing suspended sericin molecules (Figure 19) and accumulate in the apical cytoplasm. The apical border of the cell contains closely packed, long microvilli. Bundles of microfilaments protrude into the silk layer.

3.3. Anterior Silk Gland

Each anterior silk gland is a duct which is connected posteriorly to the middle silk gland. Anteriorly, the paired ducts join and connect to the spinneret (Figure 1A). Each duct is lined with a thick cuticular intima (Akai, 1960, 1973). A typical cell is drawn in Figure 20. These duct cells are characterized by deposits of glycogen and clusters of filaments (Akai, 1976a). During molts, there is a proliferation of tubular RER, Golgi complexes, and mitochondria. So the cells are more active synthetically than during intermolt periods. Many dense aggregates of fibers which are the precursors of the cuticular intima appear in the space between the apical cytoplasm and the inner surface of the cuticular intima (Figures 20, 21) during the intermolt stages and are deposited on the basal surface of the intima. During the molting stages when the intima is renewed, the fibers

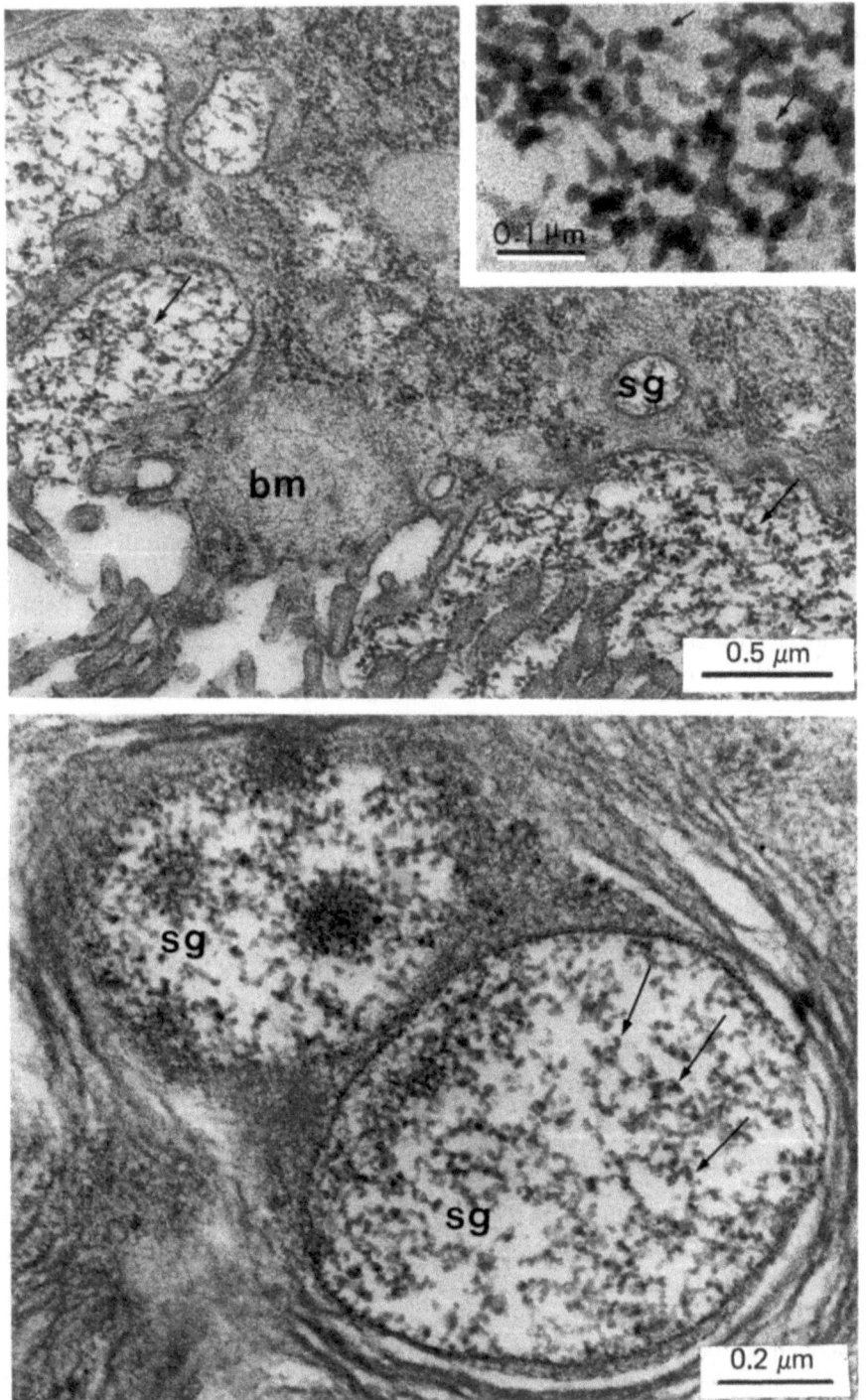

Figure 18. An apical cytoplasmic region of a cell from the anterior part of the middle division of the middle silk gland. Sericin globules (sg) are formed in Golgi complexes, and the sericin fibers are released into the silk layer (arrows). (Inset) Sericin fibers (arrows) at higher magnification.
Figure 19. Two globules in the apical cytoplasm of a cell from the anterior division of the middle silk gland. These sericin globules (sg) contain numerous sericin fibers (arrows) that will form the outer layer sericin in the lumen. From Akai (1976a).

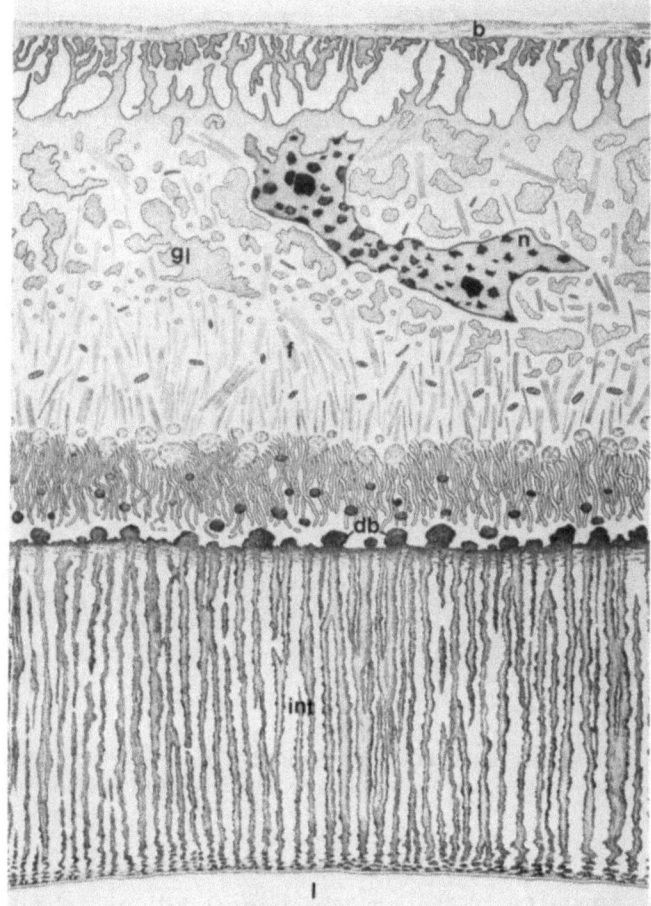

Figure 20. A schematic drawing of a cross section of a cell from the anterior silk gland. b, basement membrane; db, dense body; f, filamentous materials; gl, glycogen deposit; int, intima; l, lumen; n, nucleus. From Akai (1976a).

produced by the duct cells seem to be transferred directly into the new intima without the formation of aggregates (Akai, 1976a).

The intima shows a pattern of vertical opaque and transparent stripes (Figure 20). The opaque stripes are composed of tightly packed fibers, each 15-20 nm in diameter. The surface of the intima facing the lumen is bordered by a thin, compact layer which resembles the epicuticular layer of the larval cuticle.

3.4. Filippi's Gland

Helm (1876) suggested that these glands secreted a substance that caused the liquid silk to coagulate. Blanc (1889) proposed that the glands secrete a lubricant that facilitated the movement of the liquid silk through the spinneret. On the other hand, Tichomiroff (1879) believed that Filippi's glands were merely secondary silk glands, and later cytologists suggested that the glands were vestigial structures (Imms, 1925; Oka, 1930).

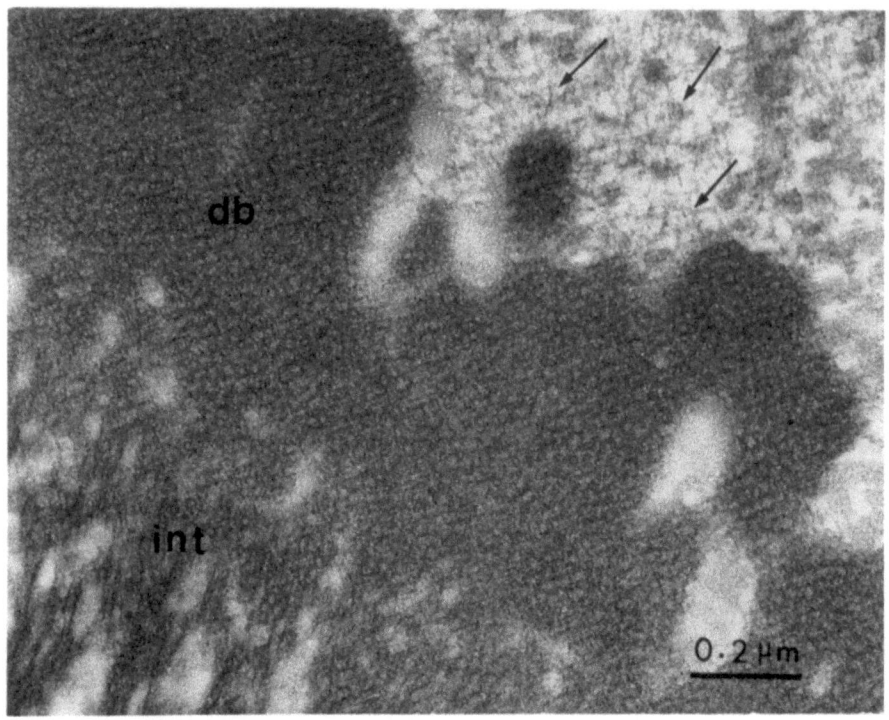

Figure 21. The basal surface of the intima (see Figure 20). Dense bodies (db) are formed by the accumulation of fibers (arrows), and they are then deposited on the surface of the cuticular intima. From Akai (1976a).

Machida (1965) found that the secretion of Filippi's gland contained proteins of completely different amino acid composition from fibroin, and he concluded that the gland could not be considered to be a secretory silk gland. Waku and Sumimoto (1974) have suggested that the gland may function in the exchange of small molecules (such as water) and ions, rather than having a secretory role.

A cell from Filippi's gland is characterized by a ramified nucleus, the presence of complicated canaliculi bearing microvilli on their inner surface, large numbers of mitochondria, and a remarkably convoluted basal plasma membrane (Figure 22). The cell lacks a well-developed cytoplasmic membrane system, such as rough or smooth ER and Golgi complexes. Free ribosomes and microtubules are plentiful (Akai, 1976a).

3.5. Liquid Silk and the Cocoon Filament

Using macroautoradiography, Fukuda and Florkin (1959) demonstrated that the fibroin synthesized in the posterior silk gland cells moves in an orderly fashion through the inside of the gland toward the anterior duct (Fukuda *et al.*, 1960). By microautoradiography, Akai (1963a) made clear the relationships between the newly secreted fibroin in different regions of the posterior silk gland

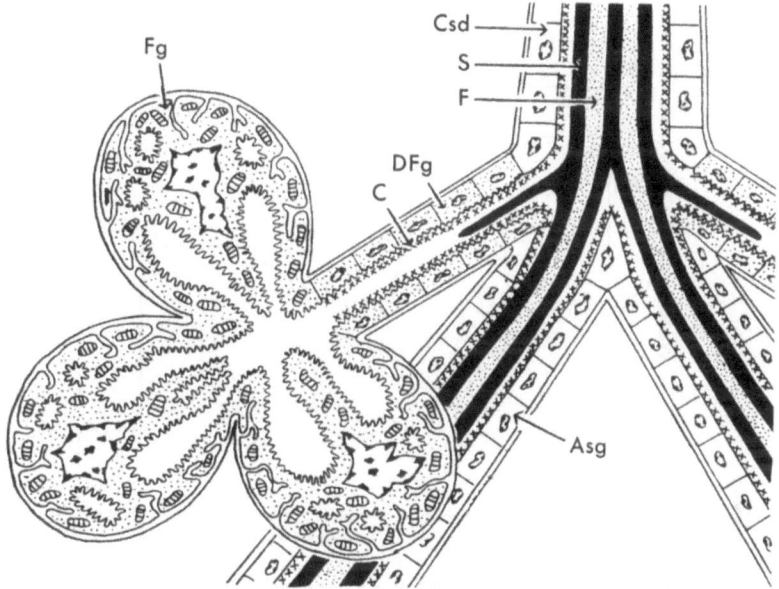

Figure 22. A diagram of Filippi's gland and the anterior silk gland. Asg, anterior silk gland; C, cuticle; Csd, common spinning duct; DFg, duct of Filippi's gland; F, fibroin; Fg, Filippi's gland; S, sericin. From Waku and Sumimoto (1974).

and plotted the subsequent routes taken in the lumen. The liquid fibroin secreted from the cells of the hind parts of the posterior silk gland takes a central course and moves more rapidly than that secreted from more anterior parts (Figure 23).

The current of liquid fibroin in the lumen is due to the movement of the spherical masses of fibroin (Figure 24). In fourth-instar larvae, about 200 spherical masses can be counted in a cross-section through the posterior silk gland (Kataoka and Akai, 1979). Each sphere within the posterior silk gland measured about 1.6 μm in diameter, but this size is reduced to 1 μm as they pass down the gland. This shrinkage seems to be due to dehydration which takes place in the middle silk gland (Shibukawa, 1959). Kataoka (1981) investigated the

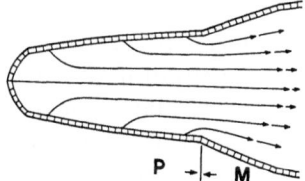

Figure 23. A diagram illustrating movement of liquid fibroin in the silk gland lumen. The arrows indicate the progressive routes taken by secreted masses of fibroin, and are proportional to the time it takes the silk to reach a given position. M, middle silk gland; P, posterior silk gland. From Akai (1976a).

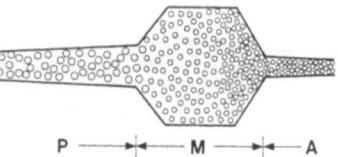

Figure 24. A diagram showing the size and the density of spherical fibroin masses in the silk gland lumen. A, anterior silk gland; M, middle silk gland; P, posterior silk gland. From Kataoka and Akai (1979).

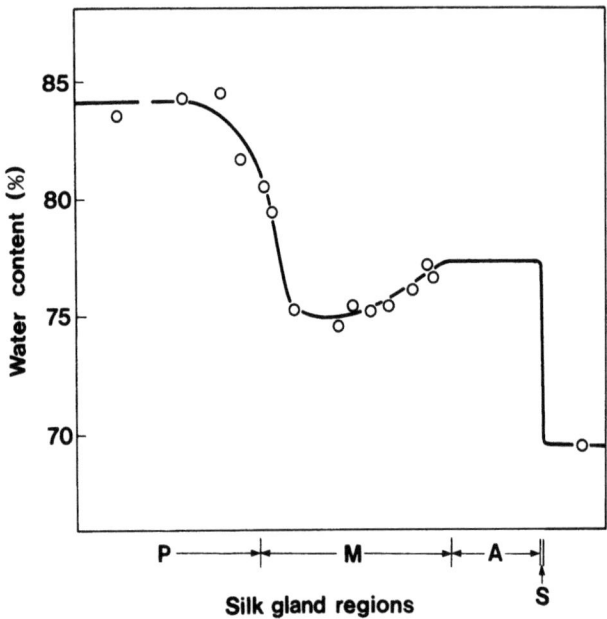

Figure 25. Changes in the water content of the liquid silk and the spun cocoon fiber. A, anterior silk gland; M, middle silk gland; P, posterior silk gland; S, spinneret. From Kataoka (1981).

water content of silk from various regions of the silk gland (Figure 25). The first change occurred as the silk moved from the posterior to the middle silk gland, and the second occurred in the spinneret.

The spherical masses of fibroin are tightly packed in the lumen of the anterior silk gland, but the elementary fibers inside the spheres are not oriented. However, they become aligned when passing through the thread press of the spinneret (Figure 26) (Akai, 1983).

Yokoyama (1951) noticed that when the subesophageal ganglion was severed, a larva never spun cocoons, and he concluded that spinning muscles innervated by the subesophageal ganglion controlled the spinning of silk.

The outer and internal structure of cocoon filaments have been observed with the scanning electron microscope (Akai, 1976b; Hayashiya and Tagawa, 1982). Mercer (1952) made an electron microscopic examination of fibrillar fragments produced by the enzymatic disintegration of silk fibroin and found fine microfibrils about 10 nm in diameter aligned parallel to the length of the fiber axis. Matsumura (1980) reported that a single filament of spun fibroin was composed of approximately 2000 fibrils, each with diameters between 0.1 and 0.4 μm, while Minagawa (1980) observed bundles composed of between 900 and 1400

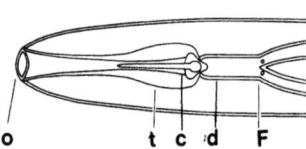

Figure 26. A schematic drawing of the spinneret of a *Bombyx* larva. Liquid silk flowing from the pair of anterior silk glands forms the cocoon filament when the silk passes through the thread press. c, chitin plate; d, common duct joining the pair of anterior silk glands; F, orifices of Filippi's glands; o, orifice of spinneret; t, thread press. From Akai (1976a).

fibrils which ranged from 0.2 to 0.4 μm in diameter. The fibrils were formed from filaments ranging between 10 and 15 nm in diameter.

4. The Hormonal Control of Silk Gland Development

The *Bombyx* silk gland is one of the most responsive organs to insect hormones, especially, ecdysone and juvenile hormone (JH). When the corpora allata which produce JH were removed from early fourth-instar larvae, these larvae become precocious pupae without going through the fourth molt (Fukuda, 1942). During this precocious development, the silk gland cells show hypertrophic development which closely resembles the behavior of the silk gland during the fifth larval instar (Fukuda, 1942; Akai and Kiguchi, 1980). Also, when larvae are ligated during the middle of the fourth instar at the level between the head and the prothorax, they become the precocious pupae, and the silk gland shows a hypertrophic development that resembles allatectomized larvae (Nishimura, 1957; Akai, 1965).

Fukuda (1962) observed that prothetelic, sixth-instar larvae (produced by the implantation of prothoracic glands early in the fifth instar) and operated fifth-instar larvae (which had been implanted during the fourth instar with corpora allata from fifth-instar larvae) could not secrete silk and never spun cocoons. The silk glands of these larvae were abnormal to various degrees. In most severely affected cases, the silk glands atrophied and transformed into threadlike structures entangled with numerous tracheae.

The development of the posterior silk glands of fourth-instar larvae allatectomized just after the third ecdysis was studied with the electron microscope (Akai and Kiguchi, 1980). The nucleolus grew enormously during the period between 24 and 72 hr after the allatectomy, and large numbers of nucleolonemata associated with numerous RNP granules form from the nucleolus (Figure 27). So the nucleolus resembles one from a fifth-instar larva during the most active period of RNA synthesis. Both RER and Golgi complexes proliferate beginning 4 hr after allatectomy. The cisternae of the RER enlarge and elementary fibroin fibers are deposited in Golgi vacuoles during the period between 24 and 96 hr after allatectomy. A silk layer between the apical surface of the gland cell and the lumen increases both in thickness and in the concentration of elementary fibroin fibers during the same period. Thus, reducing the level of JH activates the silk gland cells to synthesize a large amount of fibroin.

Conversely, the administration of ecdysone at 96 hr of the fifth-instar causes the silk gland cells to undergo a premature histolysis (Akai, 1976a). Within a few hours after ecdysone administration, the granular ER is concentrated cytoplasmic islands (Figure 28), and these are enclosed by a separate membrane to become autophagosomes (Figures 29, 30). A similar transformation occurs in silk gland cells during the normal larval–pupal metamorphosis.

Akai *et al.* (1971) showed that the JH-induced prolongation of the larval period also resulted in increased silk production. A single administration of 20 μg JH prevented metamorphosis, including the maturing of larvae and

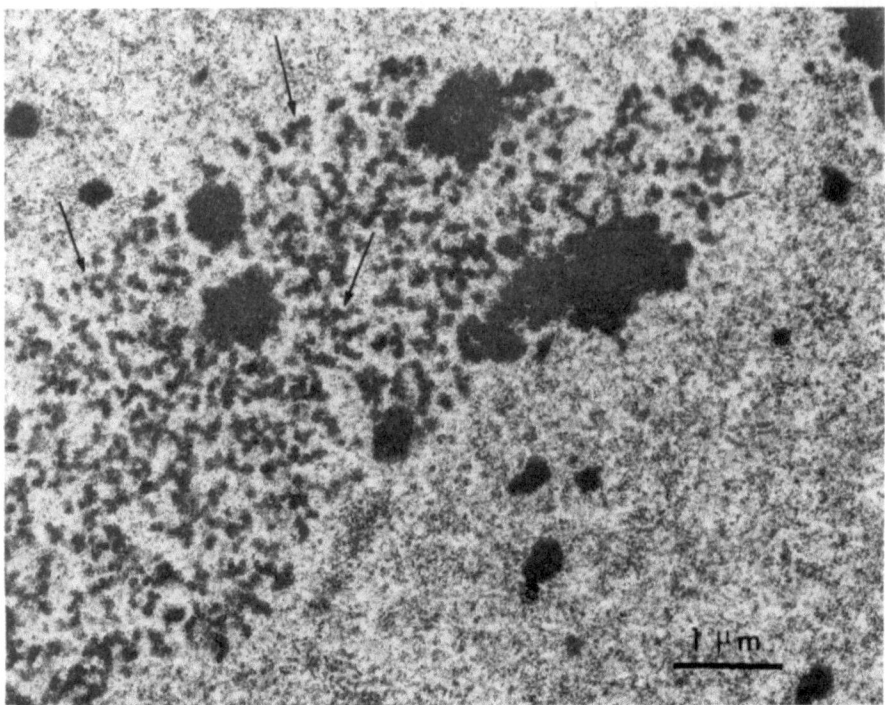

Figure 27. A nucleolus containing numerous nucleolonemata (arrows) from a posterior silk gland cell of a fourth-instar larva allatectomized after the third ecdysis. From Akai and Kiguchi (1980).

spinning. Treated larvae survive over a month (Akai and Kobayashi, 1971) and are called *dauer larvae*. The silk glands of dauer larvae contain large amounts of liquid silk, and the cells are well developed, without any signs of histolysis (Akai *et al.*, 1973). In normal fifth-instar larvae, DNA synthesis in the silk gland cells stops within 6 days of molt. In the case of the dauer larvae, DNA synthesis declines at the same time as in control animals, but synthesis resumes, showing peaks at days 6 and 10 (Figure 31), and the nuclei increase their ploidy levels. Similar studies were made by Kurata and Daillie (1978) and by Daillie (1979), who found that the quantity of DNA in JH-treated animals was nearly twice that of the controls.

In dauer larvae, fibroin synthesis continues an additional 20 days, and the fibroin stored in the lumen becomes extremely concentrated. In the gland cells, numerous globules, 1–2 μm in diameter, appear in the cytoplasm (Figure 32) (Akai, 1973, 1976a). These globules resemble the "hunger spherules" observed in several tissues of starved larvae.

5. Mutations Influencing Silk Production

A number of mutations are known in *B. mori* that are characterized by the absence of cocoons or the production of abnormal cocoons (Suzuki, 1977). Four

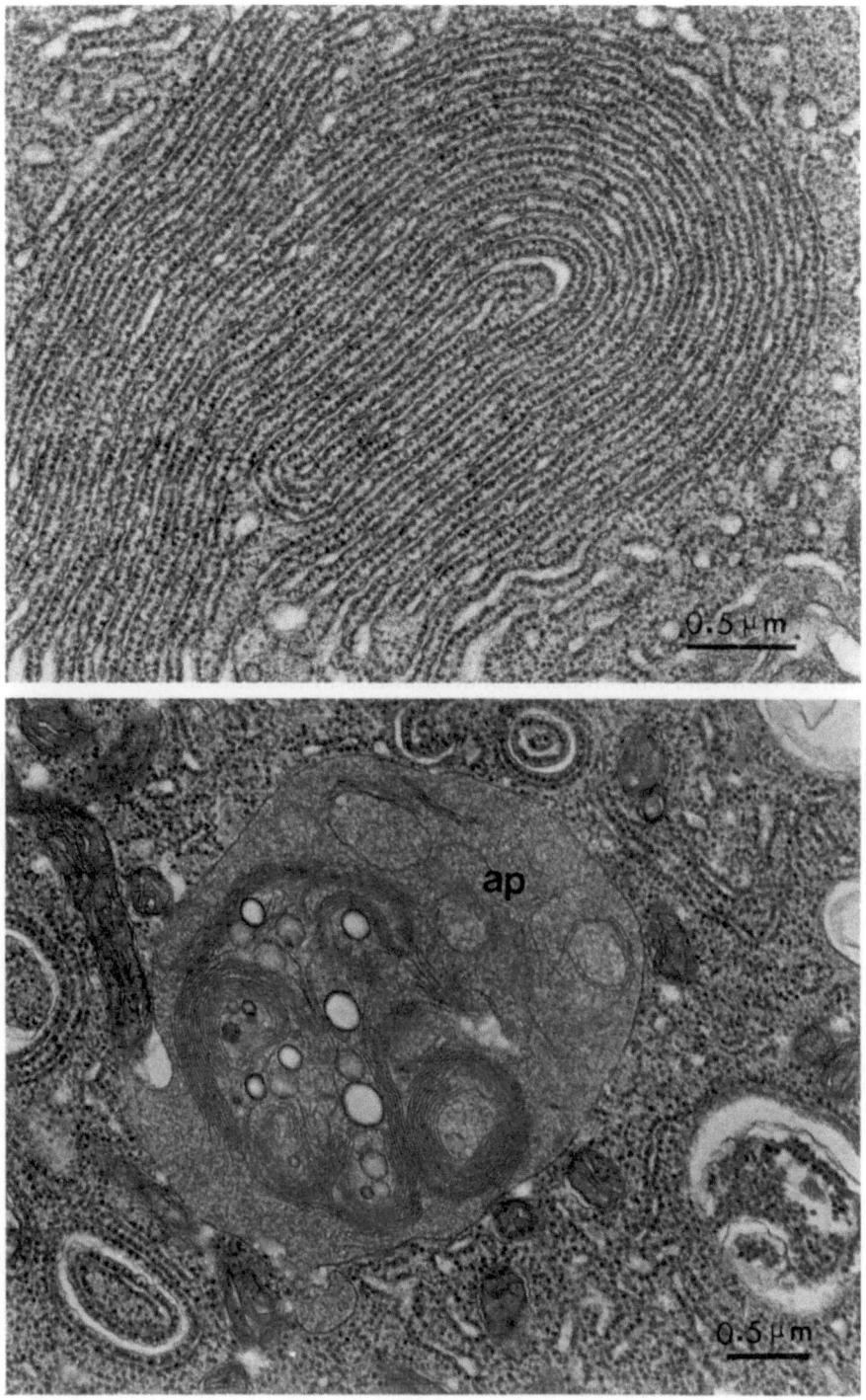

Figure 28. A lamellated body composed of RER in a cell from the posterior silk gland. These organelles formed about 6 hr after ecdysterone administration at the 96th hr of the fifth instar. From Akai (1976a).

Figure 29. An autophagosome (ap) formed in a posterior silk gland cell 24 hr after the administration of ecdysterone.

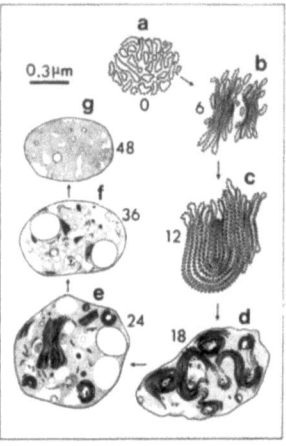

Figure 30. A diagram illustrating the intracellular digestion of RER in a cell from the posterior division of the silk gland after ecdysterone administration 96 hr after the fourth ecdysis. (a) Normal RER; (b) lamellarization of RER during 6-12 hr after administration; (c-g) lamellarization and conversion of RER to an atuophagosome. The numbers give the time in hours after ecdysterone treatment. From Akai (1976a).

mutant alleles of the *naked pupa* (*Nd*) gene have been isolated [*Nd*(1), *Nd*(2), *Nd*1, and *Nd*2]. Homozygotes produce cocoons consisting mainly of sericin or fail to produce any cocoon. The posterior silk glands are markedly inhibited in their growth, and fibroin does not accumulate in the lumen.

Fibroin mRNA levels are only about 1% the levels found in wild-type posterior silk glands. Heterozygotes (*Nd*/+) have mRNA levels 3-4% the normal levels (Suzuki and Suzuki, 1974) and show the mutant cocoon phenotype. *Nd* is located on chromosome 23 together with *Fib*, the structural gene for fibroin (Hyodo et al., 1982). Machida (1970, 1972) observed the histological characteristics of the silk glands of *Nd* larvae. Iijima (1972) studied the ultrastructure of the posterior silk gland of *Nd* larvae. Since the cisternae of the RER were remarkably

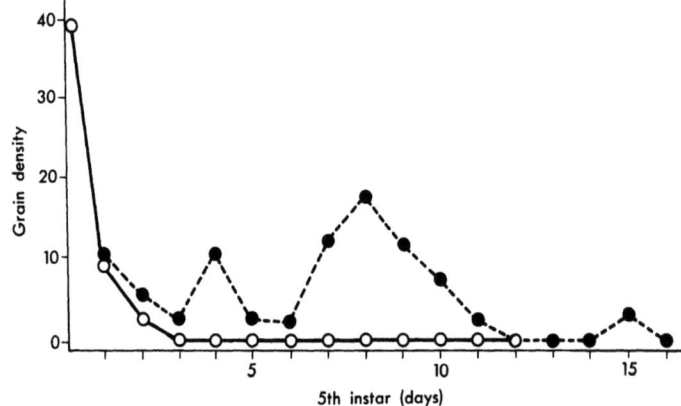

Figure 31. DNA synthesis in the silk gland cells of JH-induced dauer larvae. Each animal received 20 µg JH/g body weight 3 days after the fourth ecdysis. The synthesis of DNA in silk gland cells labeled with tritiated thymidine was estimated by quantitative autoradiography. Dotted line, JH-treated larvae; solid line, control. In controls, the silk glands degenerate between days 13 and 14. From Akai et al. (1973).

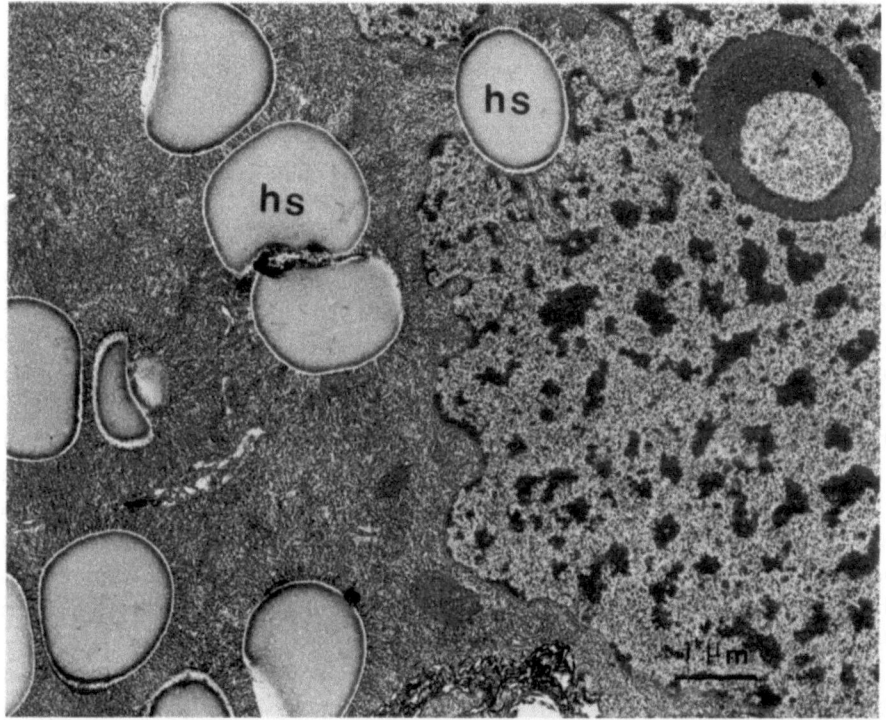

Figure 32. A region of a posterior silk gland cell of a dauer larva. The silkworm had received JH 25 days earlier. Hunger spherules (hs) are present. These appear shortly before the death of the cells. From Akai (1976a).

distended and the Golgi complexes were poorly developed, he concluded that the intracellular transport and exocytosis of the synthesized fibroin was defective.

Another mutation, *Nd-s*, is known which produces a phenotype similar to the alleles mentioned in the previous paragraph. However, *Nd-s* resides on a different chromosome, number 14 (Horiuchi and Ohi, 1966). *Nd-s* homozygotes have fibroin mRNA levels about 1% of normal, as in the case of *Nd* (2). However, the heterozygote (*Nd-s/+*) has an intermediate mRNA level; about 17% of normal (Suzuki and Suzuki, 1974). Thus, in the heterozygous condition, both *Nd* and *Nd-s* seem to suppress the functioning of the normal allele, but to different degrees. There is another major difference between *Nd* and *Nd-s* silkworms. While *Nd* sericins seem normal, the *Nd-s* silk gland produces a large quantity of a sericin with abnormal electrophoretic properties.

Doira (1970) discovered a recessive mutation (*flimsy cocoon*) residing on chromosome 3 which produces cocoons about one-third of normal weight. Cocoons from *flc* homozygotes contain both fibroin and sericin. The development of the posterior silk gland cells of *flc/flc* larvae was studied by Adachi-Yamashita *et al.* (1980). The cells seem to store fibroin globules, indicating a defect in the secretory mechanism.

The *Src-2* gene which codes for the sericin-2 polypeptide resides on chromosome 11 (Gamo, 1982).

6. Summary

The silk gland is a typical exocrine gland designed to secrete large amounts of fibroin and sericins. The gland consists of three divisions, each characterized by different secretory functions.

The posterior silk gland cells undergo 18-19 cycles of endomitosis during the larval period, and this results in highly polyploid cells. The DNA content of the average cell increases 250,000-fold, while the volume increases about 86,000-fold. These cells synthesize fibroin, and 8% of the RNA transcribed is fibroin mRNA. The fibroin molecule contains repetitive Gly-Ala-Gly-Ala-Gly-Ser sequences, and the fibroin gene resembles a satellite DNA that is transcribed. The fibroin gene, which resides on chromosome 23, is present in one copy per genome. The extremely convoluted nucleus contains numerous nucleoli, and their morphology changes with changes in RNA synthesis. The cytoplasm of mature cells contains highly concentrated RER, Golgi complexes, and fibroin globules. Nascent silk fibroin polypeptides can be observed on polysomes isolated from the posterior silk gland cells. The fibroin molecules synthesized on polysomes attached to the RER are transferred into the cisternal lumen, pass through the RER to Golgi complexes, fibroin globules, and enter the silk layer in the gland lumen.

Three types of sericin are secreted from different parts of the middle region of the silk gland. The ultrastructural characteristics of cells from the middle region resemble those of the posterior gland. Three types of sericin fibers were observed in Golgi vacuoles, sericin globules, and silk layers in different segments of the middle silk gland. The anterior silk glands serve as ducts and do not contribute to the secretion of silk.

The current of liquid fibroin moves through the lumen of the gland toward the spinneret. The liquid fibroin column is composed of spherical masses of fibers, each of these containing bundles of fibroin molecules.

Several mutations are known that influence silk production. Some (*Fib* and *Src-2*) represent mutations in the cistrons coding for fibroin and sericin molecules. Others (*Nd*, *Nd-s*, and *flc*) seem to have defects in the intracellular transport and secretion of fibroin.

The silk gland is a target tissue for ecdysone and JH. Allatectomy when performed early during the fourth instar causes hypertrophic development of the silk gland cells. High doses of JH in the early fifth instar prevent metamorphosis. However, moderate doses result in a lengthening of the larval period and an increase in silk production. Such hormonal treatments are now used to improve commercial silk production.

ACKNOWLEDGMENTS

The author is grateful to Professor Robert C. King, Northwestern University, Dr. Hideaki Maekawa, National Institute of Health (Tokyo), and Dr. Toshiki Tamura, Sericultural Experiment Station, for their help during the preparation of this review.

References

Adachi-Yamashita, N., Sakaguchi, B., and Chikushi, H., 1980, Fibroin secretion in the posterior silk gland cells of a *flimsy cocoon* mutant of *Bombyx mori*, *Cell Struct. Funct.* **5**:105-108.
Akai, H., 1960, Fine structure of the anterior division of the silk gland in the silkworm, *Bombyx mori* L., *J. Sericult. Sci. Jpn* **29**:405-410 (Japanese).
Akai, H., 1963a, On the orderly progression of fibroin in the lumen of the silk gland of the silkworm, *Bombyx mori* L., *Bull. Sericult. Exp. Stn.* **18**:191-197 (Japanese with English summary).
Akai, H., 1963b, Electron microscopical observations on the fibroin formation in the silk gland of the silkworm, *Bombyx mori*, *Bull. Sericult. Exp. Stn.* **18**:271-282 (Japanese with English summary).
Akai, H., 1964, Micromorphological changes of the glandular cells in the posterior division of the silk gland during the 5th larval instar of the silkworm, *Bombyx mori*, *Bull. Sericult. Exp. Stn.* **18**:475-511 (Japanese with English summary).
Akai, H., 1965, Studies on the ultrastructure of the silk gland in the silkworm, *Bombyx mori* L. IV, *Bull. Sericult. Exp. Stn.* **19**:375-483 (Japanese with English summary).
Akai, H., 1970, Ultrastructural localization of adenosine triphosphatase in the silk gland of the silkworm, *Bombyx mori* L. (Lepidoptera: Bombycidae), *Appl. Entomol. Zool.* **5**:112-117.
Akai, H., 1971a, Ultrastructure of fibroin in the silk gland of larval *Bombyx mori*, *Exp. Cell Res.* **69**:219-223.
Akai, H., 1971b, An ultrastructural study of changes during molting in the silk gland of larvae of the Eri-silkworm, *Philosamia cynthia ricini* Boisdval (Lepidoptera: Saturniidae), *Appl. Entomol. Zool.* **6**:27-39.
Akai, H., 1973, Ultrastructure and function of the silk gland cells, *Cell* **5**:2-17 (Japanese).
Akai, H., 1976a, *Ultrastructural Morphology of Insects*, University of Tokyo Press, Tokyo (Japanese).
Akai, H., 1976b, *Surface Structure of Insects*, University of Tokyo Press, Tokyo (Japanese with English explanation of figures).
Akai, H., 1980, Ultrastructure of the silk gland of silkworm. In *Structure of the Silk Fiber II*, edited by N. Hojo, pp. 81-100, Shinkyo Publishing, Nagano (Japanese with English summary).
Akai, H., 1981, Ultrastructure and function of silk gland, *Denken* **16**:19-24 (Japanese).
Akai, H., 1982, The effects of JH treatment on the structure and synthetic activity of the silk gland of *Bombyx mori*. In *Ultrastructure and Functioning of Insect Cells*, edited by H. Akai, R. C. King, and S. Morohoshi, Society for Insect Cells, Tokyo.
Akai, H., 1983, The structure and ultrastructure of the silk gland. In *The Physiology and Biology of Spinning in Bombyx mori*, edited by K. Shimura, *Experimentia* **39**:443-449.
Akai, H., and Kataoka, K., 1978, Fine structure of liquid fibroin in the posterior silk gland of *Bombyx* larvae, *J. Sericult. Sci. Jpn* **47**:273-278 (Japanese with English summary).
Akai, H., and Kiguchi, K., 1980, Ultrastructural changes of the posterior silk gland cells in allatectomized *Bombyx mori*, *Bull. Sericult. Exp. Stn.* **28**:1-14 (Japanese with English summary).
Akai, H., and Kobayashi, M., 1965a, Incorporation of labelled thymidine into the silk gland of the silkworm, *Nature (London)* **206**:847-848.
Akai, H., and Kobayashi, M., 1965b, Sites of fibroin formation in the silk gland in *Bombyx mori*, *Nature (London)* **206**:529.
Akai, H., and Kobayashi, M., 1966, Cytological studies on silk protein and nucleic acid synthesis in the silk gland of the silkworm, *Bombyx mori*, *Symp. Cell Chem.* **17**:131-138 (Japanese with English summary).
Akai, H., and Kobayashi, M., 1971, Induction of prolonged larval instars by the juvenile hormone in *Bombyx mori* L. (Lepidoptera: Bombycidae), *Appl. Entomol. Zool.* **6**:138-139.
Akai, H., Kiguchi, K., and Mori, K., 1971, Increased accumulation of silk protein accompanying JH-induced prolongation of larval life in *Bombyx mori* L. (Lepidoptera: Bombycidae), *Appl. Entomol. Zool.* **6**:218-220.
Akai, H., Kiguchi, K., and Mori, K., 1973, The influence of juvenile hormone on the growth and metamorphosis of *Bombyx* larvae, *Bull. Sericult. Exp. Stn.* **25**:287-305 (Japanese with English summary).
Beyer, A. L., McKnight, S., and Miller, O. L., 1979, Transcriptional units in eukaryotic

chromosomes. In *Molecular Genetics*, part III, *Chromosome Structure*, edited by J. H. Taylor, pp. 117-175, Academic Press, New York.

Blanc, L. M., 1889, Étude sur la sécrétion de la soie et la structure du brin et de la bave dans le *Bombyx mori*, *Rep. Lab. Etude Sie Lyon* **4**:54-98.

Chevallier, A., and Garel, J. P., 1979, Studies on tRNA adaptation, tRNA turnover, precursor tRNA and tRNA gene distribution in *Bombyx mori* by using 2-dimensional polyacrylamide gel electrophoresis, *Biochimie* **61**:245-262.

Chikushi, H. (ed.), 1972, *Genes and Genetical Stocks of the Silkworm*, Keigaku, Tokyo.

Daillie, J., 1979, Juvenile hormone modifies larvae and silk gland development in *Bombyx mori*, *Biochimie* **61**:275-281.

Doira, H., 1970, Genetics of the flimsy cocoon mutant, *Proc. Sericult. Sci. Kyushu* **1**:5 (Japanese).

Fournier, A., 1979, Quantitative data on the *Bombyx mori* L. silkworm: A review, *Biochimie* **61**:283-320.

Fukuda, S., 1942, Precocious development of the silk gland following ablation of the corpora allata in the silkworm, *Zool. Mag.* **54**:11-13 (Japanese with English summary).

Fukuda, S., 1962, Changes in the silkglands characteristic of prothetelic silkworm larvae, *Annot. Zool. Jpn.* **36**:139-149.

Fukuda, T., and Florkin, M., 1959, Contribution to silkworm biochemistry. VII. Ordered progression of fibroinogen in the reservoir of the silkgland during the 5th instar, *Arch. Int. Physiol. Biochim.* **67**:314-221.

Fukuda, T., Kirimura, J., Matsuda, M., and Suzuki, T., 1955, Biochemical studies on the formation of the silkprotein. I. The biosynthesis of the silkprotein, *J. Biochem.* **42**:341-346.

Fukuda, T., Suto, M., and Nakagawa, Y., 1960, Biochemical studies on the formation of the silk protein. X. Ordered progression of fibroin in the inside of the silk gland during the fifth instar, *Bull. Agric. Chem. Soc. Jpn.* **24**:501-505.

Gage, L. P., 1974, Polyploidization of the silk gland of *Bombyx mori*, *J. Mol. Biol.* **86**:97-108.

Gage, L. P., and Manning, R. F., 1976, Determination of the multiplicity of the silk fibroin gene and detection of fibroin gene-related DNA in the genome of *Bombyx mori*, *J. Mol. Biol.* **101**:327-348.

Gage, L. P., and Manning, R. F., 1980, Internal structure of the silk fibroin gene of *Bombyx mori*. I. The fibroin gene consists of a homogeneous alternating array of repetitious crystalline and amorphous coding sequences, *J. Biol. Chem.* **255**:9444-9450.

Gamo, T., 1982, Genetic variants of the *Bombyx mori* silkworm encoding sericin proteins of different lengths, *Biochem. Genet.* **20**:165-177.

Gamo, T., Inokuchi, T., and Laufer, H., 1977, Polypeptides of fibroin and sericins of the silk gland in *Bombyx mori*, *Insect Biochem.* **7**:285-295.

Garel, J. P., 1982, The silkworm, a model for molecular and cellular biologists, *Trends Biochem. Sci.* **7**:105-108.

Garel, J. P., and Keith, G., 1977, Nucleotide sequence of *Bombyx mori* L. rRNA$_1^{gly}$, *Nature (London)* **269**:350-352.

Garel, J. P., Garber, R. L., and Siddiqui, M. A., 1977, Transfer RNA in posterior silk gland of *Bombyx mori:* Polyacrylamide gel mapping of mature transfer RNA, identification and partial structural characterization of major isoacceptor species, *Biochemistry* **16**:3618-3624.

Gillot, S., 1968, Hétérogénéités fonctionelles dans l'ADN de noyaux géants: Étude autoradiographique sur la glande séricigène de *Bombyx mori* L., *Exp. Cell Res.* **50**:388-402.

Gillot, S., and Daillie, J., 1968, Rapport entre la mue et la synthese d'ADN dans la glande séricigène du ver a soie, *C.R. Acad. Sci. Ser. D* **266**:2295-2298.

Green, R. A., Morgan, M., Shatkin, A. J., and Gage, L. P., 1975, Translation of silk fibroin messenger RNA in an Ehrlich ascites cell extract, *J. Biol. Chem.* **250**:5114-5121.

Harizuka, M., 1953, Physiological genetics of the carotenoids in *Bombyx mori*, with special reference to the pink cocoon, *Bull. Sericult. Exp. Stn.* **14**:141-156 (Japanese with English summary).

Hayashiya, K., and Tagawa, T., 1982, Silk fibers. In *Illustrated Forms of Fibers*, edited by H. Kawai and T. Tagawa, pp. 106-121, Asakura Book Store, Tokyo.

Hayashiya, K., Kato, M., and Hamamura, Y., 1964, Amino acid composition of silk proteins resulting from artificial diet-fed *Bombyx mori*, *Proc. Jpn. Acad.* **40**:349-350.

Helm, F. E., 1876, Uber die Spindrüsen der Lepidopteren, *Z. Wiss. Zool. Abt. A* **26**:434-469.

Hentzen, D., Garel, J. P., and Keith, G., 1976, Anticodon loop sequences of tranfer RNA$_{CGA}^{Ser}$ and transfer $\text{RNA}^{\text{Ser}}_{\text{IGA}}$ from the posterior silkgland of Bombyx mori L., Biochem. Biophys. Res. Commun. **71:**241-248.

Horiuchi, Y., and Ohi, H., 1966, Genetics of the sericin cocoon mutant Thanghpre-S, J. Sericult. Sci. Jpn. **35:**228 (Japanese).

Hyodo, A., and Shimura, K., 1980, The occurrence of a hereditary variant of the fibroin large subunit in the silkworm, Bombyx mori, Jpn. J. Genet. **55:**203-209.

Hyodo, A., Gamo, T., and Shimura, K., 1980, Linkage analysis of the fibroin gene in the silkworm, Bombyx mori, Jpn. J. Genet. **55:**297-300.

Hyodo, A., Ueda, H., Takei, F., Kimura, K., and Shimura, K., 1982, Gene expression of two fibroin alleles in the hybrid silkworm, J-131/Nd(2), Jpn. J. Genet. **57:**551-560.

Ichimura, S., Mita, K., Zama, M., and Numata, M., 1982, Isolation of nuclei and nuclear proteins from the posterior silk gland of Bombyx mori. In *Ultrastructure and Functioning of Insect Cells*, edited by H. Akai, R. C. King, and S. Morohoshi, pp. 37-40, Society for Insect Cells, Tokyo.

Iijima, T., 1971a, Cytological changes of the fine structure of the silk gland in Bombyx larva under *in vitro* conditions, J. Sericult. Sci. Jpn. **40:**181-191.

Iijima, T., 1971b, Ultrastructure of silk gland cells in the silkworm, Bombyx mori L., injected with actinomycin D, J. Sericult. Sci. Jpn. **40:**357-367.

Iijima, T., 1972, Ultrastructure of the posterior silk gland of the "naked pupa" silkworm, Bombyx mori, J. Insect Physiol. **18:**2055-2063.

Imms, A. D., 1925, *A General Textbook of Entomology*, Methuen, New York.

Ito, H., 1915, On the metamorphosis of the silk glands of Bombyx mori, Bull. Imp. Tokyo Sericult. Coll. **1:**19-43.

Ito, S., 1977, Cytological studies on the chromosomes of silk gland cells of the silkworm with special reference to the structure and behavior of the sex chromosomes, Jpn. J. Genet. **52:**327-340.

Ito, T., 1981, Nutrition and artificial diet. In *Recent Advances of Entomology*, edited by S. Ishii, pp. 497-510, University of Tokyo Press (Japanese).

Ito, T., and Kobayashi, M., 1978, Rearing of the silkworm. in *The Silkworm: An Important Laboratory Tool*, edited by Y. Tazima, pp. 83-102, Kodansha, Tokyo.

Kataoka, K., 1981, Water content in the liquid silk and silk fibers during the spinning of larvae of the silkworm, Bombyx mori, J. Sericult. Sci. Jpn. **50:**478-483 (Japanese with English summary).

Kataoka, K., and Akai, H., 1979, Histological changes of liquid fibroin in the silk gland, J. Sericult. Sci. Jpn. **48:**171-176 (Japanese with English summary).

Kawai, N., 1976, Hormonal effects on carotenoid uptake by the silk gland in the silkworm, Bombyx mori, J. Insect Physiol.**22:**207-216.

Kawai, N., 1978, Degenerative changes and carotenoid uptake in the silk gland of the silkworm, Bombyx mori, J. Insect Physiol. **24:**17-24.

Kawakami, M., and Shimura, K., 1973, Fractionation of glycine-, alanine- and serine-transfer ribonucleic acids from the silk glands of silkworm,s J. Biochem. **74:**33-40.

Kawakami, M., Nishio, K., and Takemura, S., 1978, Nucleotide sequence of $\text{tRNA}^{\text{Gly}}_2$ from the posterior silk glands of Bombyx mori, FEBS Lett. **87:**288-290.

Kirimura, J., 1962, Studies on amino acid composition and chemical structure of silk protein by microbiological determination, Bull. Sericult. Exp. Stn. **17:**447-522 (Japanese with English summary).

Kirimura, J., and Suzuki, T., 1962, Studies on amino acid composition of silk protein by microbiological determination. III. Amino acid compositioin of silk protein and silk substance, Bull. Agric. Chem. Jpn. **36:**265-268 (Japanese).

Komatsu, K., 1975, Studies on dissolution behaviours and structural characteristics of silk sericin, Bull. Sericult. Exp. Stn. **26:**135-256 (Japanese with English summary).

Komatsu, K., 1982, Secretion and structure of liquid silk, Hikaku Kagaku **27:**193-208 (Japanese with English summary).

Komatsu, K., and Yamada, M.,, 1975, Chemical studies on sericin. II. Amino acid composition of wild and domestic cocoon sericins, J. Sericult. Sci. Jpn. **44:**105-110 (Japanese with English summary).

Kurata, K., and Daillie, J., 1978, Effect of exogenous juvenoid on the growth of the silk gland and the synthesis of nucleic acid and silk protein by the silk gland of Bombyx mori, Bull. Sericult. Exp. Stn. **27:**507-530 (Japanese with English summary).

Kurata, K., Takeshita, H., Shigematsu, H., and Sakate, S., 1974, Quantitative relationship between nucleic acids in the posterior division of silk gland and the silk in the cocoon found in various genotypes concerned with silk formation in the silkworm, Bombyx mori, *J. Sericult. Sci. Jpn.* **43**:277-282 (Japanese with English summary).

Latreille, P. A., 1802, *Histoire Naturelle, Générale et Particulière des Crustacés et des Insectes*, Dufart, Paris.

Lizardi, P. M., 1975, The length of the fibroin gene in the Bombyx mori genome, *Cell* **4**:207-215.

Lizardi, P. M., 1976, The size of pulse-labeled fibroin messenger RNA, *Cell* **7**:239-245.

Lizardi, P. M., 1979, Genetic polymorphism of silk fibroin studied by two-dimensional translation pause fingerprints, *Cell* **18**:581-589.

Lizardi, P. M., 1980, Isolation of giant silk fibroin polysomes and fibroin mRNA particles using a novel ribonuclease inhibitor, hydroxystilbamidine, *J. Cell Biol.* **87**:292-296.

Lucas, F., Shaw, J. T. B., and Smith, S. G., 1957, The amino acid sequence in a fraction of the fibroin of Bombyx mori, *Biochem. J.* **66**:468-479.

Lucas, F., Shaw, J. T. B., and Smith, S. G., 1962, Some amino acid sequences in the amorphous fraction of the fibroin of Bombyx mori, *Biochem. J.* **83**:164-171.

Machida, J., 1926, Studies on the silk substances secreted by Bombyx mori, *Bull. Sericult. Exp. Stn.* **7**:241-262 (Japanese with English summary).

Machida, Y., 1965, Studies on the silk glands in the silkworm, Bombyx mori L. 1. Morphological and functional studies of Filippi's glands in the silkworm, *Sci. Bull. Fac. Agric. Kyushu Univ.* **22**:95-108 (Japanese with English summary).

Machida, Y., 1970, Studies on the silkglands of silkworms, Bombyx mori L. II. The singularity of the silk glands in hereditary trait, naked pupae (Nd), in the silkworm (1), *Fukuoka Women's Junior College, Studies* **3**:1-21 (Japanese with English summary).

Machida, Y., 1972, Studies on the silk glands in the silkworm, Bombyx mori L. III. The singularity of the silk glands in hereditary trait, naked pupae (Nd), in the silkworm (2), *Fukuoka Women's Junior College, Studies* **10**:39-49 (Japanese with English summary).

McKnight, S. L., Sullivan, N. L., and Miller, O. L., Jr., 1976, Visualization of the silk fibroin transcription unit and nascent silk fibroin molecules on polyribosomes of Bombyx mori, *Prog. Nucleic Acid Res. Mol. Biol.* **19**:313-318

Maekawa, H., and Suzuki, Y., 1980, Repeated turn-off and turn-on of fibroin gene transcription during silk gland development of Bombyx mori, *Dev. Biol.* **78**:394-406.

Manning, R. F., and Gage, L. P., 1980, Internal structure of the silk fibroin gene of Bombyx mori. II. Remarkable polymorphism of the organization of crystalline and amorphous coding sequences, *J. Biol. Chem.* **255**:9451-9457.

Matsumura, H., 1980, Scanning electron microscopic observation of the fine structure of fibroin fibers. In *Structure of Silk Fiber II*, edited by N. Hojo, pp. 175-185, Shinkyo Publishing, Nagano (Japanese with English summary).

Matsuura, S., and Tashiro, Y., 1976, Cup-shaped mitochondria in the posterior silk gland of Bombyx mori in the prepupal stadium, *Cell Struct. Funct.* **1**:137-145.

Matsuura, S., Morimoto, T., Nagata, S., and Tashiro, Y., 1968, Studies on the posterior silk gland of the silkworm, Bombyx mori. II. Cytolytic processes in posterior silk gland cells during metamorphosis from larva to pupa, *J. Cell Biol.* **38**:589-603.

Matsuura, S., Shimadzu, T., and Tashiro, Y., 1976, Lysosomes and related structures in the posterior silk gland cells of Bombyx mori. II. In prepupal and early pupal stadium, *Cell Struct. Funct.* **1**:223-235.

Matsuzaki, K., 1963, The incorporation of C^{14}-glycine into the soluble RNA of the posterior silkgland, *J. Biochem.* **53**:326.

Matsuzaki, K., 1966, Fractionation of amino acid-specific s-RNA from the silk gland by methylated albumin column chromatography, *Biochim. Biophys. Acta* **114**:222-226.

Matsuzaki, K., 1970, The studies on specificity of silk gland tRNA from the silkworm, *Bull. Sericult. Exp. Stn.* **24**:443-478.

Maziarski, S., 1911, Recherches cytologiques sur les phenomenes secretoires dans les glandes filieres des larves des Lepidopteres, *Arch. Zellforsch.* **6**:397-442.

Mercer, E. H., 1952, The fine structure and biosynthesis of silk fibroin, *Aust. J. Sci. Res. Ser. B* **5**:366-373.

Meves, F., 1897, Zur Structur der Kerne in den Spinndrüsen der Raupe, *Arch. Mikrosk. Anat. Entwicklungsmech.* **48**:573-579.

Minagawa, M., 1980, Fine structure of silk fibers and lousiness fibers. In *Structure of Silk Fiber II*, edited by N. Hojo, pp. 187-208, Shinkyo Publishing, Nagano (Japanese with English summary).

Morimoto, T., Matsuura, S., Nagata, S., and Tashiro, Y., 1968, Studies of the posterior silk gland of the silkworm, *Bombyx mori*. III. Ultrastructural changes of posterior silk gland cells in the fourth larval instar, *J. Cell Biol.* **38**:604-614.

Nakagawa, Y., 1949, On the distribution types of the trachea and forms of the silk glands of lepidopterous larvae, *Bull. Takarazuka Insectarium* **63**:1-14 (Japanese).

Nakagawa, Y., 1970, On the evolution of the silk glands in the Lepidoptera, *Vertebr. Sci.* **20**:69-84 (Japanese).

Nakahara, W., 1917, On the physiology of the nucleoli as seen in the silk gland cells of certain insects, *J. Morphol.* **29**:55-73.

Nakajima, S., 1941, Studies on the spinnerets of lepidopterous larvae, *Bull. Miyazaki Coll. Agric. Forest.* **12**:269-338 (Japanese with English summary).

Nishimura, H., 1957, Studies on the growth of silkgland in the silkworm, *Bombyx mori*. In particular the developmental mechanism of the silkgland in ligated 4th instar larvae, *Saitama Sericult. Exp. Stn.* **31**:1-51 (Japanese).

Nunome, J., 1937, Development of the silk gland in *Bombyx mori*, *Bull. Appl. Zool. Jpn.* **9**:68-92 (Japanese with English summary).

Oba, H., 1950, Boundary between middle and hind sections of silk glands, and subdivisions of the middle section distinguished by difference of section in *Bombyx mori*, *J. Sericult. Sci. Jpn.* **19**:239-246 (Japanese).

Oba, H., 1957, Studies on the secretion and the current of liquid silk during spinning period in the silkworm, *Bombyx mori* L. II. Observation on the silk gland of insect in relation to the racial difference in lousiness quality, *Bull. Nagano Sericult. Exp. Stn.* **10**:400-405 (Japanese).

Ohmachi, T., and Shimura, K., 1981, Initiation of fibroin biosynthesis. I. Isolation of nascent fibroin peptides from the posterior silk gland of *Bombyx mori*, *J. Biochem.* **89**:531-541.

Ohmachi, T., Nagayama, H., and Shimura, K., 1982, The isolation of a messenger RNA coding for the small subunit of fibroin from the posterior silk gland of the silkworm, *Bombyx mori*, *FEBS Lett.* **146**:385-388.

Ohshima, Y., and Suzuki, Y., 1977, Cloning of the silk fibroin gene and its flanking sequences, *Proc. Natl. Acad. Sci. USA* **74**:5363-5367.

Oka, H., 1930, Untersuchungen über die Speicheldrüse der Libellen, *Z. Morphol. Oekol. Tiere* **17**:275-301.

Okabe, K., Koyanagi, R., and Koga, K., 1975, RNA in the degenerating silk gland of *Bombyx mori*, *J. Insect Physiol.* **21**:1305-1309.

Okamoto, H., Ishikawa, E., and Suzuki, Y., 1982, Structure analysis of sericin genes: Homologies with fibroin gene in the 5' flanking nucleotide sequences, *J. Biol. Chem.* **257**:15192-15199.

Ono, M., 1942, Über die Zahl von den Seidendrüssenzellen der Seidenraupe, *Bull. Imp. Kagoshima Agric. Coll.* **14**:123-160.

Ono, M., 1951, Studies on the growth of silk gland cell in silkworm larvae, *Bull. Sericult. Exp. Stn.* **13**:247-303 (Japanese with English summary).

Perdrix-Gillot, S., 1979, DNA synthesis and endomitoses in the giant nuclei of the silk gland of *Bombyx mori*, *Biochimie* **61**:171-204.

Prudhomme, J. C., and Couble, P., 1979, The adaptation of the silkgland cell to the production of fibroin in *Bombyx mori* L., *Biochimie* **61**:215-227.

Rabinovitch, M., and Vugman, I., 1959, Autoradiographic observation on the silk glands of *Bombyx mori*, *J. Biochem. Cytol.* **6**:293-294.

Rasch, E. M., 1974, The DNA content of sperm and hemocyte nuclei of the silkworm *Bombyx mori* L., *Chromosoma* **45**:1-26.

Sasaki, S., 1977, Microtubules systems and microtubular crystals in the posterior silk gland cells of *Bombyx mori*, *J. Electron Microsc.* **26**:121-127.

Sasaki, S., and Nakagaki, I., 1980, Secretory mechanism of fibroin, a silk protein, in the posterior silk gland cells of *Bombyx mori*, *Membr. Biochem.* **3**:37-47.

Sasaki, S., and Tashiro, Y., 1976, Studies on the posterior silk gland of the silkworm *Bombyx mori*. VI. Distribution of microtubules in the posterior silk gland cells, *J. Cell Biol.* **71**:565-574.

Sasaki, S., Nakajima, E., Fujii-Kuriyama, Y., and Tashiro, Y., 1981, Intracellular transport and secretion of fibroin in the posterior silk gland of the silkworm *Bombyx mori*, *J. Cell Sci.* **50**:19-44.

Sasaki, S., Nakagaki, I., Imai, Y., and Kitano, M., 1982, Electrophysiological and electron microprobe studies of the posterior silk gland cells of *Bombyx mori*. In *Ultrastructure and Functioning of Insect Cells*, edited by H. Akai, R. C. King, and S. Morohoshi, pp. 153-156, Society of Insect Cells, Tokyo.

Sasaki, T., and Noda, H., 1973, Studies on silk fibroin of *Bombyx mori* directly extracted from the silk gland. I. Molecular weight determination in guanidine hydrochloride or urea solutions, *Biochim. Biophys. Acta* **310**:76-90.

Shibukawa, A., 1959, Studies on the silk-substance within the silk gland in the silkworm, *Bombyx mori* L., *Bull. Sericult. Exp. Stn.* **15**:383-401 (Japanese with English summary).

Shigematsu, H., and Takeshita, H., 1966, Distribution of fibroin in subcellular fractions of posterior silk gland of the silkworm, *Bombyx mori* L., *J. Biochem.* **59**:223-229.

Shigematsu, H., and Takeshita, H., 1968, Effects of gamma-irradiations on development of silk glands and silk formation of the silkworm, *Bombyx mori* L., *Bull. Sericult. Exp. Stn.* **23**:121-148 (Japanese with English summary).

Shigematsu, H., Takeshita, H., and Onodera, S., 1965, Effect of actinomycin and mitomycin on fibroin synthesis in the posterior silk gland of *Bombyx mori*, *J. Biochem.* **58**:604-606.

Shigematsu, H., Takeshita, H., and Onodera, S., 1966a, Protein synthesis by ribosomal fraction of the posterior silkgland of the silkworm, *Bombyx mori*, *Symp. Cell Chem.* **17**:125-130 (Japanese with English summary).

Shigematsu, H., Takeshita, H., and Onodera, S., 1966b, Incorporation of glycine-C^{14} into fibroin in subcellular fractions of the posterior silk gland of the silkworm, *Bombyx mori* L., *J. Biochem.* **60**:140-146.

Shigematsu, H., Kurata, K., and Takeshita, H., 1978, Nucleic acid accumulation of silk gland of *Bombyx mori* in relation to silk protein, *Comp. Biochem. Physiol. B* **61**:237-242.

Shimizu, M., 1941, Eine rontgenographische Untersuchung des Sericins, *Bull. Sericult. Exp. Stn.* **10**:441-474 (Japanese with German summary).

Shimizu, S., and Horiuchi, Y., 1952, Relation between the number of silk gland cells and the amount of silk secretion in the silkworm, *Bombyx mori* L., *J. Sericult. Sci. Jpn.* **21**:37-38 (Japanese with English summary).

Shimura, K., 1978, Synthesis of silk proteins. In *The Silkworm: An Important Laboratory Tool*, edited by Y. Tazima, pp. 189-212, Kodansha, Tokyo.

Shimura, K., Kikuchi, A., Ohtomo, K., Katagata, Y., and Hyodo, A., 1976, Studies on silk fibroin of *Bombyx mori*. I. Fractionation of fibroin prepared from the posterior silk gland, *J. Biochem.* **80**:693-702.

Shimura, K., Kikuchi, A., Katagata, Y., and Ohtomo, K., 1982, The occurrence of small component proteins in the cocoon fibroin of *Bombyx mori*, *J. Sericult. Sci. Jpn.* **51**:20-26.

Sprague, K. U., 1975, The *Bombyx mori* silk proteins: Characterization of large polypeptides, *Biochemistry* **14**:925-931.

Sprague, K. U., Hagenbuchle, O., and Zuniga, M. C., 1977, The nucleotide sequence of two silk gland alanine tRNAs: Implications for fibroin synthesis and for initiator tRNA structure, *Cell* **11**:561-570.

Sprague, K. U., Roth, M. B., Manning, R. F., and Gage, L. P., 1979, Alleles of the fibroin gene coding for proteins of different lengths, *Cell* **17**:407-413.

Suzuki, Y., 1975, Fibroin messenger RNA and its genes, *Adv. Biophys.* **8**:83-114.

Suzuki, Y., 1977, Differentiation of the silk gland: A model system for the study of differential gene action. In *Results and Problems in Cell Differentiation*, edited by W. Beerman, vol. 8, pp. 1-44, Springer-Verlag, Berlin.

Suzuki, Y., 1982, Studies on fibroin gene transcription by *in vitro* genetics. In *Embryonic Development*, Part A, *Genetic Aspects*, edited by M. M. Burger and R. Weber, pp. 305-325, Liss, New York.

Suzuki, Y., and Brown, D. D., 1972, Isolation and identification of the messenger RNA for silk fibroin from *Bombyx mori*, *J. Mol. Biol.* **63**:409-429.

Suzuki, Y., and Giza, P. E., 1976, Accentuated expression of silk fibroin genes *in vivo* and *in vitro*, *J. Mol. Biol.* **107**:183-206.

Suzuki, Y., and Ohshima, Y., 1978, Isolation and characterization of the silk fibroin gene with its flanking sequences, *Cold Spring Harbor Symp. Quant. Biol.* **17**:947-957.

Suzuki, Y., and Suzuki, E., 1974, Quantitative measurements of fibroin messenger RNA synthesis in the posterior silk gland of normal and mutant *Bombyx mori*, *J. Mol. Biol.* **88**:393-407.

Suzuki, Y., Gage, L. P., and Brown, D. D., 1972, The genes for silk fibroin in *Bombyx mori*, *J. Mol. Biol.* **70**:637-649.

Tamura, T., Akai, H., and Sakate, S., 1982, Ultrastructure of the silk glands and isolation of the fibroin mRNA from *Antheraea yamamai* and other non-mulberry silkworms. In *Ultrastructure and Functioning of Insect Cells*, edited by H. Akai, R. C. King, and S. Morohoshi, pp. 139-142, Society of Insect Cells, Tokyo.

Tanaka, Y., 1911, Studies on the anatomy and physiology of the silk-producing insects. 1. On the structure of the silk gland and silk formation in *Bombyx mori*, *Coll. Agric. Tohoku Imp. Univ.* **4**:145-172.

Tashiro, Y., and Otsuki, E., 1970a, Studies on the posterior silk gland of the silkworm *Bombyx mori*. IV. Ultracentrifugal analysis of native silk proteins, especially fibroin extracted from the middle silk gland of the mature silkworm, *J. Cell Biol.* **46**:1-16.

Tashiro, Y., and Otsuki, E., 1970b, Dissociation of native fibroin by sulfhydryl compounds, *Biochim. Biophys. Acta* **214**:265-271.

Tashiro, Y., Matsuura, S., Morimoto, T., and Nagata, S., 1968a, Extrusion of nuclear materials into the posterior silk glands of silkworm, *Bombyx mori*, *J. Cell Biol.* **36**:C5-C10.

Tashiro, Y., Moromoto, T., Matsuura, S., and Nagata, S., 1968b, Studies on the posterior silk gland of the silkworm, *Bombyx mori*. 1. Growth of posterior silk gland cells and biosynthesis of fibroin during the fifth larval instar, *J. Cell Biol.* **38**:574-588.

Tashiro, Y., Otsuki, E., and Shimadzu, T., 1972, Sedimentation analyses of native silk fibroin in urea and guanidine HCl, *Biochim. Biophys. Acta* **257**:198-209.

Tashiro, Y., Shimadzu, T., and Matsuura, S., 1976, Lysosomes and related structures in the posterior silk gland cells of *Bombyx mori*. 1. In late larval stadium, *Cell Struct. Funct.* **1**:205-222.

Tashiro, Y., Matsuura, S., and Sasaki, S., 1982, Roles of the cytoskeletal systems in the biosynthesis, intracellular transport and secretion of fibroin in the posterior silk gland cells of *Bombyx mori*. In *Ultrastructure and Functioning of Insect Cells*, edited by H. Akai, R. C. King, and S. Morohoshi, pp. 149-152, Society of Insect Cells, Tokyo.

Tazima, Y., 1964, *The Genetics of the Silkworm*, Logos Press, London.

Tazima, Y., Doira, H., and Akai, H., 1977, The domesticated silkmoth, *Bombyx mori*. In *Handbook of Genetics*, edited by R. C. King, vol. 3, pp. 63-124, Plenum Press, New York.

Tichomiroff, A., 1879, Über die Entwicklungs-geschichte des Seidenwurms, *Zool. Anz.* **2**:64-67.

Tsuda, M., and Suzuki, Y., 1981, Faithful transcription initiation of fibroin gene in a homologous cell-free system reveals an enhancing effect of 5' flanking sequence far upstream, *Cell* **27**:175-182.

Tsujimoto, Y., and Suzuki, Y., 1979a, Structural analysis of the fibroin gene at the 5' end and its surrounding regions, *Cell* **16**:425-436.

Tsujimoto, Y., and Suzuki, Y., 1979b, The DNA sequence of *Bombyx mori* fibroin gene including the 5' flanking, mRNA coding, entire intervening and fibroin coating protein coding regions, *Cell* **18**:591-600.

Tsujimoto, Y., Hirose, S., Tsuda, M., and Suzuki, Y., 1981, Promoter sequence of fibroin gene assigned by *in vitro* transcription system, *Proc. Natl. Acad. Sci. USA* **78**:4838-4842.

Voigt, W.-H., 1965a, Zur Funktionellen der Fibroin-und Sericin-Sekretion der Seidendrüse von *Bombyx mori* L. I. Der proximal Abschnitt der Seidendrüse, *Z. Zellforsch. Mikrosk. Anat.* **66**:548-570.

Voigt, W.-H., 1965b, Zur Funktionellen Morphologie der Fibroin-und Sericin-Sekretion der Seidendrüse von *Bombyx mori* L. II. Mitteilung der Mediale Abschnitt der Seidendrüse, *Z. Zellforsch. Mikrosk. Anat.* **66**:571-582.

Waku, Y., and Sumimoto, K., 1974, Ultrastructure of Lyonet's gland in the silkworm (*Bombyx mori* L.), *J. Morphol.* **132**:165-186.

Yamanouchi, M., 1922, Morphologische Beobachtung über die Seidensekretion bei der Seidenraupe, *J. Coll. Agric. Sapporo* **10**:1-49.

Yokoyama, T., 1951, Studies on the cocoon-formation of the silkworm, *Bombyx mori* L., *Bull. Sericult. Exp. Stn.* **13**:183-246 (Japanese with English summary).

Zuniga, M. C., and Steitz, J. A., 1977, The nucleotide sequence of a major glycine transfer RNA from the posterior silk gland of *Bombyx mori* L., *Nucleic Acids Res.* **4**:4175-4196.

10

Structure and Development of Male Accessory Glands in Insects

GEORGE M. HAPP

1. Introduction

Reproductive accessory glands of male insects facilitate transfer of sperm to the females. The heterogeneous products of these glands include both the seminal fluids and paraseminal structures. The secretions vary widely in physical properties, biochemical composition, and physiological function. Secretions of low viscosity bathe the sperm. Within that fluid mixture are components which promote sperm maturation, which provide nourishment for stored sperm, which contribute nutrients for investment into egg yolk, or which modulate the behavior or the physiology of the female (Leopold, 1976). Secretions of higher viscosity solidify to become copulatory plugs (Bishop, 1920; Bairate and Perotti, 1970) or complex spermatophores (Tuzet, 1977). These organized secretory masses are composed of several distinct zones that form as the glandular products flow over one another in extracellular space.

In the past few decades, most research on the reproductive physiology of insects has concentrated on the phenomenology of reproductive maturation and the endocrine control of that maturation in the female sex (e.g., Engelmann, 1970; Kafatos *et al.*, 1977; Hagedorn and Kunkel, 1979; Telfer *et al.*, 1982). Although much less studied, male reproductive physiology is no less complex (e.g., Davey, 1965, 1967). At the ultrastructural level, we have many stunning descriptions of insect spermatozoa (Baccetti, 1972). The processes of spermiogenesis and spermatogenesis provide superb examples of the profound morphological changes that can occur during terminal differentiation of specialized cells (Phillips, 1974). This differentiative sequence takes place in preadult

---

GEORGE M. HAPP • Department of Zoology, University of Vermont, Burlington, Vermont 05405, USA.

stages for most insects. Spermatogenesis is modulated by the rise and fall of the concentrations of juvenile hormones and ecdysteroids that simultaneously regulate metamorphic growth (Dumser, 1980).

By the time an insect has reached the adult stage, there are so many spermatozoa present that numbers alone rarely constitute limiting factors for population growth. Delivery is another matter. Delivery of sperm to the female requires a vehicle which is produced by the accessory glands. The nature of that vehicle and its elaboration from the glands are poorly understood in most insect species. For males which mate repeatedly, the glands must go through recurrent secretory cycles to produce the charge of semen and paraseminal material for each copulation. The maturation of accessory glands is regulated by endocrine and neuroendocrine factors in almost all groups. These glands offer attractive models in which to study hormone action and secretory kinetics.

2. Accessory Glands and Their Secretions

In insects as in mammals (Mann, 1964), semen and its accompaniments are an aggregation of biochemical constituents derived form morphologically diverse and complex glands. Some of these glandular products are proteins; included among them are structural components of the wall of the spermatophore, enzymes of uncertain function such as esterase 6 (Gilbert, 1982), and molecules of known physiological significance such as mating refusal substances of *Drosophila* (Garcia-Bellido, 1964). Smaller molecules, including sugars and lipids, are reported in the semen of many phyla (Mann, 1964), including such insects as honeybees (Blum et al., 1962, 1967). The enormous diversity of reproductive strategies among species and the biochemical heterogeneity within the semen of every species are widely acknowledged. We are quite ignorant of the biochemical nature and functional significance of most seminal constituents.

The male reproductive tract is a muscular tube which runs from the testes to the gonopore. In insects, the proximal portion of the tract is mesodermal in origin while the distal portion is ectodermal. In all species, some secretions are produced by cells along the wall of the tract. In many groups, there are also distinct glands attached to the reproductive tract. These accessory glands differ in size, shape, number, anatomical placement, and embryological origin (Adiyodi and Adiyodi, 1975; Grassé, 1977).

The accessory secretions may mingle with the sperm, may precede the sperm mass, may enclose it, or may follow along afterward. Seminal vesicles, phallic glands, mushroom glands, conglobate glands, accessory glands, secretory segments of ejaculatory ducts, paragonia, ectadenia, mesadenia, simplex, and duplex are just a few of the terms which describe the organs that produce seminal or paraseminal secretions. I will employ the common terminology for each group, with the general caveat that the use of a similar trivial name does not necessarily connote homology. I will not attempt to review the morphological peculiarities of various groups nor to consider in detail the evolutionary or embryological origins of accessory glands. Rather, I will emphasize common

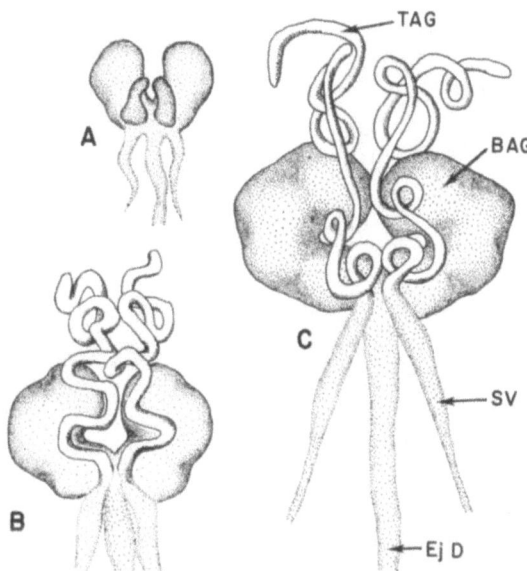

Figure 1. The accessory glands of *Tenebrio molitor* in the newly ecdysed pupa (A), the newly eclosed adult (B), and the reproductively mature adult (C). BAG, bean-shaped accessory gland; Ej D, ejacutory duct; SV, seminal vesicle; TAG, tubular accessory gland.

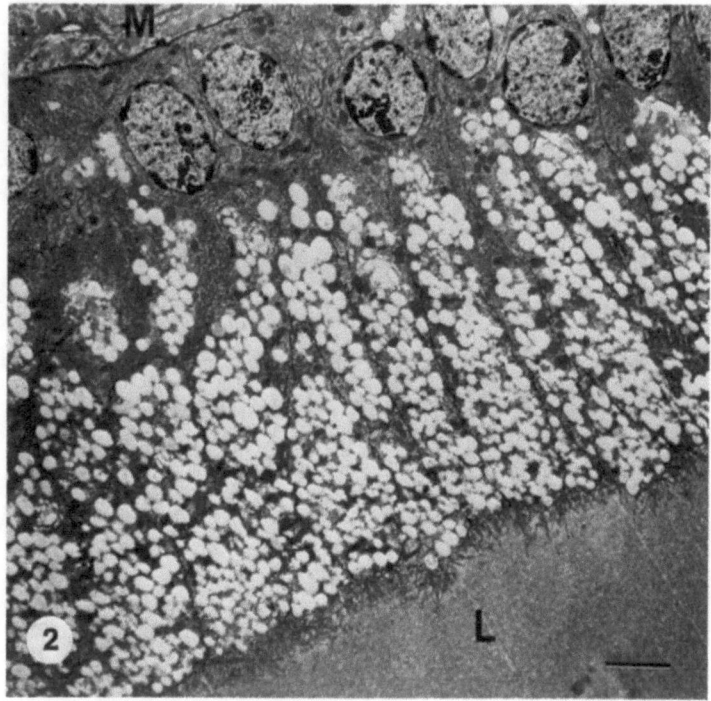

Figure 2. The secretory epithelium in TAG of a mature male. The muscle layer (M) is at the upper left and the lumen (L) of the gland at lower right. Note the basally situated nuclei and the many large secretory vesicles. (Bar = 5 μm.)

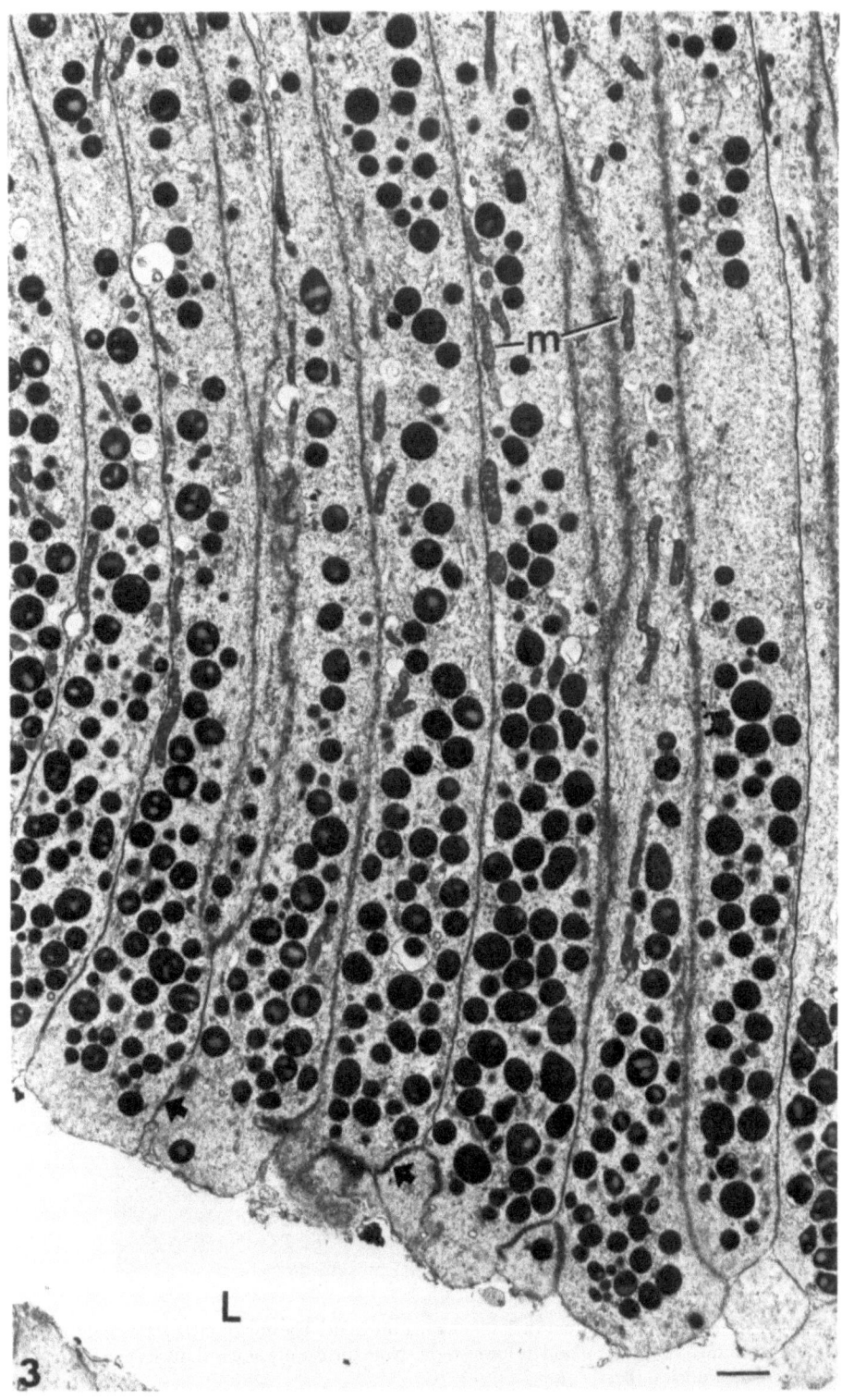

structural features of secretory epithelia, some characteristics of their products, and the process of their differentiation.

The morphology of accessory glands will be illustrated by micrographs of the tubular accessory glands (TAGs) (Figure 2) and of the bean-shaped accessory glands (BAGs) (Figure 3) of mealworm beetles (*Tenebrio molitor* L.) which have been studied over the past decade in my laboratory. For reference, the shapes and positions of these glands in the male reproductive tracts in pupae and adults are shown in Figure 1.

3. Muscular Coats and Basement Membranes

Most accessory glands consist of a secretory epithelium surrounded by a muscular sheath. The exceptions are certain ectodermal accessory glands, such as the phallic gland of cockroaches, which export their products via cuticular plumbing systems and lack any clear muscle coat. When present, the muscles typically have long sarcomeres (6-10 μm) and a poorly developed T system. The sarcoplasm contains a few free ribosomes and scattered profiles of ER. Mitochondria are usually tubular and often aligned in parallel with the muscle filaments. Within any one layer, the muscle fibers are tightly attached to each other (end-to-end) by the intercellular matrix. For most glands, there are two or more layers of muscle cells which spiral obliquely to the right or to the left, around the generally cylindrical gland. The thickness of the muscle coat seems to be correlated with the consistency of the product of each gland. In *Tenebrio*, the tubular gland secretes a fluid product and is surrounded by two or three layers of muscle cells (Figure 4) while the bean-shaped gland produces a semisolid plastic plug which is forced out by the contraction of six to eight layers of muscle cells (Figure 5).

Basement membranes surround the muscle layer and also coat the basal surface of the secretory cells (Figure 4). These extracellular matrices appear as a loose fibrous network which fills in many of the interstices between muscle cells (Figure 4). Tracheoles, enclosed in their cellular coats, run between the muscle cells and deep into the secretory epithelium. The nuclei of the tracheolar end cells may sometimes be found squeezed between the secretory cells.

4. The Secretory Epithelia

The secretory epithelia elaborate a battery of seminal and paraseminal components. In female insects, massive amounts of vitellogenic proteins are manufactured by the fat body, carried by the hemolymph, and transported by follicular cells to form the yolk (Telfer, *et al.*, 1982). With rare exceptions (Friedel and Gillot, 1976), secretory proteins are not made in the fat body of the male and

---

Figure 3. The apical zone of several cells of the secretory epithelium in the BAG. The many secretory vesicles (type 5) are packed above the zone of apical desmosomes (arrows). m, mitochondrion; L, lumen. (Bar = 1 μm.)

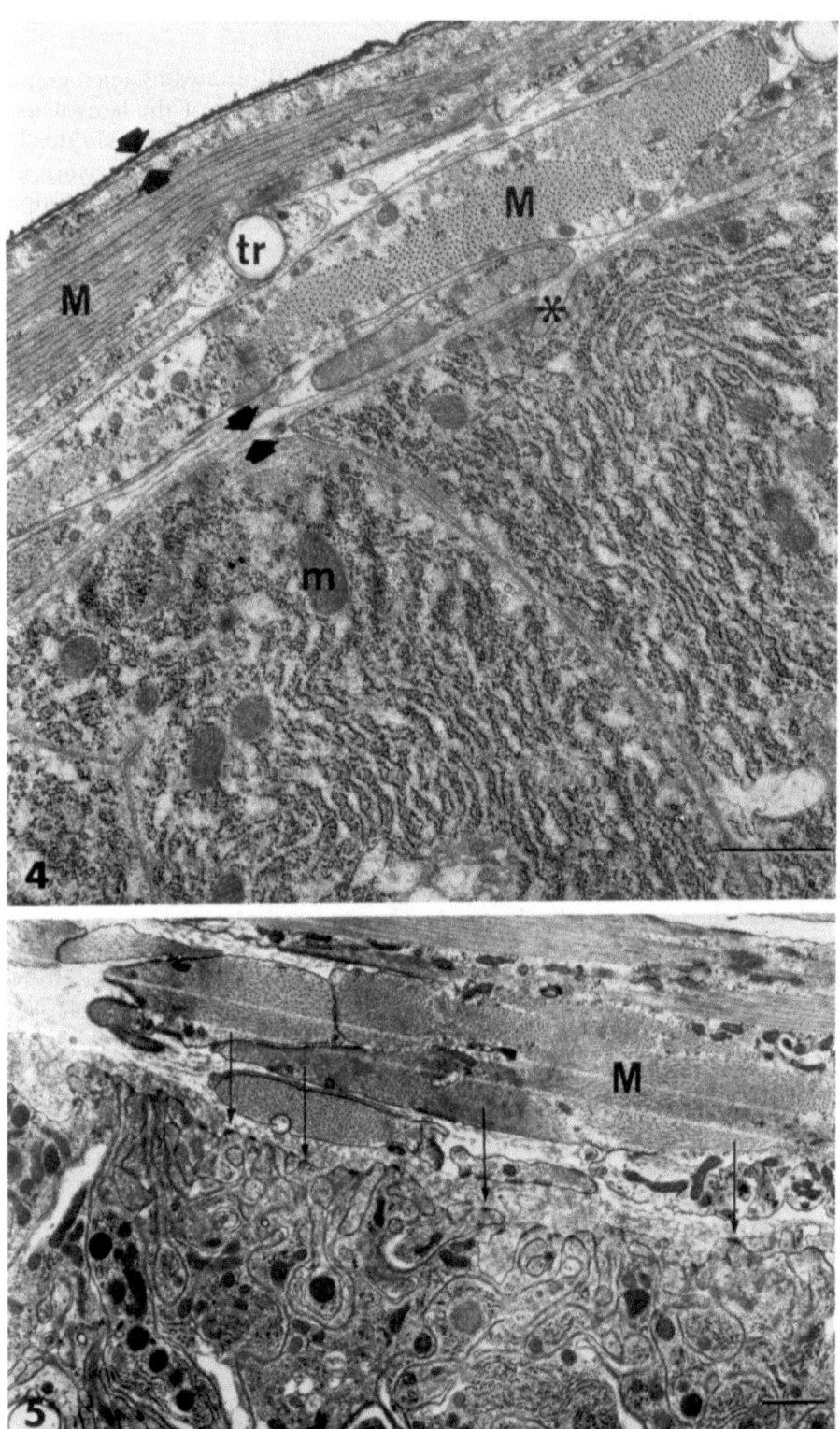

taken up by accessory glands. Secreted proteins appear to be manufactured *de novo* within the epithelium. The relative unimportance of other organs is illustrated by the fact that the accessory glands of *Tenebrio* (Happ et al., 1977) and *Rhodnius prolixus* (Barker and Davey, 1982) continue to make cell-specific proteins during organ culture *in vitro*.

Secretory epithelia of accessory glands, like most insect epithelia, are histologically simple, i.e., they are monolayers (Figure 2). There have been reports of stratification in some epithelia of accessory glands (Ohdiambo, 1969), but we suspect that the multilayered appearance is an artifact due to the plane of section and interdigitation of the cells. The individual cells may be quite long; in the bean-shaped gland of *Tenebrio*, single cells reach 300–500 µm from the basement membrane to the lumen of the gland (Dailey et al., 1980).

Each secretory cell is an independent factory for manufacture, storage, and export of secretory products. At its basal surface, precursors are absorbed from the hemolymph. Throughout most of the length of the cell, rough or smooth ER and Golgi zones are engaged in manufacture, packaging, and transport of products in membrane-bound vesicles. At the apex of each cell, secretions are expelled into the lumen.

As the male becomes reproductively mature, the lumina of the ducts and the accessory glands fill with products. During copulation, a large portion of these secreted materials is lost in the ejaculate, and subsequently they must be replenished from the secretory cells. Within a short period after mating (usually 10 hr at the most), the lumen is recharged with new products. Thus, the secretory process is discontinuous; it proceeds in episodes which are triggered by the occasional emptying of the lumina of the tract and the glands.

4.1. Intercellular Junctions

The cells of the secretory epithelium are appressed tightly to one another along their lateral surfaces. Two intercellular links, septate junctions and apical desmosomes, are major factors in maintaining the integrity of the epithelial sheet in most accessory glands.

In the bean-shaped glands of *Tenebrio*, the apical desmosomes form a belt, 0.6–1 µm in width, around each secretory cell (Figures 4, 17–19). Adjacent cells are separated by a 23- to 25-nm zone which contains flocculent extracellular material that is not well resolved in our transmission micrographs (Figure 12). Dense plaques, 30–40 nm in thickness, lie on the inner cytoplasmic surface (Figure 12). As in other insect desmosomes (Lane and Skaer, 1980), these plaques lack tonofilaments which run toward the center of the cell. Microtubules, which

Figure 4. The muscle coat and basement membranes (between arrows) surround the secretory epithelium of the TAG. The filaments in the two muscle cells (M) run at right angles to each other. A process of the inner basement membrane (asterisk) indents the basal edge of one secretory cell. The secretory cells contain scattered mitochondria (m) and much granular ER. tr, tracheole. (Bar = 1 µm.)

Figure 5. A portion of the muscle coat and the basal zone of secretory cells of the mature BAG. Note the infoldings of the basal plasma membrane and the hemidesmosome plaques (arrows) which anchor the cells to the inner basement membrane. (Bar = 1 µm.)

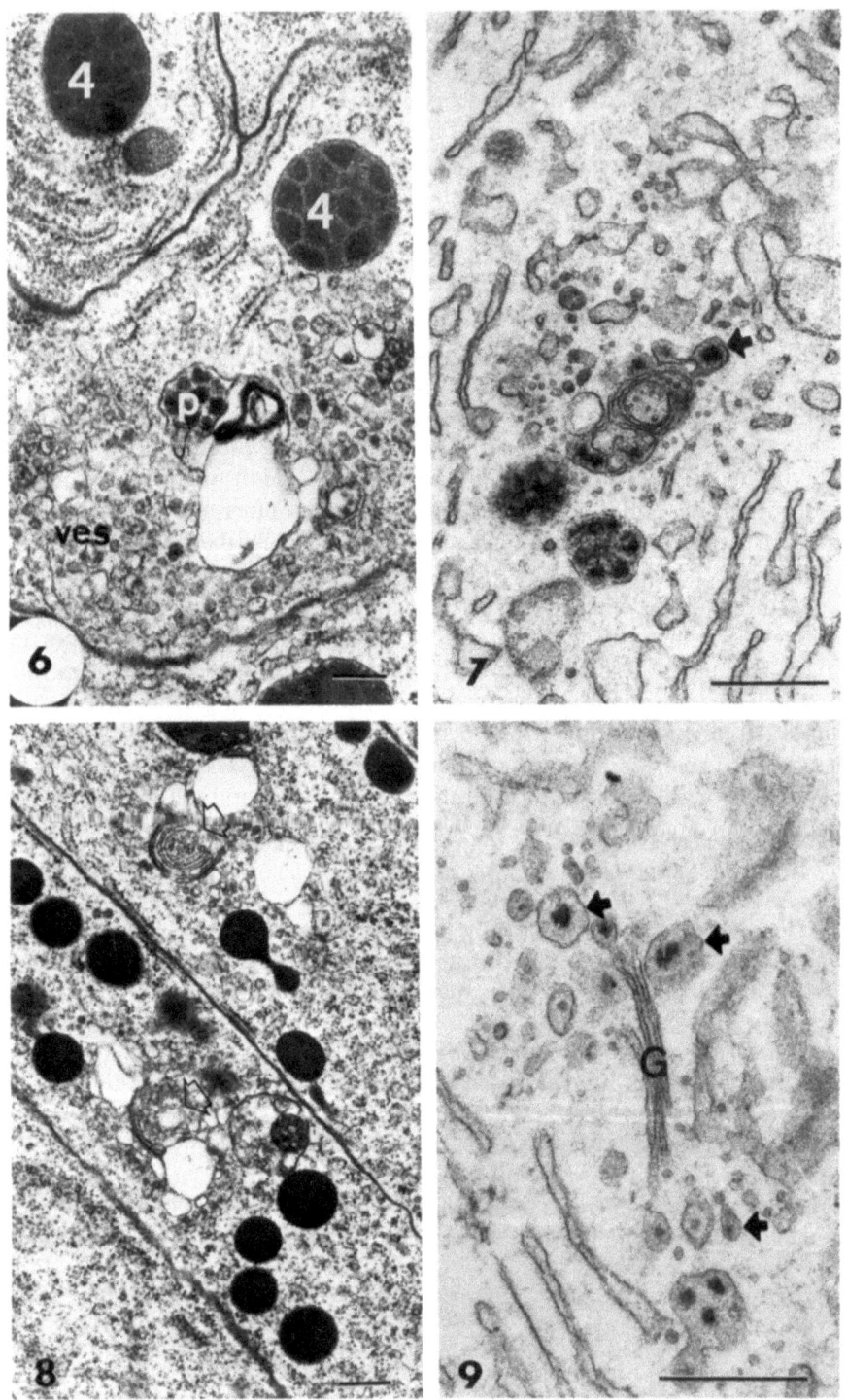

Figure 6. The Golgi region of cell type 4 of the mature BAG. Small vesicles (ves) are found throughout the Golgi zone. A precursor (p) of the definitive secretory vesicle (4) is seen at the edge of the Golgi. (Bar = 0.2 μm.)

Figure 7. Golgi zone of a secretory cell of the mature BAG. A precursor granule is indicated by the arrow. Potassium permanganate fixation. (Bar = 1 μm.)

run parallel to the long axis of the secretory cell, generally end at the level of the apical band (Figure 20).

Above the apical band, cells are held together by septate junctions. The septa are more difficult to resolve (Figure 12) as is typical of the smooth septate or continuous junctions (Lane and Skaer, 1980). The junctions usually extend 50 μm or so above the apical desmosomes. In the tall cells within thick epithelia, there is a zone above the desmosomes where the membranes of adjacent cells are separated into no more than 40-50 μm, and wisps of material (extracellular matrix?) can be seen in the gap. Finally, the cells are anchored to the basement membrane by hemidesmosome plaques (Figure 5). These structures keep the fabric of the epithelium intact during powerful contractions that occur during ejaculation.

4.2 Absorption of Precursors

Precursor molecules must percolate through basement membranes and the interstices between the muscles to reach the secretory cells. The movement and incorporation are rapid. Within 10 min of injection of tritiated amino acids into the hemocoel, we have detected radioactive secretory proteins in the accessory glands of *Tenebrio*. In the basal regions of secretory cells in the utriculi majores of the mushroom-shaped gland of male *Periplaneta americana* (Adiyodi and Adiyodi, 1974), in the simplex of *Calopodes ethlius* (Lai Fook, 1982b), in the ejaculatory duct of *Stomoxys calcitrans* (Meola, 1982), and sometimes in the basal edges of the bean-shaped gland of *Tenebrio* (Figure 5), the plasma membrane may be deeply infolded, presumably to increase the absorptive surface. No coated pits or coated vesicles have been reported at the basal surfaces of accessory glands. Coated pits usually indicate receptor-mediated endocytosis of large molecules (Steinman *et al.*, 1983). The absence of coated pits or vesicles is consistent with absorption of small molecules either by active transport, facilitated diffusion, or passive diffusion or by fluid-phase endocytosis (pinocytosis). Fluid-phase endocytosis is difficult to capture in an electron micrograph since the half-life of endocytotic vesicles is probably no more than a few seconds (Steinman *et al.*, 1983).

4.3. Biosynthetic Machinery

In active secretory cells, the cytoplasm is packed with profiles of ER, Golgi complexes, microtubules, and secretory vesicles. The membranes of the ER may be smooth (Figure 13) or ribosome-studded (Figure 10). Secretory granules accumulate in the apical half of the cell, sometimes to the exclusion of other organelles, with the result that most of the cytomembranes become confined to the basal portion of the cell (Figures 2, 4).

Figure 8. Golgi zones of a secretory cell in a BAG, 39 hr after adult ecdysis. Secretory vesicles are beginning to form. Note the central precursor granule (hollow arrow) in each Golgi zone. (Bar = 1 μm.)

Figure 9. Golgi zones of a secretory cell of a mature BAG. Note the three flattened saccules (G) and the inflations with dense central condensations (arrows). Potassium permanganate fixation. (Bar = 1 μm.)

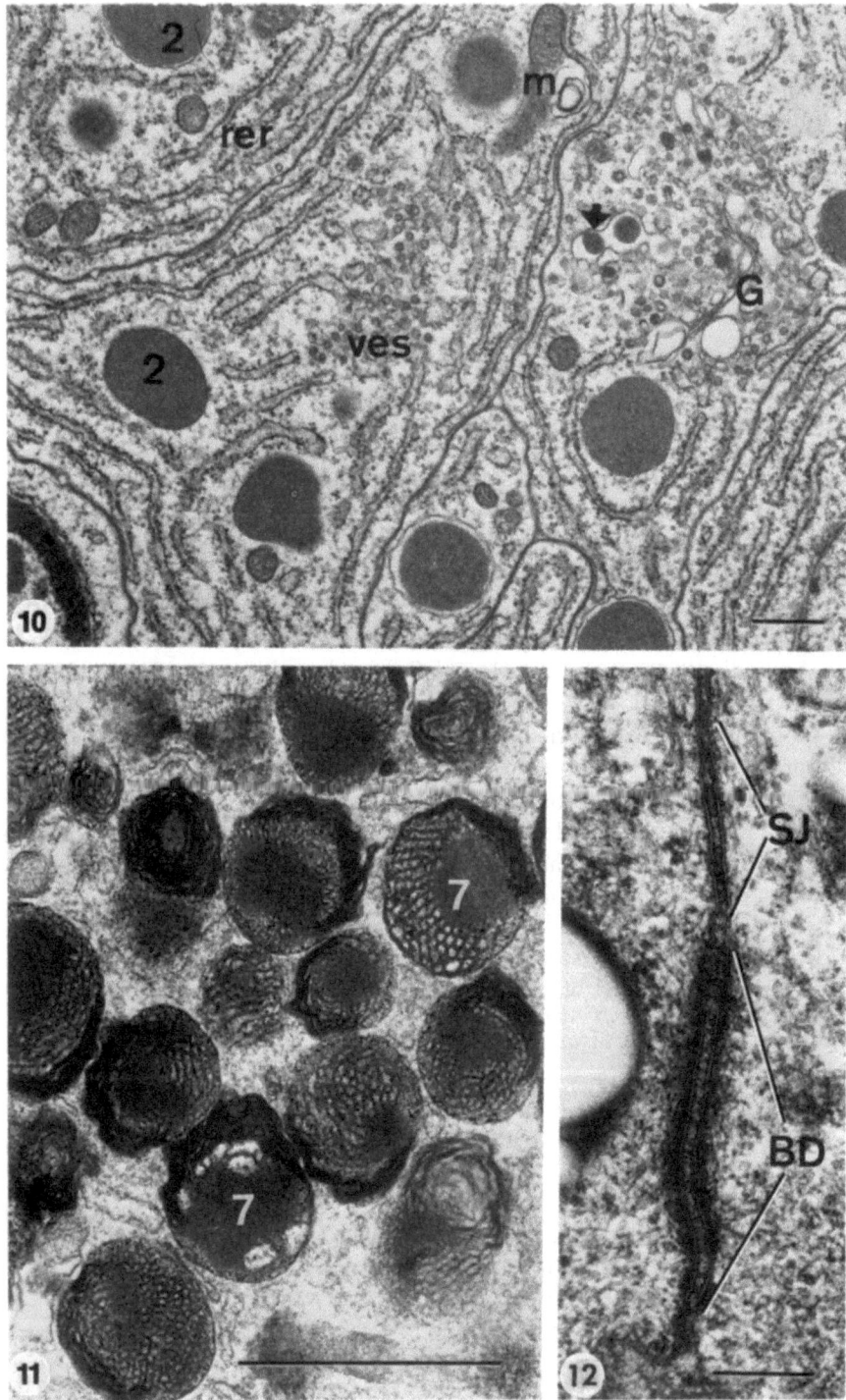

Golgi complexes lie among the cisternae. In these glands as in many other insect tissues, it is often difficult to resolve the individual components of the Golgi apparatus. Smooth-surfaced irregular vesicles (100–200 μm in diameter) appear to be budding off the membranes of the ER (Figure 10) in the vicinity of large numbers of small (50–70 μm in diameter) fuzzy vesicles which may be coated with clathrin (Pearse and Bretscher, 1981). These coated vesicles are present in large numbers throughout the Golgi zones (Figure 10). Such coated vesicles have been reported near the Golgi of the vas deferens of a rat (Friend and Fahrquhar, 1967). We presume that in the BAG, one or both of these classes of membranous sacs are transition vesicles, moving proteinaceous products from the RER to the forming face (the convex surface) of the Golgi. In favorable sections, the Golgi apparatus has at its center three (occasionally four) saccules (Figure 9) of smooth membranes which curve around sharply to form a deep cut at the maturing face (Figure 8).

The edges of the Golgi saccules are inflated. Within swollen tips of many saccules are electron-dense granules surrounded by an electron-transparent zone (Figures 7, 9, 10). We presume that these dense-core granules are secretory precursors. Often, the precursor granule lies at the center of the concave face of the membrane stack (Figure 8). A similar concentric architecture of the Golgi has been reported in the seminal vesicle of *Locusta migratoria migratorioides* (Cantacuzène, 1972). The central precursor granule(s) apparently grows in size, as if it were a nucleating site, until the electron-lucent cortex is filled with dense product. The secretory vesicles of the cells vary from homogeneous masses to multiparticle bodies. When the definitive secretory vesicle is of a complex type (see Figure 22), a cluster of presecretory granules may collect together in the membrane-bound cavity (Figure 6).

The membranes of the secretory vesicles may have surface specializations. In the tubular gland of *Acanthoscelides obtectus* (a bruchid beetle), there is a fringe of short filaments around the newly formed secretory granule (Cassier and Huignard, 1979). Surface specializations are also found in the BAG of *Tenebrio*, and the membrane may be modified as the granule matures. When a type 3 vesicle is first formed, it has scattered plaques of electron-dense material in its cortical zone (Figure 13), but by the time it moves away form the Golgi zone, it has acquired an unusual and characteristic cortex (Figure 14). In type 5 vesicles of the BAG, the newly formed vesicles have a surface fringe and approach no closer than about 100 nm to each other (Figure 15), but as these type 5 vesicles move toward the apical cell surface, they become more tightly packed together (Figure 16), perhaps due to modification of that surface coat.

Large quantities of product are present within secretory cells of active

⟵―――――――――――――――――――――――――――――――――――――

Figure 10. Type 2 cells of mature BAG. Note the saccules of the Golgi zone and the precursor granules (arrow). The dense cluster of small coated vesicles (ves) appears to be associated with inflations of cisternae of the RER. m, mitochondrion; 2, type 2 secretory vesicle. (Bar = 1 μm.)
Figure 11. Type 7 secretory vesicles of the BAG. (Bar = 1 μm.)
Figure 12. Smooth septate junction (SJ) and belt desmosome (BD) near the apical edge of the secretory epithelium of the BAG. Note the fibrous mat on the cytoplasmic face of the desmosome and the coarse granular material in the intercellular space. (Bar = 0.2 μm.)

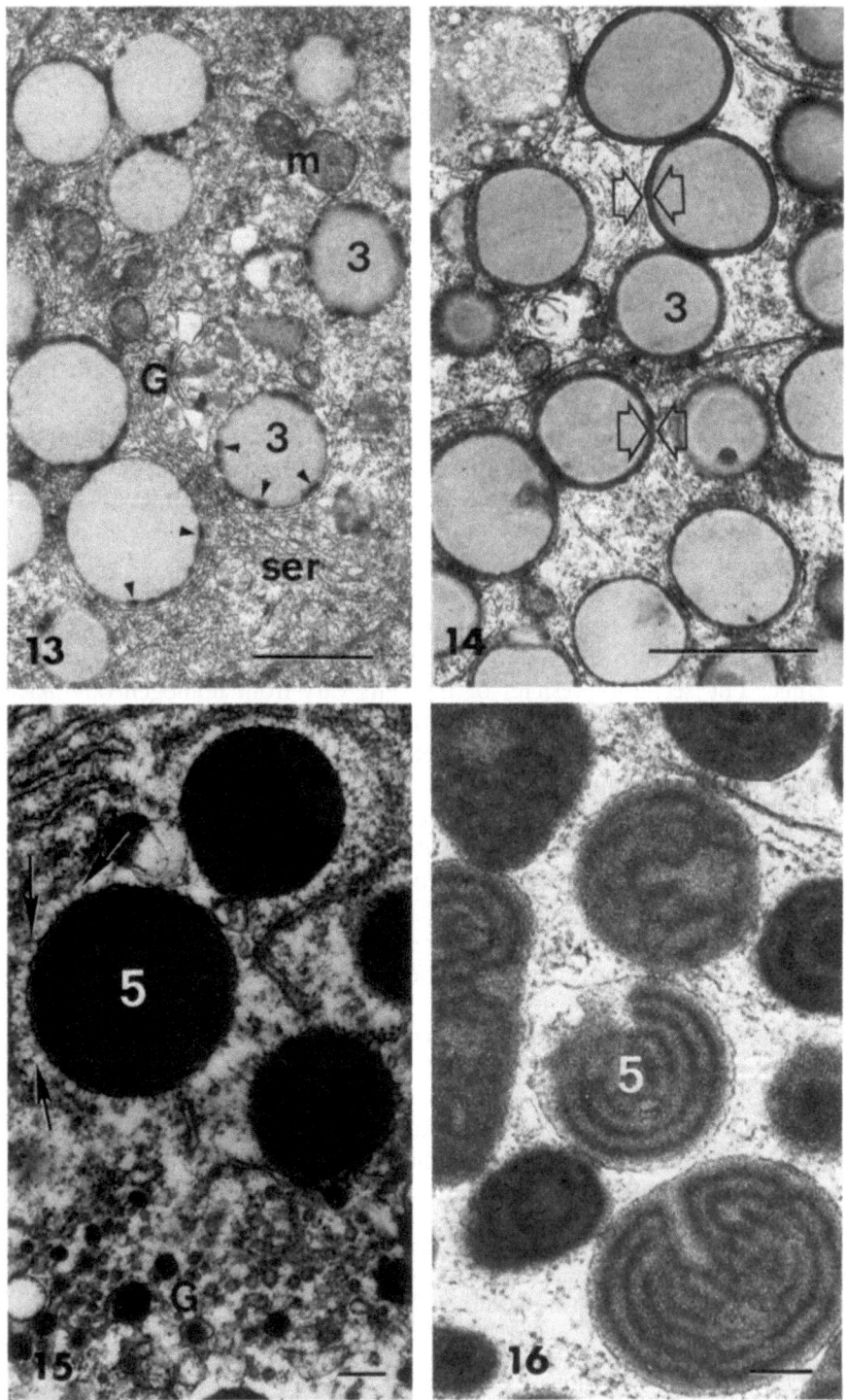

accessory glands. In cell types 1 and 2 of *Schistocerca gregaria,* Ohdiambo (1969) reports that the cisternae of the RER are much dilated but there are relatively few Golgi-derived vesicles. In these cells, storage appears to be in the cisternae, upstream from the Golgi. In most other secretory cells, storage is downstream from the Golgi. The individual secretory vesicles may remain discrete and collect in large numbers, as in the granular cells of *Anagasta kuhniella* (formerly *Ephestia*) (Reimann and Thorson, 1979b) and in the BAG of *Tenebrio* (Figure 4) (Dailey *et al.,* 1980). In the TAG, the smooth-surfaced vesicles derived from the Golgi do not remain distinct. They pack tightly together and appear to fuse forming large vesicles in the apical cytoplasm (Figure 17) (Gadzama *et al.,* 1977). Such large fusion vesicles are also seen in the middle segment of the seminal vesicle of *Locusta* (Cantacuzène, 1972).

4.4. Export of Product

As the glands mature, their lumina become filled with product. During copulation, most of the products are lost in the ejaculate, and subsequently they must be replenished from the secretory cells. After mating, the lumen is rapidly recharged with new products.

All variety of secretory devices—merocrine, holocine, and apocine—are found in accessory glands and other secretory tissues of the male tract. The paragonia (accessory glands) of *Drosophila melanogaster* liberate their product by holocrine secretion and dramatic cell death (Perotti, 1971). Holocine secretion has also been reported in the secretory cells of the ejaculatory duct of *Musca domestica* (Reimann, 1973). But in most glands the mechanisms are either merocrine or apocrine.

Merocrine secretion is seen in the vasa deferentia of *Calpodes* (Lai-Fook, 1982a) and in the TAGs of *Tenebrio* (Figure 17). The apical surfaces of these cells support irregular folds and microvilli which we presume are generated by the continual addition of membrane from the vesicles.

In the BAGs, clusters of microtubules run longitudinally along the outer zones of the cell and surround masses of secretory vesicles which become tightly packed just above the level of the desmosomal belt (Figure 20). There is a network of intracellular filaments at that level, which often seem to run across the cell, and the desmosomal membrane is frequently infolded (Figures 20, 21). A broad apical cellular process, which is poor in mitochondria and other

←

Figure 13. Newly formed secretory vesicles (3) near the Golgi (G) of a type 3 secretory cell of the BAG. Dense irregular plaques (arrowheads) are scattered about at the surface of the vesicle. m, mitochondrion. (Bar = 1 μm.)

Figure 14. Mature secretory vesicles (3) of type 3 cells of the BAG. Note the thick peripheral structure, between the hollow arrows. (Bar = 1 μm.)

Figure 15. Newly formed secretory vesicles (5) near the Golgi (G) of a type 5 secretory cell of the BAG. The arrows indicate a surface coat on each vesicle. The print has been overexposed to make the surface coat more obvious. (Bar = 0.2 μm.)

Figure 16. Mature secretory vesicles of type 5 cells contain whorls or concentric spheres of secretion of moderate electron density. The surface coat has been lost, and the vesicles are packed tightly together. (Bar = 0.2 μm.)

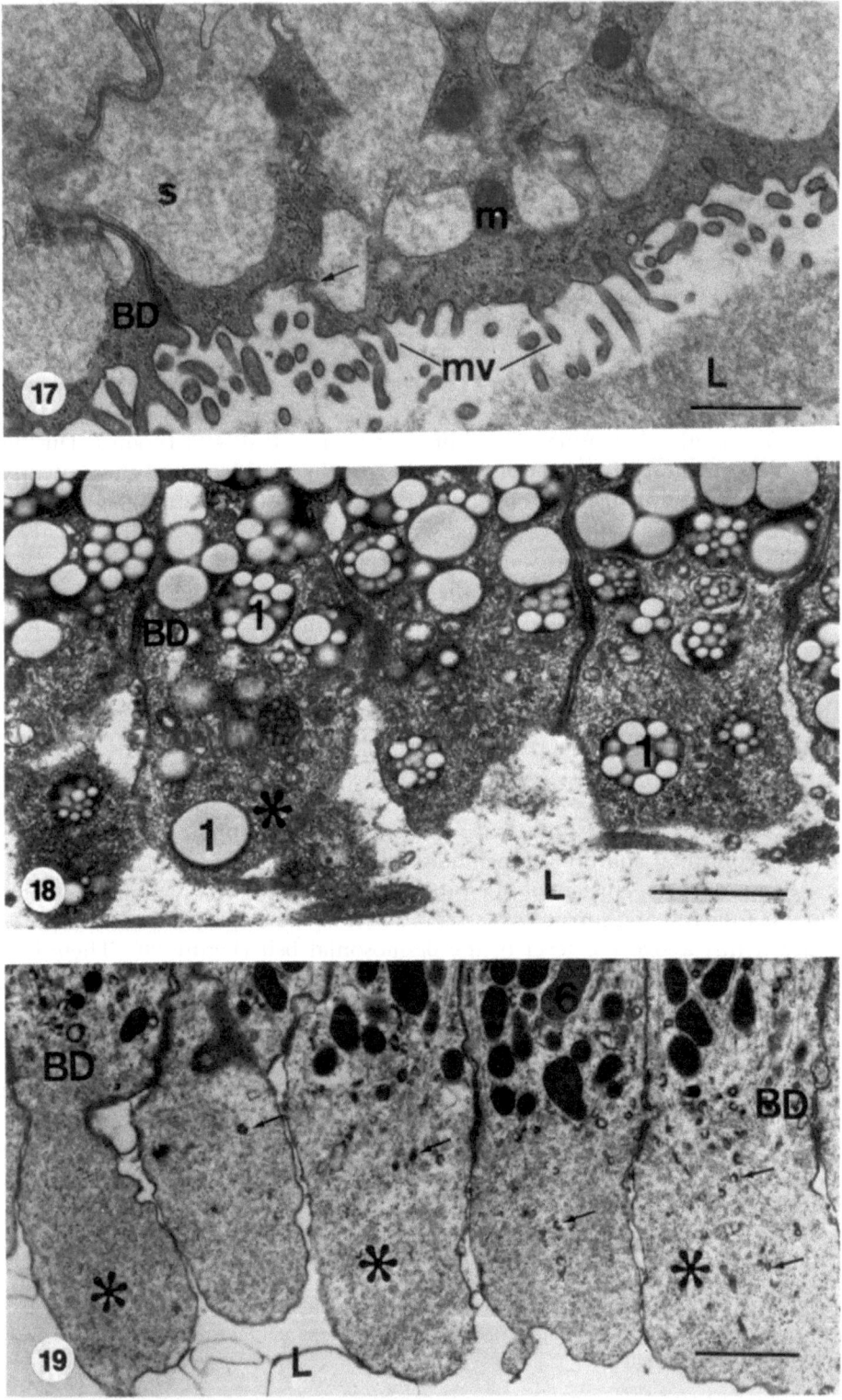

organelles, projects beyond the desmosomal belt (Figures 19, 20). In immature glands where secretion is at a low level, microtubules can be seen in that process. In this young material, the apical plasma membrane may invaginate to form a sort of funnel which reaches back up to the desmosomal level and apparently carries products to the lumen (Figure 21). In older tissues, one or more secretory vesicles can be seen in the apical process (Figures 3, 18), and in favorable sections we have seen the vesicles approaching the membrane as if about to empty their contents by exocytosis (Figure 18).

I suspect that traffic in secretory vesicles is regulated at the level of the desmosomal belt. Microtubules stream down to that belt and secretory vesicles accumulate above it. The infolded plasma membrane and network of transverse filaments might be viewed as a gate which limits the access of secretory vesicles to the apical tip.

One frequently sees coated vesicles in these apical zones (Ohdiambo, 1969; Lai-Fook, 1982a; Figures 19, 21). These vesicles may reflect recovery of membrane for transport upstream to the Golgi zones for subsequent recycling (Oliver, 1982).

In apocrine secretion, cells shed their apical process and thus export secretory vesicles. Such a mode of secretion has been reported in accessory glands of many species, e.g., gland 2 of *Schistocerca* (Ohdiambo, 1969), the seminal vesicle of castrate *Locusta* (Cantacuzène, 1972), the accessory glands of *Leptinotarsa decemlineata* (De Loof and Lagasse, 1972), and the vasa deferentia of *Anagasta* (Reimann and Thorson, 1976a).

An extreme modification for apocrine secretion is seen in the foliate cells of the simplex (ejaculatory duct) and accessory glands of *Anagasta* (Reimann and Thorson, 1979a,b) and in similar cells of *Calpodes* (Lai-Fook, 1982b). The foliate cells have elongate apical extensions, which extend well out into the lumen of the gland. These extensions arise at a narrow neck or petiole. Large numbers of microtubules run from the basal portion of the cell to end at the petiole. Distal to the petiole, the apical process lacks the ER and Golgi zones, but contains many whorls of membranes and very few microtubules and mitochondria. The apical processes are shed into the lumen and become a part of the ejaculate, and then they are rapidly regenerated within the next 7-8 hr (Reimann and Thornson, 1979a,b). Even in this special case of apocrine secretion, the desmosomal belt is the demarcation between the basal portion of the cell and the disposable apical zone.

←──────────────────────────────

Figure 17. The apical edges of two secretory cells of the TAG. The large irregular secretory vesicles (s) readily fuse with one another. The small arrow may indicate a vesicle approaching an invagination of the plasma membrane. mv, microvilli; L, lumen; BD, belt desmosome. (Bar = 1 µm.)

Figure 18. The apical edge of type 1 secretory cells of the BAG. Apical processes (asterisk) extend below the zone of belt desmosomes (BD). Type 1 secretory granules are packed tightly above the zone of desmosomes, and a few have entered the apical processes. These vesicles contain one or more electron-transparent "bubbles." (Bar = 1 µm.)

Figure 19. The apical zone of the secretory epithelium of type 6 cells of the BAG. The apical processes (asterisks) are devoid of the secretory vesicles (6) but contain small dense membrane profiles that may be coated vesicles (arrows). (Bar = 1 µm.)

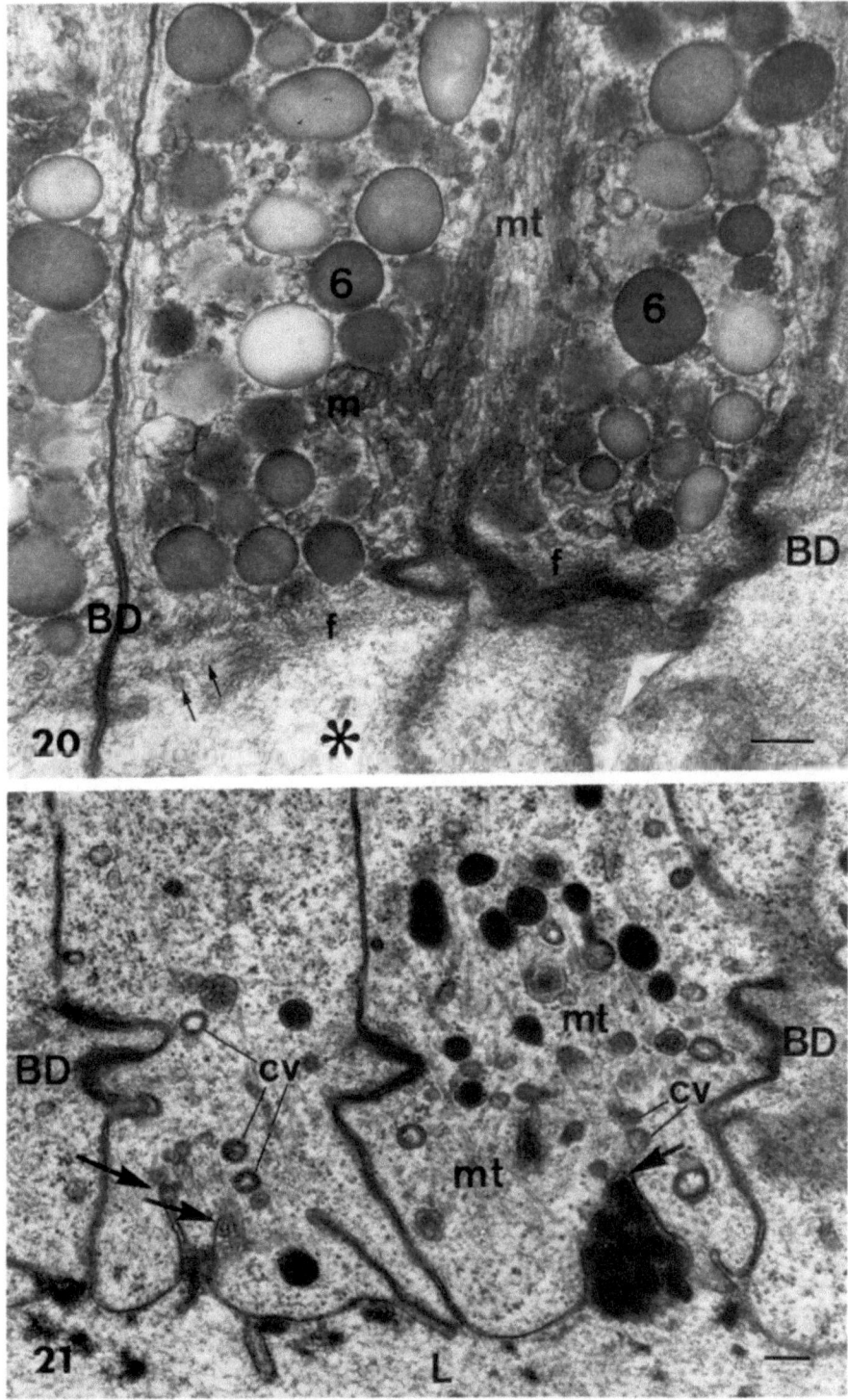

5. Semisolid Secretory Products

In addition to soluble secretions, accessory glands make spermatophores. Like the cuticle and the chorion, the wall of the spermatophore is an ordered structure formed in the extracellular space. The cuticle is a multilayered structure which is laid down by epidermal cells. As the epidermal cell adjusts the secretory blend over time, the various layers of the cuticle are deposited in succession. For production of the chorion, the follicle cells play the analogous part, temporally shifting their patterns of biosynthesis to lay down the components of the eggshell in serial order. But for a spermatophore, the various layers and zones are produced by several cell types, arranged in a parallel array of cell types which simultaneously secrete a corresponding array of products. Each of the cell types contributes to a particular layer or zone of the spermatophore.

5.1. Organized Secretion Masses

Formation of the spermatophore requires that the various secretory products be delivered or mixed in a fixed temporal order. The general strategy is as follows. Several cell types produce the heterogeneous secretory products. Each set of products is stored in a discrete site in the male tract. At mating, several precursors are mixed together in a fixed order. The order of export of the components is partly dictated by the anatomy of the male tract. During and after incorporation into the spermatophore, the components may undergo changes in physical state and arrangement.

The secretory masses within accessory glands often include insoluble structures. In the paragonia of *Drosophila*, the secretion contains many hollow filaments, with dimensions rather like microtubules (Tandler *et al.*, 1968; Perotti, 1971; Beaulaton and Perrin-Waldemer, 1975). Histochemical stains and enzyme digests indicate that these arrays of tubules have protein and glycoprotein subunits (Beaulaton and Perrin-Waldemer, 1975). The dimensions of the tubules and their ordering relative to each other change after they are expelled into the lumen. These changes may reflect reorganization of the subunits in the extracellular cavity (Perotti, 1971).

On histochemical evidence, Huignard (1975) reported that the secretion of the external median accessory gland of *Acanthoscelides* consists of a central

---

Figure 20. The apical zone of the secretory epithelium of type 6 cells of the BAG. Note the sinuous infolding of the some of the belt desmosomes (BD). Intracellular material, both irregular membranous tubes (arrows) and filaments (f), run across the cell and appear to impede the passage of the secretory vesicles into the apical projection (asterisk). mt, microtubules; m, mitochondrion. (Bar = 0.2 μm.)

Figure 21. The apical zone of secretory cells in a maturing BAG just after adult ecdysis. As in Figure 20, the belt desmosomes (BD) are infolded. Microtubules (mt) run through the zone of belt desmosomes and into the apical processes. Invaginations of the apical plasma membrane (arrows) appear to be associated with coated vesicles (cv). The invagination on the right is filled with secretory products en route to the lumen (L). (Bar = 0.2 μm.)

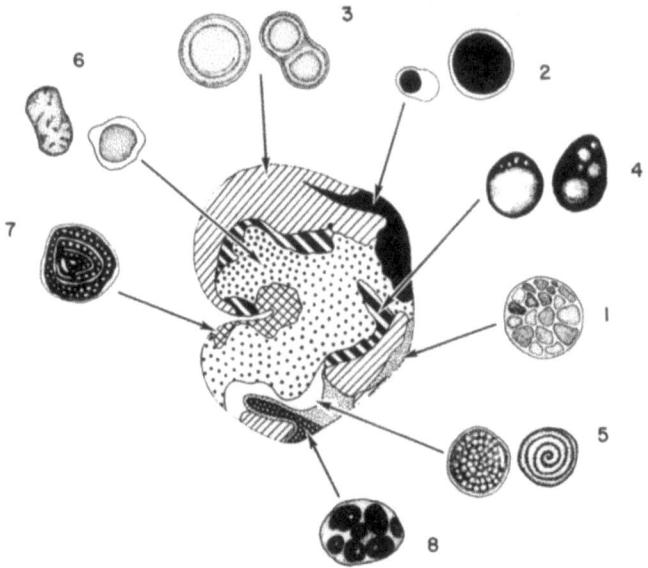

Figure 22. A diagram of the pattern of secretory cell types in the BAG and drawings of representative vesicles of each cell type.

protein mass surrounded by an acid mucopolysaccharide cortex. In the lumen of the accessory glands of *Calpodes*, there are a variety of single fibers, parallel bundles of fibers, glycogen, cellular debris, diverse electron-dense granules, and a multitude of small particles (Lai-Fook, 1982c). The contents of the accessory glands of *Anagasta* are heterogeneous, and elastic. Reimann and Thorson (1979b) have observed that the secretion of the apical section of the accessory gland will snap back to its original length even after it has "been drawn out several fold." In the BAG of *Tenebrio*, the secretory mass is semisolid, springy, and composed of several distinct zones (Figure 24). In electron micrographs, the complexity of each layer and the heterogeneity among them are quite marked (Figure 23).

5.2. Secretion into Separate Compartments Arranged in Series

In many species, the primary secretory products are kept separate from one another. In the accessory glands of *Schistocerca*, Ohdiambo (1969) recognized eight or nine different cell types based on ultrastructural and histochemical criteria: these cell types are segregated from one another so that each is confined to one or the other of the 16 tubular accessory glands. Such regional specialization also exists in the tubules of the adult and late-nymphal accessory glands of *Acheta domesticus* (Kaulenas, 1976; Kaulenas *et al.*, 1979).

In the reproductive tract of male Lepidoptera, secretions are produced by the vas deferens, accessory glands, and ejaculatory duct (composed of duplex and simplex segments). Because of differences in the morphology of the cells and their secretions, this tract can be divided into distinct regions, each composed of a band of secretory cells. In the accessory glands of *Anagasta*, there are five regions, many of which contain two cell types (foliate cells and granular cells) (Reimann and Thorson, 1979b). In *Calpodes*, there are at least 10 different regions in the tract as a whole (Lai-Fook, 1982b). Between many of the secretory regions are boundary zones which are constricted in such a way that inward extensions of the epithelial cells completely occlude the lumen. Similar constrictions were reported in *Anagasta* by Reimann and Thorson (1976b). These constrictions effectively divide the reproductive tract into many compartments arranged in series, each of which is sealed away from the others unless the circular muscle sheath relaxes. As a result, the various products are segregated from one another. Shortly after copulation begins, peristalsis squeezes out the posterior secretions. In its later stages, all the longitudinal muscles of the surrounding tract act together to shorten its length and to inject all the luminal contents including the seminal fluids (Lai-Fook, 1982d).

5.3. Secretion in Parallel to a Common Lumen

In mosquitoes and tsetse flies, there are several cell types which pass their secretory granules to a single or bipartite lumen of the accessory gland (Tongu *et al.*, 1972; Ramalingam and Craig, 1978; Kokwaro, 1982). In the BAG of *Tenebrio*, eight cell types secrete in parallel to produce the layered secretory plug. The materials exported from any one cell type form one layer of the ordered plug. To establish the distribution of the secretory cell types over the epithelium, we combined the information from stained whole glands, light histology, and transmission electron microscopy (Dailey *et al.*, 1980). The pattern is shown in Figure 22. The secretory vesicles of cell type 1 are globular, with one or more electron-transparent "bubbles" in an electron-dense cortex (Figure 18). Once the secretion from type 1 cells is expelled, it appears frothy (Figure 23). A similar secretory granule in the tubular gland of *Acanthoscelides* appears to be composed of glycoprotein (Cassier and Huignard, 1979). Secretory vesicles of cell type 3 have a complex peripheral structure (Figure 14). Secretory vesicles of cell types 4 and 8 appear faceted in cross-section (Figure 6). Those of cell types 5 (Figure 16) and 7 (Figure 11) enclose ordered networks of electron-dense granules or membranes, whereas those of cell type 6 are fairly homogeneous (Figure 20). See Dailey *et al.* (1980) for further descriptions. All cells pass their products onto the lumen of the BAG. The single secretory mass in the lumen is the ordered aggregate of the contribution from each cell type (Figure 24). Except for rudimentary histochemistry (indicating proteins, lipids, etc.) we know nothing about the biochemical composition products of these eight cell types.

Each of the eight cell types of the BAG is confined to a particular patch of the epithelium and each patch contains only one cell type. The shapes of these patches are irregular and appear random at first glance. However, the pattern is the same in every individual, and left and right glands have mirror-image patterns. In fact, the pattern appears to be an adaptation to allow the secretory products to be precisely positioned in the organized plug. Each layer of the aggregate mass is the product of cells nearby (Figure 24). The pattern of the cells

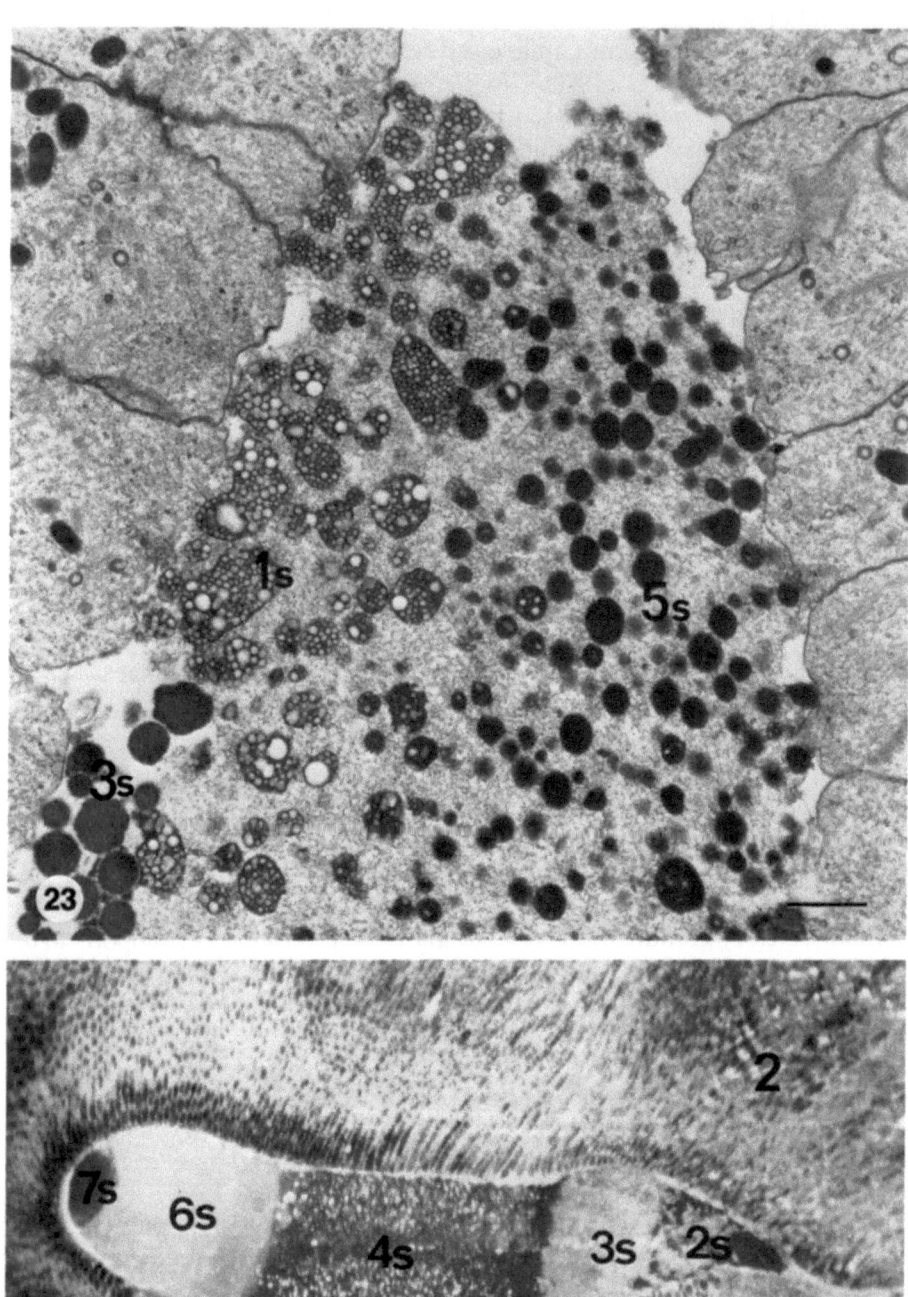

is mapped into the pattern of the plug. Not all cell types are equally common; for example, cell type 3 and cell type 6 each occupy about one-third of the total surface area, while type 5 occupies only a few percent of the total epithelium (Figure 22). The area of each patch of cells is reflected in the size of the corresponding layer in the plug. Thus, the predominant cell types (3 and 6) give rise to the thickest layers of the plug. It seems likely that in this gland, as in many others, all cell types in the epithelium are triggered to secrete after the lumen has been emptied by ejaculation. If such is the case, morphology alone, i.e., placement and size of the patch of each cell type, can explain placement and thickness of the layers in the plug. In addition, it is possible that some cell patches might secrete before others, but we have no evidence for such temporal patterning.

6. Formation of the Spermatophore

Spermatophores are constructed within a brief period which ranges from a few minutes to several hours (Tuzet, 1977). The spermatophore may be formed outside the animal and deposited on a substrate for a female to retrieve (see Schaller, 1971), it may be laid down, layer by layer, within the bursa of the female (e.g., Gerber et al., 1971), or it may be assembled during the exit of the various secretions from the male tract (e.g., Linley, 1981). Each spermatophore is composed of jellylike and waxy components which vary in viscosity, opacity, and elasticity. Some layers are visible in brightfield microscopy of unfixed spermatophores. Others can be detected by interference contrast (Linley, 1981, for *Culicoides*), by histochemistry (Gadzama and Happ, 1974, for *Tenebrio*; Gerber et al., 1971, for *Lytta*), or by transmission electron microscopy (Gadzama and Happ, 1974; Kokwaro and Ohdiambo, 1981, for *Glossina*). The complexity of the wall structure in the spermatophore of *Tenebrio* is shown in Figure 25. Comparison with the secretory plug of the BAG (Figure 23) demonstrates the extensive alterations which occur during spermatophore formation.

The organization of the spermatophore depends on the architecture of the male tract. Secretory masses are passed into the distal ejaculatory duct and then through the aedeagus in a fixed order. Thus, for example, the foliate cells of *Anagasta*, which are found toward the anterior tip of the accessory gland, contribute their product to the last-formed (outermost) layer of the spermatophore (Reimann and Thorson, 1979b).

The various components must be present in balanced quantities. The amount of each is partly defined by the volume of the lumina within the male tract, and perhaps also the female bursa. The aliquots of secretion must be delivered in an appropriate order. Some of that order is inherent in the

←——————————————————————————————

Figure 23. The edge of the secretory plug of the BAG. Secretions from cell type 1 (1s), cell type 3 (3s), and cell type 5 (5s) are shown. (Bar = 1 μm.)

Figure 24. A light micrograph of the secretory plug of the BAG *in situ* stained with toluidine blue. The numbers indicate the cell types (2, 3, 4, 6, 7) and their respective secretions in the plug (2s, 3s, 4s, 6s, 7s). (Bar = 50 μm.)

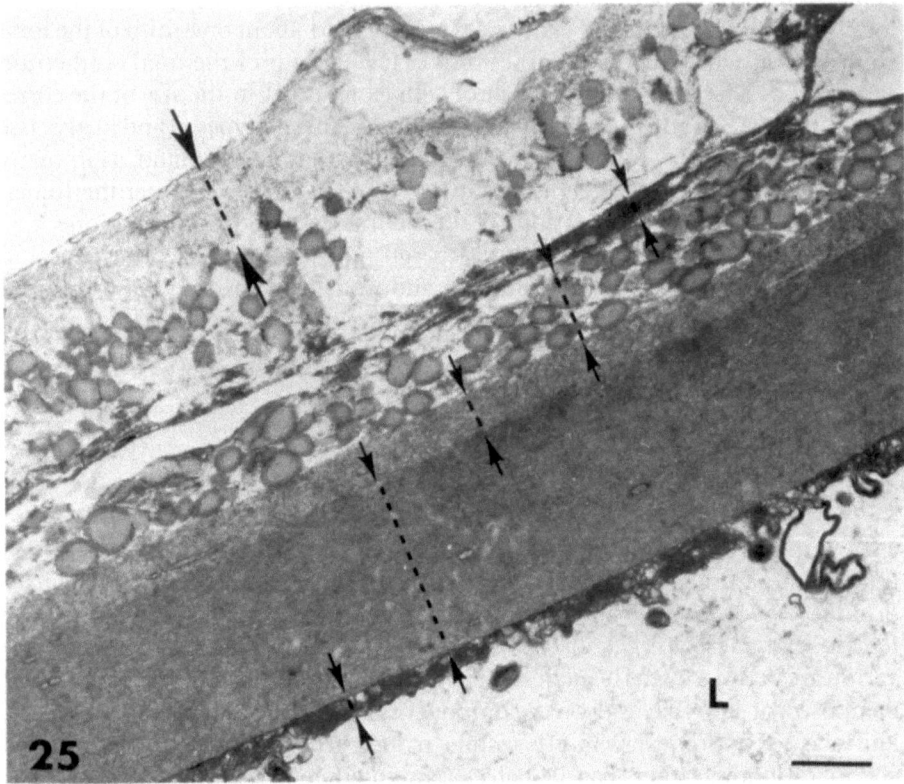

Figure 25. A section through the wall of the spermatophore of *Tenebrio*. The lumen, which contains the sperm (not shown), is at the lower right, and the outer surface is toward the upper left. The arrows bracket the layers in this wall. Comparison with Figure 23 shows that profound transformation of the secretion takes place when the spermatophore is formed. (Bar = 1 µm.)

sequential arrangement of the secretory lumina as in *Calpodes*—or in the architecture of the secretory plug as in *Tenebrio*. But nervous coordination is also important; thus, induction of ejaculation by a massive eserine injection in *Tenebrio* produces a malformed spermatophore (Happ, unpublished). For the accessory secretions from the dozen or more converging accessory glands of a cricket (*Teleogryllus commodus*) to be properly mixed together, the brain appears to be required for the coordination of the secretory sequence (Loher, 1974).

7. Evacuation of the Spermatophore

The sperm must somehow be delivered from the spermatophore to the spermatheca of the female. In some staphlinid beetles, the male aedeagus has a narrow chitinous process which forms a "guide rail" which causes the tip of the spermatophore to slip into the spermatheca (Peschke, 1978). Even when placement is not so precise, the sperm tend to leave the spermatophore near the spermathecal opening. In order for the sperm to escape from the multilayered secretory mass, the spermatophore undergoes dramatic rearrangements and contortions. The driving forces underlying spermatophore expansion and sperm expulsion are not well understood in any species. Osmotic flow of water has been invoked to explain sperm expulsion in *Acheta domesticus* (Khalifa, 1949). Gaseous CO_2 production apparently accounts for movement of sperm into the bursa of a tick (Feldman-Muhsam *et al.*, 1973). Neither explanation seems sufficient to explain the programmed sequence of evaginations which leads to the rupture of the spermatophore of *Tenebrio* (Gadzama and Happ, 1974). In this spermatophore, it appears that the outpocketings swell and then snap out, as if the walls are recoiling from previous strain. The compression and the shear forces which act on the viscous materials as they are forcibly ejected through the long narrow ejaculatory duct and the aedeagus must be of considerable magnitude. Expansion of the spermatophore of *Tenebrio* might be due in part to relaxation from shear-induced deformations during formation of the overall structure. A similar suggestion has been implied for *Culicoides melleus* by Linley (1981), who suggests "that the spermatophore behaves so as to minimize an intrinsic elastic energy."

If the molecular components of the spermatophore are actively involved in the expansion and rupture, then one might expect adaptations to prevent premature mixing or to avoid spoilage. As Lai-Fook (1982d) has suggested, the constrictions in the male tract of Lepidoptera should prevent such mixing. In *Tenebrio*, where there is considerable potential for mixing in the lumen of the BAG, the secretion is not liquid but semisolid, and the plug is regularly discarded, even if no females are encountered. Isolated males ejaculate a spermatophore onto the substrate each day (P. J. Dailey, personal communication). It may be that this "casting of seed upon the ground" is an adaptation to avoid deterioration of the wall components during protracted storage in the gland lumen. These and other intriguing possibilities can be investigated only when we know much more of the biochemistry of the secretions from the accessory glands, the rheology of the viscous materials as they surge through the male tract, and the molecular architecture of the spermatophore itself.

8. Development of Accessory Glands

Primary organogenesis of accessory reproductive organs occurs in preadult stages. Significant growth and early differentiation occur late in the preimaginal instar, either in the last larval stage of Hemimetabola or in the pupal stage of Holometabola. In those insects which mate shortly after ecdysis, the glands are fully differentiated at eclosion. For species which mate some days thereafter, there is a postecdysial period of peak differentiation.

In *Tenebrio*, the two pairs of male accessory glands are derived from a common mesodermal pouch (Huet, 1966; Poels, 1972), and the four glands are anatomically distinct at pupation (Figure 1A). At that stage, both gland pairs consist of an epithelium surrounded by a poorly differentiated muscle coat.

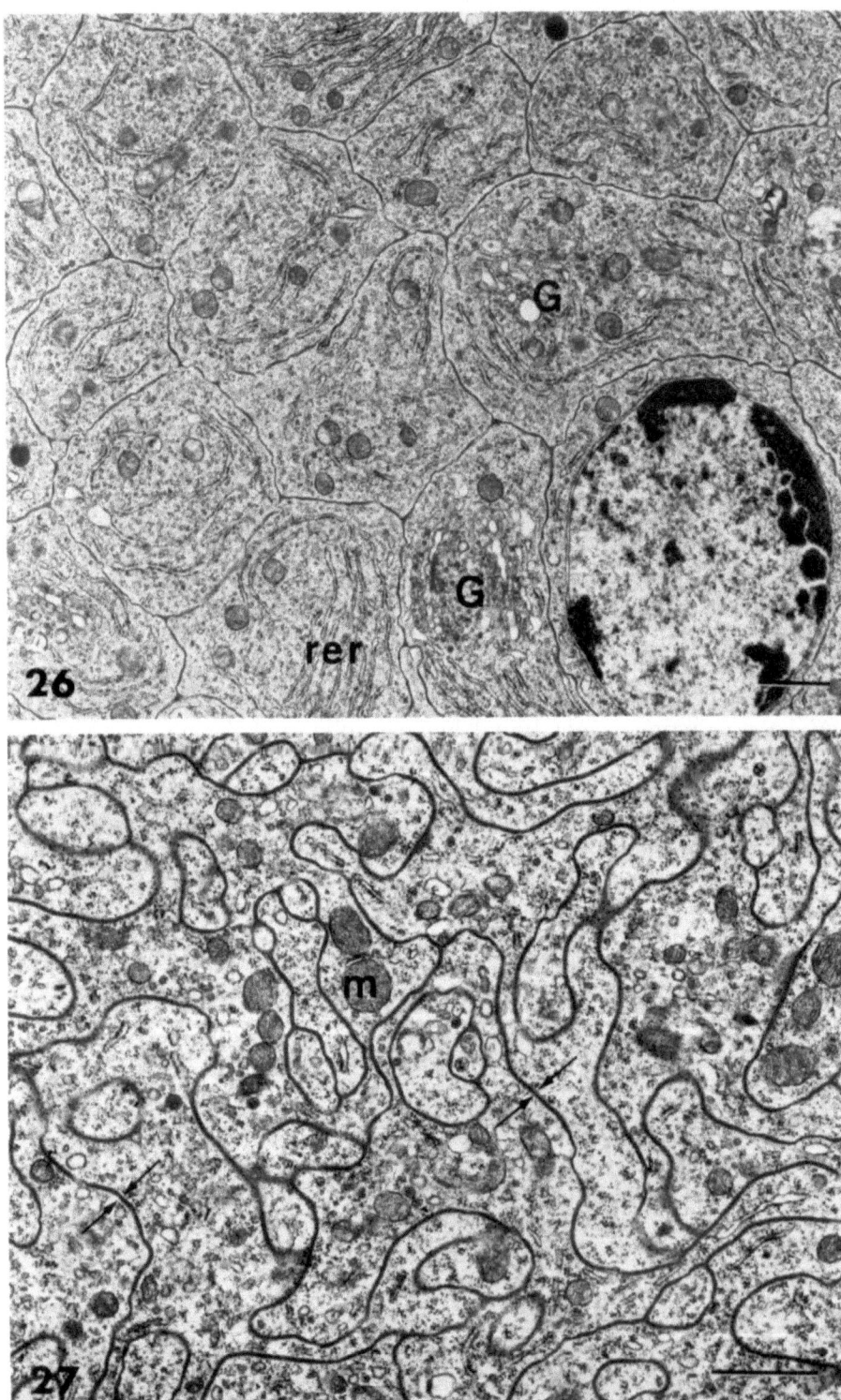

The tubular gland is like a small thumb on the surface of the mitten-shaped BAG. The closely packed columnar cells of the secretory epithelium contain sparse membranes of the ER and scattered mitochondria.

During the first 6-7 days of the 9-day pupal stage in *Tenebrio*, the secretory cells undergo cycles of division. As is common in vertebrate epithelia, the mitotic figures occur near the apical surface of the epithelium. In both BAG and TAG, the daughter cells remain linked by spindle-remnant bridges for a day or more after karyokinesis is complete. These bridges resemble fusomes reported from insect ovaries (Mandelbaum, 1980; King *et al.*, 1982) except for the fact that they are usually "plugged" with microtubules. The spindle-remnant bridges are found only near the apical surface, at about the level where apical desmosomes will eventually arise. During the later part of the pupal stage (days 4-8), adjacent cells in the BAG communicate via fused-membrane bridges which form in the midregion of the epithelium (Grimes and Happ, 1980; Happ and Happ, 1982). Once division ceases, more cytomembrane profiles appear, Golgi zones become larger, and the cells grow in volume (Figure 26). The closely appressed membranes become infolded (Figure 27), apparently in preparation for the rapid postecdysial hypertrophy. By day 8, a few secretory vesicles are seen in the TAGs (Happ and Happ, 1982) and in some of the cell types of the BAGs (Dailey and Happ, 1983). At this stage, the glands are smaller versions of the mature adult organs (Figures 1B, C). Adult-specific secretory antigens are present in TAGs from late pupae (Black *et al.*, 1982). An analogous developmental schedule has been reported for the accessory glands of *Acheta* in which Kaulenas *et al.* (1979) find low levels of adult antigens in late nymphal glands.

Growth and peak differentiation follow soon after adult ecdysis. The growth is due to hypertrophy rather than hyperplasia: In the BAG and the TAG, cells increase in height and cross-sectional area; together, these account for a 5- to 10-fold increase in gland volume (Happ and Happ, 1982; Happ *et al.*, 1982). Similar increases in cell volume occur in three of the four accessory glands of *Acanthoscelides* (Cassier and Huignard, 1979).

By 2 days after adult ecdysis, all eight cell types of the BAG contain secretory granules of the definitive adult morphology (Dailey and Happ, 1983). During this time of granule maturation, the epithelial pattern is consolidated. Some cells on the boundary lines between dissimilar patches contain transitional granules—granules intermediate in morphology between those of their two different neighbors. These transitional granules persist for only a day or two and then are destroyed as the boundary cells conform to one or the other of the adjacent definitive phenotypes (Dailey and Happ, 1983). In the TAG, maturation follows a similar time course until secretory vesicles fill most of the cell by 5-6 days after ecdysis (Gadzama *et al.*, 1977). Rapid postecdysial maturation of the ER and Golgi zones has also been reported in the male accessory

---

Figure 26. A cross-section through the midregion of the secretory cells of the TAG in a 7-day pupa. Parallel cisternae of rough endoplasmic reticulum (rer) and Golgi zones (G) are common. (Bar = 1 μm.)

Figure 27. A cross-section through the secretory cells of 9-day pupal TAG, just above the zone of apical desmosomes. The cells are linked by smooth septate junctions (between arrows) and the membranes are extensively infolded. (Bar = 1 μm.)

gland of *Acanthoscelides* (Cassier and Huignard, 1979) and *Acheta* (Kaulens *et al.*, 1979). In contrast, the secretory machinery of the paragonia of *Drosophila* is fully organized by the time of ecdysis (Federer and Chen, 1980).

Concomitant with the ultrastructural differentiation, the accessory glands increase their production of adult proteins. Leucine incorporation into a few adult-specific protein spots can be detected in the TAG and BAG just after eclosion (Black *et al.*, 1982; Happ *et al.*, 1982). The biosynthetic emphasis shifts rapidly until at 5 days later over 50% of the leucine incorporated by the TAG goes into four protein bands (Happ *et al.*, 1977; Black *et al.*, 1982).

9. Endocrine Control of Accessory Gland Development

Both ecdysteroids and juvenile hormones play roles in the development and modulation of accessory glands, but their precise actions have been defined in only a few species. Ecdysteroids are important during pupal development. In *Samia cynthia*, the spermiduct consists of secretory cells surrounded by muscle. When spermiducts from diapausing pupae are cultured *in vitro*, the secretory epithelium differentiates only when ecdysteroids are added (Szöllösi and Landureau, 1977).

In the BAG and TAG of *Tenebrio*, there are two peaks of mitotic activity in the 9-day pupal instar (Grimes and Happ, 1980; Happ and Happ, 1982). The first of these mitotic bouts (0-3 days) proceeds *in vitro* in Landureau's medium, but the second bout does not occur without addition of ecdysteroids (Happ, 1982). This second bout is correlated with a single large ecdysteroid peak in the pupa (Delbecque *et al.*, 1978). The pupal ecdysteroid peak is also correlated with a change in competence which permits the glands to synthesize adult-specific proteins. Glands from young pupae will not differentiate when implanted in adults unless they have been exposed to a large ecdysteroid concentration *in vivo* or *in vitro* (Happ, 1982). Some time ago, Ohdiambo (1966) showed that the accessory glands of late-last-instar larvae of *Schistocerca* would mature when implanted in an adult, but glands from early last-instar larvae would not do so when similarly implanted. Apparently for *Schistocerca*, some hormone (perhaps ecdysteroid) or other humoral feature of the larval environment was necessary for acquisition of competence to mature. The sequence of events which depend on the last preimaginal ecdysteroid peak deserves further study in more species.

In many hemimetabolous insects, postecdysial development of male accessory glands is dependent on juvenile hormone [e.g., *Rhodnius* (Wigglesworth, 1936), *Leucophaea maderae* (Scharrer, 1946), and *Schistocerca* (Ohdiambo, 1966)]. In intact *Schistocerca*, massive growth of the ER and Golgi begins at the fifth day when the corpora allata become active. No such expansion of the secretory organelles occurs in glands of allatectomized animals. Juvenile hormone is required for the maintenance of normal secretory morphology in the accessory glands of some beetles (*Leptinotarsa*). Within 3 days after allatectomy, the secretory cells of the accessory glands of *Leptinotarsa* begin to degenerate by sloughing off their apical halves, and in some cases the process continues to cell death (De Loof and Lagasse, 1972).

In other species, such as *Calliphora erthrocephala* (Thomsen, 1942) and *Tenebrio*, it has not been possible to show that juvenile hormones play roles in postecdysial maturation. In *Tenebrio*, I have seen normal gland development in abdomens which were isolated at adult ecdysis, but I have not been able to support vigorous development *in vitro*. The glands grow and differentiate when implanted into adults of either sex, so neither innervation nor good tracheation seems to be required. Recent experiments of Barker and Davey (1983) have shown that for *Rhodnius*, both neuroendocrine factors from the brain and juvenile hormone are necessary for full development of male accessory glands. Perhaps neuroendocrine factors are also required in other species. The regulation of peak differentiation and the recurrent secretory cycles in male accessory glands needs further study.

10. Summary and Prospects

In insects as in vertebrates, seminal and paraseminal secretions are derived from several glandular tissues. The anatomical details of the male tract are enormously varied but include adaptations for timely delivery of products during the short period of copulation. The literature on the ultrastructure of male accessory glands is somewhat sparse, but there are notable sets of papers on *Schistocerca* (Ohdiambo), *Locusta* (Cantacuzène), *Acheta* (Kaulenas), *Anagasta* (Reimann and Thorson), *Calpodes* (Lai-Fook), *Glossina morsitans morsitans* (Kokwaro and Ohdiambo), *Acanthoscelides* (Huignard), and *Tenebrio* (Happ and co-workers). The work on *Tenebrio* has received particular attention in this review.

Accessory glands are usually surrounded by muscles which provide the thrust for ejaculation. Desmosomes and septate junctions link the secretory cells tightly to each other and thus maintain the integrity of the epithelium during the pressure surges which accompany sperm transfer.

Precursors for manufacture of the secretion percolate through the overlying basement membranes and muscles to be absorbed by the secretory cells. The cells are rich in ER and in Golgi zones. As the animals become reproductively mature, the cells become turgid with stored secretions. The secretory vesicles are enormously diverse in fine structure. In *Schistocerca, Anagasta, Calpodes, Glossina,* and *Tenebrio,* published reports indicate at least 5–10 kinds of secretory cells, each of which has a product of characteristic morphology. We cannot yet associate a particular morphology with a particular class of biochemical constituents.

In most accessory glands, the secretory vesicles are transported toward the apical end of the cell and exported by apocrine or merocrine mechanisms. Secretory products fill the gland lumen and are ejaculated during copulation. Thereafter, a new charge of secretion rapidly accumulates. We know little of the ways that secretory episodes are turned on and off, about the kinetics of replenishment, or about the detailed mechanisms of transport and export.

The functions of accessory glands are diverse, and most aspects of their physiological impact on the female and their importance for sperm maturation have been reviewed elsewhere (Leopold, 1976). I have concentrated particularly on the role of many accessory glands in the formation of a spermatophore. Spermatophore structure, assembly, and evacuation are receiving increasing attention in several laboratories. The molecular components of spermatophores are the secretory products of accessory glands. The storage of these products takes place in regionally segregated compartments or in semisolid masses, so that there seems to be limited mixing of secretions until the time of assembly of the sperm sac. Biochemical, immunochemical, and histochemical analyses of spermatophores are needed to complement the morphological investigations. Information on many more species is required in order to obtain comparative information on the molecular and microscopic architecture of spermatophores. From a descriptive base, studies can proceed to analysis of the mechanisms for stabilization of the spermatophore and for its subsequent contortions which expel the sperm.

Accessory glands mature rapidly in late preadult stages and reach full secretory activity in young adults. Cell division, morphogenesis, and differentiation take place as the glands develop. These glands, which produce a small set of secretory products and have a distinct adult morphology, are attractive models for developmental biologists. They are particularly useful for questions about hormone action, about the importance of mitoses for ongoing differentiation, about pattern formation in a secretory (often mesodermal) tissue, and about the molecular rate-controlling mechanisms in progressive growth and terminal differentiation.

The ultrastructural studies of male accessory glands which have accumulated over the last two decades are intriguing examples of elaborate specializations of secretory cells and tissues. The present morphological information is sufficient to provide a springboard for biochemical analyses in experiments on the endocrine control of reproductive physiology and maturation.

ACKNOWLEDGMENTS

I thank Connie S. Bricker, Patrick J. Dailey, Njidda M. Gadzama, Margaret J. Grimes, and Christine M. Happ for their collaboration in the investigations on *Tenebrio* which have been summarized in this chapter, Christine Yuncker for drawing Figures 1 and 22, Ruth Goodridge for secretarial assistance, and the National Institutes of Health (AL-15662 and GM-26140) for financial support.

References

Adiyodi, K. G., and Adiyodi, R. G., 1975, Morphology and cytology of the accessory sex glands in invertebrates, *Int. Rev. Cytol.* **43**:353-398.

Adiyodi, R. G., and Adiyodi, K. G., 1974, Ultrastructure of the utriculi majores in the mushroom-shaped male accessory gland of *Periplaneta americana*, *Z. Zellforsch. Mikrosk. Anat.* **147**:433-440.

Baccetti, B., 1972, Insect sperm cells, *Adv. Insect Physiol.* **9**:315-397.

Bairati, A., and Perotti, M. E., 1970, Occurrence of a compact plug in the genital duct of *Drosophila* females after mating, *Drosophila Inf. Serv.* **45**:67-68.

Barker, J. F., and Davey, K. G., 1982, Intraglandular synthesis of protein in the transparent accessory reproductive gland in the male of *Rhodnius prolixus, Insect Biochem.* **12**:157-159.

Barker, J. F., and Davey, K. G., 1983, A polypeptide from the brain and corpus cardiacum of male *Rhodnius prolixus* which stimulates *in vitro* protein synthesis in the transparent accessory reproductive gland, *Insect Biochem.* **13**:7-10.

Beaulaton, J., and Perrin-Waldemer, C., 1975, Contribution à l'étude de la sécrétion des paragonies de *Drosophila melonogaster* Meig.: Ultrastructure et cytochimie des grains a microtubules, *J. Microsc. (Paris)* **24**:91-104.

Bishop, G. H., 1920, Fertilization in the honey-bee. I. The male sexual organs: Their histological structure and physiological functioning, *J. Exp. Zool.* **31**:225-265.

Black, P. N., Landers, M. H., and Happ, G. M., 1982, Cytodifferentiation in the accessory glands of *Tenebrio molitor*. VII. Crossed immunoelectrophoretic analysis of terminal differentiation in the post-ecdysial tubular accessory glands, *Dev. Biol.* **94**:106-115.

Blum, M. S., Glowska, Z., and Tauber, S., III, 1962, Chemistry of the drone honey bee reproductive system. II. Carbohydrates in the reproductive organs and semen, *Ann. Entomol. Soc. Am.* **55**:135-139.

Blum, M. S., Bumgarner, J. E., and Tauber S., III, 1967, Composition and possible significance of fatty acids in the lipid classes in honey bee serum, *J. Insect Physiol.* **13**:1301-1308.

Cantacuzène, A.-M., 1972, Recherches morphologiques et physiologiques sur les glandes annexes males des orthoptères. IV. Ultrastructure de la vésicule séminale de *Locusta migratoria migratorioides* L., *Ann. Sci. Nat. Zool. Biol. Anim.* **14**:389-410.

Cassier, P., and Huignard, J., 1979, Etude ultrastructurale des glandes annexes de l'appareil genital male chez *Acanthoscelides obtectus* Say (Coleoptera: Bruchidae), *Int. J. Insect Morphol. Embryol.* **8**:183-201.

Dailey, P. J., and Happ., G. M., 1983, Cytodifferentiation in the accessory glands of *Tenebrio molitor*. XI. Transitional cell types during establishment of patterns. *J. Morphol.* **178**:139-154.

Dailey, P. J., Gadzama, N. M., and Happ, G. M., 1980, Cytodifferentiation in the accessory glands of *Tenebrio molitor*. VI. A congruent map of cells and their secretions in the layered elastic product of the male bean-shaped gland, *J. Morphol.* **166**:289-322.

Davey, K. G., 1965, *Reproduction in the Insects*, Oliver & Boyd, Edinburgh.

Davey, K. G., 1967, The physiology of reproduction: Some lessons from insects. In *Insects and Physiology*, edited by J. W. L. Beament and J. E. Treherne, pp. 351-364, American Elsevier, New York.

Delbecque, J.-P., Hirn, M., Delachambre, J., and De Reggi, M., 1978, Cuticular cycle and molting hormone levels during the metamorphosis of *Tenebrio molitor* (Insecta, Coleoptera), *Dev. Biol.* **64**:11-30.

De Loof, A., and Lagasse, A., 1972, The ultrastructure of the male accessory reproductive glands of the Colorado beetle, *Z. Zellforsch. Mikrosk. Anat.* **130**:545-552.

Dumser, J. B., 1980, The regulation of spermatogenesis in insects, *Annu. Rev. Entomol.* **25**:341-369.

Engelmann, F., 1970, *The Physiology of Insect Reproduction*, Pergamon Press, Elmsford, N.Y.

Federer, H., and Chen, P. S., 1980, Ultrastruktur und Funktion der Paragonien von *Drosophila funebris, Rev. Suisse. Zool.* **87**:875-880.

Feldman-Muhsam, B., Borut, S., Saliternick-Givant, S., and Eden, C., 1973, On the evacuation of sperm from the spermatophore of the tick, *Onithodorus savignyi, J. Insect Physiol.* **19**:951-962.

Friedel, T., and Gillot, C., 1976, Extraglandular synthesis of accessory reproductive gland components in male *Melanoplus sanguinipes, J. Insect Physiol.* **22**:1309-1314.

Friend, D. S., and Farquhar, M. G., 1967, Functions of coated vesicles during protein absorption in the rat vas deferens, *J. Cell. Biol.* **35**:357-376.

Gadzama, N. M., and Happ, G. M., 1974, The structure and evacuation of the spermatophore of *Tenebrio molitor* L. (Coleoptera: Tenebrionidae), *Tissue Cell* **6**:95-108.

Gadzama, N. M., Happ., C. M., and Happ, G. M., 1977, Cytodifferentiation in the accessory glands of *Tenebrio* molitor. I. Ultrastructure of the tubular gland in the post-ecdysial adult male, *J. Exp. Zool.* **200**:211-222.

Garcia-Bellido, A., 1964, Das Sekret der Paragonien als Stimulus der Fekundität bei Weibchen von *Drosophila melanogaster, Z. Naturforsch. Teil B* **19**:491-495.

Gerber, G. H., Church, N. S., and Rempel, J. G., 1971, The structure, formation, histochemistry, fate, and functions of the spermatophore of *Lytta nuttalli* Say (Coleoptera: Meloidae), *Can. J. Zool.* **49**:1595-1610.

Gilbert, D. G., 1982, Ejaculate esterase-6 and initial sperm use by female *Drosophila melanogaster*, *J. Insect Physiol.* **27**:641-650.

Grassé, P.-P., 1977, Organes géntiaux mâles, *Traité Zool.* **8**(5A):125-137.

Grimes, M. G., and Happ, G. M., 1980, Fine structure of the bean-shaped accessory gland in the male pupa of *Tenebrio molitor* L. (Coleoptera: Tenebrionidae), *Int. J. Insect Morphol. Embryol.* **9**:281-296.

Hagedorn, H. H., and Kunkel, J. G., 1979, Vitellogenin and vitellin in insects, *Annu. Rev. Entomol.* **24**:475-505.

Happ, G. M., 1982, Control of cell differentiation in the accessory reproductive glands of mealworm beetles. In *The Ultrastructure and Functioning of Insect Cells*, edited by H. Akai, R. C. King, and S. Morohoshi, pp. 83-86, Society for Insect Cells, Tokyo.

Happ, G. M., and Happ, C. M., 1982, Cytodifferentiation in the accessory glands of *Tenebrio molitor*. X. Ultrastructure of the tubular gland in the male pupa, *J. Morphol.* **172**:97-112.

Happ, G. M., Yuncker, C., and Huffmire, S. A., 1977, Cytodifferentiation in the accessory glands of *Tenebrio molitor*. II. Patterns of leucine incorporation in the tubular gland of post-ecdysial adult males, *J. Exp. Zool.* **200**:223-236.

Happ, G. M., Yuncker, C., and Dailey, P. G., 1982, Cytodifferentiation in the accessory glands of *Tenebrio molitor*. VII. Patterns of leucine incorporation by the bean-shaped glands of males, *J. Exp. Zool.* **220**:81-92.

Huet, C., 1966, Etude expérimentale du développement de l'appareil génital mâle de *Tenebrio molitor* (Coléoptère: Ténébrionide), *C.R. Soc. Biol.* **160**:2021-2025.

Huignard, J., 1975, Anatomie et histologie des glandes annexes males au cours de la vie imaginale chez *Acanthoscelides obtectus* Say (Coleoptera: Bruchidae), *Int. J. Insect Morphol. Embryol.* **4**:77-88.

Kafatos, F. C., Regier, J. C., Mazur, G. D., Nadel, M. R., Blau, H. M., Petri, W. H., Wyman, A. R., Gelinas, R. E., Moore, P. B., Paul, M., Efstratiadis, A., Vournakis, J. N., Goldsmith, M. R., Hunsley, J. R., Baker, B., Nardi, J., and Koehler, M., 1977, The eggshell of insects: Differentiation-specific proteins and the control of their synthesis and accumulation during development. In *Results and Problems in Cell Differentiation*, vol. 8, edited by W. Beermann, pp. 45-145., Springer Verlag, Berlin.

Kaulenas, M. S., 1976, Regional specialization for export protein synthesis in the male cricket accessory gland, *J. Exp. Zool.* **195**:81-96.

Kaulenas, M. S., Potswald, H. E., Burns, A. L., and Yenofsky, R. L., 1979, Development of structural and functional specializations for export protein synthesis by the accessory gland of the male cricket, *Acheta domesticus* L. (Orthoptera Gryllidae), *Int. J. Insect Morphol. Embryol.* **8**:33-49.

Khalifa, A., 1949, The mechanism of insemination and mode of action of the spermatophore in *Gryllus domesticus*, *Q. J. Microsc. Sci.* **90**:281-292.

King, R. C., Cassidy, J. D., and Rousset, A., 1982, The formation of clones of interconnected cells during gametogenesis in insects. In *Insect Ultrastructure*, edited by R. C. King and H. Akai, vol. 1, pp. 3-31, Plenum Press, New York.

Kokwaro, E. D., 1982, Ultrastructure of the male accessory reproductive glands, spermatophore and spermatheca of the tsetse, *Glossina morsitans morsitans* Westwood. In *The Ultrastructure and Functioning of Insect Cells*, edited by H. Akai, R. C. King, and S. Morohoshi, pp. 53-56, Society for Insect Cells, Tokyo.

Kokwaro, E. D., and Ohdiambo, T. R., 1981, Spermatophore of the tsetse, *Glossina morsitans morsitans* Westwood: An ultrastructural study, *Insect Sci. Appl.* **1**:185-190.

Lai-Fook, J., 1982a, The vasa deferentia of the male reproductive system of *Calpodes ethlius* (Hesperidae, Lepidoptera), *Can. J. Zool.* **60**:1172-1183.

Lai-Fook, J., 1982b, Structure of the noncuticular simplex of the internal male reproductive tract of *Calpodes ethlius* (Hesperidae, Lepidoptera), *Can. J. Zool.* **60**:1184-1201.

Lai-Fook, J., 1982c, Structure of the accessory glands and duplex of the internal male reproductive system of *Calpodes ethlius* (Hesperidae, Lepidoptera), *Can. J. Zool.* **60**:1202-1215.

Lai-Fook, J., 1982d, Structure, function, and possible evolutionary significance of the constrictions in the male reproductive system of *Calpodes ethlius* (Herperidae, Lepidoptera), *Can. J. Zool.* **60**:1828-1836.

Lane, N. J., and Skaer, H. leB., 1980, Intercellular junctions in insect tissues, *Adv. Insect Physiol.* 15:35-213.

Leopold, R. A., 1976, The role of male accessory glands in insect reproduction, *Annu. Rev. Entomol.* 21:199-221.

Linley, J. R., 1981, Ejaculation and spermatophore formation in *Culicoides melleus* (Coq.) (Diptera: Ceratopogonidae), *Can. J. Zool.* 59:332-346.

Loher, W., 1974, Circadian control of spermatophore formation in the cricket *Teleogryllus commodus* Walker, *J. Insect Physiol.* 20:1155-1172.

Mandelbaum, I., 1980, Intercellular bridges and the fusome in the germ cells of the cecropia moth, *J. Morphol.* 166:37-50.

Mann, T., 1964, *The Biochemistry of Semen and of the Male Reproductive Tract*, Methuen, London.

Meola, S. M., 1982, Morphology of the region of the ejaculatory duct producing the male accessory gland material in the stable fly, *Stomoxys calcitrans* L. (Diptera: Muscidae), *Int. J. Insect Morphol. Embryol.* 14:69-77.

Ohdiambo, T. R., 1966, Growth and the hormonal control of sexual maturation in the male desert locust, *Schistocerca gregaria* (Forsköl), *Trans. R. Entomol. Soc. London* 118:393-412.

Ohdiambo, T. R., 1969, The architecture of the accessory reproductive glands of the desert locust. IV. Fine structure of the glandular epithelium, *Philos. Trans. R. Soc. London Ser. B* 256:85-114.

Oliver, C., 1982, Endocytotic pathways at the lateral and basal cell surfaces of exocrine acinar cells, *J. Cell. Biol.* 95:154-161.

Pearse, B. M. F., and Bretscher, M. S., 1981, Membrane recycling by coated vesicles, *Ann. Rev. Biochem.* 50:85-101.

Perotti, M. E., 1971, Microtubules as components of *Drosophila* male paragonia secretion: An electron microscopic study, with enzymatic tests, *J. Submicrosc. Cytol.* 3:255-282.

Peschke, K., 1978, Funktionsmorphologische Untersuchungen zur Kopulation von *Aleochara curtula* Goeze (Coleoptera, Staphylinidae), *Zoomorphologie* 89:157-184.

Phillips, D. M., 1974, *Spermiogenesis*, Academic Press, New York.

Poels, A., 1972, Histophysiologie des voies génitales mâles de *Tenebrio molitor* L. (Coléoptère: Tenebrionidae), *Ann. Soc. R. Zool. Belg.* 102:199-234.

Ramalingam, S., and Craig, G. B., Jr., 1978, Fine structure of the male accessory glands in *Aedes triseriatus*, *J. Insect Physiol.* 24:251-259.

Reimann, J. G., 1973, Ultrastructure of the ejaculatory duct region producing the male housefly accessory material, *J. Insect Physiol.* 19:213-223.

Reimann, J. G., and Thorson, B. J., 1976a, Ultrastructure of the vasa deferentia of the Mediterranean flour moth, *J. Morphol.* 149:483-505.

Reimann, J. G., and Thorson, B. J., 1976b, Ultrastructure of the ductus ejaculatorius duplex of the Mediterranean flour moth, *Anagasta kühniella* (Zeller) (Lepidoptera Pyralidae), *Int. J. Insect Morphol. Embryol.* 5:227-240.

Reimann, J. G., and Thorson, B. J., 1979a, Foliate and granule secreting cells in the ejaculatory duct (simplex) of the Mediterranean flour moth, *J. Ultrastruct. Res.* 66:1-10.

Reimann, J. G., and Thorson, B. J., 1979b, Ultrastructure of the accessory glands of the Mediterranean flour moth, *J. Morphol.* 159:355-391.

Schaller, F., 1971, Indirect sperm transfer by soil arthropods, *Annu. Rev. Entomol.*, 16:407-466.

Scharrer, B., 1946, The relationship between corpora allata and reproductive organs in adult *Leucophaea maderae* (Orthoptera), *Endocrinology* 38:46-55.

Steinman, R. M., Mellman, I. S., Muller, W. A., and Cohn, Z. A., 1983, Endocytosis and the recycling of plasma membrane, *J. Cell Biol.* 96:1-27.

Szöllösi, A., and Landureau, J.-C., 1977, Imaginal cell differentiation in the spermiduct of *Samia cynthia* (Lepidoptera): Responses *in vitro* to ecdysone and ecdysterone, *Biol. Cell.* 28:23-36.

Tandler, B., Williamson, D. L., and Ehrman, L., 1968, Unusual filamentous structures in the paragonia of male *Drosophila paulistorum*, *J. Cell. Biol.* 38:329-336.

Telfer, W. H., Huebner, E., and Smith, D. S., 1982, The cell biology of vitellogenic follicles in *Hyalophora* and *Rhodnius*. In *Insect Ultrastructure*, edited by R. C. King and H. Akai, vol. 1, pp. 118-149, Plenum Press, New York.

Thomsen, E., 1942, An experimental and anatomical study of the corpus allatum in the blowfly

Calliphora erythrocephala Meig., *Vidensk. Medd. Dan. Naturhist. Foren. Khobenhavn* **106:**319-415.

Tongu, Y., Suguri, S., Sakumoto, D., Itano, K., and Inatomi, S., 1972, The ultrastructure of mosquitoes. 6. Male accessory gland of *Culex pipiens pallens, Jpn. J. Sanit. Zool.* **23:**129-139 (Japanese with English summary).

Tuzet, O., 1977, Les spermatophores des insectes, *Traité Zool.* **8**(5A):277-330.

Wigglesworth, V. B., 1936, The function of the corpus allatum in the growth and reproduction of *Rhodnius prolixus* (Hemiptera), *Q. J. Microsc. Sci.* **79:**91-122.

11

The Photoreceptor Cells

STANLEY D. CARLSON, RICHARD L. SAINT MARIE, AND CHE CHI

1. Introduction

Insect photoreceptor cells are usually slender, cylindrical, and aggregated into various species-specific combinations beneath the corneal lens of a compound eye, ocellus, or larval stemmata. Individual cells are so sensitive to light that some are able to signal catches of a single photon (e.g., Lillywhite, 1977). This exquisite sensitivity is founded on the presence of an incompletely characterized glycoprotein (rhodopsin). Rhodopsin is the functional molecule of the photoreceptor cell, and is the major constituent of the plasma membrane of the rhabdomeric microvilli. This photopigment-containing organelle (rhabdomere) is found in the distal region of the cell (Section 5). In its more proximal reaches, the cell becomes a functional neuron with an axon (Section 7) conveying graded depolarizations to chemical and electrical synapses (Section 8).

Because of the dichotomy of structure and function in these two parts of the photoreceptor cell, it is now suspected or becoming apparent that each requires a different ionic microenvironment. To that end, crucial partitions are present (effected by membrane specializations) which form a complex barrier system between the retina and the first optic neuropil (Section 7). This is a story that is just beginning to unfold.

Because of all of this beautifully adapted ultrastructure, photoreceptor cells have the miraculous ability to transduce light energy from the environment into electrical energy meaningful to the central nervous system (CNS). Thus, the visual system reconstructs such elements of the photic environment as wavelength (UV to red), electric vector, light intensity shifts, and other temporal

---

STANLEY D. CARLSON • Department of Entomology, Neuroscience Training Program, University of Wisconsin, Madison, Wisconsin 53706, USA.   RICHARD L. SAINT MARIE • Department of Anatomy, School of Medicine, Boston University, Boston Massachusetts 02118, USA.   CHE CHI • Biomedical Division, Nicolet Instrument Corporation, Madison, Wisconsin 53711, USA.

and spatial features. If the resulting retinal mosaic is coarse and unfocused by vertebrate standards, other attributes of the image are sufficient to make vision a key factor in the survival of this preeminent group of animals.

An important part of our total understanding of insect vision comes from the knowledge of the photoreceptor cell's ultrastructure and its intermembranal relationship with its glial neighbors. (A detailed account of these glial cells can be found in Chapter 12 of this volume.)

Our goal in the present chapter is to explore the fine structural features of insect photoreceptor cells and briefly to relate these, where possible, to function. For the purpose of our discussion, the photoreceptor cell will be divided into three parts: receptor (soma), axon, and synaptic terminal. We will concentrate on the photoreceptor cells of the compound eye of Diptera and there is a ready rationale for this specialization. Diptera possess the best understood of all insect eyes and it is the one that we are the most familiar with. Further, the fly eye is a functional composite or hybrid of an apposition eye and a superposition eye. The best optical features of both types of eye are present in the fly; thus, its "neural superposition" eye (Kirschfeld, 1967) operates efficiently during the fly's diurnal and nocturnal forays and enhances both sensitivity and acuity.

2. Historical Record

The compound eyes of many insects are among their most obvious external features. It is not clear when man first intuitively understood that the two multifaceted hemispheres on the insect head were eyes. Late in the 17th century, van Leeuwenhoek had observed the corneal lens of a beetle under his glass lenses and found "that each lenslet forms a reduced inverted image located in the eye close to the cornea" (cited in Bernhard, 1966). With the help of better glass lenses, insect eyes were soon being dissected and sketched. In 1737-1738, Swammerdam (see Bernhard, 1966) provided useful depictions of the excised honeybee eye. But it was not until the 19th century that two landmark concepts of insect optics were developed, the issues and implications of which are still being studied. In 1826, Johannes Müller put forth the mosaic theory of insect vision which was modified and expanded for clear-zone eyes by Exner in 1891.

From visual function the torch was passed to structure and in 1909 the great neuroanatomist, Ramón y Cajal, published a histological treatise on the comparative anatomy of the retinas, one of which was the optic neuropil of the housefly. Using the technique of his contemporary and scientific adversary (Camillo Golgi), Ramón y Cajal clearly visualized photoreceptor cells, many kinds of visual interneurons and their accompanying glial cells. The neuronal inventory in insect eyes was further expanded upon by Ramón y Cajal and Sanchez (1915). During the same period (1910-1913), von Frisch began to obtain evidence that honeybees possessed color vision, and by 1927 Kühn had found that bees could distinguish the UV from other spectral areas. However, it was not until 1949 that von Frisch showed that honeybee vision encompassed polarization sensitivity. For a far greater wealth of historical detail on insect photoreceptors, the reader should consult reviews by Goldsmith and Bernard (1974) and Burkhardt (1977).

A truly spectacular skein of discoveries has issued forth in more recent times. The following major topics concerning insect vision are listed below together with key references. These subjects are not treated to any extent in this chapter, but they are necessary background if one is to fully appreciate and understand insect photoreceptors. Our apologies go to those many excellent workers whose papers could not be cited in the catalog that follows because of constraints on space. (1) The electrical response of single photoreceptor cells to known quantal fluxes, wavelengths, adaptation regimes, and electric vectors (Burkhardt and Autrum, 1960; Autrum and von Zwehl, 1964; Scholes, 1964; Shaw, 1969; Zettler and Järvilehto, 1971; Järvilehto and Zettler, 1971, 1973; Horridge, 1975a,b; Wehner, 1976; papers in Zettler and Weiler, 1976); (2) neuroanatomical specifics if visual neurons and pathways (Braitenberg, 1966; Kirschfeld, 1967; Ribi, 1975; Strausfeld, 1976; Strausfeld and Campos-Ortega, 1977; Strausfeld and Nässel, 1981; Shaw, 1981); (3) the visualization and quantification of insect rhodopsin (Langer and Thorell, 1966; Paulsen and Schwemer, 1972; Hamdorf *et al.*, 1973; Goldsmith and Wehner, 1977); (4) the cybernetics of insect vision (Reichardt, 1970); (5) waveguide theory and insect optics in general (Snyder, 1974; Snyder *et al.*, 1977); (6) the genetics of insect vision (Pak and Pinto, 1976; Harris and Stark, 1977; Heisenberg, 1979); (7) the development of the eye and optic lobe (Meinertzhagen, 1973; Trujillo-Cenoz and Melamed, 1978; Shelton, 1976); and (8) the blood–eye barriers (Shaw, 1977, 1978; Lane and Skaer, 1980).

3. Ommatidial Morphology

It is perhaps necessary to acquaint the general reader with some specialized terms for the insect visual system and to discuss several basic structural patterns present in various insect eyes. With regard to terminology, we are concerned mainly with the aggregate of photoreceptor cells (*ommatidium*) under each facet (Figures 1, 2). The word *retinular* is a synonym for *photoreceptor* and, on occasion, an ommatidium is called a *retinula*. The collection of microvilli on the medial border of each photoreceptor cell (Figures 1-5) is called the *rhabdomere* and the rhabdomeres of each ommatidim constitute the *rhabdom*. The rhabdomere is the photoreceptor organelle of the retinular cell.

The number of photoreceptor cells in an ommatidium in each species is usually constant, but that number varies among taxa. Most often the number is even, in the range of 6 to 12. For example, dipterans have 8, whereas honeybees have 9, and the tiger beetle *Cincindela tranquebarica* has 7 (Kuster, 1980). However, 8 appears to be the most common number. Most insects have two compound eyes, discounting cave species, some ants, and certain termite castes which have none, and whirligig beetles belonging to the genus *Dineutes*, which possess a dorsal and ventral eye on each side.

Of great interest to functional morphologists are the various kinds of photoreceptor cells in an ommatidium for therein may lie clues to the division of

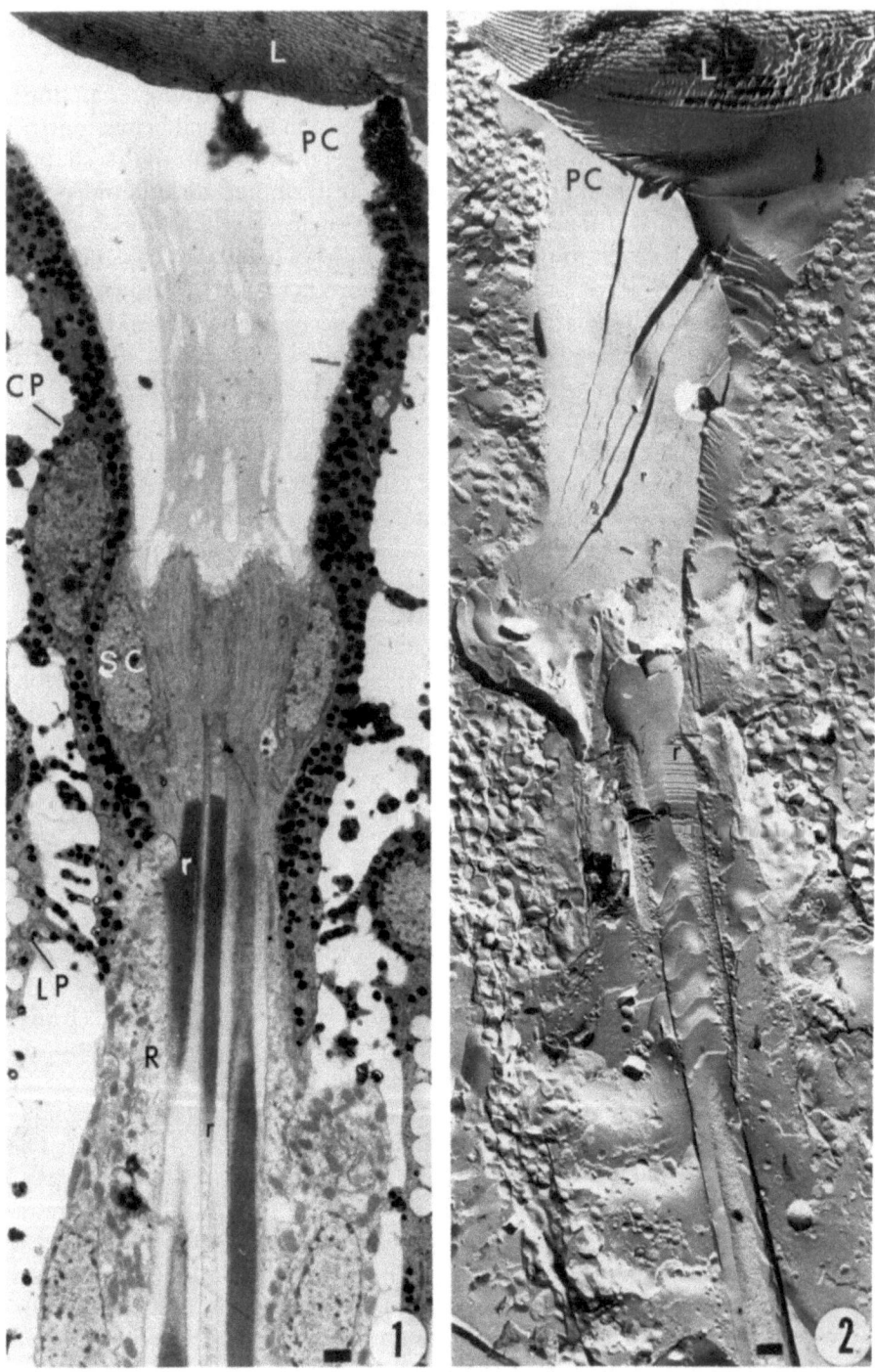

Figure 1. Thick section viewed with the high-voltage electron microscope showing the distal retinula of the housefly. Dioptric apparatus [lens (L) and pseudocone (PC)] surmount four Semper cells (SC) which directly overlie the eight photoreceptor (R) cells, some of whose rhabdomeres (r) are in this plane of section. Laterally enclosing the pseudocone are two corneal pigment cells (CP). Large pigment cells (LP) are seen on the sides (but in three dimensions encircle the retinula). (Bar = 1 μm.)
Figure 2. A freeze-fracture replica counterpart to the previous figure. (Bar = 1 μm.)

labor in the visual system. To that end in the ommatidia of given species, one finds apical cells, basal cells, cells without microvilli, central cells, peripheral cells, and tiered retinulae in which there may be two or three collections of cells in the same ommatidium, with each group forming a distinctive rhabdom at a given level. For excellent detail on the different rhabdomal patterns of insects, consult Paulus (1979). The kind of rhabdom in an ommatidium is a basic characteristic and bears elaboration.

Rhabdoms are either open (Figure 3) or fused (Figures 4, 5), meaning simply that individual rhabdomeres are contiguous or not. This characteristic has functional significance when one considers the waveguide properties of rhabdoms and the filtering and screening of light that is possible by passage through contiguous rhabdomeres or through rhabdomeres which are unfused but whose optical axes are superimposed. We will discuss in a later section (5.1) the value of twisted rhabdomeres, orthogonal sets of superimposed rhabdomeres, and short vs. long rhabdomeres in the analysis of plane-polarized light. In at least one case, however, ultrastructure has not been related to photoreceptor optics. A cross-sectioned mayfly rhabdom resembles a candelabra. Horridge (1975a) remarked that he had spent 5 years pondering the significance of this pattern "without generating a single sensible idea."

A final and basic distinction between compound eyes has allowed a classification into the photopic (apposition) and scotopic (superposition) types. The parenthesized words, in order, refer to the ommatidium being directly apposed to the dioptric apparatus (in the case of diurnal species) or being connected to the dioptric apparatus via a crystalline thread (in nocturnal-crepuscular species). The latter condition produces a sizeable "clear zone" of crystalline tracts which in most cases results in the production of superposition images. The superposition/apposition comparison extends also to screening pigment kinetics. Large pupil movements occur in the superposition eye during light and dark adaptation, while screening pigment movements are far more subtle in apposition eyes. With the advent of electron microscopy to elucidate the ultrastructural organization of insect photoreceptors, correlations between natural history and photoreceptor structure are now an essential and sometimes primary consideration.

4. Photoreceptor Fine Structure

Following the commercial availability of transmission electron microscopes (TEM) in the late 1940s, over a half-decade elapsed before Fernandez-Moran (1956) gave the world its first glimpse of thin-sectioned insect photoreceptor cells. They were from the housefly eye. In quick succession, Goldsmith and Philpott (1957), Wolken *et al.* (1957), and Fernandez-Moran (1958) provided more views of fly eyes, as well as those of lepidopterans. Despite low resolution, inadequate fixation, indifferent ultramicrotomy, and mediocre contrast (by today's standards), these early micrographs provided excitement by showing that the rhabdom was far more than merely a "refractile rod." Rather, it was made up of a linear and very regular array of microvilli. The discovery of the

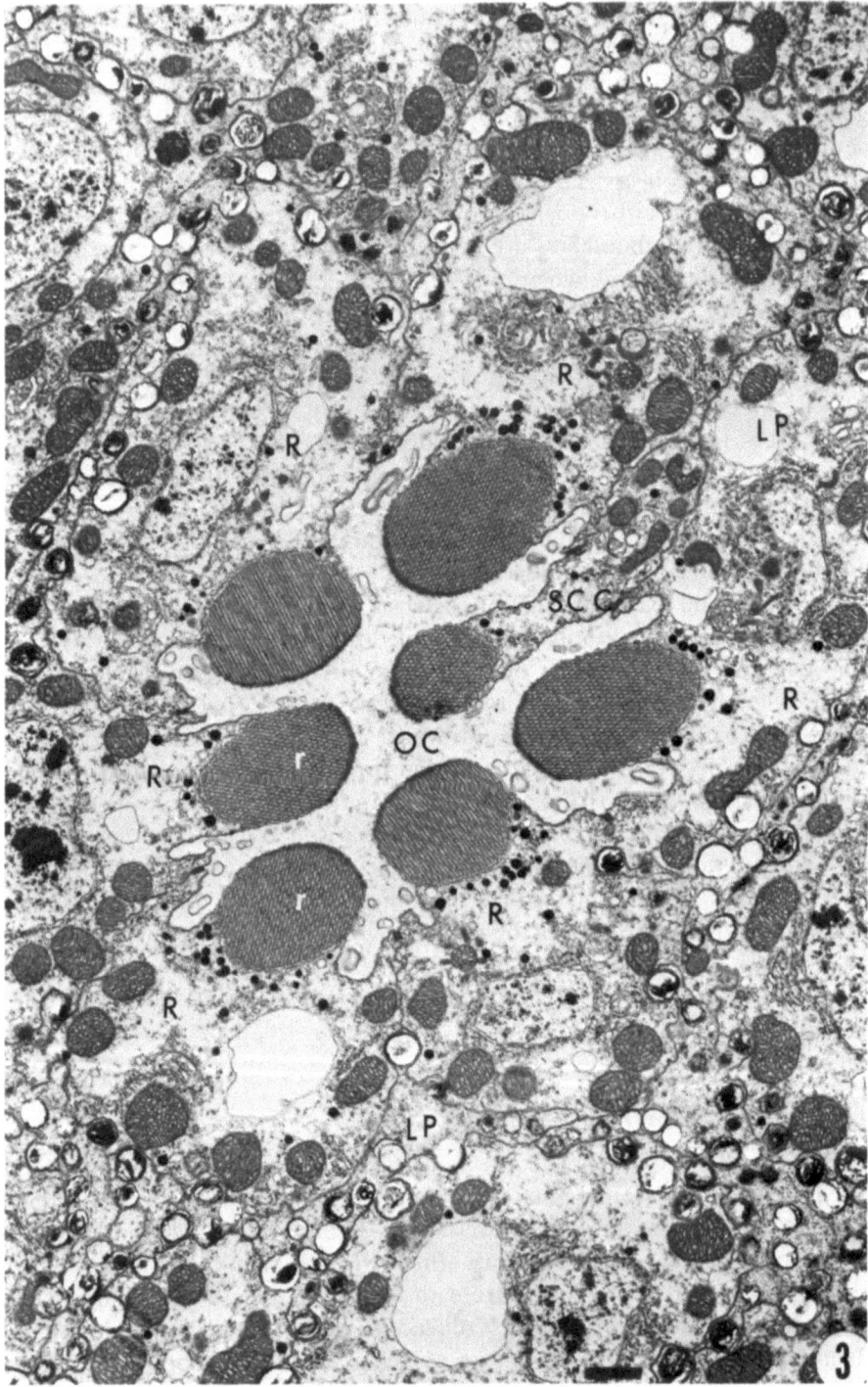

Figure 3. A cross section of a distal retinula with surrounding pigmented glia of *Drosophilia melanogaster*. Seven photoreceptor (R) cells are present at this level. One proximally situated photoreceptor cell lies beneath the superior central cell (SCC), and both cells share a common optical axis. At this distal level, the large pigment cells (LP) that enclose the retinulae are all contiguous, but

membranous nature of the photoreceptor organelle (the rhabdomere) was an extremely important addition to knowledge. At about the same time, the tracheation of the compound eye was established, and this sighting provided some understanding about how structural colors could be produced in the eye and how the spectral quality of the light entering the photoreceptors could be influenced. The latter topic will be discussed further in Section 6.

By the late 1960s, glutaraldehyde fixation was so routine that preservation and visualization of other cellular organelles (especially microtubules) was possible. Better electron optics by that time permitted resolution sufficient to distinguish gap from tight junctions, and in the past decade and a half there has been a marvelous flowering of ultrastructural findings relative to membrane specializations. The interested student of retinular fine structure will want to refer to earlier treatises by Smith (1968), Boschek (1971), Eakin (1972), Trujillo-Cenoz (1972), Goldsmith and Bernard (1974), and Carlson and Chi (1979).

From the nearly 30 years of study using TEM (including freeze-fracture), it can be stated that many of the organelles common to most cells are also present in insect photoreceptor cells. These will be described and illustrated with special emphasis on organelles and structures unique to or specially adapted to visual function, i.e., the rhabdomere, pigment granules, and certain membrane specializations.

5. Ultrastructure of the Cell Soma

5.1. Rhabdom

Eguchi et al. (1962) found (for *Bombyx mori*) that the first electrical response to light by the developing eye did not come until the rhabdom was formed. Therefore, the rhabdom is the *sine qua non* of the insect photoreceptor cell. This "photoreceptor organelle" is an absolute trimuph of bioengineering. The medial side of the cell membrane has developed into an elongate, closely packed, honeycomb array of microvilli (e.g., Figures 6, 9). Thus aggregated, the microvilli constitute a grossly increased cell surface housing a high concentration of visual pigment—a necessary and sufficient condition for an efficient photon trap.

The microvillar membrane is largely composed of the visual pigment rhodopsin (Boschek and Hamdorf, 1976; Harris et al., 1977), and its chromophore (*cis*-vitamin A aldehyde) is bent, twisted, and extremely unstable. When a photon of light collides with it, it will instantly isomerize to the all-*trans* form of the molecule (Wald, 1968).

By its construction and composition, the rhabdom has a higher index of refraction (ca. 1.35) than that of the cytoplasm or interommatidial matrix (Varela and Wiitanen, 1970). This property permits rhabdomeres to act as waveguides. Light rays directed to the rhabdom by the dioptric apparatus are thus trapped

---

at the more proximal reaches these long and slender glial cells loosely shroud each retinula. Each photoreceptor cell bears a tuft of microvilli which is separated from the rhabdomeres (r) of the neighboring cells by the fluid of the ommatidial cavity (OC). (Bar = 1 μm.)

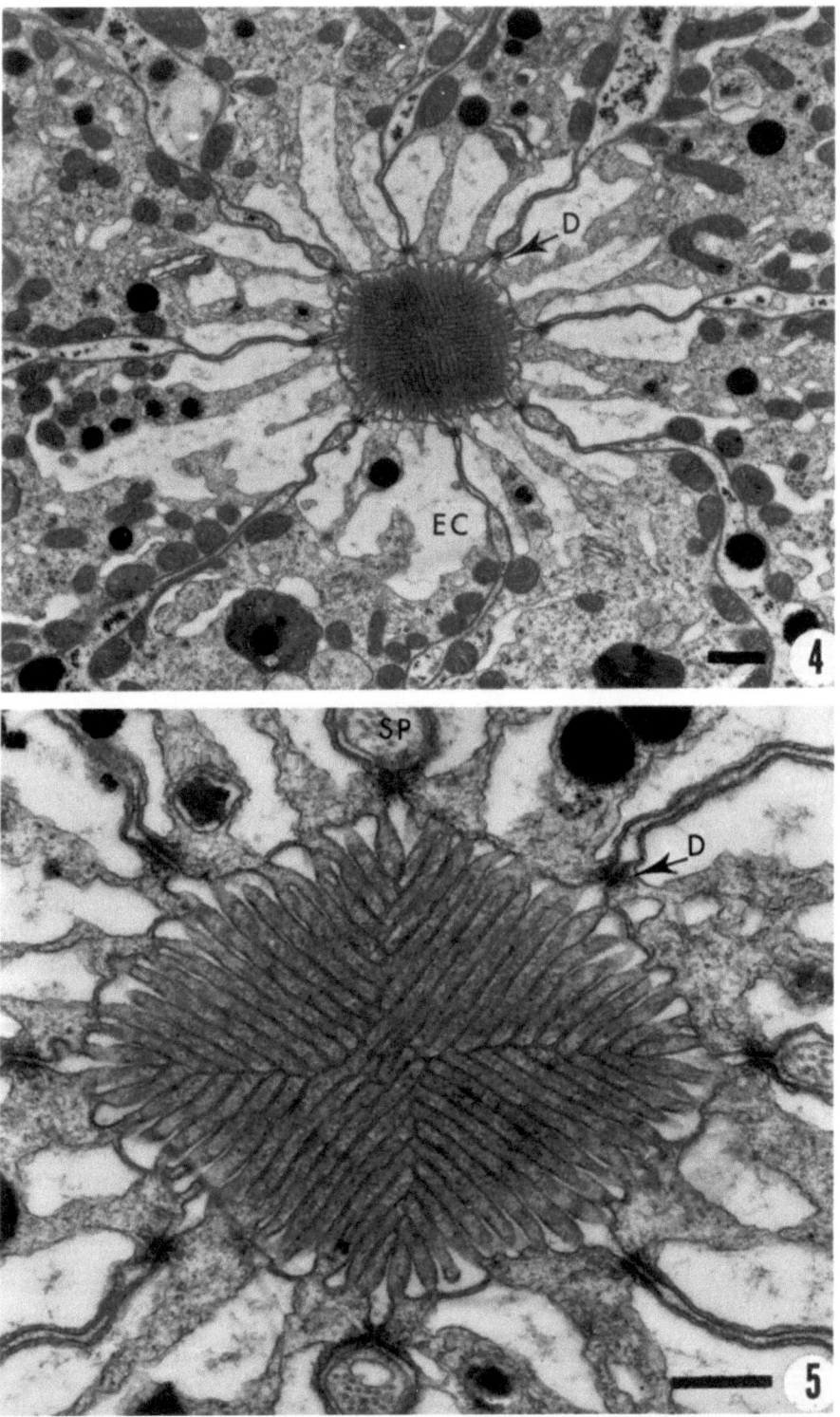

and forced to propagate down the rhabdom, where they are absorbed by the visual pigment molecules of the microvillar membranes.

Freeze-fracture studies (Boschek and Hamdorf, 1976; Nickel and Menzel, 1976; Harris *et al.*, 1977; Schinz *et al.*, 1978, 1982; Chi and Carlson, 1979) have shown that in insects the rhodopsin molecules lie semiembedded and randomly oriented in the P face of the microvillar membrane (see Figures 7, 8). Chi and Carlson (1979) have reviewed this work giving dimensions, shape characteristics, and densities of the P-face particles. Nickel and Menzel (1976) and Chi and Carlson (1979) also find ordered E-face particles that are several times larger than the P-face particles (Figure 7). Various speculations as to the functions of these intriguing intramembranous particles have been made (Chi and Carlson, 1979). They may be ion channels, a "template" imposing order on the P-face particles, precursors to rhodopsin, a UV "sensitizing" pigment, or a companion photopigment.

The ultrastructural geometry of the microvilli, relative to the light path, is important for optimal light absorption and polarization sensitivity. Microvilli in almost all insects are set at right angles to incoming quanta. The chromophore of the visual pigment molecule is dichroic, and it absorbs best along its long axis which lies in the plane of the long axis of the microvillus (e.g., Goldsmith and Bernard, 1974). This positioning is facilitated by the tight curvature of the microvilli and the elongate-elliptical shape of rhodopsin. Thus, chromophore and microvillus are orthogonal to the magnetic vector of light and also parallel to the electric (E) vector which causes the light and E vector (plane polarization) absorption to be maximized.

Another ultrastructural consideration of rhabdomeric microvilli is their assumed ability to twist along the cell's long axis. Whether or not there is an axial twist of the rhabdomeres is a subject of heated controversy. Such a deformation would necessarily reduce the summed effect of any ordering of the rhodopsin molecules, which in turn would lessen the collective ability of these molecules to maximally absorb a given E vector. More specifically, if a twist of about $1°/\mu m$ occurs, the polarization sensitivity (PS) would be more diminished in long cells

---

Figure 4. A rhabdom of the honeybee, *Apis mellifera*. The rhabdomeres of eight photoreceptor cells are contiguous at this level and throughout the length of the rhabdom. This is a fused rhabdom plan. At this low magnification, most of each photoreceptor cell can be seen. Note the enlarged cisternae (EC) of ER that radiate in spokelike fashion from the rhabdom. At this magnification, it is difficult to determine each cell's rhabdomeric contribution to the rhabdom because of the homogeneity of the collective microvilli. At the periphery of the fused rhabdom are eight belt desmosomes (D), and these enhance the adhesion of the eight cells. A ninth cell is present at a much more proximal level. Kindly provided by Dr. W. A. Ribi, Max-Planck Institut für Biologische Kybernetik, Tübingen, Federal Republic of Germany. (Bar = 1 μm.)

Figure 5. This is a higher-magnification electron micrograph of the fused rhabdom. The rhabdom presents a tetrapartite appearance and its microvilli are in two (orthogonal) orientations. The microvilli of each quadrant are contributed by two cells. The belt desmosomes (D) are better seen at this magnification as are the Semper cell processes (SP). Note the extensive space around the bases of the microvilli. This volume is confluent with the extracellular space and electron-dense tracers applied outside the cell gain access to this area and are found suspended around each microvillus (Perelet and Baumann, 1969). Kindly provided by Dr. W. A. Ribi, Max Planck Institut für Biologische Kybernetik, Tübingen, Federal Republic of Germany. (Bar = 0.5 μm.)

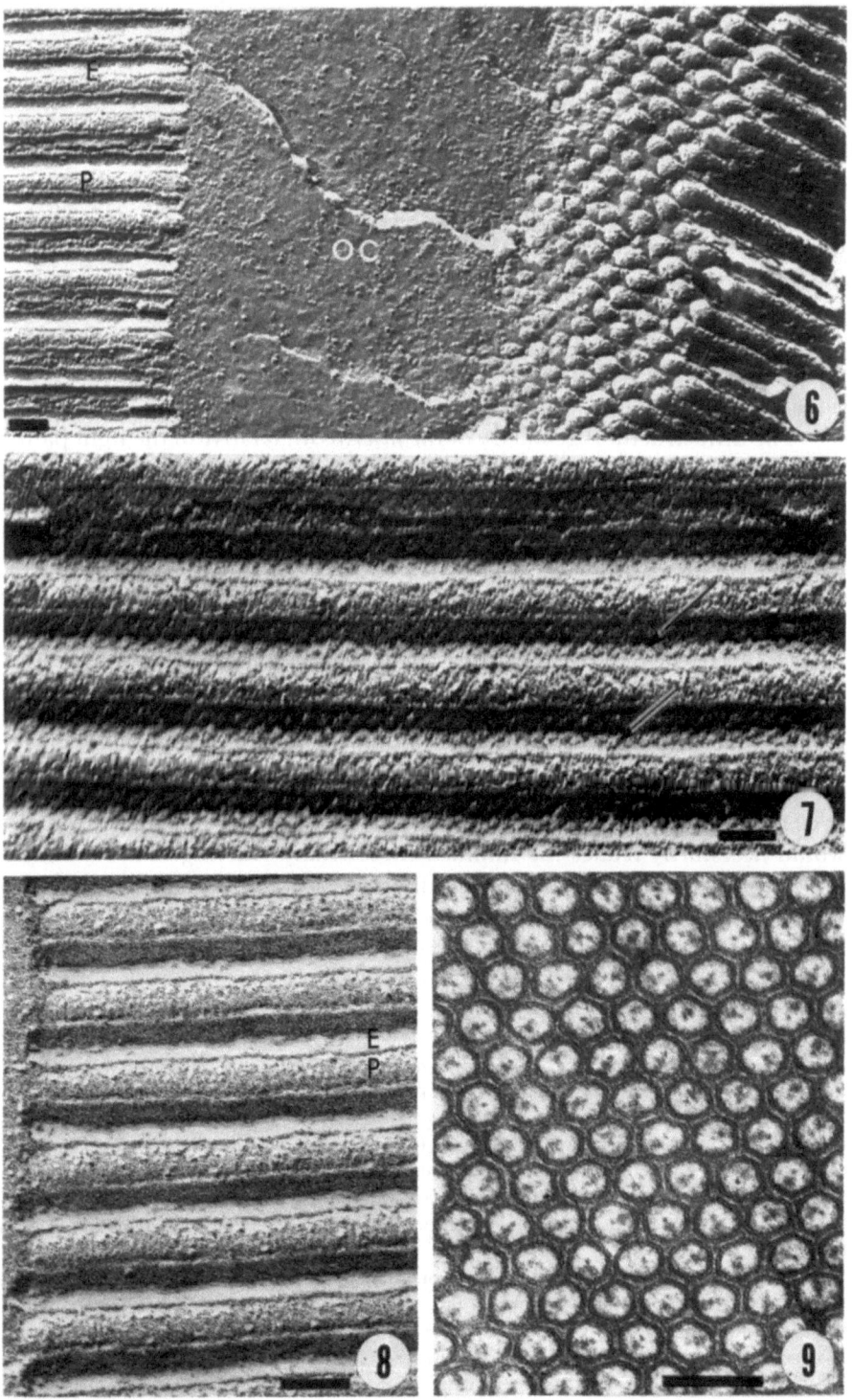

than in short ones (such as the ninth cell of the honeybee ommatidium). In this way, PS might be based on a signal ratio of two long UV cells versus that from a short UV cell. This model of the three-cell-PS unit was advanced by Wehner *et al.* in 1975. Other protagonists of twist are Menzel and Blakers (1975), Smola and Tscharntke (1979), and Wehner and Meyer (1981). Ribi (1979b, 1980) contends, however, that the ommatidia do not twist. If Ribi is correct, a new hypothesis of polarization analaysis in honeybee neurons must be set forth.

In most insects, the rhabdomere of a photoreceptor cell is a relatively straight tuft of microvilli. If the microvilli of adjacent cells in an ommatidium are contiguous, the resulting rhabdom is called "closed." The contrasting case is the "open" rhabdom in which each rhabdomere of an ommatidium is well separated from that of its neighbor. Microvilli on one cell are usually set at some angle (with respect to the horizontal) which is different from that of its neighbors. In an intriguing example, Meyer-Rochow (1974) has studied a staphylinid beetle whose rhabdom is interleafed. The microvilli of each cell emerge in a series of regularly spaced clusters which are set over or under and at right angles to similar clusters from the neighboring cell. Such an intimate orthogonal arrangement would seem to be an ideal analyzer of linearly polarized light since decapod crustaceans which possess similar rhabdoms have excellent PS. Surprisingly, however, this beetle has a low PS, and Meyer-Rochow suspects that there is electrical coupling between cells thus negating any rhabdomeric winnowing of the E vector.

The quantity and confluency of the extracellular fluid around each microvillus has a bearing on the shunting and coupling of electrical currents in excitation and inhibition. Where this relationship has been studied, no blood-eye barrier has been found to exist between extracellular space of the eye capsule and rhabdomeric microvilli. When used as an electron-dense tracer, lanthanum finds its way into the interommatidial cavity of the housefly after penetrating the R-cell junctions and belt dimensions, and these molecules are ultimately seen between each microvillus from the tip to base (Chi and Carlson, 1981). Similar findings were reported by Perelet and Baumann (1969) for the honeybee and by Shaw (1978) for the locust. Thus, there appears to be ionic continuity between the

---

Figure 6-8. Freeze-fracture replicas of the photoreceptor cells of houseflies.

Figure 6. The cleavage plane reveals rhabdomeric microvilli (r) in longitudinal (left) and oblique (right) presentations. A featureless interommatidial matrix (OC) surrounds the microvilli on their medial and lateral surfaces. Intramembranous particles seen on the P face (P) of both sets of microvilli may be rhodopsin molecules. The E face (E) of the microvilli contains fewer particles. (Bar = 0.1 μm.)

Figure 7. A higher magnification of several longitudinally cleaved microvilli. Note the prominent rows of membrane particles (arrows) semiembedded on the E face of the microvillar membrane. (Bar = 0.1 μm.)

Figure 8. The particles on the P face of the rhabdomeric microvilli are assumed to be rhodopsin molecules. Their orientation is random, and they are densely packed together. (Bar = 0.1 μm.)

Figure 9. A thin section of rhabdomeric microvilli cut transversely. Each microvillus is surrounded by a discernible extracellular space. Filamentous material is present in the center of each microvillus. Close observation reveals the contiguity of membrane from one microvillus to its neighbor in some cases. (Bar = 0.1 μm.)

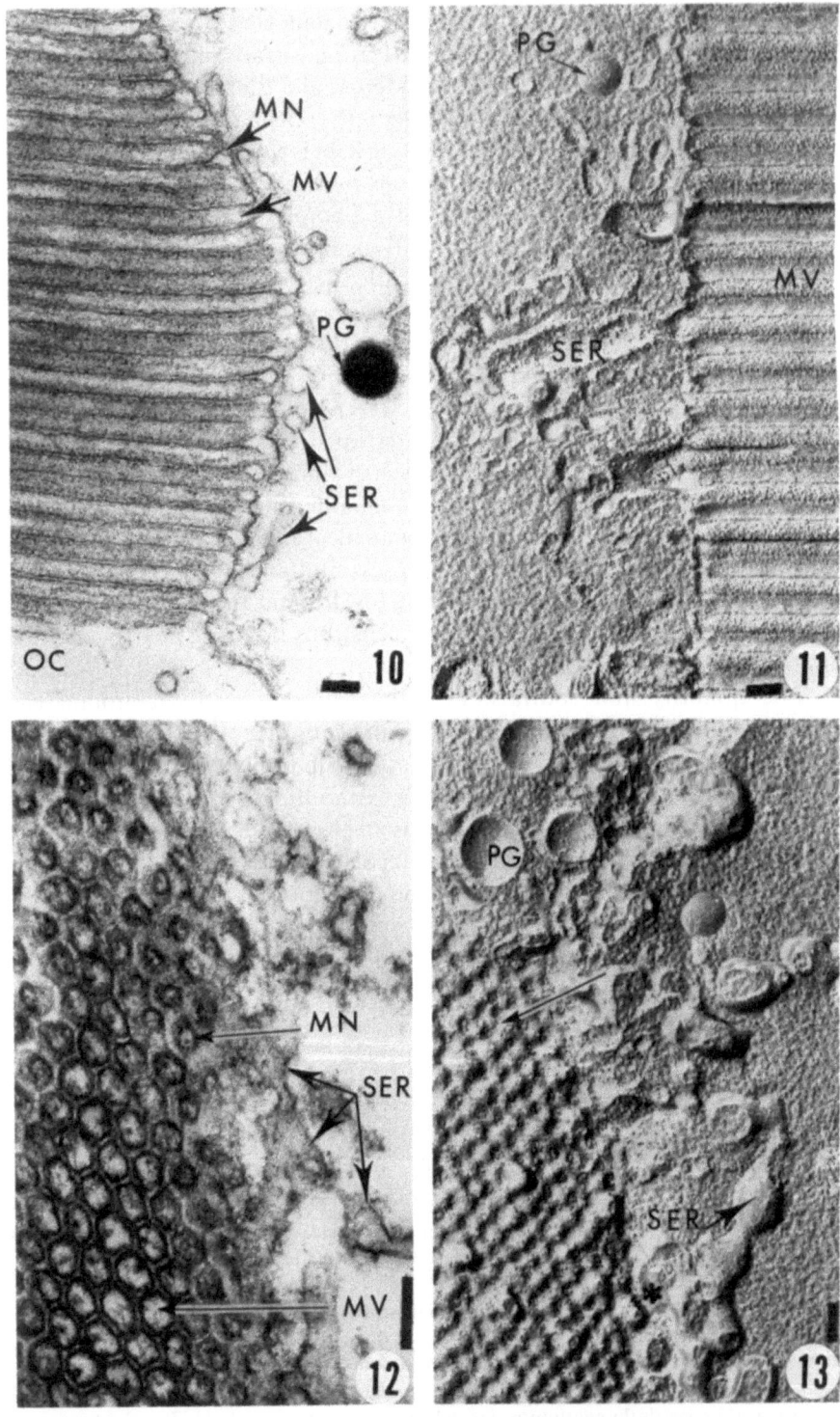

extracellular fluid bathing the outside of each ommatidium and the fluid that percolates in the miniscule spaces between and around each microvillus (Figure 16). These ionic confluency data raise more questions than they answer, particularly when one assumes that the electrophysiological characteristics of specialized (rhabdomal) and nonspecialized membranes of the photoreceptor cells are different and that adjacent cells in the same ommatidium may have dissimilar spectral sensitivity functions.

The photoreceptor cells of some homopterans lack microvilli. Duelli (1978) has described the cells from the ocelli of a male scale insect which have stacks of membranes rather than cylindrical microvilli, and his calculations show that the stacks contain no more photosensitive membrane per unit volume than rhabdomeres of housefly ocelli. Therefore, stacks may be as efficient as cylinders (microvilli), and the reader is reminded of the fine functioning stacks of double-membrane disks in the outer segments of the rods and cones of the human eye.

5.2. Nucleus

There is nothing unusual about the fine structure of the photoreceptor nucleus. In thin sections (Figure 3), a spherical nucleus is seen within which resides a nucleolus. Chromatin appears in clumps scattered throughout the nucleoplasm. A perinuclear envelope is always present and the nucleus may be nearly as voluminous as the cell itself, so that rather little cytoplasm exists in the perikaryal region.

5.3. Endoplasmic Reticulum

Both smooth (agranular) and rough (granular) variants of ER are known for insect photoreceptor cells (Figures 4, 5, 10-14). Extensive, slender, agranular cisternae exist at the base of the rhabdomeric microvilli (Figures 12-15). Trujillo-Cenoz (1972) states that in dipterns these "long channels communicate with the ommatidial central cavity and penetrate deeply in the cytoplasm." Figures 12

Figure 10. An electron micrograph of longitudinally thin-sectioned microvilli and their bases from a housefly eye. The microvillar neck (MN) constricts but does not occlude the microvillus (MV). Note the proximity of the cisternae of the smooth endoplasmic reticulum (SER). Pigment granules (PG) are scattered throughout the cytoplasm of the photoreceptor cell, and the interommatidial matrix (OC) contains an occasional vesicle. (Bar = 0.1 μm.)

Figure 11. The freeze-fracture replica counterpart to the previous figure. Note regions where the membrane of the SER is continuous with that of the micrivilli (MV). This suggests that the lumen of SER and the extracellular space are confluent. (Bar = 0.1 μm.)

Figure 12. A thin section of rhabdomeric microvilli (in cross-section). The relative diameters of the transversely sectioned microvillus at its neck (MN) and at more distal levels (MV) are seen. Note the extent and geometry of the SER. (Bar = 0.1 μm.)

Figure 13. The freeze-fracture replica counterpart to the preceding figure. Note anastomosing cisternae of the SER. The long arrow indicates the E face of the microvillar base. The asterisk points to cross-fractured ER in very close proximity to the microvilli. PG, pigment granule. (Bar = 0.1 μm.)

and 13 show that in the eye of *Musca domestica*, these cisternae are interconnected to form a network.

Dark and light adaptation of the eye may cause these cisternae to alter their morphology and volume. Generally, dark adaptation promotes an aggregation and expansion of the SER to a position around the base of the rhabdomeric microvilli, as found for honeybees (Kolb and Autrum, 1972) and locusts (Horridge and Barnard, 1965). Thus, a membranous "overcoat" of ER ensheathing the rhabdom may serve an optical function. The rhabdom cylinder plus the adjacent ER volume would make a larger-diameter "light pipe," thus permitting more light to be propagated directly through the rhabdom. This ER, when it is phototropic, is called a *palisade* (Figures 4, 5). When the eye is light adapted, the ER channels become more fragmented and retreat to the periphery of the cell. RER does not exhibit unusual motility and is found throughout the somal region, often in proximity to mitochondria (Figure 14).

Rather little is known directly about the usual metabolic roles of the ER in insect photoreceptor cells, i.e., as the site of lipid or protein biosynthesis or as a cellular storage and transport system. The ribosomes and ER likely play a sizeable role in the synthesis of rhodopsin and other membrane constituents. Pepe and Baumann (1972) and Perelet (1972) have shown this to be the case, although the synthesis site was not close to the microvilli and a transport system for rhodopsin had to be postulated. Synthesis may be cyclical given the increased anabolic activities associated with photopic conditions and the daily reconstructions of the rhabdom. Coated vesicles occur at the junction between the microvillar base and the SER. This association may indicate that membrane components (perhaps even rhodopsin) may gain immediate access to, or be recovered from, rhabdomeric microvilli via the SER. Thus, renewal of receptor membrane (with its chromophore) may depend on such putative transport capacities.

5.4. Screening Pigment Granules

In many insect species, these small, spherical, electron-dense organelles are found throughout the length of the photoreceptor cells. In the housefly, pigment granules are present mainly in the soma (Figures 3, 10), but they are sometimes seen in the axon (Figures 19, 29). This seems strange because no light screening function would be performed at this level. Granules also accumulate near the axonal constriction at the basement membrane (Figure 19).

The movement of pigment granules in the rhabdomeric portion of the cell in *Musca* has been studied by Kirschfeld and Franceschini (1969), Stavenga *et al.* (1973), and Franceschini and Kirschfeld (1976). These granules migrate to the bases of the rhabdomeres during light adaptation, but under dark conditions the granules are randomly dispersed throughout the cytoplasm. Thus, pigment movement acts as an iris to attenuate light scattering and increase acuity. Interestingly, this light reaction was not seen for the superior central cell (and presumably for its underlying counterpart, the inferior central cell) until the light levels were further increased by two log units. This was additional evidence that the two central cells constitute the photopic channel, while the peripheral cells formed a more scotopic channel.

The physical forces that effect movement of pigment granules within the cytoplasm are not known, nor have ultrastructural studies revealed any fibers or other mechanical contrivances that would enable movement. Franceschini (1975) reports that the electrical activity which results from the absorption of light in the visual pigment of the rhabdomere is in some way responsible for the uniform and rapid translocation of pigment granules to the rhabdomere base. One is tempted to speculate that pigment granules are given a "free ride" on some hitherto unseen (unstained) contractile elements of the microtrabecular lattice of the cell, and it is some constituent of the latter ground substance that is motivated by light. One can see some parallels between the screening pigment movements in insect retinular cells and the Ca^{2+}/ATP-dependent system of granule dispersion along microtubules in squirrel fish erythrophores (Luby and Porter, 1980). The lack of finding some motile element connected to pigment granules may relate to the conventional fixation techniques employed.

Finally, it should be remembered that screening pigments not only attenuate light but they preferentially absorb particular wavelengths and thus tint incoming light. This feature may enhance the spectral sensitivity of the receptors if the "window" in screening pigment absorption spans the λ_{max} of the insect's rhodopsin. Papers by Höglund (1966) and Höglund and Struve (1970) give these data, and reviews by Langer (1975) and Goldsmith and Bernard (1974) provide information on the composition of these granules.

5.5. Mitochondria

Insect photoreceptor cells contain considerable numbers of mitochondria, which is not surprising considering this cell's high energy requirements. Mitochondria are rod-shaped with a simple, lamellate pattern of cristae (Figures 3, 14). Moving about in the cytoplasm, mitochondria are probably quite flexible, and severely bent forms have been seen.

Mitochondria are not always randomly dispersed in the cytoplasm of photoreceptors, and in the fly there is a definite tendency for them to be located near the plasma membrane. Thus, one often sees a double row of mitochondria apposing each other across the intercellular cleft between two neighboring photoreceptor cells (Figure 14). In some cases there appears to be actual contact between mitochondrion and plasma membrane of the photoreceptor cell (Chi *et al.*, 1979). Earlier, Gribakin (1975) mentioned this juxtaposition in honeybee retinular cells. Between such mitochondria-lined cell membranes in the fly is an intercellular specialization which we (Chi *et al.*, 1979) have christened *the R-cell junction* (Figures 14, 15, 17).

This membrane specialization is somewhat similar to the scalariform junctions found in the transporting epithelial cells of the insect gut (see Lane and Skaer, 1980; Lane, 1982). More details about this junction as it pertains to the photoreceptor axons are presented in Section 7. The mitochondria of insect photoreceptor cells probably provide (among other things) the energy for active

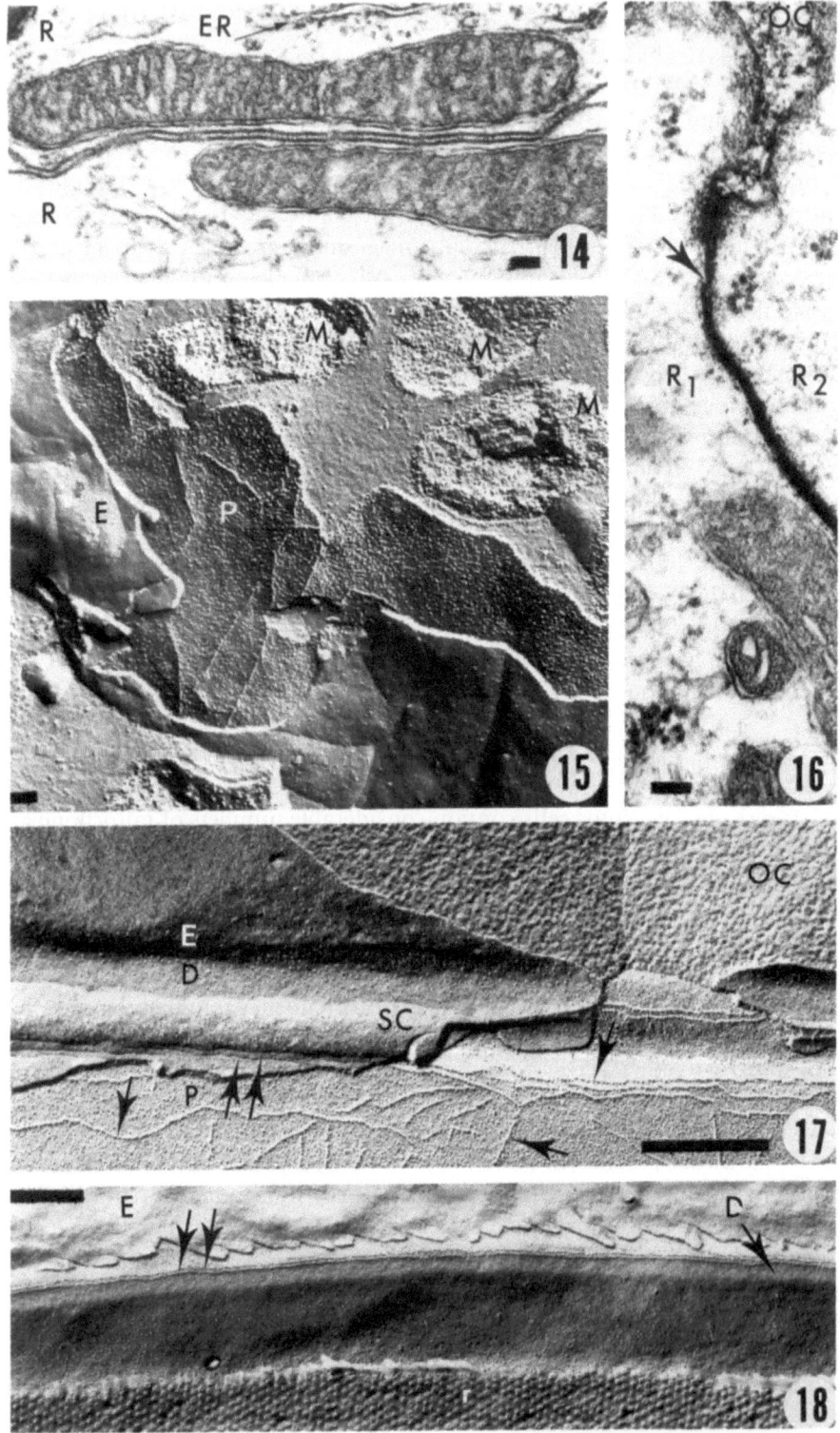

transport across the cell membrane in the immediate area of this peculiar junction. There may be a relationship between the aforementioned mitochondrial positioning near the cell membrane and Horridge's (1966) report that mitochondrial migration in *Locusta* is induced by changes in light adaptation. With light-on, the cell is depolarized or "turned on" and overall energy requirements are heightened. Mitochondrial movement to ATP-dependent sites would be advantageous in order to facilitate energy transfer.

5.6. Golgi Material

We have rarely encountered Golgi complexes in photoreceptor cells although these are commonly seen in the second-order cells [monopolar interneurons (Carlson *et al.*, 1983)]. When seen, these organelles are close to the nucleus with their forming face near the nuclear envelope.

5.7. Ciliary Structures

The nonmotile cilium with its centrioles and roots is a prominent fine structural feature in several kinds of cuticular sensillae of insects (chemoreceptors, mechanoreceptors, hygroreceptors). This composite structure, with axoneme, is a connecting piece between the outer and inner segments of the dendrite of the bipolar sense cell. In insect photoreceptors, such structures are rarely encountered, and they have not been seen in the housefly. However, Home

⬅

Figure 14. Contiguity of two photoreceptor cells (R) of the housefly. The intercellular area contains indistinct septa. The thin-section manifestation of this intercellular junction is seen in Figure 21. This specialization may be a variant of the scalariform junction. The mitochondrial alignment and proximity are characteristic of this intercellular specialization. Rough endoplasmic reticulum (ER) is adjacent to the mitochondrion. (Bar = 0.1 µm.)

Figure 15. The freeze-fracture face of the R-cell junction. The P face (P) of this membrane specialization is characterized by numerous intramembranous particles randomly distributed or organized into short, linear ridges that often extend longitudinally along the photoreceptor cell. Grooves are the counterpart of these ridges in the E face (E) of the adjacent R cell. Nearby mitochondria (M) are cross-fractured. (Bar = 0.1 µm.)

Figure 16. Lanthanum, used as an extracellular tracer, penetrates the intracellular cleft between two retinular cells (R_1, R_2) and enters the interommatidial matrix (OC). The arrow points to an apparent focal tight junction which interrupted the penetration of tracer. (Bar = 0.1 µm.)

Figure 17. A freeze-fracture replica of the junction between two photoreceptor cells. The Semper cell process (SC) extends from the Semper cells (distal to the photoreceptor cells) along much of the length of the receptor cells. The belt desmosome (D) is also present and projects in parallel with the Semper cell process. The former "staples" together the two adjacent photoreceptor cells. The correspondence of P-face ridges (single arrow) and E-face grooves (double arrow) is of particular interest, since these potential tight junctions occur consistently to either side of the Semper cell process. Other P-face particle ridges (single arrows) are seen aligned in various orientations. (Bar = 0.5 µm.)

Figure 18. A freeze-fracture replica of a large pigment cell [with exposed E face (E)] and a photoreceptor cell bearing rhabdomeric microvilli (r). Note the continuous, double rows of fused P-face particles (double arrow) just proximal to the belt desmosome (D). (Bar = 0.5 µm.)

(1972, 1975, 1976) finds ciliary structures in the photoreceptor cells of over a dozen beetle species. Home (1976) suggests that these ciliary structures play a mechanical role in maintaining the structural integrity of the rhabdom.

5.8. Other Intracellular Structures

In well-fixed photoreceptor tissue, thin sections usually reveal a wide variety of organelles other than those given separate mention in this chapter. In the housefly and most other insects examined, one sees: glycogen granules (Figure 16), vacuoles (Figure 3), dense bodies, free ribosomes, myeloid bodies, pinosomes, liposomes, and a variety of lysosomes. Meyer-Rochow (1972) has described a microsporidian that invades the retinula cells of a staphylinid beetle, and bacteria have been found in the retinula cells of the meal moth by Horridge and Giddings (1971).

Myeloid bodies and lysosomes develop extravagantly in certain genetically induced retinal pathologies (Stark and Carlson, 1982). Lysosomes are also indicative of the short duration but intense synthetic and degradative activity involved in the daily cycling of the rhabdom in insect photoreceptor cells (Blest, 1980).

6. Cells Associated with Photoreceptor Cells

No eye consists only of a mass of photoreceptor cells. Additional cellular and acellular components are invariably present in the compound eye, namely, pigmented glia, Semper cells, tracheae and tracheoblast cells, corneal lens, cones, and basement membrane (see Figure 19). All these are present and indispensible accessories of insect photoreceptor cells.

In the eye, the pigmented glial make up the most extensive extrareceptor tissue. These are discussed in some detail in Chapter 12. In the fly's peripheral retina, there are three distinct forms of these heavily pigmented cells: corneal (primary), large (secondary), and small pigment cells (Figure 19). The large and corneal pigment cells embrace the distal retinula, and a cluster of four small pigment cells are near the base of the rhabdomeres. These cells optically isolate individual ommatidia from each other, and presumably have a metabolic and ionic interdependency with the adjacent photoreceptor cells. With regard to the latter functions, Figure 18 shows the permeable scalariform junction between a large pigment cell and the neighboring retinular cell.

The corneal lens (Figure 19) is a layered cuticle which is more or less transparent. This cuticular cornea is divided into lens facets whose *en face* form is hexagonal, square, or round, and unfaceted corneas are known. Each facet exhibits both radial and axial variations in index of refractions (Kunze, 1979). On the corneal episurface, there may be arrays of discrete nipples which are too small to be seen by light microscopy (see Bernhard *et al.*, 1965). The housefly has a labyrinthine pattern of truncated nipples, while the surface of the grasshopper lens facet is perfectly smooth. Nipples are an antireflection and an impedance-

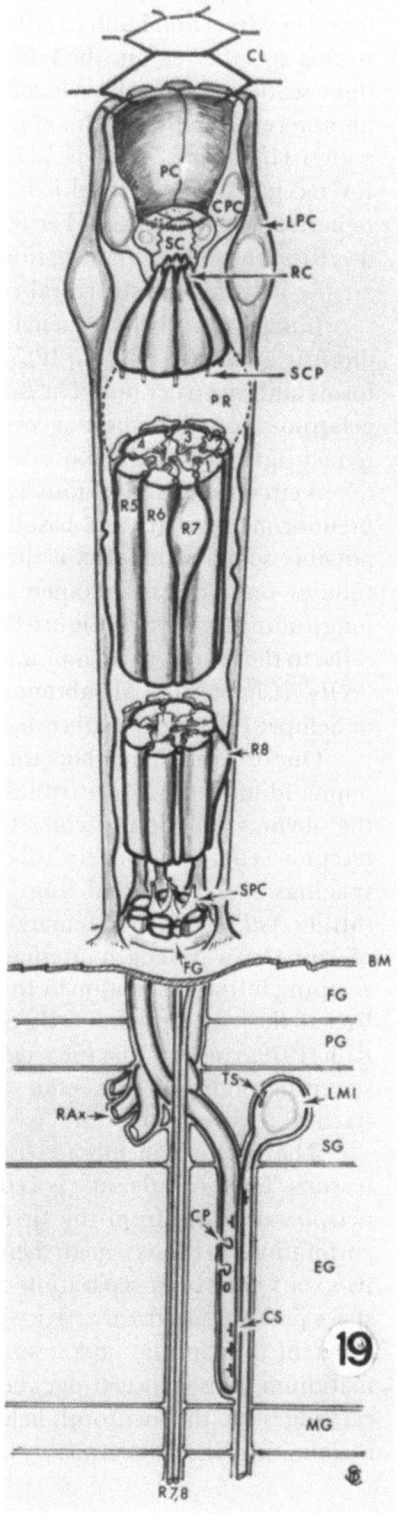

Figure 19. Semidiagrammatic representation of the dipteran ommatidium. Distally, the corneal lens (CL) overlies the clear, gelatinous pseudocone (PC) which is laterally enclosed by the corneal pigment cells (CPC). The base of the pseudocone is composed of four joined Semper cells (SC) whose long processes (SCP) descend to the basement membrane (BM) where they expand into four bulbular cell-like processes which are called *small pigment cells* (SPC). Large pigment cells (LPC) extend from each inner lens margin to the basement membrane. Seven linear photoreceptor cells form a trapezoidal array (*en face*), each terminating distally in a rhabdomere cap (RC). At a slightly lower level, each of the peripheral retinular cells (R1-6) shows a bulging perikaryal region (PR). A central cell (R7) intrudes its rhabdomere into the interommatidial cavity at this level, and its perikaryon is immediately below those of R1-6. R8 is the inferior central cell, and its rhabdomere is short and beneath that of R7. Both rhabdomeres share the same optical alignment. Just above the basement membrane, all photoreceptor cells take on the appearance of (and function as) axons (RAx). For a short distance, the axons maintain their ommatidial cohesiveness in pseudocartridge formation. Then each of the R1-6 axons twists 180° and projects to a different optic cartridge containing five lamina monopolar interneurons (LMI), one of which is seen here. Several of these relay interneurons are chemically postsynaptic (CS) to the retinal axons. Occasionally, the retinal axons are found to be postsynaptic to the monopolar interneurons. In passing from the peripheral retina to the first optic neuropil to synapse, the R axons encounter, successively, four strata of glial cells; glia of the fenestrated area (FG), glia of the pseudocartridge area (PG), satellite glia (SG), and then epithelial glia (EG). Satellite glia form trophospongia (TS) into the perikaryal region of the monopolar neurons. Capitate projections (CP) are made by epithelial glia into retinal axons. The R7, R8 axon pair from each ommatidium join below the basement membrane and project directly to the second optic neuropil without synapsing. Thus, they pass through (and are surrounded by) the marginal glia en route to the medulla externa (second optic neuropil).

matching device, enhancing transmission of light through the cornea. An example of the corneal contribution to photoreceptor function was recently found by Meyer and Labhart (1981) in the dorsal rim area of the honeybee cornea. In this specific region, the lenses have numerous pore canals which enhance light scattering, thereby widening the visual fields of the directly underlying photoreceptor cells which are primarily responsible for detecting polarization patterns in the sky. Wide visual fields are highly advantageous to these particular UV receptors. The corneal lens also screens out harmful middle and near-UV radiation (Goldsmith and Fernandez, 1968), and in some dipterans (notably the deerflies) the corneal laminations constitute quarter-wavelength filters which are the basis for the structural colors of the eye (Bernard and Miller, 1968).

Immediately below each lens facet is a cone—the basal partner of the dioptric apparatus (Figure 19). This part of the lens cylinder takes a variety of forms and constructions. The housefly has a pseudocone (Figures 1, 2) which is gelatinous, optically homogeneous and, in concert with its distal lenslet, both refract light rays to the most distal plane of the rhabdomeres.

A circular cushion of four Semper cells (Figure 1, 2, 19) form and support the pseudocone laterally and basally. Light passes through the Semper cells and possibly some of this flux is directed through longitudinally coursing microtubules onto the rhabdomere caps. These cells often have very elongate, longitudinal processes (Figure 17) that extend along and between photoreceptor cells, to the basement membrane where they enlarge to form the small pigment "cells" (Chapter 13). Membrane specializations and other ultrastructural details of Semper cells are described by Chi et al. (1979).

One or several large-bore tracheae usually project distally up the side of each ommatidium from larger trunks beneath the basement membrane. Apart from the obvious function of conveying respiratory gases to and from the photoreceptor cells, these silvery tubes in some species form a tapetum. When the spacings of the taenidial rings of certain moths and skippers were examined (Miller *et al.*, 1968), their remarkably consistent periodicity seemed to be the basis of a quarter-wavelength interference filter system. Thus, incoming white light, escaping initial absorption by the rhabdom, would reflect a particular waveband back to the photoreceptor cells. Such, however, is not the case with the housefly. Ribi (1979a) reports that the red eye glow in pierid butterflies emanates from red screening pigments that reside within the photoreceptor cells, and there is no tracheal involvement.

The basement membrane (Figure 19, 20) is a noteworthy structure for several reasons. This acellular mat is a convenient anatomical landmark, separating the peripheral retina from the first optic neuropil, and, most importantly, the perforations in the basement membrane through which each ommatidium sends its axons enforces a separation of each ommatidium from its neighbors. This space prevents electrical activity between the lateral-lying photoreceptor cells of different ommatidia, and it surrounds the photoreceptor axons of one ommatidium (a pseudocartridge) before their twisting departure to different optic cartridges in the neuropil below the basement membrane. The basement membrane of the fly is probably permeable to many ions and solutes. The area of high electrical resistance found near the basement membrane is, in all likelihood, due to membrane specializations between the lamina glia of that area and photoreceptor axons (Saint Marie and Carlson, 1982b; Chapter 12 in this volume). Odselius and Elofsson (1981) have written a useful review on the morphology and cellular derivation of the basement membrane.

7. The Axon

Three main aspects of the morphology of axons must be considered: their general form, their organelles, and their membrane specializations. Chi and Carlson (1976) have shown for the housefly that retinal axons at the region of the cartridge neck branch into each other to the point that one axon or axonal process was completely engulfed by the axoplasm of its neighbor. Axons also appeared to be herniating into each other, again, only at the level of the cartridge neck.

The organelles of the axon can be quickly enumerated. Immediately below the last rank of rhabdomeric microvilli, one notices mitochondria, longitudinally directed microtubules, a few scattered pigment granules, ribosomes, and cisternae of smooth and rough ER (Figure 20). In the plexiform layer, hordes of synaptic vesicles, numerous presynaptic structures, and capitate projections are found. These are discussed in Section 8.

Information is becoming available on the intercellular associations of photoreceptor axons, their interneurons, and the accompanying glia. Membrane specializations are the bases for chemical and electrical synapses, intercellular cohesion, ion transport and barrier systems. Knowledge of these has been instrumental to our current understanding of the functional organization of the insect visual system. What follows is a synopsis of the various junctions that the photoreceptor axons makes with its neighboring cells, be they neurons or glia, along with some thoughts as to their functional significance. Details of the fine structure and position of the various glia in the lamina can be found in Chapter 12. Specializations of the axolemma that represent electrical or chemical synapses are described in Section 8.

In the introduction of this chapter, it was stressed that the photoreceptor cell could be morphologically and physiologically bisected. This means the distal soma portion containing microvilli specializes in reception, while the proximal axonal region effects transmission of the graded receptor potentials. For functional reasons, it is suspected that each sector of the cell requires its own ionic milieu (Shaw, 1977), necessitating the peripheral retina to be ionically "walled off" from the first optic neuropil. In Figure 21-29, some of these partitions in the form of membrane specializations are seen. Figure 21 shows the P-face particle ridges that continue the R-cell junction from the distal retinula to below the basement membrane at the level of the fenestrated glia. Some polygonal enclosures are noted (Figure 22) but, for the most part, the ridges and grooves are roughly parallel to the axon's long axis. P-face ridges and E-face grooves (Figure 22) correspond to and can be correlated with creases in the

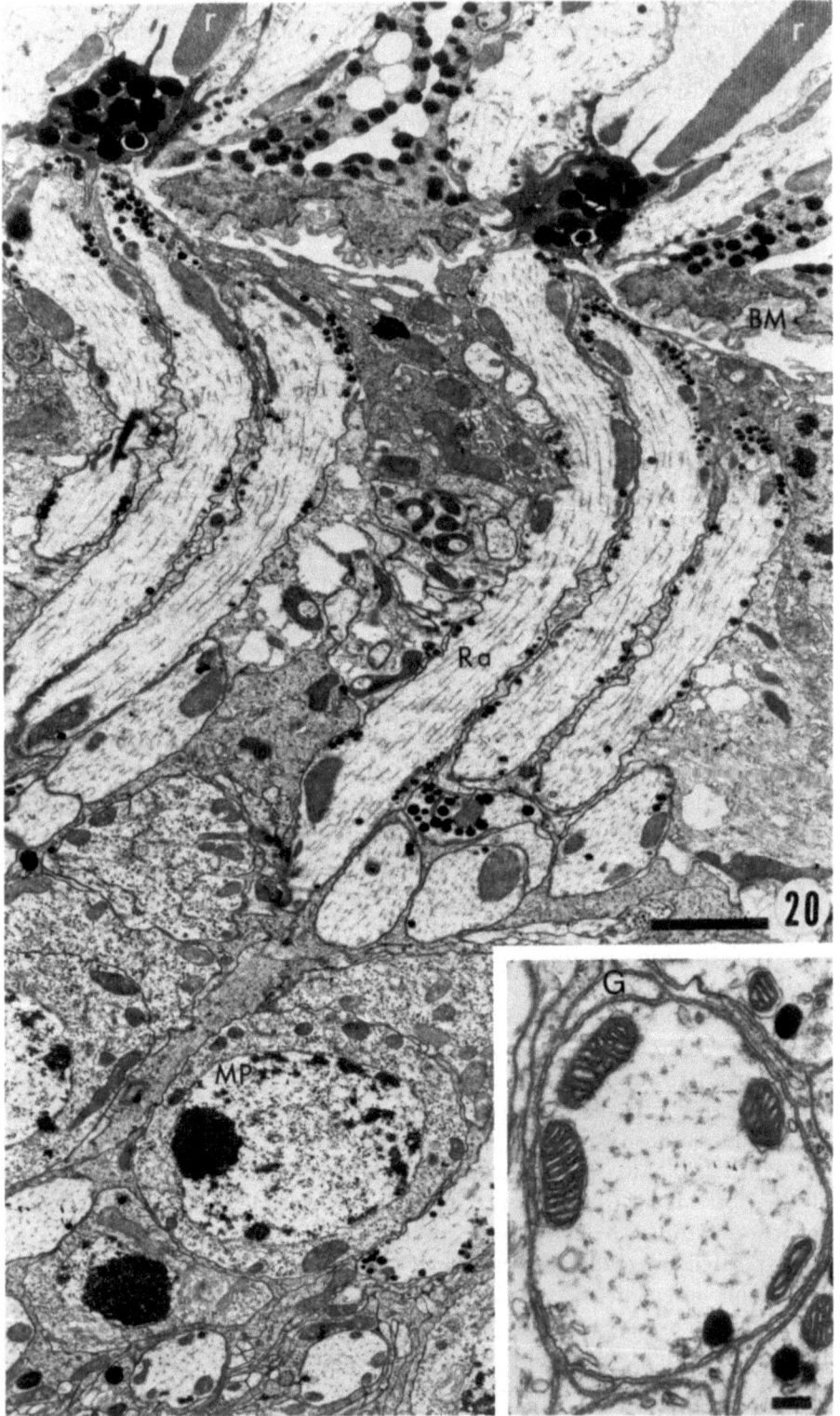

axolemma, but Figure 21 clearly indicates that the R axons do not make tight junctions with the glia in the fenestrated layer. The contours of the glial cell fit well with those of the apposing R axons, but the membranes never fuse.

Numerous and unordered intramembranous particles are seen between the fused particle ridges and in some sectors the particles form into fine, subtle ranks (Figure 22). The thin-section correlate of these superficial series of linear striae is seen in Figure 23. At the junction between the glial cell and the retinular axon, a zipperlike structure is seen. Saint Marie (1981) reported that this junction was penetrated by lanthanum and that the striations were then revealed as fine, regularly spaced, 5-nm filaments which bridged the intercellular cleft and were aligned into multiple parallel rows. This junction is rather similar to the previously mentioned R-cell membrane specialization (Section 5.5), and it also resembles the scalariform junction found in insect transporting epithelia (see Lane, 1982).

At the next (more proximal) level, the retinal axons abut the pseudocartridge glia (Chapter 12). Electron micrographs of thin-sectioned (Figure 24) and lanthanum-infiltrated cells (Figure 26) corroborate the presence of extensive septate junctions at the axoglial appositions. The rivetlike appearance of P-face particles (Figure 25) is the freeze-fracture evidence of this specialization.

As retinal axons descend to the nuclear layer (Figure 20), wrapped by the distinctly different glia of that stratum, both septate and tight junctions girdle the axon (Figure 28). Near the base of the nuclear stratum, only tight junctions are seen. A correlative thin section (Figure 27) demonstrates focal tight junctions between axon and glial cell, and part of a scalariform junction (between axon and glial cell) with the usual mitochondrion. E-face views of the axon at this level (Figure 35) and in the fenestrated layer often reveal faint parallel furrows thought to be analogous to the ordered ranks of particles found on the P face (Figure 22). The repeat distance of these linear P- and E-face arrays corresponds to the spacing between the faint striations (Figure 29) of the thin-sectioned, scalariform junctions (Saint Marie, 1981).

Projecting proximally and leaving the nuclear layer, axons of the peripheral cells enter the plexiform layer where synapses are made. At this level, neurons become stouter and are invaginated by myriads of small capitate projections from the surrounding epithelial glial cells. In freeze-fracture replicas, bowtie-shaped aggregations of ordered P-face particles (Figure 33) are found on the

⬅

Figure 20. Survey (montage) of the boundary area between the peripheral retina and the first optic neuropil. The acellular basement membrane (BM) separates these regions as well as the two different structural regions of the photoreceptor cells, i.e., above this fenestrated mat the cell posseses rhabdomeric microvilli (r). Below the basement membrane, the photoreceptor cells take on an axonal character, and each axon (Ra) of a peripheral cell of an ommatidium twists 180° and enters a given, but different, optic cartridge. Each cartridge is a columnar organization of receptor axons synapsing with interneurons. The nuclei of monopolar neurons (MP) are prominent. These twisting but undissociated axons (each bundle from a single retinula) are forming a pseudocartridge. (Bar = 1 μm.) (Inset) Enlargement of a cross-sectioned retinular axon surrounded by a satellite glial process (G). Pigment granules, mitochondria, and transected microtubules are principal organelles at this level but some smooth and rough ER are also present. (Bar = 0.1 μm.)

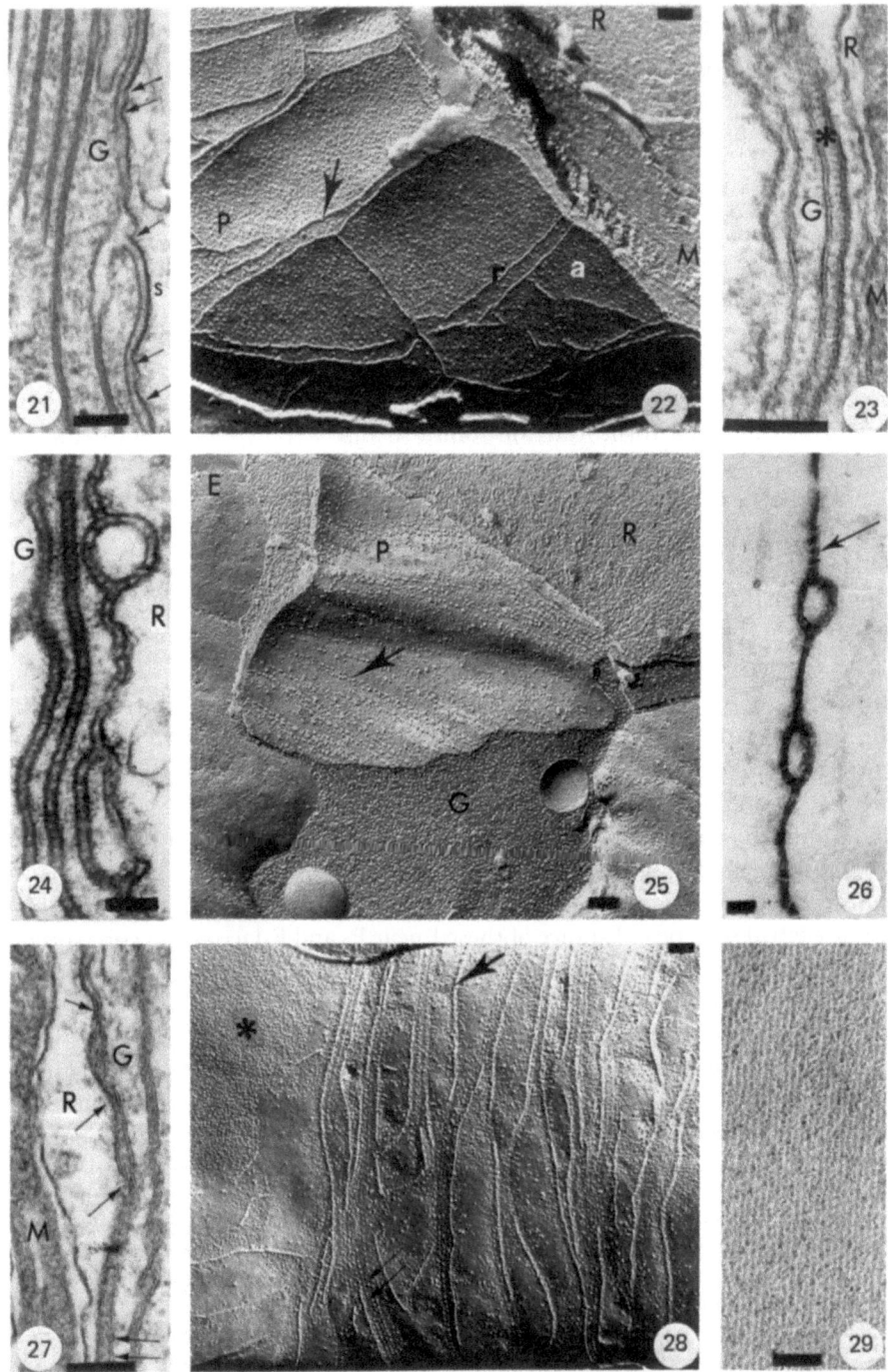

Figure 21. Intercellular junctions between a photoreceptor axon and a glial cell (G) of the fenestrated layer of the housefly. Creases and scallops (arrows) of the plasma membrane of the axon are seen on either side of an apparent scalariform junction made with the glial cell. (Bar = 0.1 μm.)

Figure 22. A freeze-fracture replica of the P face (P) of a photoreceptor axon (R) adjacent to a glial cell process, at the same level (the fenestrated layer) as in the preceding figure. Many intramembranal particles are noted on the P face, and some of these are fused into prominent ridges that are mainly

axolemma. Those formations mark the numerous presynaptic sites. Such specializations are covered in more detail in Section 8.

In contrast to the aforementioned short visual fibers from the peripheral retinular cells, the long visual fibers (from central cells) do not change in size, show any branching or herniation, and are not invaded by capitate projections from epithelial glial cells. Those axons are bonded to the glia of the fenestrated layer by scalariform junctions and by septate junctions to the pseudocartridge glia. Scalariform, septate, and tight junctions are present between the long visual fibers and the satellite glia. More complete tabulations of the membrane specializations of the retinal axons with the five types of glia that accompany them are presented by Chi and Carlson (1980a,b) and Saint Marie and Carlson (1983).

While it is useful to compile lists of membrane specializations and their loci, can these structures be related to visual function? It seems so. The plexiform layer where most of the retinal axons terminate is the inner part of a sandwich composed of the distal satellite glia and the proximal marginal glia (see Chapter 12). Certainly the tight junctions and, to some lesser extent, the septate junctions within these two glial layers retard or prevent ion movement from the adjacent tissues and hemolymph to the photoreceptor terminals and interneurons in the nonvascular lamina. These ultrastructural findings may be the basis for the sizeable standing potential found in the lamina in the dark (Burtt and Catton,

---

organized longitudinally, but may also form polygonal enclosures. The scalloped appearance of the plasma membrane of the retinal axon (seen in Figure 20) relates directly to the position and general location of the particle ridges (arrow). The remainder of the P face is packed with intramembranal particles some of which are aligned into closely packed parallel rows (a). Note cross-fracture of nearby mitochondria (M). (Bar = 0.1 μm.)

Figure 23. A rather extensive scalariform junction (∗) between a glial cell of the fenestrated layer (G) and a photoreceptor axon (R). Again note the nearby mitochondrion (M). (Bar = 0.1 μm.)

Figure 24. A septate junction between a photoreceptor axon (R) and a glial cell of the pseudocartridge stratum (G). Fingerlike processes of the glial cell alternate with more flattened, leaflike extensions of this cell, and both forms enclose the retinal axons. Thick intercellular striae are present at all cell appositions in this region. (Bar = 0.1 μm.)

Figure 25. A freeze-fracture replica counterpart to the previous figure. Aligned P-face (P) particles (arrow) are the intramembranous evidence of septate junctions. E and P faces are leaflet surfaces of glial cell membranes apposed to retinular axons (R). (Bar = 0.1 μm.)

Figure 26. The extracellular tracer lanthanum penetrates septate junctions at the level of the pseudocartridges and outlines this specialization as the tracer remains in the areas between the septa (arrow). (Bar = 0.1 μm.)

Figure 27. Membrane specializations in the nuclear layer of the first optic neuropil. Single arrows denote focal tight junctions between a retinal axon (R) and a glial cell (G) of this stratum. Below, a scalariform junction (double arrow) is seen with the usual nearby mitochondrion (M). (Bar = 0.1 μm.)

Figure 28. A freeze-fracture replica of a photoreceptor axon in the nuclear layer. On the P face are particle ridges (tight junctions, single arrow); aligned rows of particles (septate junctions, double arrow); and a particle-rich region (∗) which may be a scalariform junction. (Bar = 0.1 μm.)

Figure 29. A freeze-fracture replica of the nuclear layer of the lamina ganglionaris. The membrane of a photoreceptor axon shows parallel E-face furrows that are closely set and with low relief. This formation is thought to be the intramembranous manifestation of the scalariform junction. (Bar = 0.1 μm.)

1964; Zimmerman, 1978). Both septate and tight junctions likely impede paracellular diffusion and as such could effect an appreciable increase in tissue resistance. Not only are the upper and lower surfaces of the optic cartridges thus isolated, but the glial tight junctions that are found in each cartridge's glial envelope also form a circumferential barrier. Thus, each cartridge (Figure 30) may be thought to be ionically insulated in three dimensions (Saint Marie and Carlson, 1982b). It is not within the scope of this brief chapter to relate in detail the anatomical correlates of the lamina's electrical activity. Suffice it to say that our present understanding of glial-glial, glial-neuronal, and neuronal-neuronal membrane specializations and the synaptic polarities of the 19 odd neurons of an optic cartridge, permits one to make quite reasonable hypotheses about the mechanism of noise reduction, light adaptation, feedforward, feedback, lateral inhibition, and graded, nondecremental axonal conduction (see Shaw, 1979b, 1981; Laughlin, 1981; Saint Marie and Carlson, 1983).

8. The Synapse

8.1. Chemical Synapses

Over 3000 optic cartridges make up the lamina ganglionaris of the housefly, and each contains six photoreceptor axons (Figure 30) which make numerous electrical and chemical synapses. Electrical synapses (via gap junctions) connect neighboring axons of their own kind (Chi and Carlson, 1976; Ribi, 1978; Shaw and Stowe, 1982), and chemical synapses are made with a variety of interneurons in the lamina ganglionaris. In a few instances, the photoreceptor axons of the housefly are postsynaptic (Strausfeld and Campos-Ortega, 1977; Shaw, 1981). Obviously, a great amount of electrical activity is generated by the 18,000 photoreceptor axons, which represent but a fraction of the neurons in this neuropil. The terminals of these axons contain a high density of tiny, pleomorphic, electron-lucent synaptic vesicles (Figure 31). These vesicles are rather homogeneously distributed, and each measures about 30-40 nm in diameter. The presynaptic structure of note is a T-shaped density (Figures 31, 32) around which many of the vesicles aggregate. This T bar is a cross-sectioned view of a bowtie-shaped density over which lies an elliptical plate. Depending on the plane of section, this presynaptic density may take on a variety of appearances (Burkhardt and Braitenberg, 1976). Figures 31 and 32 show the T-bar and bowtie aspects of this structure.

In freeze-fracture replicas, presynaptic "active zones" are present within small circular elevations of the presynaptic P face. These axolemmal mesas are the freeze-fracture analogs of the glial windows, areas devoid of glial processes where neurons can appose each other directly.

Each mesa contains a "bowtie" cluster of particles and these match the presynaptic density in form and distribution (Saint Marie and Carlson, 1982a; Shaw and Stowe, 1982; Figures 32, 34). We have determined the frequency of these presynaptic structures from thin sections and freeze-fracture replicas and

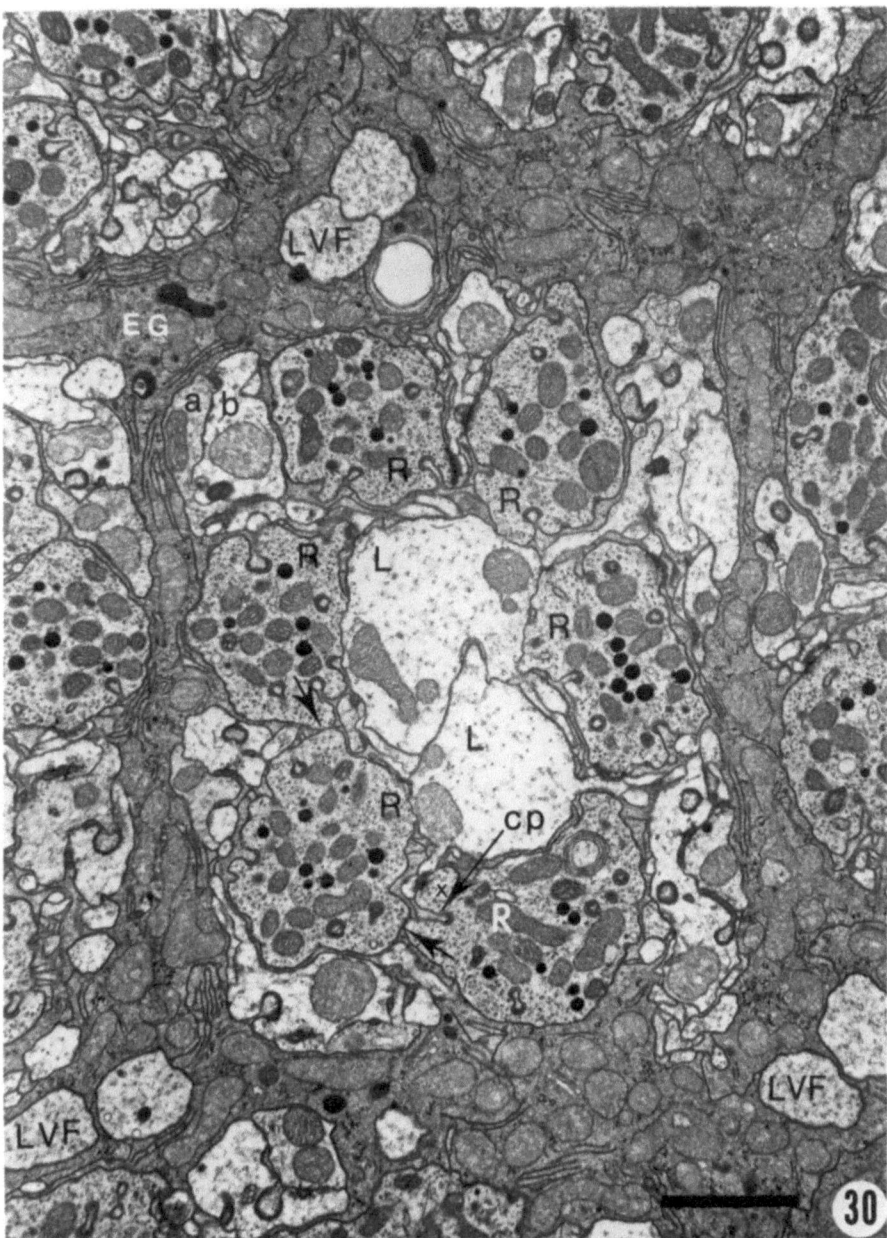

Figure 30. A cross-sectioned optic cartridge. The neuronal profiles are framed by the electron-dense (ribosome rich) cells of the epithelial glia (EG). Six moderately electron-dense (because of the numerous synaptic vesicles) retinal (R) axons encircle two electron-lucent monopolar interneurons (L). At the outer interstitial areas between retinal axons are pairs of intertwinning interneurons, the α (a) and β (b) fibers from an intrinsic (amacrine) neuron and a neuron of the second optic neuropil, respectively. Adjacent retinal axons make close appositions (arrow) which are foci for electrical coupling. The small, round, dense spots with a lucent center within the retinal axons are capitate projections (cp). These are invaginations of the epithelial glial cell into that axon. An X denotes a T-bar chemical synapse between a retinal axon and a monopolar cell process. Small dense pigment granules are also seen within each axon terminal. Pairs of long visual fibers (LVF) lie within the epithelial glial envelope of each cartridge. (Bar = 0.1 μm.)

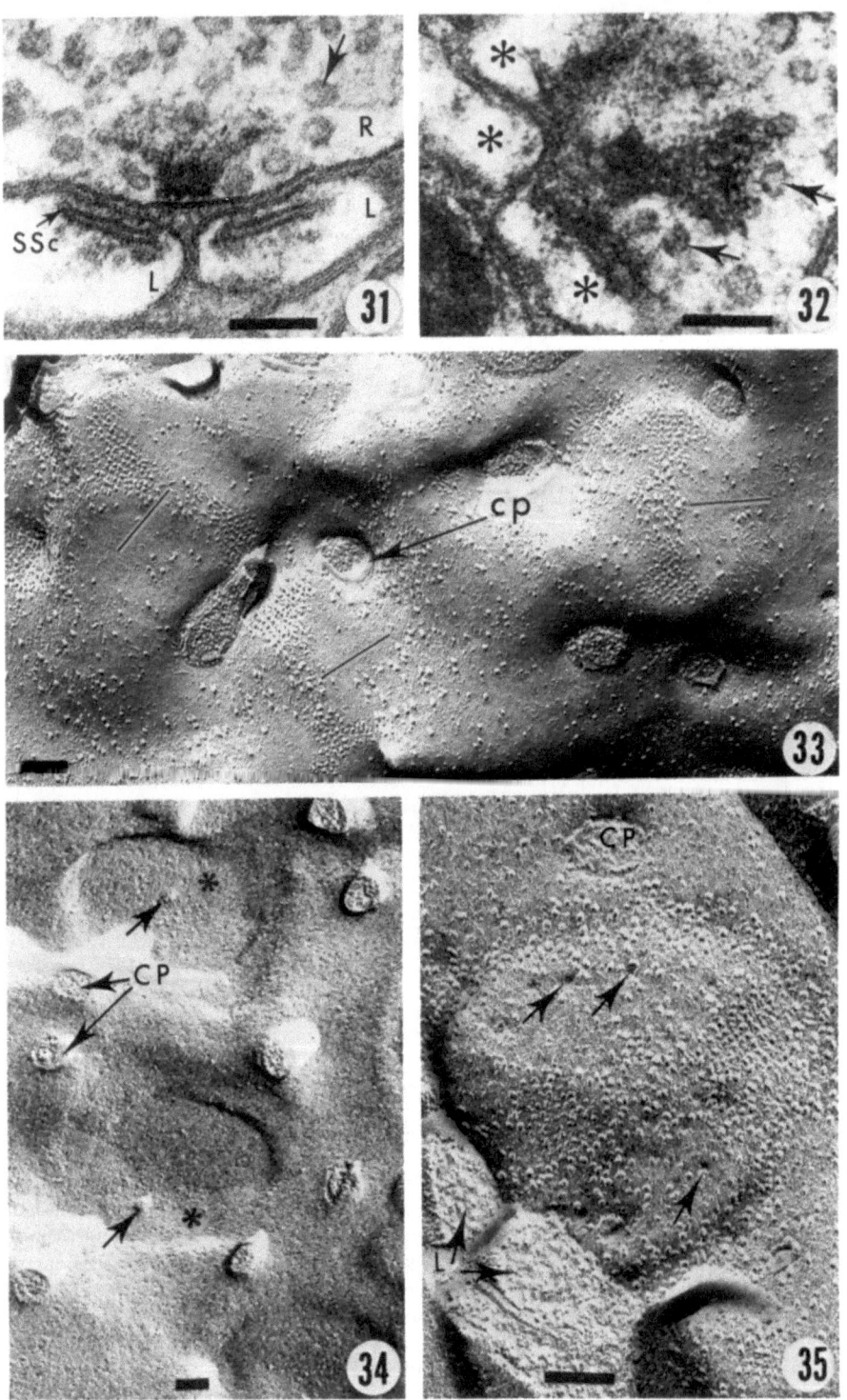

estimate that each axon terminal in the housefly forms a minimum of 170 such synapses (Saint Marie, 1981; Saint Marie and Carlson, 1982a). Similar estimates have been obtained by Nicol and Meinertzhagen (1982) from serial sections of housefly terminals (200 ± 40 synapses/terminal) and by Shaw and Stowe (1982) from freeze-fracture replicas of terminals of the blowflies *Lucilia cuprina* and *L. sericata* (190 synapses/terminal). The freeze-fracture appearance of the bowtie is that of two bows containing orthogonal rows of P-face particles and a central "knot" composed of a circular aggregate of unordered 9- to 12-nm particles (Figures 32, 34). The presence of intramembranous particles at the presynaptic site suggests that the presynaptic density is firmly anchored to the presynaptic membrane.

Freeze-fracture data also indicate that the perimeter of this particle aggregate serves as the "active zone" along which synaptic vesicles fuse with the membrane and discharge their contents. That critical event is extraordinarily evanescent and has eluded morphologists for a variety of technical reasons. Recently we were able to obtain freeze-fracture replicas of these synapses caught in the act of exocytosing neurotransmitter onto laminar interneurons in the housefly (Figure 35). This illustration of stimulation-dependent vesicular release was the first demonstration for any known inhibitory synapse. The photoreceptor axons were synaptically driven by ambient light during fixation (Figure 35) and compared with similar axons fixed in the dark (Figures 33, 34).

Figure 33 shows the E face of the unstimulated photoreceptor terminal, and this leaflet is largely unperturbed except for a very few vesicle fusions, and those "few" probably represent a low-level, dark noise. This condition is markedly different from light-stimulated terminals (Figure 35). Here numerous punctations are seen in the P face, as one would expect, and these holes occur near the

⬅──

Figure 31. The chemical synapse between the retinal axon (R) and two processes of a monopolar interneuron (L) in the first optic neuropil of the housefly. The presynaptic T bar is surrounded with synaptic vesicles (arrow). In both postsynaptic processes (L), the subsynaptic cisternae (SSc) are prominent. (Bar = 0.1 μm.)

Figure 32. A partial *en face* view of the T bar, a presynaptic structure which, in outline, resembles a bowtie (only the knot and one of the bows are seen). The perimeter of this bowtie outlines the active zone to which synaptic vesicles (arrows) migrate, line up, and discharge their contents into the synaptic cleft. Postsynaptic processes are marked (*). (Bar = 0.1 μm.)

Figure 33. Freeze-fracture replica of an unstimulated photoreceptor axon terminal. Bowtie-shaped patches of aligned particles on the P face demark chemical synaptic sites. Long bars have been placed at right angles to the long axes of the bowties. These show no particular orientation. Cross-fractured capitate projections (cp) are diagnostic for the retinal axon membrane. (Bar = 0.1 μm.)

Figure 34. Freeze-fracture replica of a retinal axon terminal (E face) which was unstimulated. The two circular depressions (surrounded by the intruding glial cell) are synaptic regions (*). Cross-fractured capitate projections (CP) from the glial cell mark the axon terminals. Short arrows mark the very few vesicle fusion sites found in the unstimulated condition. (Bar = 0.1 μm.)

Figure 35. Freeze-fracture replica of a retinal axon terminal (P face) which was light-stimulated. Punctations of the membrane (arrows) near the particle bowtie probably are the exocytotic fusions of synaptic vesicles with the presynaptic membrane. Postsynaptic processes from a laminar (L) monopolar interneuron are present adjacent to the presynaptic portion of this synapse. A capitate projection (CP) from the surrounding glial cell that invaginated into this terminal has been sheared off. (Bar = 0.1 μm.)

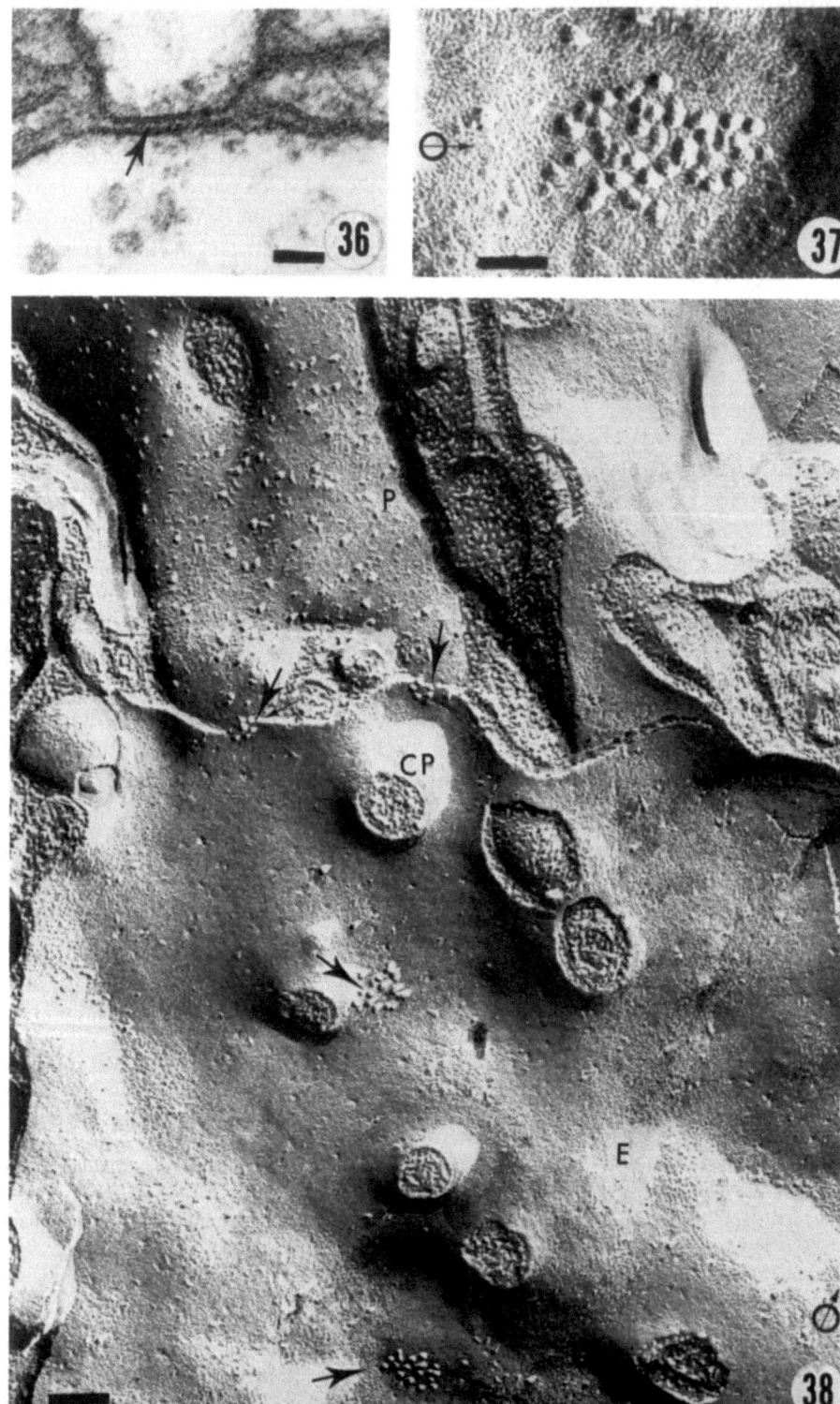

margins of the bowtie pattern of intramembranous particles. These dimples mark the exocytotic fusions of synaptic vesicles with the presynaptic membrane during discharge of the neurotransmitter. The full account of this work is in Saint Marie and Carlson (1983).

In some sections (Figure 31), the presynaptic T bar is aligned to equally cover the two underlying postsynaptic processes (dyads). Triad and tetrad configurations are also known for the retinular axon (chemical) synapse. It is the tetrad synapse that is the usual (or perhaps only) conformation found in the adult housefly (Frölich and Meinertzhagen, 1982). Nontetrad "forms" may be seen when the older postsynaptic fibers are not in the plane of section. Frölich and Meinertzhagen identify two of the postsynaptic processes as dendrites of the L1 and L2 monopolar relay interneurons, respectively, and the other pair are α-processes from an amacrine cell which is intrinsic to the lamina. In rare cases, an epithelial glial cell process intrudes into this tight-knit quartet of postsynaptic processes. Frölich and Meinertzhagen (1982) term this aggregation a *pentad*.

8.2. Electrical Synapses

Electrical synapses (gap junctions) between insect photoreceptor terminals have been verified by dye passage and microelectrode recordings (Shaw, 1979a), thin sections (Chi and Carlson, 1976; Ribi, 1978; Figure 35), and freeze-fracture techniques (Shaw and Stowe, 1982; Figures 36, 37). These electrical coupling sites between adjacent photoreceptor axons occur throughout the plexiform layer and tend to be very small (see also Shaw and Stowe, 1982). Circular patches of unordered E-face particles are the rule, and only 15–30 particles are usually found (Figure 36) in each macula. Based on similar fracture replicas of housefly lamina, Shaw and Stowe (1982) record an average of 39 particles per gap junction, and 55–62 gap junctions per terminal. It is possible that these cell-coupling junctions are a basis for the advantageous signal-to-noise ratio found by housefly photoreceptor axons by Gemperlein and Smola (1972) and to sum the attenuated depolarizations by these axons (Scholes, 1969; Shaw, 1981).

9. Concluding Remarks

Photoreceptor cells of insects are remarkable in several ways. The possession by these cells of profuse and highly organized microvilli, whose membranes are

---

Figure 36. Close apposition between photoreceptor axon terminals in the first optic neuropil of the housefly. Faint striations (arrow) are seen in the intercellular cleft between apposed axolemmas. (Bar = 50 nm.)

Figure 37. A freeze-fracture replica of a gap junction on a photoreceptor axon terminal. Unordered E-face particles are aggregated into a small circular patch. (Bar = 50 nm.)

Figure 38. A freeze-fracture replica of an optic cartridge. Two photoreceptor axons are in this field, and both P and E faces are seen. Cross-fractured capitate projections (CP) pockmark the retinal axons. Four macular gap junctions are indicated (arrows), two of which occur at the transition between the P face of one axon and the E face of the other. (Bar = 0.1 μm.)

rich in rhodopsin, permits sensitivity to such extraordinarily low light fluxes as one to several photons. It is also clear now that the receptor portion of that cell which contains rhabdomeric microvilli probably resides in an ionic environment that is different, or at least partitioned by tight and septate junctions from the extracellular milieu bathing the proximal part of the cell, namely the axon terminals. The ultrastructural geometry of the rhabdom and the cell's overall functional versatility cause it to be an excellent model system for such studies as: color vision, polarization sensitivity, neurophysiology, neuroanatomy, physiological optics, blood-brain barriers, membrane biochemistry, and neuronal-glial relationships. These statements pertain to the photoreceptor cells of the compound eye, but they also hold true for very similar cells of ocelli and the stemmata or simple eyes of larvae. These organs merit chapters of their own.

ACKNOWLEDGMENTS

The authors gratefully acknowledge financial support from Hatch Grant 2100 and additional support from the University of Wisconsin Graduate School, Project 101451. Some of the research was sponsored by a grant from the National Eye Institute, NIH (EYO-1686). Earlier drafts of this chapter were carefully reviewed by Dr. William S. Stark, Division of Biological Sciences, University of Missouri, Columbia, and we are appreciative of his critical commentary. Mr. Martin B. Garment provided able darkroom assistance.

References

Autrum, H., and von Zwehl, V., 1964, Die spektrale Empfindlichkeit einzelner Sehzellen des Bienenauges, *Z. vgl. Physiol.* **48:**357-384.
Bernard, G. D., and Miller, W. H., 1968, Interference filters in the corneas of Diptera, *Invest. Ophthalmol.* **7:**416-434.
Bernhard, C. G. (ed.), 1966, Opening address. In *The Functional Organization of the Compound Eye*, pp. 1-11, Pergamon Press, Elmsford, N.Y.
Bernhard, C. G., Miller, W. H., and Møller, A. R., 1965, The insect corneal nipple array, *Acta Physiol. Scand.* **63**(Suppl. 23):1-79.
Blest, A. D., 1980, Photoreceptor membrane turnover in arthropods: Comparative studies of breakdown processes and their implications. In *The Effects of Constant Light on Visual Processes*, edited by T. P. Williams and B. N. Baker, pp. 217-245, Plenum Press, New York.
Boschek, C. B., 1971, On the fine structure of the peripheral retina and lamina ganglionaris of the fly, *Musca domestica*, *Z. Zellforsch. mikrosk. Anat.* **118:**369-409.
Boschek, C. B., and Hamdorf, K., 1976, Rhodopsin particles in the photoreceptor membrane of an insect, *Z. Naturforsch. Teil C* **31:**763.
Braitenberg, V., 1966, Unsymmetrische Projektion der Retinulazellen auf die Lamina ganglionaris bei der Fliege *Musca domestica*, *Z. vgl. Physiol.* **52:**212-214.
Burkhardt, D., 1977, On the vision of insects, *J. Comp. Physiol.* **120:**33-50.
Burkhardt, D., and Autrum, H., 1960, Die Belichtungspotentiale einzelner Sehzellen von *Calliphora erythrocephala* Meig., *Z. Naturforsch. Teil B* **15:**612-616.
Burkhardt, W., and Braitenberg, V., 1976, Some peculiar synaptic complexes in the first visual ganglion of the fly, *Musca domestica*, *Cell Tissue Res.* **173:**287-308.

Burtt, E. T., and Catton, W. T., 1964, The potential profile of the insect compound eye and optic lobe, *J. Insect Physiol.* **10**:689-710.

Carlson, S. D., and Chi, C., 1979, Functional morphology of the insect photoreceptor, *Annu. Rev. Entomol.* **24**:379-416.

Carlson, S. D., Saint Marie, R. L., and Chi, C., 1983, Freeze fracture replication and the interpretation of glial and neuronal structures in insect nervous tissue. In *Functional Neuroanatomy*, edited by N. J. Strausfeld, pp. 339-375, Springer-Verlag, Berlin.

Chi, C., and Carlson, S. D., 1976, Close apposition of photoreceptor cell axons in the housefly, *J. Insect Physiol.* **22**:1153-1156.

Chi, C., and Carlson, S. D., 1979, Ordered membrane particles in rhabdomeric microvilli of the housefly (*Musca domestica* L.), *J. Morphol.* **161**:309-321.

Chi, C., and Carlson, S. D., 1980a, Membrane specializations of the first optic neuropile of the housefly, *Musca domestica*, I. Junctions between neurons, *J. Neurocytol.* **9**:429-449.

Chi, C., and Carlson, S. D., 1980b, Membrane specializations of the first optic neuropile of the housefly, *Musca domestica*. II. Junctions between glial cells, *J. Neurocytol.* **9**:451-469.

Chi, C., and Carlson, S. D., 1981, Lanthanum and freeze fracture studies on the retinular cell junction in the compound eye of the housefly, *Cell Tissue Res.* **214**:541-552.

Chi, C., Carlson, S. D., and Saint Marie, R. L., 1979, Membrane specializations in the peripheral retina of the housefly *Musca domestica* L., *Cell Tissue Res.* **198**:501-520.

Duelli, P., 1978, An insect retina without microvilli in the male scale insect, *Eriococcus* sp. (Eriococcidae, Homoptera), *Cell Tissue Res.* **187**:417-427.

Eakin, R. M., 1972, Structure of invertebrate photoreceptors. In *Handbook of Sensory Physiology*, edited by H. J. Dartnall, vol. VII/1, pp. 625-684, Springer-Verlag, Berlin.

Eguchi, E., Naka, K. I., and Kuwabara, M., 1962, The development of the rhabdom and the appearance of the electrical responses in the insect eye, *J. Gen. Physiol.* **46**:143-157.

Exner, S., 1891, *Die Physiologie der fazetterten Augen von Krebsen and Insecten*, Franz Deuticke, Leipzig.

Fernandez-Moran, H., 1956, Fine structure of the insect retinula as revealed by electron microscopy, *Nature (London)* **177**:742-743.

Fernandez-Moran, H., 1958, Fine structure of the light receptors in the compound eyes of insects, *Exp. Cell Res. Suppl.* 5, 586-644.

Franceschini, N., 1975, Sampling of the visual environment by the compound eye of the fly: Fundamentals and applications. In *Photoreceptor Optics*, edited by A. Snyder and R. Menzel, pp. 98-125, Springer-Verlag, Berlin.

Franceschini, N., and Kirschfeld, K., 1976, Le controle automatique du flux lumineux daus l'oeil composé des Diptères, *Biol. Cybern.* **21**:181-203.

Frölich, A., and Meinertzhagen, I. A., 1982, Synaptogenesis in the first optic neuropile of the fly's visual system, *J. Neurocytol.* **11**:159-180.

Gemperlein, R., and Smola, U., 1972, Übertragungseigenschaften der Sehzelle der Schmeissfliege *Calliphora erythrocephala*-3. Verbesserung des Signal-Stórungs-Vehaltnisses durch präsynaptische Summation in der Lamina ganglionaris, *J. Comp. Physiol.* **79**:393-409.

Goldsmith, T. H., and Bernard, G. E., 1974, The visual system of insects. In *The Physiology of Insecta*, edited by M. Rockstein, pp. 165-272, Academic Press, New York.

Goldsmith, T. H., and Fernandez, H. R., 1968, The sensitivity of housefly photoreceptors in the mid-ultraviolet and the limits of the visible spectrum, *J. Exp. Biol.* **49**:669-677.

Goldsmith, T. H., and Philpott, D. E., 1957, The microstructure of the compound eyes of insects, *J. Biophys. Biochem. Cytol.* **3**:429-440.

Goldsmith, T. H., and Wehner, R., 1977, Restrictions on rotational and translational defusion of pigment in the membranes of a rhabdomeric photoreceptor, *J. Gen. Physiol.* **70**:453-490.

Gribakin, F. G., 1975, Functional morphology of the compound eye of the bee. In *The Compound Eye and Vision of Insects*, edited by G. A. Horridge, pp. 154-176, Oxford University Press (Clarendon), London.

Hamdorf, K., Paulsen, R., and Schwemer, J., 1973, Photoregeneration and sensitivity control of photoreceptors of invertebrates. In *Biochemistry and Physiology of Visual Pigments*, edited by H. Langer, pp. 155-166, Springer-Verlag, Berlin.

Harris, W. A., and Stark, W. S., 1977, Hereditary retinal degeneration in *Drosophila melanogaster*, *J. Gen. Physiol.* **68**:261-291.

Harris, W. A. Ready, D. F., Lipson, E., Hudspeth, A. J., and Stark, W. S., 1977, Vitamin A deprivation and *Drosophila* photopigments, *Nature (London)* **266**:648-650.

Heisenberg, M., 1979, Genetic approach to a visual system. In *Handbook of Sensory Physiology*, vol. VII/6A, edited by H. Autrum, pp. 665-679, Springer-Verlag, Berlin.

Höglund, G., 1966, Pigment migration, light screening and receptor sensitivity in the compound eye of nocturnal Lepidoptera, *Acta Physiol. Scand.* **69**(Suppl. 282):1-56.

Höglund, G., and Struve, G., 1970, Pigment migration and spectral sensitivity in the compound eye of moths, *Z. vgl. Physiol.* **67**:229-237.

Home, E. M., 1972, Centrioles and associated structures in the retinula cells of insect eyes, *Tissue Cell* **4**:227-234.

Home, E. M., 1975, Ultrastructural studies of development and light-dark adaptation of the eye of *Coccinella septempunctata* L., with particular reference to ciliary structures, *Tissue Cell* **7**:703-722.

Home, E. M., 1976, The fine structure of some carabid beetle eyes, with particular reference to ciliary structures in the retinula cells, *Tissue Cell* **8**:311-333.

Horridge, G. A., 1966, The retina of the locust. In *The Functional Organization of the Compound Eye*, edited by C. G. Bernhard, pp. 513-541, Pergamon Press, Elmsford, N.Y.

Horridge, G. A., 1975a, Arthropod receptor optics. In *Photoreceptor Optics*, edited by A. W. Snyder and R. Menzel, pp. 459-478, Springer-Verlag, Berlin.

Horridge, G. A. (ed.), 1975b, *The Compound Eye and Vision of Insects*, Oxford University Press, London.

Horridge, G. A., and Barnard, P. B. T., 1965, Movement of pallisade in locust retinula cells when illuminated, *Q. J. Microsc. Sci.* **106**:131-135.

Horridge, G. A., and Giddings, C., 1971, The retina of *Ephestia* (Lepidoptera), *Proc. R. Soc. London Ser. B* **179**:87-95.

Järvilehto, M., and Zettler, F., 1971, Localized intracellular potentials from pre- and postsynaptic components in the external plexiform layer of an insect retina, *Z. vgl. Physiol.* **75**:422-440.

Järvilehto, M., and Zettler, F., 1973, Electrophysiological-histological studies on some functional properties of an insect retina, *Z. Zellforsch. mikrosk. Anat.* **136**:291-306.

Kirschfeld, K., 1967, Die Projektion der optischen Umwelt von *Musca*, *Exp. Brain Rex.* **3**:248-270.

Kirschfeld, K., and Franceschini, N., 1969, Ein Mechanismus zur Steuerung des Lichtflusses in den Rhabdomeren des Komplexauges von *Musca*, *Kybernetik* **6**:13-22.

Kolb, G., and Autrum, H., 1972, Die Feinstruktur im Auge der Biene bei Hell-und Dunkeladaptation, *J. Comp. Physiol.* **77**:113-125.

Kühn, A., 1927, Über den Farbensinn der Biene, *Z. vgl. Physiol.* **5**:762-800.

Kunze, P., 1979, Apposition and superposition eyes. In *Handbook of Sensory Physiology*, vol. VII/6A, edited by H. Autrum, pp. 441-502, Springer-Verlag, Berlin.

Kuster, J. E., 1980, Fine structure of the compound eyes and interfacetal mechanoreceptors of *Cincindela tranquebarica* Herbst (Coleoptera: Cincindelidae), *Cell Tissue Res.* **206**:123-138.

Lane, N. J., 1982, Insect intercellular junctions: Their structure and development. In *Insect Ultrastructure*, edited by R. C. King and H. Akai, vol. 1, pp. 402-433, Plenum Press, New York.

Lane, N. J., and Skaer, H. L., 1980, Intercellular junctions in insect tissues, *Adv. Insect Physiol.* **15**:35-213.

Langer, H., 1975, Properties and functions of screening pigments in insect eyes. In *Photoreceptor Optics*, edited by A. W. Snyder and R. Menzel, pp. 429-455, Springer-Verlag, Berlin.

Langer, H., and Thorell, B., 1966, Microspectrophotometry of single rhabdomeres in the insect eye, *Exp. Cell Res.* **41**:673-677.

Laughlin, S., 1981, Neural principles in the peripheral visual systems of invertebrates. In *Handbook of Sensory Physiology*, vol. VII/6B, edited by H. Autrum, pp. 133-280, Springer-Verlag, Berlin.

Lillywhite, P. G., 1977, Single photon signals and transduction in an insect eye, *J. Comp. Physiol.* **122**:189-200.

Luby, K. J., and Porter, K., 1980, Control of pigment migration in isolated erythrophores of *Holocentrus ascensionus* (Osbeck). I. Energy requirements, *Cell* **21**:13-23.

Meinertzhagen, I. A., 1973, Development of the compound eye and optic lobe of insects. In *Developmental Neurobiology of Arthropods*, edited by D. Young, pp. 51-104, Cambridge University Press, London.
Menzel, R., and Blakers, M., 1975, Functional organization of an insect ommatidium with fused rhabdom, *Cytobiology* 11:279-298.
Meyer, E. P., and Labhart, T., 1981, Pore canals in the cornea of a functionally specialized area of the honeybee's compound eye, *Cell Tissue Res.* 216:491-501.
Meyer-Rochow, V. B., 1972, An intracellular microsporidian parasite from the compound eye of *Coeophilus erythrocephalus* (Staphylinidae: Coleoptera), *Z. Parasitenkd.* 38:174-182.
Meyer-Rochow, V. B., 1974, Fine structural changes in dark-light adaptation in relation to unit studies of an insect compound eye with a crustacean-like rhabdom, *J. Insect Physiol.* 20:573-589.
Miller, W. H., Bernard, G. D., and Allen, J., 1968, The optics of insect compound eyes, *Science* 162:760-767.
Müller, J., 1826, *Zur vergleichenden Physiologie des Gesichtssinnes*, Cnobloch, Leipzig.
Nickel, E., and Menzel, R., 1976, Insect UV and green photoreceptor membrane studies by the freeze-fracture technique, *Cell Tissue Res.* 175:367-368.
Nicol, D., and Meinertzhagen, I. A., 1982, An analysis of the number and composition of the synaptic populations formed by photoreceptors of the fly, *J. Comp. Neurol.* 297:29-44.
Odselius, R., and Elofsson, R., 1981, The basement membrane of the insect and crustacean compound eye: Definition, fine structure, and comparative morphology, *Cell Tissue Res.* 216:205-214.
Pak, W. L., and Pinto, L. H., 1976, Genetic approach to the study of the nervous system, *Annu. Rev. Biophys. Bioeng.* 5:397-448.
Paulsen, R., and Schwemer, J., 1972, Studies on the insect visual pigment sensitive to ultraviolet light: Retinal as the chromophoric group, *Biochim. Biophys. Acta* 283:520-529.
Paulus, H. F., 1979, Eye structure and the monophyly of the arthropoda. In *Arthropod Phylogeny*, edited by A. P. Gupta, pp. 299-383, Van Nostrand-Reinhold, Princeton, N.J.
Pepe, I. M., and Baumann, F., 1972, Incorporation of ^{3}H-labelled leucine into the protein fraction in the retina of the honeybee drone, *J. Neurochem.* 19:507-512.
Perelet, A., 1972, Protein synthesis in the visual cells of the honeybee drone as studied with electron microscopic radioautography, *J. Cell Biol.* 55:595-605.
Perelet, A., and Baumann, F., 1969, Evidence for extracellular space in the rhabdome of the honeybee drone eye, *J. Cell Biol.* 40:825-830.
Ramón, Y., Cajal, S., 1909, Nota sobre la estructura de la retina de la mosca (*M. vomitoria* L.), *Trab. Lab. Invest. Biol. Univ. Madrid* 7:217-257.
Ramón, Y., Cajal, S., and Sanchez, D., 1915, Contribucion al conocimiento de los centros nerviosos de los insectos. Parte 1. Retina y centros opticos, *Trab. Lab Invest. Bio. Univ. Madrid* 13:1-168.
Reichardt, W., 1970, The insect eye as a model for analysis of uptake, transduction, and processing of optical data in the nervous system. In *The Neurosciences, Second Study Program*, edited by F. O. Schmitt, Rockefeller University Press, New York.
Ribi, W. A., 1975, The first optic ganglion of the bee. I. Correlations between visual cell types and their terminals in the lamina and medulla, *Cell Tissue Res.* 165:103-111.
Ribi, W. A., 1978, Gap junctions coupling photoreceptor axons in the first optic ganglion of the fly, *Cell Tissue Res.* 195:299-308.
Ribi, W. A., 1979a, Coloured screening pigments cause red eye glow hue in pierid butterflies, *J. Comp. Physiol.* 132:1-9.
Ribi, W. A., 1979b, Do the rhabdomeric structures in bees and flies really twist?, *J. Comp. Physiol.* 134:109-112.
Ribi, W. A., 1980, New aspects of polarized light detection in the bee in view of non-twisting rhabdomeric structures, *J. Comp. Physiol.* 137:281-285.
Saint Marie, R. L., 1981, A thin section and freeze fracture study of intercellular junctions and synaptic vesicle activity in the first optic neuropil of the housefly compound eye, Ph.D. thesis, University of Wisconsin, Madison.
Saint Marie, R. L., and Carlson, S. D., 1982, Synaptic vesicle activity in stimulated and unstimulated photoreceptor axons in the housefly: A freeze fracture study, *J. Neurocytol.* 11:747-761.

Saint Marie, R. L., and Carlson, S. D., 1983, Glial membrane specializations and the compartmentalization of the lamina ganglionaris of the housefly compound eye, *J. Neurocytol.* **12**:243-275.
Schinz, R. H., Lo, M. V. C., and Pak, W. L., 1978, Comparison of rhabdomeric and non-rhabdomeric plasma membrane particles in freeze-fractured *Drosophila* photoreceptors, Assoc. Res. Vision Ophthalmol., Spring meetings, Sarasota, Florida, p. 236 (Abstract).
Schinz, R. H., Lo, M. V. C., Larrivee, D. C., and Pak. W. L., 1982, Freeze-fracture study of the *Drosophila* photoreceptor membrane: Mutations affecting membrane particle density, *J. Cell Biol.* **93**:961-969.
Scholes, J. H., 1964, Discrete subthreshold potentials from the dimly lit insect eye, *Nature (London)* **202**:572-573.
Scholes, J., 1969, The electrical responses of the retinal receptors and the lamina in the visual system of the fly, *Musca*, *Kybernetik* **6**:149-162.
Shaw, S. R., 1969, Interreceptor coupling in ommatidia of drone honey bee and locust compound eyes, *Vision Res.* **9**:999-1029.
Shaw, S. R., 1977, Restricted diffusion and extracellular space in the insect retina, *J. Comp. Physiol.* **113**:257-282.
Shaw, S. R., 1978, The extracellular space and blood-eye barrier in an insect retina: An ultrastructural study, *Cell Tissue Res.* **188**:35-61.
Shaw, S. R., 1979a, Photoreceptor interaction at the lamina synapse of the fly's compound eye, Assoc. Res. Vision Ophthalmol., Spring meetings, Sarasota, Florida, p. 6 (Abstract).
Shaw, S. R., 1979b, Signal transmission by graded slow potentials in the arthropod visual system. In *The Neurosciences: Fourth Study Program*, edited by F. D. Schmitt and F. G. Worden, pp. 275-295, MIT Press, Cambridge, Massachusetts.
Shaw, S. R., 1981, Anatomy and physiology of identified non-spiking cells in the photoreceptor-lamina complex of the compound eye of insects, especially Diptera. In *Neurons without Impulses: Their Significance for Vertebrate and Invertebrate Nervous Systems*, edited by A. Roberts and B. M. H. Bush, pp. 61-115, Cambridge University Press, London.
Shaw, S. R., and Stowe, S., 1982, Freeze-fracture evidence for gap junctions connecting axon terminals of dipteran photoreceptors, *J. Cell Sci.* **53**:115-141.
Shelton, P. M. J., 1976, The development of the insect compound eye. In *Insect Development*, edited by P. A. Lawrence, pp. 152-169, Halsted Press, New York.
Smith, D. S., 1968, *Insect Cells: Their Structure and Function*, Oliver & Boyd, Edinburgh.
Smola, V., and Tscharntke, H., 1979, Twisted rhabdomeres in the dipteran eye, *J. Comp. Physiol.* **133**:291-297.
Snyder, A. W., 1974, Light absorption in visual photoreceptors, *J. Opt. Soc. Am.* **64**:216-230.
Snyder, A. W., Laughlin, S. B., and Stavenga, D. G., 1977, Information capacity of eyes, *Vision Res.* **17**:1163-1175.
Stark, W. S., and Carlson, S. D., 1982, Ultrastructural pathology of the compound eye and optic neuropiles of the retinal degeneration mutant (w rdgB^{KS222}) *Drosophila melanogaster*, *Cell Tissue Res.* **255**:11-22.
Stavenga, D. G., Zantema, A., and Kuiper, J. W., 1973, Rhodopsin processes and the function of the pupil mechanism in flies. In *Biochemistry and Physiology of Visual Pigments*, edited by H. Langer, pp. 175-180, Springer-Verlag, Berlin.
Strausfeld, N. J., 1976, *Atlas of an Insect Brain*, Springer-Verlag, Berlin.
Strausfeld, N. J., and Campos-Ortega, J. A., 1977, Vision in insects: Pathways possibly underlying neural adaptation and lateral inhibition, *Science* **195**:894-897.
Strausfeld, N. J., and Nässel, D. R., 1981, Neuroarchitecture of brain regions that subserve the compound eye of crustaceae and insects. In *Handbook of Sensory Physiology*, vol. VII/6B, edited by H. Autrum, pp. 1-132, Springer-Verlag, Berlin.
Trujillo-Cenoz, O., 1972, The structural organization of the compound eye in insects. In *Handbook of Sensory Physiology*, vol. VI/2, edited by M. G. F. Fuortes, pp. 5-61, Springer-Verlag, Berlin.
Trujillo-Cenoz, O., and Melamed, J., 1978, Development of photoreceptor patterns in the compound eyes of muscoid flies, *J. Ultrastruct. Res.* **64**:46-62.

Varela, F. G., and Wiitanen, W., 1970, The optics of the compound eye of the honeybee (*Apis mellifera*), *J. Gen. Physiol.* **55**:336-358.
von Frisch, K., 1949, Die polurisation des Himmelslichtes als orientierender Faktor bei den Tänzen der Bienen, *Experientia* **5**:142-148.
Wald, G., 1968, The molecular basis of visual excitation, *Les Prix Nobel*, pp. 260-280, The Nobel Foundation, Stockholm.
Wehner, R., 1976, Polarized-light navigation by insects, *Sci. Am.* **235**:106-115.
Wehner, R., and Meyer, E., 1981, Rhabdomeric twist in bees—Artefact or *in vivo* structure?, *J. Comp. Physiol.* **142**:1-17.
Wehner, R., Bernard, G. D., and Geiger, E., 1975, Twisted and non-twisted rhabdoms and their significance for polarization detected in the bee, *J. Comp. Physiol.* **104**:225-245.
Wolken, J. J., Capenos, J., and Turano, A., 1957, Photoreceptor structures. III. *Drosphila melanogaster*, *J. Biophys. Biochem. Cytol.* **3**:441-448.
Zettler, F., and Järvilehto, M., 1971, Decrement-free conduction of graded potentials along the axon of a monopolar neuron, *Z. vgl. Physiol.* **75**:402-421.
Zettler, F., and Weiler, R. (eds.), 1976, *Neural Principles in Vision*, Springer-Verlag, Berlin.
Zimmerman, R. P., 1978, Field potential analysis and the physiology of second-order neurons in the visual system of the fly, *J. Comp. Physiol.* **126**:297-306.

12

The Glial Cells of Insects

RICHARD L. SAINT MARIE, STANLEY D. CARLSON, AND CHE CHI

1. Introduction

To state that neuroglial cells are those ubiquitous companions of neurons is a generality not too helpful in characterizing insect neuroglia. We now know that in insect nervous systems, glial cells are not "ubiquitous" and that naked axons do exist (e.g., Trujillo-Cenóz, 1962). In addition, some types of insect glia are so specialized or evolved that they do not directly contact the neuron, but rather overlie another kind of glial cell, the latter being the true contiguous associate of the neuron.

As is typical for all neuroglia, such cells in the insect are nonexcitable, although their unique ion permeability, transport, and occluding properties often profoundly influence many aspects of nerve function. As in vertebrates, insect glial cells frequently outnumber neurons; in the stick insect by a ratio of 1.5 to 1 (Becker, 1965) and in the adult cricket by a ratio of 8.1 to 1 (Gymer and Edwards, 1967). A wealth of ultrastructural differences exists between insect glial types, and this cornucopia is both blessing and bane when functional assessments of glial–neuronal associations are to be made.

1.1. Historical Concepts of Neuroglia

The discovery and naming of neuroglia (literally, "nerve cement") is attributed to Rudolf Virchow (1860), the great 19th century pathologist. Varon

---

RICHARD L. SAINT MARIE • Department of Anatomy, School of Medicine, Boston University, Boston, Massachusetts 02118, USA.   STANLEY D. CARLSON • Department of Entomology, Neuroscience Training Program, University of Wisconsin, Madison, Wisconsin 53706, USA.   CHE CHI • Biomedical Division, Nicolet Instrument Corporation, Madison, Wisconsin 53711, USA.

and Somjen (1979) point out, however, that Virchow's premier discernment of glia was inspired by a fallacious argument and based on faulty observations. The argument was that "all tissues have connective tissue, and therefore the brain must also," and the observations were made on alcohol-fixed glial cells which contained many artifacts. A century and a quarter later, cytotechnology has advanced to the point that fixation artifacts are minimized and resolution has been enhanced by well over two orders of magnitude. Thus, a more realistic perception is emerging of the glial cell as a vital, structural, and functional associate of neurons.

1.2. Insect Neuroglia: General Concepts

Insect neuroglia were little studied or understood until Ramón y Cajal's (1913, 1916) discovery that they could be selectively impregnated with gold, staining them in a fashion similar to neurons visualized by the Golgi technique. Ramón y Cajal and Sanchez's early (1915) and farsighted appreciation of the intricacy of the insect nervous system brought forth the first visualization of isolated and more-or-less intact insect glial cells. The Spaniards were thus able to illustrate, name, and anatomically categorize various insect glial cells. Wigglesworth (1959, 1960) has since categorized the ganglionic glia of insects into four classes (including the perineurial cells) each in successive strata in the neuropil.

In a recent revision of this scheme, Strausfeld (1976) describes four classes of glia in addition to perineurial cells. Class 1 neuroglia are multipolar cells that extend from the perineurial layer and emit numerous veillike (velate) processes into the perikaryal rind. These glia form an apical cover for the neuronal cell bodies and often invaginate the latter to form trophospongia. Class 2 glia form the basal portion of the rind. These partially ensheath the neuronal perikarya and invest the neurites that project toward the neuropil. Class 2 glia thus compose an "interface" between the rind and the neuropil. Directly beneath the interface glia lie the class 3 neuroglia. These include all glia that invade the synaptic fields. Class 4 glia are sheath glia that wrap axons individually or in sets within the neuronal tracts and connectives that project between ganglia.

For the interested reader, more general information on neuroglia can be found in Glees (1955), Kuffler and Nicholls (1966), Orkand (1977), Schoffeniels *et al.* (1978), Varon and Somjen (1979), and Sears (1982). With regard to insect neuroglia, particularly their membrane specializations and role as a blood–brain barrier, comprehensive reviews are available from Lane and colleagues (Lane, 1974, 1978, 1981a, 1982; Land and Skaer, 1980; Lane and Treherne, 1980). Bullock and Horridge (1965), Roots (1978), and Radojcic and Pentreath (1979) are recommended as other instructive reviews on invertebrate glia.

1.3. The Insect Eye

The neuroglia of the peripheral and intermediate retina of the housefly have recently been delineated (Saint Marie and Carlson, 1983a,b) and are presented in this chapter to illustrate many of the fine structural features common to insect glia. To aid and orient the reader, we include the following summary of the distal visual apparatus of the insect compound eye.

Beneath the finely faceted, cuticular dome of the compound eye lies a remarkable cytoarchitecture of neurons and glia. The complexity, diversity, and interdependency of both cell types make for a functional organization that is at least as complex an organ as that of the vertebrate eye. This functional organization is detailed in our review on photoreceptor cells (Chapter 11) and is described briefly here.

Tightly packed in a fanlike array, photoreceptor cells are first-order afferents, the sole neurons of the peripheral retina. These slender, linear cells form species-specific clusters of 6 to 12 called *retinulae*. Each retinula is surrounded and optically shielded from its neighbor by a retinue of pigmented glial cells which extend from the very distal reaches of the retina to the basement membrane, below which the photoreceptor cell takes on the appearance (and function) of an axon.

Photoreceptor cell axons pass through the fenestrations of the basement membrane and immediately come into contact with the glia of the intermediate retina (lamina ganglionaris). (Figure 1 illustrates the stratification and nomenclature of glia in the fly lamina.) Saint Marie and Carlson (1983a) have shown that two distinct glial cell types (*fenestrated layer glia* and *pseudocartridge glia*) lie one atop the other and above a third type of glia (*satellite glia*) which enshroud the perikaryal rind of the intermediate retina. The upper tier of satellite glia, to accommodate Strausfeld's (1976) classification, are class 1 or rind neuroglia and these engage the neuronal somata with trophospongia. The lower tier of satellite glia invests the neurites of the monopolar neurons and form the "interface" between the rind and neuropil as is typical of class 2 neuroglia. Synaptic fields below the monopolar perikarya are enmeshed in *epithelial glia* (class 3, or neuropil glia). Axons of passage entering or departing from the base of this first optic neuropil pass through the *marginal glia* which, along with the fenestrated layer glia and pseudocartridge glia, correspond to class 4 or tract neuroglia.

Externally, a perineurium covers the insect optic lobe but bears no direct anatomical contact with the visual neurons which pass well below and parallel to the perineurial layer. These electron-dense, modified glial cells are overlaid by a thin, noncellular mat of collagenous microfibrils called the *neural lamella*. Tight junctions between perineurial cells account for the blood-eye barrier surrounding the ganglia and connectives of dipterans (Lane *et al.*, 1977; Chi and Carlson, 1981b).

Above, within, and below the first optic neuropil of the fly are transversely and longitudinally drawn ion barriers effected by glial tight (and perhaps septate) junctions. Emerging data on extracellular resistance barriers and on glial-glial and glial-neuronal occluding junctions may relate directly to the electrical activity of the intermediate retina, particularly to the dark-standing and light-stimulated field potentials; the graded, nondecremental potentials of lamina monopolar fibers; and the several forms of electrical inhibition observed in this neuropil. These structure–function correlations are discussed in Section

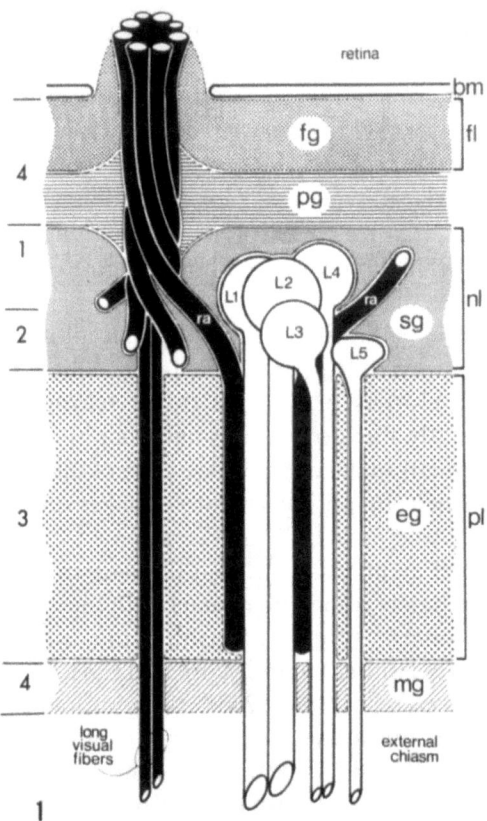

Figure 1. A diagram of a longitudinal section through the base of the peripheral retina of the housefly The section passes through the intermediate retina (lamina ganglionaris), and into the external chiasm (from Saint Marie and Carlson, 1983a). Eight retinal axons (ra) from an ommatidium pass through the fenestrated basement membrane (bm), turn and twist, projecting to different optic cartridges. The bare essentials of one optic cartridge are drawn with the five monopolar interneurons (L1-L5) seen with their axons projecting toward the second optic neuropil (not shown). Two long visual fibers of the original eight photoreceptor axons project through the lamina without synapsing. The other retinal axons are presynaptic to L1, L2, L3, and they also make electrical synapses with each other. Five ultrastructurally distinct layers of glia are found in and around the first optic neuropil. They are, from distal (above) to proximal (below): fenestrated layer glia (fg); pseudocartridge glia (pg); satellite glia (sg); epithelial glia (eg); marginal glia (mg). Given on the far right are the designations for the various strata of the intermediate retina, viz., fenestrated layer (fl); nuclear (perikaryal) layer (nl); plexiform (neuropil) layer (pl). Numbers to the left of the figure indicate the glial classification scheme of Straussfeld (1976) as applied by us to this optic ganglion.

5.6 and specifics on glial–glial and the glial–neuronal intercellular junctions are detailed in Sections 2.4 and 4.

2. Ultrastructure of Insect Neuroglia

In the past there has been little interest in classifying insect neuroglia by their fine structural characteristics though such a description can provide important functional insights. Also, it is no longer useful merely to provide a general ultrastructural description that purports to cover all insect neuroglia. Because some glial cells are restricted to particular regions of the neuron (nucleus, dendritic arbor, or axon), it is useful to consider the functional capacities and ultrastructure of a particular glial type in relation to its accompanying neuron.

In the following sections (2.1-2.4), we provide descriptions of glial cell organelles and qualify these as to cell type whenever possible. Using Figure 1, the reader can then correlate ultrastructural features of a particular glial cell to that portion of the neuron so covered. Additional details on these glia are in Saint Marie and Carlson (1983a,b), and an aid to interpretation of freeze-fractured neurons and glia can be found in Carlson et al. (1983).

2.1. The Nucleus

As a general rule, glial nuclei can be distinguished from those of neurons by their often irregular shape and the more uniform distribution of their chromatin. In Figure 2, the nucleus of a monopolar neuron in the intermediate retina of the housefly is nearly spherical, with most of its chromatin occurring in clumps. The remainder of the nucleus thus appears pale by contrast. The glial nuclei in Figure 2, on the other hand, are oval or fusiform, and although some of the chromatin is peripherally clumped, much of it is dispersed throughout the nucleoplasm giving the nucleus a greater electron opacity.

The outer membrane of the glial nuclear envelope may be completely covered with attached ribosomes (Figure 3) and the envelope itself may evaginate into the cytoplasm and appear continuous with the RER (Figure 4). Prominent nucleoli are observed infrequently in our material.

The shape of glial nuclei varies considerably and is often dictated by the morphology of the cell. Elongated cells such as the fenestrated layer glia, the pseudocartridge and epithelial glia of the fly intermediate retina tend to have oval or fusiform nuclei, whereas squamouslike cells such as the marginal glia and perineurial cells have flattened, disk-shaped nuclei. The satellite glia, which are tightly packed among the monopolar neuron somata, often have more appressed, pyramid-shaped nuclei (Saint Marie and Carlson, 1983a).

2.2. Cytoplasmic Organelles

Though neuroglial cells in the intermediate retina contain the usual complement of cellular organelles, some of these same organelles may be so elaborated as to suggest a high degree of functional specialization. The epithelial glia, for instance, are especially rich in an as yet unspecified ground substance which gives this cell type a characteristic dark appearance (Figures 2, 5). Other glial cells, such as the fenestrated layer glia (Figures 6, 7), perineurial cells (Figure 2), and satellite glia (Figure 4), though not as electron-opaque as the epithelial glia, are moderately dense when compared to adjacent neuronal profiles. Only the pseudocartridge glia (Figure 7) and marginal glia (Figure 3) appear to be as electron-lucent as the neurons.

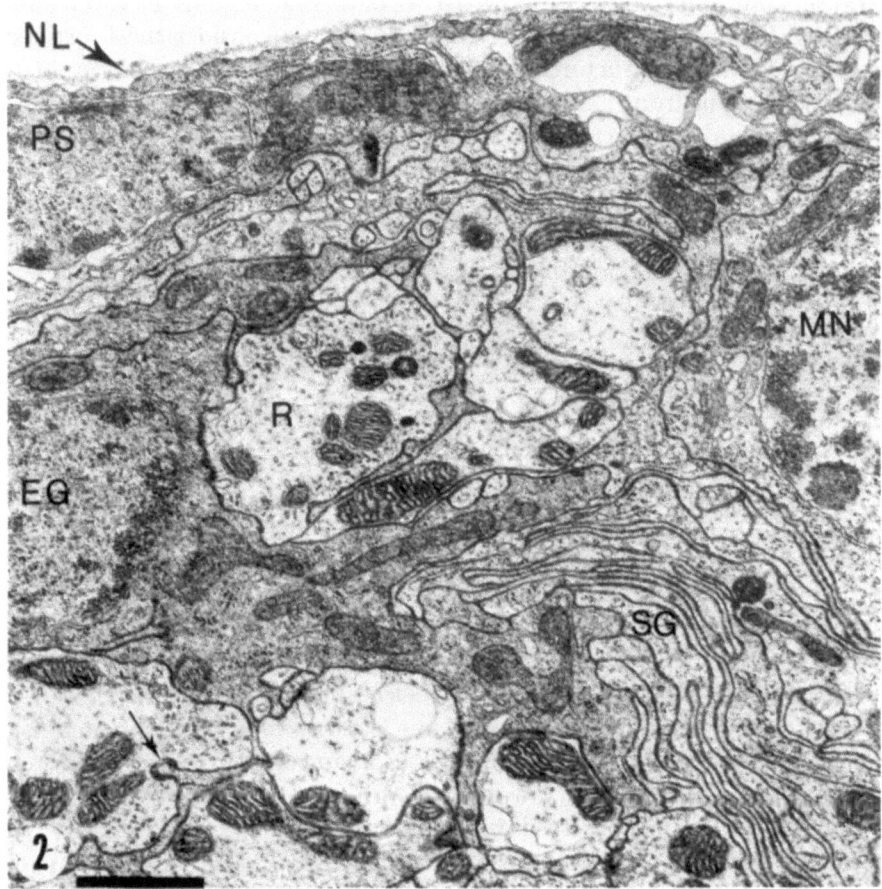

Figure 2. The lateral border of the intermediate retina at the boundary of the nuclear and plexiform layers. The perineurial cell sheath (PS) isolates the neuropil from the circulating hemolymph and lies beneath a thin neural lamella (NL). Portions of the nuclei from a monopolar neuron (MN) and an epithelial glial cell (EG) are visible. Complex interdigitating processes from the satellite glia (SG) demark the proximal limits of the nuclear layer. A discrete, punctate invagination (capitate projection) by an epithelial glial cell into a retinal axon is noted (arrow). Synaptic terminations of retinular axons (R) are found in the plexiform layer. (Bar = 1 μm.)

Unlike vertebrate neuroglia, which often contain filaments or fibrils (Peters *et al.*, 1976), the only structural elements common to insect glia are microtubules (Roots, 1978). Microtubules are abundant in the glia of the intermediate retina and can be found in even the narrowest of glial processes (Figure 6). As a general rule, microtubules occur randomly throughout the glioplasm and usually follow the contours of the many fine processes. At the distal (pseudocartridge) and proximal (marginal) limits of the optic ganglion, however, microtubules are highly specialized and arranged into vast parallel arrays. In the pseudocartridge glia (Figures 7, 8), these microtubules fill the elongate cellular extensions and run parallel to the equatorial (mediolateral) plane of the

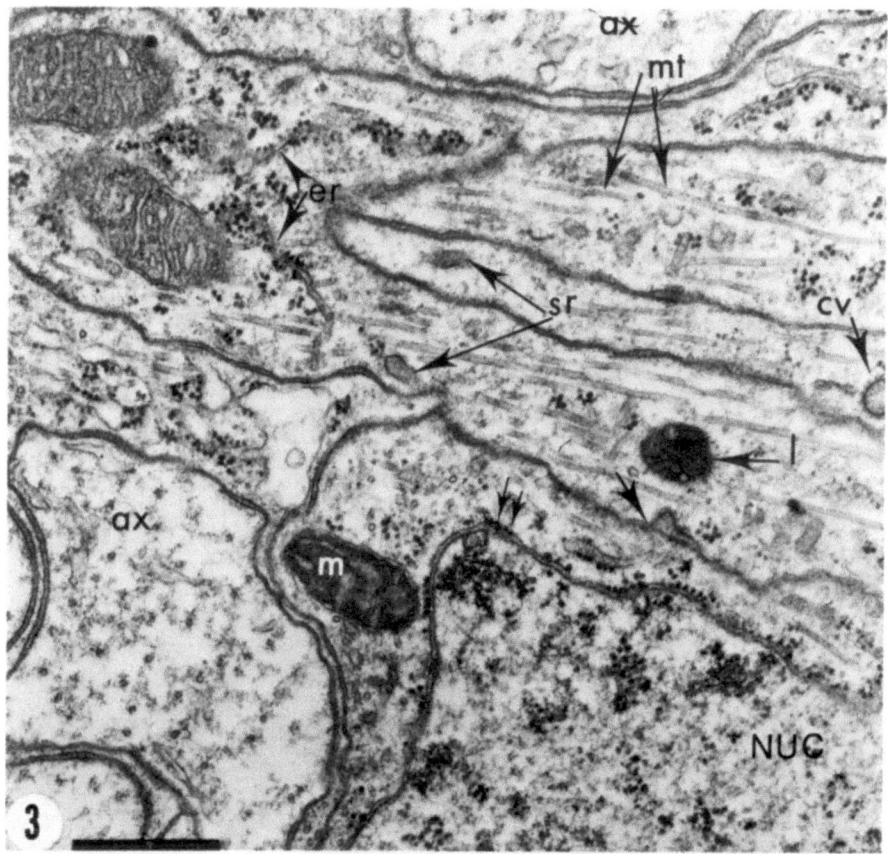

Figure 3. The proximal border of the intermediate retina. The nucleus (NUC) is from a marginal glial cell which envelopes monopolar cell axons (ax) as they project into the external chiasm. Ribosomes and polyribosomes are evident as part of the rough endoplasmic reticulum (er) or attached to the outer membrane of the nuclear envelope (double arrow). Microtubules are randomly oriented near the nucleus and in the narrow glial processes which separate the axons. Elsewhere, microtubules (mt) lie parallel to each other. Also present are: smooth endoplasmic reticulum (sr), mitochondria (m), lysosomal body (l), coated vesicle (cv), and a pinocytotic figure (arrow). (Bar = 0.5 μm.)

eye. The predominant direction of microtubules in the marginal glia, on the other hand, is dorsoventral (Figure 3). Such highly structured microtubular arrays may be related to the vital transport requirements of the occluded and avascular neuropil and this will be discussed further in Section 5.4.

Ribosomes can be found in all glia being considered here. They occur free within the cytoplasm, bound in polyribosomal clusters, or attached to the diffuse system of RER which permeates all of the glial types (Figures 3-6). As mentioned above (Section 2.1), ribosomes also attach to the nuclear envelope. SER is likewise ubiquitous, but is prominent only in the pseudocartridge glia where numerous, long, tubular cisternae extend parallel to the microtubular arrays (Figure 8). Small, membrane-bound lipid deposits are present in each of

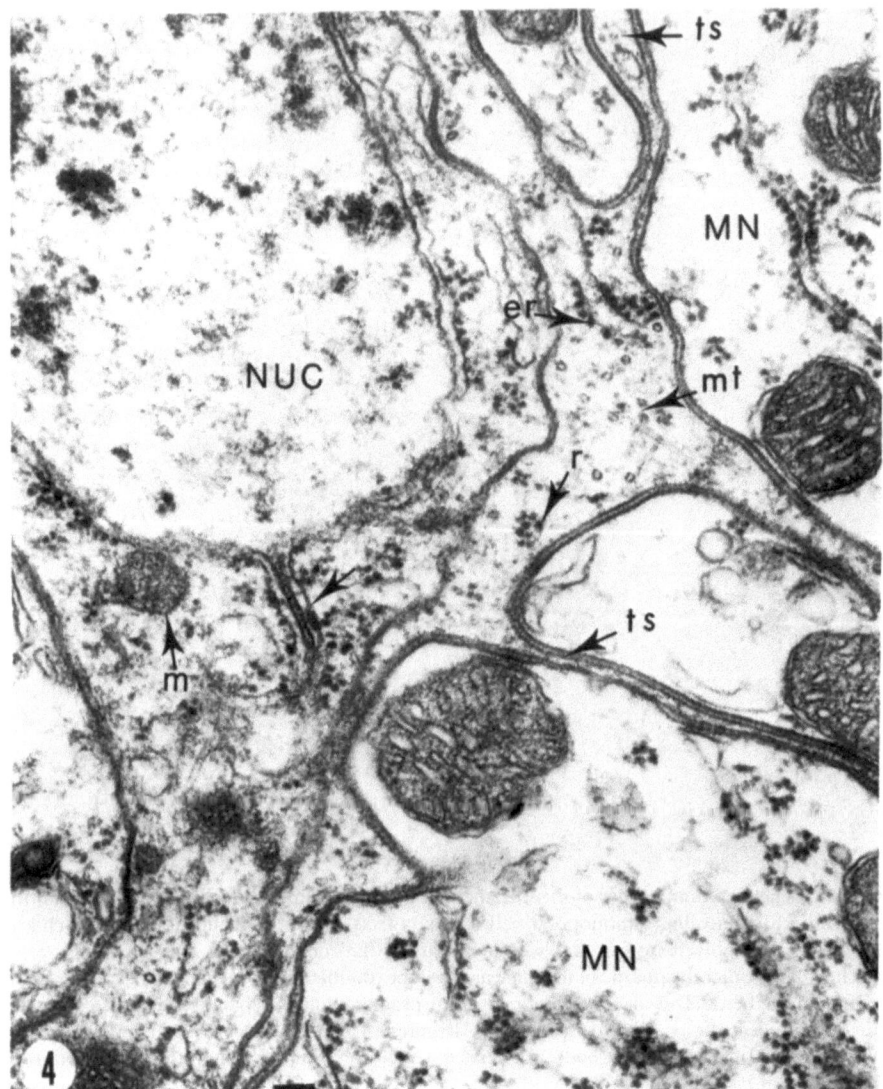

Figure 4. A nucleus (NUC) of a satellite glial cell lies adjacent to the somata of two monopolar neurons (MN). The outer membrane of the nuclear envelope is studded with ribosomes and evaginates into the cytoplasm at one point (arrow). The cytoplasm contains rough endoplasmic reticulum (er), many free ribosomes and polyribosomes (r), small mitochondria (m), and microtubules (mt). The cytoplasmic ground substance renders glia slightly more opaque than that of the adjacent neurons. Narrow glial processes, called *trophospongia* (ts), project deeply into neuronal somata. (Bar = 0.1 μm.)

the outer three glial layers, but large, spherical lipid droplets are a distinct feature only of the "interface" layer of satellite glia (Figure 9). Such droplets are commonly associated with the membranous components of the cell, i.e., the plasma membrane, mitochondria, and SER (Figure 10), and may, on occasion, appear confluent with the cisternae of the RER (Figure 11). Mitochondria are

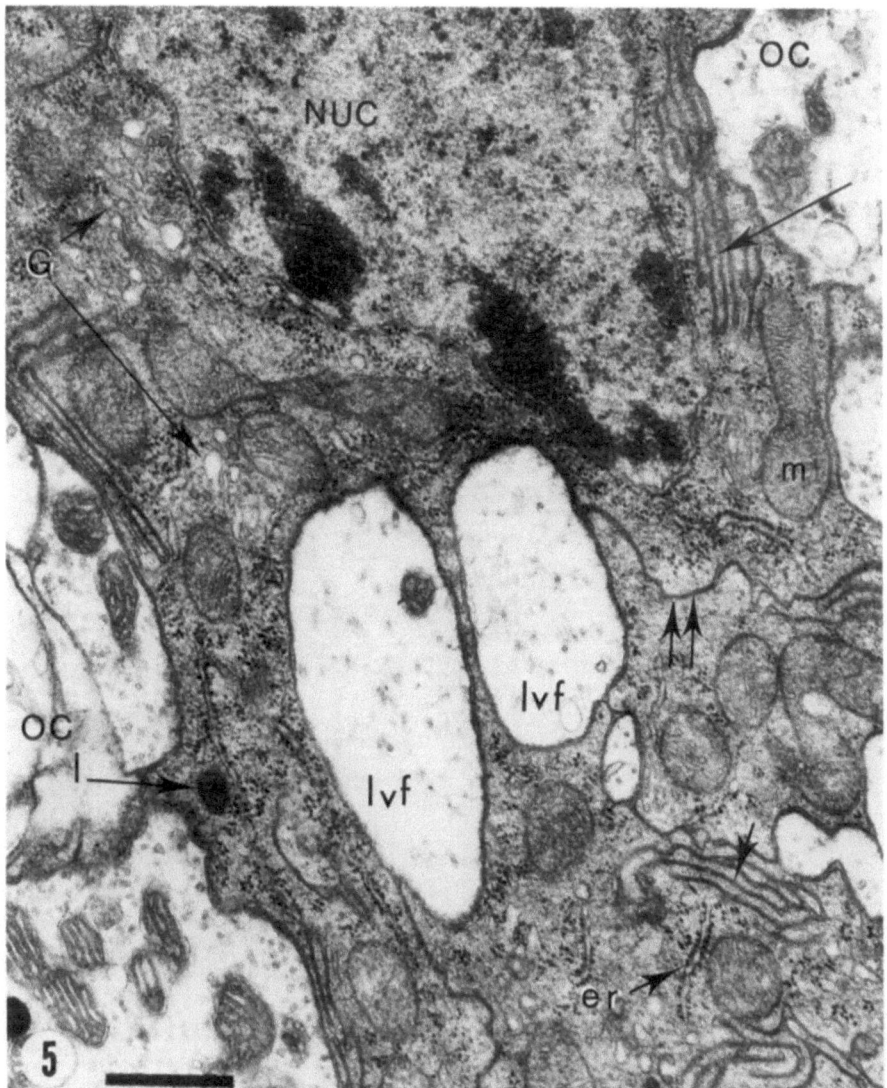

Figure 5. A cross-section through the plexiform layer. Neuronal profiles are lucent in comparison to the dense cytoplasm of the epithelial glia. A portion of an epithelial glial nucleus (NUC) is evident in the glioplasm as are small Golgi bodies (G), numerous mitochondria (m), ribosomes, rough endoplasmic reticulum (er), lysosomes (l), and stacked arrays of glial membrane (arrows). A pair of long visual fibers (lvf) and a small unidentified process pass between the optic cartridges (OC) enveloped by three epithelial glial cells. Double arrow denotes the glial appositions. (Bar = 0.5 μm.)

plentiful and can be found everywhere but in the narrowest glial processes. They are rod-shaped, moderately electron-dense, and their cristae are usually less distinct than their neuronal counterparts (Figures 2-5, 9-11). Glial Golgi bodies are small but plentiful in the vicinity of the nucleus (Figure 5). Other common but less frequently encountered organelles include coated transfer vesicles

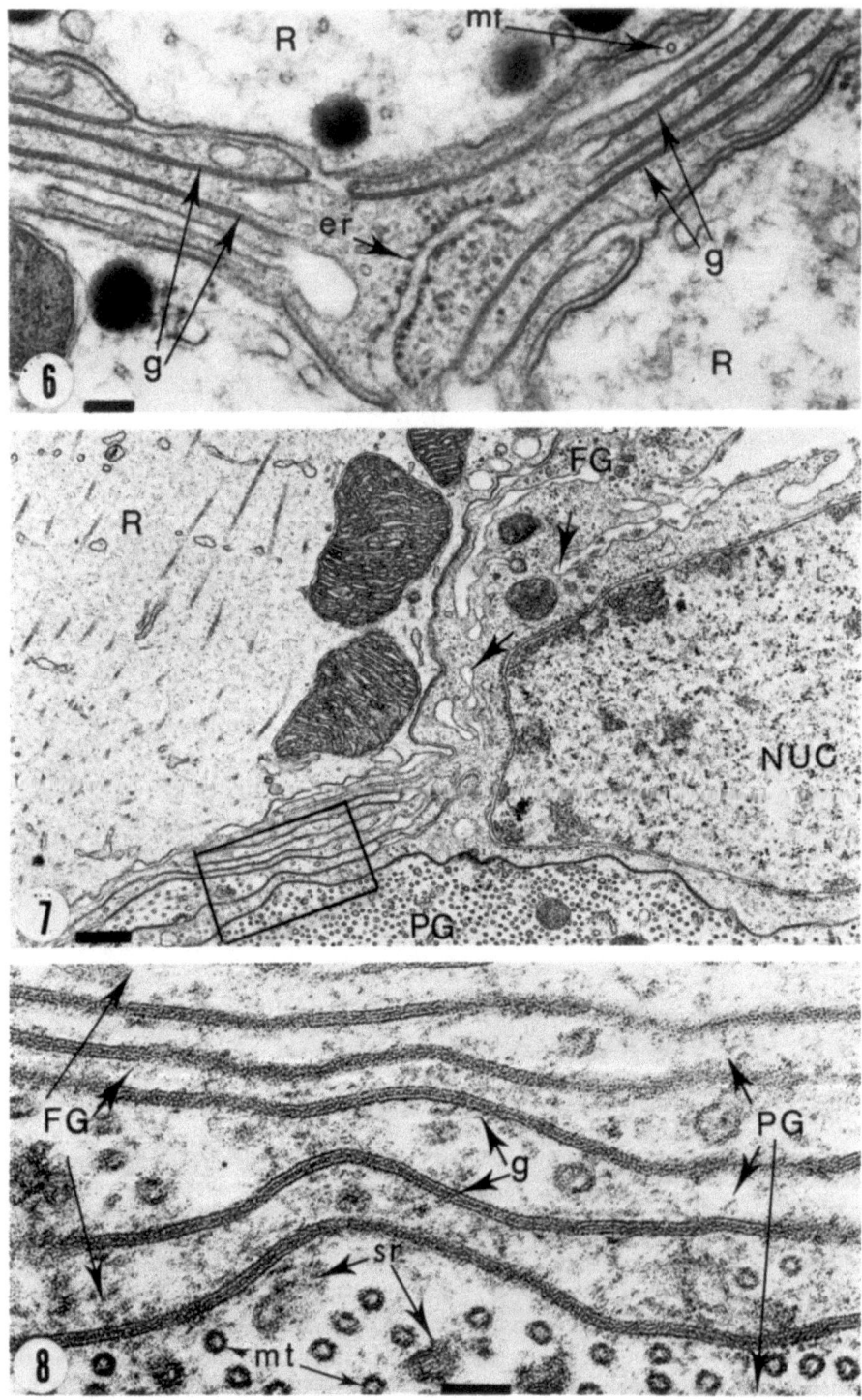

(Figure 3) and small, spherical, dense bodies, probably lysosomes (Figures 3, 5). In most of the glia of the peripheral and intermediate retina, we fail to find appreciable glycogen stores; the exceptions are perineurial cells and some pigmented glia.

2.3. Elaborations of the Plasma Membrane

The plasmalemma of insect neuroglia is a dynamic structure which exhibits a wide range of surface elaborations. This range is manifested clearly in the insect intermediate retina. For example, the surface of glia in the fenestrated layer is highly invaginated by a complex system of deep, tortuous channels which permeate the cytoplasm (Figure 7). In thin section, such surface invaginations could easily be mistaken for cisternae of SER but for the fact that they readily fill with extracellularly applied tracer dyes (Saint Marie and Carlson, 1983a) and are the site of active pinocytosis (Figure 13). Pinocytotic figures are a common feature of all glia in the intermediate retina (see also Figure 3). Another highly invaginated cell is the epithelial glial cell of the neuropil layer. In this case, however, the infolded membrane forms closely apposed laminations deep within the cell (Figure 5). Gap junctions are a common feature between apposed membranes in these stacks (see Figure 18).

Other surface modifications include complex interdigitations which may occur between homologous glial cells or at the boundary of two glial layers. Most interdigitations take the form of thin, overlapping sheets such as those found in the fenestrated layer between glial processes extending between retinular axons (Figure 6), at the border of the fenestrated layer and pseudocartridge glia (Figures 7, 8), and between processes of the satellite glia at the interface of the nuclear and neuropil layers (Figure 2). In the latter case, concentric glial processes often form a loosely wound sheath around the neurites projecting toward the plexiform layer (Figure 9). In another quite characteristic form of interdigitation, projections of the pseudocartridge glia envelop the photoreceptor axons in a manner resembling the interlocked fingers of two tightly clasped hands (Figure 12).

Figure 6. Thin glial processes in the fenestrated layer project between retinular axons (R). Interdigitating glial processes engage each other in gap junctions (g). A microtubule (mt) is clearly evident within the narrow glial process and rough endoplasmic reticulum (er) is nearby. Moderately dense ground substance pervades glial appendages. (Bar = 0.1 μm.)

Figure 7. An area in the distal lamina ganglionaris where the fenestrated layer glia (FG) overlie the pseudocartridge glia (PG). Both cell types interdigitate at their common border (boxed area, see Figure 8). Note the moderate electron density of the fenestrated layer glia and its deeply infolded channels (arrows). A retinal axon (R) projecting to an optic cartridge is seen on the left. The nucleus (NUC) of a fenestrated layer glial cell is on the right. (Bar = 0.5 μm.)

Figure 8. The boxed area in the previous figure shown at a higher magnification. The alternating processes of both glial cell types (FG, PG) are linked by extensive gap junctions (g). Parallel arrays of smooth endoplasmic reticulum (sr) and microtubules (mt) are cross-sectioned. (Bar = 0.1 μm.)

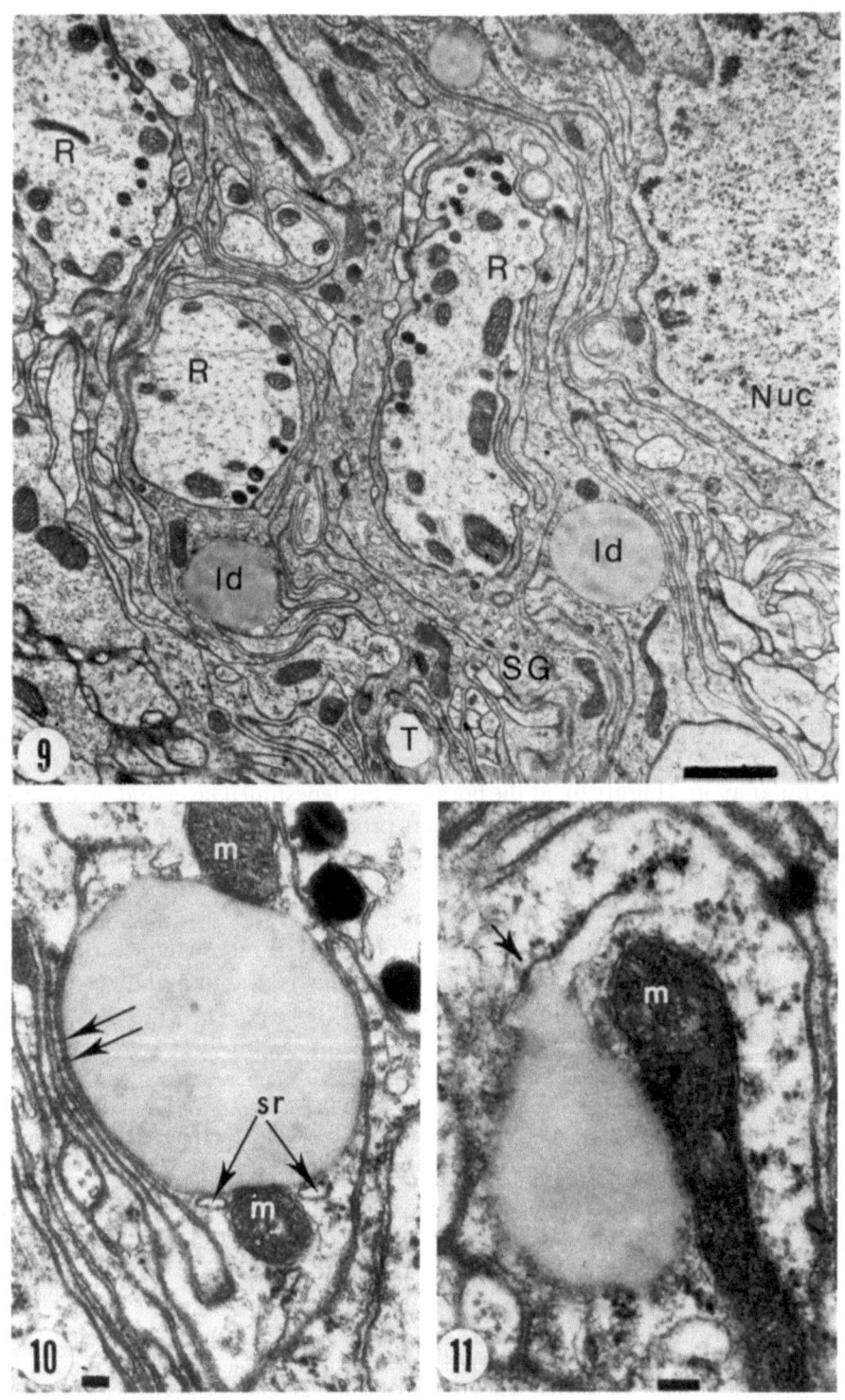

The glial membrane is also the site of some very distinct evaginations which project into neighboring neurons. Narrow veillike processes of the (class 1) satellite glia (Figure 4) project as *trophospongia* deep within the cell bodies of type I monopolar neurons (Boschek, 1971). Such invaginations of the monopolar somata continue as thin extracellular cisternae which terminate in close proximity to the neuronal RER (Figure 14) (see also Saint Marie and Carlson, 1983a). Other characteristic glial evaginations are the capitate projections and gnarls found in the neuropil layer. Capitate projections (Trujillo-Cenóz, 1965) are club-shaped extensions of the epithelial glia which stud the surface of the photoreceptor axon terminals (Figure 15; see also Figures 30, 33–35, 38 in Chapter 11). Gnarls are tuberous projections also from the epithelial glia which invaginate into the β fibers of the T1 cell which extends upward into the optic cartridges of the lamina ganglionaris from the medulla externa. Gnarls also mark the site of α–β synapses (Figure 16; see also Burkhardt and Braitenburg, 1976).

2.4. Interglial Junctions

These glial–glial associations will be discussed briefly as to type, frequency, and distribution. For a more comprehensive treatment of this topic, the reader is referred to Chi and Carlson (1980) and Saint Marie and Carlson (1983b). Details on junctional substructure and development are in Lane and Skaer (1980) and Lane (1982).

Gap junction. The gap junction is the most prevalent interglial membrane specialization in the housefly lamina and is found between all apposed neuroglia, homologous and heterologous. This finding implies an intimate metabolic and ionic interdependence among these cells. Though ubiquitous, gap junctions do vary in size and frequency within and between the various glial layers. Along the lateral borders of apposed fenestrated layer glia, gap junctions occur in roughly circular plaques of moderate size, up to 1 μm in diameter (Figure 17). In other glial layers, the plaques may be smaller and fewer (Figures 12, 26, 28).

Gap junctions are prominent between membranes of the tightly stacked glial processes that extend between retinular axons in the fenestrated layer (Figure 6) and those that invaginate into the epithelial glia (Figure 18). In both

---

Figure 9. Proximal portion of the nuclear layer. Numerous laminations, composed of processes from satellite glia (SG), envelop the retinular axons (R). In this cross-section, these laminations appear to form a loosely wound sheath around the axon. A prominent lipid deposit (ld) lies within the satellite glial cell, and the nucleus (Nuc) of the latter cell is seen. (Bar = 1 μm.)

Figure 10. A higher-magnification electron micrograph of a lipid droplet in one of the satellite glia. Note also the proximity of the plasmalemma (double arrow) to the lipid. Smooth ER (sr) is nearby and on both sides of a mitochondrion (m). (Bar = 0.1 μm.)

Figure 11. This lipid droplet in a satellite glial cell appears confluent with the cisternae of rough ER (arrow). Also note the proximity to a mitochondrion (m). (Bar = 0.1 μm.)

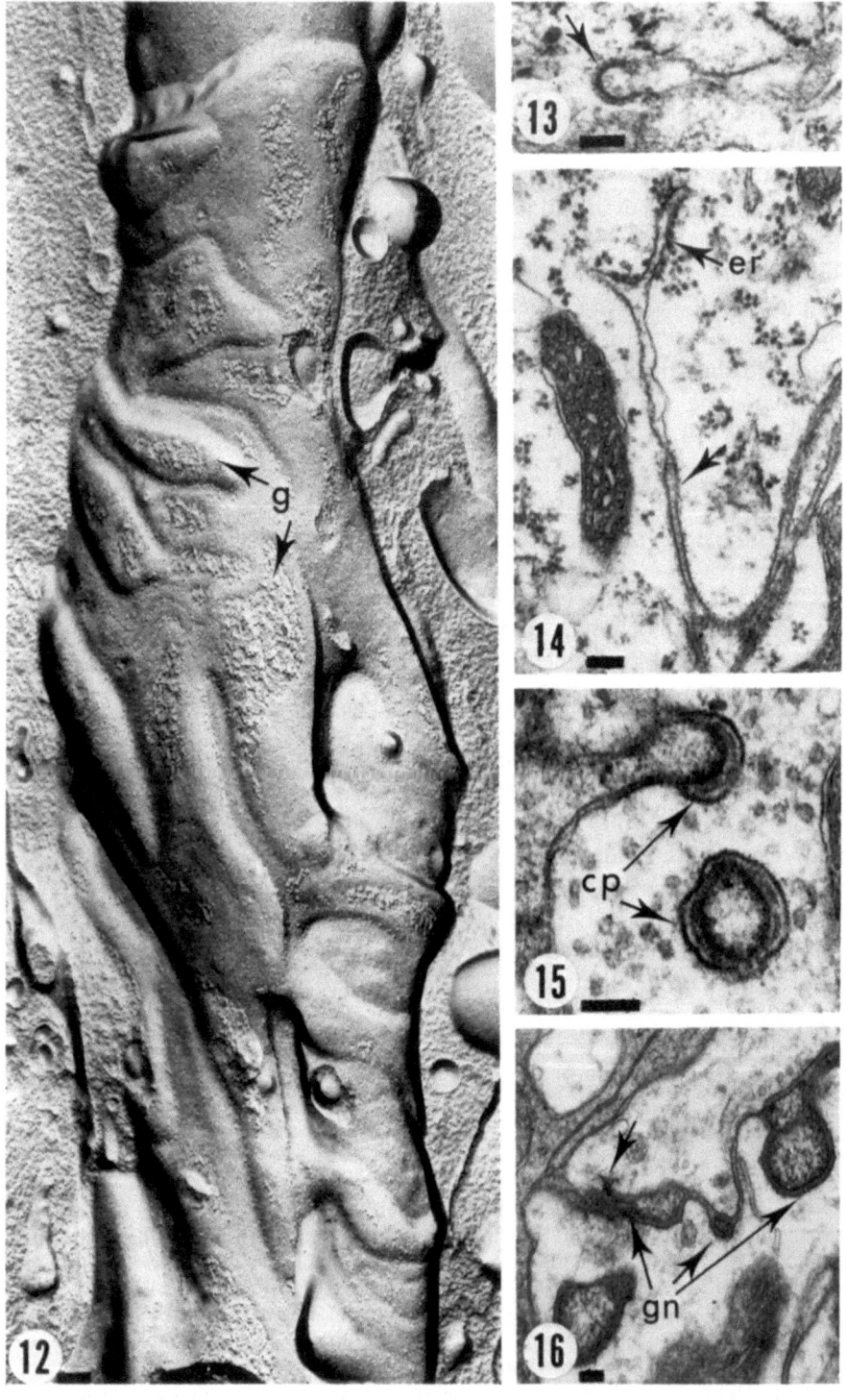

cases, gap junctions can be found between processes of the same parent cell. Similar associations in the blowfly have been dubbed *reflexive gap junctions* (Lane and Swales, 1978b). At other glial boundaries, gap junctions may engage nearly all of the surface between apposed cells (Figure 19). Gap junctions are also abundant between perineurial cells covering the peripheral and intermediate retina (Chi and Carlson, 1981b) and between specialized supporting cells in the peripheral retina (Section 3.1).

Desmosomes and hemidesmosomes. Desmosomes are found at many glial appositions in the peripheral (Section 3.1) and intermediate retina. In the intermediate retina, maculae adhaerentes measure 0.20-0.35 μm in diameter and are seen in the form of circular plaques (Figure 20). Hemidesmosomes in the fly optic lobe are noted with less frequency. These specializations "staple" perineurial cells to the neural lamella (Chi and Carlson, 1981b), and they link the basement membrane to the glia of the fenestrated layer and to the large pigmented glial cells of the peripheral retina (Figure 21). Both forms of the desmosome appear to possess adhesive properties and are freely penetrated by extracellular tracers.

Septate junctions. This intercellular association is found between glia in the distal portion of the intermediate retina, among pseudocartridge glial cells and among the satellite glia. In the former instance, the extent of the septate junctions is remarkable, occupying nearly all of the available intercellular space (Figure 22). This is most apparent in fracture replicas of this region (Figure 24). Both pleated-sheat (Figure 23) and honeycomb appearances of the septate junction have been noted. Septate junctions are also widespread between the various pigmented glia and Semper cells (Figures 33, 34) of the peripheral retina (Section 3.1), between perineurial cells of the optic lobe, as well as between the latter and the immediately underlying glia (see Chi and Carlson, 1981b).

From this checklist, it would appear that glial cells of the peripheral and intermediate retinas are a well-knit lot, largely through the adhesive properties of septate junctions. It has also been suggested that septate junctions may retard the passage of solutes through extracellular space. Lane and Skaer (1980) reviewed the literature on this topic and found that evidence of extracellular

---

Figure 12. Freeze-fracture above the nuclear layer of the housefly lamina ganglionaris reveals the interdigitating processes of the pseudocartridge glia as they ensheath a photoreceptor axon. Also note the small irregular plaques of E-face particles which represent glial–glial gap junctions (g). (Bar = 0.1 μm.)

Figure 13. A coated pinocytotic vesicle (arrow) attached to one of the deep extracellular channels that permeate the glia of the fenestrated layer. (Bar = 0.1 μm.)

Figure 14. Tropnospongium terminates into a narrow extracellular channel (arrow) in close proximity to a cisterna of rough endoplasmic reticulum (er) within a cell body of a monopolar neuron. (Bar = 0.1 μm.)

Figure 15. A capitate projection (cp) in longitudinal and cross section. These invaginations into the retinular axons are from the epithelial glia. Note synaptic vesicles in the axoplasm of a retinular axon (Bar = 0.1 μm.)

Figure 16. Gnarls (gn) are evaginations of the epithelial glia into the axoplasm of β-fibers of the optic cartridge. Note T-shaped synaptic ribbon (arrow) nearby. (Bar = 0.1 μm.)

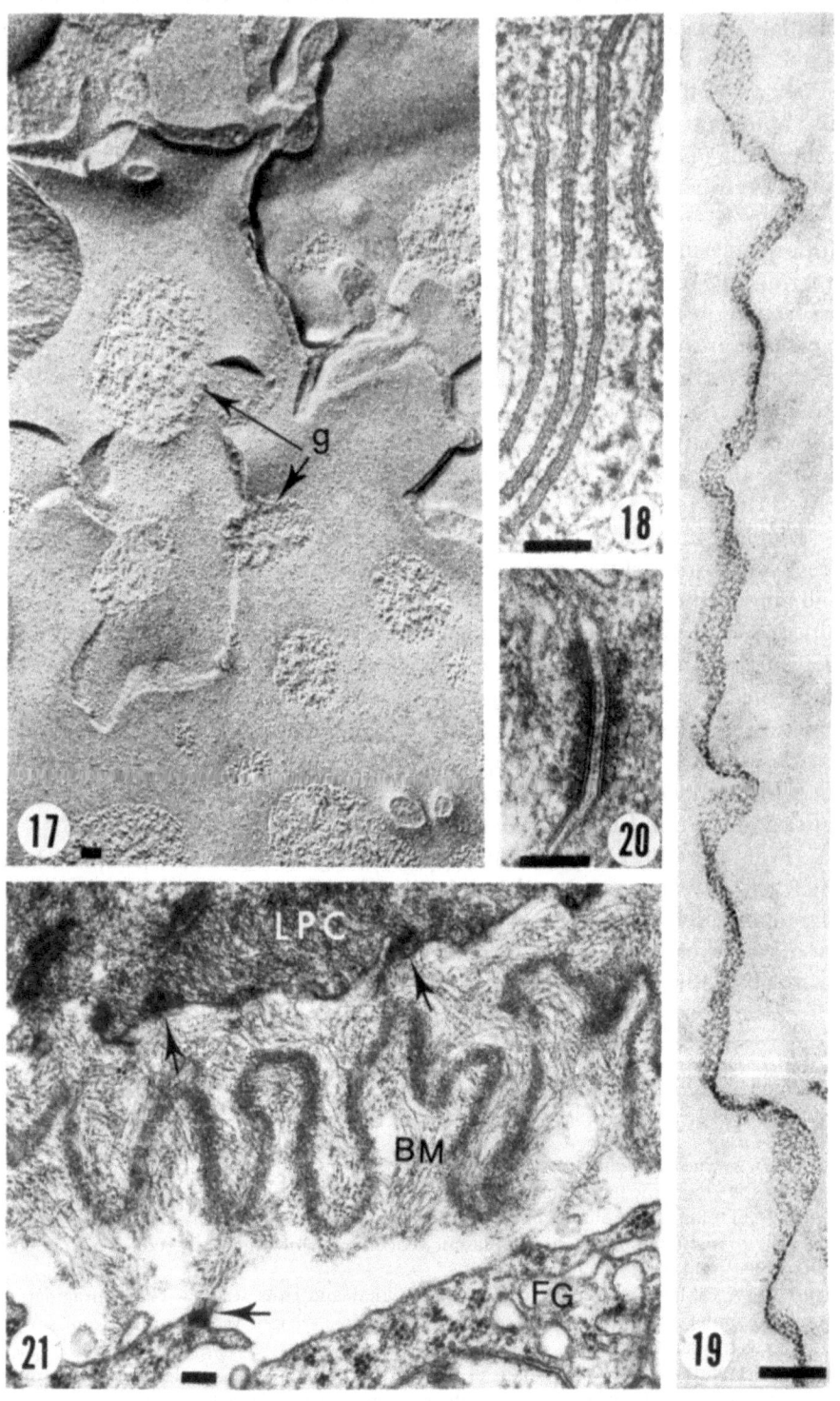

occlusion by septate junctions was largely wanting. Nonetheless, evidence for this role continues to accumulate (see Saint Marie and Carlson, 1983b).

Tight junctions. Throughout most of the 1970s, the existence of tight junctions (zonulae occludentes) in insects was largely overlooked by many cytologists. Lane and co-workers (summarized in Lane, 1981a) have since rectified this disparity with abundant evidence that intercellular fusion of membranes does occur in insects, though not on the grand scale seen in vertebrates.

Tight junctions have recently been reported in the insect eye (Saint Marie and Carlson, 1980, 1983b; Lane, 1981b; Saint Marie, 1981). In the peripheral retina, junctions resembling tight junctions are noted between retinular (photoreceptor) cells sharing the same ommatidium (see Figures 16-18 in Chapter 11). These usually occur as a few aligned strands adjacent to the belt desmosome. The leakiness of these junctions to extracellular dyes (Chi and Carlson, 1981a), however, raises doubts as to their effectiveness as an occluding barrier at this site. More important are the zonulae occludentes found in the intermediate retina. Here the zonulae are quite extensive, often involving scores of intramembranous fusions between apposed glial cells (Figures 25-28) and between glia and neurons (Section 4). Though simpler in geometry than the anastomosing tight junctions of vertebrates, the insect counterparts bear all of the essential characteristics of true tight junctions, i.e., the punctate fusion of apposed membranes in thin section (Figure 27) and the correspondence of P-face ridges and E-face grooves on apposed membranes in freeze-fracture replicas (Figure 28).

At the proximal boundary of the intermediate retina, tight junctions are a common feature, occupying nearly all of the available space between marginal glia (Figure 25), thus sealing the base of the neuropil. Within the neuropil, tight junctions between epithelial glia (Figure 26) circumferentially isolate each optic cartridge from its neighbors. The satellite glia of the interface layer represent the "cap" of the junctional "bottle" that encloses each optic cartridge. Here, arrays of tight junctions mingle with septate junctions (Figure 27). A more complete account of this system of paracellular barriers and their possible roles in visual signaling is given in Saint Marie and Carlson (1983b).

Figure 17. The freeze-fracture replica surface at the apposition of two fenestrated layer glial cells of the housefly. This field illustrates the frequency, size, and shape of E-face particle plaques, which represent gap junctions (g). (Bar = 0.1 μm.)

Figure 18. Invaginated membrane of an epithelial glial cell. The plasmalemma of this cell folds into membranous stacks which engage in reflexive gap junctions. (Bar = 0.1 μm.)

Figure 19. An apposition of a fenestrated layer glial cell and a pseudocartridge glial cell. The extensive gap junction has been infiltrated with lanthanum. (Bar = 0.1 μm.)

Figure 20. A demosome between two perineural cells at the lateral border of the intermediate retina. (Bar = 0.1 μm.)

Figure 21. The basement membrane (BM) is interposed between a large pigment cell (LPC) and glia of the fenestrated layer (FG). Arrows point to hemidesmosomes. (Bar = 0.1 μm.)

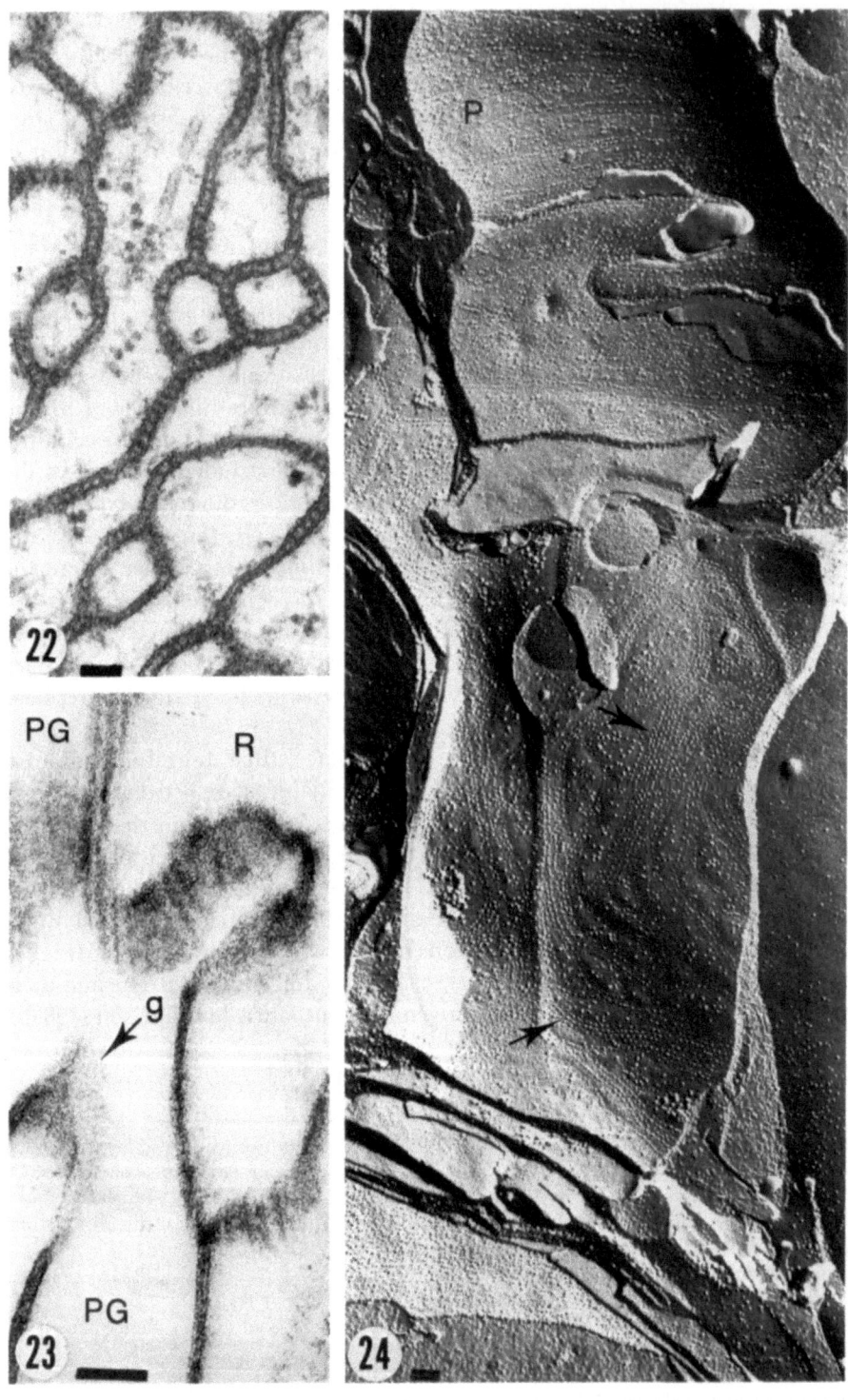

3. Specialized Supporting Cells of the Insect Eye

While photoreceptor cells are the only neuronal elements of the peripheral retina, several forms of pigmented glia and tracheal cells are also part of its *dramatis personae*.

3.1. Pigmented Glia

In many classifications of glial cells, the retinal pigment cells have not been mentioned. Based on their intimate structural association with receptor cells, a case can be made that pigment cells be included in the family of glia. It was primarily because of the sequential depolarization of first the photoreceptors and then the pigmented glia in response to light that Baumann (1975) was prompted to conclude that pigmented cells should "be considered as the equivalent of glial cells in other nervous structures." This in-tandem electrical activity takes place without direct coupling between photoreceptor and pigment cell types. Depolarization of the photoreceptor cells leads to increased extracellular K^+ which, in turn, transiently decreases the pigment cell potential (Coles and Tsacopoulos, 1979). The latter effect can be at once widespread among the pigmented glia when gap junctions connect these pigment cells as in the honeybee (Perrelet, 1970).

Pigmented glia are found in the compound eyes and ocelli of all insects. However, it is in the compound eye that the pigmented glia exhibit their greatest diversity of cellular morphology and ultrastructure, and this finding probably relates to several functions that have been ascribed to these cells; e.g., optical screening (Langer, 1975), mechanical support for each ommatidium (Chi and Carlson, 1976), and metabolic intercourse with the retinal cells (Tsacopoulos *et al.*, 1981).

There are three distinct types of pigmented glia in the insect eye: primary (corneal), secondary (large), and basal (small) pigment cells. Each kind maintains varying degrees of contact with the adjoining retinular cells.

Large pigment cells. In the fly, a cluster of large (secondary) pigment cells (LPCs) suspends the entire ommatidium. This is accomplished through a distal anchorage into the base of the corneal lenslet. These slender, elongate cells then extend (in parallel with the even more elongate retinular cells) to another firm embedment (via hemidesmosomes) in the basement membrane (Figure 21). A

Figure 22. Interdigitating processes of pseudocartridge glial cells which are totally bonded by septate junctions. (Bar = 0.1 μm.)

Figure 23. An unstained correlate to the previous figure. Lanthanum has filled the septate junctions between two pseudocartridge glial cells (PG) and a retinular axon (R). This tracer reveals the pleated-sheet form. Note also the glial-glial gap junction (g). (Bar = 0.1 μm.)

Figure 24. A freeze-fracture replica revealing the P pace of the pseudocartridge glia. Note the undulating rows of particles which represent interglial septate junctions (arrows). P face (P) of the glial cell also lies beneath a retinular axon, and the P-face particles here are oriented at right angles to the neuronal projection. (Bar = 0.1 μm.)

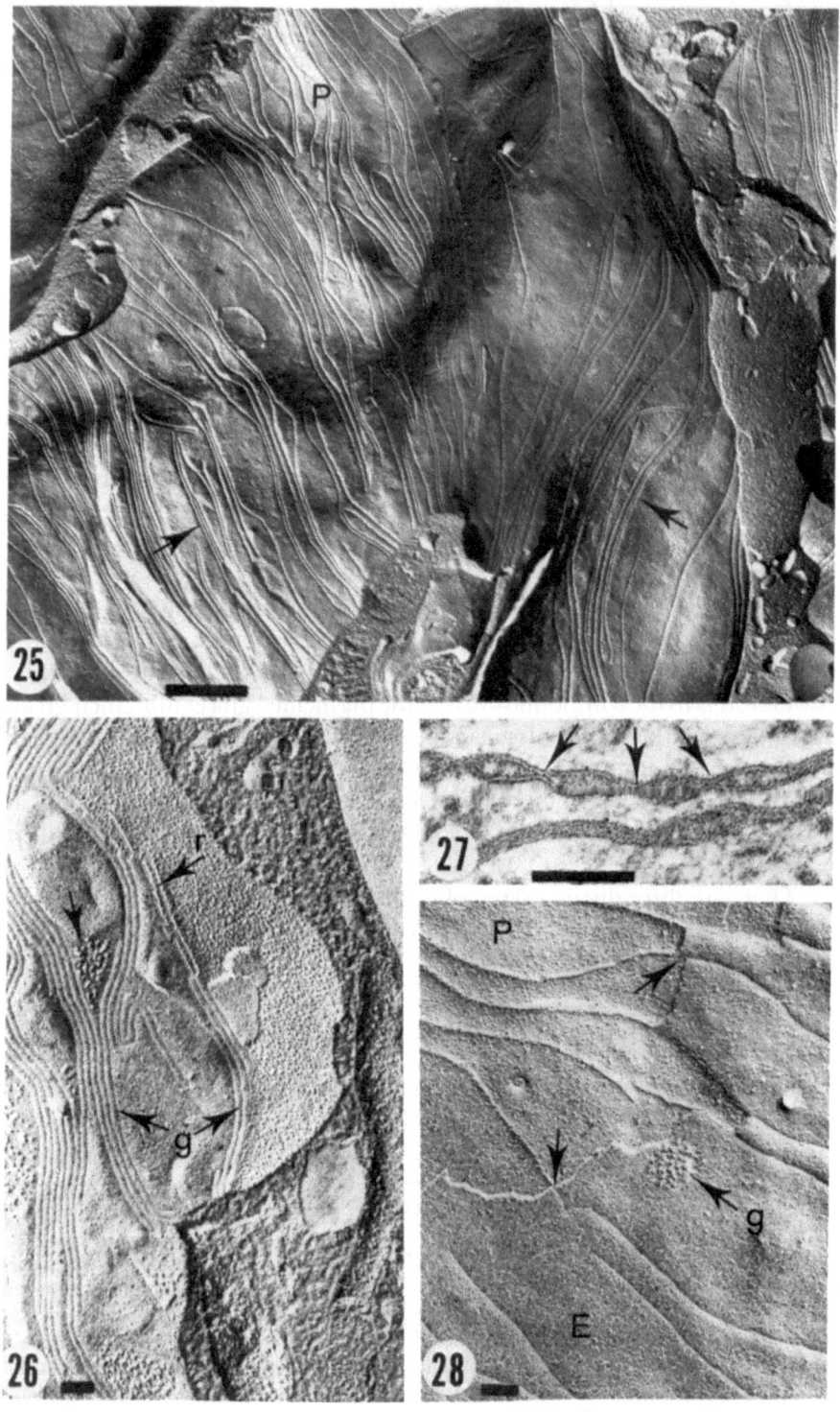

few micrometers below the corneal lens, the function of surrounding each pseudocone is taken over by two corneal pigment cells. Throughout most of the length of the peripheral retina (from pseudocone base to basement membrane), the LPCs maintain a loose contact with the photoreceptor cells via slender processes that wedge between the retinular cells (Figure 3 in Chapter 11).

Corneal pigment cells. These cells (two per ommatidium in the fly) provide most of the lateral wall and structural integrity of the pseudocone. Proximally, corneal (primary) pigment cells (CPCs) enclose the four Semper cells and then embrace the distal cap of the retinular cells. In the distal pseudocone area, the LPCs are sandwiched between CPCs of the neighboring ommatidia. This particular LPC-CPC construction is noteworthy, particularly in the interstice region between four contiguous pseudocones. Here the LPCs take on a blocklike character and fill (and presumably reinforce) these critical corner areas. Septate junctions abound between LPCs, between CPCs (Figure 33), and between LPCs and CPCs (Chi and Carlson, 1976; Chi *et al.*, 1979). Gap junctions also occur at each of these interfaces (Figure 30; see also Chi *et al.*, 1979).

Apart from their obvious light-filtering activity, a function almost never ascribed to "more conventional" glial cells, the copious glycogen and lipid stores in these pigmented cells (Ribi, 1978) point to the more usual glial role, that of metabolic and nutritional support for the neighboring photoreceptor cells. Presumably, glycogen and lipid in the pigmented glia are made available to the photoreceptors whose energy requirements for ion pumping and the synthesis of proteins, especially opsins, are probably high.

Semper cells (SCs) and small (basal) pigment cells (SPCs). The presumed glial lineage of this cell group is based on: (1) their distal and proximal placements (at all times contiguous with the photoreceptor cell/axon); and (2) their transfer of metabolites via gap junctions with the corneal pigment cells (Figure 34). It is now apparent that in the fly like the mosquito (Brammer, 1970), SCs and SPCs represent a single cell type. The anucleate SPCs are proximal bulbous extensions of SCs. A cluster of four SCs underlies each pseudocone. From the base of each SC issues several processes, one of which extends along the entire length of the peripheral retinal between a pair of retinular cells (see Figure 17, Chapter 11). Just above the basement membrane, processes of the four SCs terminate, each as a bulbous ending filled with pigment granules (Figure 29).

Figure 25. A freeze-fracture replica of a marginal glial cell demonstrating its P face (P). This plane illustrates the pervasive nature of glial tight junctions in this glial layer. An arrow denotes P-face ridges. (Bar = 0.5 μm.)

Figure 26. A freeze-fracture surface showing intramembranal areas between two epithelial glial cells. E-face grooves (g) correspond with P-face ridges (r). Note the tight parallel configuration of these ridges/grooves which run parallel to the neuronal projection. Also note the small gap junction on the glial E face (arrow). (Bar = 0.1 μm.)

Figure 27. A thin section showing satellite glia processes. Apposed glial membranes fuse (arrows) into focal tight junctions which alternate with intercellular septa. (Bar = 0.1 μm.)

Figure 28. P-face (P) ridges of one marginal glial cell align (arrows) with the E-face (E) grooves of an adjacent cell, as is typical of tight junctions. A small E-face gap (g) junction is also present. (Bar = 0.1 μm.)

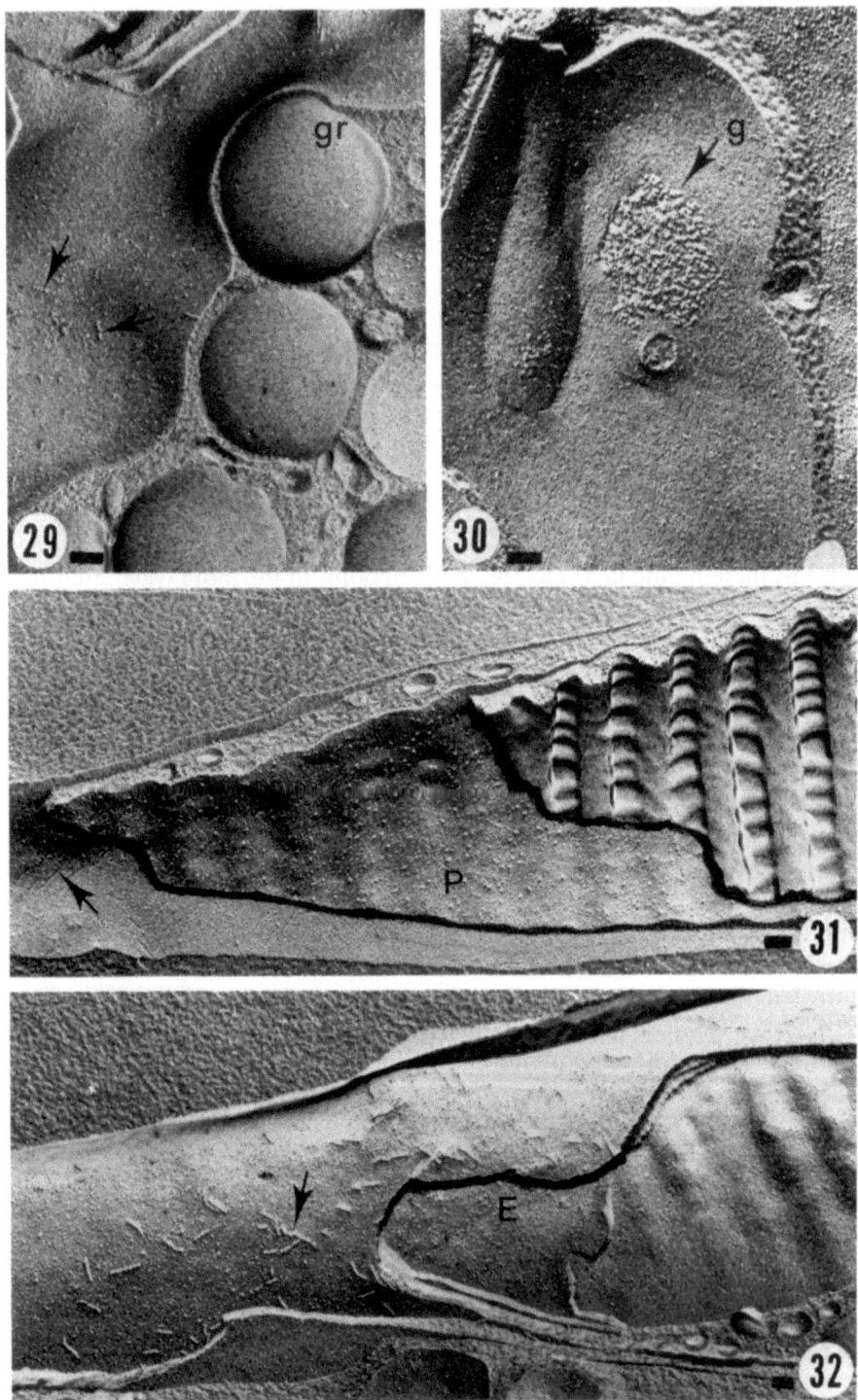

This cluster is nestled medial to the axons of an ommatidium at the base of the rhabdomeres. Here the four pigment-filled endings (SPCs) are linked to each other by gap junctions and extensive desmosomes (Figure 35).

The distal aspect of each SC contains the nucleus and an abundance of vertically oriented microtubules and SER. They contain little or no stored material or synthetic organelles (Golgi, ribosomes, RER, or mitochondria) (Boschek, 1971), an advantageous condition considering that they are in the light path. The apical surface of the SC contains numerous microvilli which contact the pseudocone; and their lateral surfaces, where they appose each other, are covered with intercellular junctions (septate junctions, gap junctions, and desmosomes) (Chi et al., 1979). The former two specializations are also present between the SCs and the CPCs (Figure 34).

Short, intramembranous particle ridges (P face) or grooves (E face) occur frequently along freeze-fractured SPCs (Figure 29), SCs and their interretinular processes, LPCs and CPCs. Such ridges and grooves are also evident along tracheal cells (Figures 31, 32) in the housefly eye, along the glial–axonal apposition of the long visual fibers in the first optic neuropil (Saint Marie and Carlson, 1983b) and surrounding the postsynaptic membrane at the synapses between retinular cells and monopolar cells (Carlson et al., 1983). Similar short, moniliform ridges have been reported in a variety of insect tissues (summarized in Lane, 1979; Lane and Skaer, 1980) including nervous tissue (neurons and glia), gut, trachea, and muscle. Based on correlative thin sections, Lane and Skaer (1980) believe that such specializations represent focal tight junctions. Their scanty and discontinuous nature, however, is not compatible with an occluding function, and therefore these authors (Lane, 1979; Lane and Skaer, 1980) propose that such short particle arrays may function as enzymatic receptor or adhesion or contact-guidance recognition sites.

3.2. Tracheal Cells

Each ommatidium in the peripheral retina usually has one or two large-bore tracheae ascending distally in parallel with the photoreceptor cells (see Chi et al., 1979). These tracheae are branches of an extensive bed of air tubes which

Figure 29. A freeze-fracture view of the small pigment cells from the housefly retina. Large pigment granules (gr) reside in their cytoplasm. Numerous short, P-face ridges (arrows) are found within the membrane. (Bar = 0.1 μm.)

Figure 30. A freeze-fracture plane showing a macular patch of unordered, E-face particles which denote a gap junction (g) between two large pigment cells in the peripheral retina. (Bar = 0.1 μm.)

Figure 31. A freeze-fracture replica revealing intramembranal surfaces of a tracheal cell. An E-face groove on the outer tracheal cell membrane is denoted by the arrow. The P-face (P) leaflet of the inner tracheal cell membrane is also seen. The toothed taenidial rings of the tracheal cell are conspicuous on the right. (Bar = 0.1 μm.)

Figure 32. A freeze-fracture replica showing numerous short, P-face ridges (arrow) on the P face of the outer membrane of the tracheal cell. Note the annulated relief of the chitinous tracheal surface (far right). E = E face of the inner membrane of the tracheal cell. (Bar = 0.1 μm.)

extend across the top of the intermediate retina in the domain of the fenestrated layer glia. These tracheae have their origin in a widespread tracheal system in the external chiasm (between first and second optic neuropils). Tracheoblast cell bodies in this chiasm also project fine tracheoles distally into the optic cartridges.

In fracture replicas, the two membrane surfaces of the tracheal cell are easily distinguished. The inner membrane surrounding the tracheal tube is unremarkable, exhibiting a sparse but uniform distribution of P-face particles (Figure 31) and a smooth E face (Figure 32). By contrast, the outer membrane of the cell contains many short, P-face particle ridges (Figure 32) and E-face grooves (Figure 31). The significance of these membrane specializations is discussed above in Section 3.1 and in Lane (1979) and Lane and Skaer (1980). A comprehensive treatment of tracheae is set forth in Volume 1 of this series by Noirot and Noirot-Timothée (1982).

4. Neuro-glial Junctions

Knowledge of membrane associations between neurons and glia provides us with critical insights into mutual structural relations and, perhaps more importantly, into metabolic and other functional interdependencies that exist between these two very different kinds of cells. For background, it is instructive to consider how the plasma membranes of neurons and glia associate with each other. Previously we have mentioned trophospongia, capitate projections, and gnarls (Section 2.3). Now the five known intercellular junctions present between glia and neurons of the insect eye will be discussed.

4.1. Desmosomes

Neuro-glial desmosomes are rarely encountered in the housefly eye and never in the intermediate retina. In the peripheral retina, desmosomes secure the distal cap of the photoreceptor cell to the base of the SCs (Saint Marie, unpublished), and the proximal end of the retinula cell (just beneath the end of the rhabdomere) to the SPCs (Figure 35). These junctions thus should be considered as an appreciable factor in the tautness of the retinular cells and in any axial twisting of the ommatidium, should the latter occur.

4.2. Septate Junctions

Both retinular axons and laminar monopolar interneurons make septate junctions with glia. Specifically, pseudocartridge glia and retinal axons are extensively bound to each other via septate junctions (Figures 23, 24; see also Figures 24–26 in Chapter 11). In turn, the next more proximal glial layer, the satellite glia, binds the retinal axons (Figure 28 in Chapter 11) as well as the neurite neck of the monopolar interneurons (Figure 36) with tight and septate junctions. Glial–neuronal septate junctions are not found in the synaptic neuropil of the lamina ganglionaris, nor below it in the external chiasm.

Freeze-cleaved preparations of the pseudocartridge glia show rows of 10-nm, P-face particles, aligned in parallel. In Figure 24, we see the difference in orientation of the P-face particles (of the septate junction) between a glial–glial apposition versus that of a glial–neuronal juxtaposition. Particle rows encircle the axon and are thus perpendicular to the axon's long axis. When adjacent glial membranes are cleaved, on the other hand, the parallel particle rows are flexuous to the point of forming enclosed loops.

Though a few rows of septa may easily be penetrated by extracellular tracers (Figure 23), the extensive zonulae (50–100 rows) above the nuclear rind of the intermediate retina probably effect a partial and selective barrier to extracellular solutes (Saint Marie and Carlson, 1983b) and may contribute to the increased extracellular resistance found in this region (Shaw, 1975; Zimmerman, 1978). Below the nuclear rind (Figure 36), tight as well as septate junctions would form the ultimate sealing qualities of that barrier system (Section 4.4). Other glial–neuronal septate junctions are listed in Lane and Skaer (1980).

4.3. Scalariform Junctions

Scalariform junctions, like septate junctions, exhibit a ladderlike appearance when sectioned transversely. Differentiation between the two is not difficult as the scalariform junction shows indistinct subtle septa (Figure 22 in Chapter 11) in contrast to the bolder, stouter, and far better resolved septa of the septate junction (Figure 22). In addition, scalariform junctions are often associated with closely apposed mitochondria which is not the case for septate junctions. Mitochondria may line up along the plasma membrane with the mitochondrial long axes parallel to and often within 90 Å of the membrane. One obvious example of the scalariform junction in the insect eye is that which links the peripheral photoreceptor cells (Chi et al., 1979; Chi and Carlson, 1981a). This intercellular association is also found in the intermediate retina between the fenestrated layer glia and photoreceptor axons (see Figures 20–22 in Chapter 11) and between satellite glia and those same axons (Figures 26–28 in Chapter 11).

Scalariform junctions are found in a wide variety of insects, most often between cells of transporting epithelia. Knowing that, it becomes more understandable why mitochondria might be nearby if energy for active transport at that site is required; and why the junction is freely permeable to tracers if such pumping activity is facilitated by access to ion pools of the greater extracellular space. Conversely, it is not completely clear what function the scalariform junction has in nervous tissue. Lane et al. (1977), Lane and Swales (1978a, b), and Lane and Skaer (1980) tentatively assign a role in ion transport to scalariform junctions. Such a concept would seem to be compatible for the housefly lamina as well, given the disposition of scalariform junctions immediately distal and proximal to the retina–lamina diffusion barrier (Saint Marie, 1981).

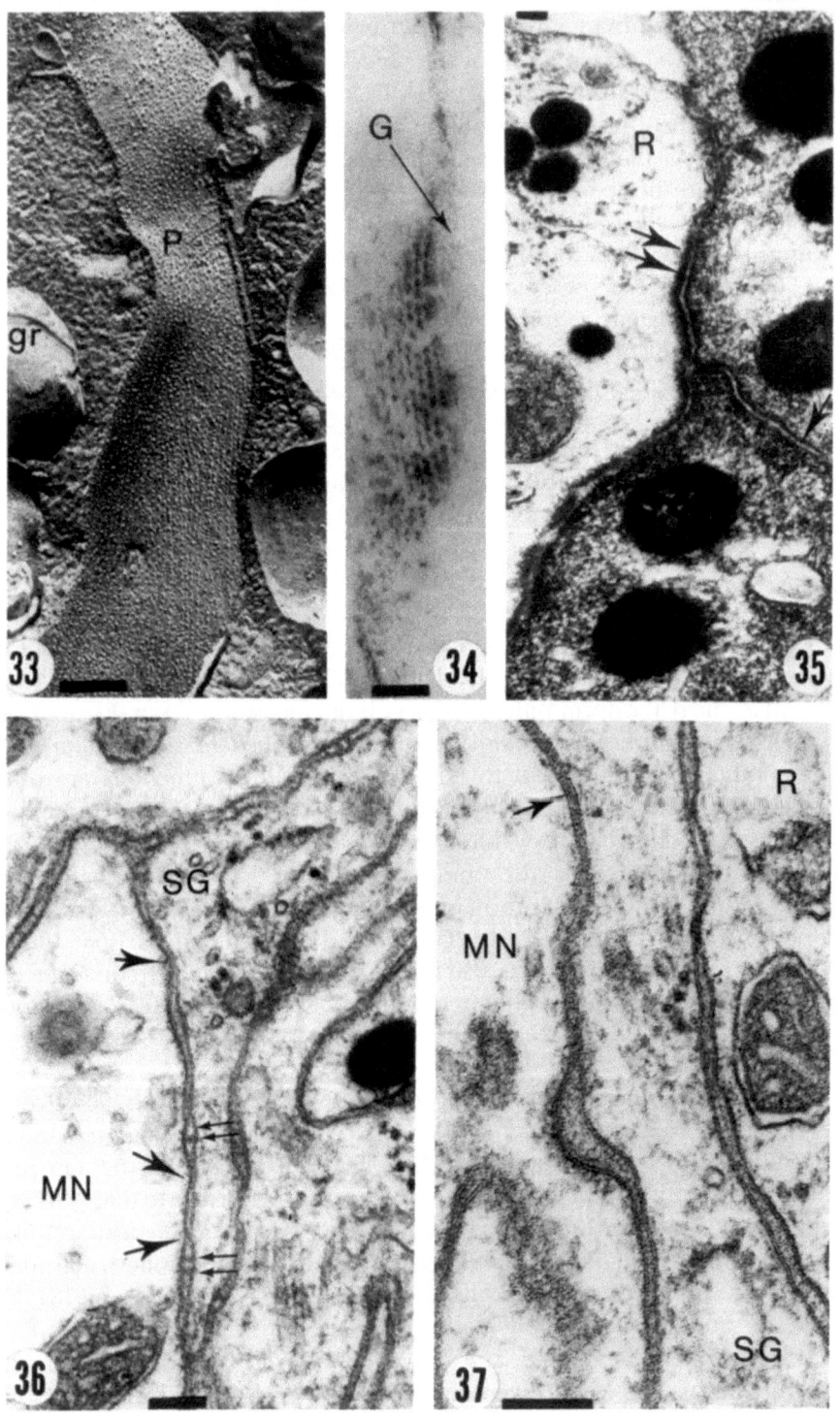

4.4. Tight Junctions

Fused membranes between neurons and glia totally occlude intercellular spaces and thus form the basis of the system of paracellular ion-barriers. Several regions of high resistance are known for the insect eye (Shaw, 1975; Zimmerman, 1978) based on analyses of field and DC potentials. This electrical compartmentalization will be discussed more fully in Section 5.6. For the present, a checklist of neuro-glial tight junctions is a prerequisite for better understanding the electrical activity of the lamina.

Tight junctions occur not only between glia in the intermediate retina (Section 2.4), but also between glia and neuron in the distal and proximal portions of the first optic neuropil. Distally, satellite glia form tight junctions between retinal axons (Figures 32, 33 in Chapter 11) and monopolar neurites (Figure 36), and at the base of the lamina the marginal glia make tight junctions with axons entering and exiting from the external chiasm. More details about this fore and aft perilamina barrier are available in Saint Marie and Carlson (1983b). In this study, typical tight junctions were revealed from an analysis of thin sections and freeze-fracture replicas.

4.5. Gap Junctions

The gap junction (macula communicans or nexus) is present between a variety of cell types in many insect tissues and is the site of electrical and metabolic coupling between cells so linked. A complete roster of insect cells forming gap junctions can be gleaned from Lane (Chapter 14 in Volume 1 of this series, and Lane, 1978, 1981a). This membrane specialization is the structural correlate of the electrical synapse. Very few of these low-resistance pathways have been identified in insects. If this type of cell-to-cell communication is rare between insect neurons, it is even more illusive between insect neurons and their glia. In vertebrates, we are aware of only two such examples (Morales and Duncan, 1975; Walker and Hild, 1969).

Figure 33. A freeze-fracture replica of septate junctions, appearing as parallel rows of unfused P-face (P) particles, between two corneal pigment cells of the housefly. The pigment granules (gr) are large, and closely packed. (Bar = 0.25 μm.)

Figure 34. Lanthanum infiltrates a septate junction between a Semper cell and a corneal pigment cell revealing the pleated-sheet pattern. Also note the infiltrated gap junction (G). (Bar = 0.1 μm.)

Figure 35. A field in the proximal portion of the peripheral retina. A desmosome (arrow) is seen linking two small pigment cells. Similar desmosomes (double arrow) bind small pigment cells to the retinular axon (R). Pigment granules (of a specific kind) reside in each cell type. (Bar = 0.1 μm.)

Figure 36. A thin section taken from the proximal nuclear layer. Focal tight junctions (large arrows) fuse the membrane of a satellite glia (SG) to that of the neurite of a monopolar relay neuron (MN). These tight junctions alternate with glial-neuronal septate junctions (double, small arrows). (Bar = 0.1 μm.)

Figure 37. A satellite glial process (SG) projects between a retinular axon (R) and a cell body of a monopolar neuron (MN). This glial process engages the neuronal soma by a gap junction (arrow). (Bar = 0.1 μm.)

Thin sections and freeze-fracture replicas of the perikaryal rind (nuclear layer) of the fly lamina indicate that gap junctions are common between adjacent monopolar neurons (Carlson et al., 1983) and between monopolar somata and their satellite glia (Figure 37; Saint Marie and Carlson, in preparation). Though electrical coupling between certain monopolar neurons may be anticipated based on the similarity of their input and postsynaptic response, what interpretation can be placed on apparent electrical synapses between neurons and glia?

The neuron in question, the lamina monopolar interneuron, possesses a sizeable nucleus, and its perikaryal region tapers to a narrow neurite prior to entering the neuropil. The perikaryon is outside of the path of the orthodromically directed postsynaptic potential (which goes directly from dendrite to axon). The internal resistance of the narrow neurite neck should lessen the likelihood of electrical shunting toward the large cell body. Thus, one can infer that these glial-somatic gap junctions probably have little influence on the electrical properties of the monopolar cell dendrites and axon, though one should not discount the importance of gap junctions in the exchange of metabolites (see Section 5.3) between glia and neuron, since these junctions are found in the partially occluded compartment of the nuclear layer (Saint Marie and Clarson, 1983b).

5. Structure-Function Relationships

The ultimate challenge for an anatomical treatise is that of deducing, or at least proposing functions based on ultrastructural findings. The definitive list of functions in which neurons and glia collaborate will be lengthy and truly awesome, if it is ever drawn up. The following six subjects are those for which some knowledge has accumulated. As to the blood-brain barrier facilitating ionic homeostasis, considerable information is available, and therefore this topic is the most comprehensive. Compartmentalization of neuropil, on the other hand, comes from very recent data obtained from a few laboratories, but these shortcomings in no way diminish the importance of the subject.

5.1. Structural Support

Before the first membrane specializations of insect glial cells were visualized, there was the intuitive notion that neuroglia ("nerve cement") accomplished more in overall structural integrity for nervous tissue than merely filling the voids between neurons. In the earliest times, such a belief was founded on the extravagant array of interdigitating glial processes that wrapped some insect axons. Later, glial invaginations into nerve cells such as capitate projections and gnarls were observed, and these structures seemed to provide anchoring or cohesive effects. More recently, membrane specializations have been better characterized by high-resolution capabilities and freeze-fracture studies. Intercellular adhesions, such as septate and tight junctions and desmosomes between glial cells (Section 2.4), and in some cases between glia and neuron (Section 4), offer one the impression of mechanical stability in certain neuropils. Particular adhesive membrane specializations in the insect optic lobe are illustrated and discussed in Chi and Carlson (1980), Lane (1981b), and Saint Marie and Carlson (1983b), while more general reviews of glial-glial and glial-neuronal intercellular junctions can be found in Lane and Skaer (1980) and Lane (1981a, 1982).

Apart from glial cell geometry and the various membrane specializations, one must also consider certain glial organelles as support elements. Microtubules are a prominent feature in many insect glial cells. Not only the presence but the collective and consistent orientation of microtubules are factors to be reckoned with when considering the mechanical rigidity of a neuropil. For example, in the housefly eye, both pseudocartridge and marginal glial cells are replete with microtubules (Section 2.2 and also Saint Marie and Carlson, 1983a). In both cell types, the microtubules are directed at right angles to the long axis of the passing axons, thus producing a fibrous mat which completely ensheaths the optic ganglion. Such features probably also have functional importance in the transport of vital metabolites to the neuropil (Section 5.3), but the more established cytoskeletal function of microtubules is one that should not be minimized.

One must be impressed by the asymmetry and, in some cases, the almost filigree nature of glial processes. Again, the glia of the insect eye (Section 2.3) are outstanding examples of cells which form a variety of complex interdigitations among themselves as they "fill the gaps" between the neurons they envelop. An integral part of this architectural organization is the glial function of maintaining neuron position, spacing, and orientation within the neuropil. By placing mechanical constraints through membrane specializations and enveloping processes, glia confine neurons to the appropriate slots in the axonal feltwork of the adult neuropil. This "glial guidance" will be discussed further in Section 5.5.

Glial contacts with neurons may also result in dramatic changes in neuronal dimensions. This is particularly true at the boundary of the perikaryal rind and neuropil layer in the insect lamina (Saint Marie and Carlson, 1983a). Here multiple concentric cuffs of glial processes, replete with adhering junctions, envelop the emerging neurite of the monopolar perikaryon. This results in the neurite being greatly narrowed until it enters the plexiform layer. Retinular axons also increase in diameter as they emerge from this proximal portion of the nuclear layer.

5.2. The Blood-Brain Barrier and Ionic Homeostasis

Of all the known and ascribed functions for insect glia, ionic homeostasis is the best documented. Much of this understanding comes from the work of Lane, Treherne, and colleagues who have extensively studied the permeability properties of the glia of insect ganglia. The glial barrier maintains the ionic microenvironment of the nerve cell surfaces in these avascular ganglia. It is the neuronal bathing medium that, when ionically correct, permits the initiation and propagation of neuronal signals.

The point is made (Skaer and Lane, 1974) that in many insects (particularly the phytophagous species), the inverted ratio (high K^+; low Na^+) in the hemolymph is inimical to nerve function, should hemolymph ever bathe the neurons directly. The basis for the vagaries of hemolymph composition arises from the wide variety of foods ingested by the collective Insecta. In addition, dietary habits may also change drastically from larva to adult, and this temporally alters the composition of the hemolymph. Nevertheless, in the face of this variability, a remarkable ionic homeostasis of the neuronal extracellular environment is maintained. Such compositional extremes between blood and extracellular fluid are usually not found in other animals (Abbott and Treherne, 1977). It follows that, of all the invertebrates, insects are the only group with a blood-brain barrier in insects is considerably less complex (at least structurally) understood years before the structural correlates (tight junctions) were uncovered.

There is an early literature dating from the 1950s (e.g., Hoyle, 1953; Twarog and Roeder, 1956) in which an ionic diffusion barrier between the insect nerve cell and the bathing solution was proposed. In the above cited cases, high potassium concentrations in the media failed, for some time, to eliminate axonal conduction, but impulses were quickly blocked when the nerve was desheathed. Next O'Brien and Fisher (1958) demonstrated that sheathed nerves were resistant to particular toxins and pharmacological agents. Eldefrawi and co-workers studied the permeability of fatty acids (Edlefrawi and O'Brien, 1966), quaternary ammonium compounds (Eldefrawi and O'Brien, 1967a), alcohols (Eldefrawi and O'Brien, 1967b), and polarity effects (Eldefrawi et al., 1968). In passing years, a large number of organic molecules (many of the insecticides) were shown to be blocked from entering a sheathed nerve, and therefore interest shifted to the structure of the sheath.

Ultrastructural tracer studies (Pichon et al., 1972; Lane, 1972; Skaer and Lane, 1974; Lane et al., 1977) have elucidated the fine structure of the perineurium (the layer of modified glial and perineurial cells composing the innermost nerve sheath) and the acellular neural lamella that wraps the perineurium. In addition to describing the kind and disposition of cells of this sheath, Lane and co-workers have now identified the intercellular junctions responsible for the tightness and structural cohesion of these sheath cells.

The neural lamella was seen to be a collagenous mat overlying a monolayer of perineurial cells beneath which are several strata of glial cells. Many slender meandering, intercellular clefts were found to exist between glial cells—the system of glial lacunae. All these elements are interposed between hemolymph and the neuronal surface. Somewhere in that labyrinth of nerve-encircling cells, barriers existed that exclude such vital cations as sodium and potassium. When the concentrations of these ions in the bathing medium were experimentally manipulated (Pichon et al., 1972), no change occurred in the action potentials elicited from the sheathed cord. However, the desheathed cord failed to function under the same conditions.

Electron microscopic studies (especially by Lane, 1972, 1974; Skaer and Lane, 1974; Lane et al., 1977) have shown, in thin sections and freeze-fracture replicas, that the critical obstacles to the ingress of ions, and toxins, were tight junctions that existed between the lateral, highly convoluted membranes of the perineurial cells. Using the electron-opaque tracer, lanthanum, Lane (1972) found that lanthanum could enter the neural lamella and the lacunae, the clefts and gap junctions between adjacent perineurial cells, but could penetrate no further. At that point, tight junctions in the perineurial cells held back the invading tracer. As an aside, it is interesting to remember that as late as 1973 (e.g., Satir and Gilula, 1973), tight junctions were claimed to be the exclusive property of vertebrates.

Lane and co-workers have continued the ultrastructural characterizations of this sheath in a variety of insect species and recently have charted the ultrastructural changes in the blood-brain barrier especially with regard to assembly and breakdown of gap and tight junctions in the ventral nerve cord during larval and pupal development of the fly *Calliphora erythrocephala* (Lane and Swales, 1978a,b). Other properties of the perineurial cells have been documented, e.g., sodium transport (Schofield and Treherne, 1978), permeability to aliphatic alcohols (Thomas, 1976a,b), and enzymatic activity (O'Brien, 1967; Treherne and Pichon, 1972; Houk and Beck, 1976). Thus, there is a growing body of data on the blood-brain barrier of various insect species.

In summary, there is a barrier located in the perineurial sheath, and tight junctions are the major (and perhaps exclusive) membrane specializations protecting the nerve surface from an ionically hostile hemolymph. While the blood-brain barrier in insects is considerably less complex (at least structurally) than that of animals with backbones, the barrier performs essentially the same functions in both vertebrates and invertebrates. The latter statement is a deliberate but useful oversimplification.

A model by Lane and Treherne (1980) is instructive in understanding how insect glia "fine tune" the K^+-Na^+ cation pools in the extracellular fluid bathing nerve cells. These authors illustrate Na^+ entry into perineurial cells by free diffusion and by a carrier-mediated process. Simultaneously, Na^+ is transported to hemolymph via an ATP-dependent K^+/Na^+ pump. These events cause Na^+ to be concentrated in the fluid sector and K^+ to be sequestered intracellularly by the glia. Also, there is free K^+ and Na^+ diffusion intracellularly (between glial and perineurial cells) via gap junctions. This scheme yields a K^+-poor, Na^+-rich extracellular fluid advantageous for nerve function, i.e., adequate Na^+ for entry into the nerve membrane in depolarization. K^+, effluxed by the depolarized neuron, is taken up by the contiguous glial cells.

A good example of this model in action is that of the honeybee drone eye. Coles and Tsacopoulos (1979) find that it is likely that the glia (in this case the pigmented glia) surrounding the drone retinulae "take up potassium during photostimulation . . . [and] . . . they simultaneously lose sodium." "These ion movements can be summarized as tending to maintain homeostasis of a K^+ and Na^+ in extracellular space." Our studies have revealed that the housefly perineurium overlying the first optic neuropil is somewhat leaky, in that the extracellular tracer lanthanum gains entry through the glial sheath (Chi and Carlson, 1981b) and is deposited in the extracellular space between glial cell and neuron. It can be argued that tightness of the barrier need not be a paramount consideration in the adult *Musca* because there is a "good" Na/K ratio on the order of 11–14 : 1 (Florkin and Jeuniaux, 1974; Zimmerman, 1978).

5.3. Metabolic Commerce

There seem to be numerous ultrastructural clues that indicate exchange of metabolites and ions between homologous and heterologous insect glia, as well as between neurons and glia. One can readily believe in the transfer of small molecules and ions between glial cells because of the extensive gap junctions they make with each other. On the other hand, glial–neuronal gap junctions are not widely known or well documented, and the cell-to-cell communication this specialization effects has been discussed in Section 4.5.

A quarter century ago, Wigglesworth (1960) noted that glycogen reserves of the cockroach perineurial cells were depleted in starvation. It was then hypothesized that neurons (particularly their neuronal somata) acquired carbohydrates through the invading glial trophospongia. It is suggestive that trophospongia frequently terminate very near the ER of the neuron in the fly lamina (see Section 2.3 and Saint Marie and Carlson, 1983a). Treherne (1960) found that trehalose and glucose molecules in the circulating hemolymph were quickly assimilated by ventral cord ganglia of *Periplaneta americana*, and much of that fraction was consigned to amino acid metabolism, notably glutamic acid and glutamine. To gain access to the ganglionic neurons, those previously mentioned saccharides must have passed transmembranally through the perineurial cells (not intercellularly, because of intervening tight junctions between those cells) and then permeated through glial gap junctions and/or percolated through the glial lacunar system to reach the neuronal surface. There is no literature bearing on lipid and protein transfer from insect glia to neurons, but this absence of data is most likely due to experimental inertia. The giant axon of the squid obtains proteins directly from its ensheathing glial cells (Lasek *et al.*, 1974).

Appropriate to this discussion is the insect eye, and starting distally, we see abundant stores of glycogen and lipid (with few mitochondria) in the LPCs that circumferentially shroud the retinular cells from lens to basement membrane (Chi and Carlson, 1976). The retinular cells show essentially no reserves, but many mitochondria, so the food stores of the pigmented glia may ultimately become a major substrate for the sensory cell's metabolism. At a more proximal level in the fly retina, the glia of the intermediate retina have numerous coated transfer vesicles, abundant free and membrane-bound ribosomes, and well-developed systems of microtubules. These structures probably facilitate the uptake, synthesis, and/or transport of essential metabolites across the lamina barrier systems. Lipid droplets in the satellite glia and the pervasive arrays of SER within the pseudocartridge glia suggest that lipids and proteins are stored in these cells. The predominant lateral organization of microtubules in the pseudocartridge and marginal glia suggests how a long-range (interglial) transport of materials from perineurial cells might be accomplished. Since the fly does not have a vascular supply of hemocoel channels in its neuropils (as do other insects), neuroglial microtubules may be of considerable importance in both the long-range and local dispersion of metabolites.

The glial transfer of any specific material to the photoreceptor axons or their adjacent interneurons has not been found to date. Campos-Ortega (1974) determined that tritiated γ-aminobutyric acid injected into the hemolymph was readily sequestered by the fly's marginal glia and, to a lesser extent, by epithelial glia (these cells reside within the synaptic field of the retinal axons), but did not appear in the adjacent neurons. Tsacopoulos *et al.* (1981) concluded from the arrangement of mitochondria in the retinulae versus reserve food material in the neighboring glial cells that there was a compartmentalization of metabolism. The latter paper and one by Evequoz *et al.* (1978) show that photostimulation of the insect retinular cells increases glycogen turnover in the adjacent glial cells— a metabolism that fuels the activated photoreceptors. Apart from potential exchange of metabolites between neurons and glia, the latter also synthesize and secrete enzymes whose target is the nerve cells in the immediate vicinity. It is known for *Melanoplus* (Lane, 1968) and *Musca* and *Calliphora* (Griffiths, 1979) that phosphatases, especially acid phosphatases, are synthesized in the SER of the glial cell and secreted into the extracellular space. Griffiths (1979) proposed that microtubules may effect transport of this enzyme. A light-induced pathology of glial cells was ultrastructurally charted in a visual mutant of *Drosophila* ($wrdgB^{KS222}$) by Stark and Carlson (1982). In *Drosophila* mutants ($wrdgB^{KS222}$) having the gene $Acph-1^{nll}$ (lacking acid phosphatase), there is no light-induced, glial-initiated pathology (Harris and Stark, 1977). Thus, the acid phosphatase gene is transcriptionally active in the glial cell.

5.4. Phagocytosis

Most, if not all, glial cells have the capacity to remove, break down, and phagocytize degenerating neurons and muscles that they contact. In the adult insect, glial cells are believed to be able to divide (Radojcic and Pentreath, 1979). Thus, potential voids created by phagocytized and internalized cellular debris are avoided as glial cells multiply and add glioplasmic volume in that area. It should be remembered that while glial cells have the capacity for phagocytosis, they are not phagocytes per se. Varon and Somjen (1979) state: "Palay has raised the distinction between a cell's '. . . being a phagocyte' and 'being capable of phagocytosis. . . .' Speaking broadly, a phagocyte is a professional; but other cells, whose main calling is some other activity may also be amateurs at phagocytosis."

The engulfing activity of true phagocytes and fat body cells is well known in insect tissue other than nerve and muscle (Wigglesworth, 1972). Since the central nervous system is avascular, there is no opportunity for hemocytes to enter neuropil and release lytic enzymes to digest injured or senescent neurons and endocytotically absorb the debris. That function would be best, if not solely, accomplished by the glia. Such a role for glia may also be important in muscle tissue. Because of the extensive and intimate glial invaginations into insect muscle (see Rheuben and Reese, 1978), we are led to speculate that muscle glia may have a phagocytic function. In some instances (e.g., *Calliphora*), it is known that nonglial phagocytes provide lysosomes for the programmed muscle histolysis that occurs during metamorphosis, but in other species ". . . phagocytes do not play any part in the process" (Finlayson, 1975). In the muscle degeneration–regeneration literature (reviewed by Finlayson, 1975), the autolytic changes in muscles are minutely detailed, but almost no mention is made of any glial contribution, real or potential, to the phagocytic cleanup operation. Close observation of glial cells during metamorphosis is absolutely essential, if we are to assess the contributions of glia to this important function. Insect neuroanatomists would do well to emulate the studies on molluscs where viruses, ferritin, and fragments of dying neurons have been shown to be internalized by glia (Reinecke, 1976; Borovyagin *et al.*, 1972).

There is some information in this regard from studies of some visual mutants of *Drosophila*. In one, $wrdgB^{KS222}$, photoreceptor cell death occurs when the newly emerged adult is exposed to light (Harris and Stark, 1977). Stark and Carlson (1982) have described the ultrastructural pathology of the retinal axons. A multiplication of glial cells (with an increase in glioplasm) is coincidental with the death of axons from R1-6. Griffiths (1979) had shown earlier that (after cutting the photoreceptor cell perikarya) the demise of photoreceptor axons in *Musca* and *Calliphora* came about from the production of at least two different acid phosphatases from the "satellite glia" of the lamina. "Satellite glia" could mean any or all of three distinct glial cell types: fenestrated glia, pseudocartridge glia, and "satellite" glia (as delineated by Saint Marie and Carlson, 1983a). The enzymatic and cytochemical specifics of the resulting lysosomal activity by these glia are given by Griffiths (1979). This glial cell reaction occurs within a few minutes of the operation. In the $wrdgB^{KS222}$ mutant, a similar gliosis with lysosomal activity (directed against the adjacent photoreceptor axons) was observed within 3 days after emergence.

In earlier studies, the interaction of insect glia with adjacent severed or degenerating axons is equivocal. Boulton (1969) failed to find a consistent glial response to deteriorating axons in *Schistocerca gregaria*, since normal and pathological glia were associated with both nondegenerating and degenerating axons. However, Hess (1960) found for *Periplaneta* that the glial sheath hypertrophied above severed axons which subsequently degenerated.

5.5. Glia in Neurogenesis

Glial guidance of neuronal cell movements and control of the ultimate location of neurons in postembryonic insects have been studied only to a limited extent. Pipa and co-workers (Pipa and Woolever, 1965; Pipa, 1967, 1973; Tung and Pipa, 1971, 1972) have demonstrated that glial cells in the wax moth *Galleria* can exert "intrinsic tractive forces" on interganglionic connectives of the wax moth so that the connectives are greatly shortened by the time the adult condition is attained. These tractive forces are so great that the interganglionic axons are caused to loop in the first stage of shortening. Interestingly, such startling mechanical energetics by the glial cells seem to take a toll on these cells. A glial degeneration ensues that mimics that of the neighboring axons. Pipa (1973) concludes that this glial autolysis is a factor responsible for the scanty ensheathment of adult axons which is in stark contrast with the glial covering of "nearly every axon" in the larva.

As retinal axons of Diptera mature during the late pupal stage, each pseudocartridge bundle of axons twists 180° so that each axon "spins off" to a different location than that of its neighbor and into a locus in a separate optic cartridge. Hanson *et al.* (1972) studied this postembryonic neuronal development and concluded that the force responsible for the twisting of the axons could come from either an interaction between retinular and second-order axons or between retinular growth cones and an unknown constituent in the neuropil. What more likely candidate for this "unknown constituent" than glia? Saint Marie and Carlson (1983b) provide considerable detail about septate and tight junctions between retinular axons and the pseudocartridge and satellite glia of the optic neuropil. These intercellular adhesions which probably play a considerable role in enforcing neuronal geometries in the adult may also play a vital role in their development.

5.6. Compartmentalization of Neuropil

In Sections 2.4 and 4.1-4.5 we showed that each optic cartridge is encapsulated by tight junctions: distally, basally, and circumferentially. The satellite glia form a distal cup over the neuropil which fits into the proximal cap of marginal glia. Both glial covers come together as (conceptually) hollow hemispheres over the core of neuropil. The barrier system bisects the photoreceptor cell environment into a rhabdomal portion and a synaptic domain. In like manner, the monopolar dendrites (in the optic cartridges of the lamina ganglionaris) are isolated from their axonal termination in the medulla externa. Laterally, the epithelial glia isolate each optic cartridge, and the long visual fibers, which pass through the lamina en route to the second optic neuropil, are virtually sealed off from the laminar cartridges by the occluding junctions (as are the tracheae). How do all of these neuro-glial barrier systems affect nerve transmission and integration in the first optic neuropil?

First, the perilamina barrier may account for the dark, standing potential (-30 to -100 mV) of the intermediate retina (Zettler and Järvilehto, 1971; Zimmerman, 1978). This sizeable potential difference is thought to be due to the circumferential layer of neuroglia forming an extracellular resistance barrier around the lamina ganglionaris of the fly (Zimmerman, 1978) as well as the locust (Shaw, 1975). As Laughlin (1975) suggests, current electrotonically conducted down the photoreceptor axons results in the extracellular depolarization of the neuropil. These light-stimulated depolarizations [lamina depolarizations (LDs)] have long been known (e.g., Burtt and Catton, 1964; Shaw, 1968; Zimmerman, 1978) to appreciably depolarize the dark, standing potential.

The glial barrier system also modifies the visual signals from the receptors vis-à-vis the lamina field potentials. Laughlin (1974) contends that the depolarization of the retinal axon results ultimately in the depolarization of nearby extracellular spaces which increases the potential difference across the membrane and diminishes neurotransmitter release. This field effect should impinge on the membrane potentials of all neurons in a cartridge. It is now thought that the LDs are involved in several forms of electrical inhibition: light adaptation of the photoreceptors or feedback inhibition (Laughlin, 1974), feedforward inhibition of laminar monopolar cells (Shaw, 1968; Autrum et al., 1970), and lateral inhibition at the photoreceptor–monopolar cell synapse (Shaw, 1975).

The large field potentials in the lamina cartridges should have little effect on the long visual fibers (R7, R8) which are walled off from the cartridges by glial tight junctions. Color and E-vector information directly conveyed by these long, slender fibers to the medulla externa would otherwise be swamped by the large LDs of the nearby cartridges.

Finally, the glial barrier isolates the dendritic fields of the monopolar interneurons (in the lamina) from their axon terminals (in medulla externa). This ion barrier between the two ends of the monopolar cell effects an increase in extracellular resistance which is in series with the resistance of the axonal membrane. Such geometry may account for the apparent nondecremental conduction of the graded potentials in these axons. Thus, the basis for this surprising and unusual conduction may be due to the passive properties of the tissue. The "active mechanism" proposed by Zettler and Järvilheto (1973) now seems less likely.

In summary, we see barrier systems set out by the neuroglia bisecting the distal and proximal ends of the photoreceptors and all of the laminar interneurons. In addition, the massed tight junctions between the epithelial glia laterally isolate each optic cartridge. The precise location of these ion barriers can be correlated with areas of high electrical resistance. These factors probably modify the electrical activity of the various lamina neurons relative to: signal decrement, background intensity shifting, particular forms of intracartridge electrical inhibition, and the elimination of cross-talk between cartridges and between the short and long visual channels. Thus, knowledge of the kind, location, and extent of the membrane specializations of each glial stratum allows a greater understanding of the role neuroglia play in neural integration.

6. Concluding Remarks

Insect neuroglia occupy more than half the volume of brain and accompanying ventral nerve cord. This chapter summarizes some of the functions performed by glial–neuronal associations and hopefully will suggest further studies involving toxicology, embryology, cytology, and cytochemistry. We have studied the glia of the peripheral and intermediate retina of the housefly. These cells comprise all previously classified (Strausfeld, 1976) glia, plus modified glia such as: perineurial cells, Semper calls, and various pigmented glia. The cellular cooperation with, and importance of glia to, neurons is often seen in glial morphology and submicroscopic cytology and in the elaborations and specializations of their plasma membranes. The fine structural details can be related to functions as diverse as: blood-brain barrier, metabolic interactions, neuropilar partitioning, phagocytosis, and neurogenesis. Lastly, we propose the neuroglial cell as important in its own right. Homologous and heterologous glia are linked via extensive gap junctions, a finding that shows glial cells (taken collectively) are very interdependent and capable of concerted metabolic actions. The old adage that "in numbers there is strength" is apt for neuroglia. One understands that insect neuroglia represent great numbers of cells, a variety of cellular form and kind, and, very likely, a diversity of functions.

References

Abbott, N. J., and Treherne, J. E., 1977, Homeostasis of the brain microenvironment: A comparative account. In *Transport of Ions and Water in Animals*, edited by B. L. Gupta, R. B. Moreton, J. L. Oschman, and B. J. Wall, pp. 481-510, Academic Press, New York.

Autrum, H., Zettler, F., and Järvilheto, M., 1970, Postsynaptic potentials from a single monopolar neuron of the ganglion opticum I of the blowfly *Calliphora*, *Z. vgl. Physiol.* 70:414-424.

Baumann, F., 1975, Electrophysiological properties of the honey bee retina. In *The Compound Eye and Vision of Insects*, edited by G. A. Horridge, pp. 53-74, Oxford University Press (Clarendon), London.

Becker, H. W., 1965, The number of neurons, glial and perineurium cells in an insect ganglion, *Experientia* 21:719.

Borovyagin, V. L., Salanki, J., and Zs.-Nagy, I., 1972, Ultrastructural alterations in the cerebrum ganglion of *Anadonta cygnea* L. (Mollusca, Pelecypoda) induced by transection of the cerebrovisceral connective, *Acta Biol. Hung.* 23:31-45.

Boschek, C. B., 1971. On the fine structure of the peripheral retina and lamina ganglionaris of the fly, *Musca domestica*, *Z. Zellforsch. mikrosk. Anat.* 118:369-409.

Boulton, R. S., 1969, Degeneration and regeneration in the insect central nervous system, *Z. Zellforsch. Mikrosk. Anat.* 101:98-118.

Brammer, J. D., 1970, The ultrastructure of the compound eye of a mosquito *Aedes egypti* L., *J. Exp. Zool.* 175:181-195.

Bullock, T. H., and Horridge, G. A., 1965, *Structure and Function in the Nervous Systems of Invertebrates*, Freeman Press, San Francisco.

Burkhardt, W., and Braitenberg, V., 1976, Some peculiar synaptic complexes in the first visual ganglion of the fly, *Musca domestica*, *Cell Tissue Res.* 173:287-308.

Burtt, E. T., and Catton, W. T., 1964, The potential profile of the insect compound eye and optic lobe, *J. Insect Physiol.* 10:689-710.

Campos-Ortega, J. A., 1974, Autoradiographic localization of n-γ-aminobutyric acid uptake in the lamina ganglionaris of *Musca* and *Drosophila*, *Z. Zellforsch. mikrosk. Anat.* 147:415-31.

Carlson, S. D., Saint Marie, R. L., and Chi, C., 1983, Freeze-fracture replication and the interpretation of glial and neuronal structures in insect nervous tissue. In *Functional Neuroaviation*, edited by N. J. Strausfeld, pp. 339-375, Springer-Verlag, Berlin.

Chi, C., and Carlson, S. D., 1976, The large pigment cell of the compound eye of the housefly *Musca domestica*, *Cell Tissue Res.* 170:77-88.

Chi, C., and Carlson, S. D., 1980, Membrane specializations in the first optic neuropil of the housefly *Musca domestica* L. II. Junctions between glial cells, *J. Neurocytol.* 9:451-469.

Chi, C., and Carlson, S. D., 1981a, Lanthanum and freeze-fracture studies on the retinular cell junction in the compound eye of the housefly, *Cell Tissue Res.* 214:541-552.

Chi, C., and Carlson, S. D., 1981b, The adult housefly perineurium: Ultrastructure and permeability to lanthanum, *Cell Tissue Res.* **217**:373-386.

Chi, C., Carlson, S. D., and Saint Marie, R. L., 1979, Membrane specializations in the peripheral retina of the housefly *Musca domestica*, *Cell Tissue Res.* **198**:501-520.

Coles, J. A., and Tsacopoulos, M., 1979, Potassium activity in photoreceptors, glial cells and extracellular space in the drone retina: Changes during photostimulation, *J. Physiol. (London)* **290**:525-549.

Eldefawi, M. E., and O'Brien, R. D., 1966, Permeability of the abdominal nerve cord of the American cockroach to fatty acids, *J. Insect Physiol.* **12**:1133-1142.

Eldefrawi, M. E., and O'Brien, R. D., 1967a, Permeability of the abdominal nerve cord of the cockroach, *Periplaneta americana* L., to quaternary ammonium salts, *J. Exp. Biol.*, **41**:1-12.

Eldefrawi, M. E., and O'Brien, R. D., 1967b, Permeability of the abdominal nerve cord of the American cockraoch, *Periplaneta americana* (L.), to aliphatic alcohols, *J. Insect Physiol.* **13**:691-398.

Eldefrawi, M. E., Toppozada, A., Salpeter, M. M., and O'Brien, R. D., 1968, The location of penetration barriers in the ganglia of the American cockroach, *Periplaneta americana*, *J. Exp. Biol.* **48**:325-338.

Evequoz, V., Deshusses, J., and Tsacopoulos, M., 1978, The effect of photostimulation on glycogen turnover in the retina of the honeybee drone, *Experientia* **34**:897.

Finlayson, L. H., 1975, Development and degeneration. In *Insect Muscle*, edited by P. N. R. Usherwood, pp. 75-149, Academic Press, New York.

Florkin, M., and Jeuniaux, C., 1974, Hemolymph composition. In *Physiology of Insecta*, vol. 5, edited by M. Rockstein, pp. 255-307, Academic Press, New York.

Glees, P., 1955, *Neuroglia, Morphology and Function*, Thomas, Springfield, Illinois.

Griffiths, G. W., 1979, Transport of glial cell acid phosphatase by endoplasmic reticulum into damaged axons, *J. Cell Sci.* **36**:361-389.

Gymer, A., and Edwards, J. S., 1967, The development of the insect nervous system. I. An analysis of postembryonic growth in the terminal ganglion of *Acheta domesticus*, *J. Morphol.* **123**:191-197.

Hanson, T. E., Jiang, Y. H., and Lee, J. Y., 1972, Growth cone dynamics in the lamina of *Drosophila*, *Cal. Tech. Biol. Annu. Rep.* **38**:41-42.

Harris, W. A., and Stark, W. S., 1977, Hereditary retinal degeneration in *Drosophila melanogaster*. A mutant defect associated with the transduction process, *J. Gen. Physiol.* **69**:261-291.

Hess, A., 1960, The fine structure of degenerating nerve fibers, their sheaths, and their terminations in the central nerve cord of the cockroach *(Periplaneta americana)*, *Biophys. Biochem. Cytol.* **7**:339-344.

Houk, E. J., and Beck, S. D., 1976, Comparative ultrastructure and the blood-brain barrier in diapause and nondiapause larvae of the European corn borer *Ostrinia nibilalis* (Hübner), *Cell Tissue Res.* **162**:449-510.

Hoyle, G., 1953, Potassium and insect nerve muscle, *J. Exp. Biol.* **30**:121-135.

Kuffler, S. W., and Nicholls, J. G., 1966, The physiology of neuroglia cells, *Ergeb. Physiol. Biol. Chem. Exp. Pharmakol.* **57**:1-90.

Lane, N. J., 1968, The thoracic ganglia of the grasshopper, *Melanoplus differentialis:* Fine structure of the perineurium and neuroglia with special reference to the intracellular distribution of phosphates, *Z. Zellforsch. mikrosk. Anat.* **86**:293-312.

Lane, N. J., 1972, Fine structure of a lipidopteran nervous system and its accessibility to peroxidase and lanthanum, *Z. Zellforsch. mikrosk. Anat.* **131**: 205-222.

Lane, N. J., 1974, The organization of insect nervous systems. In *Insect Neurobiology*, edited by J. E. Treherne, pp. 1-71, North-Holland, Amsterdam.

Lane, N. J., 1978, Intercellular junctions and cell contacts in invertebrates. In *Electron Microscopy 1978*, vol. 3, edited by J. M. Sturgess, pp. 673-691, Imperial Press, Toronto.

Lane, N. J., 1979, Intramembranous particles in the form of bracelets or assemblies in arthropod tissues, *Tissue Cell* **11**:1-18.

Lane, N. J., 1981a, Tight junctions in arthropod tissues, *Int. Rev. Cytol.* **73**:243-318.

Lane, N. J., 1981b, Vertebrate-like tight junctions in the insect eye, *Exp. Cell Res.* **132**: 482-488.

Lane, N. J., 1982, Insect intercellular junctions: Their structure and development. In *Insect Ultrastructure*, edited by R. C. King and H. Akai, vol. 1, pp. 402-433, Plenum Press, New York.

Lane, N. J., and Skaer, H. LeB., 1980, Intercellular junctions in insect tissues. In *Advances in Insect Physiology*, vol. 15, edited by M. J. Berridge, J. E. Treherne, and V. B. Wigglesworth, pp. 35-213, Academic Press, New York.

Lane, N. J., and Swales, L. S., 1978a, Changes in the blood-brain barrier of the central nervous system in the blowfly during development, with special references to the formation and disaggregation of gap and tight junctions. I. Larval development, *Dev. Biol.* **62**:389-414.

Lane, N. J., and Swales, L. S., 1978b, Changes in the blood-brain barrier of the central nervous system in the blowfly during development, with special references to the formation and disaggregation of gap and tight junctions. II. Pupal development and adult flies, *Dev. Biol.* **62**:415-431.

Lane, N. J., and Treherne, J. E., 1980, Functional organization of arthropod neuroglia. In *Insect Biology in the Future—VBW 80*, edited by M. Locke and D. S. Smith, pp. 765-795, Academic Press, New York.

Lane N. J., Skaer, H. LeB., and Swales, L. S., 1977, Intercellular junctions in the central nervous system of insects, *J. Cell Sci.* **26**:175-199.

Langer, H., 1975, Properties and functions of screening pigments in insect eyes. In *Photoreceptor Optics*, edited by A. W. Snyder and R. Menzel, pp. 429-455, Springer-Verlag, Berlin.

Lasek, R. J., Gainer, H., and Przybylski, R. J., 1974, Transfer of newly synthesized proteins from Schwann cells to the squid giant axon, *Proc. Natl. Acad. Sci. USA* **71**:501-523.

Laughlin, S. B., 1974, Neural integration in the first optic neuropil of dragonflies. II. Receptor signals in the lamina, *J. Comp. Physiol.* **92**:357-375.

Laughlin, S. B., 1975. The function of the lamina ganglionaris. In *Neural Principles in Vision*, edited by F. Zettler and R. Weiler, pp. 175-193, Springer-Verlag, Berlin.

Morales, R., and Duncan, P., 1975, Specialized contacts of astrocytes with astrocytes and with other cell types in the spinal cord of the cat, *Anat. Rec.* **182**:255-266.

Noirot, C., and Noirot-Timothée, C., 1982, The structure and development of the tracheal system. In *Insect Ultrastructure*, edited by R. C. King and H. Akai, vol. 1, pp. 351-381, Plenum Press, New York.

O'Brien, R. D., 1967, Barrier systems in insect ganglia and their implications for toxicology, *Fed. Proc.* **26**:1056-1061.

O'Brien, R. D., and Fisher, R. W., 1958, The relation between ionization and toxicity to insects of some neuropharmacological compounds, *J. Econ. Entomol.* **51**:169-175.

Orkland, R. K., 1977, Glial cells. In *Handbook of Physiology*, section 1, *The Nervous System*, vol. 1, *Cellular Biology of Neurons*, part 2, edited by J. M. Brookhart, V. B. Mountcastle, E. R. Krandel, and S. R. Geiger, pp. 855-875, American Physiological Society, Bethesda.

Perrelet, A., 1970, The fine structure of the retina of the honey bee drone: An electron microscopic study, *Z. Zellforsch. mikrosk. Anat.* **108**:530-562.

Peters, A., Palay, S. L., and Webster, H. deF., 1976, *The Fine Structure of the Nervous System*, 2nd ed., Saunders, Philadelphia.

Pichon, Y., Sattelle, D. B., and Lane, N. J., 1972, Conduction processes in the nerve cord of the moth *Manduca sexta* in relation to its ultrastructural and hemolymph ionic composition, *J. Exp. Biol.* **56**:717-734.

Pipa, R. L., 1967, Insect neurometamorphosis. III. Nerve cord shortening in a moth, *Galleria mellonella* (L.), may be accomplished by humoral potentiation of neuroglial motility, *J. Exp. Zool.* **164**:47-60.

Pipa, R. L., 1973, Proliferation, movement, and regression of neurons during the postembryonic development of insects. In *Developmental Neurobiology of Arthropods*, edited by D. Young, pp. 105-129, Cambridge University Press, London.

Pipa, R. L., and Woolever, P. S., 1965, Insect neurometamorphosis. II. The fine structure of perineural connective tissue, adiopohemocytes, and the shortening ventral nerve cord of a moth, *Galleria mellonella* (L.), *Z. Zellforsch. mikrosk. Anat.* **68**:80-101.

Radojcic, T., and Pentreath, V. W., 1979, Invertebrate glia, *Prog. Neurobiol.* **12**:115-179.

Ramón y Cajal, S., 1913, Sobre un nuevo proceder de impregnacion de la neuroglia y sus resultados en los centros nerviosos del hombre y animales, *Trab. Lab. Invest. Biol. Univ. Madrid* **11**:219-237.

Ramón y Cajal, S., 1916, El proceder del oro-sublimado para la coloracion de la neuroglia, *Trab. Lab. Invest. Biol. Univ. Madrid* **14**:155-162.

Ramón y Cajal, S., and Sanchez, D., 1915, Contribution al conocimiento de los centros nerviosos de los insectos, Parte 1, Retina y centros opticos, *Trab. Lab. Invest. Biol. Univ. Madrid* **13**:1-168.

Reinecke, M., 1976, The glial cells of the cerebral ganglia of *Helix pomata* L. (*Gastropoda pulmonata*). II. Uptake of ferritin and ^{3}H-glutamate, *Cell Tissue Res* **169**:361-382.

Rheuben, M. B., and Reese, T. S., 1978, Three dimensional structure and membrane specializations of moth excitatory neuromuscular synapse, *J. Ultrastruct. Res.* **65**:95-111.

Ribi, W. A., 1978, A unique hymenopteran compound eye: The retina fine structure of the digger wasp *Sphex cognatus* Smith (Hymenoptera, Spheaidae), *Zool. Jahrb. Anat. Bb.* **100**:299-342.

Roots, B. I., 1978, A phylogenetic approach to the anatomy of glia. In *Dynamic Properties of Glial Cells*, edited by E. Schoffeniels, B. Franck, L. Hertz, and D. B. Tower, pp. 45-54, Pergamon Press, Elmsford, N.Y.

Saint Marie, R. L., 1981, A thin-section and freeze-fracture study of intercellular junctions and synaptic vesicle activity in the first optic neuropil of the housefly compound eye, Ph.D. thesis, University of Wisconsin, Madison.

Saint Marie, R. L., and Carlson, S. D., 1980, Glia-glial and glia-axonal tight junctions: Possible substrate for lateral electrical inhibition at the fly photoreceptor synapse, *Invest. Opthalmol. Vis. Sci.* **11**(Suppl.):246.

Saint Marie, R. L., and Carlson, S. D., 1983a, The fine structure of glia in the laminal ganglionaris of the housefly, *Musca domestica* L., *J. Neurocytol.* **12**:243-275.

Saint Marie, R. L., and Carlson, S. D., 1983b, Glial membrane specializations and the compartmentalization of the lamina ganglionaris of the housefly, *J. Neurocytol.* **12**:243-275.

Satir, P., and Gilula, N. B., 1973, The fine structure of membranes and intercellular communication in insects, *Annu. Rev. Entomol.* **18**:143-166.

Schoffeniels, E., Franck, B., Hertz, L., and Tower, D. B. (eds.), 1978, *Dynamic Properties of Glial Cells*, Pergamon Press, Elmsford, N.Y.

Schofield, P. K., and Treherne, J. E., 1978, Kinetics of sodium and lithium movements across the blood-brain barrier of an insect, *J. Exp. Biol.* **75**:239-251.

Sears, T. A. (ed.), 1982, *Neuronal-Glial Cell Interrelationships, Dahlem Konferenzen*, Springer-Verlag, Berlin.

Shaw, S. R., 1968, Organizaton of the locust retina, *Symp. Zool. Soc. London* **23**:135-163.

Shaw, S. R., 1975, Retinal resistance barriers and electrical lateral inhibition, *Nature (London)* **255**:480-483.

Skaer, H. LeB., and Lane, N. J., 1974, Junctional complexes, perineural and glial-axonal relationships, and the ensheathing structures of the insect nervous system; a comparative study using conventional and freeze-cleaving techniques, *Tissue Cell* **6**:695-718.

Stark, W. S., and Carlson, S. D., 1982, Ultrastructural pathology of the compound eye and optic neuropiles of the retinal degeneration mutant $(wrdgB^{KS222})$ *Drosophila melanogaster*, *Cell Tissue Res.* **225**:11-22.

Strausfeld, N. J., 1976, *Atlas of an Insect Brain*, Springer-Verlag, Berlin.

Thomas, M. V., 1976a, Insect blood-brain barrier: An electrophysiological investigation of its permeability to the aliphatic alcohols, *J. Exp. Biol.* **64**:101-118.

Thomas, M. V., 1976b, Insect blood-brain barrier: A radio-isotope study of the kinetics of exchange of a liposoluble molecule (*n*-butanol), *J. Exp. Biol.* **64**:119-130.

Treherne, J. E., 1960, The nutrition of the central nervous system in the cockroach, *Periplaneta americana* L.: The exchange and metabolism of sugars, *J. Exp. Biol.* **37**:513-533.

Treherne, J. E., and Pichon, Y., 1972, The insect blood-brain barrier, *Adv. Insect Physiol.* **9**:257-313.

Trujillo-Cenóz, O., 1962, Some aspects of the structural organization of the arthropod ganglia, *Z. Zellforsch. mikrosk. Anat.* **56**:649-682.

Trujillo-Cenóz, O., 1965, Some aspects of structural organization of the intermediate retina of dipterans, *J. Ultrastruct. Res.* **13**:1-33.

Tsacopoulos, M., Poitry, A., and Borsellino, A., 1981, Diffusion and consumption of oxygen in the superfused retina of the drone (*Apis mellifera*) in darkness, *J. Gen. Physiol.* **77**:601-628.

Tung, A. S.-C., and Pipa, R. L., 1971, Fine structure of transected interganglionic connectives and degenerating axons of wax moth larvae, *J. Ultrastruct. Res.* **36**:694-707.

Tung, A. S.-C., and Pipa, R. L., 1972, Insect neurometamorphosis. V. Fine structure of axons and neuroglia in the transforming interganglionic connectives of *Galleria mellonella* (L.) (Lepidoptera), *J. Ultrastruct. Res.* **39**:556-567.

Twarog, B. M., and Roeder, K. D., 1956, Properties of the connective tissue sheath of the cockroach abdominal nerve cord, *Biol. Bull.* **111**:278-286.

Varon, S. S., and Somjen, G. G., 1979, Neuron-glia interactions, *Neurosci. Res. Progr. Bull.* **17**:1-239.

Virchow, R., 1860, *Cellular Pathology*, translated from second German edition by F. Chance, Churchill, London.

Walker, F. D., and Hild, W. J., 1969, Neuroglia electrically coupled to neurons, *Science* **165**:602-603.

Wigglesworth, V. B., 1959, The histology of nervous system of an insect, *Rhodnius prolixus* (Hemiptera). II. The central ganglia, *Q. J. Microsc. Sci.* **100**:299-313.

Wigglesworth, V. B., 1960, The nutrition of the central nervous system of the cockroach, *Periplaneta americana:* The mobilization of reserves, *J. Exp. Biol.* **37**:500-512.

Wigglesworth, V. B., 1972, *The Principles of Insect Physiology*, 7th ed., Chapman and Hall, London.

Zettler, F., and Järvilheto, M., 1971, Decrement-free conduction of graded potentials along the axon of a monopolar neuron, *Z. vgl. Physiol.* **75**:402-421.

Zettler, F., and Järvilehto, M., 1973, Active and passive axonal propagation of non-spike signals in the retina of *Calliphora, J. Comp. Physiol.* **85**:89-104.

Zimmerman, R. P., 1978, Field potential analysis and the physiology of second-order neurons in the visual system of the fly, *J. Comp. Physiol.* **126**:297-316.

13

Mechanosensitive and Olfactory Sensilla of Insects

THOMAS A. KEIL AND
R. ALEXANDER STEINBRECHT

1. Introduction

Insects are the most successful land dwellers among the invertebrates. They owe this achievement largely to their cuticle, which provides mechanical support and protects against water loss at the same time. This exoskeleton forms a barrier between the environment and the interior milieu, so that sense organs need unique adaptations to remain accessible to external stimuli.

Mechanoreceptors in soft-bodied animals can receive stimuli anywhere within the integument. In insects, the stimulating forces reach the sensory cells only at places where the cuticle is deformable. Special lever structures developed which can transmit stimuli from the outside to the inside of the body, taking advantage of the adjustable mechanical properties of cuticle.

Chemoreceptors must take care that the prevention of water loss does not interfere with the accessibility of the receptor cells for stimulating molecules. Other terrestrial animals, like mammals, hide their olfactory receptors in inverted cavities and protect the chemosensory membranes by a layer of mucus. The "everted noses" of insects represent an entirely different solution, taking advantage of the unique permeability properties of cuticle and special stimulus transport structures.

In insect mechanosensitive and olfactory sensilla, structure-function relationships are quite well understood. As compared with other types, these

---

THOMAS A. KEIL AND R. ALEXANDER STEINBRECHT • Max Planck Institute of Behavioral Sciences, D-8131 Seewiesen, Federal Republic of Germany. Dedicated to Dietrich Schneider on the occasion of his 65th birthday.

sensilla show the greatest wealth of structural adaptations. In this chapter, we survey the ultrastructure of a few well-studied examples rather than giving a comprehensive review. We shall first describe the "Bauplan" common to all insect sensilla and then lay special emphasis on specific stimulus-transmitting structures. Physiological results are incorporated wherever necessary for the interpretation of structure in functional terms.

2. General Organization and Morphogenesis of Sensilla

A typical insect sensillum (Figure 1) consists of a hairlike structure protruding beyond the cuticle surface, one or several sensory cells, and three auxiliary cells, the thecogen, trichogen, and tormogen cells. In addition, there are one or several glial cells. The bipolar sensory neurons send their receptive dendrites toward the hair, their axons to the CNS. The auxiliary cells form a large

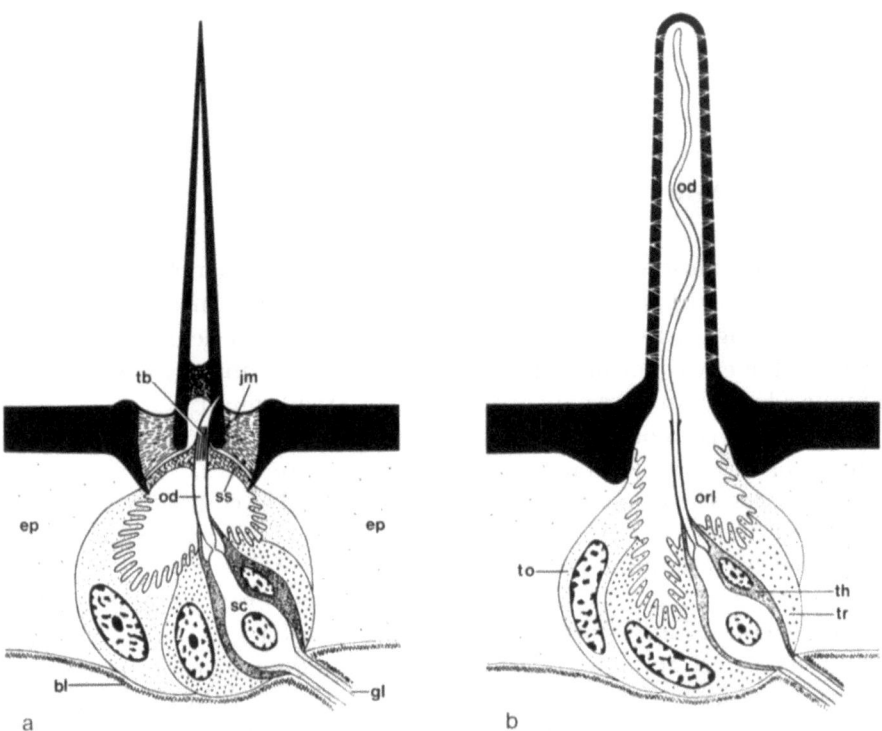

Figure 1. General organization of insect mechanosensory (a) and olfactory (b) sensilla. The sensory cell (sc) is enveloped by the thecogen (th), trichogen (tr), and tormogen cell (to). The axon is enveloped by the glia cell (gl), and the whole epithelium (ep) is underlain by the basal lamina (bl). The outer dendritic segment (od), which is enclosed by the dendrite sheath, runs through the outer receptorlymph cavity (orl). In mechanoreceptors, its tip, which contains the tubular body (tb), is supported by the socket septum (ss) and attached to the hair shaft, which is suspended in the joint membrane (jm). In olfactory sensilla, the outer dendritic segment runs up to the tip of the hair, the cuticular wall of which is perforated by numerous pores.

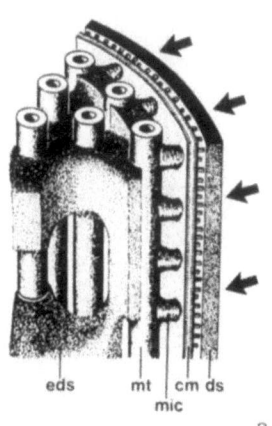 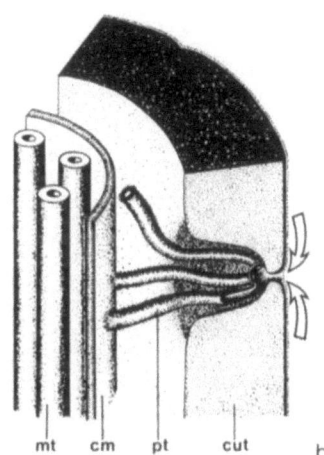

Figure 2. Idealized drawing of the structures mainly involved in stimulus transmission. (a) Microtubule-membrane complex in mechanoreceptors. The stimulating forces (arrows) are transmitted by cuticular lever structures (omitted here; see Figures 1a, 8a, 10a) to the dendrite sheath (ds) and via small bridge structures to the dendritic membrane (cm) and the membrane-integrated cones (mic). These are backed by the tubular body, consisting of microtubules (mt) and an electron-dense substance (eds). (b) Pore-tubule systems of olfactory sensilla. Odor molecules are presumed to enter pores (arrows) in the cuticular hair wall (cut), diffuse along the pore tubules (pt), and reach the dendritic membrane (cm) either via direct contacts or via the receptorlymph.

subcuticular cavity, the receptorlymph space, which is confluent with the hair lumen and surrounds the dendrites. The differentiations specific for the sensory modalities are located in the cuticular structures of the hairs and in the outer segments of the dendrites (Figure 2).

For insect sensilla, the term *Kleinorgan* (Henke, 1953) or *organule* (Lawrence, 1966) has been proposed; their cellular organization can be understood best if we look at their development. Beginning in the late 1930s, there have been many light microscopic investigations, mainly by the groups of Kühn and Henke (for review see Henke, 1953), but electron microscopic studies are still rare. Figure 3 outlines the developmental steps summarizing the observations of Wigglesworth (1953), Peters (1965), Lawrence (1966), and Ernst (1972). Like in noninnervated hairs and scales, epidermal glands, and ommatidia, differentiation starts with a single epidermal cell, the stem or mother cell. This bristle mother cell 1 (Lawrence, 1966) divides into a bristle mother cell 2 and an accessory cell (Figure 3a) (at this stage, further development can be inhibited by application of mitomycin C; Sanes and Hildebrand, 1976). The bristle mother cell 2 divides again (Figure 3b) into the tormogen and trichogen cell, whereas the accessory cell undergoes two successive divisions (Figures 3b, c), finally yielding the "neurilemma cell" (Lawrence, 1966) and the sensory cell (Figure 3d). At this stage, the tormogen and trichogen cells start endomitotic activity which can be inhibited if the sensory cell is killed before; in this case, development stops (Clever, 1958, 1960). If the sensory cell dies after start of those endomitoses, normal development follows, yielding a noninnervated hair. This

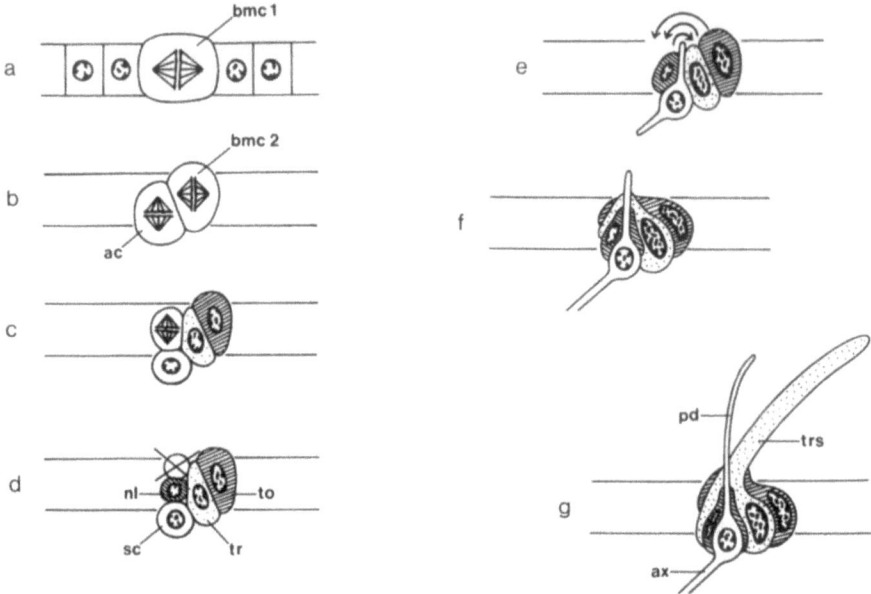

Figure 3. Synoptic presentation of sensillum development. The bristle mother cell 1 (bmc 1) undergoes sequential and unequal mitoses, finally yielding the sensory cell (sc), neurilemma (nl) = thecogen cell, trichogen (tr), and tormogen cell (to). ac, accessory cell; ax, axon; pd, primary dendrite; trs, trichogen cell sprout (for further explanation, see text).

is what happens in lepidopteran scales. The neurilemma cell represents the thecogen cell of the fully differentiated sensillum.

About the time of endomitotic activities, the sensory cell forms two outgrowths (Figure 3e). The apical process, the dendrite, grows beyond the epidermis; the basal process, the axon, grows toward the CNS. Contact with the CNS, however, is not essential for sensillum development: if the peripheral nerve is cut (Wigglesworth, 1953; Clever, 1960), or even if the brain is removed (Sanes *et al.*, 1976), the axons still continue growing and normal sensilla result. At the same time (Figures 3e, f), the three auxiliary cells grow concentrically around each other with the dendrite and the sensory cell body in the center. The thecogen cell secretes an electron-dense sheath around the outgrowing dendrite. The trichogen cell forms a long, slender process, growing beyond the epidermis. The basis of this process is enveloped by the tormogen cell (Figure 3g). Both cells define the final shape of the later hair shaft and socket.

The cuticle of the hair shaft is secreted by the trichogen cell. The tormogen cell secretes the basal parts of the hair shaft and the socket structures. In the final stages of morphogenesis, both cells retract from the cuticle and form a large subcuticular space, the receptorlymph cavity. It is remarkable that in mechanosensitive and olfactory sensilla, the process of the trichogen cell does not follow the primary dendrite as in gustatory sensilla (see Gnatzy and Schmidt, 1972;

De Kramer and van der Molen, 1984), but runs separate through the exuvial space (Figure 3g; Ernst, 1972; Sanes and Hildebrand, 1976; Gnatzy, 1978). This free part of the dendrite degenerates as the cuticle is secreted. In mechanoreceptors, the tubular body develops in the tip of the remaining dendrite. In olfactory receptors, the stump of the dendrite grows out a second time into the hair shaft after retraction of the trichogen cell.

3. Fine Structure of the Cellular Components of Sensilla

3.1. The Sensory Neurons

The sensory cells have a more electron-lucent cytoplasm than the other sensillar cell types. The *perikarya* are rich in mitochondria and free polyribosomes, but contain few cisternae of RER. Some lysosomes, dense bodies, and multivesicular bodies are usually present (Figure 4). Some organelles, however, appear to be less abundant in mechanoreceptor cells than in olfactory receptors, e.g., the Golgi apparatus, SER, and vesicles (for mechanoreceptors see Moran *et al.*, 1971; Gnatzy, 1976, 1978; Keil, 1978; for olfactory receptors see Ernst, 1969; Steinbrecht, 1980).

The dendrites are subdivided into an inner and an outer segment by a short ciliary structure. The *inner segment* originates at the perikaryon and contains all the typical organelles of the latter; sometimes there is an accumulation of mitochondria, while vesicles are generally abundant. At the level of the *ciliary segment*, the dendrites leave the complex of enveloping cells and enter an *inner receptorlymph cavity* formed by the thecogen cell (see Section 3.4). Here they narrow and assume the structure of cilia (Gray, 1960; Slifer and Sekhon, 1960) which, however, always lack the central microtubule doublet (Figure 5; see also Eakin, 1965; Thurm, 1970). Two basal bodies, which are mostly arranged in tandem and consist of the typical microtubule triplets, are found immediately beneath the cilium (see Gaffal and Bassemir, 1974). There is a ciliary rootlet which originates in the distal basal body, encloses the proximal one, and runs for several micrometers through the inner dendritic segment in a proximal direction.

The cilium soon loses its typical structure and thickens to some extent, forming the *outer dendritic segment*, which is completely devoid of organelles other than microtubules, 10-nm filaments, and vesicles (Marshall, 1973), sometimes multivesicular bodies. The nine microtubule doublets can be followed quite far up into the outer dendritic segment; numerous single microtubules are added along its length (see Section 4.6). The dendrite sheath encloses the dendritic outer segments up to the hair base. In mechanoreceptors, the sheath is attached to the hair base where the dendrite ends (Figure 1a; see Section 4). In olfactory sensilla, the sheath ends at the hair base while the dendrite(s) runs into the hair lumen well up to the tip (Figure 1b).

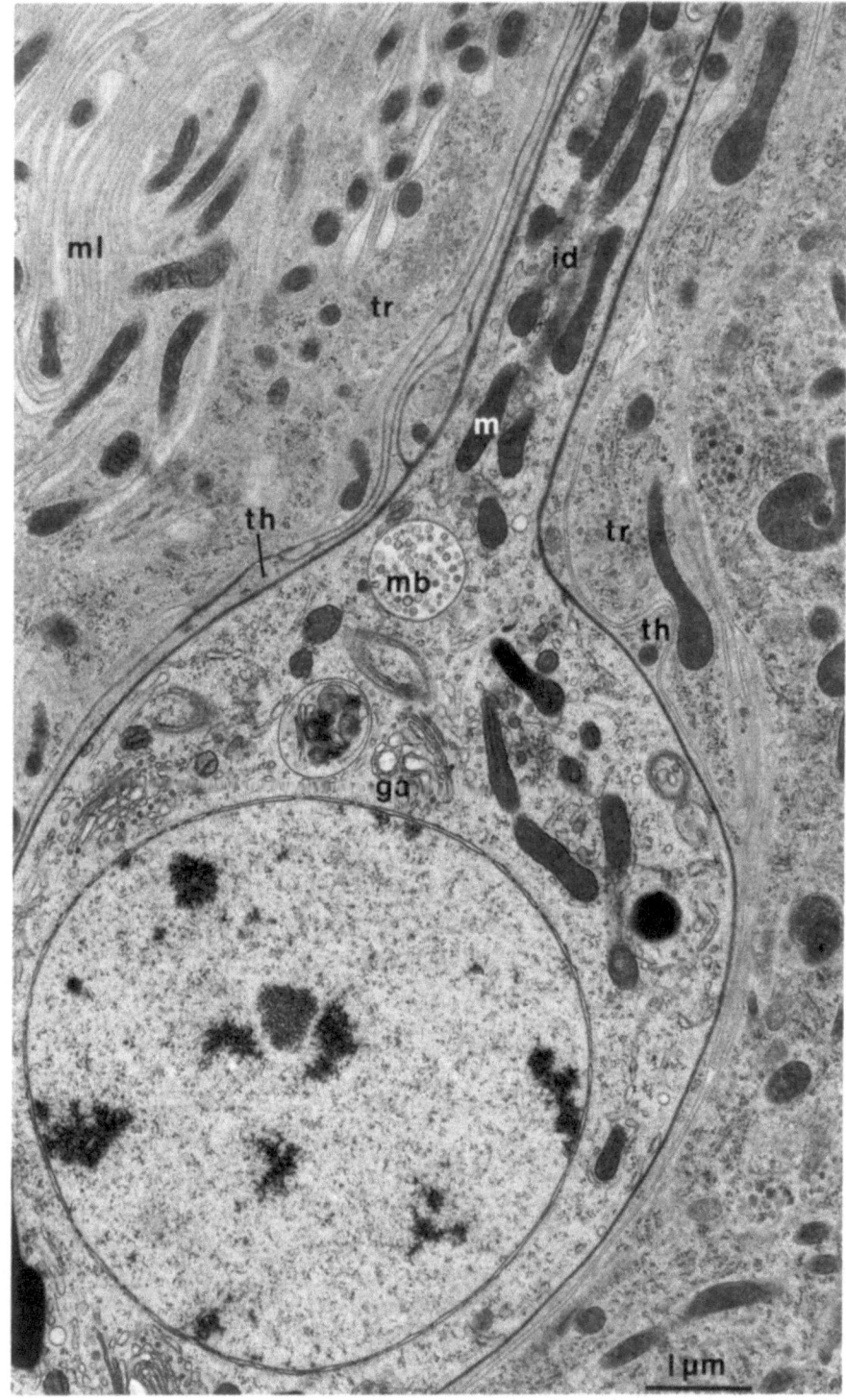

3.2. The Auxiliary Cells

As in all insect epidermal cells, the main function of the auxiliary cells during morphogenesis is secretion and shaping of cuticle, which is reflected by their ultrastructural organization. In the fully developed sensillum, these cells exhibit the typical features of ion-transporting cells (for review see Berridge and Oschman, 1972; Thurm and Küppers, 1980). This structural and functional change during the insect's life cycle is an interesting aspect of sensillum biology and has been studied in detail with insect mechanoreceptors (Gnatzy, 1978; Keil, 1978; Gnatzy and Romer, 1980).

3.2.1. The Tormogen Cell

This is the outermost cell of the sensillum unit. During morphogenesis, it secretes the basal part of the hair shaft, the inner part of the hair socket, and in mechanoreceptors the joint membrane and parts of the socket septum. Its polyploid nucleus is much larger than that of epidermal cells. During the secretion of the cuticle, the cytoplasm is rich in granular ER, Golgi apparatus, secretory granules and vesicles, also in mitochondria and irregularly arranged microtubules. The apical membrane displays small microvilli with electron-dense plaques (Locke, 1976; Locke and Huie, 1979).

When the animal is due to molt, the appearance of the cell changes. The cytoplasm is still rich in mitochondria and Golgi apparatus, but the granular ER is replaced by numerous free ribosomes. There are also multivesicular and dense bodies as well as isolation bodies (Locke and Collins, 1965). The number of microtubules is drastically reduced. The most important change, however, occurs with the apical membrane. It retracts from the cuticle, remaining attached to the hair socket only with a narrow rim, and develops a large subcuticular cavity, the outer receptorlymph space. The apical membrane now forms numerous long slender microvilli with central filaments and apical plaques which protrude into the cavity. In the final stage, the microvilli are transformed to a large extent into microlamellae. The cytoplasmic leaflet becomes studded with 8-nm particles (see Section 3.4), and numerous elongate mitochondria can be found associated with the apical membrane. The granular ER reappears, and a basal labyrinth is formed by the cell (Keil, 1978).

Figure 4. Sensory cell of an olfactory sensillum in longitudinal section (*Bombyx mori* ♂, pheromone receptor). The perikaryon contains a conspicuous multivesicular body (mb), several Golgi apparatus (ga), and numerous vesicles and free ribosomes. The inner segment of the dendrite (id) shows an especially high density of mitochondria (m). The neuron is surrounded by a thin, but complete envelope of the thecogen cell (th); note the special junction between these two cells as shown by the very dense contour; parts of the trichogen cell (tr) are also shown with RER, mitochondria, and elaborate microlamellae (ml) of the plasma membrane where it faces the receptorlymph cavity. Freeze-substituted specimen.

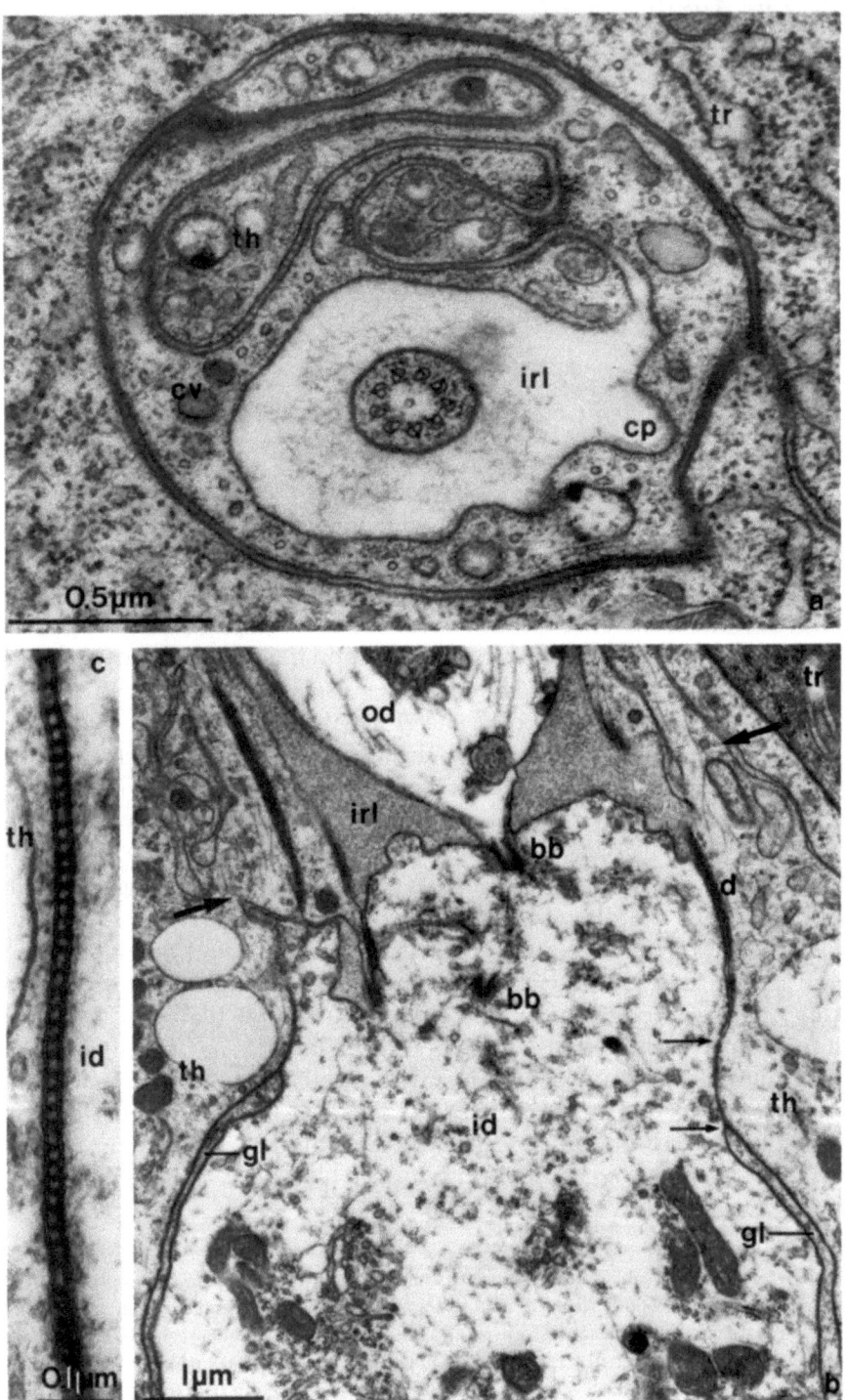

3.2.2. The Trichogen Cell

During morphogenesis, the trichogen cell carries the bulk of secretory activity in most cases and secretes the hair shaft, parts of the socket septum (in mechanoreceptors), as well as the wall-pore systems in olfactory hairs. The cytoplasmic organization is similar to that of the tormogen cell, but the most conspicuous structures are the densely packed, parallel microtubules in the hair-forming apical process (Lawrence, 1966). After deposition of the hair wall, the cell retracts completely from the cuticle, forming the outer receptorlymph cavity together with the tormogen cell, and assuming a similar ultrastructural organization. Like the tormogen cell, this cell can secrete macromolecules into the receptorlymph cavity (Phillips and Vande Berg, 1976).

In the thoracic macrochaete of certain flies (*Calliphora erythrocephala*, *C. vicina*, and *Sarcophaga barbata*), the trichogen cells are extremely large, and some of their nuclei contain as much as 8192 times the haploid amount of DNA. The chromosomes have undergone cycles of endoreduplication to generate giant polytene chromosomes (Ribbert, 1967; Trepte, 1976), and the nuclear envelopes contain a very high concentration of annuli (Figure 6). But after termination of morphogenesis, this cell degenerates and disappears completely (Keil, 1978).

3.2.3. The Thecogen Cell and the Dendrite Sheath

The thecogen cell is the innermost auxiliary cell and, in most cases, directly envelops the inner dendritic segment as well as the sensory cell body. During development, it secretes the dendrite sheath (Schmidt and Gnatzy, 1971; Ernst, 1972). During the secretory phase, the cell is rich in granular ER and secretory vesicles (Gnatzy, 1978). Afterwards, its most conspicuous organelles are parallel microtubules associated with desmosomes and hemidesmosomes toward the dendrite and dendrite sheath, respectively. The thecogen cell encloses the neuron forming a thin but continuous layer. In the region of the ciliary segment, the intercellular cleft widens and forms the inner receptorlymph cavity

---

Figure 5. Filiform sensillum of *Acheta* cercus. (a) Cross-section of ciliary structure, showing nine microtubule doublets with side arms; and a central structure which has been observed in many sensilla, but certainly is no microtubule. Note meandering course of thecogen (th) cell mesaxon, and coated pits (cp) and vesicles (cv). The pleated septate junction between thecogen and trichogen (tr) cell is marked by the regularly arranged dots on the cytoplasmic membrane leaflets. (b) Transitory region from inner (id) to outer (od) dendritic segment. Note microtubules and vesicles in outer segment. The inner segment contains two basal bodies (bb, which are not arranged exactly in tandem in these sensilla) as well as mitochondria, Golgi fields, polyribosomes, microtubules, and vesicles. The zone of septate junctions (small arrows) lies distally to the glia cell (gl). Spot desmosomes (d) between dendrite and thecogen cell (th) are very extended. The inner receptorlymph cavity (irl) is filled with flocculent material. Large arrows indicate sectioning plane of (a). (c) Pleated septate junction between inner dendritic segment (id) and thecogen cell (th). Preparation as in Figure 12.

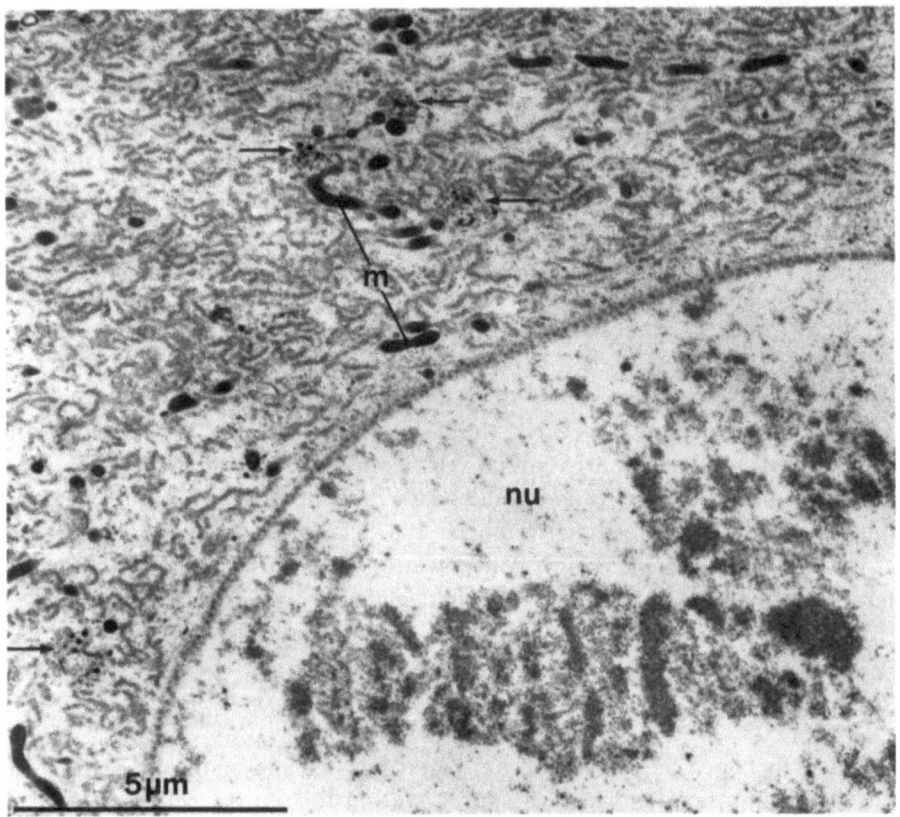

Figure 6. Trichogen cell of *Calliphora* macrochaeta on day 6 of pupation, when cuticle secretion is under way. Note large polytene chromosome in nucleus (nu) as well as nucleopores in obliquely sectioned membrane. The cytoplasm is densely filled with granular ER and contains many electron-dense mitochondria (m) and Golgi fields (arrows). The many irregularly running microtubules are barely visible. Bloc-stained with uranyl acetate.

(see Section 3.4). The electron-dense dendrite sheath surrounds the outer dendritic segments up to the hair base. It is apically attached to the hair wall in many sensilla, especially in mechanoreceptors, where it can form a very complex unit with the socket septum (see Section 4.6). Generally, the dendrite sheath is more massive in mechanosensitive than in olfactory sensilla, where it sometimes appears rather porous.

3.3. Membrane Contact Structures

The distribution of intercellular junctions reflects the epithelial nature of sensilla (Thurm and Küppers, 1980). The arrangement of belt desmosomes, septate junctions, and gap junctions, which is typical for the insect epidermis, can also be found with the three auxiliary cells (Figure 5), which thus form part of the "functional syncytium" (Caveney and Berdan, 1982). The sensory cells, however, are exceptional, because typical gap junctions have never been observed between neuron and thecogen cell. In olfactory sensilla of the silkmoth, after freeze substitution, the intercellular cleft between these cells is reduced to 6 nm over the entire neuronal surface (Steinbrecht, 1980; see also Figure 4). The inner dendritic segments form massive spot desmosomes backed by microtubules (Ashhurst, 1970) with the thecogen cell about 1 μm proximal of the ciliary region. Proximal of the spot desmosomes, a band of pleated septate junctions extends over several micrometers.

Experiments with ionic lanthanum (Keil and Thurm, 1979) suggest that the septate junctions in the apical region of a sensillum form a diffusion barrier between the hemolymph and the receptorlymph. The question whether the paracellular diffusion barrier across insect epithelia is formed by septate or tight junctions is not yet finally settled (see Lane, 1982). It might depend on the type of tissue. In the electrically "tight" sensory epithelium, the septate junctions are prominent (Figure 5c), but tight junctions could not be found. For a discussion of the junctions between sensillar and glial cells, see Section 3.5.

It must be pointed out that a very tight contact is formed between the lateral apical membrane of the tormogen cell and the cuticle of the hair socket, sealing the receptorlymph cavity from the general subcuticular space. This has been found in thoracic macrochaetae of *Calliphora*, cercal mechanoreceptors of *Acheta*, and antennal olfactory hairs of *Antheraea* and *Bombyx*. The electrophysiological correlate of this structure is a high resistance between neighboring sensilla (Thurm and Küppers, 1980; De Kramer et al., 1984).

3.4. The Receptorlymph Spaces

The large subcuticular space (Figures 1,7) which is formed by the tormogen and trichogen cell has been named *receptorlymph cavity* by Nicklaus et al. (1967) and *Sensillenliquorraum 2* by Ernst (1969). It is the "vacuole" of the earlier investigators (see Slifer et al., 1957; Slifer and Sekhon, 1963). There is also an *inner receptorlymph cavity* (*Sensillenliquorraum 1* by Ernst, 1969) which is formed between the thecogen cell and the ciliary region of the dendrites (cf. Section 3.2.3). According to tracer studies, both spaces are connected in mechanoreceptors (Keil and Thurm, 1979), and the same is supposed for olfactory sensilla. In the following, only the outer, subcuticular, receptorlymph cavity will be referred to.

The receptorlymph cavity is bordered by the apical membrane of the tormogen and trichogen cells, mainly by the tormogen in mechanoreceptors and mainly by the trichogen in olfactory receptors (Figure 1). The most conspicuous ultrastructural differentiations are the numerous microvilli and microlamellae of the apical cell membranes, which are studded on their cytoplasmic side with 8-nm particles, and which considerably enlarge the surface area (Keil, 1978; Steinbrecht and Gnatzy, 1984; Gnatzy et al., 1984). Smith (1969) has been the first to point out the strong structural resemblance to ion-transporting epithelial cells of insects (reviewed by Berridge and Oschman,

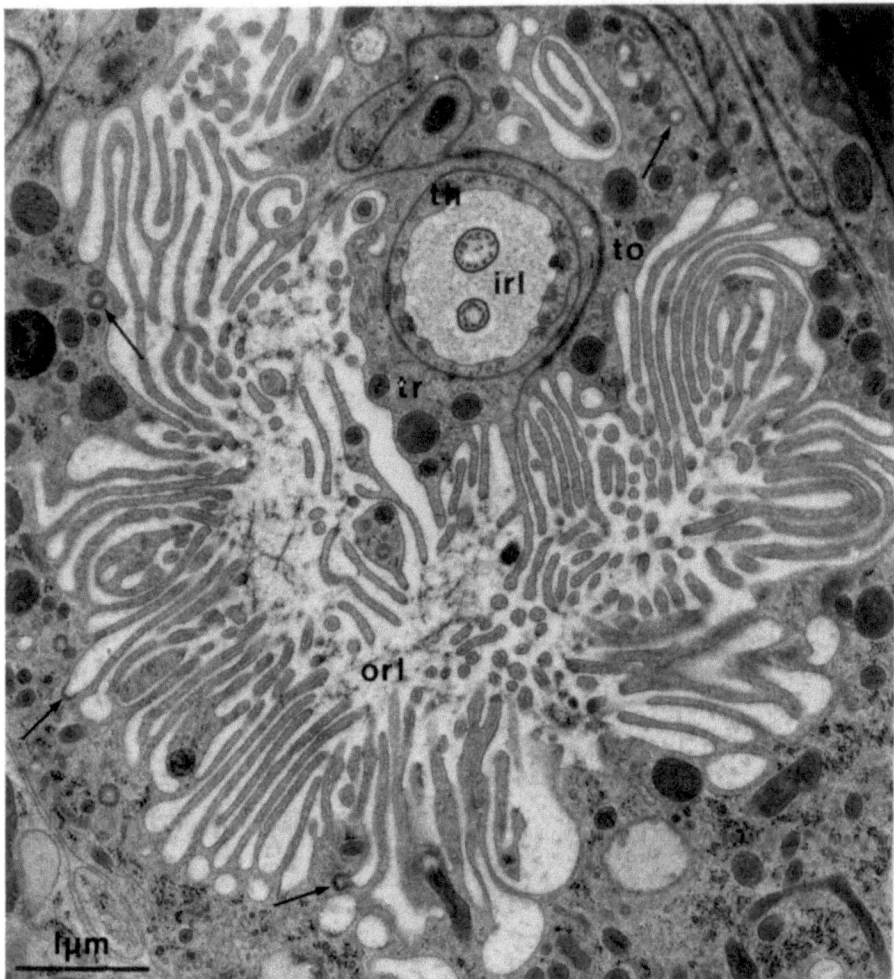

Figure 7. Two olfactory dendrites cross-sectioned at the level of the cilia are comprised by the thecogen cell (th) within the inner receptorlymph cavity (irl). The trichogen cell (tr) and tormogen cell (to) form concentric wraps. At this level, the large outer receptorlymph cavity (orl) is mainly bordered by microvilli and microlamellae of the tormogen cell; coated pits and vesicles (arrows) are frequent. *Bombyx mori* ♂, sensillum trichodeum, freeze-substituted specimen.

1972); the 8-nm particles probably being identical to Harvey's (1980) *portasomes*. Besides numerous mitochondria, there are also coated pits and vesicles at the base of the membrane folds.

In campaniform sensillum fields of the haltere of *Musca*, Thurm (1970) observed a transepithelial voltage (TEV) across the sensory epithelium which had a marked influence on the receptor potential. This TEV has also been found in other insect sensilla (Thurm and Wessel, 1979; review by Thurm and Küppers, 1980). In hemimetabolous insects, the TEV disappears while the receptorlymph space is reduced during the molting cycle (W. Gnatzy, unpublished, quoted in Thurm, 1974). However, Moran et al. (1976) and Gnatzy and Tautz (1977) could show that sensilla do not lose their excitability during this period. Erler and Thurm (1978) reported that the influence of the TEV on the receptor potential differs markedly in different insect mechanoreceptors.

It has been shown by measurements with ion-sensitive electrodes, by flame photometry, and X-ray microanalysis that the receptorlymph is rich in potassium as well as sulfur, but poor in sodium, chlorine, and phosphorus; just contrary to the hemolymph (Küppers, 1974; Kaissling and Thorson, 1980; Steinbrecht and Zierold, 1982). Histochemistry showed that the receptorlymph contains acid mucopolysaccharides, namely hyaluronic acid (Gnatzy and Weber, 1978; Keil, 1979), similar to the axonal environment in the CNS (Ashhurst and Costin, 1971).

There is strong evidence that active K^+-transport via the tormogen and/or trichogen cells into the receptorlymph space is a common feature of insect sensilla (Thurm and Küppers, 1980). But to maintain the different ionic composition of receptorlymph and hemolymph, the diffusion of ions through the intercellular clefts of the epithelium has to be inhibited, similar to the blood–brain barrier (Lane and Treherne, 1980). In insect sensilla, this diffusion barrier is provided by the septate junctions, as described in Section 3.3.

Acid mucopolysaccharides are biopolymers with most interesting ion- and water-binding properties (reviews by Manery, 1966; Laurent, 1970). At physiological pH, their carboxyl groups are dissociated, thus yielding a very high number of negative charges, which bind cations. These, in turn, can cross-link macromolecules like hyaluronic acid, causing them to form aggregates of high complexity; they also can bind water molecules to the chains (Winter and Arnott, 1977). Dunstone (1962) claims that cations can remain more or less free and able to move if merely attracted by weak electrostatic forces.

The functions of the receptorlymph can be summarized as follows:

1. It provides an aqueous environment for the dendrites and contains the cations essential for the generation of biopotentials.
2. It contains macromolecules which are capable of water- and ion-binding, which are probably secreted by the tormogen and trichogen cells.
3. It may play a role in stimulus transport and inactivation in olfactory sensilla (see Section 5.2.1).

3.5. Glia Cells and Neuron–Glia Interrelations

Glia cells envelop the axons and often the sensory cell bodies in most insect sensilla, but there are cases of naked sensory axons in blood-sucking insects (Steinbrecht and Müller, 1976). The ultrastructure of sensillar glia cells is similar to the general description given in Chapter 12, though they do not fit perfectly into one of the different groups described there.

In some mechanoreceptors of Orthoptera, the glia cells form extensive trophospongia with the neurons (Gnatzy, 1976, 1978). Submembrane cisternae are frequently facing each other in both cell types (Keil and Steinbrecht, 1983). Mostly, the glia joins the thecogen cell in the region where the axon leaves the sensory cell body, as in olfactory hairs of *Antheraea* and *Bombyx*. In mechanoreceptors of flies, glia processes extend into the cleft between sensory and thecogen cell. In certain mechanoreceptors of Orthoptera, there is even a separate glialike cell which fills the cleft between sensory and thecogen cell (Gnatzy and Schmidt, 1971; Gnatzy, 1976) and adjoins the axonal glia exactly at the point of axon origin. At least in mechanoreceptors of *Calliphora*, septate junctions have been found to seal the mesaxon from the axon origin to its entry into the peripheral nerves.

In very few cases it is known how the axonal glia joins the thecogen cell and the sensory cell body. In olfactory receptors of *Bombyx* and *Antherea*, there are extended complexes of septate junctions which seal the gap between sensory cell, axon, thecogen cell, and glia cell (Steinbrecht and Gnatzy, 1984; Keil and Steinbrecht, 1983). Such junctions are definitely absent in the region of the axon origin in cercal mechanoreceptors of *Acheta* and thoracic mechanoreceptors of *Calliphora*; in the latter, ionic lanthanum has free access to the cleft between thecogen and sensory cell (Keil and Thurm, 1979). However, it is not yet possible to settle the question whether this difference is a fundamental one between mechano- and olfactory receptors, or merely between different insect orders. There are functional implications for electrical circuit models (Thurm and Küppers, 1980; Kaissling and Thorson, 1980).

4. Fine Structure of Mechanosensitive Sensilla

4.1. General Remarks

Insect mechanoreceptors can be roughly subdivided into four types: two types with and two without external hair structures.

The first type is mainly represented by touch receptors where the hair is moved by direct contact with the substrate. The hair is suspended in the body cuticle by means of a thick, elastic, fibrillar membrane, the joint or socket membrane. Also, multimodal sensilla belong to this type, e.g., most taste hairs, where several gustatory neurons are associated with one mechanosensitive neuron.

The second type is represented by extremely sensitive receptors which respond to airstreams, gravity, sound, or vibration, e.g., trichobothria. The easily movable hair is connected to the body cuticle by a thin cuticular lamella only.

The third type is the *campaniform sensilla*, in which the outer cuticular parts are transformed into a cupola-shaped structure. They are proprioreceptors which monitor mechanical deformations of the cuticle, and, therefore, are located in body regions where such deformations can be expected, viz. on legs, wings, halteres, predominantly near the joints.

The fourth type is represented by the internal *chordotonal* and *scolopidial receptors*. These are mainly tension receptors and often associated with auditory organs. They consist of the same cellular parts, but their connection with the cuticle differs fundamentally from the other three types.

In this chapter, we shall discuss the properties of the first three types in general (Sections 4.2-4.4) and give special attention to well-investigated representatives of the first two groups (Section 4.5 and 4.6). For comprehensive reviews on mechanoreceptors in general, see McIver (1975), and on chordotonal organs, see Moulins (1976).

4.2. Cuticular Parts

The hair, bristle, or cupola are the externally visible cuticular structures of mechanoreceptors. They consist mainly of exocuticular material which shows no substructure in thin sections as well as in cross-fractures.

The structures which are most important for mechano-transduction are located in the joint region of the hair shaft where it interacts with the dendrite tip, namely the *joint membrane* and the *socket septum* (Gaffal et al., 1975) (Figure 1a).

As will be shown in Sections 4.5 and 4.6, the structure and mechanical properties of the joint membrane vary markedly depending on the type of mechanoreceptor. The socket septum is a fibrillar structure which primarily acts as a support holding the dendrite in its proper place. It probably provides the counterpressure when the hair base is moved against the dendrite tip. The socket septum, dendrite sheath, and hair base can form a very complex unit (Figure 10).

The mechanical construction of the hair socket, which in some cases allows deflection only in certain directions, is the structural basis of the directional sensitivity of many mechanoreceptors. The directional sensitivity may, however, also be caused by the shape of the dendrite tip which often is flattened in a way that it presents its broad side to the stimulating force, or a combination of both features (for references, see Gaffal et al., 1975; Gnatzy and Tautz, 1980).

4.3. Cellular Parts

The most conspicuous feature of insect mechanoreceptors is the *tubular body* (Thurm, 1964), which is situated in the tip of the dendrite where hair base and socket septum interact (Figure 1a, 2a). There is a large variability in the construction of the tubular body. In a few receptors, its cross-section is circular [e.g., tibial thread hairs of *Acheta* (Gaffal and Theiss, 1978)], mostly, it is more or less flattened [e.g., campaniform sensilla on blowfly halteres (Smith, 1969), cockroach legs (Moran et al., 1971) filiform hairs of cricket cerci (Gnatzy and Tautz, 1980); see also Figures 8b, 12a].

The tubular body proper is an aggregation of densely packed microtubules, which are interconnected by bridge structures or by an electron-dense matrix, sometimes with 10-nm filaments. The number of microtubules varies from about 30 (tsetse fly mechanoreceptor, Figure 2 in Rice et al., 1973) to well over 1000 (cercal filiform hairs of cockroaches, Figure 2 in Nicklaus et al., 1967). The electron-dense intermediate substance can be very massive with few electron-lucent spots (Figure 12), arranged in regular layers (Gaffal and Hansen, 1972),

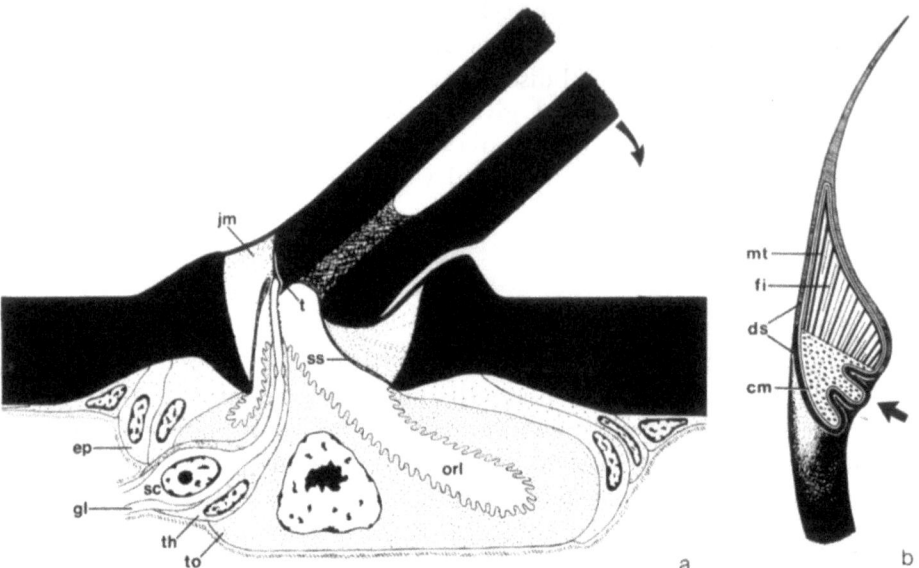

Figure 8. Organization of the thoracic macrocheata of *Calliphora* (a). Note the large tormogen cell (to) and absence of trichogen cell, as well as the massive joint membrane (jm) and the connection of the dendrite tip with the tubular body (shown enlarged in b) with socket septum (ss) and hair base. ep, epidermal cell; gl, glia cell; orl, outer receptorlymph cavity; sc, sensory cell; t, tongue; th, thecogen cell. Arrow indicates deflection of hair shaft. (b) Tip of the dendrite with dendrite sheath (ds), cell membrane (cm), and the tubular body consisting mainly of microtubules (mt) and 10-nm filaments (fi). Arrow indicates direction of stimulating forces.

or reduced to intertubular bridge structures (Smith, 1969; Keil, 1978; Völker, 1982).

There is a distinct peripheral layer of microtubules, which in some very small or flattened tubular bodies are the only microtubules present (Smith, 1969; Rice *et al.*, 1973). These are connected with the dendritic membrane via fine bridge structures (Figure 12b), which appear to be general features of insect mechanoreceptors (Gaffal and Hansen, 1972; Gnatzy and Tautz, 1980; Völker, 1982). The bridges have been named *membrane-integrated cones* (MIC) by Thurm (1981). They are most probably the structures which are responsible for the depolarization of the dendritic membrane when they receive mechanical pressure. The stimulus-transducing mechanism will be discussed in the next section.

The tubular body can be attached to supporting structures like folds or ribs of the dendrite sheath, but also to the free microtubules of the outer dendritic segment (see Section 4.6).

The microtubules were suggested to play a central role in mechanosensory transduction, a notion mainly based on degradation experiments (Moran and Varela, 1971). Erler (1983), however, showed (1) that vinblastine degrades the proximal tubules of mechanoreceptive dendrites in *Acheta*, but not those of the tubular body proper, and (2) that this does not necessarily destroy the electrical activity of the dendrite. Thus, it seems that the microtubules act in the more passive role of a dendritic cytoskeleton, and that the tubular body supports the dendritic membrane just in the place where it is deformed by the extracellular structures.

4.4. Stimulus Uptake and Transmission

Insect mechanoreceptors respond to stimuli of very different nature and strength. The way in which these are transmitted to the tubular body depends on the architecture and mechanical properties of the cuticular and subcuticular components of the sensilla (Figures 1a, 8a, 10).

The hair shaft forms a first-order lever which can turn around an assumed pivot point close to its base. As the lower lever arm is only short, a large deflection of the hair tip is transformed into a very small, oppositely directed movement of the hair base which now can press against the tip of the dendrite. It depends on the mechanical properties of the socket septum how much the dendritic tip is indented and how much it can give way. Thurm (1965) showed that lateral compression of the dendritic tip is the adequate stimulus. However, quantification of this compression so far is based on light microscopic observations only [Thurm (1965) claims 0.1 μm at maximal hair deflection]. Attempts to fix a mechanoreceptor in a "stimulated" (deflected) state never gave reasonable results, because the shrinkage during preparation might be larger than the possible deformation (Matsumoto and Farley, 1978; Gnatzy and Tautz, 1980). Calculations based on sensillum geometry are uncertain because of the unknown mechanical properties of the basal structures.

The MICs now are thought to be the structures directly involved in mechano-electric transduction (Thurm, 1982). Movement of the hair base results in a pressure against the dendrite sheath (Figure 2a). Sheath and membrane are connected by fine filaments (Keil, 1978; Völker, 1982; Gnatzy and Tautz, 1980). The membrane in turn is pressed against the cones, which then might cause the opening of ion channels (Thurm, 1982).

4.5. The Macrochaetae of Calliphora

These sensilla are presented as examples for the first group of mechanoreceptors (see Section 4.1). They are large bristles up to 2 mm in length and 30 μm in diameter, which are located on head and thorax of certain Diptera. They have been thoroughly investigated by Ribbert (1967) and Trepte (1976, cytogenetics), Keil (1978, ultrastructure), and Theiss (1979, physiology). Their biological function, however, is not clear.

A remarkable feature of these sensilla is their morphogenesis. The giant trichogen cell, which has a polytene nucleus, forms the hair and degenerates thereafter (Section 3.2.2; Figure 6). The tormogen cell is also large and polytene but to a lesser degree (Trepte, 1976); it forms a very large receptorlymph cavity in the imago (Keil, 1978).

The size of the neuron is comparable to those of other sensilla and does not reflect the extraordinary size of all other parts of the sensillum. The ciliary region of the dendrite is transformed into the "fibrillar body," as is the case in all known mechanoreceptors of true flies (Smith, 1969; Keil, 1978). The tip of the dendrite sheath is attached to the hair base. The tubular body lies at the point where the hair shaft forms a tongue; it is inserted between this tongue and the apex of the dome-shaped socket septum (Figure 8a). The tubular body is roof-shaped on one side (facing the socket septum) and has deep invaginations on the other (facing the tongue); the arrangement of microtubules and neurofilaments is shown in Figure 8b. The movements leading to stimulation can be deduced from the geometry of the cuticular parts (Gaffal et al., 1975; Keil, 1978; Theiss, 1979): if the hair shaft is deflected "downwards," the tongue moves "upwards" and presses the tubular body against the socket septum.

The macrochaeta of *Calliphora* deserves special attention, because due to its extraordinary size the structures of the joint could be studied in detail. The mechano-elastic properties of the joint membrane have been investigated by Theiss (1979), who came to the conclusion that it could not consist of pure resilin as proposed by Thurm (1965) for a honeybee mechanoreceptor. In the light microscope, the membrane is transparent and stains with methylene blue. It becomes brittle if dry, but regains its elasticity under water. If excited with short-wave UV, it emits a blue fluorescence with a spectrum similar to that of resilin (for the properties of resilin, see Andersen and Weis-Fogh, 1964). After boiling in KOH, the membrane is only partly degraded, but the fluorescence disappears almost completely. In the electron microscope, it can be seen that the membrane is composed of fibers, which are stained by OsO_4 and uranyl acetate as well as phosphotungstic acid, are embedded in an electron-lucent matrix, and are anchored in the hair as well as in the body cuticle (Figure 9).

We can thus conclude that the joint membrane consists of protein (which stains with phosphotungstic acid and is degraded by KOH) and chitin (which is not degraded by KOH), and that the material has some of the properties of resilin. Following the results of Blackwell and Weih (1980) and Hillerton (1980), we may propose as the molecular arrangement that fibrils consisting of a chitin core and a protein sheath are anchored in the sclerotized cuticle of body and hair and run in the direction of the tension. These fibrils are embedded in a resilin matrix with water-containing spaces (see the two-component model of insect cuticle, as discussed by Neville, 1975).

4.6. Filiform and Clavate Hairs on the Cerci of Acheta

Just as the fly macrochaetae are characterized by their massive joint membrane, so the filiform and clavate hairs on the cricket cercus are characterized by the extreme reduction of this structure. They are presented here as examples for the second group of mechanoreceptors (see Section 4.1). These sensilla are among the best studied insect sensory organs as regards their ultrastructure, development, and physiology, as well as their central nervous projections and neuronal processing (e.g., Gnatzy and Schmidt, 1971; Edwards and Palka, 1974; Bischof, 1975; Gnatzy, 1978; Tobias and Murphey, 1979, Gnatzy and Tautz, 1980; Murphey, 1981).

The *filiform hairs* are up to 1.5 mm long and are arranged in rows along the cerci. The geometry of the hair base allows deflection in only one exactly defined plane. Hairs of each row can move either parallel or normal (L and T hairs) to the long axis of the cercus. They respond to air streams and low sound frequencies (Gnatzy and Tautz, 1980). The direction of the stimuli is monitored by the different hair rows and somatotopically transmitted to the last abdominal ganglion (Murphey, 1981). In *clavate hairs,* which are gravity receptors (Bischof, 1975), the shaft has been transformed into a fluid-filled club, but otherwise they are structurally similar to the filiform hairs. Clavate sensilla are arranged in two clusters at the base of the cerci, facing each other. The base of each sensillum is enclosed by a wide cup which limits the possible hair deflection. The cup is surrounded by several campaniform sensilla, which are stimulated if the hair is deflected beyond its limiting angle so that the whole cup is moved (Dumpert and Gnatzy, 1977).

The filiform sensilla are remarkable for their very complex stimulus-transmitting apparatus (see Gnatzy and Tautz, 1980). Figure 10 shows a three-dimensional reconstruction of the noncellular parts of this apparatus, which only allows exactly defined movements. The base of the hair shaft forms an oval-shaped plate with two slightly rounded lateral projections which fit into two lateral bearings of the ring-shaped socket. The lateral projections can thus act as trunnions, allowing the hair to move only in a plane normal to their axis. The basal plate is bulged downwards, forming a shovel-shaped projection, the *stimulating edge,* which lies about 1-2 μm below the trunnions. The joint membrane is reduced to a thin cuticular lamella, which forms a very sharp fold between socket and basal plate. This lamella most probably suspends the hair in a way that it is not entirely resting on the trunnions; it might also displace the axis of pivoting distal to the trunnion axis as discussed by Thurm (1982, see Figures 15, 32a). The dendrite sheath, which ends in the ecdysial canal, has a sharp incision on a level with the stimulating edge. Just opposite the edge, a crescent-shaped plate is "welded" onto the dendrite sheath which is supported by a cushion on this side (Gnatzy, 1978). The cushion consists of an electron-dense scaffolding and a fibrillar intermediate substance and is spanned by a dome-shaped, fibrillar structure which ends at the rear of the crescent-shaped plate. The whole apparatus is suspended in a cuticular framework which is attached to the hair socket.

The dendrite sheath tightly encloses the outer dendritic segment. The number of microtubules within the dendrite increases from 18 in the ciliary segment to about 100 (small filiform hairs) or about 1000 (large filiform hairs) in the region of the tubular body.

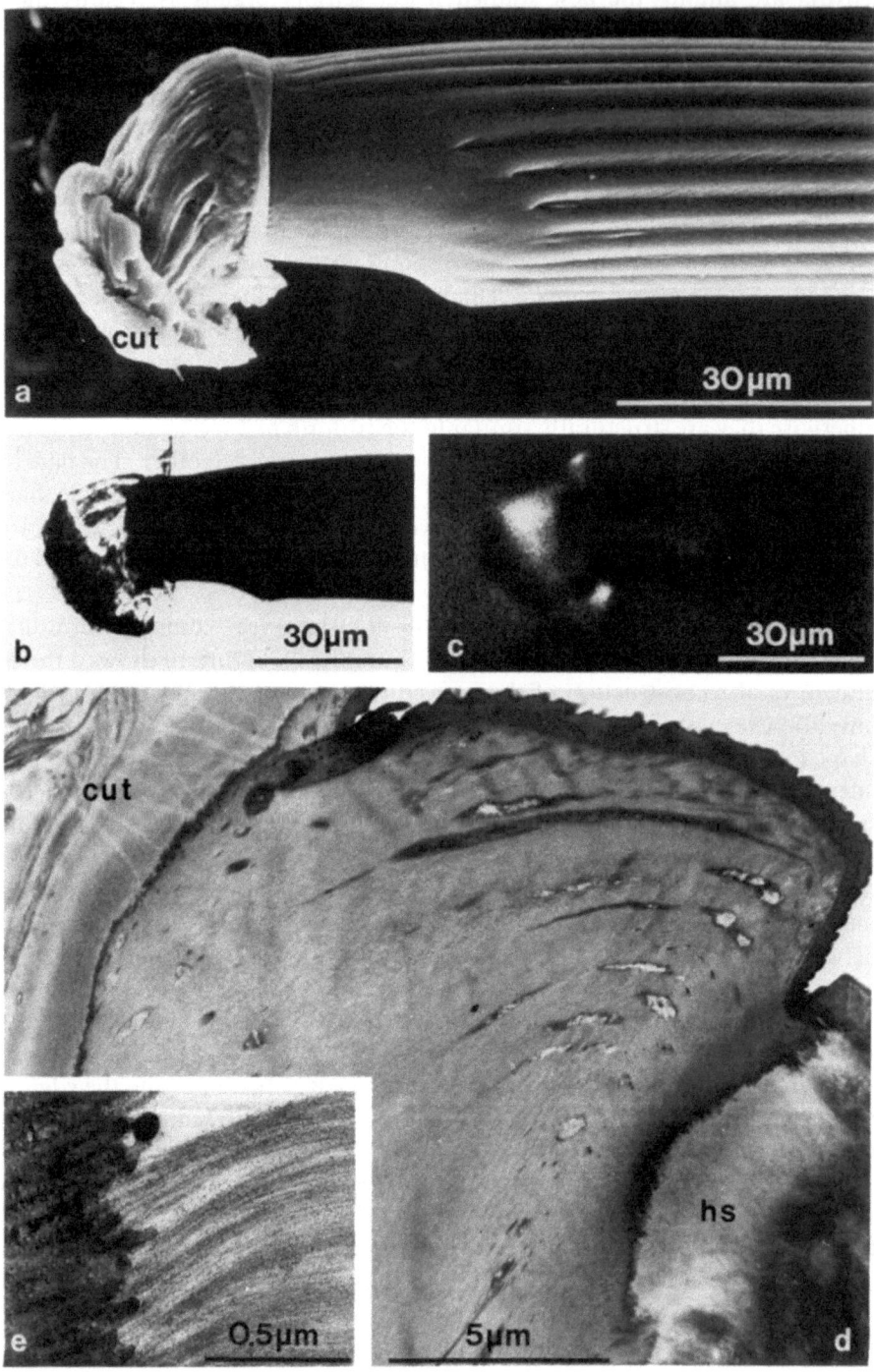

The increase of microtubule number has been quantified in a series of cross-sections of the dendrite represented by Figure 12a. On a level with the inward-projecting ribs of the dendrite sheath (Figures 10, 11), 650 free microtubules have been counted. In this region, many tubules gather to form bundles (Figure 11). These tubules are connected via hemidesmosome-like structures to the dendrite membrane and the ribs. Additional microtubules originate here which gradually join the bundles. These bundles form the tubular body which consists of 700 densely packed microtubules associated with an electron-dense intermediate substance. This tubular body proper is surrounded by a dendritic cytoskeleton formed by 370 free microtubules. The incision is finally entered by only 520 tubules of the tubular body, and only in this narrow region are the MICs found. Most microtubules end immediately above this level (Figure 12c); only very few run into the dendrite tip which ends in the ecdysial canal.

Figure 10 may help to understand the mode of stimulus transmission. The axis of pivoting is thought to run slightly above the hair's basal plate (see above). A deflection of the hair shaft of less than 0.5° already yields a nearly maximal amplitude of the receptor potential (Thurm et al., 1983). Such a deflection is thought to cause a movement of the proximal lever arm of 20 nm at most. The stimulating edge presses the dendrite membrane against the MICs which are backed by the tubular body. If the deflection of the hair becomes larger, the whole apparatus should be able to give way in a graded manner, as can be deduced from its biomechanical construction. Thus, the tubular body possibly is not compressed at all. It is most probable that a mechanical interaction between MICs and membrane is sufficient to yield a membrane depolarization.

The MICs are candidates for being the essential stimulus-transducing elements for several reasons (Thurm, 1982). First, they are found just in the region of the maximal mechanical impact. Second, their connection with the receptor membrane is well compatible with the short receptor latencies (0.1 msec) which should require close association of the stimulus-receiving structures and the ion channels in the membrane. Third, they would serve for concentrating the low energetic input of the mechanical stimulus on a few discrete structures, e.g., 10^{-19} Wsec to 10^{-16} Wsec/1000 MICs.

Figure 9. Socket structures of the *Calliphora* macrochaeta. (a) Scanning electron micrograph of macrochaeta which has been torn out of its socket with parts of the socket cuticle (cut) adhering. (b) Light micrograph (bright field) of an isolated macrochaeta, showing the transparent joint membrane with adhering parts of the cuticular socket (dark). (c) Same hair as in (b) shown in dark field excited with UV light. The socket membrane is fluorescing with a blue color. (d) Tangential section of the joint membrane, showing the course of its fibers between cuticular socket (cut) and hair shaft (hs). Bloc-stained with phosphotungstic acid (Silverman and Glick, 1969), dehydration with ethylene glycol (Pease, 1966). (e) Fibers (right) attached to the cuticular socket (left). Bloc-stained with uranyl acetate (Locke et al., 1971). Micrograph (a) taken by Mr. C. Göcke, Münster; (b, c) taken in cooperation with Professor K. Hamdorf, Bochum.

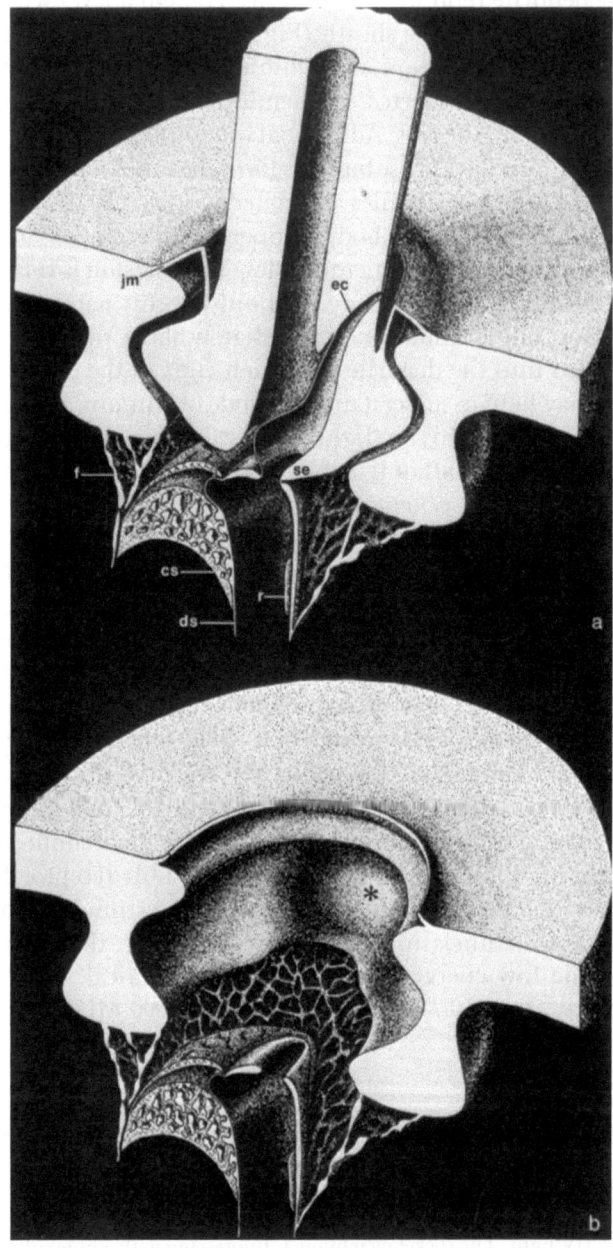

Figure 10. The cuticular parts of the stimulus-transmitting apparatus of the cercal filiform hair of *Acheta*, shown with the hair in its proper place (a) and removed (b). The possible movements of the hair shaft are defined by the cuticular bearings (asterisk in b) and the reduced joint membrane (jm). Deflection of the hair shaft to the right causes a movement of the stimulating edge (se) to the left which presses against the dendrite tip (compare Figure 12a). The supporting structures are provided by the dendrite sheath (ds) reinforced opposite to the stimulating edge, the cushion (cs), and the cuticular framework (f). ec, ecdysial canal; r, ribs of the dendrite sheath bearing the tubular body (compare Figure 11).

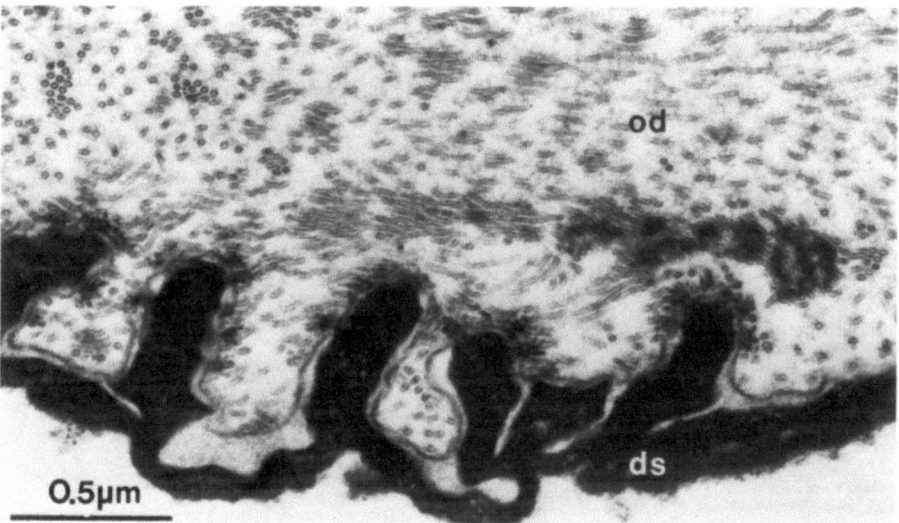

Figure 11. Cross-section of the outer dendritic segment just below the tubular body. The dendrite sheath (ds) forms extensions into the dendrite (od): many microtubules are attached to the cell membrane in this region. Other microtubules start to form clusters with intercalated electron-dense substance; additional microtubules originate here as well. Preparation as in Figure 12.

5. Fine Structure of Olfactory Sensilla

5.1. General Remarks

There is but one feature in insect olfactory sensilla which can be regarded as a modality-specific structure. This is the presence of pores in the wall of the cuticular hair, which could not be detected before the introduction of electron microscopy (Richards, 1952; Slifer *et al.*, 1959). A great number of investigations using both, electron microscopy and electrophysiology, has since been carried out (for review see Slifer, 1970; Altner and Prillinger, 1980; Zacharuk, 1980; Chapman, 1982; Steinbrecht, 1984). These studies showed without exception that all sensilla with a multiporous wall have an olfactory function. Two fundamentally different types of wall pores have been recognized by Steinbrecht (1969) and are now taken for classification of olfactory sensilla (Altner, 1977):

1. Wall pores with *pore tubules* are present in all *single-walled olfactory sensilla*.
2. Wall pores with *spoke channels* are found exclusively in *double-walled sensilla*.

In addition to these cuticular structures, we find, however, no conspicuous cellular feature typical for olfactory dendrites like the tubular body in mechanoreceptors. The structures responsible for stimulus recognition and the subsequent electrical events must be comprised in the molecular organization of the receptor membrane and, therefore, are not yet accessible to fine structural analysis (for current hypotheses see Ritter, 1979; Wright, 1982).

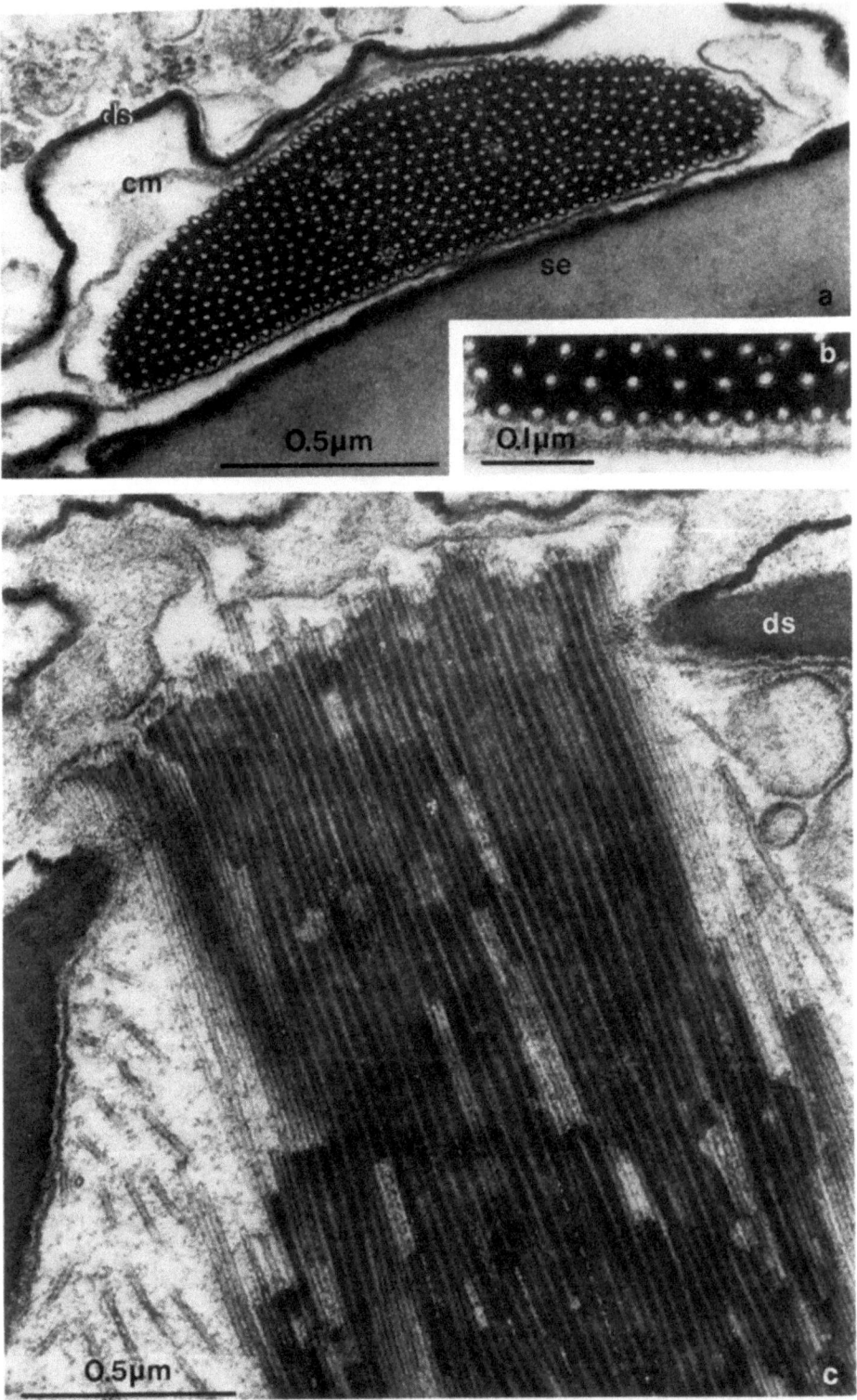

In the following sections, typical examples for olfactory sensilla will be described with emphasis on the fine structure of their cuticular parts and their outer dendritic segments.

5.2. Single-Walled Olfactory Sensilla with Pore Tubules

5.2.1. The Pheromone-Sensitive Sensilla of Bombyx

These sensilla, known as the long sensilla trichodea, besides belonging to the first insect chemoreceptors studied by electrophysiological methods, are especially well-known systems in terms of sensillum fine structure, receptor cell specificity and sensitivity, and the behavioral aspects of orientation and intraspecific communication (for references see Steinbrecht and Schneider, 1980; Schneider, 1980).

On the average, 17,000 long sensilla trichodea are present on a single antenna of a male *Bombyx mori* (Steinbrecht, 1970). They are arranged in a regular array so that the bipectinate antennal flagellum obtains a sievelike appearance (Figure 13a). In fact, about 30% of the odor molecules contained in a passing airstream are filtered out by this "odor sieve" (Kaissling, 1971). Moreover, theoretic calculations (Adam and Delbrück, 1968), and measurements with tritium-labeled pheromone (Steinbrecht and Kasang, 1972) established that more than 80% of the adsorbed stimulus molecules are caught by the long (100 μm) and slender (1-3 μm) sensilla trichodea, although the hair surfaces total less than 13% of the whole antennal surface. The cuticular hair wall of the sensilla trichodea is fairly thick and perforated by pores in relatively low density (2-7 μm^2; Figures 13b,c). Nevertheless, behavioral and electrophysiological experiments lead to the conclusion that the stimulus transport from the hair surface to the receptor membrane of the dendrites within the hair lumen is highly efficient. At threshold intensities, the majority of adsorbed molecules stimulates the receptor cells (Kaissling and Priesner, 1970).

The number, distribution, and fine structure of the pores in *Bombyx* have been described in detail by Steinbrecht (1973; see also Figures 2b, 14a,b):

A narrow *pore canal* leads into several *pore tubules* running within a wider *liquor channel*. The lumen of the pore canal is filled with an electron-lucent, chloroform-insoluble material which also covers the outer hair surface (layer L2, Steinbrecht and Kasang, 1972). Some pore tubules contact the dendrites, but most are seen to end free in the hair lumen. The frequency of pore tubule-dendrite contacts depends on the fixation procedure (*B. mori:* Steinbrecht and

---

Figure 12. The tubular body of the cercal filiform hair of *Acheta*. (a) Cross-section of tubular body immediately above incision. Note peripheral microtubule layer, connected to the dendritic membrane (cm) by fine bridges, opposing the stimulating edge (se) of the hair. The number of microtubules is 520. ds, dendrite sheath. (b) Tubule–membrane complex opposite of stimulating edge. The MICs between the peripheral microtubule layer and dendritic membrane are clearly visible. Note microtubule subunits. (c) Grazing section of tubular body, showing parallel course of microtubules with electron-dense intermediate substance. Most of these tubules end immediately above the dendritic incision. The free microtubules (left) end just below. Fixed in glutaraldehyde with tannic acid; bloc-stained with uranyl acetate.

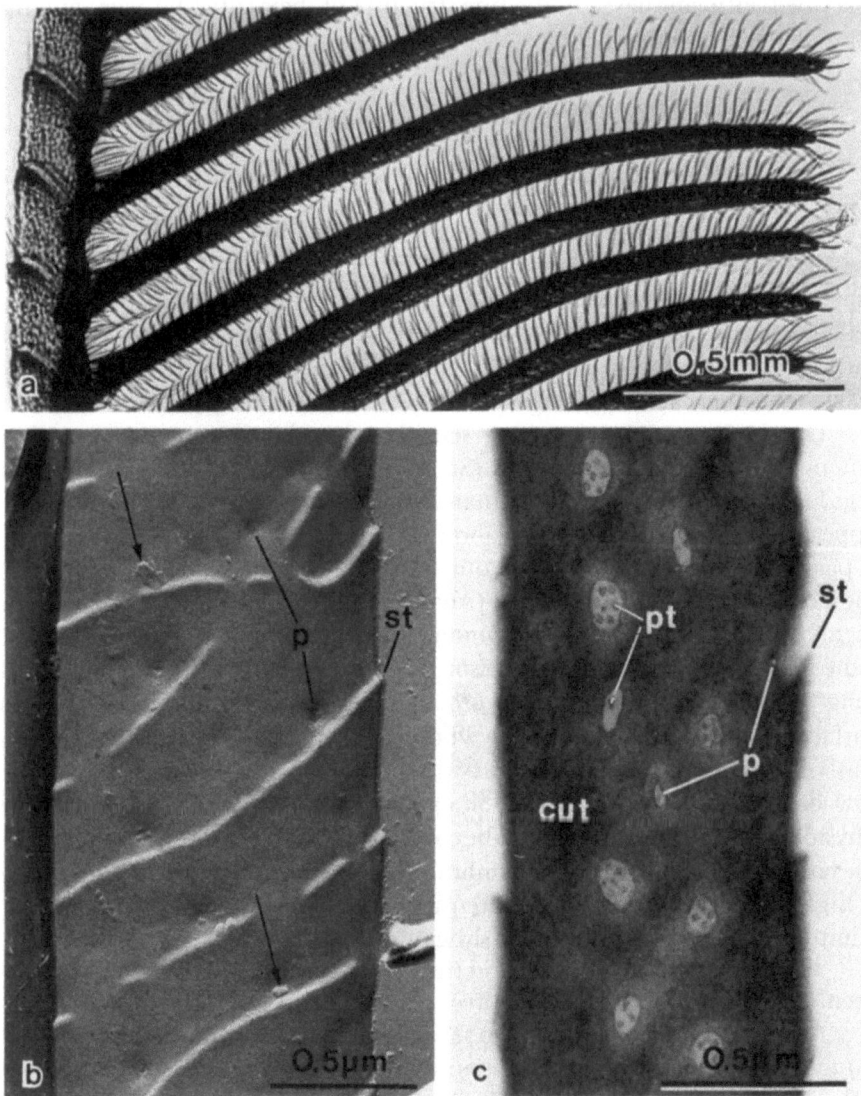

Figure 13. Bombyx mori ♂, pheromone-sensitive sensilla trichodea. (a) The regular array of sensilla trichodea on the branched antennal flagellum forms an effective "odor sieve" (see text). (b) Surface of a sensory hair after freeze-fracturing; surface steps (st), and pores (p) are evident. The outermost epicuticular layers have been removed by the fracturing, except for a few remnants (arrows). (c) Tangential section through cuticle (cut) of hair wall. Up to seven pore tubules (pt) are found in connection with each pore (p) comprised within a liquor channel (compare Figures 2b, 14b). (a) from Steinbrecht (1970); (b, c) from Steinbrecht (1973)).

Müller, 1971; *Antheraea polyphemus:* Keil, 1982). Little can be said at present about the chemical nature of the pore tubules; they probably consist of lipid-insoluble as well as lipid-soluble material (Hawke and Farley, 1971b; Steinbrecht, 1980).

When the pore tubule systems were discovered by Slifer *et al.* (1959) in thin-walled sensilla of a grasshopper, they were thought to be fine extensions of the dendrites directly probing the ambient air. Later, Ernst (1969) showed that the pore tubules in olfactory sensilla of *Necrophorus* are extracellular structures formed by the trichogen cell. Therefore, any contact between the pore tubules and dendrites can only occur secondarily after the trichogen cell has retracted from the hair lumen. A possible mechanism for establishing this contact has recently been proposed by Keil (1982), who observed a fuzzy surface coat on the dendrites and pore tubules, but not on the inner surface of the hair wall.

The pore tubule systems have been intuitively taken as the pathways of odor molecules by most investigators, but the evidence in favor of this notion is only indirect:

1. Whenever sensilla with typical wall pores and pore tubules (Figures 14, 15) have been studied by electrophysiology, olfactory function was demonstrated.
2. Tracer substances applied from outside could be shown to penetrate into the pores and pore tubules of some sensilla (Ernst, 1969; Chu and Axtell, 1971; Hawke and Farley, 1971b; Foelix, 1972).
3. Structures similar to the pore tubules, the wax canal filaments, are widespread among insect cuticles. Because of their abundance in secreting or resorbing cuticles (for references see Steinbrecht, 1973), transport function has been proposed for these structures too (Locke, 1965).

Experiments with tritium-labeled pheromone indicated the diffusion coefficient of bombykol along the hair to be at least 5×10^{-7} cm^2/sec (Steinbrecht and Kasang, 1972). Using this value for both, the diffusion on the hair surface as well as along the pore tubules, and substituting the morphometric data of the sensillum, Steinbrecht (1973) calculated the diffusion time of a stimulus molecule from its first impact on the hair surface until it reaches the end of a pore tubule as shorter than 5 msec in most hair regions. Thus, stimulus transport by diffusion would be fast enough if most pore tubules would contact the dendrite *in vivo*; the comparatively long latency of the receptors (>50 msec; Kaissling, 1974) then would be caused by other rate-limiting processes in stimulus transduction.

If, on the other hand, the low frequency of contacts between pore tubules and dendrites (see above) is not merely a preparatory artifact, two possible pathways of stimulus transport have to be considered: (1) the pheromone molecules arriving at the dead end of a pore tubule move back and invade other pore tubules or (2) they pass into the receptorlymph to reach the dendrites (Steinbrecht, 1973). At first sight, crossing the phase border between the lipophilic pore tubule and the aqueous receptorlymph appears difficult for a lipophilic molecule such as bombykol, but this might be facilitated by the macromolecules contained within the receptorlymph (see Section 3.4). The pheromone-binding protein recently discovered in the receptorlymph of silkmoths and supposed to be involved in stimulus inactivation (Vogt and Riddiford, 1981) might even work as a carrier for pheromone molecules (R. Vogt, personal communication). At present, there is no direct evidence to discriminate between the two hypothetical pathways. It is clear that in either case the actual stimulus transport time will be a multiple of the diffusion time as

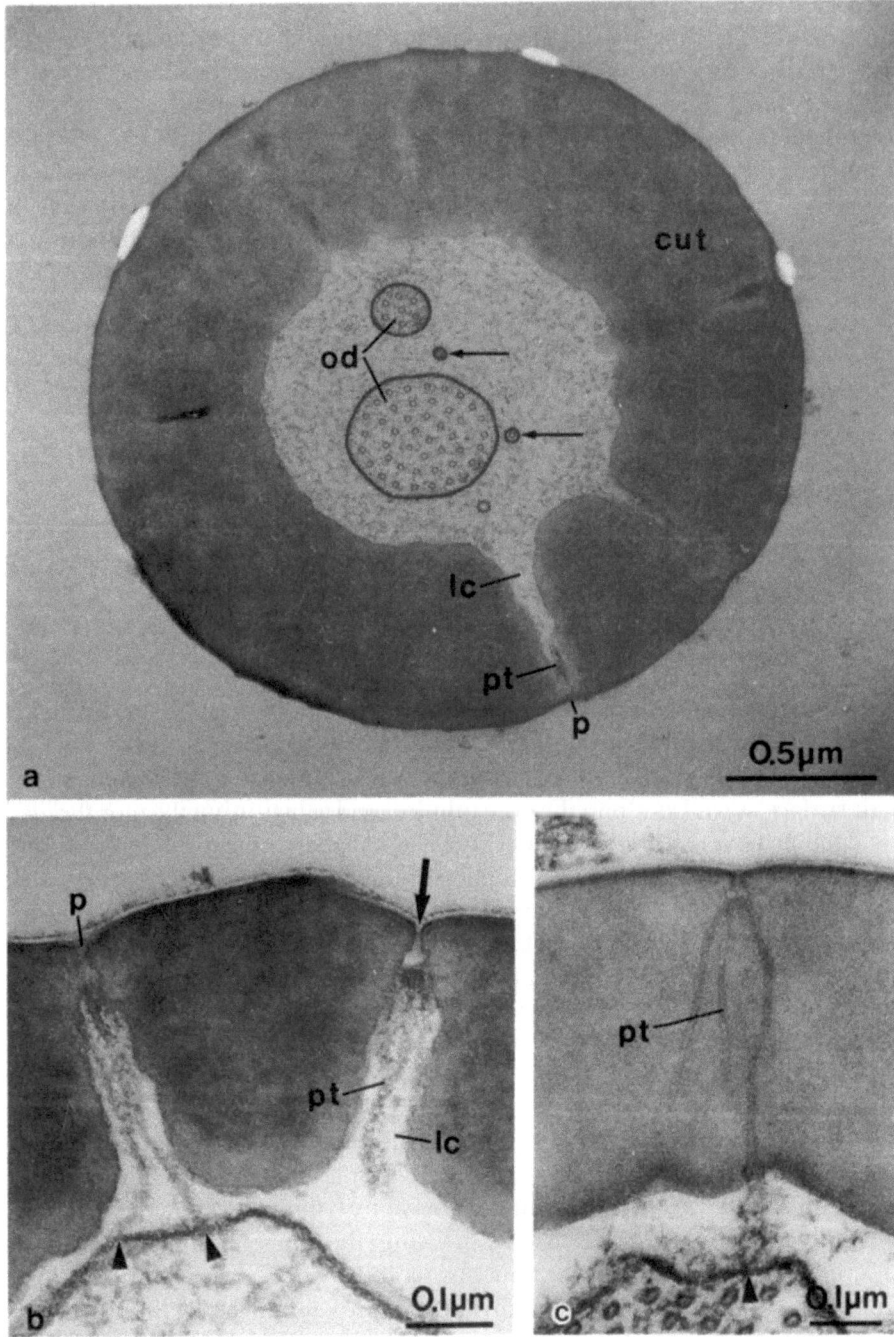

Figure 14. Thick-walled olfactory hairs in cross-section. (a, b) Sensillum trichodeum of *Bombyx mori* ♂. The outer dendritic segments (od) of the two pheromone-sensitive receptor cells have different caliber; arrows point to short side branches containing but one microtubule. The cuticular wall (cut) is perforated by narrow pores (p) each leading into several pore tubules (pt) which are surrounded by a wider liquor channel (lc) filled with receptorlymph, like the hair lumen. An electron-lucent layer

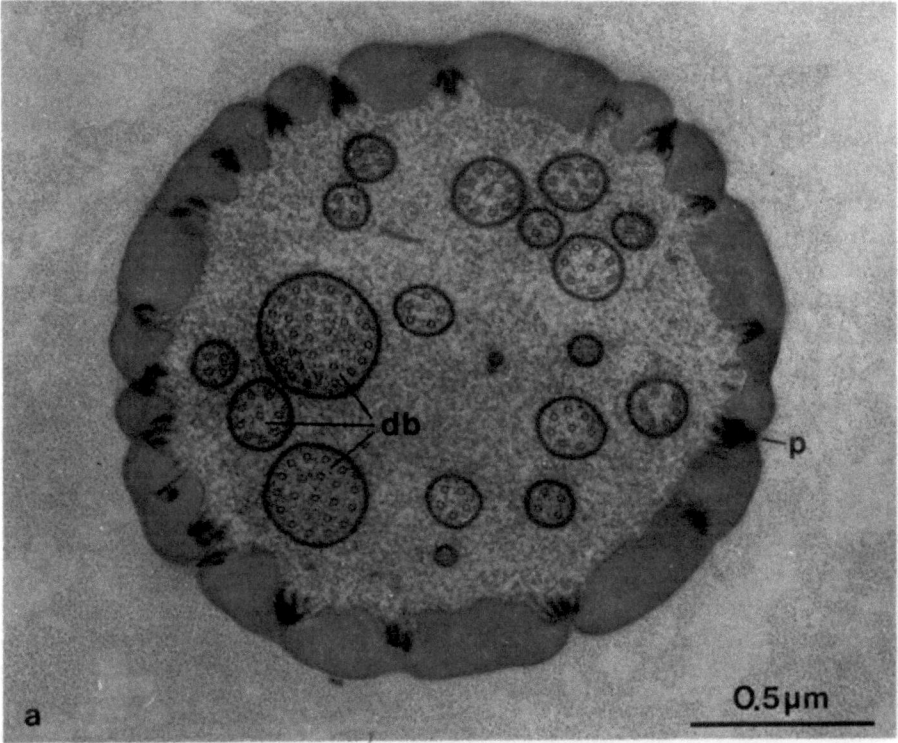

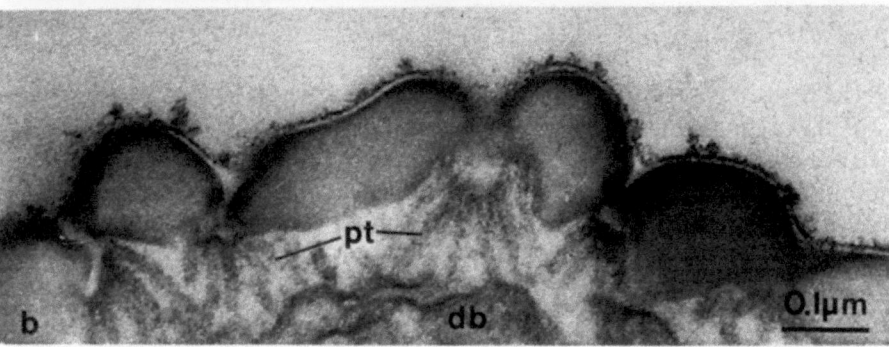

Figure 15. Thin-walled olfactory hair in cross-section (sensillum basiconicum of *Bombyx mori*) displaying a high density of pores each leading into numerous pore tubules. The outer segments of three olfactory receptor cells form many dendritic branches (db) which frequently get in contact with pore tubules as shown in (b). Note the outermost epicuticular layers in (b). (a) Freeze-substituted specimen; (b) OsO₄ fixation. (b) from Steinbrecht (1973).

covers the hair surface and fills the pores (arrow). The inner part of the pore tubules is usually not preserved by freeze-substitution in acetone (a), but after chemical fixation with Dalton's mixture (b), pore tubules are found protruding into the hair lumen and sometimes ending in contact with the dendritic membrane (arrowheads). (c) *Antheraea polyphemus* ♂, pheromone-sensitive sensillum trichodeum; pore tubules (pt) traverse the cuticle without a liquor channel and may end in contact (arrowhead) with the dendrite. Preparation as in Figure 12. (b) from Steinbrecht (1973); (c) from Keil (1982).

calculated above and could well account for the long receptor latencies with weak stimuli. The pore tubules finally could play a role in protecting the stimulus molecules on their way to the receptor membrane from degradating esterases found in the receptorlymph (Vogt and Riddiford, 1981).

The dendrites of the sensilla trichodea of *Bombyx* and many other moths are unbranched, except for occasional, extremely thin side branches. The two dendritic outer segments contain a different number of microtubules and have a characteristic difference in diameter, which is, however, often masked by quasi-periodic thickenings of the dendrites along the hair (Steinbrecht, 1973). The caliber difference of the two dendrites can be related to the consistently observed difference in the amplitude of the action potential recorded extracellularly in response to the two pheromone components of *Bombyx* females, bombykol (= E(10),Z(12)-hexadecadien-1-ol: large spikes) and bombykal (= E(10),Z(12)-hexadecadien-1-al: small spikes) (Kaissling et al., 1978).

In freeze-etch preparations of sensilla trichodea of *Bombyx*, cross-fractured dendrites show up perfectly smooth and round, thus corroborating the results of freeze-substituted preparations. The rarely obtained *en face* views of freeze-fractured outer dendritic membranes reveal a very high density of intramembrane particles. This indicates a high protein content of the receptor membrane. The ciliary region of these dendrites, on the other hand, has a very low density of intramembrane particles (Steinbrecht, 1980). Comparable results have been reported for olfactory receptors of vertebrates and also of an insect (Menco and van der Wolk, 1982).

5.2.2. Other Single-Walled Olfactory Sensilla

There are also other olfactory sensilla on the antenna of *B. mori*. Besides the sensilla trichodea, there are the sensilla basiconica which respond to various fatty acids and alcohols, but also to extracts of mulberry leaves (E. Priesner, unpublished, quoted in Schneider and Steinbrecht, 1968). Although basically belonging to the same morphological type, these sensilla differ markedly from the sensilla trichodea by: (1) their small size (hair length 10-20 μm), (2) the thin cuticular wall ($\sim$0.15 μm), (3) the high density of pores (20/μm^2), (4) the high number of pore tubules per pore ($\sim$17), and (5) the great number of dendritic branches filling the hair lumen (Steinbrecht, 1973; see also Figure 15).

Similar sensilla are found in nearly every insect species investigated. Famous examples are the thin-walled sensilla basiconica of the grasshopper *Melanoplus* (Slifer et al., 1959; Slifer and Sekhon, 1964b), the flesh fly *Sarcophaga* (Slifer and Sekhon, 1964a), the carrion beetle *Necrophorus* (Ernst, 1969), and the cockroach *Periplaneta* (Altner et al., 1977; Toh, 1977). The receptors in these sensilla usually have broad, sometimes overlapping reaction spectra; often a function in food finding and selection is observed (*Necrophorus*: Boeckh, 1962; *Calliphora*: Kaib, 1974; *Periplaneta*: Sass, 1978).

Although the distribution and shape of the pores as well as the length and number of the pore tubules per pore vary considerably from species to species, the diameter of the pore tubules remains remarkably uniform (Schneider and

Steinbrecht, 1968; Meinecke, 1975; Altner and Prillinger, 1980; Zacharuk, 1980). This, however, does not imply their chemical identity, for differential resistance of pore tubules to solvents has been shown in two sensillum types of the cockroach *Arenivaga* (Hawke and Farley, 1971b).

Contacts between pore tubules and dendrites have been observed more frequently in sensilla basiconica than in sensilla trichodea, possibly due to the dense packing of dendritic branches in the hair lumen (Figure 15b; for references see Keil, 1982). Nevertheless, the functional significance of the high numbers of pore tubules and of the dendritic branching remains obscure due to the lack of relevant experimental data.

Single-walled olfactory sensilla are extremely variable in size, shape, and arrangement. Certainly, optimizing stimulus uptake and transport is not the only adaptive pressure on antennal and sensillar structure. Resistance to mechanical stress and water loss, compatibility with flight aerodynamics, as well as structural limitations imposed by the need for grooming, may suppress the formation of odor sieves and also change the form and distribution of sensilla (see Kaissling, 1971, for examples). Thus, the cuticular hair may even be totally transformed into pore plates more or less flush with the antennal surface as in Hymenoptera (for references see Stepper *et al.*, 1983; Schmidt and Kuhbandner, 1983), Homoptera (for references see Lewis and Marshall, 1970; Bromley *et al.*, 1979), or lamellicorn beetles. In the latter group, Meinecke (1975) studied the inventory of olfactory sensilla of 42 species and found an impressive variety of cuticular structures showing morphological transitions in parallel with the presumed evolutionary lineage.

5.3. Double-Walled Olfactory Sensilla with Spoke Channels

The double-walled olfactory sensilla have a remarkably complex cuticular structure. An inner cylindrical cuticular wall shrouds the dendrites and is basally continuous with the dendrite sheath. An outer cuticular wall concentrically surrounds the inner wall being smooth and devoid of pores in the basal regions, but longitudinally grooved in the distal parts of the peg. These grooves are connected with the inner cylinder by hollow cuticular spokes, so that receptorlymph from the innervated lumen may reach the bottom of the grooves. Thus, the radial *spoke channels* represent the wall pores of double-walled sensilla. Typical examples of this type are the coeloconic sensilla of the locust (Steinbrecht, 1969) or the silkmoth (Figure 16a) or the antennal grooved pegs of mosquitoes (McIver, 1974) or mealworm beetles (Harbach and Larsen, 1977); more are listed in the reviews by Altner and Prillinger (1980) and Zacharuk (1980). The function of double-walled olfactory sensilla has been studied most intensely in *Locusta* (Boeckh, 1967; Kafka, 1970). It appears that double-walled multiporous sensilla occur together with single-walled multiporous sensilla on the antennae of most insect species studied so far; they are, however, lacking in the whole group of Collembola (Altner and Prillinger, 1980).

In different species, the spoke channels may be quite variable in length, diameter, and shape. Sometimes they branch to form complex canal systems,

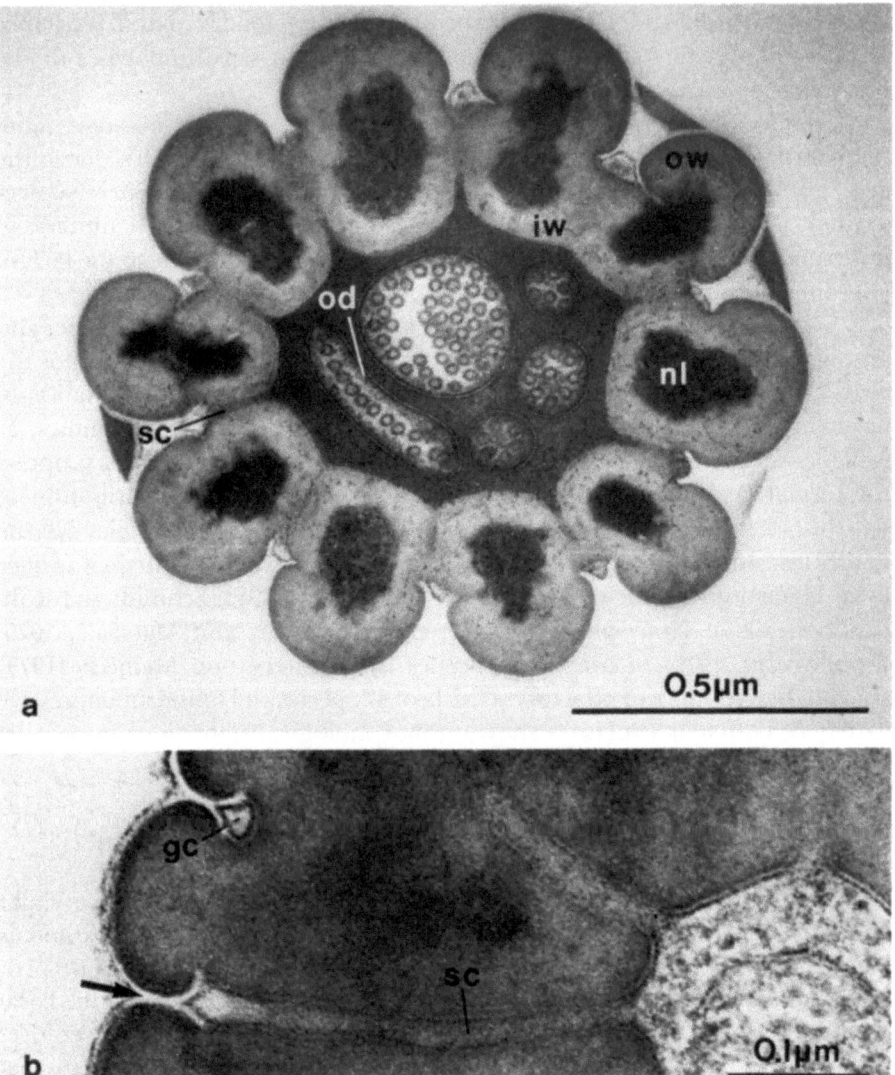

Figure 16. Examples of double-walled sensilla with wall pores. (a) *Bombyx mori*, sensillum coeloconicum, cross-section through freeze-substituted peg. Five dendrites (od) are surrounded by an inner (iw) and outer (ow) cuticular wall, the latter being longitudinally grooved. Radial spoke channels (sc) lead from the inner lumen into the longitudinal grooves. The inner lumen as well as the noninnervated lumina (nl) between the cuticular walls are filled with extremely dense receptorlymph. (b) *Cimex lectularius*, double-walled peg at high magnification. The longitudinal grooves are covered by an electron-lucent epicuticular layer (arrow) so that the receptorlymph within the spoke (sc) and groove channels (gc) is not in direct contact with ambient air. OsO_4 fixation. (b) from Steinbrecht and Müller (1976).

and sometimes they fuse to form radial slits (Meinecke, 1975). There is still some uncertainty about the outer end of the spoke channels. Meinecke (1975) claims that the receptorlymph filling the slits in the type P sensillum of *Potosia* is directly in contact with the ambient air. Hawke and Farley (1971a), on the other hand, in coeloconic sensilla of *Arenivaga* clearly demonstrate an electron-lucent epicuticular layer at the bottom of the grooves covering the spoke channels. The same is true for type C sensilla of the bedbug, *Cimex lectularius* (Steinbrecht and Müller, 1976; Figure 16b) and double-walled sensilla of the bark beetle, *Ips typographus* (Hallberg, 1982). This electron-lucent layer, although easily overlooked, is now generally known as the outer epicuticle and assumed to be of paramount importance for waterproofing (see Filshie, 1982).

As discussed by Steinbrecht and Müller (1976), stimulus diffusion along the spoke channels may follow the same principles as proposed for sensilla with pore-tubule systems. Dense material found in the lumen of the spoke channels of some species may provide a suitable milieu for diffusion (Steinbrecht, 1969; Altner et al., 1977). Altner et al., (1977) observed a different specificity range of receptors in double- and single-walled sensilla of *Periplaneta americana*. Single-walled sensilla in this species were found to respond mostly to fatty alcohols and terpenes, double-walled sensilla to fatty acids and amines. The lack of comparable studies in other species, however, does not allow us to generalize from these observations.

6. Final Remarks

It might be worthwhile before closing to have a look at sensilla other than mechanosensitive or olfactory. Gustatory sensilla (for review see Hansen, 1978; Altner and Prillinger, 1980) have a terminal pore through which the stimulus (sugar or salt solutions or even pure water) will reach the receptor endings of the dendrites. Hygroreceptors (for review see Altner and Prillinger, 1980; Steinbrecht 1984) are tightly enclosed by short aporous pegs and in the fine structure of their dendrites resemble chordotonal organs. Possibly they are modified mechanoreceptors responding to hygroscopic swelling or shrinking of the cuticular apparatus (Yokohari, 1978). Least is known about structure-function correlations in thermoreceptors (for review see Loftus, 1978).

It should also be remembered that sensilla are morphologically and morphogenetically defined units, but that the various receptor cells, despite being arranged together with their common auxiliary cells and being coupled to a nonneural auxiliary voltage (see Section 3.4), function as largely independent units, each having its own specificity. Thus, it is common that different receptor cells in the same olfactory or gustatory sensillum serve different odor or taste qualities, respectively. Even different stimulus modalities may be combined in a given sensillum. Gustatory cells are mostly accompanied by a mechanoreceptor, while hygroreceptors often are combined with a thermoreceptor (cold–moist–dry triad, e.g., Waldow, 1970); in Collembola, even receptors of presumed olfactory function are combined with a mechanoreceptor according to fine structural evidence (Altner and Ernst, 1974).

In conclusion, we can say that insect sensilla are oligocellular units of an amazingly uniform basic organization. Yet these organules show a structural and functional complexity almost comparable to large multicellular organs. Because of their easy accessibility, they are ideal systems for the experimental approach, as shown by electron microscopy, developmental biology, cytogenetics, biochemistry, and electrophysiology. The combination of these different lines of research caused a remarkable progress in our understanding of insect sense organs during the last 25 years.

ACKNOWLEDGMENTS

We are grateful to Professors Dietrich Schneider and Karl-Ernst Kaissling for critical discussion of the text, and to Heidi Hoesch, Barbara Müller, Inge Susann Rössel, and Judith Schuler for help in preparing the manuscript. This chapter contains previously unpublished work done at the University of Münster with support by the Deutsche Forschungsgemeinschaft (T.K.).

References

Adam, G., and Delbrück, M., 1968, Reduction of dimensionality in biological diffusion processes. In *Structural Chemistry and Molecular Biology*, edited by A. Rich and N. Davidson, pp. 198-215 Freeman, San Francisco.

Altner, H., 1977, Insect sensillum specificity and structure: An approach to a new typology. In *Olfaction and Taste*, vol. VI, edited by J. Le Magnen and P. MacLeod, pp. 295-303, Information Retrieval, London.

Altner, H., and Ernst, K.-D., 1974, Struktureigentümlichkeiten antennaler Sensillen bodenlebender Collembolen, *Pedobiologia* 14:118-122.

Altner, H., and Prillinger, L., 1980, Ultrastructure of invertebrate chemo-, thermo-, and hygroreceptors and its functional significance, *Int. Rev. Cytol.* 67:69-139.

Altner, H., Sass, H., and Altner, I., 1977, Relationship between structure and function of antennal chemo-, hygro-, and thermoreceptive sensilla in *Periplaneta americana*, *Cell Tissue Res.* 176:389-405.

Andersen, S. O., and Weis-Fogh, T., 1964, Resilin: A rubberlike protein in arthropod cuticle, *Adv. Insect Physiol.* 2:1-65.

Ashhurst, D. E., 1970, An insect desmosome, *J. Cell Biol.* 46:421-425.

Ashhurst, D. E., and Costin, N. M., 1971, Insect mucosubstances. II. The mucosubstances of the central nervous system, *Histochem. J.* 3:297-310.

Berridge, M. J., and Oschman, J. L., 1972, *Transporting Epithelia*. Academic Press, New York and London.

Bischof, H.-J., 1975, Die kuelenförmigen Sensillen auf den Cerci der Grille *Gryllus bimaculatus* als Schwererezeptoren, *J. Comp. Physiol.* 98:277-288.

Blackwell, J., and Weih, M. A., 1980, Structure of chitin-protein complexes: Ovipositor of the ichneumon fly *Megarhyssa*, *J. Mol. Biol.* 137:49-60.

Boeckh, J., 1962, Elektrophysiologische Untersuchungen an einzelnen Geruchsrezeptoren auf den Antennen des Totengräbers (*Necrophorus*, Coleoptera), *Z. v gl. Physiol.* 46:212-248.

Boeckh, J., 1967, Reaktionsschwelle, Arbeitsbereich und Spezifität eines Geruchsrezeptors auf der Heuschreckenantenne, *Z. v gl. Physiol.* 55:378-406.

Bromley, A. K., Dunn, J. A., and Anderson, M., 1979, Ultrastructure of the antennal sensilla of aphids. I. Coeloconic and placoid sensilla, *Cell Tissue Res.* 203:427-442.

Caveney, S., and Berdan, R., 1982, Selectivity in junctional coupling between cells of insect tissues. In *Insect Ultrastructure*, edited by R. C. King and H. Akai, vol. 1, pp. 434-465, Plenum Press, New York.

Chapman, R. F., 1982, Chemoreception: The significance of receptor numbers, *Adv. Insect Physiol.* 16:247-356.

Chu, I.-W., and Axtell, R. C., 1971, Fine structure of the dorsal organ of the house fly larva, *Musca domestica* L., *Z. Zellforsch. mikrosk. Anat.* 117:17-34.

Clever, U., 1958, Untersuchungen zur Zelldifferenzierung und Musterbildung der Sinnesorgane und des Nervensystems im Wachsmottenflügel, *Z. Morphol. Oekol. Tiere* **47**:201-248.

Clever, U., 1960, Der Einfluss der Sinneszellen auf die Borstenentwicklung bei *Galleria mellonella* L., *Wilhelm Roux Arch. Entwicklungmech. Org.* **152**:137-159.

De Kramer, J. J., and Van Der Molen, L., 1984, Development of labellar taste hairs in the blowfly *Calliphora vicina* Rob. Desv. *Zoomorphology* **104**:1-10.

De Kramer, J. J., Kaissling, K.-E., and Keil, T., 1984, Passive electrical properties of insect olfactory sensilla may produce the biphasic shape of spikes. *Chem. Senses* **8**:289-295.

Dumpert, K., and Gnatzy, W., 1977, Cricket combined mechanoreceptors and kicking response, *J. Comp. Physiol.* **122**:9-25.

Dunstone, J. R., 1962, Ion-exchange reactions between acid mucopolysaccharides and various cations, *Biochem. J.* **85**:336-351.

Eakin, R. M., 1965, Evolution of photoreceptors, *Cold Spring Harbor Symp. Quant. Biol.* **30**:363-370.

Edwards, J. S., and Palka, J., 1974, The cerci and abdominal giant fibres of the house cricket, *Acheta domesticus*. I. Anatomy and physiology of normal adults, *Proc. R. Soc. London Ser. B* **185**:83-103.

Erler, G., 1983, Sensitivity of an insect mechanoreceptor after destruction of dendritic microtubules by means of vinblastine, *Cell Tissue Res.* **229**:673-684.

Erler, G., and Thurm, U., 1978, Die Impulsantwort epithelialer Rezeptoren in Abhängigkeit von der transepithelialen Potentialdifferenz, *Verh. Dtsch. Zool. Ges.* **71**:279.

Ernst, K.-D., 1969, Die Feinstruktur von Riechsensillen auf der Antenne des Aaskäfers *Necrophorus* (Coleoptera), *Z. Zellforsch. mikrosk. Anat.* **94**:72-102.

Ernst, K.-D., 1972, Die Ontogenie der basiconischen Riechsensillen auf der Antenne von *Necrophorus* (Coleoptera), *Z. Zellforsch. mikrosk. Anat.* **129**:217-236.

Filshie, B. K., 1982, Fine structure of the cuticle of insects and other arthropods. In *Insect Ultrastructure*, edited by R. C. King and H. Akai, vol. 1, pp. 281-312, Plenum Press, New York.

Foelix, R. F., 1972, Permeability of tarsal sensilla in the tick *Amblyomma americanum* L. (Acarina, Ixodidae), *Tissue Cell* **4**:130-135.

Gaffal, K.-P., and Bassemir, U., 1974, Vergleichende Untersuchung modifizierter Cilienstrukturen in den Dendriten mechano- und chemosensitiver Rezeptorzellen der Baumwollwanze *Dysdercus* und der Libelle *Agrion*, *Protoplasma* **82**:177-202.

Gaffal, K.-P., and Hansen, K., 1972, Mechanorezeptive Strukturen der antennalen Haarsensillen der Baumwollwanze *Dysdercus intermedius* Dist., *Z. Zellforsch. mikrosk. Anat.* **132**:79-94.

Gaffal, K.-P., and Theiss, J., 1978, The tibial threadhairs of *Acheta domesticus* L. (Saltatoria, Gryllidae). The dependence of stimulus transmission and mechanical properties on the anatomical characteristics of the socket apparatus. *Zoomorphologie* **90**:41-51.

Gaffal, K.-P., Tichy, H., Theiss, J., and Seelinger, G., 1975, Structural polarities in mechanosensitive sensilla and their influence on stimulus transmission (Arthropoda), *Zoomorphologie* **82**:79-103.

Gnatzy, W., 1976, The ultrastructure of the thread-hairs on the cerci of the cockroach *Periplaneta americana* L.: The intermoult phase, *J. Ultrastruct. Res.* **54**:124-134.

Gnatzy, W., 1978, Development of the filiform hairs on the cerci of *Gryllus bimaculatus* Deg. (Saltatoria, Gryllidae), *Cell Tissue Res.* **187**:1-24.

Gnatzy, W., and Romer, F., 1980, Morphogenesis of mechanoreceptor and epidermal cells of crickets during the last instar, and its relation to molting-hormone level, *Cell Tissue Res.* **213**:369-391.

Gnatzy, W., and Schmidt, K., 1971, Die Feinstruktur der Sinneshaare auf den Cerci von *Gryllus bimaculatus* Deg. (Saltatoria, Gryllidae). I. Faden- und Keulenhaare, *Z. Zellforsch. mikrosk. Anat.* **122**:190-209.

Gnatzy, W., and Schmidt, K., 1972, Die Feinstruktur der Sinneshaare auf den Cerci von *Gryllus bimaculatus* Deg. (Saltatoria, Gryllidae). IV. Die Haütung der kurzen Borstenhaare, *Z. Zellforsch. mikrosk. Anat.* **126**:223-239.

Gnatzy, W., and Tautz, J., 1977, Sensitivity of an insect mechanoreceptor during moulting, *Physiol. Entomol.* **2**:279-288.

Gnatzy, W., and Tautz, J., 1980, Ultrastructure and mechanical properties of an insect mechanoreceptor: Stimulus-transmitting structures and sensory apparatus of the cercal filiform hairs of *Gryllus*, *Cell Tissue Res.* **213**:441-463.

Gnatzy, W., and Weber, K. M., 1978, Tormogen cell and receptorlymph space in insect olfactory sensilla: Fine structure and histochemical properties in *Calliphora*, *Cell Tissue Res.* **189**:549-554.

Gnatzy, W., Mohren, W., and Steinbrecht, R. A., 1984, Pheromone receptors of *Bombyx mori* and *Antheraea pernyi*. II. Morphometric data, *Cell Tissue Res.* **235**:35-42.

Gray, E. G., 1960, The fine structure of the insect ear, *Philos. Trans. R. Soc. London Ser. B* **243**:75-94.

Hallberg, E., 1982, Sensory organs in *Ips typographus* (Insecta: Coleoptera)—Fine structure of antennal sensilla, *Protoplasma* **111**:206-214.

Hansen, K., 1978, Insect chemoreception. In *Taxis and Behavior*, edited by G. L. Hazelbauer, pp. 233-292, Chapman & Hall, London.

Harbach, R. E., and Larsen, J. R., 1977, Fine structure of antennal sensilla of the adult mealworm beetle, *Tenebrio molitor* L. (Coleoptera: Tenebrionidae), *Int. J. Insect Morphol. Embryol.* **6**:41-60.

Harvey, W. R., 1980, Water and ions in the gut. In *Insect Biology in the Future*, edited by M. Locke and D. S. Smith, pp. 105-124, Academic Press, New York.

Hawke, S. D., and Farley, R. D., 1971a, Antennal chemoreceptors of the desert burrowing cockroach *Arenivaga* sp., *Tissue Cell* **3**:649-664.

Hawke, S. D., and Farley, R. D., 1971b, The role of pore structures in the selective permeability of antennal sensilla of the desert burrowing cockroach, *Arenivaga* sp., *Tissue Cell* **3**:665-674.

Henke, K., 1953, Über Zelldifferenzierung im Integument der Insekten und ihre Bedingungen, *J. Embryol. Exp. Morphol.* **1**:217-226.

Hillerton, J. E., 1980, Electron microscopy of fibril-matrix interactions in a natural composite, insect cuticle, *J. Materials Sci.* **15**:3109-3112.

Kafka, W. A., 1970, Molekulare Wechselwirkungen bei der Erregung einzelner Riechzellen, Z. vgl. Physiol. 70:105-143.

Kaib, M., 1974, Die Fleisch- und Blumenduftrezeptoren auf der Antenne der Schmeissfliege *Calliphora vicina*, *J. Comp. Physiol.* **95**:105-121.

Kaissling, K.-E., 1971, Insect olfaction. In *Handbook of Sensory Physiology*, vol. IV-1, edited by L. M. Beidler, pp. 351-431. Springer-Verlag, Berlin.

Kaissling, K.-E., 1974, Sensory transduction in insect olfactory receptors. In *Biochemistry of Sensory Functions*, edited by L. Jaenicke, pp. 243-273, Springer-Verlag, Berlin.

Kaissling, K.-E., and Priesner, E., 1970, Die Riechschwelle des Seidenspinners, *Naturwissenschaften* **57**:23-28.

Kaissling, K.-E., and Thorson, J., 1980, Insect olfactory sensilla: Structural, chemical and electrical aspects of the functional organization. In *Receptors for Neurotransmitters, Hormones and Pheromones in Insects*, edited by D. B. Sattelle, L. M. Hall, and J. G. Hildebrand, pp. 261-282, Elsevier/North-Holland, Amsterdam.

Kaissling, K.-E., Kasang, G., Bestmann, H. J., Stransky W., and Vostrowsky, O., 1978, A new pheromone of the silkworm moth *Bombyx mori;* Sensory pathway and behavioral effect, *Naturwissenschaften* **65**:382-384.

Keil, T., 1978, Die Makrochaeten auf dem Thorax von *Calliphora vicina* Robineau-Desvoidy (Calliphoridae, Diptera): Feinstruktur und Morphogenese eines epidermalen Insekten-Mechanoreceptors, *Zoomorphologie* **90**:151-180.

Keil, T., 1979, Rutheniumrot-Färbung sensorischer Einheiten der Insekten-Epidermis, *Eur. J. Cell Biol.* **19**:78-82.

Keil, T., 1982, Contacts of pore tubules and sensory dendrites in antennal chemosensilla of a silkmoth: Demonstration of a possible pathway for olfactory molecules, *Tissue Cell* **14**:451-462.

Keil, T., and Steinbrecht, R. A., 1983, Beziehungen zwischen Sinnes-, Hüll- und Gliazellen in epidermalen Mechano- und Chemorezeptoren von Insekten. Verh. Dtsch. zool. Ges. 76:294.

Keil, T., and Thurm, U., 1979, Die Verteilung von Membrankontakten und Diffusionsbarrieren in epidermalen Sinnesorganen von Insekten, Verh. Dtsch. zool. Ges. 72:285.

Küppers, J., 1974, Measurements on the ionic milieu of the receptor terminal in mechanoreceptive sensilla of insects. In *Abhandlungen Rheinisch-Westfälische Akademie der Wissenschaften* 53,

Symposium Mechanoreception, edited by J. Schwartzkopff, pp. 387-394, Westdeutscher Verlag, Opladen.

Lane, N. J., 1982, Insect intercellular junctions: Their structure and development. In *Insect Ultrastructure,* edited by R. C. King and H. Akai, vol. 1, pp. 402-433, Plenum Press, New York.

Lane, N. J., and Treherne, J., 1980, Functional organisation of arthropod neuroglia. In *Insect Biology in the Future,* edited by M. Locke and D. S. Smith, pp. 765-795, Academic Press, New York.

Laurent, T. C., 1970, Structure of hyaluronic acid. In *Chemistry and Molecular Biology of the Intercellular Matrix,* edited by E. A. Balazs, pp. 703-732, Academic Press, New York.

Lawrence, P. A., 1966, Development and determination of hairs and bristles in the milkweed bug, *Oncopeltus fasciatus* (Lygaeidae, Hemiptera), *J. Cell Sci.* 1:475-498.

Lewis, C. T., and Marshall, A. T., 1970, The ultrastructure of the sensory plaque organs of the antennae of the Chinese lantern fly, *Pyrops candelaria* L., (Homoptera, Fulgoridae), *Tissue Cell* 2:375-385.

Locke, M., 1965, Permeability of insect cuticle to water and lipids, *Science* 147:295-298.

Locke, M., 1976, The role of plasma membrane plaques and Golgi complex vesicles in cuticle deposition during the moult/intermoult cycle. In *The Insect Integument,* edited by H. R. Hepburn, pp. 237-258, Elsevier, Amsterdam.

Locke, M., and Collins, J. V., 1965, The structure and formation of protein granules in the fat body of an insect, *J. Cell Biol.* 26:857-884.

Locke, M., and Huie, P., 1979, Apolysis and the turnover of plasma membrane plaques during cuticle formation in an insect, *Tissue Cell* 11:277-291.

Locke, M., Krishnan, N., and McMahon, J. T., 1971, A routine method for obtaining high contrast without staining sections, *J. Cell Biol.* 50:540-544.

Loftus, R., 1978, Peripheral thermal receptors. In *Sensory Ecology,* edited by M. A. Ali, pp. 439-466, Plenum Press, New York.

McIver, S. B., 1974, Fine structure of antennal grooved pegs of the mosquito, *Aedes aegypti, Cell Tissue Res.* 153:327-337.

McIver, S. B., 1975, Structure of cuticular mechanoreceptors of arthropods, *Annu. Rev. Entomol.* 20:381-397.

Manery, J. F., 1966, Connective tissue electrolytes, *Fed. Proc.* 25:1799-1803.

Marshall, A. T., 1973, Vesicular structures in the dendrites of an insect olfactory receptor, *Tissue Cell* 5:233-241.

Matsumoto, D. E., and Farley, R. D., 1978, Comparison of the ultrastructure of stimulated and unstimulated mechanoreceptors in the taste hairs of the blowfly *Phaenicia serricata, Tissue Cell* 10:63-76.

Meinecke, C.-C., 1975, Riechsensillen und Systematik der Lamellicornia (Insecta, Coleoptera), *Zoomorphologie* 82:1-42.

Menco, B. P. M., and van der Wolk, F. M., 1982, Freeze-fracture characteristics of insect gustatory and olfactory sensilla. I. A comparison with vertebrate olfactory receptor cells with special reference to ciliary components, *Cell Tissue Res.* 223:1-27.

Moran, D. T., and Varela, F. G., 1971, Microtubules and sensory transduction, *Proc. Natl. Acad. Sci. USA* 68:757-760.

Moran, D. T., Chapman, K. M., and Ellis, R. A., 1971, The fine structure of cockroach campaniform sensilla, *J. Cell Biol.* 48:155-173.

Moran, D. T., Rowley, J. C., III, Zill, S. N., and Varela, F. G., 1976, The mechanism of sensory transduction in a mechanoreceptor: Functional stages in campaniform sensilla during the molting cycle, *J. Cell Biol.* 71:832-847.

Moulins, M., 1976, Ultrastructure of chordotonal organs. In *Structure and Function of Proprioceptors in the Invertebrates,* edited by P. J. Mill, pp. 387-426, Chapman & Hall, London.

Murphey, R. K., 1981, The structure and development of a somatotopic map in crickets: The cercal afferent projection, *Dev. Biol.* 88:236-246.

Neville, A. C., 1975, *Biology of the Arthropod Cuticle,* Springer-Verlag, Berlin.

Nicklaus, R., Lundquist, P. G., and Wersäll, J., 1967, Die Übertragung des Reizes auf den distalen Fortsatz der Sinneszelle bei den Fadenhaaren von *Periplaneta americana, Verh. Dtsch. zool. Ges.* 61:578-584.

Pease, D. C., 1966, The preservation of unfixed cytological detail by dehydration with "inert" agents, *J. Ultrastruct Res.* **14**:356-378.

Peters, W., 1965, Die Sinnesorgane an den Labellen von *Calliphora erythrocephala* MG. (Diptera), *Z. Morphol. Oekol. Tiere* **55**:259-320.

Phillips, C. E., and Vande Berg, J. S., 1976, Directional flow of sensillum liquor in blowfly (*Phormia regina*) labellar chemoreceptors, *J. Insect Physiol.* **22**:425-429.

Ribbert, D., 1967, Die Polytänchromosomen der Borstenbildungszellen von *Calliphora erythrocephala* unter besonderer Berücksichtigung der geschlechtsgebundenen Strukturheterozygotie und des Puffmusters während der Metamorphose, *Chromosoma* **21**:296-344.

Rice, M. J., Galun, R., and Finlayson, L. H., 1973, Mechanotransduction in insect neurons, *Nature New Biol.* **241**:286-288.

Richards, A. G., 1952, Studies on arthropod cuticle. VIII. The antennal cuticle of honeybees, with particular reference to the sense plates, *Biol. Bull.* **103**:201-225.

Ritter, F. J. (ed.), 1979, *Chemical Ecology: Odour Communication in Animals*, Elsevier/North-Holland, Amsterdam.

Sanes, J. R., and Hildebrand, J. G., 1976, Origin and morphogenesis of sensory neurons in an insect antenna, *Dev. Biol.* **51**:300-319.

Sanes, J. R., Hildebrand, J. G., and Prescott, D. J., 1976, Differentiation of insect sensory neurons in the absence of their normal synaptic targets, *Dev. Biol.* **52**:121-127.

Sass, H., 1978, Olfactory receptors on the antenna of *Periplaneta*: Response constellations that encode food odors, *J. Comp. Physiol.* **128**:227-233.

Schmidt, K., and Gnatzy, W., 1971, Die Feinstruktur der Sinneshaare auf den Cerci von *Gryllus bimaculatus* Deg. (Saltatoria, Gryllidae). II. Die Häutung der Faden- und Keulenhaare. *Z. Zellforsch. mikrosk. Anat.* **122**:210-225.

Schmidt, K., and Kuhbandner, B., 1983, Ontogeny of the sensilla placodea on the antennae of *Aulacus striatus* Jurine (Hymenoptera: Aulacidae), *Int. J. Insect Morphol. Embryol.* **12**:43-57.

Schneider, D., 1980, Pheromone von Insekten: Produktion-Reception-Inaktivierung, *Nova Acta Leopold. N.F.* **51**:249-278.

Schneider, D., and Steinbrecht, R. A., 1968, Checklist of insect olfactory sensilla, *Symp. Zool. Soc. London* **23**:279-297

Silverman, L., and Glick, D., 1969, The reactivity and staining of tissue proteins with phosphotungstic acid, *J. Cell Biol.* **40**:761-767.

Slifer, E. H., 1970, The structure of arthropod chemoreceptors, *Annu. Rev. Entomol.* **15**:121-142.

Slifer, E. H., and Sekhon, S. S., 1960, The fine structure of the plate organs on the antenna of the honey bee, *Apis mellifera* Linnaeus, *Exp. Cell Res.* **19**:410-414.

Slifer, E. H., and Sekhon, S. S., 1963, Sense organs on the antennal flagellum of the small milkweed bug, *Lygaeus kalmii* Stal (Hemiptera, Lygaeidae), *J. Morphol.* **112**:165-193.

Slifer, E. H., and Sekhon, S. S., 1964a, Fine structure of the sense organs on the antennal flagellum of a flesh fly, *Sarcophaga argyrostoma* R.-D. (Diptera, Sarcophagidae), *J. Morphol.* **114**:185-208.

Slifer, E. H., and Sekhon, S. S., 1964b, The dendrites of the thin-walled olfactory pegs of the grasshopper (Orthoptera, Acrididae), *J. Morphol.* **114**:393-410.

Slifer, E. H., Prestage, J. J., and Beams, H. W., 1957, The fine structure of the long basiconic sensory pegs of the grasshopper (Orthoptera, Acrididae) with special reference to those on the antenna, *J. Morphol.* **101**:359-397.

Slifer, E. H., Prestage, J. J., and Beams, H. W., 1959, The chemoreceptors and other sense organs on the antennal flagellum of the grasshopper (Orthoptera, Acrididae), *J. Morphol.* **105**:145-191.

Smith, D. S., 1969, The fine structure of haltere sensilla in the blowfly, *Calliphora erythrocephala* (Meig.), with scanning electron microscopic observations on the haltere surface, *Tissue Cell* **1**:443-484.

Steinbrecht, R. A., 1969, Comparative morphology of olfactory receptors. In *Olfaction and Taste*, Vol. III, edited by C. Pfaffmann, pp. 3-21, Rockefeller University Press, New York.

Steinbrecht, R. A., 1970, Zur Morphometrie der Antenne des Seidenspinners, *Bombyx mori* L.: Zahl und Verteilung der Riechsensillen (Insecta Lepidoptera), *Z. Morphol. Tiere* **68**:93-126.

Steinbrecht, R.A., 1973, Der Feinbau olfaktorischer Sensillen des Seidenspinners (Insecta,

Lepidoptera): Rezeptorfortsätze und reizleitender Apparat, *Z. Zellforsch. mikrosk. Anat.* **139**:533–565.

Steinbrecht, R. A., 1980, Cryofixation without cryoprotectants: Freeze substitution and freeze etching of an insect olfactory receptor, *Tissue Cell* **12**:73–100.

Steinbrecht, R. A., 1984, Arthropoda: Chemo-, thermo-, and hygroreceptors. In *Biology of the Integument*, edited by J. Bereiter-Hahn, A. G. Matoltsy, and K. S. Richards, Vol. 1, pp. 523–553, Springer-Verlag, Berlin.

Steinbrecht, R. A., and Gnatzy, W., 1984, Pheromone receptors of *Bombyx mori* and *Antheraea pernyi*. I. Reconstruction of the cellular organization of the sensilla trichodea, *Cell Tissue Res.* **235**:25–34.

Steinbrecht, R. A., and Kasang, G., 1972, Capture and conveyance of odour molecules in an insect olfactory receptor. In *Olfaction and Taste*, vol. IV, edited by D. Schneider, pp. 193–199, Wissenschaftliche Verlagsgesellschaft, Stuttgart.

Steinbrecht, R. A., and Müller, B., 1971, On the stimulus conducting structures in insect olfactory receptors, *Z. Zellforsch. mikrosk. Anat.* **117**:570–575.

Steinbrecht, R. A., and Müller, B., 1976, Fine structure of the antennal receptors of the bed bug, *Cimex lectularius* L., *Tissue Cell* **8**:615–636.

Steinbrecht, R. A., and Schneider, D., 1980, Pheromone communication in moths: Sensory physiology and behaviour. In *Insect Biology in the Future*, edited by M. Locke and D.S. Smith, pp. 685–703, Academic Press, New York.

Steinbrecht, R. A., and Zierold, K., 1982, Cryo-embedding of small frozen specimens for cryo-ultramicrotomy, *Proc. 10th Int. Congr. Electron Microsc. Hamburg* **3**:183–184.

Stepper, J., Becker, C., and Schmidt, K., 1983, Feinbau und Ontogenese der Porenplatten auf den Antennen von *Pimpla turionellae* (Hymenoptera, Ichneumonidae), *Zoomorphology* **102**:11–32.

Theiss, J., 1979, Mechanoreceptive bristles on the head of the blowfly: Mechanics and electrophysiology of the macrochaetae, *J. Comp. Physiol.* **132**:55–68.

Thurm, U., 1964, Mechanoreceptors in the cuticle of the honey bee: Fine structure and stimulus mechanism, *Science* **145**:1063–1065.

Thurm, U., 1965, An insect mechanoreceptor. Part I. Fine structure and adequate stimulus, *Cold Spring Harbor Symp. Quant. Biol.* **30**:75–82.

Thurm, U., 1970, Untersuchungen zur funktionellen Organization sensorischer Zellverbände, *Verh. Dtsch. zool. Ges.* **64**:79–88.

Thurm, U., 1974, Basics of the generation of receptor potentials in epidermal mechanoreceptors of insects. In *Abhandlungen Rheinisch-Westfälische Akademie der Wissenschaften 53, Symposium Mechanoreception*, edited by J. Schwartzkopff, pp. 355–385, Westdeutscher Verlag, Opladen.

Thurm, U., 1981, Mechano-electric transduction, *Biophys. Struct. Mech.* **7**:245–246.

Thurm, U., 1982, Mechano-elektrische Transduktion. In *Biophysik*, edited by W. Hoppe, W. Lohmann, H. Markl, and H. Ziegler, pp. 691–696, Springer-Verlag, Berlin.

Thurm, U., and Küppers, J., 1980, Epithelial physiology of insect sensilla. In *Insect Biology in the Future*, edited by M. Locke and D. S. Smith, pp. 735–763, Academic Press, New York.

Thurm, U., and Wessel, G., 1979, Metabolism-dependent transepithelial potential differences at epidermal receptors of arthropods. I. Comparative data, *J. Comp. Physiol.* **134**:119–130.

Thurm, U., Erler, G., Gödde, J., Kastrup, H., Keil, T., Völker, W., and Vohwinkel, B., 1983, Cilia specialized for mechanoreception, *J. Submicrosc. Cytol.* **15**:151–155.

Tobias, M., and Murphey, R. K., 1979, The response of cercal receptors and identified interneurons in the cricket (*Acheta domesticus*) to airstreams, *J. Comp. Physiol.* **129**:51–59.

Toh, Y., 1977, Fine structure of antennal sense organs of the male cockroach, *Periplaneta americana*, *J. Ultrastruct. Res.* **60**:373–394.

Trepte, H.-H., 1976, Das Puffmuster der Borstenapparat-Chromosomen von *Sarcophaga barbata*, *Chromosoma* **55**:137–164.

Vogt, R. G., and Riddiford, L. M., 1981, Pheromone binding and inactivation by moth antennae, *Nature (London)* **293**:161–163.

Völker, W., 1982, Lebendbeobachtungen an kutikulären Reizübertragungsstrukturen campaniformer Sensillen und Hochauflösungs-Elektronenmikroskopie der reizaufnehmenden Sinneszellregion, Dissertation, Zoologisches Institut, Universität Münster.

Waldow, U., 1970, Elektrophysiologische Untersuchungen an Feuchte-, Trocken- und Kälterezeptoren auf der Antenne der Wanderheuschrecke *Locusta*, *Z. vgl. Physiol.* **69:**249-283.

Wigglesworth, V. B., 1953, The origin of sensory neurones in an insect, *Rhodnius prolixus* (Hemiptera), *Q. J. Microsc. Sci.* **94:**93-112.

Winter, W. T., and Arnott, S., 1977, Hyaluronic acid: The role of divalent cations in conformation and packing, *J. Mol. Biol.* **117:**761-784.

Wright, R. H., 1982, *The Sense of Smell*, CRC Press, Boca Raton, Florida.

Yokohari, F., 1978, Hygroreceptor mechanism in the antenna of the cockroach *Periplaneta*, *J. Comp. Physiol.* **124:**53-60.

Zacharuk, R. Y., 1980, Ultrastructure and function of insect chemosensilla, *Annu. Rev. Entomol.* **25:**27-47.

14

The Cytopathology of Baculovirus Infections in Insects

YOSHINORI TANADA AND ROBERTA T. HESS

1. Introduction

Baculoviruses are the most widely investigated of the viruses which infect insects. More than 320 of them have been isolated, mainly from insects and a few other arthropods such as mites and crustaceans. These viruses contain closed, circular, double-stranded DNA enclosed within rod-shaped, enveloped nucleocapsids. They are placed in the family Baculoviridae (Matthews, 1982) which is separated into two recognized subgroups: A, nuclear polyhedrosis viruses, and B, granulosis viruses. Two other subgroups have been proposed: C, nonoccluded, rod-shaped nuclear viruses, and D, nonoccluded nuclear viruses with a polydisperse DNA genome. In the nuclear polyhedrosis viruses (NPVs), numerous enveloped virions are occluded in occlusion bodies, called *polyhedra*, in the nuclei of infected cells (Figure 1), whereas in the granulosis viruses (GVs), the enveloped virions are occluded singly or rarely two or more in occlusion bodies called *capsules* which develop in the nucleus or cytoplasm (Figure 2). In subgroups C and D, the virions are not occluded in occlusion bodies.

All baculoviruses are presently known to infect only arthropods, and none has been found to infect vertebrates. Many of them have host ranges restricted to one or a few species, and this characteristic is a primary reason for proposing them as candidates for the microbial control of insect pests. Most apparent infections occur in the larva, but they also develop in the pupa and adult. Some of them cause systemic infections in insects and infect most of the major tissues, while others are confined to a single or only a few host tissues (Figures 3, 4).

YOSHINORI TANADA AND ROBERTA T. HESS • Department of Entomological Sciences, University of California, Berkeley, California 94720, USA.

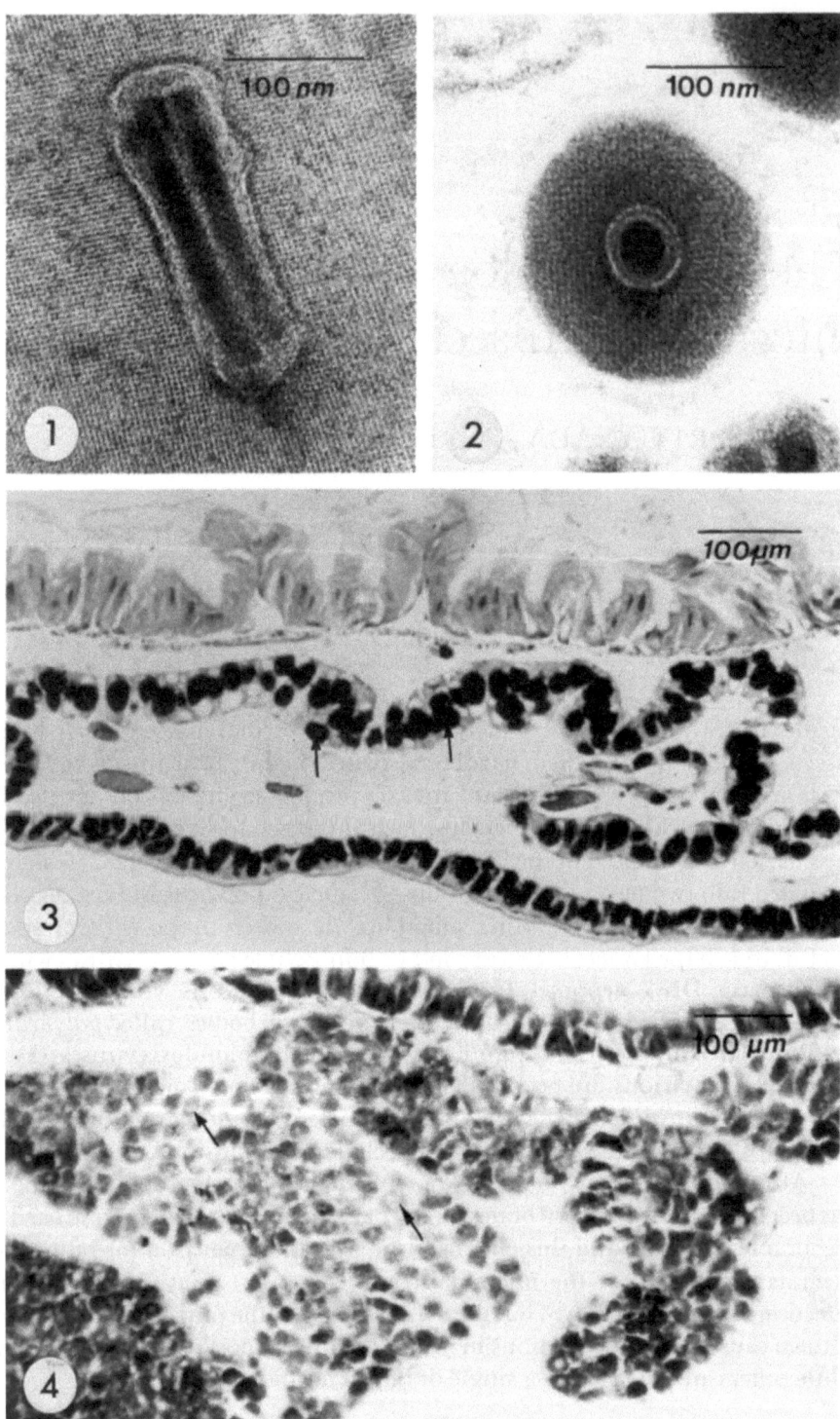

2. Mode of Virus Infection

Since the pathways of baculovirus infection and virogenesis cause physicochemical and cytological pathologies, they will be described first. These aspects have been treated in comprehensive reviews (Bergold, 1958; Aizawa, 1963; Paschke and Summers, 1975; Smith, 1976; Summers, 1977; Tinsley and Harrap, 1978; Granados, 1980; Faulkner, 1981). The pathways of infection by NPVs and GVs are, as a whole, similar, and they shall be considered together. Insects generally become infected with baculoviruses by ingesting the proteinaceous occlusion bodies (polyhedra and capsules). The occlusion bodies are dissolved in the midgut lumen through the action of the alkaline digestive juice, and the enveloped virions are liberated (Figure 5). Virus entry occurs in the midgut, but of the two major cell types, only the columnar epithelial cells are invaded by the enveloped nucleocapsids. Harrap and Robertson (1968) and Harrap (1970) were the first to observe that an NPV initially infects the midgut epithelium. The envelopes of the virions fuse with the cell membranes of the microvilli of columnar midgut cells (Kawanishi *et al.*, 1972; Tanada *et al.*, 1975; Granados, 1978) (Figure 6). The attachment appears to be dependent on the phospholipid and the ionic charge of the virus envelope (Yamamoto and Tanada, 1977, 1978a). A synergistic factor, a lipoprotein, that occurs in the capsule matrix of a GV of the armyworm, *Pseudaletia unipuncta*, enhances the attachment of the enveloped virion to the host cell membrane (Tanada *et al.*, 1975, 1980), and the phospholipid in the synergistic factor molecule is involved in the enhancement (Yamamoto and Tanada, 1978b). The site of attachment of the synergistic factor is apparently the virus receptor site.

The nucleocapsid enters the microvillus through an opening at the junction of the fusion (Summers, 1969, 1971; Harrap, 1970). The adsorption, fusion, and virus entry take place within 0.25 to 4 hr postfeeding (Kawanishi *et al.*, 1972; Granados, 1978; Granados and Lawler, 1981). Nucleocapsid movement in the microvillus into the cytoplasm appears to follow a gradient (Figure 7). Microtubules may be involved in the vectorial movement of the nucleocapsid in the cytoplasm to the nucleus (Granados, 1978; Granados and Lawler, 1981). The nucleocapsid becomes attached to a nucleopore within 1 hr and the nucleic acid (DNA) is discharged into the nucleus (Summers, 1969, 1971;

Figures 1 and 2. Paracrystalline proteinaceous occlusion bodies of baculoviruses.

Figure 1. A portion of polyhedral matrix of an NPV of *Autographa californica* propagated in the larva of *Spodoptera exigua* showing in a longitudinal section a bundle of two nucleocapsids enclosed in an envelope.

Figure 2. Cross-section of an oval-cylindrical capsule of a GV of the codling moth, *Cydia pomonella*, showing the occluded enveloped nucleocapsid.

Figures 3 and 4. Longitudinal paraffin sections of larvae of the armyworm, *Pseudaletia unipuncta*.

Figure 3. An NPV infecting major larval tissues. Hypertrophied infected nuclei are filled with polyhedra stained black with iron hematoxylin (arrows).

Figure 4. A GV infecting apparently only the fat body. Some nuclei are greatly hypertrophied and others are disrupted with no obvious boundary between the nucleus and the cytoplasm (arrows).

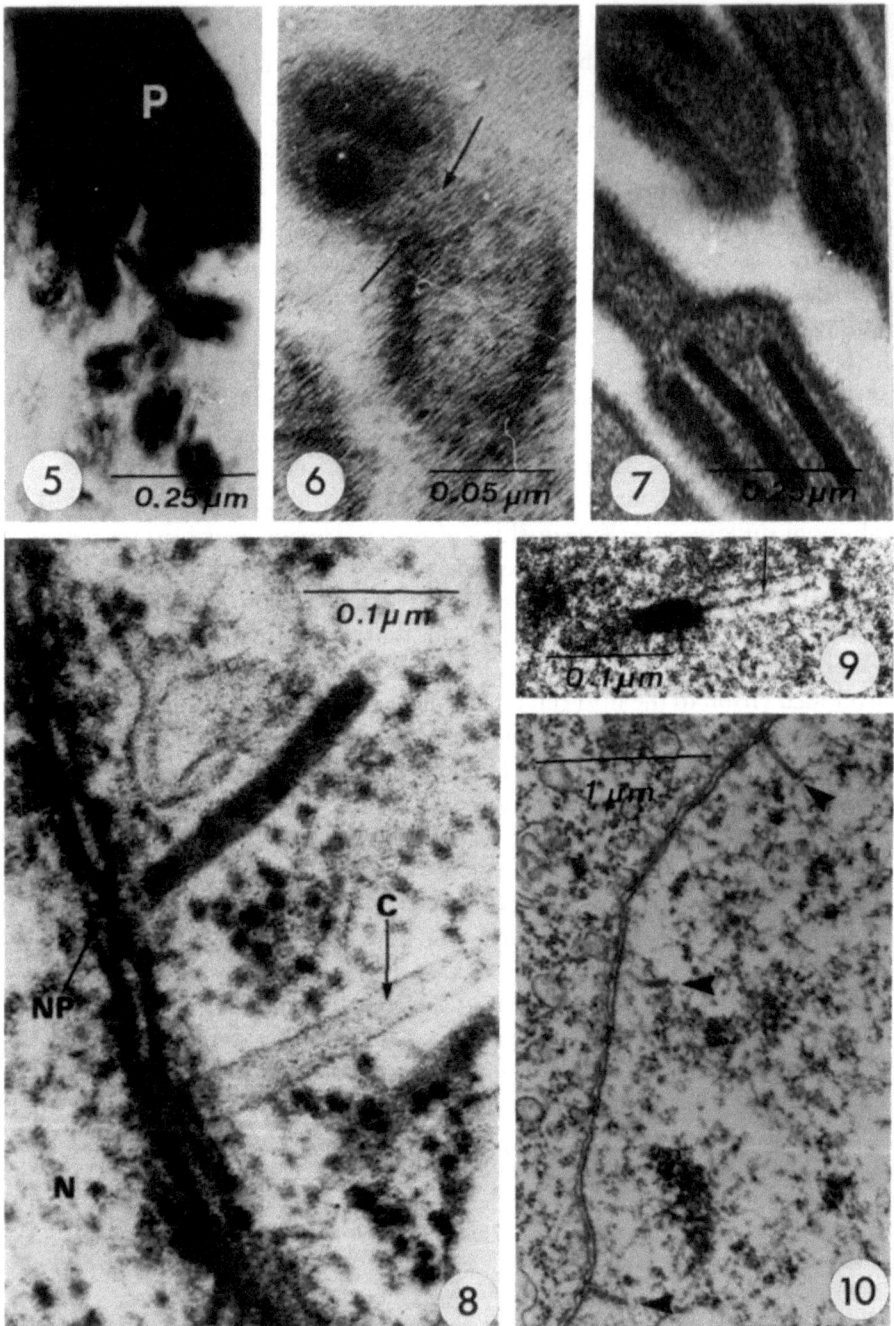

Figures 5-7. Midguts of armyworm larvae infected with an NPV. Reproduced from Tanada *et al.* (1975).

Figure 5. A partially dissolved polyhedra (P) in the midgut lumen showing the liberated enveloped nucleocapsids.

Figure 6. Attachment of an enveloped virus bundle to a microvillus of a columnar cell (arrows). Contents within the virus envelope and the cytoplasm appear to be confluent.

Kawanishi et al., 1972; Raghow and Grace, 1974; Tanada and Hess, 1976) (Figure 8), or the nucleocapsid enters through a nucleopore and is uncoated in the nucleoplasm (Hirumi et al. 1975; Granados, 1978, 1980) (Figure 9). Granados (1980) has speculated that the site of nucleocapsid uncoating separates the GVs from the NPVs; i.e., uncoating occurs at the nucleopore complex for GVs within the nucleoplasm for NPVs, and by both methods for the subgroup D baculovirus. However, Walker et al., (1982) have found in some cells the uncoating of the GV nucleocapsids in the nucleus (Figure 10), and this differs from that proposed by Summers (1971) and Granados (1980). In the codling moth, *Cydia (Laspeyresia) pomonella,* the GV nucleocapsids uncoat in the nucleoplasm (unpublished data).

The uncoating of the nucleocapsid initiates the eclipse period of virus replication, and this period is terminated with the appearance of the virogenic stroma to produce the nucleocapsids (Xeros, 1956; Huger, 1960; Huger and Krieg, 1961) (Figure 11). The nucleocapsids are enveloped either from membranes produced by *de novo* synthesis in the nucleus or by budding through cell membranes. Those that are enveloped in the nucleus are usually occluded in occlusion bodies.

At an early stage in the formation of nucleocapsids, some of them pass out through the nucleopores or openings in the nuclear envelope (Kawamoto et al., 1976; Adams et al., 1977; Hess and Falcon, 1977), while still others bud through the nuclear envelope (Injac et al., 1971; MacKinnon et al., 1974; Kawamoto et al., 1976, 1977; Tanada and Hess, 1976; Adams et al., 1977; Granados and Lawler, 1981). Those that pass through the nuclear envelope acquire two loose-fitting membranes and enter the cytoplasm (Hirumi et al., 1975; Kawamoto et al., 1976, 1977; Knudson and Harrap, 1976). Apparently, the nucleocapsids lose their envelopes in the cytoplasm (Hirumi et al., 1975; Granados and Lawler, 1981). The naked nucleocapsids acquire envelopes while passing through the basal plasma membrane to form bipolar-enveloped virions (Tanada and Hess, 1976; Adams et al., 1977; Hess and Falcon, 1977; Kawamoto et al., 1977; Granados and Lawler, 1981) (Figure 12). Enveloped and unenveloped nucleocapsids may occur in intercellular spaces (Adams and Wilcox, 1968; Tanada and Leutenegger, 1970; Injac et al., 1971; Hunter et al., 1975) (Figure 13). Recently, Granados and Lawler (1981) have observed that some virions of the NPV of *Autographa californica* enter the midgut cells, pass directly through the cytoplasm, and bud through the plasma membrane into the hemocoel as early as $\frac{1}{2}$ hr after infection.

Figure 7. Nucleocapsids within the microvilli.
Figure 8. A portion of a larval midgut nucleus of *Trichoplusia ni* showing nucleocapsids of a GV attached to nucleopores (NP) where DNA uncoating takes place. N, nucleus; C, capid. Reproduced from Summers (1971).
Figure 9. Nucleocapsids (arrow) of the NPV of *Heliothis zea* within the nucleoplasm of a columnar midgut cell shortly after infection. DNA uncoating takes place in the nucleus. Reproduced from Granados (1980).
Figure 10. Nucleocapsidlike structures (arrow heads). 2 days after treatment, found inside the nucleus of a fat body cell of *Spodoptera frugiperda* infected with a GV. Reproduced from Walker et al. (1982).

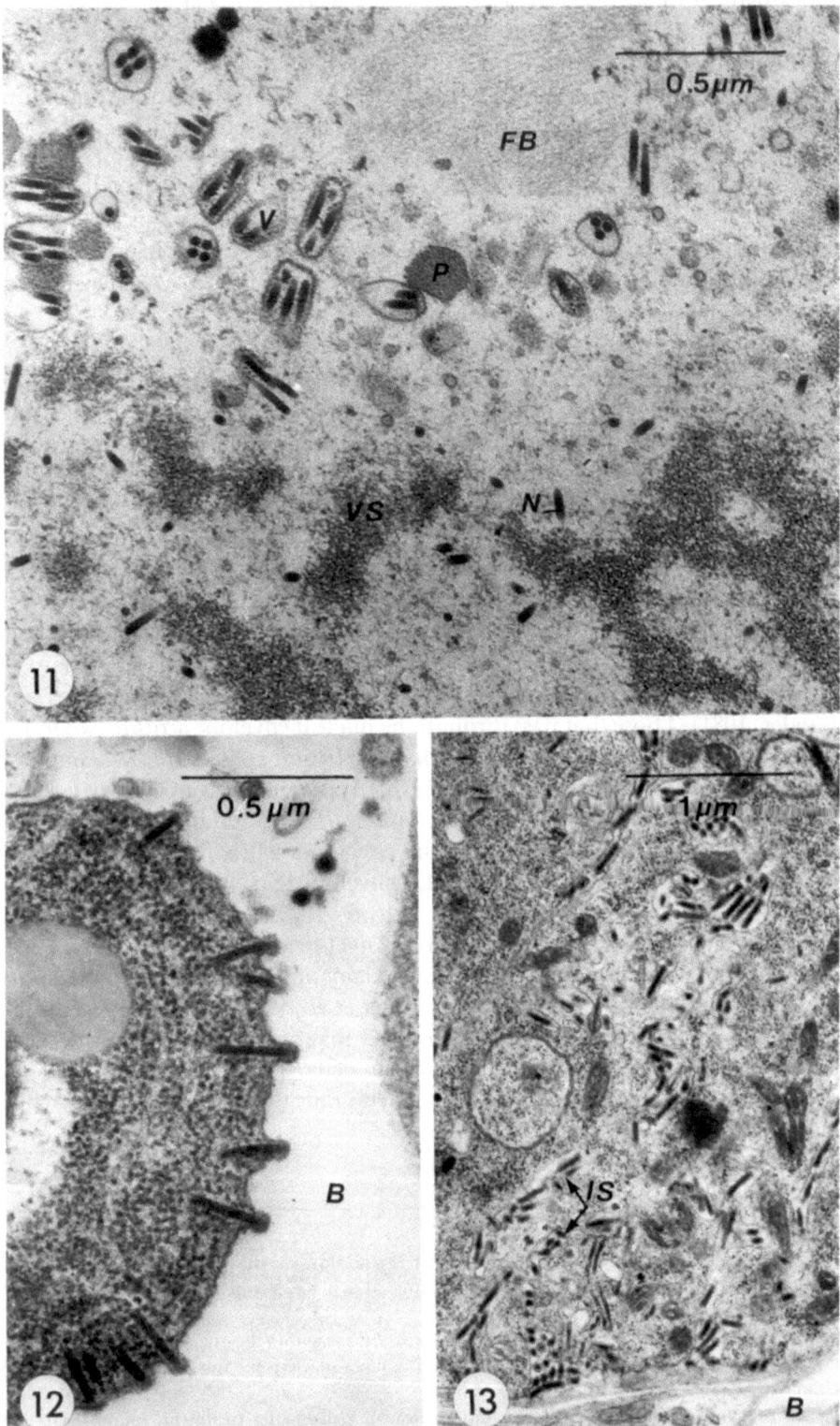

A few nucleocapsids become enveloped in the nucleus and are occluded in occlusion bodies, but occlusion bodies are infrequently formed early in the infections of the midguts of lepidopterous insects, with the possible exception of the NPV of the noctuid, *Diparopsis watersi*, in which numerous polyhedra are produced in the midgut epithelial cells (Croizier et al., 1980). This is a major difference between infections in the midgut cells from those in the hemocoelic tissues. Occlusion bodies occur, however, in abundance in baculoviruses (NPVs) that infect mainly or only the cells of the midgut epithelia of sawflies (Hymenoptera) and sciarid flies (Diptera).

Unenveloped and enveloped nucleocapsids occur in the hemolymph of virus-infected larvae. The general consensus is that the enveloped nucleocapsids that have budded through the basement plasma membrane are responsible for infecting the susceptible host cells in the hemocoel. The envelopes of the nucleocapsids have spikes or peplomers (Adams et al., 1977; Hess and Falcon, 1977; Kawamoto et al., 1977; Granados and Lawler, 1981) (Figure 14).

The enveloped virions probably invade intrahemocoelic cells by the attachment of the ends of the envelopes containing the peplomers to the cell membrane and enter the cell by viropexis (pinocytosis) (Adams et al., 1977; Hess and Falcon, 1977; Kawamoto et al., 1977) (Figure 15). Recently, we have observed the lateral attachment of the enveloped virion with the cell membrane and the subsequent fusion and invasion of the nucleocapsid into the cytoplasm (unpublished data). The nucleocapsids migrate to the nucleus and uncoat either at the nucleopore or in the nucleoplasm and virogenesis proceeds as described for the midgut cell, except that many of the nucleocapsids become enveloped by membranes within the nucleus and are subsequently occluded in occlusion bodies. In the case of NPVs, numerous singly enveloped or multiply enveloped nucleocapsids are occluded in a polyhedra (Bergold, 1958) but with GVs, usually one or rarely several enveloped nucleocapsids are occluded in a capsule (Hughes, 1952; Huger, 1963).

The emergence of the nucleocapsids from the nucleus into the cytoplasm and out through the plasma membrane of hemocoelic cells is the same as in the midgut cell (Kawamoto et al., 1976, 1977; Adams et al., 1977; Hess and Falcon, 1977). In general, the NPV infection in hemocoelic cells is much more dramatic than that of midgut cells.

The *in vitro* replication of NPVs is similar to that of hemocoelic cells (Granados, 1980). The enveloped virions enter the cell by viropexis (Hirumi et al., 1975; Adams et al., 1977; Granados et al., 1981), and the nucleocapsids penetrate and uncoat at the nucleopores or in the nucleoplasm (Raghow and Grace, 1974; Hirumi et al., 1975; Granados and Lawler, 1981), and virogenesis takes place in the nucleus. The nucleocapsids emerge from the cell by budding

---

Figure 11. Virogenic stroma (VS) in the nucleus of a midgut columnar cell of an armyworm larva infected with an NPV. FB, fibrous body; N, nucleocapsid; P, developing polyhedra; V, enveloped nucleocapsids.

Figure 12. Budding of nucleocapsids through the plasma membrane to invade the hemocoel (B) of a codling moth larva infected with a GV.

Figure 13. Nucleocapsids in the intercellular space (IS) of midgut cells of a codling moth larva. Nucleocapsids are presumed to move basally and enter the hemocoel (B).

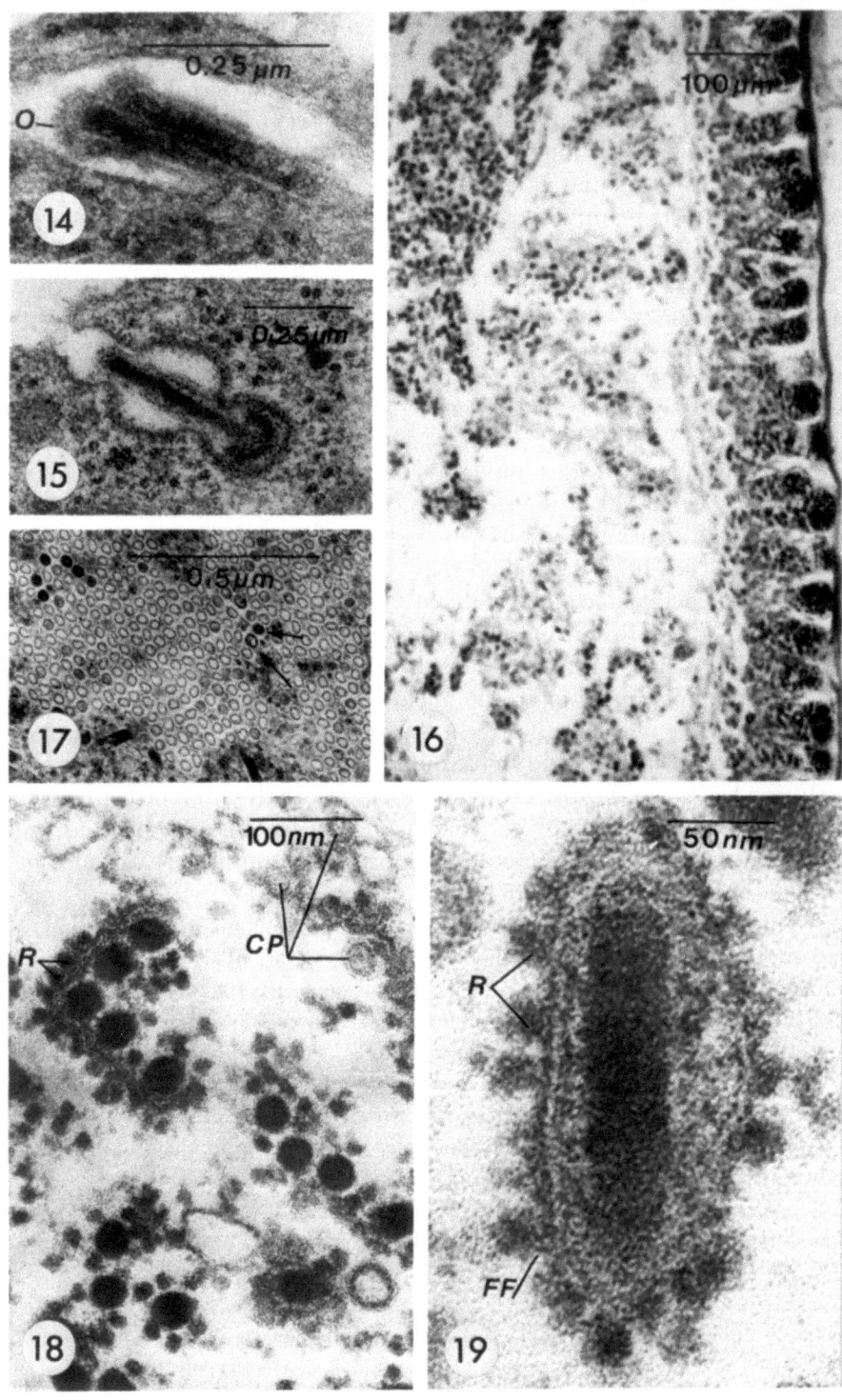

through the plasma membrane and acquire an envelope with peplomers (MacKinnon et al., 1974; Hirumi et al., 1975; Knudson and Harrap, 1976; Summers and Volkman, 1976). These infectious virions are surrounded by a fragile envelope (Henderson et al., 1974; Dougherty et al., 1975; Summers and Volkman, 1976; Hess and Falcon, 1977).

3. Gross Pathology of Nuclear Polyhedroses of Lepidoptera

The occluded NPVs infect mainly insects in the order Lepidoptera (all major families), to a lesser extent in the order Hymenoptera (sawfly families), and a few species in the orders Coleoptera, Neuroptera, Diptera, and Trichoptera (Martignoni and Iwai, 1981). Infections are most pronounced in the larval stage, but also occur in the pupal and adult stages. Signs, symptoms, and pathologies in these diverse groups of insects vary, especially with the types of tissues infected and with the stages of the insect.

In most lepidopterous larvae, there are no external signs or symptoms or changes in appetite until approximately two-thirds of the period of lethal infection has passed (Aizawa, 1963). About 4 to 7 days after infection, the integument becomes swollen and gradually changes color and luster, i.e., an increase in opaqueness, milkiness, and glossiness. The normally clear hemolymph turns cloudy and milky. The larva is increasingly more lethargic but may continue to feed up to a few days before death. Just prior to death or shortly after death, the integument, if the hypodermal cells are infected, becomes fragile and is easily torn when handled. The larva is in a wilted condition, typical of most NPVs. The body contents are a fluid mass. Dead larvae generally die attached to elevated branches or locations by means of their abdominal and caudal prolegs.

Most NPVs cause a systemic infection in lepidopterous larvae and infect the major tissues, such as the fat body, trachea, hypodermis, and blood (Figure 3). Cells of other tissues, e.g. glands, nerves, and muscles, are also susceptible in some cases. Some NPVs are tissue specific and infect only one or a few tissues of

---

Figure 14. Nucleocapsids of an NPV of *Estigmene acrea* acquiring an envelope by budding through the cell membrane of a larval midgut cell. Note the presence of glycoprotein spikes of peplomer (O) at one end of the envelope.

Figure 15. Peplomer-end of an enveloped nucleocapsid associated with the cell plasma membrane of *Euproctis subflava*. Peplomers attach to the plasma membrane, and the nucleocapsid enters by viropexis. Reproduced from Kawamoto et al. (1977).

Figure 16. Longitudinal paraffin section of the adult midgut of *Oryctes rhinoceros* infected with a baculovirus. Note discharged cells in the lumen. Extensive proliferation of midgut cells suggests a malignant carcinoma. Photograph taken from a slide provided by Dr. K. Marschall.

Figure 17. Tubular profiles that are considered to be capsids present in the nucleus of a midgut cell of an *Estigmene acraea* larva infected with the NPV of *Autographa californica*. Stages suggestive of capsid maturation are shown (arrows). Reproduced from Hess and Falcon (1981).

Figure 18. Granules (R) associated with nucleocapsids in a midgut nucleus of an armyworm larva infected with an NPV. Granules are frequently attached to fibrous filaments. CP, capsid.

Figure 19. Granules (R) and fibrous filaments (FF) associated with a developing envelope of a nucleocapsid. Figures 18 and 19 reproduced from Tanada and Hess (1976).

the lepidopterous larva. This is the case with the hypertrophy strain (HNPV) that mainly infects the trachea and hypodermis of the larva of the armyworm, *P. unipuncta*, as compared to the typical (TNPV) strain that infects most tissues (Tanada *et al.*, 1969). The intensity of infection in the various tissues may alter the typical syndrome of NPV. For example, Pavan *et al.* (1981) have reported two variants of an NPV of the velvet bean caterpillar, *Anticarsia gemmatalis*, one of which produces the wilt symptom while the other does not. The nonwilt variant produces few polyhedra in the majority of the infected mesodermal and epidermal cells. In another example, two strains (SV and BV) of the NPV of the Douglas-fir tussock moth, *Hemerocampa pseudotsugata*, differ in the fragility of the integument of the infected larva (Hughes and Addison, 1970). The integument of a larva infected with the BV strain may rupture before the larva succumbs to the virus.

In general, NPV-infected cells exhibit hypertrophy and no, or very little, hyperplasia, Petre and Ploaie (1969), however, report an unusual case of hyperplasia in the gypsy moth larva, *Lymantria dispar*, infected with an NPV. The simple epithelia of the trachea and hypodermis become stratified when infected with the virus. Petre and Ploaie (1969) point out the similarity of this hyperplasia to malignancy which produces tumors.

4. Gross Pathology of Nuclear Polyhedroses of Hymenoptera

The sawflies are susceptible to occluded NPVs (subgroup A). The initial sign in the infected sawfly larva is a faint yellow discoloration particularly on the third to fifth abdominal segments (Bergold, 1958; Aizawa, 1963). The larva gradually becomes inactive and loses its appetite. Often, it excretes a dark-brown fluid from the anus, and vomits a milky-white fluid. At death, the larva is flaccid and the skin becomes fragile and, when broken, expels the liquid body contents. The number of frass pellets excreted by the larva is reduced 48 to 60 hr after feeding on the virus (Bird and Whalen, 1953).

Sawfly NPVs are tissue specific and infect only the midgut epithelium (Bird, 1952). The infection in the midgut causes the regenerative nidal cells to proliferate and produce a tumor which often occurs in the hemocoel (Bird, 1949). Neilson and Elgee (1968), however, have shown that the tumor is the result of hemocyte encapsulation of virus-infected cells. Such tumors also occur in uninfected insects.

5. Gross Pathology of Nuclear Polyhedroses of Diptera

Several dipterous insects, e.g., mosquito, crane fly, and sciarid fly, are susceptible to occluded NPVs. Several species of mosquitoes in different genera have been found susceptible to NPVs. The nuclear polyhedrosis in the mosquito larva is similar to that of the occluded NPV of Hymenoptera in that the infection is confined to the digestive tract (Clark *et al.*, 1969; Federici and Lowe, 1972; Hall and Fish, 1974). The infected larva is sluggish and eventually stops feeding.

The pronounced symptom is the milky-white, hypertrophied cells of the digestive tract which are visible through the integument. The larva dies shortly after the rupture of the infected cells.

In the crane fly, *Tipula paludosa*, the NPV produces very unusually shaped occlusion bodies. The infected larva becomes increasingly paler as the infection progresses until it appears chalky white, as contrasted to the earthy color in the uninfected larva (Rennie, 1923; Aizawa, 1963). A milky-white liquid exudes on rupturing the integument, and many refractive crescent-shaped occlusion bodies are present in the fluid. The virus infects the blood and fat cells.

The sciarid fly, *Rhynchosciara angelae*, which has giant polytene chromosomes, is susceptible to an NPV (Diaz and Pavan, 1965). The virus causes a delay in larval development, and the larvae are smaller and paler than uninfected larvae. Some infected larvae may complete the larval stage (Pavan *et al.*, 1971). The virus infects the cells of a pair of long glands (caeca) attached to the midgut and also cells in the posterior part of the midgut. The cells hypertrophy and form small tumors.

6. Gross Pathology of Granuloses

The GVs are known to infect only members of the order Lepidoptera and are more species specific than the NPVs (Ignoffo, 1968). The gross pathologies and insect stages infected by GVs are similar to those of NPVs, but differences occur depending on the types of tissues infected (Figure 4). The first indication of infection in the larva is the loss of appetite and a progressive change, especially on the ventral side, from its usual color to a pale whitish or milky-yellow appearance (Huger, 1963). In some cases, the larva becomes mottled at an advanced stage of infection and may later have a brownish discoloration. Some GV-infected larvae, especially those with a systemic infection, are smaller at death as compared to uninfected larvae, but others in which the infection is limited mainly to the adipose tissues, may become larger and live longer than uninfected larvae. The change in color is usually accompanied by progressive weakening, sluggishness, and flaccidity. Diarrhea occurs in some cases. The blood, at an advanced stage of infection, is often milky white and turbid, owing to the presence of large numbers of capsules derived from the disintegrated, infected tissues.

Some GVs infect mainly the fat bodies, but others cause systemic infection similar to that of the typical NPVs with infections in the hypodermis, trachea, fat body, blood, etc. In a systemic infection, the larval integument, just prior to or at larval death, is fragile, as in nuclear polyhedroses, and the dead larva is wilted. When GV-infected cells are observed with the compound light microscope, they appear yellowish to light brownish in color. This is unlike the NPV-infected cells, whose hypertrophied nuclei are filled with polyhedra and appear refractive and dark.

The GV of the Western grape leaf skeletonizer, *Harrisinia brillians*, infects principally the midgut epithelium (Smith *et al.*, 1956). This appears to be unique, since most GVs infect the midgut epithelium mainly during the initial stage of infection, as in the NPVs.

7. Gross Pathology of Baculovirus of Oryctes

The baculovirus which infects the Indian rhinoceros beetle, *Oryctes rhinoceros*, is considered to be a nonoccluded virus (Matthews, 1982), but the presence of occlusion bodies has been reported by Huger (1970), Marschall (1971), and Monsarrat *et al.*, (1973). Larvae, pupae, and adults are susceptible to the virus. Symptoms of infected larvae vary, but the most obvious ones are the cessation of feeding, a gradual increase in hemolymph turbidity, a pearly appearance of the abdomen, and diarrhea (Huger, 1965, 1966). Necrotic melanized areas develop in the midgut and fat body lobes (Monsarrat *et al.*, 1973). With the successive disintegration of the fat tissues and an increase in the hemolymph volume, a "dropsied" condition results and the larva becomes translucent. At a terminal stage of infection, the larva is shiny, beige and waxen in color, especially in the broadened abdominal region. Often, the increased turgor causes the hindgut to herniate. The terminally infected larva frequently develops a whitish mottled pattern because of chalky-white bodies that forms beneath the integument. Dead larvae either undergo putrefaction to become flaccid, brownish, or successively bluish or bluish-black. They also may shrink and mummify, but this is less frequent.

Acute and chronic infections of the virus occur depending on the virus dosage and the stage (larva or adult) of the beetle (Marschall, 1972, 1973). There are differences in pathologies in infected larvae and adults. In an acute case in the larva, the infection begins in the midgut epithelium and spreads to all tissues, and the larva dies in a short time. In adults, the infection differs in that the initially infected midgut develops a severe hyperplasia that produces an enormous amount of virus. At a high level of infection, the virus infects other tissues, and the adult becomes immobilized and dies within a week. In a chronic infection, initially only the midgut epithelium is infected, and other tissues remain unaffected until the very final stage of infection. The infected adult stops feeding, becomes diarrheic, and excretes mainly virus. The infected adult not only becomes a very productive virus factory, but also serves in the dissemination and transmission of the virus (Zelazny, 1976). The extensive proliferation of the midgut epithelium suggests a malignant carcinoma produced by the virus (Figure 16).

Another baculovirus, similar to that of *O. rhinoceros*, infects a few midgut cells of the adult gyrinid beetle, *Gyrinus natator* (Gouranton, 1972). It causes a chronic infection and infected adults do not display apparent symptoms.

8. Gross Pathology of Parasitoid Baculovirus

Hymenopterous parasitoids (family Braconidae) are susceptible to baculoviruses which do not form occlusion bodies. The viruses multiply in certain epithelial calyx cells of the ovary (Poinar *et al.*, 1976; Stoltz and Vinson, 1979a).

They are excreted into the calyx lumen and are introduced into the host larva when the parasitoid oviposits in the host. They appear to be normal components of female wasps and cause no harm to the parasitoids. They may serve a beneficial function for the parasitoids in combating the immune reaction of the host lepidopterous larvae (Rotheram, 1967, 1973; Poinar et al., 1976; Stoltz and Vinson, 1979a).

The signs and symptoms in the parasitoids caused by the baculoviruses have not been clarified. The viruses, however, appear to affect the growth and weight of the host lepidopterous larvae (Stoltz and Vinson, 1979a).

9. Cytopathology of Occluded NPV in Lepidoptera

The cytopathologies of the different baculoviruses appear to be generally similar in the various types of hosts described above. In some cases, it is difficult to differentiate structures associated with the replication of the virus from those caused by cellular injury. We shall describe these structures and their association with viral development. The following descriptions are based on the nuclear polyhedrosis of lepidopterous larvae.

The first indication of cytopathology is the fusion of the enveloped virion to the microvillus cell membrane of the larval midgut epithelium. The fusion appears to result in the incorporation of the viral envelope into the cell membrane (Figure 6). At the fusion site, an opening is formed through which the naked nucleocapsids migrate into the microvillus (Figure 7). How the opening is formed in the membranes is not known, but enzymatic activity may be involved.

The infected midgut cell shows no apparent cytopathology until after the invasion of the viral DNA into the nucleus when the chromatin granules disperse to the peripheral areas. The nucleolus disappears with the concomitant appearance of a dense network (virogenic stroma) and dense bodies of various sizes (Figure 11). The nucleus increases in size. Elongate tubular profiles appear in the vicinity of the virogenic stroma (Figure 17). They are considered by most investigators to be virus capsids (Summers, 1971; Hughes, 1972; Granados, 1980). The assembly of the viral DNA is believed to occur in the stroma. Subsequently, in some cases the inner layer of the nuclear envelope shows blebbing, infolding, and cisternae (Summers and Arnott, 1969; Tanada and Hess, 1976; Falcon and Hess, 1977), but not in others (Stoltz et al., 1973; MacKinnon et al., 1974). If the layers of the nuclear envelope separate, the inner layer extending into the nuclear matrix may become trilaminar (Tanada and Hess, 1976). At a later stage, the nucleocapsids are closely associated with granules, 17.5 nm in diameter, which appear to be ribosomes and are frequently attached to fibrous filaments (F) (Figure 18). Such granule-filament complexes seem to coil around the nucleocapsids (Figure 19) and may develop into an envelope formed *de novo* through the action of the ribosomelike particles. Similar structures occur during the *de novo* envelopment of other NPVs (Kawamoto et al., 1977).

Globular structures appear, sometimes in large numbers, to form the envelopes surrounding the nucleocapsids. Most observers consider the enve-

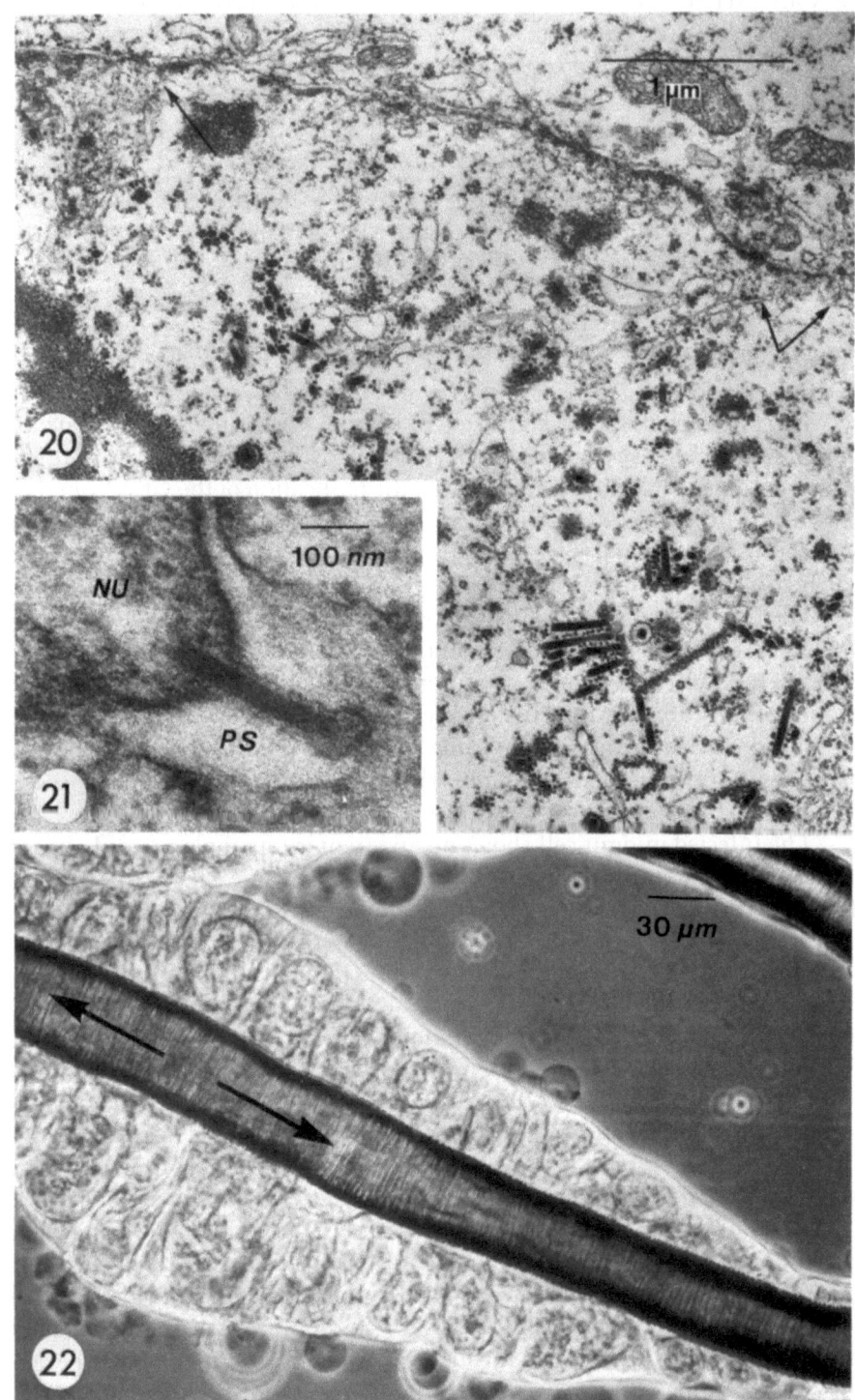

lopes to be formed *de novo* (Hughes, 1972; Stoltz *et al.*, 1973; Mackinnon *et al.*, 1974; Kawamoto *et al.*, 1976, 1977; Granados, 1980). There is the possibility, as described above, that the envelopes may originate from nuclear membranes, especially if proliferation occurs in the nuclear envelope (Figure 20).

Fibrillar masses, which also occur in infected cells of the hemocoel, appear in the nuclei and cytoplasm of infected midgut cells (Croizier *et al.*, 1980). The mitochondria are enlarged, clear, and often clustered together and appear vacuolated.

In most cases, only a few of the nucleocapsids in the midgut epithelium are enveloped and are occluded. Most nucleocapsids exit from the nucleus. Those that bud through the nuclear envelope cause the inner nuclear layer of the envelope to become thicker and denser at the point of contact with the nucleocapsid (Kawamoto *et al.*, 1976, 1977; Hess and Falcon, 1977) (Figure 21). The inner layer then wraps around the nucleocapsids, which initially lie free in the perinuclear cisternae, before entering the cytoplasm. These enveloped nucleocapsids presumably lose their envelopes before passing through the cytoplasm.

When the nucleocapsids pass through the cell plasma membrane, the junction of the cell membrane and the budding nucleocapsid is structurally modified to form a multilayered, bulbous configuration associated with a halo of fine filaments or spikes called *peplomers* (Tanada and Hess, 1976; Adams *et al.*, 1977; Hess and Falcon, 1977; Kawamoto *et al.*, 1977; Granados and Lawler, 1981) (Figure 14). Not only nucleocapsids but also other nonviral structures also acquire envelopes with peplomers when they pass through the cell plasma membrane (Kawamoto *et al.*, 1977). These spikes or peplomers are also produced in virus-infected cells in the hemocoel, e.g., fat body and tracheal cells (Summers and Volkman, 1976; Adams *et al.*, 1977; Hess and Falcon, 1977; Kawamoto *et al.*, 1977). Other investigators have reported peplomer formation in virus-infected tissue culture cells (Summers and Volkman, 1976; Adams *et al.*, 1977). The membranes acquired from the plasma membrane fit very loosely around the nucleocapsids except at the anterior end. The nucleocapsids with peplomers invade the susceptible cell through the peplomer-end facing the host cell membrane and penetrate into the cell by means of viropexis (Adams *et al.*, 1977; Kawamoto *et al.*, 1977). Dense materials appear along the inner lining of the plasma membrane at the point of contact of the spiked membrane (Figure 15).

We have investigated in depth two strains of an NPV infecting the armyworm, *P. unipuncta*. One strain, the typical NPV (TNPV), produces cytopathologies comparable to those caused by most NPVs, whereas the

Figure 20. Proliferation of the inner nuclear membrane (arrows) of a midgut cell of an armyworm larva infected with an NPV.

Figure 21. A nucleocapsid budding from the nucleus (NU) into the perinuclear space (PS). Figures 20 and 21 reproduced from Tanada and Hess (1976).

Figure 22. Light micrograph of a trachea of an armyworm larva infected with the hypertrophy strain of an NPV. Arrows indicate infection gradient. Reproduced from Ritter *et al.* (1982).

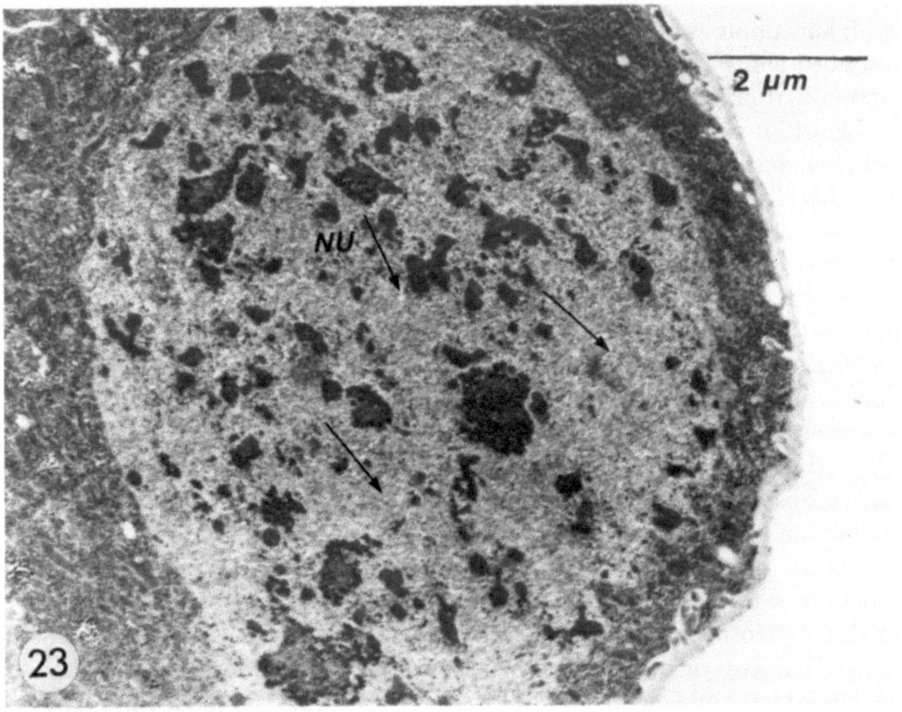

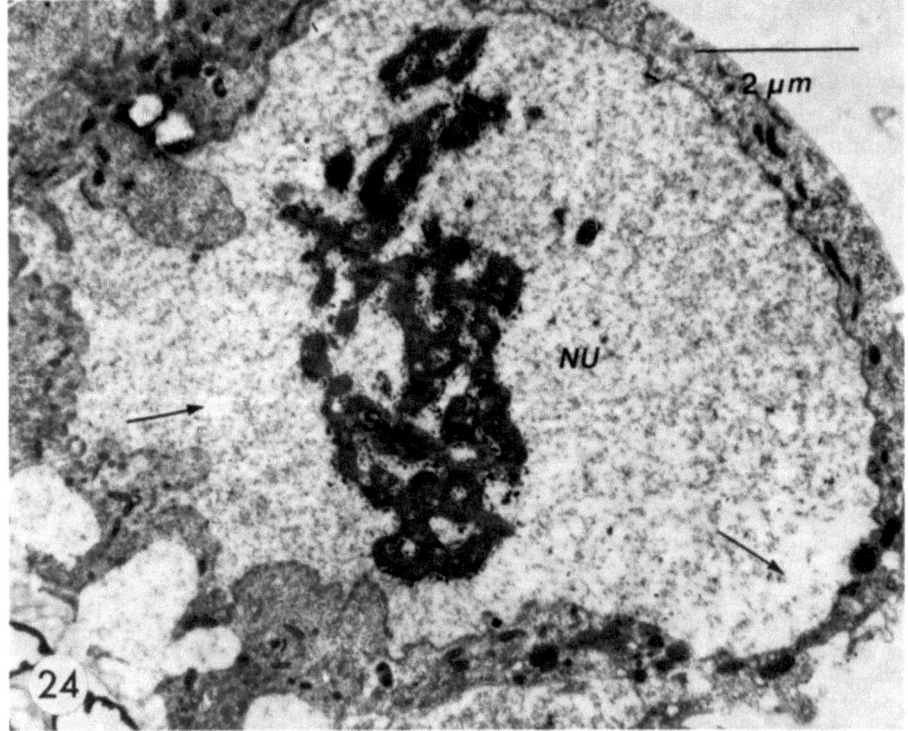

hypertrophy strain (HNPV) causes much more pronounced and distinct cellular alterations (Tanada *et al.*, 1969, 1982; Ritter *et al.*, 1982). The tracheal cells infected by HNPV form a gradient which follows the progressive infection of the virus along the trachea (Figure 22). Thus, a detailed chronological picture of virogenesis and cytopathology is observed in the gradient of infected cells. The cytopathologies caused by these two strains are presented in detail. These two viruses differ in the morphogenetic sequence of virogenesis and associated cytopathologies because of the slow virogenic development in HNPV-infected cells which show virogenic stromata after 120 hr postinoculation (hpi) as compared to TNPV-infected cells which develop stromata in 64 hpi. The nucleus of the HNPV-infected tracheal cell reacts dramatically with the uncoating of the nucleocapsid in the nucleus (Ritter *et al.*, 1982). The initial sign is a change in the density of the matrix which is more loosely constructed than in the uninfected nucleus (Figure 23). The hypertrophy of the cell is clearly visible at 5 days postinoculation (dpi). There is a loss of matrix material. The granular chromatin is more diffused and dispersed throughout the nucleoplasm and gradually disappears to form a relatively "clear" nuclear picture. In cells infected by TNPV and most other NPVs, clumps of chromatin remain up to and even after the appearance of the virogenic stroma. In HNPV-infected cells, there appear dense, compact, interconnected fibrillar strands that are densely packed together in place of the granular chromatin (Figure 24). Punctate densities, which are composed of numerous densely packed fine fibrils, are scattered at the periphery of the fibrillar strands. The punctate densities appear to have the same substructure as that of the fibrils except with electron-dense deposits on them. Subsequently, the first recognizable sign of virus infection is apparent with the appearance of the "virogenic stroma" at about 120 hpi with HNPV (Figure 25). The virogenic stroma is dispersed among dense strands of fibrils which are uniformly distributed throughout the nucleus.

The period of viral infection in the cell up to the appearance of the virogenic stroma and nucleocapsids is the eclipse period. The unique feature of the HNPV infection during the eclipse period is that the hypertrophy in the host cell is evident at a very early stage of infection before any obvious signs of virus infection are apparent. Associated with this hypertrophy are the disappearance of granular chromatin and the appearance of dense interconnected fine fibrils and punctate densities. These cellular alterations suggest that complex cytological and biochemical changes are taking place during the eclipse period.

In cells infected with TNPV and most other NPVs, the chromatin granules are usually marginated and persist along the periphery of the nucleus after the

---

Figure 23 and 24. Trachael cells of armyworm larvae infected with the hypertrophy strain of an NPV.

Figure 23. An infected nucleus (NU) showing the initial changes in the density of the nuclear matrix (arrows).

Figure 24. A nucleus (NU) containing dense, compact interconnected fibrous strands (S) not normally apparent in the typical NPV infection. Figures 23 and 24 reproduced from Ritter *et al.* (1982).

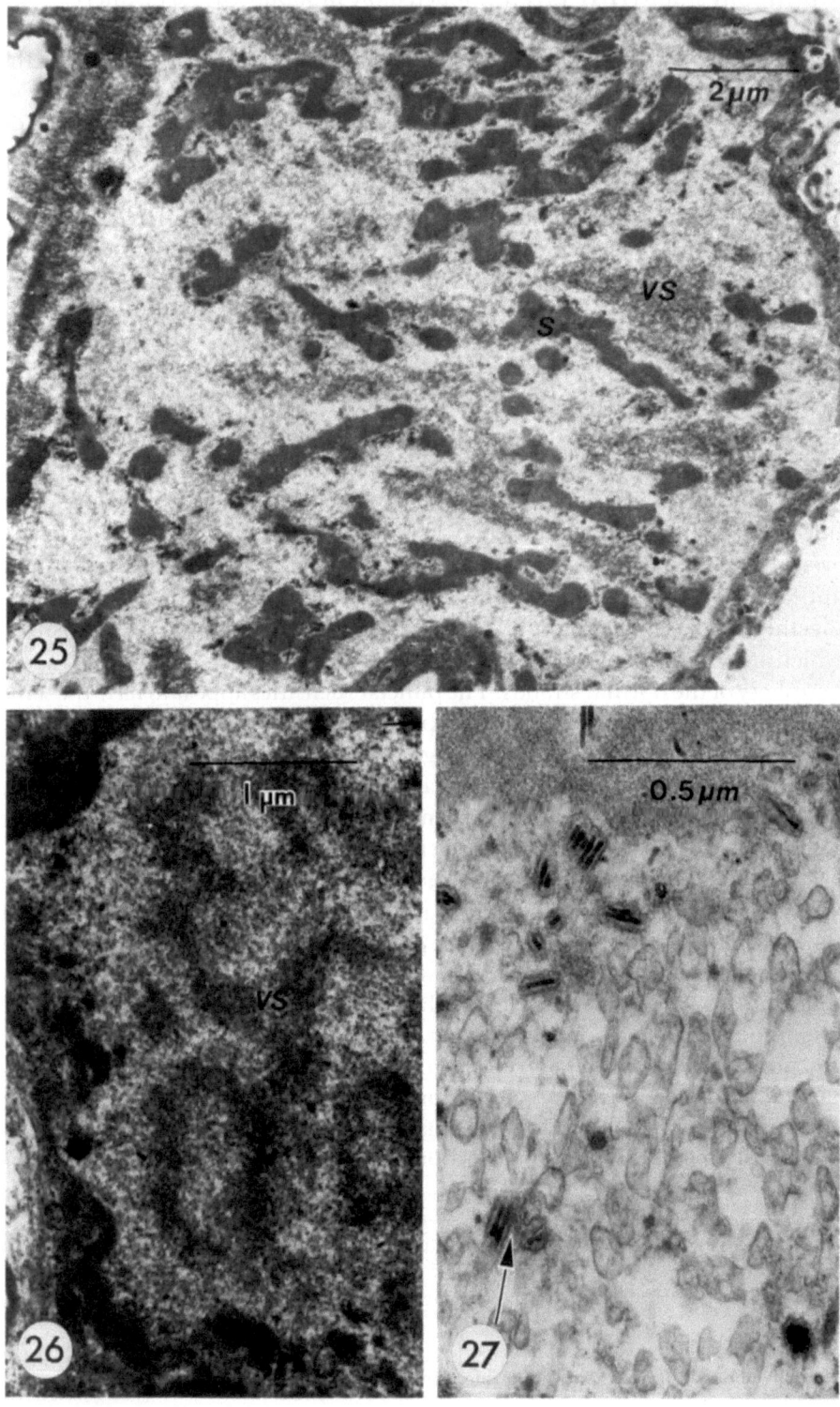

appearance of the centrally located virogenic stroma, which in TNPV infection appears about 64 hpi (Ritter et al., 1982) (Figure 26). The cellular and nuclear hypertrophies develop at about 3 dpi with TNPV. In cells infected with TNPV, the presence of dense fibrillar strands and punctate densities has not been observed, either because of the abbreviated eclipse period or because these structures are unique for HNPV.

After the appearance of the virogenic stromata in HNPV-infected cells, several sequential, morphogenetic changes develop that are comparable to those occurring in TNPV-infected cells (Tanada et al., 1969, 1982; Tanada and Hess, 1976). The changes are (1) the presence of empty profiles similar to nucleocapsids which are closely associated with the virogenic stroma, (2) electron-dense profiles suggestive of nucleocapsid maturation, and (3) the appearance of viral envelopes, dense polyhedral membranes, polyhedral occlusion bodies, and fibrous bodies. However, certain changes in HNPV-infected cells are more pronounced or do not occur in those infected with TNPV. Concomitant with the appearance of enveloped virions, extensive areas of membranous profiles appear in the nucleus (Figure 27). These membranes, in serial sections, seem to be small sacs approximately the same size as the envelopes without virions. These profiles have been observed in most NPV infections, but such abundant formation of membranous profiles is unusual. Moreover, in HNPV-infected cells, a greater number of dense granules is formed than in TNPV-infected cells. The granules are apparently formed from punctate densities in the fibrillar strands which appear during the eclipse period of HNPV-infected cells (Figure 28). The densities, which seem to be formed by electron-dense deposits in the periphery of the fibrils, occur in high concentration in some areas and also appear as aggregates (Figure 29) that are arranged in round to oval groups, similar in size and shape to the granules (Figures 30, 31). The fully developed granules are very electron-dense, about 0.5 nm in diameter, contain transparent areas, and rodlike forms are occluded within them (Figure 32). The significance of these granules and their aggregates in virogenesis is unknown. Similar granules or dark bodies have been observed in other NPV infections (Asayama et al., 1974; Falcon and Hess, 1977). At a late stage of HNPV infection, the nucleus is filled with polyhedra which occlude the enveloped virions at random (Figure 33). The cellular and nuclear hypertrophies reach their maximum, and the cell is about three times the size of a cell infected at a comparable stage with TNPV.

In the nucleus appear large fibrous bodies whose peripheries are associated with dense laminar profiles (Figure 33). These structures occur commonly in nuclear polyhedroses (Krieg and Huger, 1969; Summers and Arnott, 1969;

Figures 25-27. Tracheal cells of armyworm larvae.

Figures 25. Portion of a nucleus infected with the hypertrophy strain of an NPV showing the virogenic stroma (VS) and fibrous strands (S).

Figure 26. Portion of a nucleus infected with the typical strain of an NPV showing the marginated chromatin and the centrally located virogenic stroma (VS). Figures 25 and 26 reproduced from Ritter et al. (1982).

Figure 27. Portion of a nucleus infected with the hypertrophy strain showing extensive membrane proliferation. Some nucleocapsids are enveloped (arrow). Reproduced from Tanada et al. (1982).

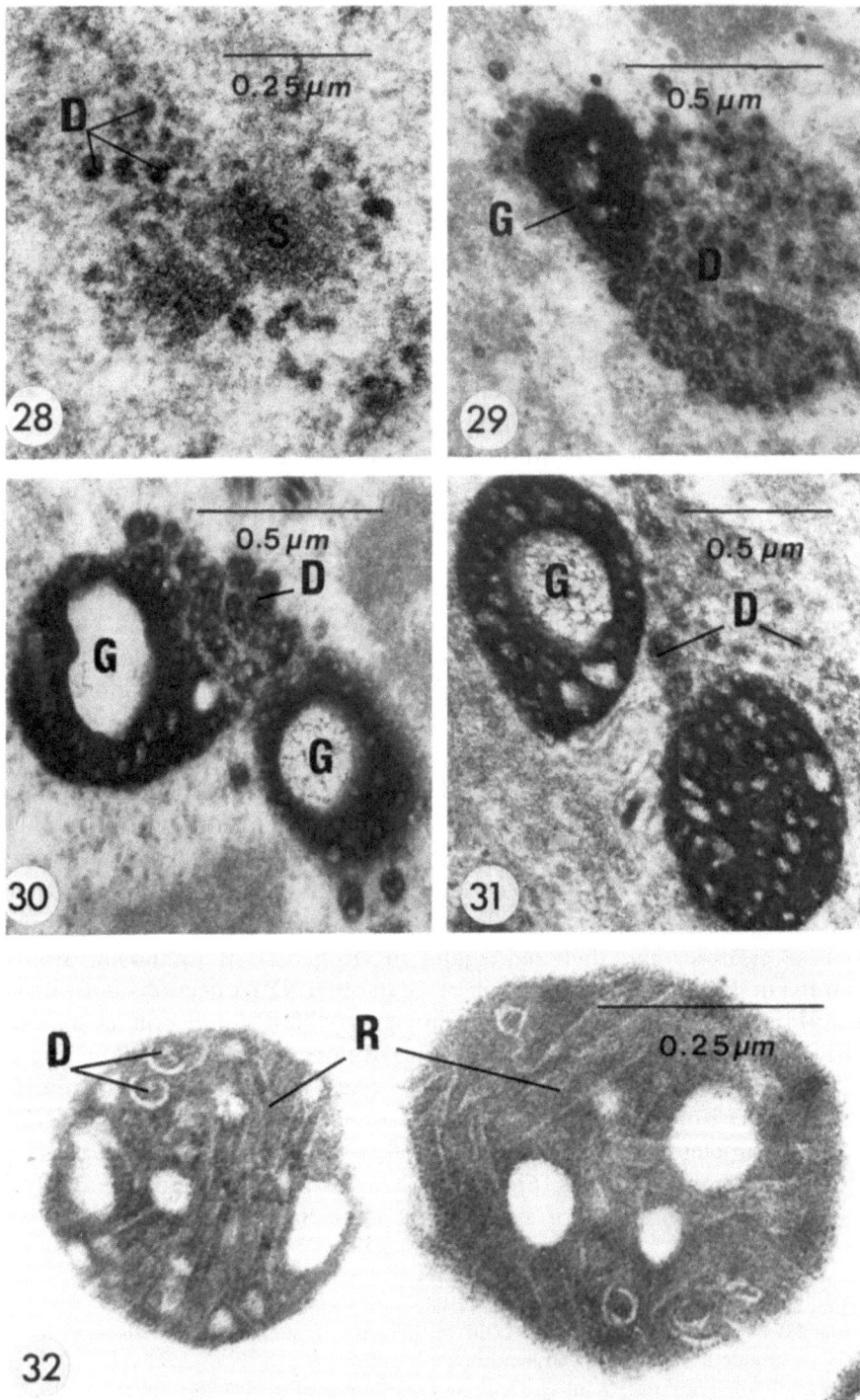

Figures 28-32. Portions of nuclei of tracheal cells of armyworm larvae infected with the hypertrophy strain of an NPV showing the development of the granules.

Harrap, 1972a,b; Injac *et al.*, 1973; Asayama *et al.*, 1974; MacKinnon *et al.*, 1974; Hirumi *et al.*, 1975; Hess and Falcon, 1977; Chung *et al.*, 1980). The fibrous material in the nucleoplasm is believed to serve as ultrastructural precursors to the macromolecular lattice of the polyhedra and the polyhedral membrane. Fibrous bodies of the same substructure are also found in large numbers in the cytoplasm of HNPV-infected cells (Figure 33). This has also been observed by others in NPV-infected cells (Krieg and Huger, 1969; Summers and Arnott, 1969; Injac *et al.*, 1973; Asayama *et al.*, 1974; Croizier and Quiot, 1975; Hudson *et al.*, 1979). The cytoplasmic fibrous bodies in HNPV-infected cells are associated with microtubules of varying densities (Figure 34). These microtubules occur as a normal component of uninfected tracheal cells.

Hudson *et al.*, (1979) have concluded from the presence of the fibrous bodies in the cytoplasm that the bodies are by-products or waste products associated with the increased metabolic activity in infected cells. The fibrous bodies, however, may be precursors to the polyhedral protein in the nucleus and cytoplasm. In the cytoplasm of HNPV—but not in that of TNPV-infected cells, there are occasional polyhedralike bodies with a paracrystalline substructure similar to that of the polyhedra in the nucleus (Figure 35). The cytoplasmic polyhedra, however, do not contain virions and have many clear areas. Proteinaceous polyhedrashaped bodies without virions occur also in the cytoplasm of tracheal cells of the almond moth larva, *Cadra cautella* (Adams and Wilcox, 1968).

Hudson *et al.*, (1979) have observed multilayered structures in the nuclei and occasionally in the cytoplasm of cells of the velvet bean caterpillar, *Anticarsia gemmatalis*, infected with an NPV. These structures are irregular in shape and size, but always consist of concentric layers that vary in thickness. In the infected nucleus, these workers also reported vacuolelike structures lined with a double membrane or with evenly spaced striations which have not been described previously in nuclear polyhedroses.

The cell and nuclear membranes of the hypertrophied cell infected with HNPV show a large number of folds which greatly increase the surface areas of the cell and nucleus (Tanada *et al.*, 1969). The mitochondria in the hypodermis increase in numbers at an early stage of infection. As the HNPV infection progresses in both the hypodermis and the trachea, the mitochondria degenerate and the cristae disappear leaving a vacuolelike structure.

Near the terminal stage of HNPV infection, the greatly hypertrophied cells

---

Figure 28. Initial granule formation associated with the appearance of punctate densities (D) on fibrous strands (S).

Figure 29. Aggregates of punctate densities that are continuous with the outer edge of a developing granule (G).

Figure 30. Granules composed of a concentration of punctate densities.

Figure 31. Fully formed granules. Granules are round to oval in shape and contain transparent areas and punctate densities.

Figure 32. An underdeveloped electron micrograph revealing the substructure of the dense granules which consists of punctate densities and rod-shaped structures (R). All figures reproduced from Tanada *et al.* (1982).

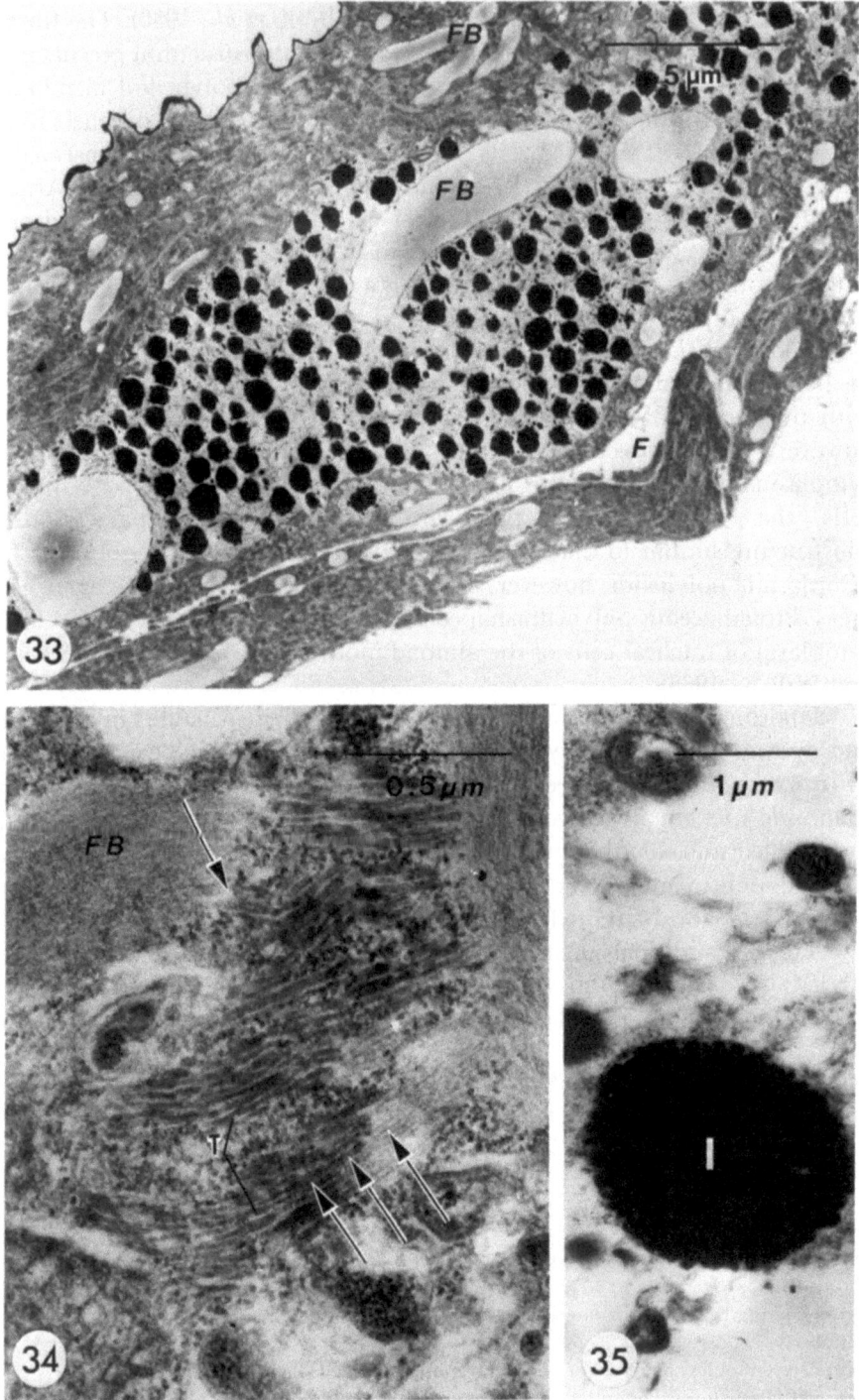

Figures 33-35. Portions of tracheal cells of armyworm larvae infected with NPVs.

exhibit only a little destruction which appears as a tearing of the mestracheon folds between the cells along with the separation of the septate desmosomes (Figure 33). In less hypertrophied cells, the integrity of the intracellular spaces is not affected, but attenuation develops in the cellular sheath between the hypertrophied cell body containing the nucleus and the adjoining cell. Injac et al., (1973) have observed, in the trachea of *Hyphantria cunea* infected with an NPV, fissures between the hypertrophied infected tracheal cell and the cuticle. In most NPVs, when the nucleus is completely or almost filled with polyhedra, the nuclear envelope breaks down and is followed by the lysis of the cell.

At a late stage of NPV infection, the virus replication occurs, in some cells, not in the nuclei but in the cytoplasm (Asayama and Kawamoto, 1975; Falcon and Hess, 1977). All stages of virogenesis including the occlusion of viruses in polyhedra are found in the cytoplasm. The nucleus of the cell undergoes chromatin margination but without further signs of virus replication. Interspersed with the mitochondria, glycogen, and rough endoplasmic reticulum (RER) are membranous profiles which occur at much higher frequency than in an uninfected nucleus. The membranous profiles are presumed to form the virus envelopes. Within the cytoplasm, fibrillar structures similar to the virogenic stroma are associated with nucleocapsids.

10. Cytopathology of Occluded NPV in Hymenoptera

Aside from the very early study by Bird and Whalen (1954), we are not aware of subsequent studies on the cytopathology of this group of interesting NPVs which are tissue specific for the midgut epithelium of the sawfly. The first symptom of infection in the midgut epithelium is the swelling of the nuclei and nucleoli and the coagulation of the chromatin. Within 48 hr, all stages of virus development are found and the nucleocapsids are associated with the chromatin mass (virogenic stroma). Particles dissociate from the chromatin mass and appear in the nuclear sap, where they are thicker and denser than those in the chromatin. The nucleoli persist even after the formation of polyhedra. Numerous spherical membranes occur in the nucleus. These appear to be the precursors of the virus envelopes. The nucleoli continue to increase in size and become more diffuse. At a late stage of infection, the nuclei are filled with polyhedra.

Figure 33. A cell infected with the hypertrophy strain showing fibrous bodies (FB) in both the nucleus and the cytoplasm. Extensive hypertrophy of the nucleus and cell causes a tear in the mestracheon folds (F). Reproduced from Tanada et al. (1982).

Figure 34. A cell infected with the typical strain showing variation in densities (arrows) of the microtubule profiles (T) in the cytoplasm. Also shown is the association of the microtubule profiles with fibrous bodies (FB). Reproduced from Tanada and Hess (1976).

Figure 35. A cell infected with the hypertrophy strain showing in the cytoplasm a polyhedral body (I) containing no virions. Reproduced from Tanada et al. (1982).

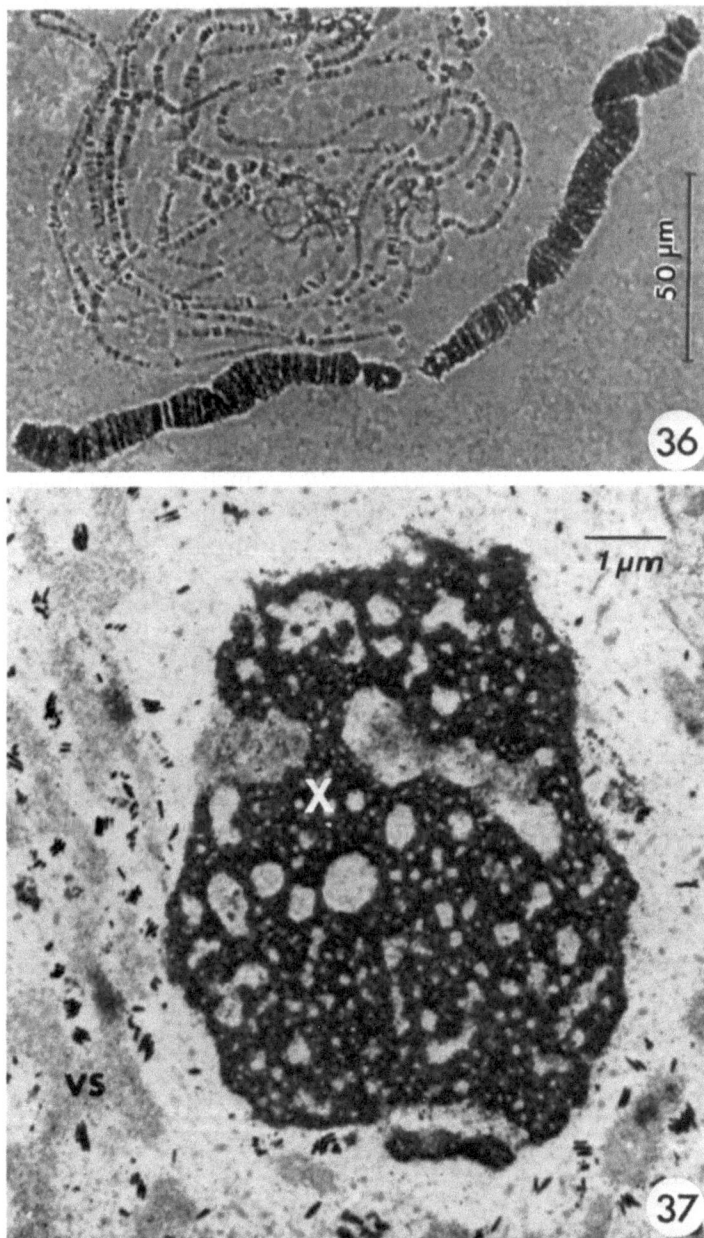

Figure 36. A light micrograph of a smear of a larval gastric-caeca cell of the sciarid fly, *Rhynchosciara angelae*, infected with an occluded NPV. Chromosome on the bottom is greatly increased in width, and the one at the top is increased in length. Reproduced from Paven et al. (1971).

Figure 37. Portion of the enlarged nucleus of a larval gastric-caeca cell of the sciarid fly infected with virus showing the enlarged chromosome (X) and the virogenic stroma (VS). Reproduced from Da Cunha et al. (1972).

11. Cytopathology of Occluded NPV in Diptera

In the mosquito larva, the NPV infects the cells of the cardia, gastric caeca, and the regenerative cells of the midgut (Clark et al., 1969; Federici and Lowe, 1972; Hall and Fish, 1974). Virogenesis is essentially similar to that of other occluded baculoviruses. The nucleus hypertrophies and the polytene chromosomes disintegrate and move toward the nuclear envelope. Long nucleocapsids are formed from the virogenic stroma, and through their cleavages produce the normal-length nucleocapsids (Federici, 1980), as in the case of the NPV of the tussock moth, H. pseudotsugata (Hughes, 1972). The viral envelopes are formed de novo from membranous structures that appear as irregular ribbons or pleomorphic vesicles and from granular components of the nucleoplasm (Federici, 1980). The enveloped virion has a slightly bulbous appearance at one end (Hall and Fish, 1974; Federici, 1980). The formation of occlusion bodies differs from that of other NPVs. The numerous bodies coalesce in the nucleus to form eventually one or a few spindle-shaped polyhedra (Federici and Lowe, 1972).

Aside from the early studies of Smith and Xeros (1954) and Smith (1955), very little is known of the cytopathology in the crane fly, *Tipula paludosa*, infected with NPV. A chromatin mass (virogenic stroma) appears in the center of the hypertrophied nucleus of an infected blood cell. Virus particles (nucleocapsids) appear near the mass. Subsequently, the particles are enclosed singly or in bundles within vesicles (envelopes) and accumulate near the nuclear membrane. Instead of the random occlusion of enveloped nucleocapsids as in Lepidoptera, the masses of particles are occluded together in polyhedral matrix protein in close association with the nuclear membrane. As the crescent-shaped polyhedra are formed, they are displaced outside of the nucleus to lie in the cytoplasm.

The occluded NPV in the sciarid fly, *R. angelae*, causes distinct pathology in the giant polytene chromosomes (Pavan et al., 1971; Da Cunha et al., 1972). The infection results in the hypertrophy of the gland and midgut epithelial cells in which the polytene chromosomes also increase in size (Figure 36). The giant chromosomes show an increase in polyteny, an accelerated RNA production manifested by specific puffs in the chromosomes, an increase in width and/or depth, and the release of many micronuclei. Subsequently, constrictions and breaks in the chromosomes appear in specific regions which vary with the virus strain (Da Cunha et al., 1972) (Figure 37). The nucleolus enlarges greatly, and the ribosomes and polyribosomes increase in numbers. The mitochondria swell and become globular, and the matrix is much less electron-dense. An abundance of short, open-ended membranous profiles, which are the source of virus envelopes, appear in the nucleus (Stoltz et al., 1973). The enveloped nucleocapsids are occluded near the inner layer of the nuclear envelope and the polyhedra are attached to the envelope (Da Cunha et al., 1972; Stoltz et al., 1973). At the terminal stage of infection, the nuclear envelope breaks down and the giant chromosomes undergo catabolic destruction.

12. Cytopathology of Granuloses

The pathology of GV infection in midgut cells resembles that of the occluded NPV with the fusion of the enveloped virion to the microvillus and the entrance of the nucleocapsid into the cytoplasm. The uncoating of the nucleocapsid takes place at the nucleospore (Summers, 1971; Granados, 1980). Walker et al. (1982), however, have observed nucleocapsidlike structures within the nucleus and they have suggested that the site of uncoating may vary with the cell types.

After a brief eclipse period of the GV infection in the midgut cell nucleus, a virogenic stroma is formed, nucleocapsids are produced, and the cell becomes hypertrophied. The nucleocapsids acquire envelopes produced by *de novo* synthesis in the nucleus or cytoplasm or by budding through cell membranes. Unlike the NPV infection, the nuclear envelope of a GV-infected cell loses its structural integrity at an early stage of virus infection (24 hpi), resulting in the intermingling of nuclear and cytoplasmic components (Huger and Krieg, 1961; Bird, 1963; Tanada and Leutenegger, 1968, 1970; Summers, 1969, 1971; Hunter et al., 1975; Pinnock and Hess, 1977; Walker et al., 1982). (Figure 38). This rapid and drastic disassembly of the nuclear envelope is a major difference in the cytopathology between GV and NPV infections.

Although virogenesis begins in the nucleus, it continues in both the nuclear and the cytoplasmic areas after the disruption of the nuclear envelope. At a late stage of nuclear membrane disruption (3-4 dpi), the cell contains virogenic stroma, newly formed nucleocapsids, and frequently myelinlike figures (Figure 39). About 8-10 dpi, the nucleocapsids are enveloped but, as in the case with NPV, they are infrequently occluded in occlusion bodies. At this late stage, there is no evidence of a distinct nuclear membrane. In some infected midgut cells, enveloped virions singly or in aggregates occur in vesicles (Summers, 1969, 1971; Tanada and Leutenegger, 1970). At 24 hpi, virus particles, mostly enveloped nucleocapsids, may occur in continuous rows (Figure 40) in intercellular spaces between midgut cells and near the basement membrane. At a late stage, many virus rods are embedded in the basement membrane or are budding through the membrane (Tanada and Leutenegger, 1970; Summers, 1971). Fibrous material found in the nucleus and cytoplasm of NPV-infected cells also occurs in GV-infected cells, but no membranous profiles are associated with them (Summers and Arnott, 1969).

Figure 38. An early GV infection in a hypodermal cell nucleus of *Archips argyrospila* larva. A few nucleocapsids (N) are visible and the nuclear membrane is undergoing lysis (arrows). Lying on the nuclear membrane are electron-dense clumps (E) peculiar to infected nuclei. Reproduced from Pinnock and Hess (1978).

Figure 39. A portion of a cell of a codling moth larva infected with a GV showing virogenic stroma (VS) and nucleocapsids (N) in the cytoplasm.

Figure 40. Budding of nucleocapsids into the intercellular space (IS) of larval midgut cells of the codling moth. Nucleocapsids are presumed to migrate basally to the hemocoel (B).

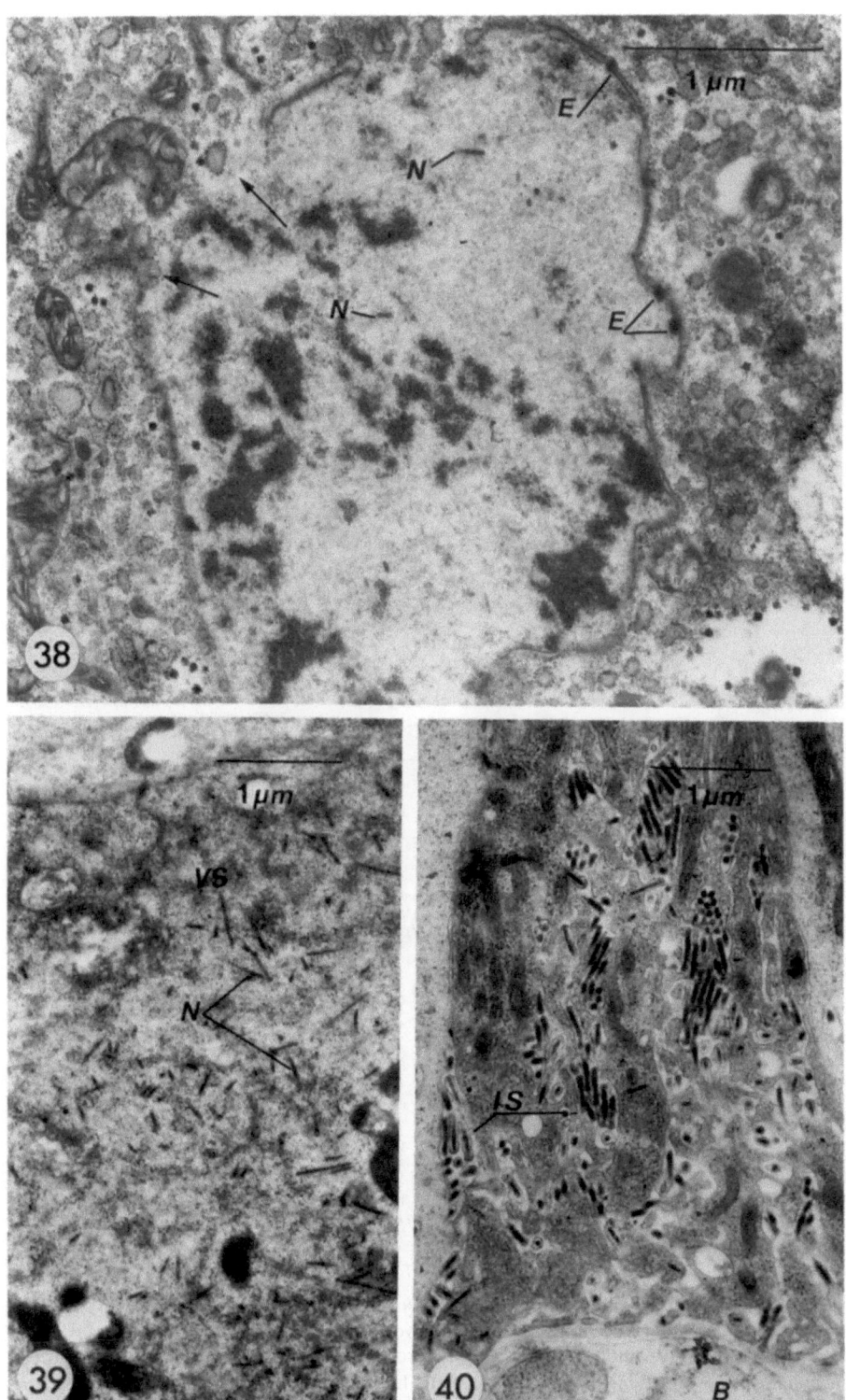

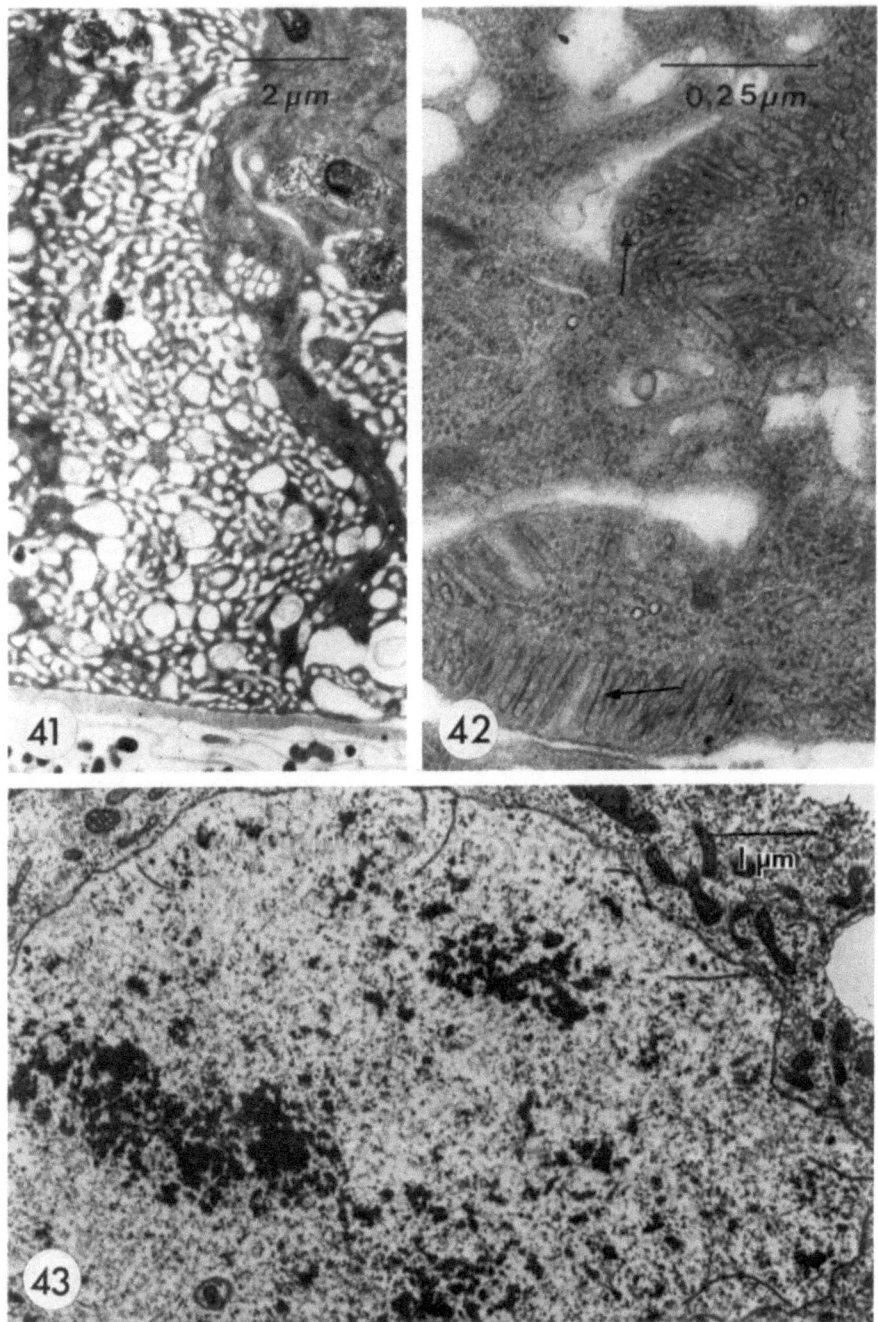

Figures 41 and 42. Portions of midgut cells of *Archips argyrospila* larvae infected with a GV.
Figure 41. Late stage of infection showing extensive vacuolation in the basal region.
Figure 42. Apparently empty nucleocapsidlike profiles (arrows) associated with the basal cell membrane. They are aligned as if to bud into the hemocoel. Figures 41 and 42 reproduced from Pinnock and Hess (1977).

An unusual cytopathology occurs in the larval midgut cells of the fruit tree leafroller, *Archips argyrospila*, infected with a GV (Pinnock and Hess, 1977). The midgut cells rarely show virus replication, and when infected, the replication in this site is partially blocked or inhibited. The cytoplasm of the cell is highly vacuolated especially in the basal cell region (Figure 41). The nucleus exhibits chromatin loss and has usually more than one large fibrillar mass of material of about the same electron density as chromatin. Interspersed upon these large masses are electron-dense particles (30–50 nm). The nucleoplasm also contains circular membranous profiles of various sizes. Nucleocapsidlike structures occur in the nucleus and cytoplasm. These structures accumulate in the basal regions of the midgut cells and are sometimes aligned as though budding through the basal membrane (Figure 42). Annulate lamellae and myelinlike membrane whorls are found in the cytoplasm. The mitochondrial matrix granules are enlarged considerably. The myelinlike whorls may be remnants of the nuclear envelope (Hunter *et al.*, 1975).

The cytopathology of GV infections in tissues of the hemocoel will be described in detail in particular with the infection in fat body cells. The first recognizable cell alteration, after the invasion of the nucleocapsid into the nucleus, is the formation of small intranuclear blebs of the inner layer (membrane) of the nuclear envelope (Walker *et al.*, 1982). Subsequently, intranuclear extensions of the inner nuclear membrane are formed (Figure 43). The nucleoplasm develops a less-consolidated heterochromatin, most of which is coalesced and displaced into an eccentric intranuclear position. The rest of the nucleus contains only small remnants of heterochromatin and has a "cleared" appearance (Figure 44). Small electron-dense clumps that are occasionally associated with nuclear pores appear along the nuclear envelope (Pinnock and Hess, 1978; Walker *et al.*, 1982) (Figure 38). Shortly thereafter, the virogenic stroma develops in the nucleus, but no nucleocapsid or other virion structure is initially associated with it. These changes are rapidly followed by the disassembly of the nuclear envelope and the formation of paired cisternae and stacks of cisternae. Ribosomes are associated with both free surfaces of the stacked cisternae (Tanada and Leutenegger, 1968; Hunter *et al.*, 1975; Asayama and Inagaki, 1975; Pinnock and Hess, 1978; Walker *et al.*, 1982).

While the nuclear alterations are developing, pathologies appear in the cytoplasm with drastic changes in the RER (Figure 45) and the mitochondria (Figure 46). There is an increase in myelin whorls and a loss in lipid and glycogen contents. The mitochondria may assume a balloon shape with fragmented cristae (Tanada and Leutenegger, 1968; Asayama and Inagaki, 1975). Cellular junctions, e.g., intercellular gap junctions, desmosomes, and hemidesmosomes, disappear, and eventually none can be found between the cells containing the GV capsules (Walker *et al.*, 1982). Such affected fat cells detach from the surrounding cells and basal lamina. Uninfected cells appear to encroach the areas vacated by the infected cells.

Figure 43. Intrannuclear extensions of the nuclear membrane (annulate lamellae) in a fat cell of *Spodoptera frugiperda* larva infected with a GV. Reproduced from Walker *et al.* (1982).

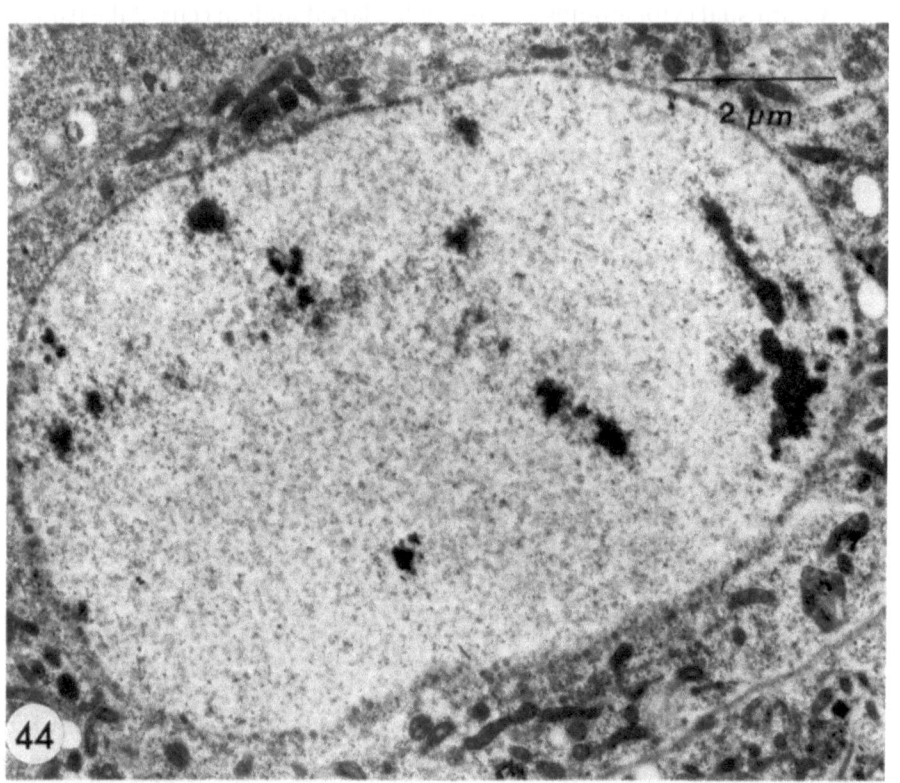

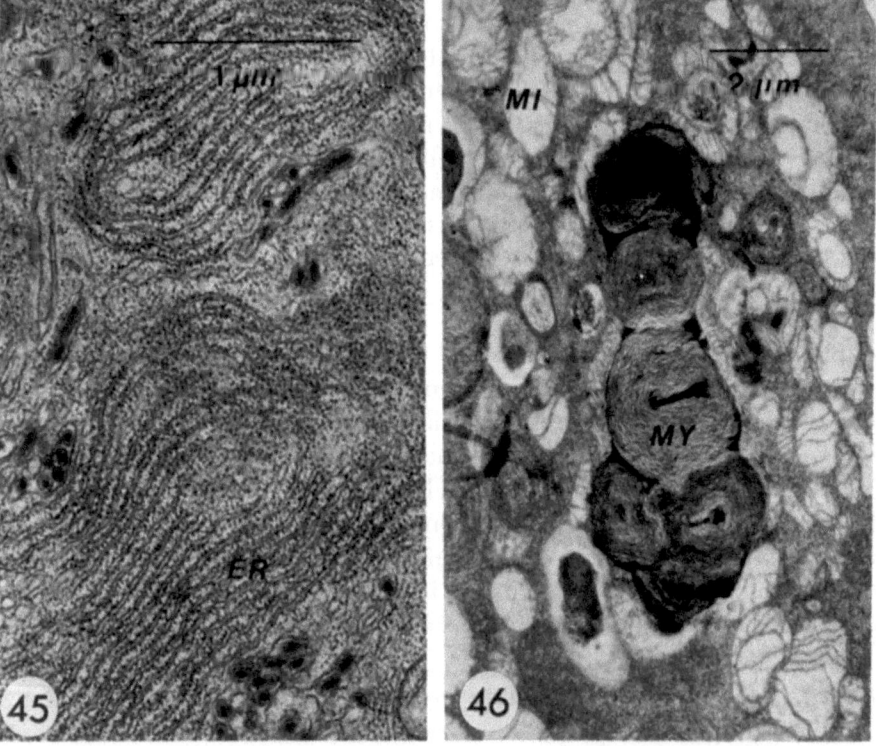

Annulate lamellae, which are characteristic of GV infections, are formed in the cytoplasm (Figure 47). Virus capsules appear by 3 dpi with massive virus formation by 4 days and an overproduction of capsule constituents by 7 days. With the breakdown of the nuclear envelope, the only recognizable nucleoplasm is the nucleoluslike material. Masses of coiled filaments are interspersed among the capsules (Tanada and Leutenegger, 1968) (Figure 48). Similar masses have been observed in other GV infections (Arnott and Smith, 1969; Walker *et al.*, 1982). Hexagonally packed, tubular membrane arrays and membranous profiles appear late in infection. In moderately to highly infected hypertrophied fat bodies, there is an increase in mitotic activity, frequently in apparently uninfected fat cells at the edges of the fat lobes.

In some cases, the GV replication occurs in the cytoplasm, although most of the replication occurs in the nuclei of tissue cells in the hemocoel (Huger, 1963; Pinnock and Hess, 1978). During the early stages of virogenesis, the nucleus appears to be unaffected, but there is no detailed study on the cytopathology in the nucleus when virogenesis occurs in the cytoplasm.

The enveloped nucleocapsids are occluded by matrix proteins usually starting at one end and extending toward the other end until the entire nucleocapsid is occluded to form the capsule (Hughes, 1952; Huger and Krieg, 1961; Arnott and Smith, 1968). Not infrequently, aberrant capsule formation occurs in the fat and other tissues of the hemocoel (Arnott and Smith, 1968; Hunter and Hoffmann, 1972; Pinnock and Hess, 1978; Walker *et al.*, 1982).

13. Cytopathology of Nonoccluded NPV in Oryctes rhinoceros

In the infection of the nonoccluded NPV in *O. rhinoceros*, only the cytopathology in infected fat body cells in the larva has been studied in detail (Huger, 1966; Monsarrat *et al.*, 1973). The infected cells show extensive vacuolation, followed by necrosis. The fat globules, albuminoid spheres, and other contents disintegrate and are reabsorbed. At an advanced stage, considerable areas of the fat body are reduced to a more or less loose network consisting of connective tissue membranes, cell membranes, and cytoplasmic remnants. The nuclei are pycnotic or hypertrophied. The disintegrated parts of the fat body form a highly viscous mass. Some fat cells, which are associated with the infected areas of the fat body, proliferate extensively and separate into single cells which float free in the hemolymph. Virus replication also occurs during the proliferation of the fat cells. The chalky-white bodies, visible beneath the

---

Figure 44. A fat cell nucleus of a codling moth larva infected with a GV showing the margination of the chromatin and the clearing of the nuclear matrix.

Figure 45. Extensive proliferation of the endoplasmic reticulum (ER) in a midgut cell of a *Trichoplusia ni* larvae infected with a GV. Such proliferation occurs commonly in GV infections. Reproduced from Tanada and Leutenegger (1970).

Figure 46. Swollen mitochondria (MI) with disrupted cristae in a fat cell of a codling moth larva infected with a GV. Myelin whorls (MY) are very conspicuous.

integument of larvae at the terminal stage of the disease, consist of clusters of disintegrating fat cells containing a dense network of strands and aggregates of needlelike crystals.

In ultrathin sections of a fat cell at an early stage, the hypertrophied nucleus contains, in some cases, an abundance of spherical and fibrillar membranous profiles (Huger, 1966; Monsarrat et al., 1973) (Figure 49). Virogenic stroma and nucleocapsids appear near the margins of the nucleus. At the final phase of virogenesis, the cytoplasm breaks down and the nuclei are densely packed with enveloped nucleocapsids and spherical profiles. Long tubular filaments, possibly nucleocapsids arranged in an irregular fashion or in parallel arrays, occur in the nucleus. The nuclear envelope increases in thickness and becomes disorganized. Virogenic stroma may occur also in the cytoplasm, accompanied by multilaminar membranes of the myelin type that disintegrate as virogenesis progresses. Polyhedra are rarely formed in such tissues or in organs as the fat body, midgut, ovarian sheath, and the internal wall of the spermatheca (Huger, 1970; Marschall, 1971; Monsarrat et al., 1973). Virus rods occur in the polyhedra (Figure 50).

The cytopathology and replication of this NPV have also been studied *in vitro* in the cell line of *O. rhinoceros*. The replication of the virus in the nucleus is similar to that in the insect (Quiot et al., 1973). This is also the case in the cell line of *Spodoptera frugiperda*, but in that of *Aedes albopictus*, the multiplication occurs in the cytoplasm (Kelly, 1976). No occlusion bodies are formed. The cytoplasm of the cultured cell is highly vacuolated and disorganized, and the mitochondria are degenerated.

The most dramatic result of infection occurs in the midgut of chronically infected adult beetles (Huger, 1970; Marschall, 1971, 1973). The infected midgut of the beetle undergoes intensive proliferation of the replicative nidi which form cells that are infected by the virus (Figure 16). The enormous accumulation of infected cells results in massive amounts of virus being excreted by the chronically infected adult beetles.

14. Cytopathology of Nonoccluded NPV in Gyrinus natator

The nonoccluded NPV of *G. natator* infects only a portion of the midgut cells and causes no apparent symptoms in the adult beetle (Gouranton, 1972).

Figure 47. Annulate lamellae (AL) in a hypodermal cell of an *Archips argyrospila* larva infected with a GV. Annulate lamellae, which are characteristic of infections, appear to be associated with the endoplasmic reticulum (ER). Reproduced from Pinnock and Hess (1978).

Figure 48. A portion of a larval fat cell of the codling moth infected with a GV. Capsules are frequently associated with islands of coiled filaments (CF).

Figure 49. Proliferation of membranous profiles (M) in a fat cell nucleus of an *Oryctes rhinoceros* adult infected with a baculovirus.

Figure 50. Polyhedra (P), occluding enveloped nucleocapsids, in the cytoplasm of the ovarian sheath cell of an *Oryctes rhinoceros* adult infected with a baculovirus. Reproduced from Monsarrat et al. (1973).

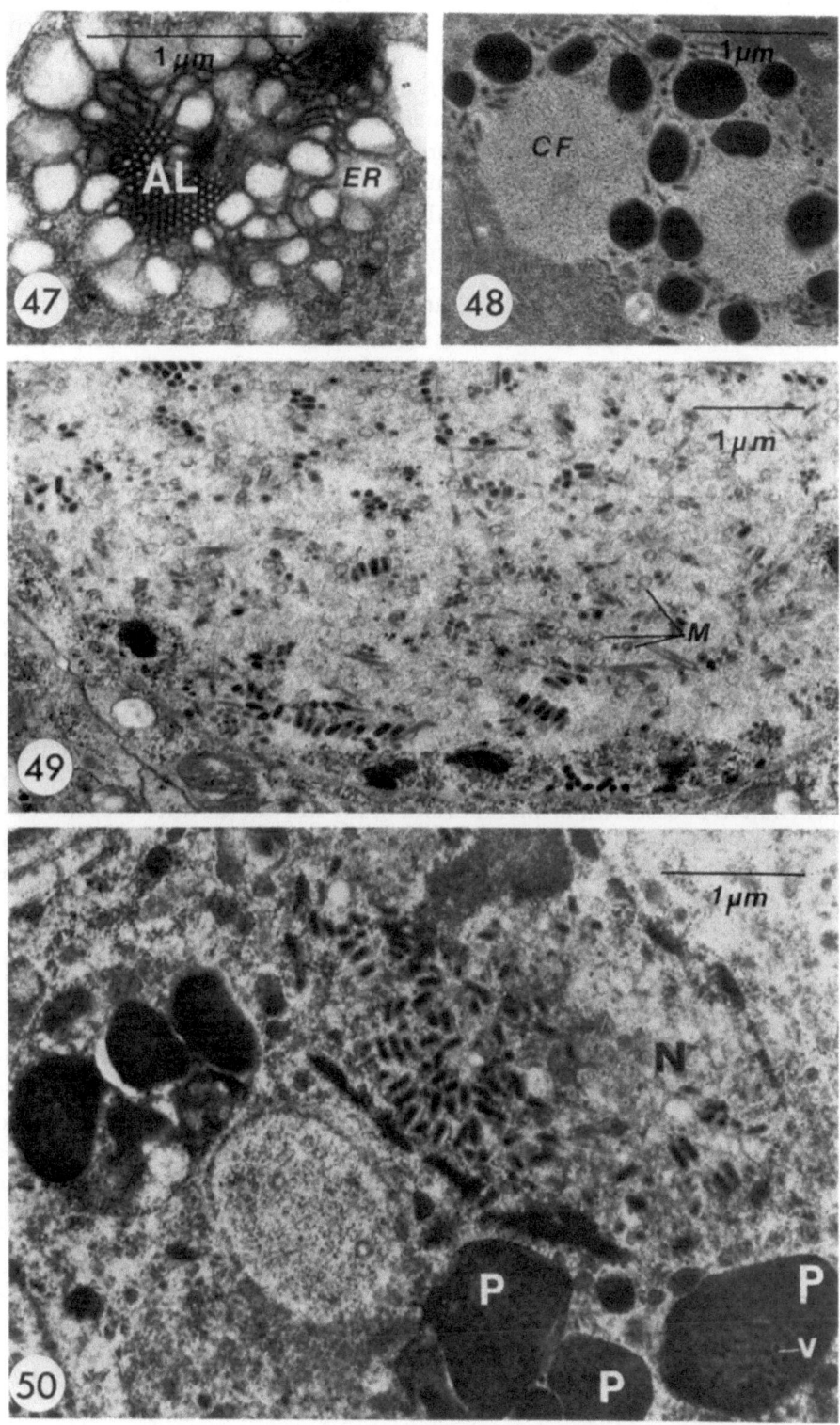

The first signs are the characteristic network and dense bodies. The network is probably the virogenic stroma. At an early stage of infection, fibrils occur abundantly in the nucleus, and since they disappear as viral particles are formed, they are considered to be DNA fibrils. Gouranton (1972) has observed only enveloped nucleocapsids and no naked nucleocapsids. Within the nucleus are microtubules, cylindrical structures, and cup-shaped membranes which form the virus envelopes. The inner nuclear membrane occasionally proliferates. This virus, unlike most other NPVs, does not cause a lethal infection.

15. Cytopathology of Nonoccluded NPV of Parasitoids

The nonoccluded NPVs of parasitoids (mainly family Braconidae) replicate in the ovary (calyx) of wasps and are transmitted to host larvae during the parasitoids' oviposition (Poinar et al., 1976; Stoltz et al., 1976; Stoltz and Vinson, 1977, 1979a) (Figure 51). The virus may be taken up by cells of the host larva, but no virus replication has been observed in these cells (Stoltz and Vinson, 1979b) (Figure 52).

In the nuclei of the calyx epithelial cells of the parasitoid, the first visible manifestation of virogenesis is the appearance of open-ended membrane segments, presumably envelope precursors formed *de novo,* and the appearance of a virogenic stroma (Stoltz and Vinson, 1977). The virus particles may be concentrated in membrane-bound vesicles in the cytoplasm and nucleus (Poinar et al., 1976; Hess et al., 1979). At times, the vesicles contain many whorls of membranes of densely staining bodies. The virus possesses an unusual protrusion of the viral envelope which is involved either or both in the penetration of basement membranes and in the entry of the nucleocapsids into the parasitoid's host cell (Stoltz and Vinson, 1979b). The virus particles emerge from the calyx cells by budding through the basement membrane or after the lysis of the cells.

16. Summary

The cytopathologies of cells infected by the various subgroups of baculoviruses are essentially similar. The initial injury is the invasion of the nucleocapsid through the host cell membrane. The infected cell exhibits no evident cytopathology until the virus uncoats in the nucleoplasm. With the initiation of virogenesis, differences in cytopathic effects appear among the different types of baculoviruses. The most pronounced difference is the early

---

Figure 51. Presence of virus particles, which are produced in the nucleus (NU) of a calyx cell, in the lumen (L) of the calyx of the parasitoid braconid, *Phanerotoma flavitestacea.*
Figure 52. Presence of parasitoid virus particles in the larval hemocoel (B) and tissues (H) of the host, *Paramyelois transitella.* Figures reproduced from Poinar et al. (1976).

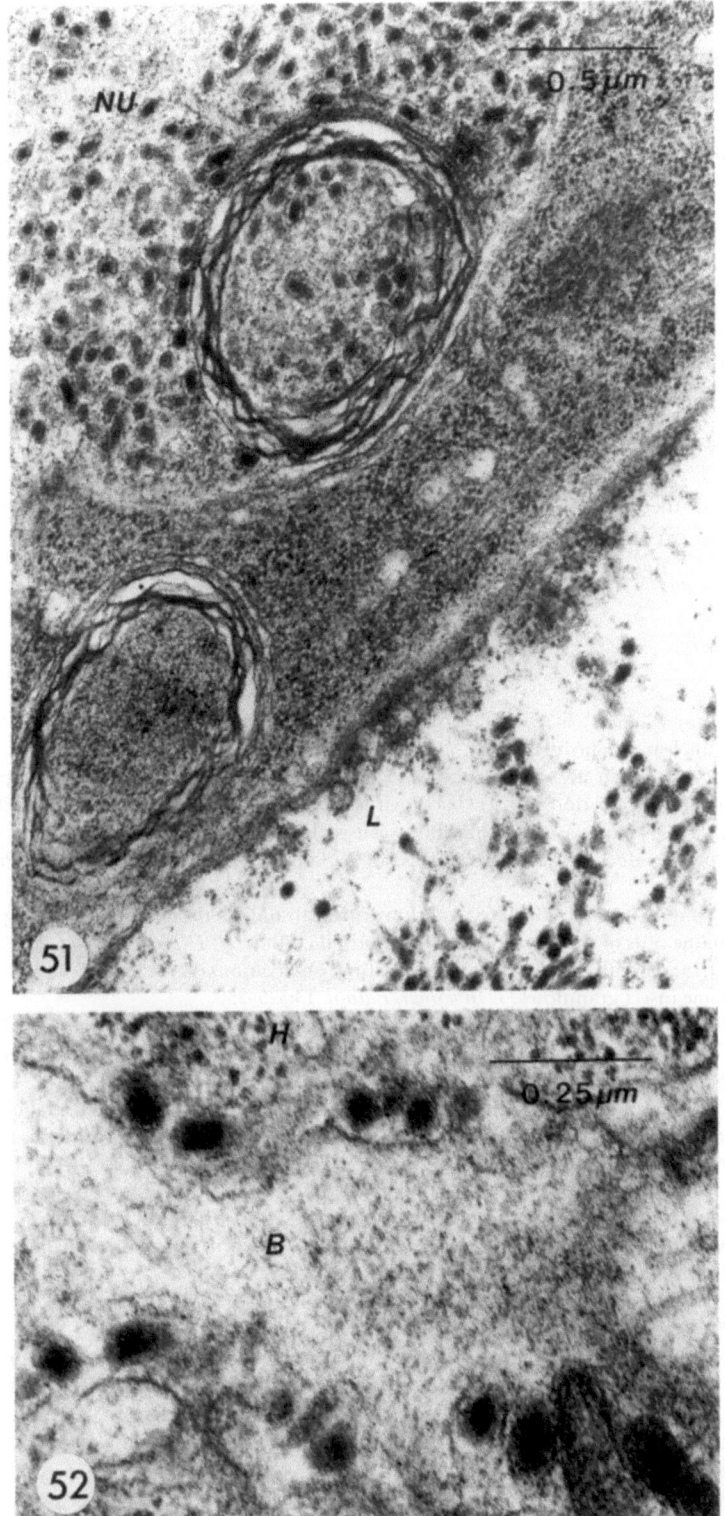

and rapid disassembly of the nuclear envelope in the granulosis as compared to the nuclear polyhedrosis and other baculoviruses. The origin, function, and fate of most structures appearing during infection and their association with virogenesis have not been established. Structures that have been defined are the virogenic stroma, nucleocapsid, viral envelope, and occlusion bodies; the nature of other structures is still unresolved.

Most studies are concerned principally with the nucleus since it is the site of virogenesis and is transformed drastically. Only limited information is available on cytoplasmic injuries and malformations. Greater emphasis on the pathology of the entire cell, i.e., including the cells of different tissues, would result in a better understanding of the biochemical, molecular, and cytological mechanisms underlying baculovirus infections at the cellular level.

ACKNOWLEDGMENTS

We thank the authors for the use of their figures. Portions of this review were based on studies supported by National Science Foundation Grant PCM-8201247.

References

Adams, J. R., and Wilcox, T. A., 1968, Histopathology of the almond moth, *Cadra cautella*, infected with a nuclear polyhedrosis virus, *J. Invertebr. Pathol.* **12**:269-274.

Adams, J. R., Goodwin, R. H., and Wilcox, T. A., 1977, Electron microscopic investigations on invasion and replication of insect baculoviruses *in vivo* and *in vitro*, *Biol. Cell.* **28**:261-268.

Aizawa, K., 1963, The Nature of infections caused by nuclear-polyhedrosis viruses. In *Insect Pathology, An Advanced Treatise*, edited by E. A. Steinhaus, pp. 381-412, Academic Press, New York.

Arnott, H. J., and Smith, K. M., 1968, An ultrastructural study of the development of a granulosis virus in the cells of the moth *Plodia interpunctella* (Hbn.), *J. Ultrastruct. Res.* **21**:251-268.

Arnott, H. J., and Smith, K. M., 1969, Ultrastructural observations on the branched rods associated with some insect granuloses, *J. Invertebr. Pathol.* **13**:345-350.

Asayama, T., and Inagaki, I., 1975, Cell alteration caused by the infection with the granulosis virus in the diamondback moth, *Plutella xylostella*, and the site of appearance of nucleocapsid, *Jpn. J. Appl. Entomol. Zool.* **19**:79-84.

Asayama, T., and Kawamoto, F., 1975, An electron microscope observation on the fat body cell of the brown tail moth, *Euproctis similis* Fuessly, infected with a nucleopolyhedrosis virus, *Jpn. J. Appl. Entomol. Zool.* **19**:1-9.

Asayama, T., Inagaki, I., Kawamoto, F., and Suto, C., 1974, Electron microscope observations on the maturation process of the nucleopolyhedrosis viruses of the Oriental tussock moth, *Euproctis subflava* Bremer, and the Japanese giant silkworm, *Dictyoploca japonica* Butler, *Jpn. J. Appl. Entomol. Zool.* **18**:189-197.

Bergold, G. H., 1958, Viruses of insects. In *Handbuch der Virusforschung*, edited by C. Hallauer and K. F. Meyer, Vol. 4, pp. 60-142, Springer, Berlin.

Bird, F. T., 1949, Tumors associated with a virus infection in an insect, *Nature (London)* **163**:777.

Bird, F. T., 1952, On the multiplication of an insect virus, *Biochim. Biophys. Acta* **8**:360-368.

Bird, F. T., 1963, On the development of granulosis virus, *J. Insect Pathol.* **5**:368-376.

Bird, F. T., and Whalen, M. M., 1953, A virus disease of the European pine sawfly, *Neodiprion sertifer* (Geoffr.), *Can. Entomol.* **85**:433-437.

Bird, F. T., and Whalen, M. M., 1954, Stages in the development of two insect viruses, *Can. J. Microbiol.* **1**:170-174.

Chung, K. L., Brown, M., and Faulkner, P., 1980, Studies on the morphogenesis of polyhedral inclusion bodies of a baculovirus *Autographa californica* NPV, *J. Gen. Virol.* **46**:335-347.
Clark, T. B., Chapman, H. C., and Fukuda, T., 1969, Nuclear-polyhedrosis and cytoplasmic-polyhedrosis virus infections in Louisiana mosquitoes, *J. Invertebr. Pathol.* **14**:284-286.
Croizier, G., and Quiot, J.-M., 1975. Etude en microscopie électronique des structures cellulaires réticulées induites chez les Lépidoptères par des Baculovirus, *C.R. Acad. Sci. Ser. D* **281**:1055-1057.
Croizier, G., Amargier, A., Godse, D.-B., Jacquemard, P., and Duthoit, J.-L., 1980, Un virus de polyédrose nucléaire découvert chez le lépidoptère Noctuidae *Diparopsis watersi* (Roth.) nouveau variant du *Baculovirus* d'*Autographa californica* (Speyer), *Coton Fibres Trop.* **35**:414-423.
Da Cunha, A. B., Pavan, C., Biesele, J. J., Riess, R. W., and Simoes, L. C. G., 1972, III. An ultrastructural study of the development of a nuclear polyhedrosis with effects on giant polytene chromosomes, *Stud. Genet.* **7**:117-143.
Diaz, M., and Pavan, C., 1965, Changes in chromosomes induced by microorganism infection, *Proc. Natl. Acad. Sci. USA* **54**:1321-1327.
Dougherty, E. M., Vaughn, J. L., and Reichelderfer, C. F., 1975, Characteristics of the non-occluded form of a nuclear polyhedrosis virus, *Intervirology* **5**:109-121.
Falcon, L. A., and Hess, R. T., 1977, Electron microscope study on the replication of *Autographa* nuclear polyhedrosis virus and *Spodoptera* nuclear polyhedrosis virus in *Spodoptera exigua, J. Invertebr. Pathol.* **29**:36-43.
Faulkner, P., 1981, Baculovirus. In *Pathogenesis of Invertebrate Microbial Diseases*, edited by E. W. Davidson, pp. 3-37, Allanheld, Osmum, New Jersey.
Federici, B. A., 1980, Mosquito baculovirus: Sequence of morphogenesis and ultrastructure of the virion, *Virology* **100**:1-9.
Frederici, B. A., and Lowe, R. E., 1972, Studies on the pathology of a baculovirus in *Aedes triseriatus, J. Invertebr. Pathol.* **20**:14-21.
Gouranton, J., 1972, Development of an intranuclear nonoccluded rod-shaped virus in some midgut cells of an adult insect, *Gyrinus natator* L. (Coleoptera), *J. Ultrastruct. Res.* **39**:281-294.
Granados, R. R., 1978, Early events in the infection of *Heliothis zea* midgut cells by a baculovirus, *Virology* **90**:170-174.
Granados, R. R., 1980, Infectivity and mode of action of baculoviruses, *Biotechnol. Bioeng.* **22**:1377-1405.
Granados, R. R., and Lawler, K. A., 1981, *In vivo* pathway of *Autographa californica* baculovirus invasion and infection, *Virology* **108**:297-308.
Granados, R. R., Lawler, K. A., and Burand, J. P., 1981, Replication of *Heliothis zea* baculovirus in an insect cell line, *Intervirology* **16**:71-79.
Hall, D. W., and Fish, D. D., 1974, A *Baculovirus* from the mosquito *Wyeomyia smithii, J. Invertebr. Pathol.* **23**:383-388.
Harrap, K. A., 1970, Cell infection by a nuclear polyhedrosis virus, *Virology* **42**:311-318.
Harrap, K. A., 1972a, The structure of nuclear polyhedrosis viruses. I. The inclusion body, *Virology* **50**:114-123.
Harrap, K. A., 1972b, The structure of nuclear polyhedrosis viruses. III. Virus assembly, *Virology* **50**:133-139.
Harrap, K. A., and Robertson, J. S., 1968, A possible infection pathway in the development of a nuclear polyhedrosis virus, *J. Gen. Virol.* **3**:221-225.
Henderson, J. F., Faulkner, P., and MacKinnon, E. A., 1974, Some biophysical properties of virus present in tissue cultures infected with the nuclear polyhedrosis virus of *Trichoplusia ni, J. Gen. Virol.* **22**:143-146.
Hess, R. T., and Falcon, L. A., 1977, Observations on the interaction of baculoviruses with the plasma membrane, *J. Gen. Virol.* **36**:525-530.
Hess, R. T., and Falcon, L. A., 1981, Electron microscope observations of *Autographa californica* (Noctuidae) nuclear polyhedrosis virus replication in the midgut of the saltmarsh caterpillar, *Estigmene acraea* (Arctiidae), *J. Invertebr. Pathol.* **37**:86-90.
Hess, R. T., Poinar, G. O., Jr., and Caltagirone, L. E., 1979, DNA-containing particles in the calyx of *Phanerotoma flavitestacea* (Hymenoptera: Braconidae), *J. Invertebr. Pathol.* **33**:129-132.

Hirumi, H., Hirumi, K., and McIntosh, A. H., 1975, Morphogenesis of a nuclear polyhedrosis virus of the alfalfa looper in a continuous cabbage looper cell line, *Ann. N.Y. Acad. Sci.* **266:**302-326.

Hudson, J. S., Carner, G. R., and Barnett, O. W., 1979, Ultrastructure of fat body cells of the velvetbean caterpillar infected with a nuclear polyhedrosis virus, *J. Invertebr. Pathol.* **33:**31-39.

Huger, A., 1960, Über die Natur des Fadenwerkes bei der granulose von *Choristoneura fumiferana* (Hbn.) (Lepidoptera, Tortricidae), *Naturwissenschaften* **15:**385-389.

Huger, A., 1963, Granuloses of insects. In *Insect Pathology, An Advanced Treatise,* edited by E. A. Steinhaus, pp. 531-575, Academic Press, New York.

Huger, A., 1965, Ein neuer Typ von Insektenviren aus malaiischen Populationen von *Oryctes rhinoceros* (L.) (Col., Scarabaeidae), *Naturwissenschaften* **52:**542.

Huger, A. M., 1966, A virus disease of the Indian rhinoceros beetle, *Oryctes rhinoceros* (Linnaeus), caused by a new type of insect virus, *Rhabdionvirus oryctes* gen. n., sp. n., *J. Invertebr. Pathol.* **8:**38-51.

Huger, A. M., 1970, Report on the activities of the "Institut fur Biologische Schadlingsbekampfung," Darmstadt, Germany: Further diagnostic and histopathological studies on *Oryctes rhinoceros* and two of its predators. UNDP/SPC project for research on the control of the coconut palm rhinoceros beetle. Semi-Annu. Rep. Proj. Manager, Nov. 1969 to May 1970, pp. 16-19.

Huger, A., and Krieg, A., 1961, Electron microscope investigations on the virogenesis of the granulosis of *Choristoneura murinana* (Hübner), *J. Insect Pathol.* **3:**183-196.

Hughes, K. M., 1952, Development of the inclusion bodies of a granulosis virus, *J. Bacteriol.* **64:**375-380.

Hughes, K. M., 1972, Fine structure and development of two polyhedrosis viruses, *J. Invertebr. Pathol.* **19:**198-207.

Hughes, K. M., and Addison, R. B., 1970, Two nuclear polyhedrosis viruses of the Douglas-fir tussock moth, *J. Invertebr. Pathol.* **16:**196-204.

Hunter, D. K., and Hoffmann, D. F., 1972, Cross infection of a granulosis virus of *Cadra cautella,* with observations on its ultrastructure in infected cells of *Plodia interpunctella, J. Invertebr. Pathol.* **20:**4-10.

Hunter, D. K., Hoffmann, D. F., and Collier, S. J., 1975, Observations on a granulosis virus of the potato tuberworm, *Phthorimaea operculella, J. Invertebr. Pathol.* **26:**397-400.

Ignoffo, C. M., 1968, Specificity of insect viruses, *Bull. Entomol. Soc. Am.* **14:**265-276.

Injac, M., Vago, C., Duthoit, J.-L., and Veyrunes, J.-C., 1971. Libération (Release) des virions dans les polyédroses nucléaires, *C.R. Acad. Sci. Ser. D.* **273:**439-441.

Injac, M., Duthoit, J.-L., and Amargier, A., 1973, Étude histo- et cytopathologique d'une polyèdrose nuclèaire de lècaille fileuse (*Hyphantria cunea* Drury) Lepidoptera, Arctiidae, *Ann. Zool. Ecol. Anim.* **5:**99-110.

Kawamoto, F., Asayama, T., and Kobayashi, M., 1976, Acquisition of the envelope of nuclear polyhedrosis viruses in the Chinese oak silkworm, *Antheraea pernyi* Guer-Min, and the Japanese giant silkworm, *Dictyoploca japonica* Butler, *Appl. Entomol. Zool.* **11:**59-69.

Kawamoto, F., Suto, C., Kumada, N., and Kobayashi, M., 1977, Cytoplasmic budding of a nuclear polyhedrosis virus and comparative ultrastructural studies of envelopes, *Microbiol. Immunol.* **21:**255-265.

Kawanishi, C. Y., Summers, M. D., Stoltz, D. B., and Arnott, H. J., 1972, Entry of an insect virus *in vivo* by fusion of viral envelope and microvillus membrane, *J. Invertebr. Pathol.* **20:**104-108.

Kelly, D. C., 1976, "*Oryctes*" virus replication: Electron microscopic observations on infected moth and mosquito cells, *Virology* **69:**596-606.

Knudson, D. L., and Harrap, K. A., 1976, Replication of a nuclear polyhedrosis virus in a continuous cell culture of *Spodoptera frugiperda:* Microscopy study of the sequence of events of the virus infection, *J. Virol.* **17:**254-268.

Krieg, A., and Huger, A. M., 1969, New ultracytological findings in insect nuclear polyhedroses, *J. Invertebr. Pathol.* **13:**272-279.

MacKinnon, E. A., Henderson, J. F., Stoltz, D. B., and Faulkner, P., 1974, Morphogenesis of nuclear polyhedrosis virus under conditions of prolonged passage *in vitro, J. Ultrastruct. Res.* **49:**419-435.

Marschall, K. J., 1971, Report of K. J. Marschall (Insect Pathologist). UNDP/SPC project for research on the control of the coconut palm rhinoceros beetle. Rep. Proj. Manager, June 1970 to May 1971, pp. 166-191.

Marschall, K. J., 1972, Report of K. J. Marschall (Insect Pathologist). UNDP/SPC project for research on the control of the coconut palm rhinoceros beetle. Rep. Proj. Manager, June 1971 to June 1972.

Marschall, K. J., 1973, Histopathology of *Rhabdionvirus* infections. UNDP/FAO project for research on the control of the coconut palm rhinoceros beetle (phase II). Rep. Proj. Manager, July 1972 to June 1973.

Martignoni, M. E., and Iwai, P. J., 1981, A catalogue of viral diseases of insects, mites and ticks. In *Microbial Control of Pests and Plant Diseases 1970-1980*, edited by H. D. Burges, pp. 897-911, Academic Press, New York.

Matthews, R. E. F., 1982, Classification and nomenclature of viruses, *Intervirology* 17:1-199.

Monsarrat, P., Meynadier, G., Croizier, G., and Vago, C., 1973, Recherches cytopathologiques sur une maladie virale du Coléoptère *Oryctes rhinoceros* L., *C.R. Acad. Sci. Ser. D* 276:2077-2080.

Neilson, M. M., and Elgee, D. E., 1968, Tumorlike bodies in virus-infected and noninfected adults of the spruce sawfly, *Diprion hercyniae*, *J. Invertebr. Pathol.* 10:70-75.

Paschke, J. D., and Summers, M. D., 1975, Early events in the infection of the arthropod gut by pathogenic insect viruses. In *Invertebrate Immunity*, edited by K. I. Maramorosch and R. E. Shope, pp. 75-112, Academic Press, New York.

Pavan C., Da Cunha, and Morsoletto, C., 1971, Virus-chromosome relationships in cells of *Rhynchosciara* (Diptera, Sciaridae), *Caryologia* 24:371-389.

Pavan, O. H., Boucias, D. G., and Pendland, J. C., 1981, The effects of serial passage of a nuclearpolyhedrosis virus through an alternate host system, *Entomophaga* 26:99-108.

Petre, A., and Ploaie, P., 1969, Cell proliferations in *Lymantria dispar* L. larvae infected with the nuclear polyhedrosis virus, *Experientia* 25:842-844.

Pinnock, D. E., and Hess, R. T., 1977, Electron microscope observations on granulosis virus replication in the fruit tree leaf roller, *Archips argyrospila:* Infection of the midgut, *J. Invertebr. Pathol.* 30:354-361.

Pinnock, D. E., and Hess, R. T., 1978, Morphological variations in the cytopathology associated with granulosis virus in the fruit-tree leaf foller, *Archips argyrospila*, *J. Ultrastruct. Res.* 63:252-260.

Poinar, G. O., Jr., Hess, R., and Caltagirone, L. E., 1976, Virus-like particles in the calyx of *Phanerotoma flavitestacea* (Hymenoptera: Braconidae) and their transfer into host tissues, *Acta Zool.* 59:161-165.

Quiot, J.-M., Monsarrat, P., Meynadier, G., Croizier, G., and Vago, C., 1973, Infection des culture cellulaires de Coléoptères par le "virus *Oryctes*" *C.R. Acad. Sci. Ser. D* 276:3229-3231.

Raghow, R., and Grace, T. D. C., 1974, Studies on a nuclear polyhedrosis virus in *Bombyx mori* cells *in vitro*. 1. Multiplication kinetics and ultrastructural studies, *J. Ultrastruct. Res.* 47:384-399.

Rennie, J., 1923, Polyhedral disease in *Tipula paludosa* (Meigen), *Proc. R. Soc. Edinburgh Ser. A* 20:265-267.

Ritter, K. S., Tanada, Y., Hess, R. T., and Omi, E. M., 1982, Eclipse period of baculovirus infection in larvae of the armyworm, *Pseudaletia unipuncta*, *J. Invertebr. Pathol.* 39:203-209.

Rotheram, S., 1967, Immune surface of eggs of a parasitic insect, *Nature (London)* 214:700.

Rotheram, S., 1973, The surface of the egg of a parasitic insect. II. The ultrastructure of the particulate coat on the egg of *Nemeritus*, *Proc. R. Soc. London Ser. B* 183:195-204.

Smith, K. M., 1955, Intranuclear changes in the polyhedrosis of *Tipula paludosa* (Meig.), *Parasitology* 45:482-487.

Smith, K. M., 1976, *Virus-Insect Relationships*, Longman Group, London.

Smith, K. M., and Xeros, N., 1954, An unusual virus disease of a dipterous larva, *Nature (London)* 173:866-867.

Smith, O. J., Hughes, K. M., Dunn, P. H., and Hall, I. M., 1956, A granulosis virus disease of the Western grape leaf skeletonizer and its transmission, *Can Entomol.* 88:507-515.

Stoltz, D. B., and Vinson, S. B., 1977, Baculovirus-like particles in the reproductive tracts of female parasitoid wasps. II. The genus *Apanteles*, *Can. J. Microbiol.* 23:28-37.

Stoltz, D. B., and Vinson, S. B., 1979a, Viruses and parasitism in insects, *Adv. Virus Res.* 24:125-176.

Stoltz, D. B., and Vinson, S. B., 1979b, Penetration into caterpillar cells of virus-like particles injected during oviposition by parasitoid ichneumonid wasps, *Can. J. Microbiol.* **25:**207-216.

Stoltz, D. B., Pavan, C., and Da Cunha, A. B., 1973, Nuclear polyhedrosis virus: A possible example of *de novo* intranuclear membrane morphogenesis, *J. Gen. Virol.* **19:**145-150.

Stoltz, D. B., Vinson, S. B., and MacKinnon, E. A., 1976, Baculovirus-like particles in the reproductive tracts of female parasitoid wasps, *Can. J. Microbiol.* **22:**1013-1023.

Summers, M. D., 1969, Apparent *in vivo* pathway of granulosis virus invasion and infection, *J. Virol.* **4:**188-190.

Summers, M. D., 1971, Electron microscopic observations on granulosis virus entry, uncoating and replication processes during infection of the midgut cells of *Trichoplusia ni*, *J. Ultrastruct. Res.* **35:**606-625.

Summers, M. D., 1977, Baculoviruses (Baculoviridae). In *The Atlas of Insect and Plant Viruses*, edited by K. Maramorosch, pp. 3-28, Academic Press, New York.

Summers, M. D., and Arnott, H. J., 1969, Ultrastructural studies on inclusion formation and virus occlusion in nuclear polyhedrosis and granulosis virus-infected cells of *Trichoplusia ni* (Hübner), *J. Ultrastruct. Res.* **28:**462-480.

Summers, M. D., and Volkman, L. E., 1976, Comparison of biophysical and morphological properties of occluded and extracellular nonoccluded baculovirus from *in vivo* and *in vitro* host systems, *J. Virol.* **17:**962-972.

Tanada, Y., and Hess, R. T., 1976, Development of a nuclear polyhedrosis virus armyworm, *Pseudaletia unipuncta*, *J. Invertebr. Pathol.* **28:**67-76.

Tanada, Y., and Leutenegger, R., 1968, Histopathology of a granulosis-virus disease of the codling moth, *Carpocapsa pomonella*, *J. Invertebr. Pathol.* **10:**39-47.

Tanada, Y., and Leutenegger, R., 1970, Multiplication of a granulosis virus in larval midgut cells of *Trichoplusia ni* and possible pathways of invasion into the hemocoel, *J. Ultrastruct. Res.* **30:**589-600.

Tanada, Y., Hukuhara, T., and Chang, G. Y., 1969, A strain of nuclear-polyhedrosis virus causing extensive cellular hypertrophy, *J. Invertebr. Pathol.* **13:**394-409.

Tanada, Y., Hess, R. T., and Omi, E. M., 1975, Invasion of a nuclear polyhedrosis virus in midgut of the armyworm, *Pseudaletia unipuncta*, and the enhancement of a synergistic enzyme, *J. Invertebr. Pathol.* **26:**99-104.

Tanada, Y., Inoue, H., Hess, R. T., and Omi, E. M., 1980, Site of action of synergistic factor of a granulosis virus of the armyworm, *Pseudaletia unipuncta*, *J. Invertebr. Pathol.* **34:**249-255.

Tanada, Y., Hess, R. T., and Omi, E. M., 1982, Unique virus morphogenesis and cytopathology of a baculovirus (hypertrophy strain) in larva of the armyworm, *Pseudaletia unipuncta*, *J. Invertebr. Pathol.* **40:**197-204.

Tinsley, T. W., and Harrap, K. A., Viruses of invertebrates. In *Comprehensive Virology*, vol. 12, edited by M. Fraenkel-Conrat and R. R. Wagner, pp. 1-101, Plenum Press, New York.

Walker, S., Kawanishi, C. Y., and Hamm, J. J., 1982, Cellular pathology of a granulosis virus infection, *J. Ultrastruct. Res.* **80:**163-177.

Xeros, N., 1956, The virogenic stroma in nuclear and cytoplasmic polyhedrosis, *Nature (London)* **178:**412-413.

Yamamoto, T., and Tanada, Y., 1977, Possible involvement of phospholipids in the infectivity of baculoviruses, *J. Invertebr. Pathol.* **30:**279-281.

Yamamoto, T., and Tanada, Y., 1978a, Biochemical properties of viral envelopes of insect baculoviruses and their role in infectivity, *J. Invertebr. Pathol.* **32:**202-211.

Yamamoto, T., and Tanada, Y., 1978b, Phospholipid, an enhancing component in the synergistic factor of a granulosis virus of the armyworm, *Pseudaletia unipuncta*, *J. Invertebr. Pathol.* **31:**48-56.

Zelazny, B., 1976, Transmission of a baculovirus in populations of *Oryctes rhinoceros*, *J. Invertebr. Pathol.* **27:**221-227.

III

The Ultrastructure of Cells in Pathological States

15

Comparative Ultrastructure of Wild-Type and Tumorous Cells of *Drosophila*

ELISABETH GATEFF, ROSHANA SHRESTHA, AND HIROMU AKAI

1. Introduction

In *Drosophila*, genetic factors cause malignant and benign neoplasms (Gateff, 1978a,b,c). In addition, compactly growing lethal tumors have been obtained from eye-antennal imaginal discs during serial subculture in the abdomens of female flies (Gateff, 1978a,b). The following tumor types have been found: (1) lethal-benign imaginal disc neoplasms, (2) malignant neuroblastomas, (3) malignant blood cell neoplasms, and (4) a benign gonial cell neoplasm. The fine structure of some of the above tumors has been studied and compared to the fine structure of corresponding wild-type cells. These will be discussed below.

2. Comparative Ultrastructure of Wild-Type and Tumorous Imaginal Discs

The epithelium of a wild-type imaginal disc represents a folded monolayer of cells which connect to each other with junctional complexes. The basal cell surface is in contact with a basement membrane, and the apical portion of the

---

ELISABETH GATEFF • Institute of Genetics, Johannes Gutenberg University, Mainz 6500, Federal Republic of Germany.   ROSHANA SHRESTHA • Department of Zoology, Triburan University, Kathmandu, Nepal.   HIROMU AKAI • The Sericultural Experiment Station, Yatabe, Ibaraki 305, Japan.

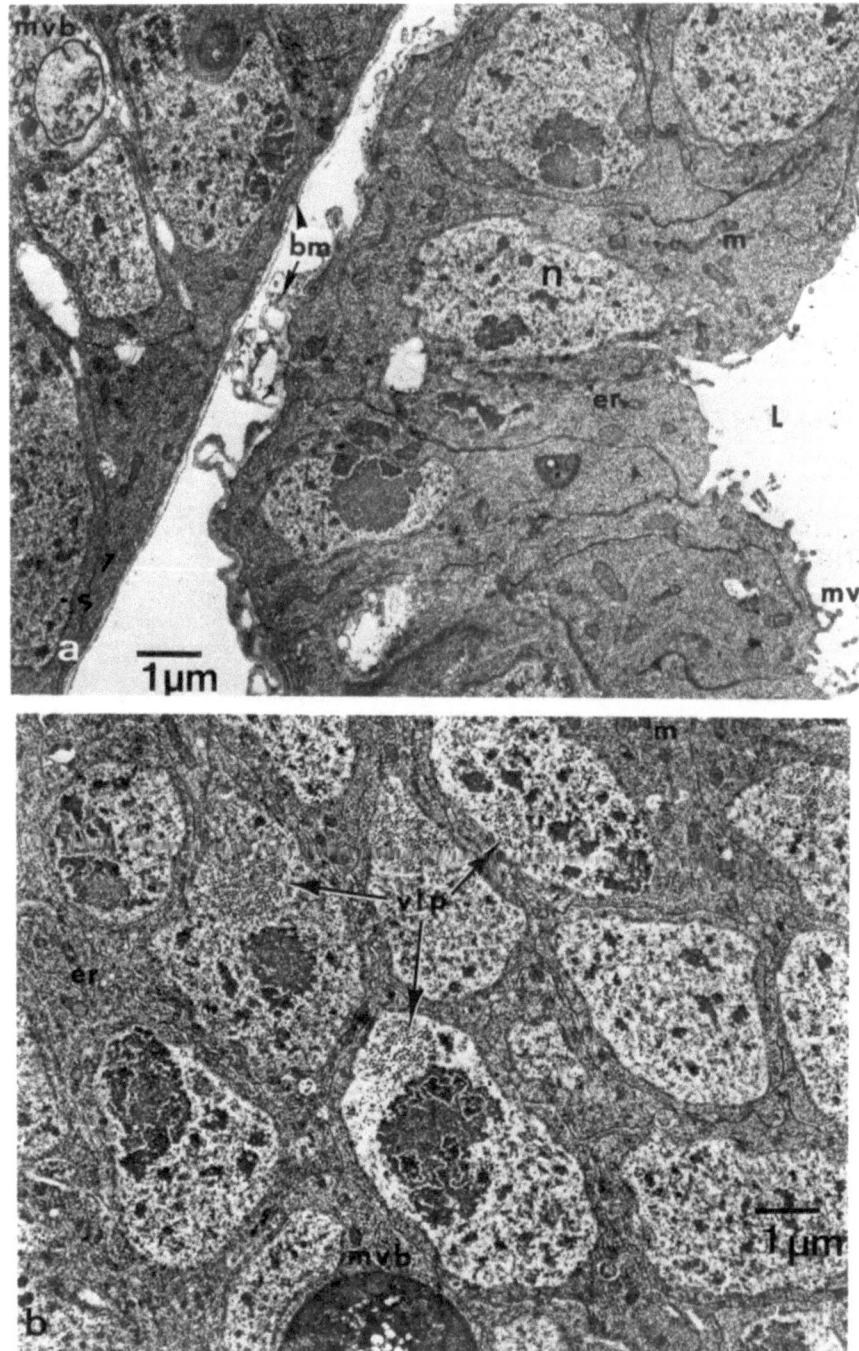

Figure 1. (a) An electron micrograph of allotypic tissue, derived from subline I_3 in the 18th transfer generation, which resembles normal imaginal disc cells. The cells are arranged in monolayers. Note the basement membrane (bm) and the microvilli (mv) on the apical surface facing the lumen (L). The basement membrane of the left cell layer is smooth, whereas that of the right cell layer is irregular and

cell shows microvilli and serves as the secretory surface for chitin. The elongated nuclei are located in the lower third of the cells. They show marginally located nucleoli and irregularly dispersed heterochromatin. The ribosome-rich cytoplasm exhibits few mitochondria, ribbons of RER, and considerable numbers of lipid droplets. For details on the fine structure of wild-type imaginal discs, consult Poodry (1980).

Imaginal disc tumors have been found in 12 nonallelic, recessive lethal mutants (Gateff, 1978a,b,c). Using Hadorn's method (1966), Gateff (1978a) has shown that tumors also originate during subculture of wild-type imaginal discs in female flies. Ultrastructural studies have been performed on imaginal discs of the mutant *lethal (2) giant larvae*^{4} [$l(2)gl^1$] and on differentiating and tumorous tissue sublines, derived from a wild-type eye-antennal disc after subculture in wild-type adult flies for various periods of time (Gateff et al., 1974).

The transition from the differentiating into the tumorous state of the above tissue sublines and the ultrastructural changes associated with the altered differentiation capacity of the cells will be described first. Up to the eighth transfer generation, the eye-antennal disc tissue grew moderately fast and differentiated into eye-antennal cuticular structures (autotypic tissue; Hadorn, 1966). The morphology, histology, and fine structure of autotypic tissue resembled closely those of wild-type imaginal discs. During the eighth transfer generation, however, the developmental capacity of the various tissue sublines changed. In some sublines, transdetermination took place (allotypic tissue; Hadorn, 1966), while in others, the capacity for differentiation was entirely lost (atelotypic tissue; Hadorn, 1966). The transdetermined, allotypic tissue sublines differentiated, up to the 22nd transfer generation, into both autotypic eye-antennal and allotypic genital cuticular patterns of normal (normotypic) appearance. After the 22nd transfer generation, the cuticular patterns into which allotypic tissue sublines metamorphosed became progressively abnormal (anormotypic tissue; Hadorn, 1966) and increasingly larger portions of the tissue became atelotypic. This gradual transformation from normotypic into anormotypic and further into atelotypic tissue, continued up to the 40th transfer generation. After this, the tissue became entirely atelotypic and, thus, represented an autonomously growing lethal-benign neoplasm.

These changes of the capacity to differentiate were correlated with increased growth rates and aberrant morphology, histology, and fine structure (Gateff et al., 1974). The tissue sublines, consisting of auto- and allotypic tissue, showed only minor differences, when their fine structure was compared to that of wild-type imaginal discs or autotypic tissue (Figure 1a). The cells were cuboidal as opposed to columnar in wild-type imaginal discs. They showed normal numbers of junctional complexes, apical microvilli, and usually straight and

---

curled (arrows). See text for further discussion. er, endoplasmic reticulum; m, mitochondrion; mvb, multivesicular body; n, nucleus. (b) An electron micrograph of a region exhibiting compactly arranged cells from anormotypic, transdetermined tissue subline I_2 in the 24th transfer generation. Notice the clusters of viruslike particles (vlp) in the nuclei of the cells (arrows) and the multivesicular bodies (mvb). m, mitochondrion. From Gateff et al. (1974).

smooth basement membranes (Figure 1a). In the cytoplasm, densely packed ribosomes, few, small mitochondria, short segments of RER, and Golgi bodies were observed. Allotypic tissue contained few lipid droplets, in contrast to mature wild-type imaginal disc cells *in situ*. Increased numbers of viruslike particles (vlp) appeared primarily in the nuclei (Akai *et al.*, 1967).

After the 22nd transfer generation, more and more cells began to lose their basal-apical polarity and clumped together (Figure 1b). Thus, the tissue represented a mosaic of cells in monolayers and cells in clusters. The basement membrane showed striking changes, such as detachment from the basal cell surface and curling. A beginning of this process also can be seen occasionally in allotypic tissue sublines (Figure 1a). In parallel with abnormalities of the basement membrane, the cells also showed poorly developed microvilli. This apparently coincides with reduced secretory activity and, thus, abnormal cuticular patterns. In regions with a compact cellular arrangement, neither basement membrane nor microvilli were formed, and thus no cuticle could be secreted. The cells in such regions exhibited fewer junctions, no lipid droplets, but considerable concentrations of multivesicular bodies and vlp (Figure 1b). Thus, after 40 transfer generations, the tissue sublines transformed gradually into atelotypic tissue exhibiting the typical characteristics of malignant or benign neoplasms, such as rapid, autonomous, growth associated with a loss of the capacity for differentiation.

As already mentioned, another portion of the tissue sublines transformed directly from the autotypic to the atelotypic (tumorous) state. The cells of these tissue sublines showed even more extreme fine structural aberrations when compared to the wild-type cells. Figure 2 shows such a directly transformed atelotypic tissue from the 21st transfer generation. The cells and their nuclei were extremely variable in size and shape. Compared to the cells in the compact regions of anormotypic tissue, the cells of directly transformed, atelotypic tissue exhibited many cytoplasmic processes and granular and membranous material in the intercellular spaces (compare Figures 1b and 2). Junctional complexes among the cells were rare. The electron density of the cytoplasm varied considerably from cell to cell, the majority of the cells being of the electron-dense type. Such cells contained more ribosomes, RER, mitochondria, and multivesicular bodies than the electron-lucent cells. Large numbers of vlp were observed in the nuclei and less often in the cytoplasm (Akai *et al.*, 1967). Degenerating cells were frequently seen. Thus, the above studies established close correlations between the capacity for cuticular pattern secretion and the morphology, histology, and fine structure of imaginal disc-derived cells subcultured *in vivo*.

Of the 12 mutants causing imaginal disc tumors, the fine structure of only $l(2)gl^4$ imaginal discs has been studied (Gateff and Schneiderman, 1974). The imaginal discs from $l(2)gl^4$ larvae represent clumps of cells which closely resemble cells from the directly transformed, atelotypic tissue sublines mentioned above. Like the cells of atelotypic tissue, $l(2)gl^4$ imaginal disc cells show no apical-basal polarity. Instead, cytoplasmic processes are distributed all over their surfaces, and the numbers of junctions among the cells are reduced.

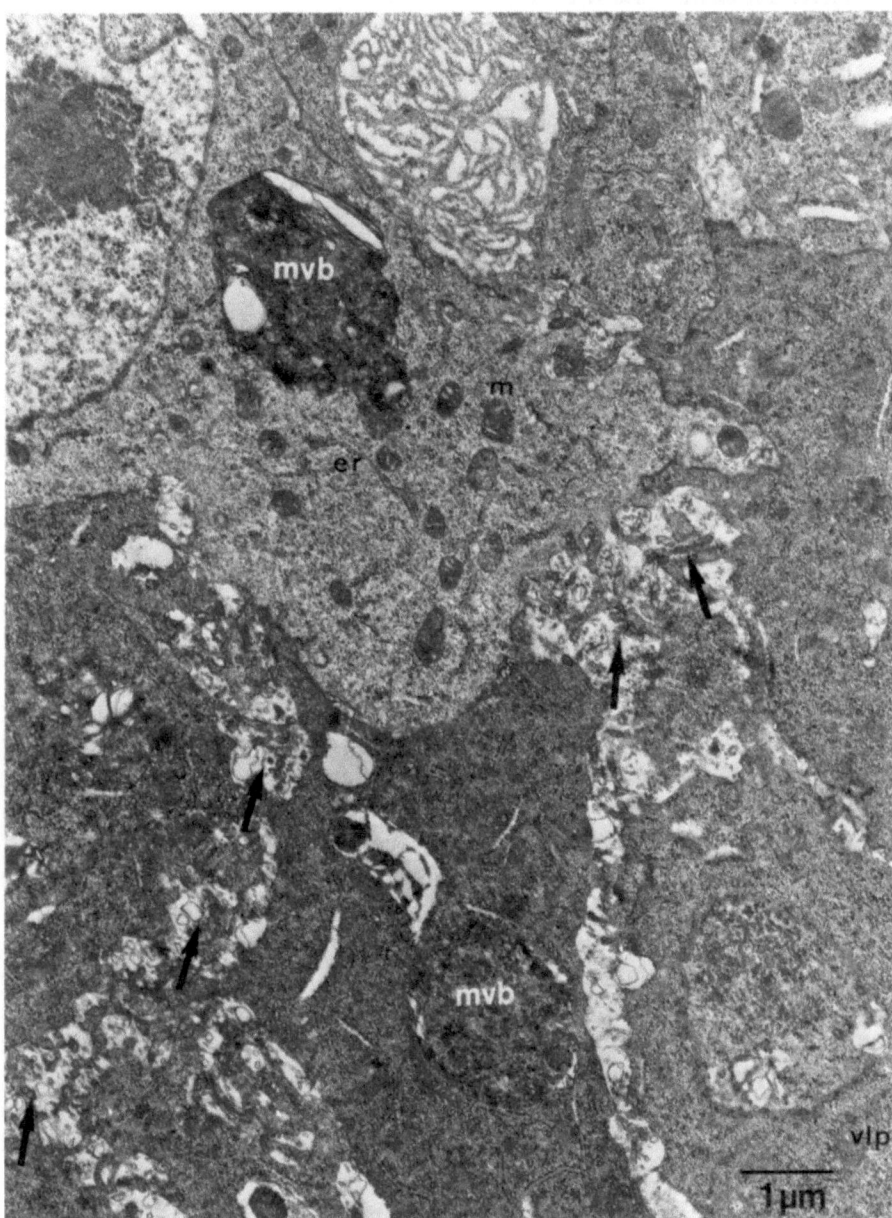

Figure 2. An electron micrograph of "directly transformed," atelotypic tissue subline I_6 in the 21st transfer generation, showing cells of various electron densities. Extensive intercellular spaces with numerous cytoplasmic processes are obvious (arrows). er, endoplasmic reticulum; m, mitochondrion; mvb, multivesicular body; vlp, viruslike particles. From Gateff *et al.* (1974).

3. Comparative Ultrastructure of Wild-Type and Malignant Optic Neuroblasts and Ganglion Mother Cells

Of the seven nonallelic mutants developing malignant neuroblastomas, only the $l(2)gl^4$ neuroblasts and ganglion mother cells were studied at the ultrastructural level. In order to understand the fine structural changes in the mutant optic neuroblasts and ganglion mother cells, a brief outline will follow of the ultrastructure of the cell types in the presumptive adult optic centers in the mature, third-instar, wild-type larval brain.

3.1. Ultrastructure of the Neuroblasts and Ganglion Mother Cells in the Mature Wild-Type Larval Brain

For details on the fine structure and development of the wild-type nervous system and the presumptive adult optic centers in the larval brain, consult Kankel *et al.* (1980) and Hofbauer (1979), respectively.

The larval nervous system is surrounded by the perilemma which consists of a monolayer of syncytial glial cells (the perineurium), and the neurilemma, a fibrous membrane secreted by the above glial cells. Beneath the perilemma is the cellular cortex made up of the following cell types: (1) neuroblasts, (2) ganglion mother cells, (3) neurons, (4) neurosecretory cells, and (5) several types of glial cells. The neuropil is located centrally (Figure 3a). The lateral portion of each brain hemisphere is occupied by the presumptive optic centers, which during adult development differentiate into the three optic glomeruli. The presumptive adult optic centers consist of large neuroblasts restricted to the outer and inner formation centers. Located in between are many layers of smaller optic ganglion mother cells and optic neurons (Figure 3a). We will be concerned here only with the fine structure of the adult optic neuroblasts and ganglion mother cells since they are the malignant cells in the $l(2)gl^4$ neuroblastoma.

The outer optic formation center is located under the perineurium, while the inner is adjacent to the neuropil (Figure 3a). The neuroblasts, found in the optic formation centers, are oval with an average size of 5×12 μm, and its nucleus measures 1.5×3 μm. Except for the differences in cell size among the optic neuroblasts and the ganglion mother cells, no striking fine structural variations were found. Both cell types show a cytoplasm of medium electron density with numerous ribosomes, moderate amounts of mitochondria, and short segments of RER (Figure 3b).

3.2. Ultrastructure of the Neuroblasts and Ganglion Mother Cells in the Mature $l(2)gl^4$ Larval Brain

Compared to the size of the brain hemispheres in a mature wild-type larva (150–170 μm in diameter), the $l(2)gl^4$ brain hemispheres are enlarged (200–250 μm). The cause for this increase in size are neuroblastomas developing in the presumptive adult optic centers. The five cell types found in the wild-type brain

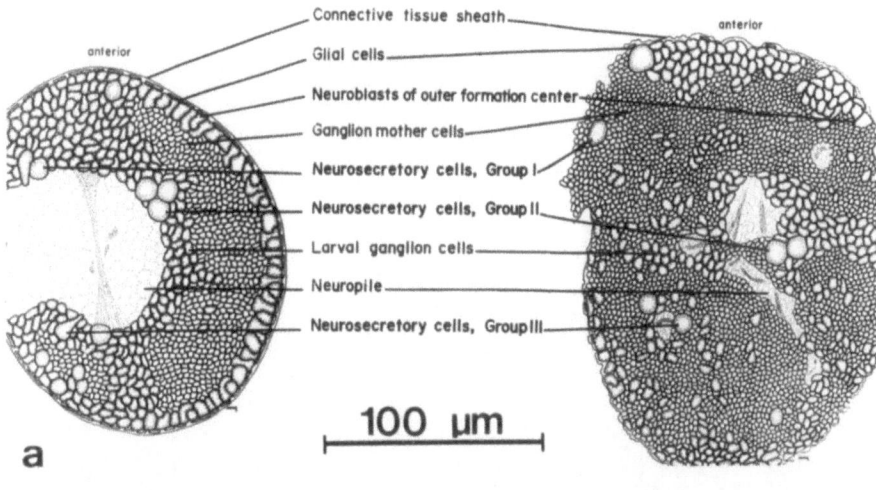
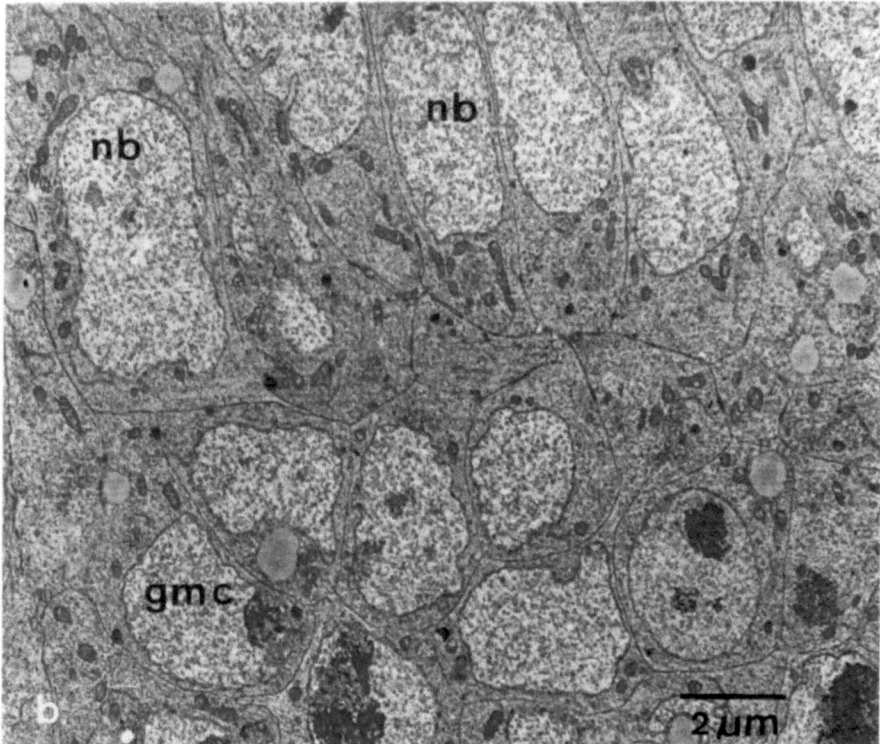

Figure 3. (a) A schematic drawing of a frontal section through a wild-type and an $l(2)gl^4$ brain hemisphere of mature third-instar larvae. The neuroblasts of the inner optic formation center, located close to the neuropil, are not shown. From Gateff and Schneiderman (1969). (b) An electron micrograph from the presumptive adult optic centers in the brain of a 118-hr-old wild-type larva. The outer formation center is represented by the row of elongated optic neuroblasts (nb). Below it are round ganglion mother cells (gmc). All cells are invested by thin glial sheaths. Courtesy of Dr. A. Hofbauer.

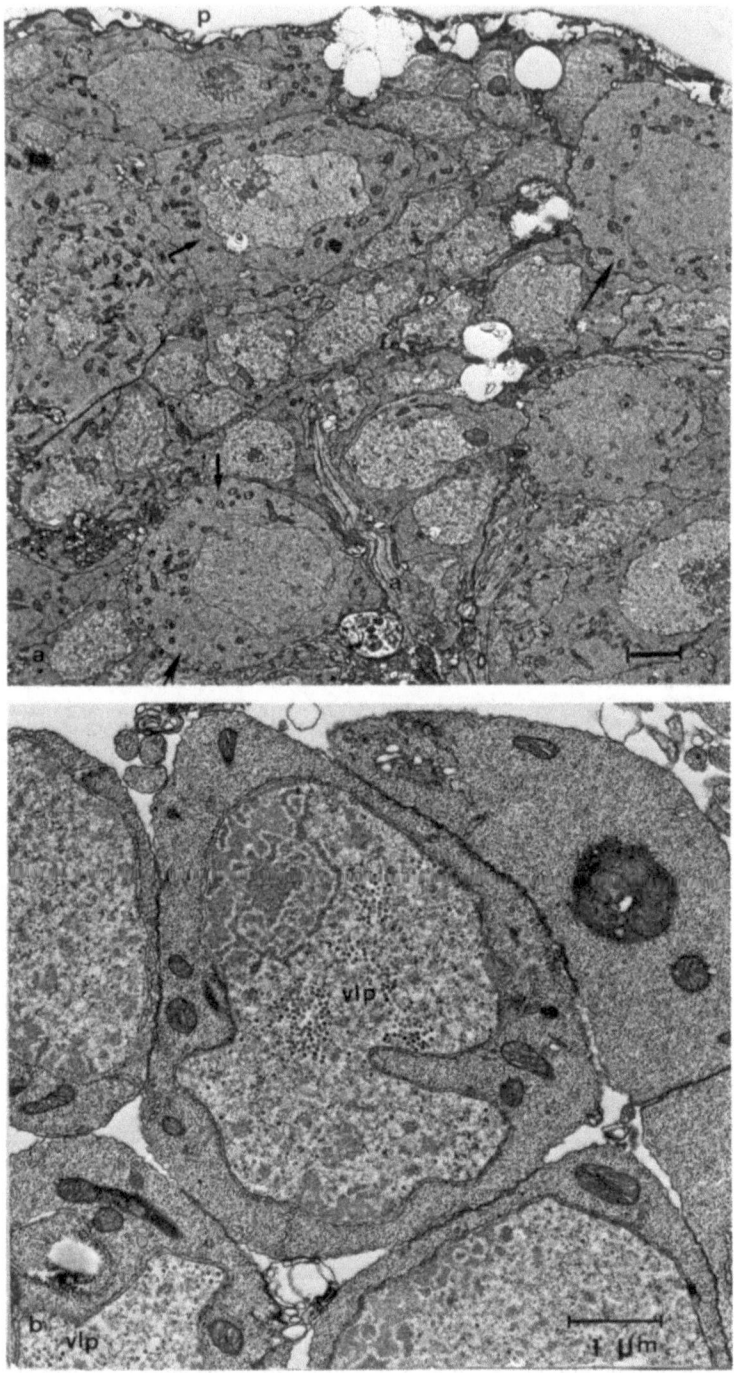

Figure 4. (a) An electron micrograph from the lateral region of an $l(2)gl^4$ brain from a mature larva. Beneath the perilemma (p), large neuroblasts (arrows) and smaller ganglion mother cells surrounded by fine glial processes can be seen. A few axons (a) are present in the lower portion of the micrograph. (b) An electron micrograph of $l(2)gl^4$ neuroblasts cultured *in vivo* in wild-type female flies for seven transfer generations. Note the abnormal nuclear shape, the viruslike particles (vlp), the absence of glia, and the lack of specialized junctions.

are also present in the mutant brain. However, optic neuroblasts and ganglion mother cells are the predominant cell types (Figure 3a). Unlike the normal brain, the different cell types are not restricted to specific areas, but are intermixed in the mutant brain (compare Figures 3 and 4a). The optic neuroblasts are not arranged into an inner and an outer formation center as in the wild-type brain, but are found interspersed between the ganglion mother cells (Figure 4a). Furthermore, neuroblasts as well as ganglion mother cells invade the healthy portion of the brain. In contrast to the wild-type brain, many degenerating cells occur throughout the mutant brain.

At the fine structural level, wild-type and tumorous neuroblasts and ganglion mother cells show minor differences. They possess more cytoplasm and mitochondria and thus are generally larger than their wild-type counterparts (Figure 4a). Neuroblasts, for instance, show an average size of 7×12.5 μm as opposed to 5×12 μm for the wild-type cells. Mutant neurosecretory cells are also larger and store increased amounts of neurosecretory droplets. This may result from the interruption or destruction of the axons by the invading tumor cells (Akai, 1978). All cells are invested in thin glial sheaths. The increased numbers of cells in the mutant brain stretch the perilemma and sometimes cause it to break.

The malignant characteristics of the $l(2)gl^4$ neuroblasts and ganglion mother cells became even more pronounced after transplantation of small pieces of mutant brain into the abdomens of wild-type flies. Here they continued to grow in a rapid, autonomous, invasive and lethal fashion (Gateff and Schneiderman, 1969). Electron micrographs were taken of tumors from the first, second, and seventh transfer generations. These growths consisted of cells that resembled neuroblasts and ganglion mother cells (Figure 4b). Glial cells and neurosecretory cells were never observed. The cells did not form junctions with each other. The large variations of size and shape, characteristic for the malignant neuroblasts and ganglion mother cells and their nuclei *in situ*, increased still further with the time in culture *in vivo*. Many of the nuclei showed abnormal lobulations and harbored large numbers of vlp (Figure 4b; Akai *et al.*, 1967). At the ultrastructural level, there are no significant differences between the *in vivo*-subcultured neuroblasts and ganglion mother cells and the ones observed *in situ* (compare Figures 4a and b).

4. Comparative Ultrastructure of Wild-Type and Tumorous Blood Cells

In the hemolymph of a fully grown, third-instar, wild-type larva, two cell types can be distinguished. These are the plasmatocytes and the crystal cells (Rizki, 1957). At the end of larval life, plasmatocytes differentiate into podocytes and lamellocytes. The above two types of blood cells originate in the lymph glands, which are considered the larval hematopoietic organs (Stark and Marshall, 1930; Gateff, 1977; Shrestha and Gateff, 1982a). Plasmatocytes originate in the lymph gland from proplasmatocytes and the crystal cells from procrystal cells (Gateff, 1977; Shrestha and Gateff, 1982a).

Before the discussion of the malignant blood cells, the ultrastructure of the primordial blood cells in the wild-type, larval hematopoietic organs and the free blood cells in the hemolymph will be considered.

4.1. Ultrastructure of Blood Cell Precursors in the Hematopoietic Organs and of the Free Blood Cells in the Hemolymph of Wild-Type Larvae

In fine structure preparations of the hematopoietic organs from fully grown, wild-type larvae, three types of cells can be observed: the prohemocytes, the proplasmatocytes, and the procrystal cells. Prohemocytes are undifferentiated with the typical features of embryonic cells, such as a ribosome-rich cytoplasm with few mitochondria, small amounts of RER, Golgi complexes, primary lysosomes, and large nuclei. Prohemocytes differentiate into proplasmatocytes which show almost 50% more primary lysosomes than prohemocytes. Causally related to the differentiation of the primary lysosomal system is the formation of Golgi complexes and the increase of RER (Shrestha and Gateff, 1982a).

At the end of larval life, proplasmatocytes enter the hemolymph and differentiate into plasmatocytes. Plasmatocytes are round to oval and characterized by a threefold increase of the number of primary lysosomes over that of proplasmatocytes. They possess well-developed Golgi complexes and RER (Figure 5a). Phagocytotic vacuoles occur sporadically in them.

Plasmatocytes transform into podocytes which show extensive cytoplasmic processes and a still further increase in the number of primary lysosomes. The transformation of a podocyte into a lamellocyte involves a flattening to a 1-μm thickness and an increase of cell size (plasmatocyte, 8–10 μm; lamellocyte, 16–24 μm; Figure 5b). Larval lamellocytes show a 40% increase of primary lysosomes in their cytoplasm over that of plasmatocytes, while pupal lamellocytes show an increase of more than 100%. Lamellocytes are characterized further by a well-developed RER and a cytoplasmic margin devoid of cell organelles (Figure 5b). More than 20% of the lamellocytes show phagocytotic vacuoles as compared to 5% in plasmatocytes and podocytes taken together.

Thus, on the fine structural level, the differentiation of prohemocytes into proplasmatocytes within the hematopoietic organs, and further into plasmatocytes, podocytes, and lamellocytes in the hemolymph is characterized by the numerical increase of primary lysosomes which in turn is closely associated with an increase of cell organelles, such as the Golgi complex, RER, and mitochondria. The development of the primary lysosomal system is connected with the functional differentiation of the cells of the plasmatocyte line, whose main role is phagocytosis.

Figure 5. An electron micrograph of a plasmatocyte (a) and a lamellocyte (b) from the hemolymph of a mature wild-type larva. Note the organelle-free margin of the lamellocyte. A 6-μm-long cellular process is not included in the photograph. av, autophagic vacuole; G, Golgi body; ly, primary lysosome; m, mitochondrion; n, nucleus; nu, nucleolus; rER, rough endoplasmic reticulum. From Shrestha and Gateff (1982a).

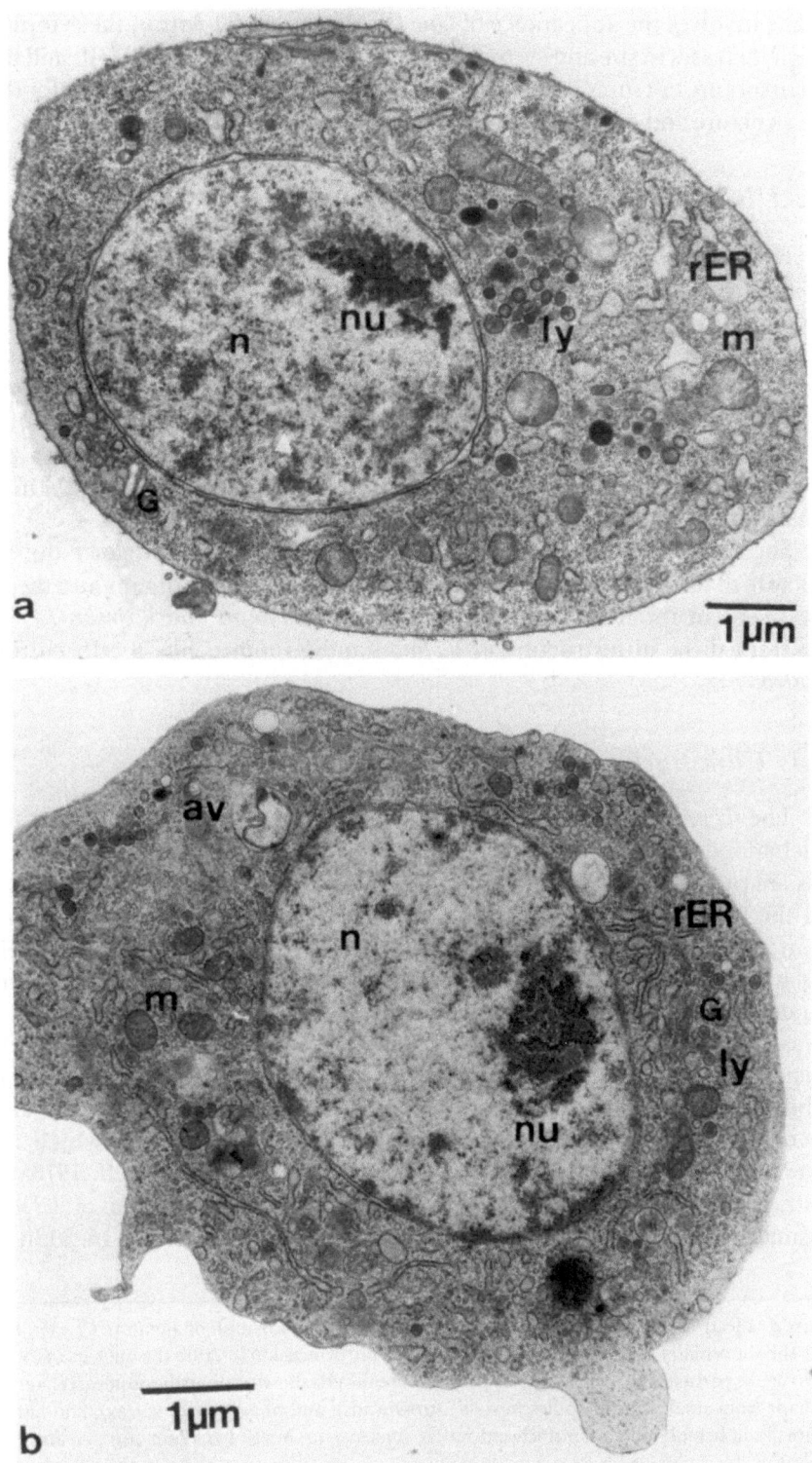

The differentiation of crystal cells from procrystal cells in the hematopoietic organs involves the appearance of fine fibrillar material within the cytoplasm which increases in size and eventually crystallizes. Since the crystal cells and their precursors are not tumorous in any of the mutants, the reader is referred for their fine structure and development to Shrestha and Gateff (1982a).

4.2. Ultrastructure of Tumorous Blood Cells

Hematopoietic malignancies have been found in five nonallelic larval, recessive lethal mutants, which fall into two classes: (1) two mutants in which blood cell differentiation takes place and (2) three mutants in which blood cell differentiation is impaired. The two mutants belonging to the first class are: *lethal (1) malignant blood neoplasm* [*l(1)mbn*] (Gateff, 1978a,b,c) and *Tumorous lethal*[1] (*Tum*[1]) (Corwin and Hanratty, 1976; Hanratty and Ryerse, 1981). The mutants lacking blood cell differentiation are: *l(2)mbn* (Gateff, 1978a,b,c), *l(3)mbn-1* (Gateff, 1978a,b,c), and *l(3)mbn-2* (Gateff, unpublished). In all mutants, the tumorous blood cell type is the plasmatocyte.

Shrestha (1979) and Shrestha and Gateff (1982b) investigated the fine structure of the primordial blood cells in the hematopoietic organs and the free blood cells in the hemolymph of the mutants *l(1)mbn* and *l(3)mbn-1*. They also studied the ultrastructure of *l(2)mbn* and *l(3)mbn-1* blood cells cultured *in vitro*.

4.2.1. Ultrastructure of the l(1)mbn Blood Cells

The *l(1)mbn* mutant belongs to the first type of blood tumor mutant in which blood cell differentiation takes place. About eight times more free blood cells are present in the hemolymph of mutant larvae than in wild-type larvae, and the hematopoietic organs are drastically enlarged. In the hematopoietic organ, one finds prohemocytes and proplasmatocytes which in their fine structure closely resemble wild-type prohemocytes and proplasmatocytes. Differentiated plasmatocytes, podocytes, and lamellocytes are also present. They are ultrastructurally similar to their free counterparts in the hemolymph, but differ considerably from wild-type plasmatocytes, podocytes, and lamellocytes. Figure 6 shows two *l(1)mbn* plasmatocytes from the hemolymph of a mature larva. Their nuclei have abnormal shapes. One of them contains vlp which have never been observed in the wild-type cells (Akai *et al.*, 1967; Gateff, 1978a). In contrast to wild-type plasmatocytes (Figure 5a), the cytoplasm of *l(1)mbn* plasmatocytes contains elliptical or round primary lysosomes. In addition,

---

Figure 6. Electron micrographs of plasmatocytes from the hemolymph of a mature *l(1)mbn* larva. Note the abnormally shaped nuclei (n), the cytoplasmic protrusion (c), into the nucleus in (b), and the viruslike particles (vlp) in (a). The cytoplasm of both cells shows primary lysosomes (ly), ranging in shape from small, round bodies to rods (arrowheads); autophagic vacuoles (av), and lamellar bodies (b). h, hemolymph; m, mitochondrion; n, nucleus; nu, nucleolus. From Shrestha and Gateff (1982b).

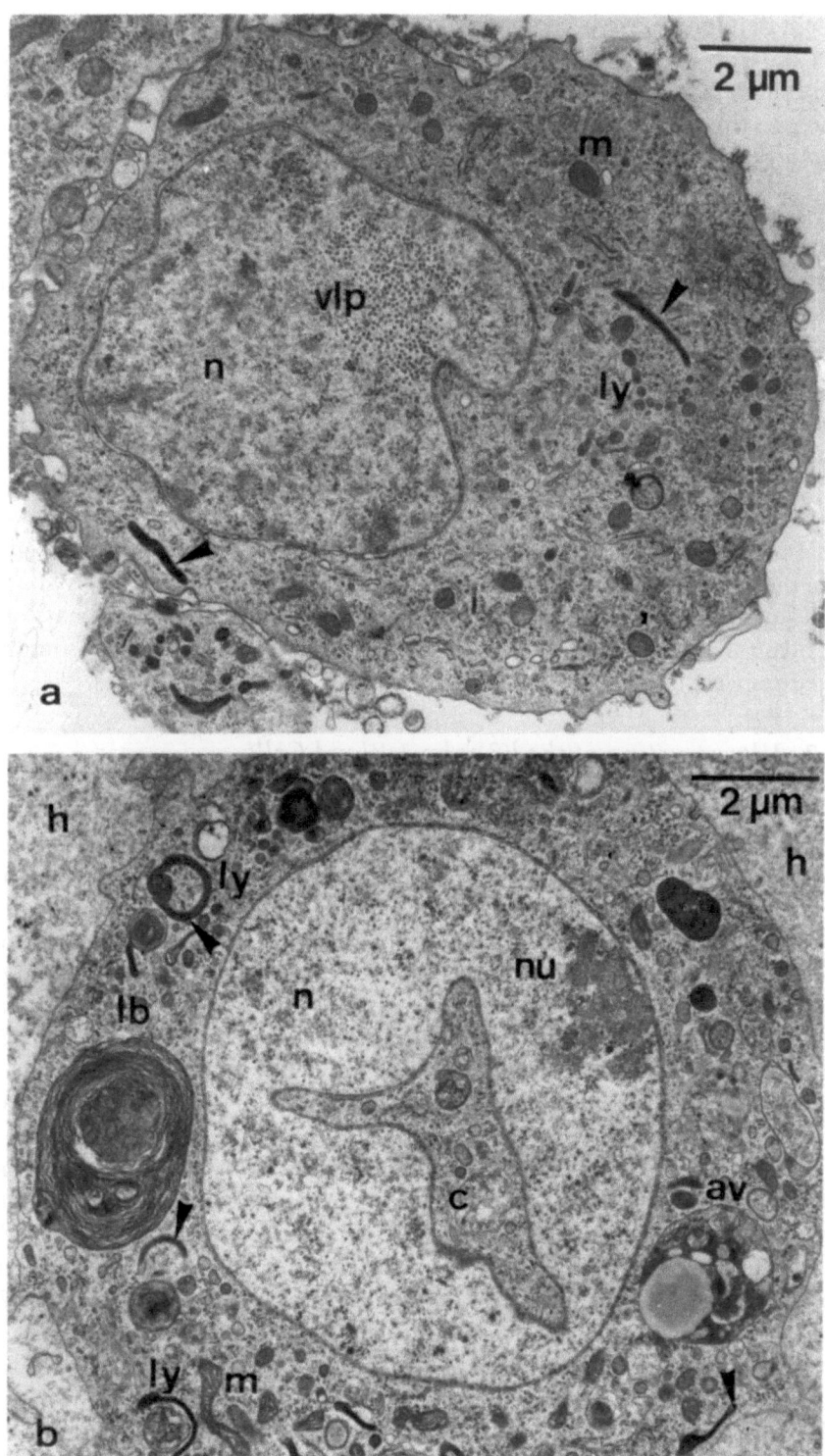

variable numbers of electron-dense structures of different shapes and sizes are present that have never been observed in wild-type larvae. Due to their positive reaction with the acid phosphatase test, they are considered lysosomelike. Podocytes as well as lamellocytes contain similarly shaped, lysosomelike structures in their cytoplasm. They reach the highest number in *l(1)mbn* lamellocytes. Compared to the wild-type cells, the number of primary lysosomes and lysosomelike structures in mutant plasmatocytes is increased. In the wild-type, with the exception of lamellocytes which show increased numbers of secondary lysosomes (autophagic-phagocytotic vacuoles and multivesicular bodies), all other cell types of the plasmatocyte line have relatively few secondary lysosomes. In contrast, the *l(1)mbn* plasmatocytes, podocytes, and lamellocytes exhibit increased amounts of secondary lysosomes. For example, mutant plasmatocytes and lamellocytes respectively possess about 6 and 10 times more secondary lysosomes in their cytoplasm than in their wild-type counterparts.

In conclusion: Prohemocytes and proplasmatocytes in the hematopoietic organs of *l(1)mbn* larvae do not show significant changes in their fine structure when compared to their wild-type counterparts. However, mutant plasmatocytes, podocytes, and lamellocytes from the hematopoietic organs, as well as from the hemolymph exhibit considerable changes, such as irregularly shaped nuclei containing vlp's and drastically increased numbers of primary and secondary lysosomes in the cytoplasm. Abnormally shaped lysosomelike structures were more abundant than the regularly shaped lysosomes.

4.2.2. Ultrastructure of the l(3)mbn-1 Blood Cells

The *l(3)mgn-1* mutation belongs to the class of mutants in which blood cell differentiation is impaired. As in *l(1)mbn*, the hematopoietic organs are enlarged and the blood cell concentration is increased: 2.5×10^6 blood cells/ml as compared to the wild-type value of 1.6×10^4 blood cells/ml. The majority of the cells in the hematopoietic organs and the hemolymph are immature, resembling either prohemocytes, proplasmatocytes, or plasmatocytes. Throughout the body cavity, secondary nests of hematopoietic cells steadily produce more immature blood cells which remain undifferentiated. However, about 5% of the free blood cells in the hemolymph grow abnormally. This results in extremely large cells which are designated as giant plasmatocytes and giant lamellocytes (Shrestha, 1979; Figure 7b, inset).

While prohemocytes, proplasmatocytes, and plasmatocytes show minor fine structural differences with their wild-type counterparts, **giant plasmatocytes and lamellocytes** diverge much more from them (compare Figures 5 and 7). Besides the differences in size, giant plasmatocytes exhibit small cytoplasmic processes and abnormally shaped, eccentrically located nuclei (Figure 7a, compare with Figure 7b, inset). Within the nucleus, vlp can sometimes be observed, and the cytoplasm shows large numbers of acid phosphatase-positive primary and secondary lysosomes, the latter of autophagic nature. Many irregularly shaped, small, single vesicles are also present. ER, mitochondria, and phagocytotic vacuoles were rarely observed. Thus, it appears that in the giant

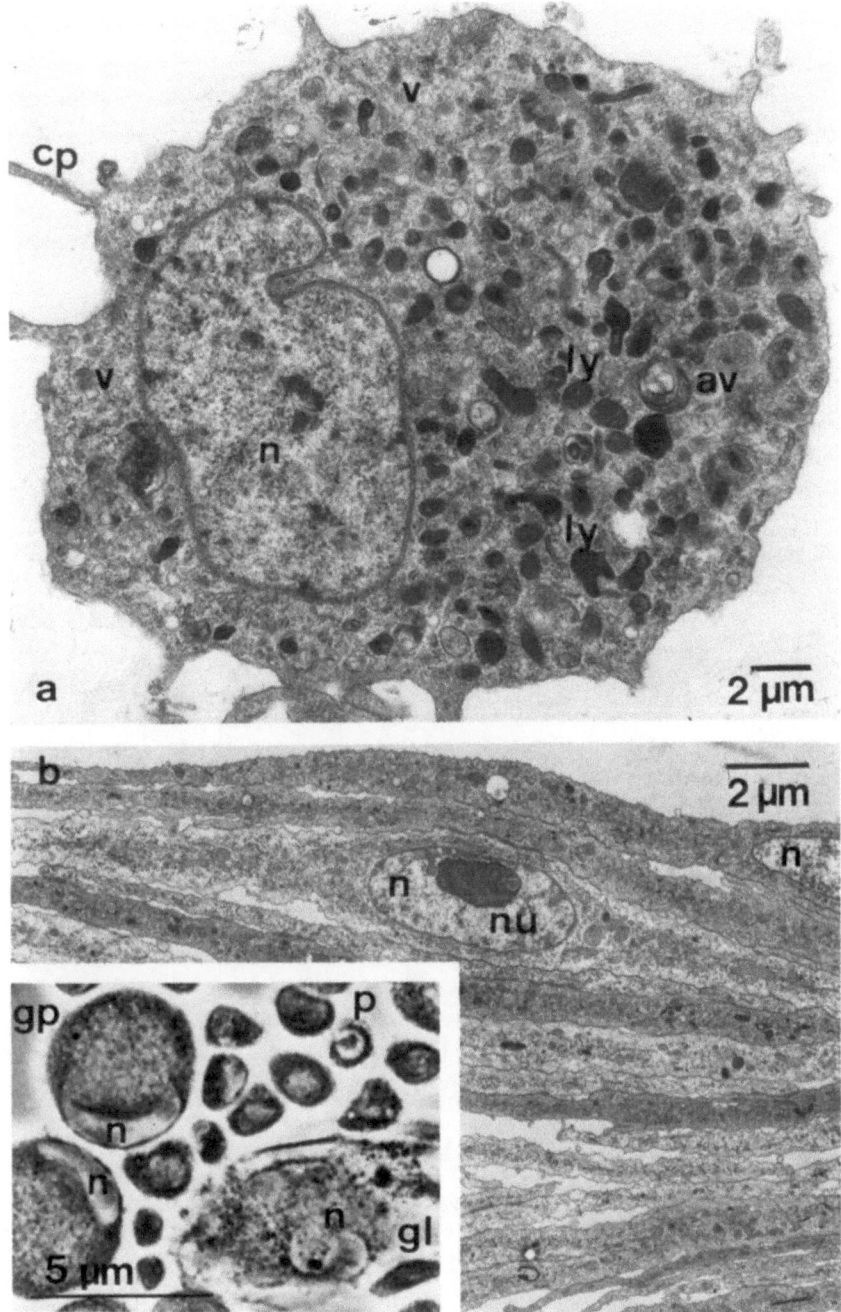

Figure 7. (a) An electron micrograph of a giant plasmatocyte from the hemolymph of a mature, third-instar *l(3)mbn* larva. Compare the sizes of the giant plasmatocytes (gp) and plasmatocytes (p) in the phase-contrast micrograph inset in (b). Note the abnormally shaped nucleus (n) and the numerous primary lysosomes (ly) within the cytoplasm. (b) Longitudinal section through a cluster of giant lamellocytes. The ends of the cells are not shown. For size comparisons, see giant lamellocyte (gl) in the inset. av, autophagic vacuole; v, irregular, small vesicles; nu, nucleolus. From Shrestha (1979).

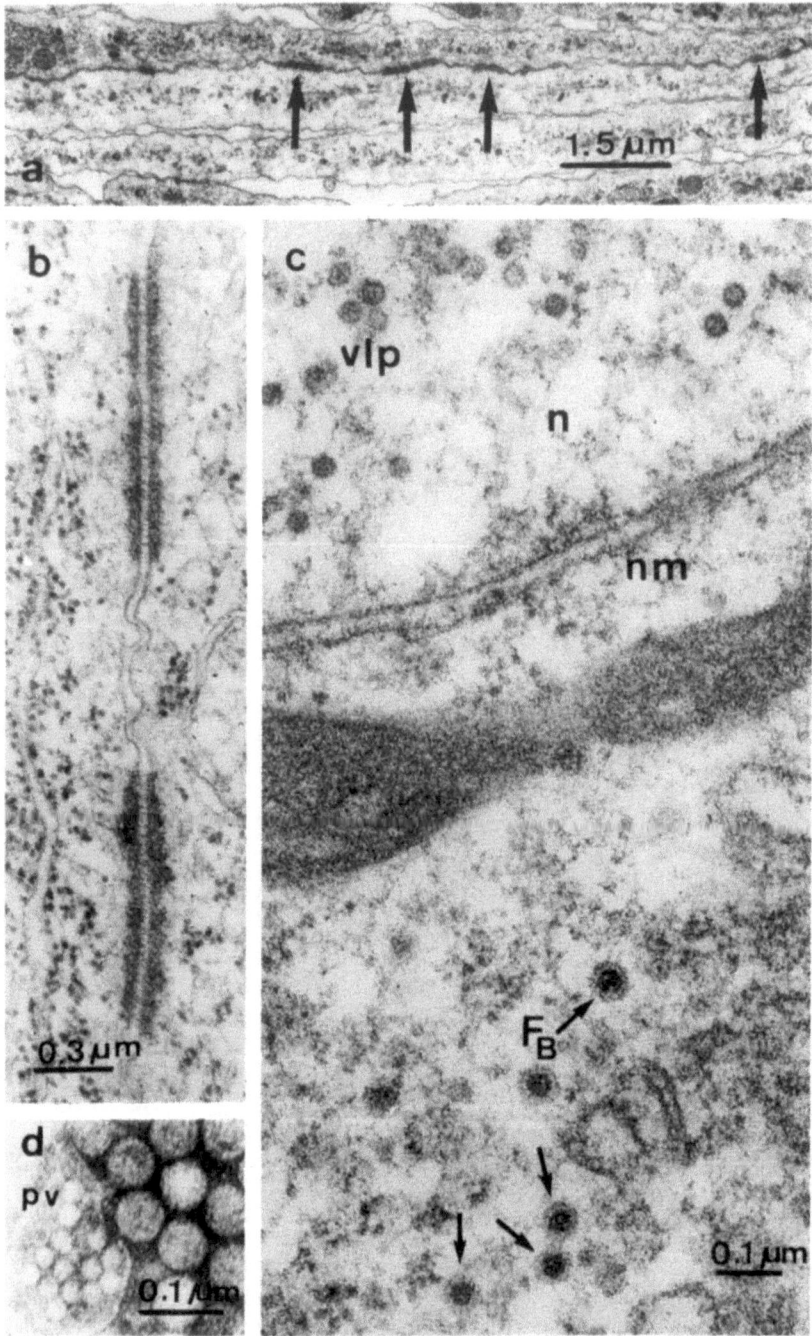

Figure 8. (a) An electron micrograph of a small portion from a longitudinal section through an *(3)mbn* giant lamellocyte (compare also with Figure 7b). Arrows show a number of desmosomes. (b) Electron micrograph of a desmosome as shown in (a). Note the dense plate apposing the inner leaflet of each cell membrane. The intracellular space, measuring approximately 29 nm, is traversed by

plasmatocyte, an abnormally high production of primary lysosomes takes place which may be the cause of their giant size.

The giant lamellocytes occur with a frequency of 2–3%. They originate from giant plasmatocytes through a flattening process and are found predominantly in clusters (Figure 7b), forming extensive desmosomes between each other (Figures 8a, b). Within a giant lamellocyte two regions can be recognized, a cortical, organelle-free region and a central region containing secondary lysosomes. However, these occur less often than in the giant plasmatocytes (Figure 7b). Microtubules running along the length of the cells were observed sporadically. Some of the giant lamellocyte bundles contained phagocytotic vacuoles. The polyphenol oxidase test demonstrated the melanization reaction in 100% of the lamellocytes, while only some of the plasmatocytes showed melanization (Shrestha, 1979).

4.2.3. Ultrastructure of Tumorous Blood Cells in Vitro

Because of the large numbers of free blood cells in the $l(2)mbn$ and $l(3)mbn-1$ tumorous larvae, it was possible to establish blood cell lines *in vitro* (Gateff *et al.*, 1980). The cultures consist of cells that resemble plasmatocytes, podocytes, and lamellocytes. In older cultures, many of the cells show phagocytotic vacuoles. The test for acid phosphatase revealed considerable numbers of primary and secondary lysosomes within the cytoplasm. Cells, in the process of lysis, had the enzyme-substrate precipitate both in the cytoplasm and along the outer cell membrane. The polyphenol oxidase test was negative for the cells of both cell lines.

In addition to large numbers of vlp's in the nuclei of these cells, two types of RNA viruses were detected: (1) a new reovirus named F_b and (2) a picornavirus designated as the *Drosophila* C-virus (DCV; Figures 8c, d; Gateff *et al.*, 1980).

5. Discussion

Foulds (1969) cautions that the classification of tumors into benign and malignant is somewhat arbitrary, since numerous intermediate stages exist between the above extremes, and this applies also to the tumors of *Drosophila*. The criteria used to assess the degree of malignancy in vertebrate tumors are: (1) cytological changes, (2) changes in the intercellular relationships, and (3) the invasion of malignant cells into surrounding, healthy tissue (Ghadially, 1980).

←

fibrils of 12.5-nm thickness with a periodicity of 15 nm. From Shrestha (1979). (c) Detail from an electron micrograph of an $l(2)mbn$ plasmatocyte in culture *in vitro*, showing vlp in the nucleus (n) and F_B virions in the cytoplasm. The vlp possess a single-layered capsid and are 35 nm in diameter. The reovirus F_B exhibits a double-layered capsid and a diameter of 60 nm. nm, nuclear membrane. From Gateff *et al.* (1980). (d) Electron micrograph of purified F_B virus with 14 C-virus particles (pv), which were a constant contaminant of the preparations. The C-virus measures 23 nm in diameter. From Gateff *et al.* (1980).

The cytological changes include deviations in the size and shape of the nucleus and the nucleolus. Further, changes are seen in the structure and the amount of the chromatin. Within the cytoplasm, abnormalities of the mitochondria, the RER, the Golgi complex, and the presence of other cytoplasmic inclusions, such as primary and secondary lysosomes, provide additional criteria for the characterization of the degree of malignancy. The most prominent changes in the intercellular relationships are alterations of the cell junctions and the cell membrane. Invasion can be assessed by light microscopy more easily than by electron microscopy.

Applying the above criteria to the *Drosophila* tumors, we find variable degrees of ultrastructural deviations from the normal state. This is demonstrated clearly during the transformations of autotypic into allotypic and further into atelotypic sublines of imaginal disc tissues in culture *in vivo* (Gateff et al., 1974). During this transformation, the imaginal disc cells lose their basal-apical polarity and their secretory activity. This process begins with abnormalities of the basement membrane, such as curling and detachment from the base of the cells (Figure 1a), and on the apical surface the number of microvilli is reduced. The cuticule secreted by such cells contains fewer hairs and bristles, and there are large regions of plain cuticle. With progress in culture *in vivo*, more and more cells lose their basal-apical polarity until finally the tissue consists almost exclusively of cluster cells. Cells in this clumped arrangement are incapable of cuticle secretion, show fewer junctions, and considerable numbers of vlp in their nuclei. In wild-type female flies, such tissue sublines grow autonomously in a benign or malignant fashion. Furthermore, the different tumorous tissue sublines showed ultrastructural abnormalities to various degrees, depending on whether they experienced transdetermination before becoming tumors or whether they transformed directly from the differentiating, autotypic state to the tumorous condition. Cells from the directly transformed tumorous tissue sublines exhibited the most extreme membrane abnormalities (Figure 2).

A similarly extreme phenotype was shown by the $l(2)gl^4$ imaginal discs *in situ* (Gateff and Schneiderman, 1969). Changes of the membrane properties are the main features of the malignant neuroblasts and ganglion mother cells in the $l(2)gl^4$ larval brain, which apparently accounts for their invasive behavior (Figure 4a). Except for the generally increased cellular and nuclear sizes of the mutant neuroblasts and ganglion mother cells, cytoplasmic and nuclear deviations from their wild-type counterparts were minimal. After subculture *in vivo*, the cellular and nuclear size variations increased still further. Many of the nuclei appeared deformed and exhibited large numbers of vlp (Akai et al., 1967; Figure 4b). The cultured neuroblasts and ganglion mother cells formed no firm junctions with each other.

The tumorous hemocytes from the *l(1)mbn* and *l(3)mbn-1* mutants show drastic cytoplasmic abnormalities, in addition to membrane changes (Shrestha, 1979; Shrestha and Gateff, 1982a,b). The hematopoiesis in the two mutants differs in one main respect. In the *l(1)mbn* mutant, blood cell maturation is not impaired, while in *l(3)mbn-1*, most blood cells in the hemolymph remain immature (Gateff, 1977; Shrestha, 1979; Shrestha and Gateff, 1982b). The immature blood cell precursors in the hematopoietic organs or the ones from the l(3)mbn-1 larval hemolymph resemble very closely in their fine structure the respective wild-type blood cell precursors in the hematopoietic organs. The mature, free blood cells in the l(1)mbn hemolymph, on the other hand, exhibit an array of ultrastructural aberrations, such as increased numbers of primary and secondary lysosomes, Golgi complexes, RER, mitochondria, and abnormally shaped nuclei containing large numbers of vlp's (Figures 6 and 8c). They also show increased phagocytotic activity when compared to the wild-type. Contrary to the wild-type situation, the tumorous plasmatocytes, podocytes, and lamellocytes invade into larval tissues and cause their destruction.

The mutant Tum^1 exhibits a phenotype very similar to the above l(1)mbn phenotype. The low-power electron micrographs of Hanratty and Ryerse (1981) demonstrate very few differences between the mutant and wild-type primordial blood cells in the hematopoietic organs. On the other hand, the plasmatocytes, podocytes, and lamellocytes in the hemolymph show increased numbers of primary and secondary lysosomes and cell membranes with many cytoplasmic processes, as do l(1)mbn plasmatocytes, podocytes, and lamellocytes.

While the majority of the plasmatocytes in the (3)mbn-1 mutant remain undifferentiated, a small percentage engage in an abnormal differentiation, resulting in the production of giant plasmatocytes, podocytes, and lamellocytes. These cells contain very large numbers of primary lysosomes (Figure 7). The undifferentiated cells in the hematopoietic organs and the hemolymph show only minor fine structural deviations from the corresponding wild-type blood cells.

In conclusion, comparing the fine structural abnormalities associated with the tumorous state in the four types of tumor cells, it becomes obvious that the changes are either of a general nature, i.e., increased numbers of autophagic vacuoles and secondary lysosomes, or they affect specific cellular structures directly involved in the functioning of the cells. The later changes include the basement membrane abnormalities, the changes of the microvillar surface in tumorous imaginal disc cells, the inability of ganglion mother cells to form axons, the incapacity of mutant, tumorous blood cells to recognize self from nonself, and the production of excessive amount of primary lysosomes.

References

Akai, H., 1978, Brain neoplasm in *Drosophila* and its neurosecretory cells, *Cell* 5:38-41.

Akai, H., Gateff, E., Davis, L., and Schneiderman, H. A., 1967, Virus-like particles in normal and tumorous tissues of *Drosophila*, *Science* 157:810-813.

Corwin, H. O., and Hanratty, W. P., 1976, Characterization of a unique lethal tumorous mutation in *Drosophila*, *Mol. Gen. Genet.* 14:345-347.

Foulds, L., 1969, *Neoplastic Development*, vol. 1, Academic Press, New York.

Gateff, E., 1977, Malignant neoplasms of the hematopoietic system in three mutants of *Drosophila melanogaster*, *Ann. Parasitol. Hum. Comp.* 52:81-83.

Gateff, E., 1978a, Malignant and benign neoplasms of *Drosophila melanogaster*. In *The Genetics and Biology of Drosophila*, vol. 2b, edited by M. Ashburner and T. R. F. Wright, pp. 182-261, Academic Press, New York.

Gateff, E., 1978b, Malignant neoplasms of genetic origin in *Drosophila melanogaster*, *Science* **200:**1448-1459.

Gateff, E., 1978c, The genetics and epigenetics of neoplasms in *Drosophila*, *Biol. Rev. Cambridge Philos. Soc.* **53:**123-168.

Gateff, E., and Schneiderman, H. A., 1969, Neoplasms in mutant and cultured wild type tissues of *Drosophila*, *Natl. Cancer Inst. Monogr.* **31:**365-397.

Gateff, E., and Schneiderman, H. A., 1974, Developmental capacities of benign and malignant neoplasms of *Drosophila*, *Wilhelm Roux Arch. Entwicklungsmech. Org.* **176:**23-65.

Gateff, E., Akai, H., and Schneiderman, H. A. 1974, Correlations between developmental capacity and structure of tissue sublines derived from the eye-antennal imaginal discs of *Drosophila melanogaster*, *Wilhelm Roux Arch. Entwicklungsmech. Org.* **176:**89-123.

Gateff, E., Gissmann, L., Shresta, R., Plus, N., Phister, H., and zur Hansen, H., 1980, Characterization of two tumorous blood cell-lines of *Drosophila melanogaster* and the viruses they contain. In *Invertebrate Systems in Vitro*, edited by E. Kurstak, K. Maramorosch, and A. Dübendörfer, pp. 517-533, Elsevier/North-Holland, Amsterdam.

Ghadially, F. N., 1980, *Diagnostic Electron Microscopy of Tumors*, Butterworths, London.

Hadorn, E., 1966, Konstanz, Wechsel und Typus der Determination und Differenzierung in Zellen aus männlichen Genitalscheiben von *Drosophila melanogaster* in Dauerkultur, *Dev. Biol.* **13:**424-509.

Hanratty, W. P., and Ryerse, J. S., 1981, A genetic melanotic neoplasm of *Drosophila melanogaster*, *Dev. Biol.* **83:**238-249.

Hofbauer, A., 1979, Die Entwicklung der optischen Ganglien bei *Drosophila melanogaster*, Dissertation, Fakultät für Biologie, Universität, Freiburg.

Kankel, D. R., Ferrus, A., Garren, S. H., Harte, P. J., and Lewis, P. E., 1980, The structure and development of the nervous system. In *The Genetics and Biology of Drosophila*, vol. 2d, edited by M. Ashburner and T. R. F. Wright, pp. 295-363, Academic Press, New York.

Poodry, C. A., 1980, Imaginal discs: Morphology and development. In *The Genetics and Biology of Drosophila*, vol. 2d, edited by M. Ashburner and T. R. F. Wright, pp. 407-441, Academic Press, New York.

Rizki, M. T. M., 1957, Alterations in the hemocyte population of *Drosophila melanogaster*, *J. Morphol.* **100:**137-459.

Shrestha, R., 1979, A comparative light and electron microscopic, developmental study of the hematopoiesis in the wild type and two blood tumor mutants of *Drosophila melanogaster*, Dissertation, Fakultät für Biologie, Albert-Ludwigs Universitat, Freiburg.

Shrestha, R., and Gateff, E., 1982a, Ultrastructure and cytochemistry of the cell types in the larval hematopoietic organs and the hemolymph of *Drosophila melanogaster*, *Dev. Growth Differ.* **24:**64-82.

Shrestha, R., and Gateff, E., 1982b, Ultrastructure and cytochemistry of the cell types in the tumorous hematopoietic organs and the hemolymph of the mutant *lethal(1) malignant blood neoplasm* (*l(1)mbn*) of *Drosophila melanogaster*, *Dev. Growth Differ.* **24:**83-98.

Stark, M. B., and Marshall, A. K., 1930, The blood forming organ of the larva of *Drosophila melanogaster*, *J. Am. Inst. Homeopathy* **23:**1204-1206.

16

The Cellular Defense System of *Drosophila melanogaster*

TAHIR M. RIZKI AND ROSE M. RIZKI

1. Introduction

Although insects lack adaptive immune systems endowed with memory elements and finely tuned discriminatory powers such as those found in the vertebrates, they do possess internal defense mechanisms for combating foreign materials (Salt, 1970). Among these are the cellular responses of phagocytosis and encapsulation used to resist parasitic and microbial infections. These processes do not involve opsonization of foreign materials (Scott, 1971; Anderson *et al.*, 1973), so it appears that the insect hemocyte surfaces themselves must play the crucial role in discriminating nonself from self. This being the case, analysis of insect defense reactions requires study of hemocyte surface receptor sites on which extracellular cues operate as well as investigation of the body's own tissue surfaces to which the hemocytes remain neutral.

This report deals primarily with the larval hemocytes of *Drosophila melanogaster*. The available mutations that influence larval hemocyte function in this species provide the minimum material necessary to analyze some of the roles of each of the hemocyte types. Our approach stresses the use of mutations to selectively block specific hemocyte functions or disrupt cellular organization to delineate events in hemocyte responses that otherwise are tightly interwoven and difficult to analyze. Such genetic analyses are not currently possible in any other insect. We refer to other species only briefly, in comparison to the *D. melanogaster* system.

TAHIR M. RIZKI AND ROSE M. RIZKI • Division of Biological Sciences, University of Michigan, Ann Arbor, Michigan 48109, USA. We respectfully dedicate this chapter to Dr. George Salt for generating enthusiasm among the students of insect cellular defense reactions through his critical and stimulating synthesis of the subject.

Defense against parasites in the hemocoel is one of the major roles ascribed to insect hemocytes, but consideration of this subject is not included in the present chapter for several reasons. As far as we are aware, ultrastructural studies of *Drosophila* hemocytes combating parasites have not been published although other aspects of host–parasite relationships have been well studied (Walker, 1959; Nappi and Streams, 1969; Carton and Kitano, 1981). These will be reviewed in a forthcoming publication (Carton *et al.*, 1984). Secondly, the ultrastructural features of cellular capsules are common to those surrounding parasites and those formed around aberrant tissues in melanotic tumor mutants. The latter are considered in Section 6. Finally, ultrastructural details of encapsulation in other insects have been described (Grimstone *et al*, 1967; Ratcliffe and Rowley, 1979).

Unfortunately, there is no uniform classification for insect hemocytes, but there are similarities among the cellular defense responses of different insects (Ratcliffe and Rowley, 1979). Whether these similarities are functional (analogous) or reflect similarities due to descent (homologous) is not clear. This distinction requires probing at the genic level. Hopefully, developmental and physiological genetic studies of *D. melanogaster* hemocytes will eventually lead us to identification of genic determinants of the cellular defense system, and homologies of these components among various insect groups can then be examined at the genic level using recombinant DNA technologies.

2. Distribution of Hemocytes in the Larva

Circulating hemocytes in the hemocoel and the heart can be observed directly through the body wall of the living larva with a microscope and appropriate illumination (Rizki, 1957a). Many hemocytes in fixed specimens are found resting against the body wall of the caudal hemocoel and on the imaginal discs and pericardial cells. Therefore, these hemocytes are conveniently handled by fixing and processing these regions of the larva for scanning electron microscopy (SEM) or transmission electron microscopy (TEM). Another convenient source of hemocytes for electron microscopy is the lymph glands which are paired structures along the heart of the larva. These organs in *Drosophila* were originally described as "blood-forming organs" by Stark and Marshall (1931) who observed hemocytes leaving the organs by diapedesis. The lymph glands of some mutant strains release their cellular contents during the larval period (Castiglioni, 1957; Röhrborn, 1961) but the function of the lymph glands as a source of blood cells has been questioned by several authors (El Shatoury, 1955; Srdic and Reinhardt, 1980), and this controversy remains unsettled. Our concern with the lymph glands in this chapter is as an organ containing cells with the characteristics of the hemocytes that are found in the hemocoel. The most convincing evidence of this similarity is the crystal cell which is unique and present both in the hemocoel and in the lymph glands (Rizki, 1960; T. Rizki and Rizki, 1980c).

Robertson (1936) noted the breakdown of the larval lymph glands in the pupal stage. This phenomenon has been accurately monitored by utilizing hot water treatment to artificially blacken crystal cells within lymph glands *in situ* (Rizki *et al.*, 1980). The cells of the lymph glands must be released prior to tanning of the puparium since aggregates of blackened cells in the lymph gland regions are no longer detectable at this stage (Rizki and Rizki, in preparation).

3. Classification of the Larval Hemocytes

The classification of the larval hemocytes of *D. melanogaster* that we proposed (Rizki, 1956, 1957a) was based on studies utilizing phase microscopy, stained blood smears, and histological sections. Since that time the hemocytes have been examined by SEM and TEM (Rizki *et al.*, 1976; Rizki and Rizki, 1978), and we have recently investigated hemocyte surfaces using plant lectins (Rizki and Rizki, 1983b) and monoclonal antibodies (unpublished observations). The later studies complemented and confirmed the original classification which we consider operational and subject to modification with further understanding of the functions of the hemocytes.

The cornerstone of the classification given in Figure 1 is the interrelationship between the cells of the plasmatocyte line and the distinction between this cell line and the crystal cells which contain paracrystalline inclusions. The plasmatocytes and their variant forms have, or differentiate, adhesive surfaces for reactions against nonself, and the crystal cells have phenol oxidase activity which is utilized in defense reactions requiring protein cross-linking and melanization (T. Rizki and Rizki, 1980c). Detailed information on these hemocyte types which has been published will not be reviewed here due to space limitations.

There is no evidence that the crystal cells and plasmatocytes are replenished from a common stem cell (prohemocyte) during larval life nor is there information to exclude this possibility. A cell heavily laden with paracrystalline inclusions has never been encountered in cell division. This does not, however, exclude the possibility that cells prior to developing paracrystalline inclusions are distinct from cells giving rise to new plasmatocytes. Hemocytes in division generally resemble small plasmatocytes (Figure 2), but the recent observation of melanized daughter cells in an embryo of the *Black cell* mutant suggests that cells with latent hemolymph phenol oxidase activity may undergo mitosis (Rizki *et al.*, 1980).

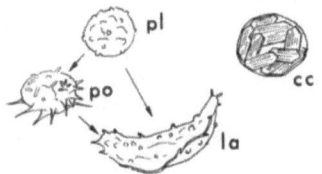

Figure 1. A diagram illustrating the relationships between the variants of the plasmatocytic line: plasmatocyte (pl), podocyte (po), and lamellocyte (la). These cells are phagocytic and are involved in encapsulation. The crystal cell (cc) contains phenol oxidases that are utilized for the cross-linking of proteins and melanization.

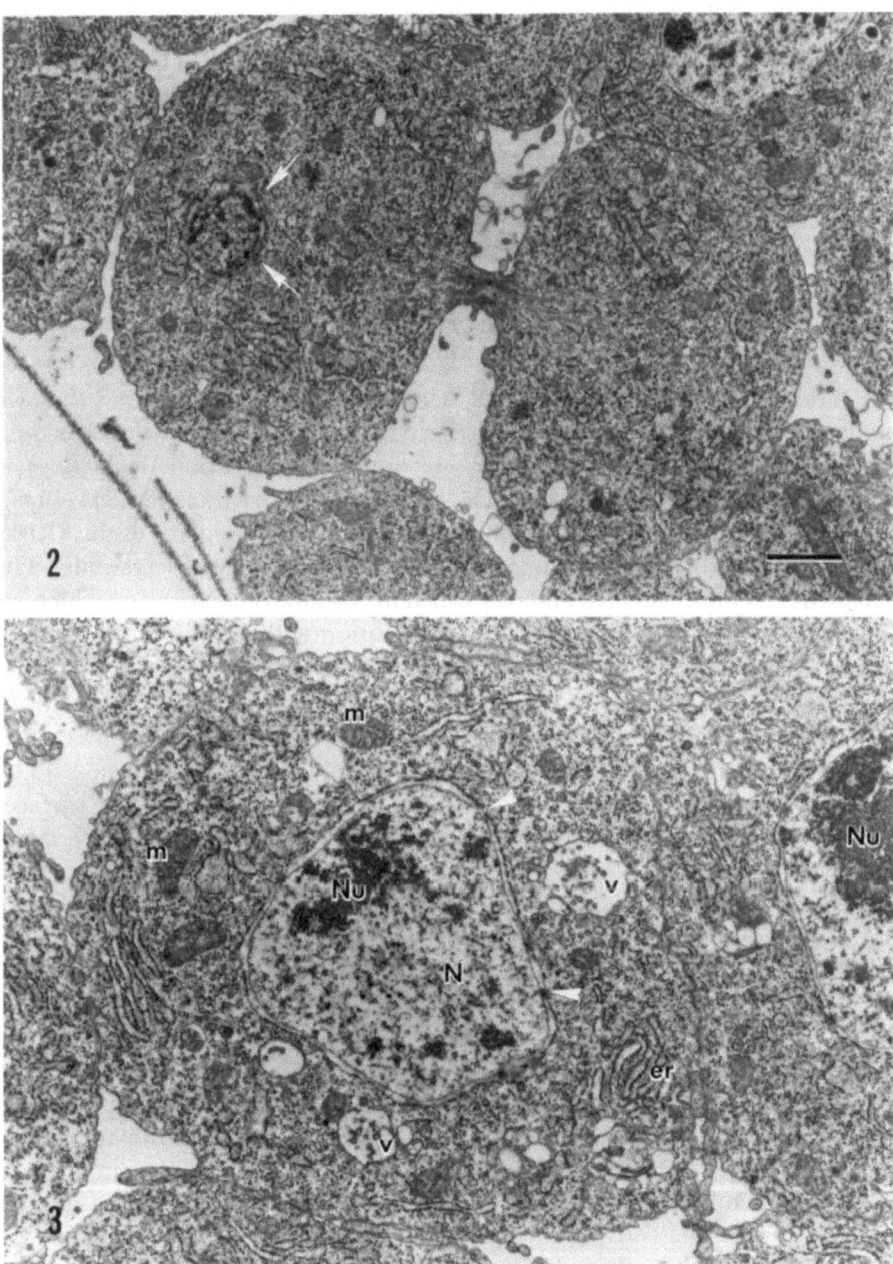

Figure 2. Sibling hemocytes in telophase from the lymph gland of an early third-instar larva from the *Oregon-R* (*Ore-R*) wild-type strain of *D. melanogaster*. The stem body containing a bundle of microtubules connects the two cells. Condensation of the nuclear membrane around the chromatin, and the forming nuclear pores (arrows) can be seen in the left cell. (Bar = 1 μm.)

Figure 3. A plasmatocyte in the lymph gland of a young third-instar *Ore-R* larva. m, mitochondrion; er, endoplasmic reticulum; N, nucleus; Nu, nucleolus; v, large vesicles containing small electron dense circular profiles of unknown nature. The arrowheads point to some of the nuclear pores. The magnification is the same as in Figure 2.

3.1. The Plasmatocyte and Its Variants

The plasmatocytes which comprise the bulk of the larval hemocyte population are phagocytic cells comparable to plasmatocytes in other insects, or cells named *amoebocytes* or *macrophages* by other workers (review in Gupta, 1979). TEM examination shows that plasmatocytes contain a variety of inclusions such as endocytotic vesicles, lysosomes, residual bodies, and lamellar bodies (myelin forms) in addition to mitochondria, RER and unattached ribosomes, Golgi bodies, and microtubules. The cells have a prominent nucleolus and most of their chromatin appears to be diffuse (Figure 3). In the SEM, the plasmatocyte surfaces show a variety of small knobs of various sizes and signs of pleating (Figure 4).

The podocyte is a plasmatocytic variant characterized by filamentous or membranous extensions of the cytoplasmic surface (Figure 5). Many of the hemocytes resting on the body wall or on tissues in the body show these surface features, and plasmatocytes also assume this appearance as they undergo differentiation to the flattened lamellocytic form. The intracellular disposition of organelles and their nature does not distinguish these cells from plasmatocytes, and thus agrees with the original classification of podocytes as a morphological variant of the plasmatocyte. Indeed, it is difficult in sectioned material to discriminate between the profiles of a pleated surface of a plasmatocyte and the filamentous surface of a podocyte, especially since the surfaces of the latter are highly variable. Whether it is necessary to adopt the specific name *podocyte* for these cells might be questioned. On the other hand, this operational term is useful, and the classification for *D. melanogaster* hemocytes has always recognized this form as a plasmatocytic variant (Rizki, 1956). It has never been accorded the status of a unique category as in studies in other insects (Gupta, 1979).

Lamellocytes form the walls of capsules to enclose foreign bodies that enter the hemocoel, and the morphology of these cells is remarkably adapted for this function. They are flattened, disk-shaped cells measuring 50-60 μm in diameter and less than 0.2 μm in thickness at their outermost borders (Figures 6-8). Their derivation from plasmatocytes was deduced on the basis of the sudden change in cell frequencies, i.e., a decrease in plasmatocytes correlated with an increase in podocyte-lamellocyte variants in the absence of any unusual mitotic activity (Rizki, 1957a). This suggestion was later borne out by direct observations of individual cells differentiating *in vitro* (Rizki, 1962). Intermediate forms between plasmatocytes, podocytes, and lamellocytes can be recognized in melanotic tumor larvae examined by SEM (Figure 9) (T. Rizki and Rizki, 1980a), but this method of sequencing is only acceptable because we have previously established the link between the spherical and the flattened forms. In the *Ore-R* wild-type strain, lamellocytes are not abundant until the end of larval life, but in melanotic tumor strains many lamellocytes are present throughout the third instar (Rizki, 1957b).

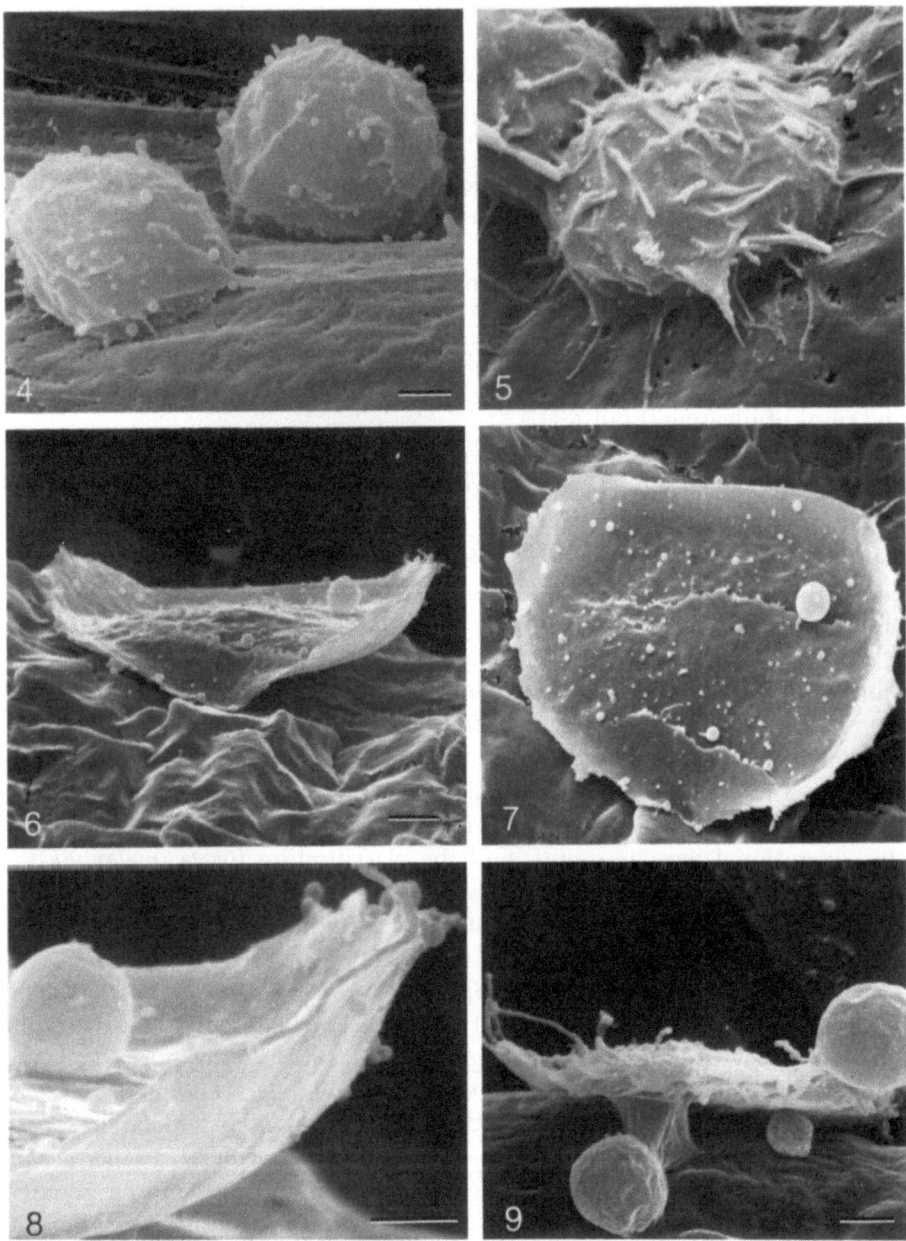

Figure 4. Two plasmatocytes within the body cavity of an *Ore-R* larva. (Bar = 1 μm.)
Figure 5. A podocyte in an *Ore-R* larva. The magnification is the same as in Figure 4.
Figures 6 and 7. Two views at the same magnification of the same lamellocyte from a *tu-W* larva. (Bar = 1 μm.)
Figure 8. A higher-magnification scanning electron micrograph of the part of the lamellocyte in Figure 6 showing the vesiculate surface. (Bar = 1 μm.)
Figure 9. A transitional form between a podocyte and a lamellocyte showing extrusion of large and small vesicles. Also note the filaments on the left side of this cell from a *tu-W* larva. (Bar = 1 μm.)
Figures 4–9 reproduced from Rizki and Rizki (1979) with permission of Springer-Verlag.

3.2. The Crystal Cell

Crystal cells are characterized by prominent, cytoplasmic paracrystalline inclusions that are rectangular or elongate with tapered ends (Figure 10). The cells are generally large and show a characteristic topology due to the underlying paracrystalline inclusions which are randomly oriented in the cytoplasm. That these surface features are sufficient to spot the crystal cells among other hemocytes was demonstrated by first selecting cells with these characteristics and then dissecting them open with a micromanipulator interfaced to the SEM to confirm the presence of paracrystalline inclusions within the selected cells (Rizki et al., 1976).

In addition to the paracrystalline inclusions and mitochondria, many free ribosomes and some electron-dense amorphous bodies are present in the cytoplasm of crystal cells. The nuclei of cells with prominent paracrystalline inclusions very often have large masses of condensed chromatin in contrast with the more diffuse chromatin and small regions of condensed chromatin usually seen in the plasmatocytes. This nuclear characteristic has been observed in the crystal cells of other species of the D. melanogaster species subgroup as well as in species lacking highly ordered inclusions of the melanogaster type (see Section 4).

The paracrystalline inclusions are not membrane bound. In some preparations, interconnecting fibers between ribosomes and the parallel arrays of the paracrystalline material can be seen. Figures 11 and 12 illustrate the stacking of parallel and cross elements. Fixation of the paracrystalline structures is better when formaldehyde is included in the primary fixative, otherwise there is considerable damage resulting in diffusion of the fibers or large empty spaces in the form of geometric outlines previously occupied by paracrystalline material (Figures 13, 14). That the crystal cells are extremely sensitive to changes in the surrounding medium is apparent by the ease with which they swell and rupture when they are removed from the hemocoel (Rizki, 1978). This sensitivity is probably related to the function of the cells in setting off the chain of reactions that results in melanization in case of injury to the larva (see Sections 8 and 9).

Since we recognize crystal cells by their characteristic paracrystalline inclusions, positive identification of young cells just beginning to differentiate as crystal cells is difficult. This problem has been tackled recently by Shrestha and Gateff (1982). An additional difficulty in recognizing the formative stages of the paracrystalline inclusions is posed by variations in paracrystalline material due to incomplete fixation of the total structure, such as illustrated in Figures 13 and 14. We consider that the aggregates of ribonucleoproteins found in some cells without paracrystalline inclusions may represent early stages in the synthesis of paracrystal proteins (Figure 16).

4. Crystal Cells in Other Drosophila Species

Of the eight known species in the D. melanogaster species subgroup (Lemeunier and Ashburner, 1976; Tsacas and Bächli, 1981), seven have

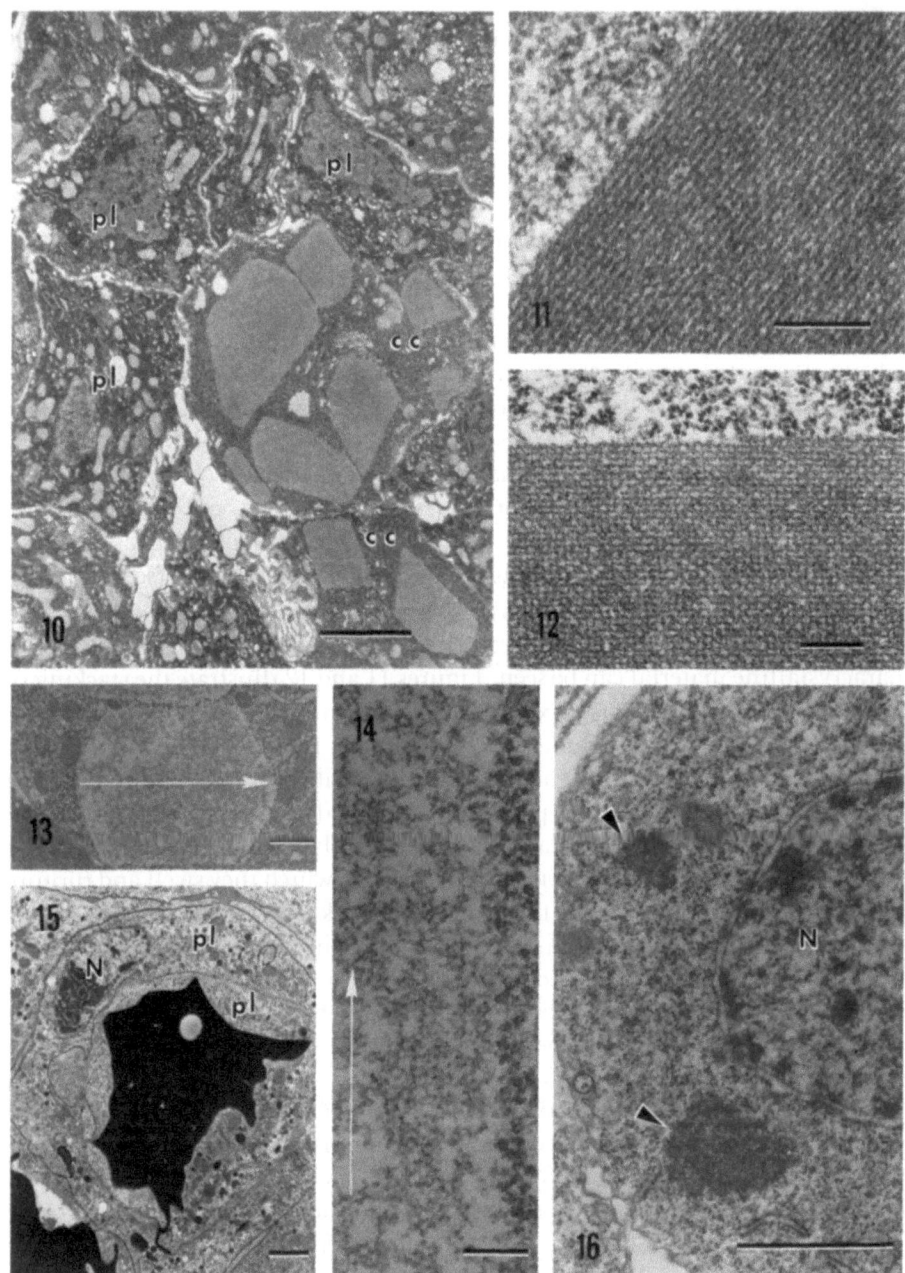

Figure 10. Two crystal cells in an *Ore-R* lymph gland. The geometric profiles of the paracrystalline inclusions depend on the plane of section. Note the marked difference between the cytoplasm of the crystal cells (cc) and the surrounding plasmatocytes (pl). (Bar = 2 μm.)

Figures 11 and 12. Orientation of paracrystalline proteins in two different planes. The crystals are not bound by membranes. Electron-dense fibers interconnect surrounding free ribosomes and elements of the paracrystalline matrix. (Bars = 0.2 μm.)

hemocytes with paracrystalline inclusions. These are, in addition to *melanogaster: simulans, mauritiana, erecta, yakuba, sechellia,* and *orena* (T. Rizki and Rizki, 1980b, and unpublished observations). TEM studies of the crystal cells have been completed for all these species except the last two. In the eighth species, *D. teissieri*, hemocytes with large spherical inclusions show the functional characteristics of the crystal cells, i.e., melanization with hot water and sensitivity to environmental change (T. Rizki and Rizki, 1980b). The inclusions in *D. teissieri* are not membrane bound. However, they lack the paracrystalline arrangement seen in *D. melanogaster,* and their fibrous components are randomly dispersed (Figures 17-19).

Drosophilids outside the *D. melanogaster* subgroup (22 have been examined) have spherical inclusions that resemble those of *D. teissieri* (Rizki *et al.*, in preparation). TEM examination of the inclusions from several of these species reveals that they lack regularity of fibril orientation. Thus, regularly oriented paracrystalline inclusions are probably unique to the *D. melanogaster* species subgroup which is of relatively recent origin (Throckmorton, 1975).

The crystal cells which are the sole source of hemolymph phenol oxidase activity confer a selective advantage on the larva by virtue of their role in wound healing and other defense reactions involving cross-linking of proteins and melanization (see Sections 8 and 9). In *D. melanogaster,* a number of nonallelic genes affecting phenol oxidase activity (Peeples *et al.*, 1969) and crystal cell organization (T. Rizki *et al.*, 1980; Rizki and Rizki, 1981a) have been identified. The evolution of a genotype which changed a nonparacrystalline phenotype to a paracrystalline phenotype, the former prevalent among drosophilids outside the *melanogaster* sibling group, must have taken place in the face of stringent selective pressure against interruption of the functional state of this cell type.

Analysis of crystal cells among the drosophilids provides evidence of the genetic relationship between cell type and homology, or similarity by descent. Melanization is a feature common to other insects, and hemocytes which contain either large globular-type cytoplasmic inclusions, rupture readily, or show a blackening reaction have been described in these species. Such cells have been named coagulocytes, spherule cells, oenocytoids, and cystocytes (Grégoire and Goffinet, 1979).

Figure 13. Illustration of partial fixation of a paracrystalline inclusion with the outlining cytoplasm sufficiently preserved to show the inclusion profile. The arrow shows the long axis of the crystal. (Bar = 1 µm.)

Figure 14. Higher magnification of the base of the inclusion in Figure 13. The coiled fibers of the matrix are clearly visible. The arrow shows the long axis of the crystal. (Bar = 0.1 µm.)

Figure 15. A fully melanized and sclerotized crystal cell in the lymph gland of a *Bc/Bc* mutant larva. The "black" cell is encapsulated by two partially flattened plasmatocytes (pl). The nucleus (N) of one of the plasmatocytes is oblong rather than spherical. (Bar = 1 µm.)

Figure 16. A young hemocyte in the lymph gland. Aggregates of ribosomes/ribonucleoproteins (arrowheads) which are not bound by membrane may represent initiation of paracrystalline material in the cytoplasm. (Bar = 1 µm.)

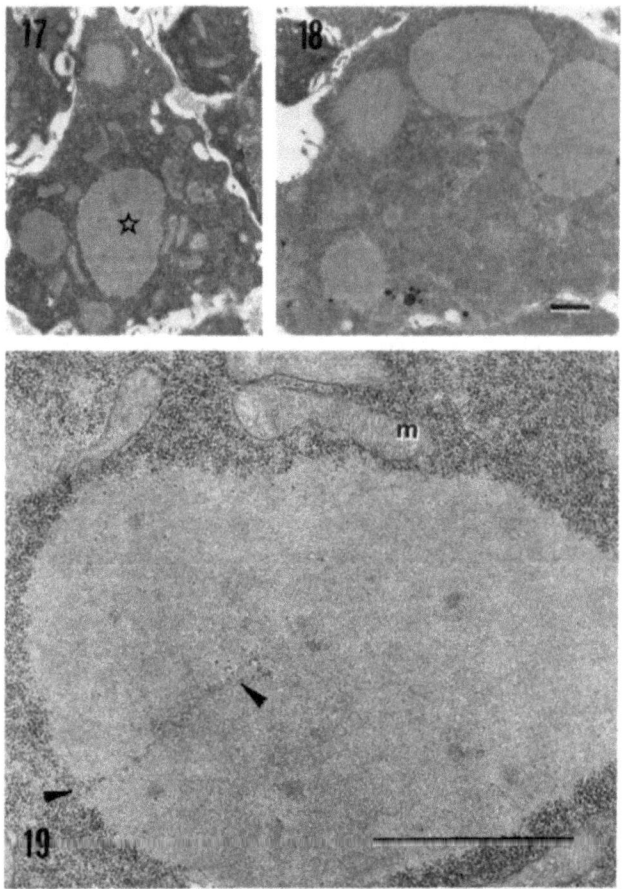

Figures 17 and 18. Two crystal cells from *D. teissieri* (Bar = 1 μm.)

Figure 19. A higher-magnification transmission electron micrograph of the starred inclusion in Figure 17. The contents are not paracrystalline. However, these inclusions, like those of *D. melanogaster*, are not membrane bound, and ribosomes are intimately connected at the periphery. In this case, ribosomes (arrowheads) also extend into the body of the inclusion. m, mitochondrion (B = 1 μm.)

5. Phagoctyosis

The phenomenon of phagocytosis can be divided into three distinct processes: (1) adhesion of a small foreign object to the hemocyte surface; (2) ingestion of the object by the hemocyte; (3) digestion or destruction of the ingested material. To determine whether all *Drosophila* hemocytes engage in phagocytosis or this function is performed by a subset of the hemocyte population, we injected *Escherichia coli* into Ore-R larvae and fixed specimens for SEM and TEM 15 min and 2–3 hr postinjection (T. Rizki and Rizki, 1980c). Figures 20–22 show bacteria adhering to plasmatocytes in the hemocoel of a

Drosophila larva. Cells having the topology of crystal cells (Rizki *et al.*, 1976) do not have bacteria attached to their surfaces. That crystal cells are not active in phagocytosis was also established by examining Epon sections in the TEM. Bacteria were readily found in plasmatocytes (Figure 26), but we never observed crystal cells with ingested bacteria (T. Rizki and Rizki, 1980c).

The differentiated lamellocyte surfaces retain the adhesive properties of the plasmatocytes. Injected bacteria in melanotic tumor larvae adhere to the surface of lamellocytes forming the outer layer of the capsule wall as well as to the surfaces of plasmatocytes in various stages of differentiation to the lamellocytic form (T. Rizki and Rizki, 1980c). Most of the bacteria are attached to the small filaments at the borders of the lamellocytes, but a few bacteria adhere to the main cell body (Figures 23, 24). Such material has not been sectioned, so we cannot establish whether fully differentiated lamellocytes are capable of ingesting bacteria. This seems doubtful considering the thinness of the cells. However, lamellocyte surfaces folding over bacteria were observed (Figure 25), and the sandwiching of adherent bacteria between lamellocytes during capsule formation can permanently trap bacteria and assist in ridding the body of circulating infective agents.

5.1. Cell Disruption as a Source of Humoral Factors

Experiments involving injection of bacterial suspensions into *Drosophila* larvae invariably result in disruption of a few hemocytes (Figures 27, 28). We have also observed ruptured hemocytes within the lymph glands under nonexperimental conditions in a melanotic tumor strain of *D. yakuba* (Figure 29). In this case, phagocytosis of the cellular debris by neighboring cells in the gland can be seen. Intracellular contents including soluble enzymes released from ruptured hemocytes may be mistaken for secreted humoral factors such as bacteriolysins or opsonins. This complication cannot be ignored in evaluating the source of humoral components.

5.2. Absence of Opsonization in Phagocytosis

Rapid assessment of the adherence, engulfment, and subsequent destruction of bacteria by *D. melanogaster* hemocytes was accomplished by exposing hemocytes to bacteria labeled with a fluorescent dye, and then examining the cells with fluorescence and phase optics. For these studies, *E. coli* (strain J5) grown on YT-broth with or without glucose were labeled with diamidino-2-phenylindole-2 HCl (DAPI), a DNA intercalating dye (T. M. Rizki and D. S. Eilender, unpublished). Hemocytes, both *in vivo* and *in vitro*, were challenged with these two groups of bacteria. Bacteria grown in the presence of glucose attached to plasmatocytic cell surfaces whereas bacteria grown in the absence of glucose did not. Thus, the nature of the bacterial surface must be important for their adhesion to *Drosophila* hemocyte surfaces, and the hemocytes must recognize a specific conformation of the bacterial cell surface rather than foreignness or nonself *per se*. Since both the bacteria and the hemocytes were

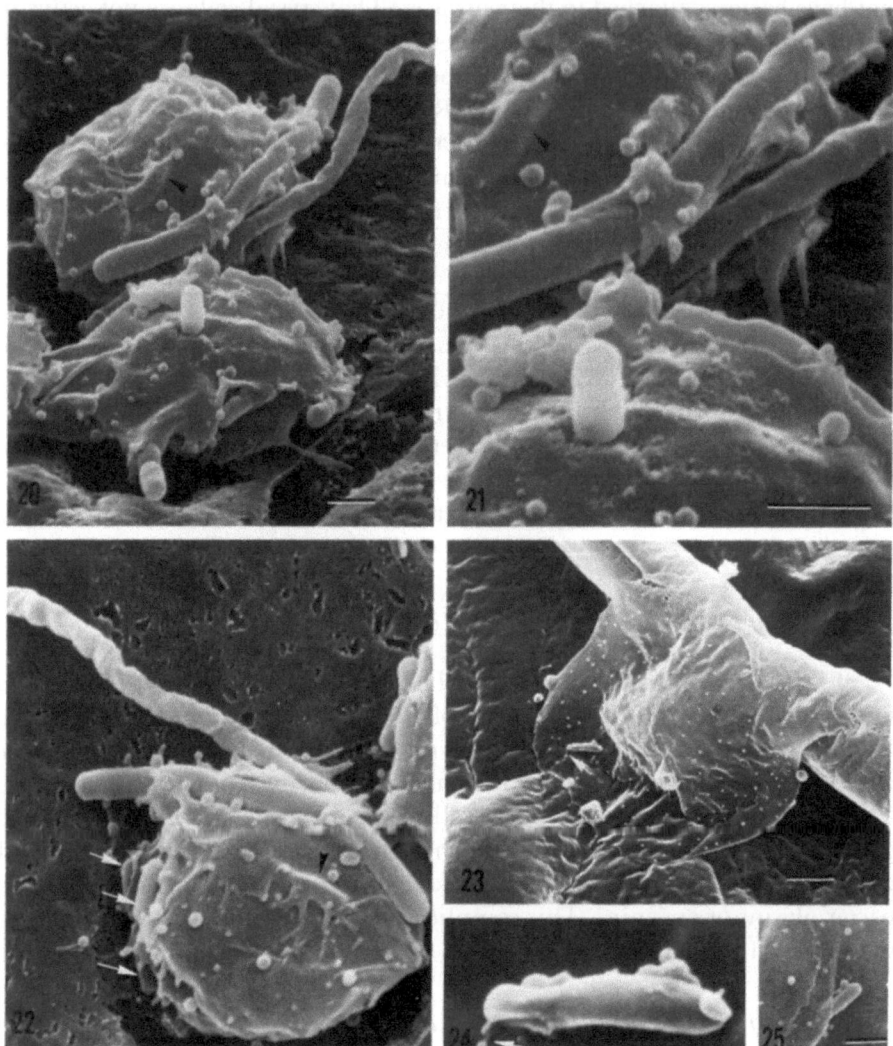

Figure 20. Phagocytosis of *E. coli* (strain HB101) by plasmatocytic cells *in vivo situ*. Some of the bacteria are partially ingested and at least one is completely covered by the folded surface membrane (arrowhead). (Bar = 1 μm.)

Figure 21. A higher-magnification scanning electron micrograph showing the details of membrane folding over adherent bacteria. The arrowhead points to the same bacterium under the surface membrane as in Figure 20. (Bar = 1 μm.)

Figure 22. Polar view of the same plasmatocyte as in Figure 20 showing more bacteria (arrows) which have been entrapped. Note the folding of the surface membrane over the same bacterium as in Figure 20 (arrowhead). The magnification is the same as in Figure 20.

Figure 23. Adhesion of a bacterium (*E. coli*, strain J5) to the surface of a lamellocyte of a *Bc/Bc* larva. (Bar = 3 μm.)

Figure 24. A higher-magnification electron micrograph of the same bacterium shown in Figure 23. The magnification is the same as in Figure 21.

Figure 25. A region of a lamellocyte surface folding over a bacterium. This lamellocyte is part of the outermost layer of a *tu-W* capsule wall. (Bar = 2 μm.)

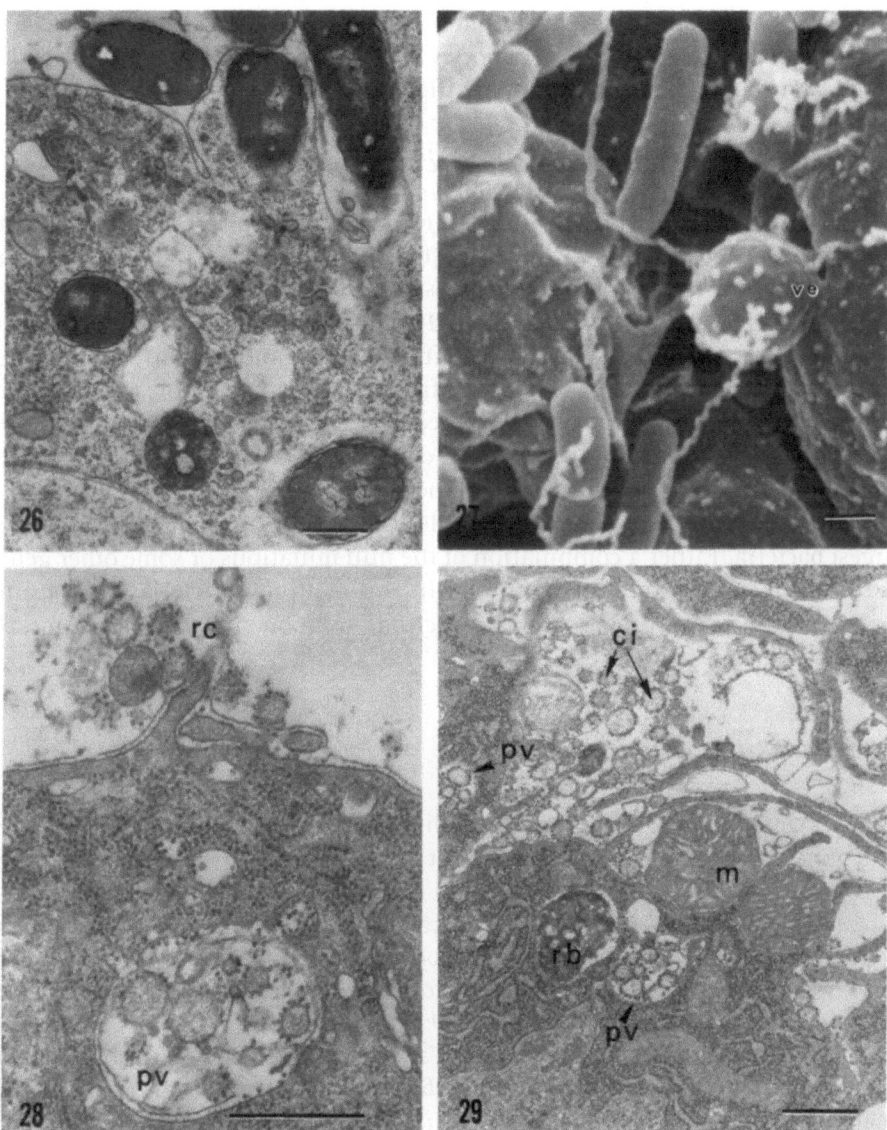

Figure 26. A transmission electron micrograph of a section passing through a phagocytic plasmatocyte showing bacteria adhering to the surface, being ingested, and in the process of digestion. (Bar = 0.5 μm.)

Figure 27. A scanning electron micrograph of phagocytic cell surfaces in a larva injected with *E. coli*. The fine threads are tubular cisternae released from ruptured hemocytes. Some of the cisternae are connected to a vesicle (ve). (Bar = 0.5 μm.)

Figure 28. A transmission electron micrograph of a section showing contents of ruptured cells (rc) in a specimen from the same experiment represented in Figure 27. Part of a hemocyte with a phagocytic vacuole (pv) containing similar profiles as seen outside the cell. (Bar = 0.5 μm.)

Figure 29. A naturally occurring cell lysis and phagocytosis within the lymph gland of a *D. yakuba* larva. Mitochondria (m), ribosome-bound cisternae (ci), and other cellular contents can be seen outside the phagocyte where some of them are in the process of being engulfed. Note the phagocytic vacuole (pv), the residual body (rb). (Bar = 0.5 μm.)

washed with physiological saline before they were mixed for the *in vitro* studies, the necessity for opsonin-mediated adhesion of bacteria to the hemocyte surface is ruled out. Using cockroach hemocytes *in vitro*, Scott (1971) and Anderson *et al.* (1973) came to this conclusion on different grounds.

5.3. A Mutation Affecting Phagocytosis

DAPI-labeled *E. coli* were also used to evaluate the role of phosphatases in phagocytosis. Labeled bacteria were injected into wild-type larvae, larvae from a strain carrying a number of third-chromosome markers including acid phosphatase (*Acph*) and alkaline phosphatase (*Aph*) null mutations (MacIntyre, 1966 a,b), and larvae from several strains with various combinations of the same third-chromosome markers. Hemocytes from these strains sampled within 4 hr postinjection all had adherent or engulfed bacteria. By 24 hr postinjection, the bacteria had been digested in all strains except in (*ru Aph e ca Acph^{n13}*). Since two of the other tested strains also had the *Acph^{n13}* allele, it is unlikely that the absence of acid phosphatase activity alone is responsible for the failure to digest bacteria. Either a combination of the null mutations or some other gene is this third chromosome affects intracellular digestion of bacteria. This strain will be useful for correlation ultrastructural components with biochemical processes during phagocytosis.

6. Encapsulation

Foreign objects too large to be phagocytosed by single hemocytes are surrounded by layers of lamellocytes that form compact capsules enclosing the extraneous materials. Cellular capsules are generally melanized and remain within the insect as inert bodies. Encapsulation of biological as well as nonbiological materials has been described in a number of insects (Salt, 1970; Ratcliffe and Rowley, 1979). In *Drosophila*, eggs of parasitic wasps are destroyed by encapsulation (Walker, 1959; Nappi and Streams, 1969; Carton and Kitano, 1981), and in one reported case, a massive bacterial infection was also contained by this mechanism (Rizki, 1969). Cellular capsules also develop in melanotic tumor mutants of *Drosophila*, but in these strains the site of encapsulation is a gene-conditioned, aberrant tissue. Several melanotic tumor mutations having high penetrance and expressivity provide excellent material for studying the phenomen of capsule formation (Oftedal, 1952; Wilson *et al.*, 1955; Sang and Burnet, 1963; Sparrow, 1978).

6.1. Capsule Formation in Melanotic Tumor Mutants

In *tu-W* (2-66.2) (Rizki and Rizki, 1981b) melanotic tumor larvae, lamellocytes are present from the beginning of the third instar. However, encapsulation does not begin until the mid third instar when the basement membrane overlying some of the caudal fat body cells disintegrates (Rizki and Rizki, 1979).

The importance of a change in tissue surfaces is realized in the tu-Sz^{ts} (1-34.3) temperature-sensitive mutant where the larvae develop melanotic tumors at 26°C but not at 18°C. Lamellocytes are present throughout the third instar at both temperatures. At higher temperature, it is only after the basement membrane overlying atypical cells of the caudal fat body becomes abnormal that capsule formation begins. At 18°C, there are no atypical cells, the basement membrane of the caudal fat body remains normal, and lamellocytes do not form capsules at this site (T. Rizki and Rizki, 1980a).

It is clear that the lamellocytes in tu-W and tu-Sz^{ts} larvae react specifically against tissues developing aberrant surfaces during larval life. The hemocyte response to defective tissues in other nonallelic melanotic tumor mutations is the same as that seen in these two mutant strains, but the tissue surface modifications that precede hemocyte activity are unique for each melanotic tumor gene. Thus, the phenomenon of melanotic tumor formation can be partitioned into two processes: (1) a gene-conditioned tissue surface modification and (2) a response by the body's cellular defense against the atypical surfaces. Since the outcome of the second step, a melanotic cellular capsule, is the same as that found when a foreign object such as a parasite enters the body, the tissue surfaces in these mutants must be altered such that they are no longer recognized as self by the hemocytes. This implies that the hemocytes constantly monitor the normality of the body's internal surface linings in the hemocoel. The surfaces of the internal organs suspended within the hemocoel and the body wall are covered by basement membrane, so this covering itself must provide signals to the hemocytes for the distinction of nonself from self.

6.2. Basement Membrane as a Factor for Recognition of Self

D. melanogaster hemocytes remain neutral to tissues covered by normal basement membrane and react against tissues that are not (R. Rizki and Rizki, 1980). Implants with these two type of surfaces were tested in tu-Sz^{ts} larvae growing at 18°C. These larvae are ideal hosts, since they have an abundance of lamellocytes that are not engaged in capsule formation. The tu-Sz^{ts} lamellocytes encapsulate *D. melanogaster* (*Ore-R*) larval fat body implants with injured surfaces and fat body implants whose basement membranes have been removed by collagenase, but do not encapsulate *Ore-R* implants with unaltered surfaces. On the other hand, undamaged heterospecific tissues, such as *D. virilis* fat bodies and imaginal discs, are encapsulated suggesting that tu-Sz^{ts} lamellocytes are capable of detecting differences between the molecular architecture of the basement membranes of *D. melanogaster* and *D. virilis*. A recently isolated monoclonal antibody that recognizes *D. melanogaster* basement membrane but not *D. virilis* basement membrane confirms that there are antigenic differences between these surface covering (Rizki *et al.*, 1983).

The lamellocytes of tu-Sz^{ts} larvae have also been used as a bioassay to monitor changes in tissue surfaces during development and aging (Rizki *et al.*, 1983). Fat body from various ages of larval donors remains unencapsulated in tu-Sz^{ts} hosts, but fat body from larvae with everted spiracles and from donors of

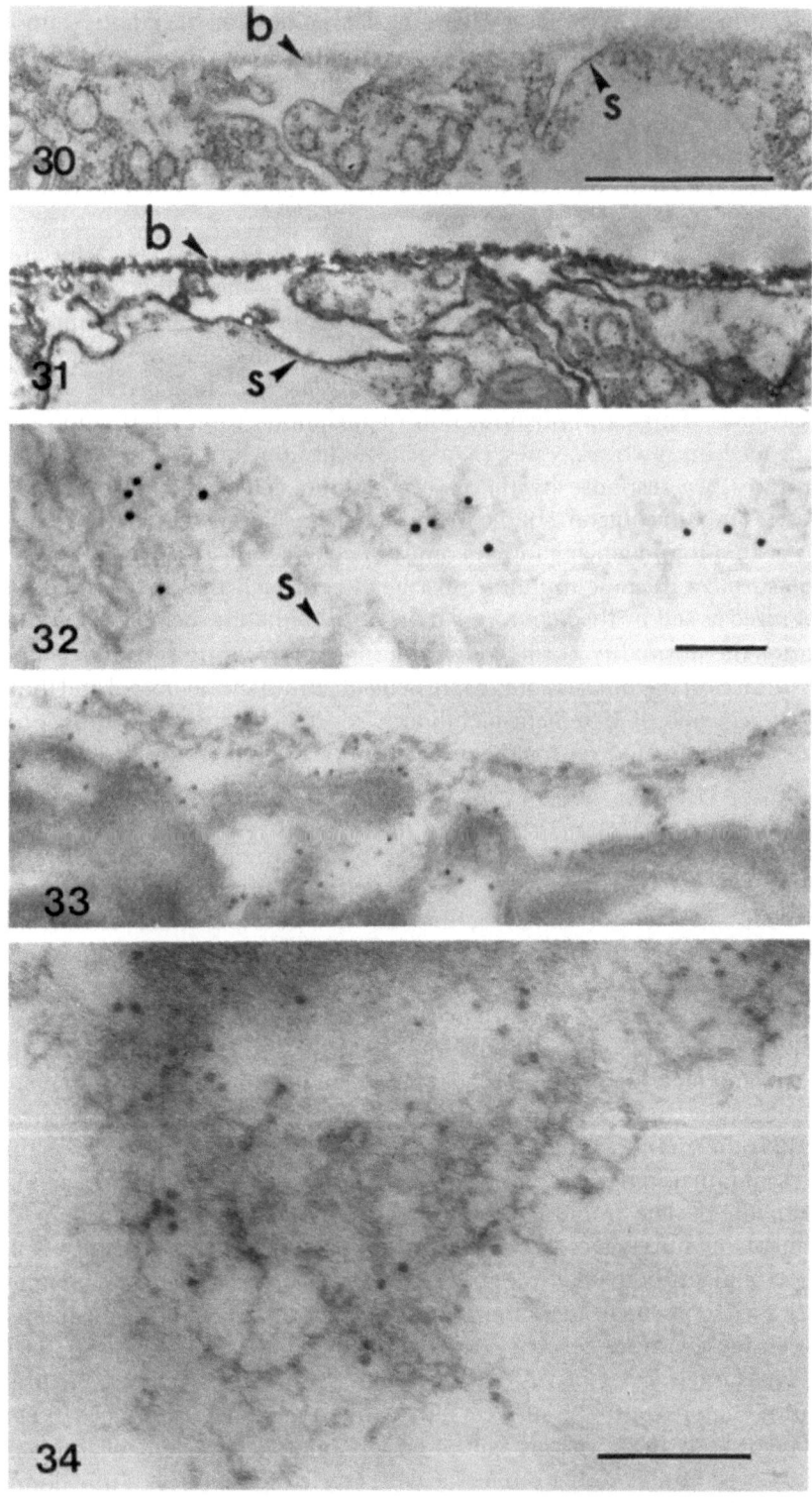

subsequent stages in pupariation and pupation are encapsulated. During pupation, the basement membrane of the larval fat body disintegrates and the individual adipose cells are dispersed in the pupal hemocoel. Since the pupal fat body cells are not covered by basement membrane, their susceptibility to encapsulation is in line with the observations described above. Moreover, it seems that the biochemical changes in the tissue surfaces that must precede the dissolution of the basement membrane may be occurring as early as the spiracle-everted stage, and the tu-Sz^{ts} lamellocytes are capable of discriminating these changes which cannot be detected at the ultrastructural level (Rizki et al., 1983).

6.3. Ultrastructural Examination of Basement Membrane

A thin layer of basement membrane (basal lamina) forms a continuous covering around the entire fat body. This surface coat consists of fine fibrils forming a network with open spaces (Figures 30-34). These fibrils also project at various angles from the surface of the basement membrane. Presumably, the open spaces in the basement membrane allow passage of materials in and out of the underlying fat body cells. The arrangement of the fibrils can be seen best in tangential sections which also reveal extensions of the fibrils to small protuberances on the fat body cell surfaces. It has been suggested that such interconnections restrict mobility of the underlying cell surface molecules so that the fat body cells remain polarized during larval life (Rizki and Rizki, 1983a).

Histochemical studies show that the basement membrane of *D. melanogaster* reacts with periodic acid–Schiff reagent after amylase treatment and binds alcian blue. Moreover, inclusion of alcian blue or ruthenium red in primary and secondary fixatives for TEM increases the electron density of basement membrane elements. Thus, *Drosophila* basement membrane contains carbohydrate moieties in association with material digestible by collagenase (see Section 6.2). Recently, Monson et al. (1982) isolated genomic DNA clones encoding collagen from *D. melanogaster* and demonstrated that the expression of these genes during development coincides with the expected pattern of basement membrane genesis (Natzle et al., 1982). Whether the carbohydrate moieties are due to glycosylation of collagen or whether there are other glycoproteins such as laminin in *Drosophila* basement membrane is not known.

◄─────────────────────────────────

Figures 30 and 31. Sections of fat body reacted with P-WGA. The control (Figure 30) was treated with P-WGA that had been preincubated with chitobiose to establish the specificity of the WGA binding illustrated in Figure 31. Note the strong peroxidase reaction of WGA bound to the basement membrane (b) and the adipose cell surfaces (s) in Figure 31. (Bar = 1 μm.)

Figure 32. Distribution of Au-WGA in the matrix of basement membrane of larval fat body. The colloidal gold particles fail to reach the underlying cell surfaces (s). (Bar = 0.1 μm.)

Figure 33. F-WGA binds to basement membrane and fat body cell surfaces. The small, electron-dense dots are ferritin. Same magnification as Figure 32.

Figure 34. A high-resolution transmission electron micrograph of a tangential section showing F-WGA binding to the fibrous components of basement membrane. (Bar = 0.1 μm.) Material illustrated in Figures 32-34 was not osmicated, and the sections were stained with vanadatomolybdate.

We have investigated carbohydrate moieties in *Drosophila* basement membrane by use of the plant lectin, wheat germ agglutinin (WGA), which binds specifically to *N*-acetyl glucosamine residues. This binding can be blocked by pretreating WGA with chitobiose. Ultrastructural localization of WGA binding sites employed WGA conjugated to ferritin (F), to peroxidase (P), and colloidal gold (Au) coated with WGA (Rizki *et al.*, 1983). Au-WGA does not penetrate the tissue surface, and its binding is restricted to basement membrane sites, where as F-WGA and P-WGA bind to basement membrane as well as the underlying cell surfaces (Figures 31-33). There is no difference in the affinity of F-WGA to the basement membranes of larval fat bodies prior to and after the spiracle-everted stage. Nor were we able to distinguish the basement membranes of these stages by using fluorescein isothiocyanate-conjugated (FITC) WGA which is useful for examining the distribution of WGA binding sites over large areas of tissue surface. However, at 26°C the atypical surfaces of tu-Sz^{ts} larval fat bodies lose their FITC-WGA binding sites (Rizki and Rizki, 1983b). The latter correlates with the changes in these surfaces at the ultrastructural level (T. Rizki and Rizki, 1980a).

7. Differentiation of Competent Lamellocytes

Plant lectins have been utilized to distinguish subpopulations of vertebrate lymphocytes (Reisner *et al.*, 1976; Irlé *et al.*, 1978; Hammarström *et al.*, 1978), and this approach has recently been adopted to study *D. melanogaster* hemocytes (Rizki and Rizki, 1983b). Two populations of lamellocytes are distinguished on the basis of FITC-WGA binding: one which lights up with speckled surfaces (spk$^+$) and the other not showing this characteristic (spk$^-$). In tu-Sz^{ts} larvae at 18°C, about 95% of the lamellocytes are spk$^-$ whereas at tumor-permissive temperature (26°C) 80% are spk$^+$. The evidence that the spk$^+$ population differentiates in response to foreign surfaces comes from transplantation experiments using *D. virilis* donor tissues in tu-Sz^{ts} hosts at 18°C. Within 24 hrs after receiving heterospecific implants, the hosts' spk$^+$ lamellocyte population rises to about 60%. Moreover, the spk$^+$ lamellocyte frequency is high in two other nonallelic melanotic tumor mutants (tu-W and tu-$bw = mt^4$) and low in the *Ore-R* wild-type (nontumorous) strain. The possibility of clonal differentiation of spk$^+$ populations is suggested by the presence of spk$^+$ plasmatocytes in larvae with spk$^+$ lamellocytes, since lamellocytic forms are differentiated from plasmatocytes.

The presence of spk$^+$ lamellocytes under capsule-forming conditions suggests that these are the capsule-competent cells. The mode of operation of spk$^+$ hemocytes in capsule formation has not been fully established, but developing capsule walls contain both spk$^+$ and spk$^-$ lamellocytes. Moreover, WGA$^+$ material on atypical tissue surfaces that had previously lost the normal WGA binding sites may originate from spk$^+$ hemocytes (see next paragraph). We propose (Rizki and Rizki, 1983b) that the spk$^+$ hemocytes extrude "xenophilic" material which is the glue for adhesion between spk$^+$ and spk$^-$

hemocytes as well as between aberrant surfaces and lamellocytes. In the absence of spk$^+$ hemocytes (e.g., in tu-Sz^{ts} larvae at 18°C), spk$^-$ lamellocytes do not adhere to one another to form capsules.

We have not as yet examined the binding of WGA to aberrant fat body cell surfaces and lamellocyte surfaces by EM. However, the material on the atypical fat body surfaces may be associated with surface materials that we observed previously by SEM and TEM (Rizki and Rizki, 1974, 1979). Figures 35 and 36 illustrate the deposits of vesicular material seen on aberrant fat body cell surfaces in regions where plasmatocytes and lamellocytes are located, and this material appears to be extruded by the hemocytes. That the materials on the hemocyte surfaces may contain carbohydrate moeities has been illustrated by including alcian blue in the EM fixative. There is a differential distribution of electron-dense material due to alcian blue staining on some hemocyte surfaces and not others. This material is also present at the contact points between lamellocyte surfaces (Figures 37, 38). Whether this material is the same as, or is related to, the materials which are WGA$^+$ is unknown at this time, but these studies suggest that ultrastructural localization of appropriately tagged molecular probes, such as ferritin-conjugated lectins and monoclonal antibodies, will be a worthwhile approach to analyze lamellocyte–lamellocyte binding and adhesion of lamellocytes to nonself surfaces.

Other remaining questions concern the nature of the stress signals which induce differentiation of competent hemocytes and whether these signals require processing by a specific subpopulation of plasmatocytes. Further, what

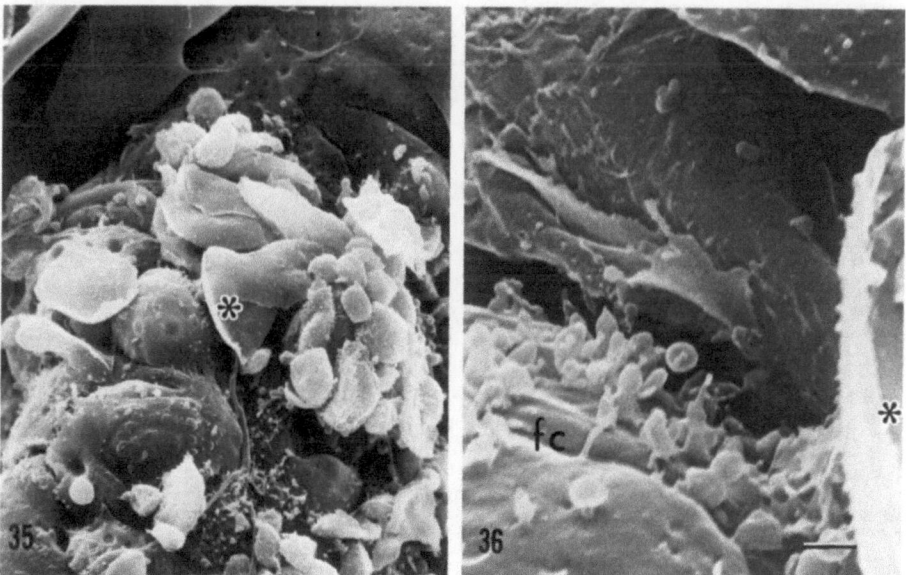

Figures 35 and 36. Lamellocytes and other hemocytes on the surface of a tu-Sz^{ts} caudal fat body. The area adjacent to the lamellocyte marked with an asterisk is magnified in Figure 36 to show the vesicles coating the fat body cells (fc). (Bar = 1 μm.)

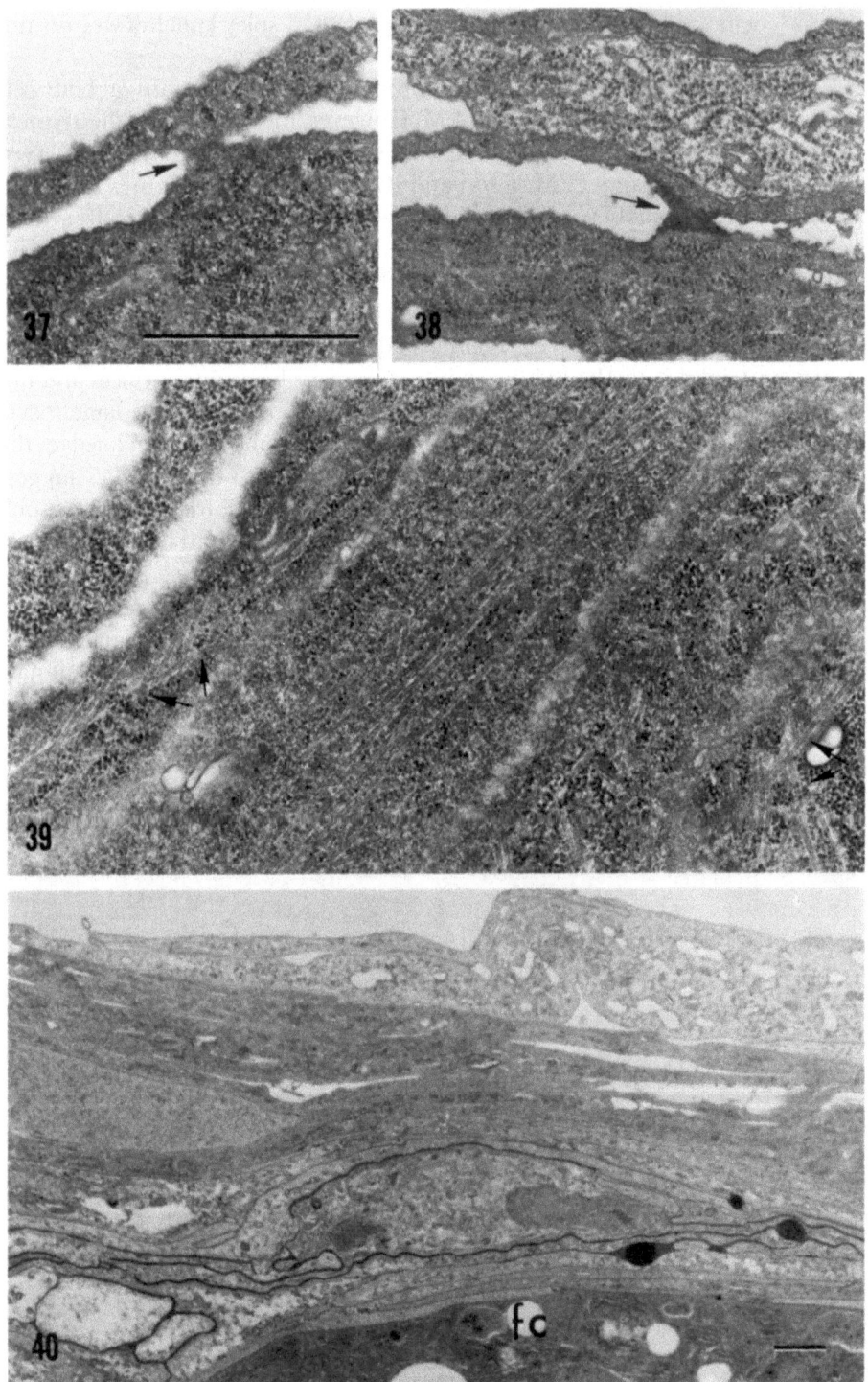

causes the differentiation of spk⁻ lamellocytes in the absence of capsule-forming conditions? Equally important is consideration of the delayed response of lamellocytes to implanted tissues. For example, encapsulation of implanted *D. virilis* tissue does not commence for 12–14 hr, whereas adherence of bacteria to hemocyte surfaces is rapid. Finally, what role is played by the microtubules (Figure 39) and other cytoskeletal structures of the lamellocytes as the latter are layered in the capsule wall (Figure 40)?

8. Contribution of Crystal Cells to Melanization Reactions

When *Drosophila* hemolymph is exposed to air, it blackens within a few minutes. This melanization is mediated by hemolymph phenol oxidases. Two mutant genes known to affect crystal cell function clearly demonstrate that the crystal cells are the sole source of larval hemolymph phenol oxidase activity (Rizki and Rizki, 1959). Hemocytes with paracrystalline inclusions are absent in *Black cell* (*Bc*; 2-80.6) larvae which contain instead black sclerotized cells in their lymph glands and hemocoel (Figure 15). The intracellular release of phenol oxidases in the mutant crystal cells results in self destruction of the enzyme complexes, due to cross-linking and tanning of proteins, and deprives the hemolymph of its phenol oxidase activity (Rizki *et al.*, 1980). Some mutant alleles of *lozenge* (*lz*; 1-27.7) lack hemolymph phenol oxidase activity (Peeples *et al.*, 1969), and these mutants do not have hemocytes with paracrystalline inclusions (Rizki and Rizki, 1981a). The absence of both phenol oxidase and paracrystalline structures among the hemocytes of these *lz* mutants strongly suggests that the paracrystalline material may include proenzyme or enzyme proteins of the phenol oxidase complex. When combined with the *Bc* mutant gene, the phenol oxidase-minus *lz* mutants act as suppressors of the black cell phenotype, since the latter depends on the presence of phenol oxidase activity.

The availability of mutant genes blocking hemolymph phenol oxidase activity made it possible to demonstrate that the source of melanization and

---

Figure 37. Contact point (arrow) between two lamellocyte surfaces that are engaged in capsule formation around the lymph gland of a *tu-bw* larva. Fixation without alcian blue. (Bar = 1 μm.)
Figure 38. This tissue was treated the same as that in Figure 37 except that the fixative contained alcian blue. Note the electron-dense material (arrow) between the lamellocyte surfaces. Same magnification as Figure 37.
Figure 39. A section passing through the capsule wall in a *tu-bw* larva showing the disposition of microtubules of the lamellocytes. Parallel arrays of microtubules in the center of the electron micrograph run diagonally and in a crisscross pattern in the regions indicated by double arrows. Same magnification as Figure 37.
Figure 40. A section passing through the capsule wall around an implanted *D. virilis* fat body recovered from a *tu-Sz^{ts}* host. There are intercellular layers of melanized material with electron-dense deposits between the lamellocytes of the inner and outer layers. fc, fat body cell. (Bar = 1 μm.)

cross-linking of cellular capsules is the crystal cell. In double mutants, such as *tu-W Bc/tu-W Bc,* lamellocytic capsules form, but due to failure of cross-linking and tanning of the capsular walls, the tumors remain unmelanized. These amelanotic tumors are digested and lysed during pupation. However, the absence of crystal cell function in *Bc* larvae neither impairs phagocytosis of ingested *E. coli* by the plasmatocytes nor affects the encapsulation response of its lamellocytes to implated *D. virilis* tissue (Rizki and Rizki, unpublished observations). Such studies strengthen the conclusion that plasmatocytic cell functions are independent of crystal cell function, and demonstrate that crystal cells do not play a direct role in recognition of foreign surfaces.

9. Wound Healing

When the body wall of the *Ore-R* wild-type larva is punctured, the hemolymph seeping through the wound becomes hardened and darkened forming a black crust over the injured cuticle. Mutant larvae with impaired crystal cell function (*Bc* and *lz* alleles; see Section 8) fail to show this healing response and often bleed to death. Those *lz* alleles that have crystal cells with paracrystalline inclusions (Rizki and Rizki, 1981a) form black crusts over an injury site. We suppose this process to be extracellular, since massive accumulations of blood cells are absent from the injured site (Rizki and Rizki, unpublished observations), and it is known that the release of enzymes from ruptured crystal cells can cause darkening and hardening of the surrounding hemolymph contents (Rizki, 1978).

The role of crystal cells in would healing and melanization of cellular capsules are the only functions of the crystal cells that have been detected thus far. However, crystal cells first appear in the middle of the egg embryonic stage and are present in larvae and early puparia (Rizki *et al.*, 1980). After using hot water to artificially blacken the crystal cells *in situ*, we find decreasing numbers of crystal cells in progressively older puparia. Crystal cells are not detectable by this method after eversion of the cephalic complex in the pupa (Rizki and Rizki, unpublished observations). We hope that the contribution of crystal cells to the developmental physiology of the larva and pupa can be ascertained by comparative studies of mutants such as *Bc* and *lz*.

10. Genetic Dissection of the Cellular Defense System

A number of genes affecting various components of the cellular defense system have been described in this chapter: melanotic tumor mutations affecting tissue surfaces so that they are no longer accepted as self; mutations inactivating crystal cell function; and a mutation (referred to as factor III) on the third chromosome causing defective phagocytosis. The nature of the interactions required for a successful encapsulation reaction can be appreciated by systematically combining these mutant genes and observing the resultant phenotype.

As described in Section 8, combination of the *Bc* gene with the *tu-W* gene results in the formation of amelanotic capsules that do not survive metamorphosis. Therefore one of the normal functions of the crystal cells in the encapsulation reaction must be the preservation of capsule walls formed by the lamellocytes. That the development of a hardened, melanotic capsule wall is important for survival of an affected individual can be demonstrated by combining the *tu-W* and *Bc* genes with factor III. The triple mutant dies in the white puparial stage, suggesting that the phagocytic function is indispensable for survival in the absence of normal crystal cell function. Presumably, the phagocytic function of the plasmatocytes is not crucial when crystal cell function is operating normally since the aberrant tissue masses are effectively confined within tightly bound melanized capsule walls.

Recognition of non self by the plasmatocytes and lamellocytes is independent of the processes involved in intracellular digestion of foreign materials, since factor III, which disrupts the digestive step in phagocytosis, does not interfere with the adhesion between plasmatocytes and foreign materials. Double mutant individuals carrying *Bc* and factor III are viable. Thus, in the absence of a tissue surface aberration (such as that caused by *tu-W*), both the crystal cell function and the phagocytic function are nonessential for survival of the individual. The latter two cell functions become integral components of the cellular defense system on demand. Mutant genes that affect these two cell types block the facultative roles of the hemocytes. Whether the obligatory functions of the blood cells are also affected, and what they are, remains to be investigated.

11. Summary and Concluding Remarks

On the basis of morphological and functional criteria, the larval hemocytes of *D. melanogaster* are separable into two distinct lines: (1) the crystal cell line which carries phenol oxidases and substrates utilized in melanization reactions and (2) the plasmatocyte and its variants—podocytes and lamellocytes—whose surfaces have adhesive properties for phagocytosis and encapsulation of foreign materials. Mutant genes causing a defect in crystal cell function do not interfere with functions of the plasmatocyte line; a mutant gene affecting the phagocytic capability of the plasmatocytes does not disrupt crystal cell function.

Subpopulations of plasmatocytes and lamellocytes have been distinguished recently using FITC-conjugated lectins, and frequency differences of the subpopulations have been correlated with capsule formation. Thus, the way has been paved for analysis of insect hemocyte surface interactions by molecular probes which can be tagged for localization at the ultrastructural level.

No consideration has been given to the possible existence of a histocompatibility system in *Drosophila* since *D. melanogaster* will accept donor tissues from sibling species. Recently, Shalev *et al* (1983) provided serologic evidence that *Drosophila* embryonic cells in culture have H-2-like and β_2-microglobulin-like antigens. This suggests that these antigens have been conserved during the course of evolution from invertebrates to vertebrates and that they are likely coded by orthologous genes. We have recently found that *D. melanogaster* DNA shares sequence homology with mouse H-2 DNA. (Lalanne *et al.*, 1982).

ACKNOWLEDGMENT

NIH Grant 01945 supported the preparation of this chapter and the unpublished studies included in some sections.

References

Anderson, R. S., Holmes, B., and Good, R. A., 1973, In vitro bactericidal capacity of *Blaberus craniifer* hemocytes, *J. Invertebr. Pathology* 22:127-135.
Carton, Y., and Kitano, H., 1981, Evolutionary relationships to parasitism by seven species of the *Drosophila melanogaster* subgroup, *Biol. J. Linn. Soc.* 16:227-241.
Carton, Y., Bouletreau, M., van Lenteren, J. C., and van Alphen, J.C.M., 1984. The *Drosophila* parasitic wasps, In *The Genetics and Biology of Drosophila*, Vol. 3, edited by M. Ashburner, H. L. Carson, and J. N. Thompson, Academic Press, New York, in press.
Castiglioni, M. C., 1957, Le cellule dell-emolinfa di *Drosophila melanogaster* in relazione al genotipo e alla produzione degli pseudotumori, *Atti III Riun, A.G.I. Ric. Sci.* Suppl. 27:51-58.
El Shatoury, H. E., 1955, The structure of the lymph glands of *Drosophila* larvae, *Wilhelm Roux Arch. Dev. Biol.* 147:489-495.
Grégoire, C. H., and Goffinet, G., 1979, Controversies about the coagulocyte. In *Insect Hemocytes*, edited by A. P. Gupta, pp. 189-229, Cambridge University Press, London.
Grimstone, A. V., Rotherham, S., and Salt, G., 1967, An electron-microscope study of capsule formation by insect blood cells, *J. Cell Sci.* 2:281-292.
Gupta, A. P. (ed.), 1979, Hemocyte types: Their structures, synonymies, interrelationships, and taxonomic significance. In *Insect Hemocytes*, pp. 85-127, Cambridge University Press, London.
Hammarström, S., Hellström, U., Dillner, M.-L., Perlmann, P., Perlmann, H., Axelsson, B., and Robertsson, E.-S., 1978, Fractionation of lymphocytes on insolubilized *Helix pomatia* A hemagglutinin and wheat germ agglutinin. In *Affinity Chromatography*, edited by O., Hoffmann-Ostenhof, pp. 273-286, Pergamon Press, Elmsford, N.Y.
Irlé, C., Piguet, P.-F., and Vassalli, P., 1978, In vitro maturation of immature thymocytes into immunocompetent T cells in the absence of direct thymic influence, *J. Exp. Med.* 148:32-45.
Lalanne, J. L., Bregegere, F., Delarbe, C., Abastado, J. P., Gachelin, G., and Kourilsky, P., 1982, Comparison of nucleotide sequences of mRNAs belonging to the mouse H-2 multigene family, *Nucleic Acids Res.* 10:1039-1049.
Lemeunier, F., and Ashburner, M., 1976, Relationships within the *melanogaster* species subgroup of the genus *Drosophila* (Sophophora). II. Phylogenetic relationships between six species based upon polytene chromosome banding sequences. *Proc. R. Soc. London Ser. B* 193:275-294.
MacIntyre, R. J., 1966a, The genetics of an acid phosphatase in *D. melanogaster* and *D. simulans*, *Genetics* 53:461-474.
MacIntyre, R. J., 1966b, Locus of the structural gene for 3rd larval instar alkaline phosphatase, *Drosophila Inf. Serv.* 41:62.
Monson, J. M., Natzle, J., Friedman, J., and McCarthy, B. J., 1982, Expression and novel structure of a collagen gene in *Drosophila*, *Proc. Natl. Acad. Sci. USA* 79:1761-1765.
Nappi, A. J., and Streams, F. A., 1969, Haemocytic reactions of *Drosophila melanogaster* to the parasites of *Pseudeucoila mellipes* and *P. bochei*, *J. Insect Physiol.* 15:1551-1566.
Natzle, J. E., Monson, J. M., and McCarthy, B. J., 1982, Cytogenetic location and expression of collagen-like genes in *Drosophila*, *Nature* 296:368-371.

Oftedal, P., 1952, Histology and histogenesis of Drosophila tumors, Science 116:392-393.
Peeples, E., Geisler, A., Whitcraft, C. J., and Oliver, C. P., 1969, Activity of phenol oxidases at the puparium formation stage in development of nineteen lozenge mutants of Drosophila melanogaster, Biochem. Genet. 3:563-569.
Ratcliffe, N. A., and Rowley, A. F., 1979, Role of hemocytes in defense against biological agents. In Insect Hemocytes, edited by A. P. Gupta, pp. 331-414, Cambridge University Press, London.
Reisner, Y., Linker-Israeli, M., and Sharon, N., 1976, Separation of mouse thymocytes into two subpopulations by the use of peanut agglutinin, Cell. Immunol. 25:129-134.
Rizki, R. M., and Rizki, T. M., 1974, Basement membrane abnormalities in melanotic tumor formation, Experientia 30:543-546.
Rizki, R. M., and Rizki, T. M., 1979, Cell interactions in the differentiation of a melanotic tumor in Drosophila, Differentiation 12:167-178.
Rizki, R. M., and Rizki, T. M., 1980, Hemocyte responses to implanted tissues in Drosophila melanogaster larvae, Wilhelm Roux Arch. Dev. Biol. 189:207-213.
Rizki, R. M., Rizki, T. M., Bebbington, C. R., and Roberts, D. B., 1983, Drosophila larval fat body surfaces: Changes in transplant compatibility during development, Wilhelm Roux Arch. Dev. Biol. 192:1-7.
Rizki, T. M., 1956, Blood cells of Drosophila as related to metamorphosis. In Physiology of Insect Development, edited by F. L. Campbell, pp. 91-94, University of Chicago Press, Chicago.
Rizki, T. M., 1957a, Alterations in the haemocyte population of Drosophila melanogaster, J. Morphol. 100:437-458.
Rizki, T. M., 1957b, Tumor formation in relation to metamorphosis in Drosophila melanogaster, J. Morphol. 100:459-472.
Rizki, T. M., 1960, Melanotic tumor formation in Drosophila, J. Morphol. 106:147-158.
Rizki, T. M., 1962, Experimental analysis of hemocyte morphology in insects, Am. Zool. 2:247-256.
Rizki, T. M., 1969, Hemocyte encapsulation of streptococci in Drosophila, J. Invertebr. Pathol. 12:339-343.
Rizki, T. M., 1978, The circulatory system and associated cells and tissues. In The Genetics and Biology of Drosophila, vol. 2b, edited by M. Ashburner and T. R. F. Wright, pp. 397-452, Academic Press, New York.
Rizki, T. M., and Rizki, R. M., 1959, Functional significance of the crystal cells in the larva of Drosophila melanogaster, J. Biophys. Biochem. Cytol. 5:235-240.
Rizki, T. M., and Rizki, R. M., 1978, The role of hemocytes in melanotic tumor formation. In Comparative Pathobiology, vol. 4, edited by L. A. Bulla, Jr., and T. C. Cheng, pp. 85-96, Plenum Press, New York.
Rizki, T. M., and Rizki, R. M., 1980a, Developmental analysis of a temperature-sensitive mutant in Drosophila melanogaster, Wilhelm Roux Arch. Dev. Biol. 189:197-206.
Rizki, T. M., and Rizki, R. M., 1980b, The direction of evolution in the Drosophila melanogaster species subgroup based on functional analysis of the crystal cells, J. Exp. Zool. 212:323-328.
Rizki, T. M., and Rizki, R. M., 1980c, Properties of the larval hemocytes of Drosophila melanogaster, Experientia 36:1223-1226.
Rizki, T. M., and Rizki, R. M., 1981a, Alleles of lz as suppressors of the Bc-phene in Drosophila melanogaster, Genetics 97:s90.
Rizki, T. M., and Rizki, R. M., 1981b, Genetics of tumor-W in Drosophila melanogaster, J. Hered. 72:78-80.
Rizki, T. M., and Rizki, R. M., 1983a, Basement membrane polarizes lectin, binding sites of Drosophila larval fat body cells, Nature 30:340-342.
Rizki, T. M., and Rizki, R. M., 1983b, Blood cell surface changes in melanotic tumor mutants of Drosophila, Science 220:73-75.
Rizki, T. M., Rizki, R. M., Allard, L. F., and Bigelow, W. C., 1976, Micromanipulation of tissues and cells of the Drosophila larva in the SEM, Scanning Electron Microscopy/1976 II, edited by O. Johari and R. P. Becker, pp. 611-618, IIT Research Institute, Chicago.
Rizki, T. M., Rizki, R. M., and Grell, E. H., 1980, A mutant affecting the crystal cells in Drosophila melanogaster, Wilhelm Roux Arch. Dev. Biol. 188:91-99.
Robertson, C. W., 1936, The metamorphosis of Drosophila melanogaster including an accurately timed account of the principal morphological changes, J. Morphol. 59:351-399.

Röhrborn, G., 1961, *Drosophila* tumors and the structure of larval lymph glands, *Experientia* **17**:507.
Salt, G., 1970, *The Cellular Defense Reactions of Insects*, Cambridge University Press, London.
Sang, J. H., and Burnet, B., 1963, Physiological genetics of melanotic tumors in *Drosophila melanogaster*. I. The effects of nutrient balance on tumor penetrance in the tu^K strain, *Genetics* **48**:235-253.
Scott, M. T., 1971, Recognition of foreignness in invertebrates. II. *In vitro* studies of cockroach phagocytic haemocytes, *Immunology* **21**:817-828.
Shalev, A., Pla, M., Ginsburger-Vogel, T., Echalier, G., Lögdberg, L, Björck, L., Colombani, J., and Segal, S., 1983, Evidence for β_2-microglobulin-like and H-2-like antigenic determinants in Drosophila, **130**:297-302.
Shrestha, R., and Gateff, E., 1982, Ultrastructure and cytochemistry of the cell types in the larval hematopoietic organs and hemolymph of *Drosophila melanogaster*, *Dev. Growth Differ.* **24**:65-82.
Sparrow, J. C., 1978, Melanotic "tumours" In *The Genetics and Biology of Drosophila*, vol. 2b, edited by M. Ashburner and T. R. F. Wright, pp. 277-313, Academic Press, New York.
Srdic, Z., and Reinhardt, C., 1980, Histolysis initiated by "lymph gland" cells of *Drosophila*, *Science* **207**:1375-1377.
Stark, M. B., and Marshall, A. K., 1931, The blood-forming organ of the larva of *Drosophila melanogaster*, *J. Am. Inst. Homeopathy* **23**:1204-1206.
Throckmorton, L. H., 1975, The phylogeny, ecology, and geography of *Drosophila*. In *Handbook of Genetics*, vol. 3, edited by R. C. King, pp. 421-469, Plenum Press, New York.
Tsacas, L., and Bächli, G., 1981, *Drosophila sechellia*, N. SP., huitième espèce du sous-groupe melanogaster des Iles Séchelles, *Rev. Fr. Entomol.* **3**:146-163.
Walker, I., 1959, Die Abwehrreaktion des Wirtes *Drosophila melanogaster* gegen die zoophage Cynipide *Pseudeucoila bochei* Weld, *Rev. Suisse Zool.* **66**:569-632.
Wilson, L. P., King, R. C., and Lowry, J. L., 1955, Studies on the *tu-W* strain of *D. melanogaster*, *Growth* **19**:215-244.

Author Index*

Abbott, N. J., 464
Abu-Hakima, R., 25, 27
Adachi-Yamashita, N., 355
Adam, G., 501
Adams, J. R., 521, 523, 531, 537
Addison, R. B., 526
Adiyodi, K. G., 366, 373
Adiyodi, R. G., 366, 373
Aggarwal, S. K., 11, 16, 23, 28, 256, 274, 276, 277, 282, 283, 285
Agui, N., 271, 272, 279, 285, 287, 291
Aizawa, K., 519, 525, 526, 527
Akai, H., 49, 50, 51, 70, 208, 323, 324, 325, 327-355, 559, 562, 567, 570, 576
Akutsu, S., 50, 51, 57
Aldrich, H. C., 131, 135, 139, 140
Aldrich, H. C., 129, 133, 137, 138
Altner, H., 499, 506, 507, 509
Alzouma, I., 35
Amos, W. B., 113
Anderson, E., 4, 5, 7, 8, 9, 11, 12, 13, 16, 17, 21, 23, 24, 25, 27-30, 33, 35, 36, 38
Anderson, F., 208
Anderson, R. S., 579, 592
Anderson, S. O., 494
Ando, H., 4, 5, 7, 8, 11, 13, 15, 16, 17, 23, 27
Andriés, J. C., 202, 206, 208, 209, 211, 223, 224, 228
Anstee, J. H., 235, 239, 241, 248
Arnott, H. J., 242, 529, 535, 537, 542, 547
Arnott, S., 489
Asayama, T., 535, 537, 539, 545
Ashburner, M., 585
Ashhurst, D. E., 119, 120, 125, 139, 142, 202, 204, 487, 489
Atkins, M. D., 306
Auber, J., 115, 117, 127
Autrum, H., 399, 410, 470
Axtell, R. C., 503

Bacetti, B., 248, 365
Bächli, G., 585
Baehr, J. C., 35, 376
Bahr, G. F., 77
Bairate, A., 365
Baker, G., 304
Ballan-Dufrançais, C., 199, 241, 245, 248, 257
Barber, S. B., 141
Barker, J. F., 371, 391
Bassemir, U., 481
Bassurmanova, O. K., 274, 276
Baumann, F., 407, 410, 453
Bayon, C., 205, 211, 221, 228
Beadsley, J. H., Jr., 304
Beament, J. W. L., 28
Beams, H. W., 11, 12, 23, 235, 239
Beaulation, J., 282, 283, 381
Beck, S. D., 174, 465
Becker, H. W., 435
Beermann, W., 76, 77, 79, 81, 87
Bell, D. M., 241
Bell, R. A., 269, 304
Belyaeva, E. S., 81
Bencze, J. L., 86
Bennett, C. E., 16, 17, 19
Berdan, R., 21, 23, 25, 247, 486
Berendes, H. D., 77, 78, 79
Berg, G. J., 67
Bergold, G. H., 519, 523, 526
Bergot, B. J., 272, 274
Berkaloff, A., 241
Berlese, E., 110
Bernhard, C. G., 398, 399, 403, 405, 410, 411, 414, 416
Berridge, M. J., 233, 242, 249, 483, 487
Berry, D. J., 4, 9, 11, 19, 21
Beyer, A. L., 331, 332, 333
Bienz-Eisler, G., 120
Bier, K., 9, 21

*This index contains only textual citations to authors.

Bignell, D. F., 228
Bilinski, S., 65
Bird, E. T., 526, 539, 542
Bischof, H.-J., 495
Bishop, G. H., 365
Bjersing, L., 281
Black, P. N., 389, 390
Blackwell, J., 494
Blakers, M., 407
Blanc, L. M., 324, 347
Bland, K. P., 235, 237, 239
Blazse, K. I., 282, 285
Blest, A. D., 414
Blomquist, G. T., 303, 305
Bloom, W., 279
Blum, M. S., 366
Bodenstein, D., 173, 270
Boeckh, J., 506, 507
Bollenbacher, W. E., 272, 282, 291
Bonhag, P. F., 4, 5, 8, 9, 11, 12, 13, 24, 28, 29, 30, 31, 33
Borg, T. K., 274, 276
Borovyagin, V. L., 468
Boschek, C. B., 403, 405, 447, 457
Boulingand, M. Y., 289
Bowers, B., 255
Bowers, W. S., 275
Bownes, M., 270
Bracegirdle, B., 112
Bradley, T., 7
Braitenberg, V., 399, 422, 447
Brammer, J. D., 455
Bretscher, M. S., 375
Bridges, C. B., 76, 77, 78, 79, 91
Bridges, P. N., 76, 82, 91, 92
Brodie, D. A., 191
Bromley, A. K., 507
Brooks, M., 173
Brower, P. T., 25
Brown, C. L., 25
Brown, D. D., 324, 331
Brown, K. S., Jr., 304, 306
Brunet, P. C. J., 176
Brunt, A., 4, 5, 11, 13, 17, 24, 27
Bryan, J. H. D., 16
Bücher, T., 142
Bullard, B., 120, 124, 125, 127, 129
Bulliere, D., 27
Bullock, T. H., 436
Büning, J., 4, 7, 8, 9, 11-17, 21, 24, 28, 29, 30, 31, 33, 35-38
Burghardt, R. C., 25
Burgos, M., 202, 204
Burkhardt, D., 399, 422
Burkholder, G. D., 78
Burnet, B., 592

Burnside, B., 127
Burtt, E. T., 421, 469
Butt, F. H., 51
Butterworth, F. M., 167

Calvez, B., 282
Campos-Ortega, J. A., 399, 422, 467
Candia-Carnevali, M. D., 121
Cantacuzène, A.-M., 375, 377, 379, 391
Capco, D., 9, 21, 27
Carlson, S. D., 397, 403, 405, 407, 413, 414, 417, 421, 422, 425, 427, 435-349, 445, 447, 449, 451, 453, 455, 457, 459, 461, 462, 463, 465-469
Carron, C. P., 23
Carton, Y., 580, 592
Case, D., 30
Cassier, P., 211, 257, 272, 375, 383, 389, 390
Castiglioni, M. C., 580
Catton, W. T., 421, 469
Cave, M. D., 11, 35
Caveney, S., 21, 23, 25, 127, 247, 486
Challice, C. E., 255
Chambers, T. C., 202
Chapman, R. F., 499
Charnley, A. K., 241
Charpin, P., 276
Charpin, S., 274, 276
Chemes, H. E., 12
Chen, P. S., 390
Cheung, W. W. K., 202, 204, 229, 249
Chevallier, A., 332
Chi, C., 397, 403, 405, 407, 411, 416, 417, 421, 422, 427, 435, 437, 447, 449, 451, 453, 455, 457, 459, 463, 465, 466
Chikushi, H., 323
Chippendale, G. M., 167, 276
Choi, W. C., 9, 11, 21, 28, 30, 33, 37
Christensen, A. K., 281
Chu, I.-W., 503
Chung, K. L., 537
Church, N. S., 60
Clark, T. B., 526, 541
Claude, A., 77
Clever, U., 479, 480
Cochrane, D. G., 134, 160, 164
Cohen, C., 113
Coles, J. A., 453, 465
Collins, J. V., 154, 167, 177, 180, 182, 191, 256, 480
Comings, D. E., 84, 87
Cone, M. V., 13
Cooper, D. P., 135
Copeland, E., 249
Corwin, H. O., 570
Costin, N. M., 489
Couble, P., 330, 334, 337
Couchman, J. R., 4, 8, 13, 15, 24

Author Index

Counce, S. J., 60
Craig, G. B., Jr., 383
Crang, R. E., 235
Crick, F., 81
Croizier, G., 523, 531, 537
Crossley, A. C., 255, 256
Crowder, L. A., 240
Cullen, M. J., 114, 115, 136, 139, 140, 141
Cymborowski, B., 271

Da Cunha, A. B., 541
Dailey, P. J., 235, 371, 383, 389
Daillie, J., 330, 352
Daly, H. V., 138
Daneholt, B., 201
Davenport, D., 9, 21, 24, 27
Davey, K. G., 25, 27, 365, 371, 391
Davies, I., 221
Davies, R. G., 306
Davis, N. T., 4, 11
Day, M. F., 110
Dean, R. L., 154, 162, 164, 167, 168, 173, 180, 189, 282, 291, 295
De Duve, C., 191
Degrugillier, M. E., 51, 60, 61
Dekel, N., 25
deKort, C. A. D., 277
De Kramer, J. J., 481
Delbecque, J.-P., 282, 289, 293, 390
Delbrück, M., 501
Deleurance, S., 274, 276
De Loof, A., 23, 25, 28, 158, 379, 390
de Priester, W., 158, 206, 211
Derksen, J., 83
Devauchelle, G., 217
Diaz, M., 526
Dietz, A., 308
Dittmann, F., 20
Doira, H., 355
Dorn, A., 275, 276, 277
Dortland, J. F., 158
Dougherty, E. M., 525
Downe, A. E. R., 167
Duelli, P., 409
Dumpert, K., 495
Dumser, J. B., 366
Duncan, P., 461
Dunlap, H. L., 11, 13
Duspiva, F., 9, 27

Eakin, R. M., 403, 481
Ebeling, W., 305
Edlefrawi, M. E., 464
Edwards, G. A., 137, 140, 254
Edwards, J. S., 435, 495

Eguchi, E., 403
Eichelberg, D., 240, 242, 249
Eilender, D. S., 589
Elder, H. Y., 112, 115, 121, 125, 129, 133, 137, 139, 142
Elgee, D. E., 526
Elliot, J. H., 13
Elliott, H. J., 274, 277, 279
Elofsson, R., 417
El Shatoury, H. E., 580
Endo, Y., 211
Engelmann, F., 4, 23, 365
Engels, W., 12, 21, 24
Eppig, J., 25
Erickson, R. P., 602
Erler, G., 489, 492
Ernst, K.-D., 479, 481, 487, 503, 506, 509
Eschenberg, K. M., 11, 13
Evans, J. J. T., 309, 316
Evans, W. G., 305
Evequoz, V., 467
Exner, S., 398

Fahrquhar, M. G., 154, 281, 375
Fain-Maurel, M. A., 206, 208, 209, 211, 213, 219, 221, 223, 225, 257
Falcon, L. A., 521, 523, 525, 529, 531, 535, 537, 539
Farley, R. D., 493, 502, 503, 507, 508
Faulkner, P., 519
Fawcett, D. W., 12, 279
Federer, H., 390
Federici, B. A., 526, 541
Feldman-Muhsam, B., 387
Fernandez, H. R., 416
Fernandez-Moran, H., 401
Ferreira, C., 211, 220, 221
Feyereisen, R., 275, 276
Fiil, A., 24
Filshie, B. K., 202, 209, 215, 219, 288, 289, 309, 509
Finlayson, L. H., 468
Fish, D. D., 526, 541
Fisher, R. W., 464
Florkin, M., 348, 466
Flower, N. E., 209
Foelix, R. F., 503
Foldi, I., 229, 303, 309, 311-318, 320
Folliot, R., 17, 215
Forbes, A. R., 202
Foulds, L., 575
Fournier, A., 325, 330, 331
Franceschini, N., 410, 411
Francois, J., 202, 205
Franzini-Armstrong, C., 131, 135

Friedel, T., 271, 277, 288, 369
Friend, D. S., 375
Frölich, A., 427
Fukuda, S., 351
Fukuda, T., 348
Fullilove, S. L., 65
Furtado, A., 28, 29, 35
Fux, T., 23

Gabe, M., 248, 255
Gadzama, N. M., 377, 385, 387, 389
Gaffal, K.-P., 481, 491, 492, 494
Gage, L. P., 330, 331, 337
Gamo, T., 325, 334, 343, 355
Garcia-Bellido, A., 366
Garel, J. P., 324, 332
Garrett, W. E., 119
Gassner, G., 67
Gateff, E., 559, 561, 563, 565, 567, 568, 570, 575, 576, 585
Gay, H., 77
Gehring, W. J., 86, 90
Gerber, G. H., 5, 385
Gerrity, R. G., 60
Gersch, M., 282, 285
Ghadially, F. M., 575
Giacomelli, F., 282, 283
Giddings, C., 411
Gilbert, L. I., 270, 276, 281, 287, 289, 290, 291, 293, 295, 296, 366
Gillot, C., 369
Gillot, S., 330, 331
Gilula, N. B., 465
Giza, P. E., 331
Glees, P., 435
Glitho, I., 281
Gnatzy, W., 480, 481, 483, 485, 487, 489–493, 495
Goffinet, G., 587
Goldsmith, M. R., 50, 399, 401, 403, 405, 411, 416
Goll, D. E., 129
Goltzene, F., 277
Gonzalez, R. H., 304
Gouranton, J., 204, 206, 213, 215, 217, 218, 229, 244, 528, 548, 550
Grace, T. D. C., 519, 523
Granados, R. R., 519, 521, 523, 529, 531, 542
Granger, N. A., 277
Grassé, P.-P., 366
Gray, E. G., 481
Green, L. B. F., 242
Green, R. A., 334
Greenawalt, G. H., 141

Greenstein, M. E., 289
Grégoire, C. H., 587
Grell, E. H., 75, 77
Gribakin, F. G., 411
Griffiths, G. W., 173, 174, 467, 468
Grimal, A., 276
Grimes, M. G., 389, 390
Grimstone, A. V., 247, 580
Gross, J., 3, 4
Gruzova, M. N., 21
Gupta, A. P., 583
Gupta, B. L., 129, 249, 254
Gymer, A., 435

Hadley, N. F., 306
Hadorn, E., 561
Hagedorn, H. H., 23, 27, 365
Hägele, K., 78, 81, 90
Hall, D. W., 526, 541
Hallberg, E., 509
Hamdorf, K., 399, 403, 405
Hammarström, S., 596
Hamon, A. B., 316
Hamon, C., 17
Hanratty, W. P., 570, 577
Hansen, K., 491, 492, 509
Hanson, J., 115
Hanson, T. E., 469
Happ, C. M., 389, 390, 391
Happ, G. M., 371, 385, 387, 389, 390
Haque, M. S., 309
Harbach, R. E., 507
Hardie, J., 121, 122, 123
Harizuka, M., 343
Harrap, K. A., 519, 521, 525, 537
Harris, W. A., 86, 399, 403, 405, 467, 468
Harrison, J. B., 202, 204
Harverson, R. C., 137, 141
Harvey, W. R., 208, 488
Hashimoto, A., 316
Hawes, C., 121, 122, 123
Hawke, S. D., 502, 503, 507, 508
Hayashi, K., 325, 350
Hayashi, T., 115, 117
Hecker, H., 202, 211, 220, 222
Hegner, R. W., 3
Heisenberg, M., 399
Helm, F. E., 332, 347
Helminen, H. J., 153, 171
Helms, T. J., 13
Heming, B. S., 4
Henderson, J. F., 525
Henke, K., 479
Henking, H., 3

Hentzen, D., 332
Herman, W. S., 272, 281, 283
Herskowitz, I. H., 77
Hess, A., 468
Hess, R. T., 521, 523, 525, 529, 531, 535, 537, 539, 542, 545, 547, 550
Hetru, C., 27
Hild, W. J., 461
Hildebrand, J. G., 479, 481
Hill, A. V., 129
Hill, S., 275
Hillerton, J. E., 494
Hiruma, K., 271
Hirumi, H., 521, 523, 525, 537
Hochman, B., 77, 81
Hofbauer, A., 564, 565
Hoffmann, D. F., 547
Hoffmann, J., 255
Hogen Esch, T., 158
Hoglund, G., 411
Holter, P., 204
Home, E. M., 413, 414
Hopkins, C. R., 249
Horiuchi, Y., 325, 355
Horridge, G. A., 399, 401, 410, 413, 414, 436
Houk, E. J., 173, 205, 222, 465
House, C. R., 235, 237, 239
Hoyle, G., 115, 120, 121, 464
Hruban, Z., 191
Hudson, J. S., 537
Huebner, E., 3, 4, 5, 7, 8, 9, 11, 12, 13, 15, 16, 17, 19, 20, 21, 23, 24, 25, 27-31, 33, 35-38
Huger, A., 521, 523, 527, 528, 535, 537, 542, 547, 548
Hughes, K. M., 523, 526, 529, 531, 541
Huie, P., 171, 191, 483
Huet, C., 35. 387
Huignard, J., 375, 381, 383, 389, 390, 391
Hulmes, D., 120
Hunter, D. K., 521, 542, 545, 547
Huxley, H. E., 115, 131
Hyams, J. S., 13, 16, 17
Hyodo, A., 332, 354

Ichimura, S., 325
Ignoffo, C. M., 527
Iijima, T., 327, 329, 333, 354
Ikeda, M., 255
Imms, A. D., 347
Inagaki, I., 545
Injac, M., 521, 537, 539
Injeyan, H. S., 25, 27
Irlé, C., 596
Ito, S., 331
Ito, T., 324, 342

Iwai, P. J., 525
Iwanaga, T., 211

Jäckle, H., 9
Jackson, L. L., 305
Jacobson, A. G., 65
Jaffe, L. F., 19
Jamrich, M., 81, 83
Janzen, H. G., 234
Jarial, H. S., 241, 249
Järvilehto, M., 399, 469, 470
Jeannel, R., 5
Jeantet, A. Y., 208, 213, 214, 215, 223, 224, 241, 244, 254
Jeffrey, W., 9, 27
Jeuniaux, C., 466
Joachim, F. G., 269
Johannsen, O. A., 51
Johnson, R. A., 275
Jorgensen, W. K., 121
Josephson, R. K., 137, 141, 142
Judd, B. H., 81, 89
Junquera, P., 25

Kadiri, Z., 225
Kafatos, F. C., 24, 28, 365
Kafka, W. A., 507
Kaib, M., 506
Kaissling, K.-E., 489, 490, 501, 503, 506, 507
Kalisch, W. E., 78, 81, 90
Kanda, T., 50, 51
Kankel, D. R., 564
Kasang, G., 501, 503
Kataoka, K., 327, 334, 337, 340, 349, 350
Kathirithamby, J., 140
Katsuno, S., 59
Kaufman, T. C., 81
Kaulenas, M. S., 382, 389, 390, 391
Kavenoff, R., 81
Kawai, M., 343
Kawai, S., 316
Kawakami, M., 332
Kawamoto, F., 521, 523, 529, 531, 539
Kawamura, N., 61
Kawanishi, C. Y., 519, 521
Keil, T., 477, 481, 483, 485, 487, 489, 490, 492, 493, 494, 502, 503, 507
Keino, H., 50, 63
Keith, G., 332
Kelly, D. C., 548
Kelly, D. E., 135
Kelly, T. J., 27
Kendall, M. D., 235, 239
Kerkis, A. J., 79

Kessel, R. G., 23, 235, 239, 255
Khalifa, A., 387
Khan, T. R., 274
Kiguchi, K., 351, 352
Kilby, B. A., 167
King, D. S., 281
King, P. E., 4, 8, 13, 15, 24
King, R. C., 4, 9, 24, 25, 28, 70, 76, 77, 87, 255, 256, 274, 276, 277, 279, 281, 282, 283, 285, 356, 389
Kirimura, J., 325
Kirschfeld, K., 398, 399, 410
Kitajima, W. E., 202
Kitano, H., 580, 592
Kitaoka, S., 316
Kleine-Schonnefeld, H., 21
Kloc, M., 9, 11, 12, 15, 16, 21, 28, 29, 30, 33, 37, 38
Kloetzel, J. A., 240
Knudson, D. L., 521, 525
Kobayashi, M., 324, 327, 330, 338, 339, 352
Koch, E. A., 25, 27
Koeppe, J. K., 24
Köhler, A., 3
Kokwaro, E. D., 383, 385, 391
Kolb, G., 410
Komatsu, K., 325, 342, 343
Korfsmeier, K. H., 12
Korschelt, E., 3
Kosztarab, M., 305, 306, 316
Kramer, S. J., 167
Krieg, A., 521, 535, 537, 542, 547
Krishnan, N., 289, 309
Kristensen, N. P., 4
Krogh, I. M., 277
Ksiażkiewicz, M., 4, 13, 15
Kuda, A. M., 135
Kuffler, S. W., 436
Kuhbandner, B., 507
Kühn, A., 398
Kummel, G., 249
Kunkel, J. G., 23, 365
Kunkel, J., 154
Kunze, P., 411
Küppers, J., 483, 486-490
Kurata, K., 330, 352
Kurihara, M., 11, 21
Kuster, J. E., 399

Labhart, T., 416
Labour, G., 158
Lafont, R., 282
Lagasse, A., 25, 379, 390
Lagnez, M., 27

Lai-Fook, J., 127, 373, 377, 379, 382, 383, 387, 391
Laird, C. D., 78, 81, 84, 86, 90
Lalanne, J. L., 602
Landureau, J.-C., 390
Lane, N. J., 201, 202, 204, 251, 371, 373, 399, 411, 419, 436, 437, 447, 449, 457, 458, 459, 461, 463, 464, 465, 467, 487, 489
Langer, H., 399, 411, 453
Lanzrein, B., 274
Larsen, J. R., 507
Larsen, W. J., 152, 160, 164, 168, 176, 190
Lasek, R. J., 466
Latreille, P. A., 323
Laufer, H., 240
Laughlin, S., 422, 469, 470
Laurent, T. C., 489
Lauverjat, S., 167, 168, 235, 237, 239
Laverdure, A. M., 27, 35
Lawler, K. A., 519, 521, 523, 531
Lawrence, P. A., 479, 485
Lee, W. M., 141
Lefevre, G., Jr., 76, 77, 86, 87, 89
Lefevre, H. M., 79
Leffrey, W., 21
Lehane, M. J., 208, 211, 219
Lemeunier, F., 585
Leopold, R. A., 51, 60, 61, 365, 391
Leslie, R. A., 234
Leuchtenberger, C., 11
Leutenegger, R., 521, 542, 545, 547
Lewis, C. T., 507
Lezzi, M., 81
Lhonoré, J., 241
Liechty, L., 275
Lillywhite, P. G., 397
Lindsay, K. I., 229
Lindsley, D. L., 75, 77
Linley, J. R., 385, 387
Littau, N. C., 244
Lizardi, P. M., 331, 334
Locke, M., 151, 153, 154, 160, 162, 164, 167, 171, 173, 174, 177, 180, 182, 184, 189, 190, 191, 256, 282, 288, 289, 291, 295, 305, 308, 309, 316, 317, 318, 483, 503
Lockey, K. H., 304
Loftus, R., 509
Loher, W., 386
Longs, F. J., 23
Lossinsky, A. S., 79
Loughton, B. G., 167
Lowe, R. E., 526, 541
Luby, K. J., 411
Lucas, F., 337
Lutz, D., 9, 12, 16, 20, 28, 29, 30, 35, 36, 37, 38
Lyonet, P., 109

Author Index

MacGregor, H. C., 9, 12, 13, 16, 17
Machida, J., 324, 325
Machida, Y., 348, 354
MacIntyre, R. J., 592
Mackinnon, E. A., 521, 525, 529, 531, 537
Maekawa, H., 331, 356
Magakyan, Y. A., 21
Mahowald, A. P., 4, 9
Maillet, P. L., 244
Mala, J., 285
Manabe, Y., 311, 313–316, 318
Mandaron, P., 291
Mandelbaum, I., 389
Manery, J. F., 489
Manning, R. F., 331, 337
Mann, T., 366
Margaritis, L. H., 25, 28
Marschall, K. J., 528, 548
Marshall, A. K., 567
Marshall, A. T., 202, 204, 229, 249, 309, 316, 318, 481, 507
Martignoni, M. E., 525
Marty, F., 154
Masner, P., 4, 5, 9, 13, 27, 28, 29, 35, 38
Matile, P., 154
Matsumoto, D. E., 493
Matsumura, H., 350
Matsuura, S., 327, 330, 333, 338, 342
Matsuzaki, K., 332
Matsuzaki, M., 4, 5, 7, 8, 11, 13, 15, 16, 17, 23, 27, 29, 50, 51
Matthews, R. E. F., 517, 528
Matuszewski, B., 9, 12, 15, 16, 21, 28, 30, 33, 37, 38
Mays, U., 9, 11, 12, 21, 27, 38
Maziarski, S., 324, 332
Mazur, G. D., 28, 50
McCann, F. V., 121
McClintock, J., 171, 173, 180, 184
McDermid, H., 153, 154, 160, 162, 173, 184
McGhee, J. D., 80
McIver, S. B., 507
McKnight, S. L., 332, 334, 336
McIntosh, J. R. 113
McMahon, J. T., 164, 189
McNeill. P. A., 113
Meats, M., 27
Meinecke, C.-C., 507, 508
Meinertzhagen, I. A., 399, 425, 427
Melamed, J., 399
Mellanby, H., 30
Melnikova, F. J., 276
Menco, B. P. M., 506
Mendelson, M., 137
Menzel, R., 405, 407

Meola, S. M., 24, 373
Mercer, E. H., 350
Meredith, J., 249
Meves, F., 324
Meyer, E., 407, 416
Meyer-Rochow, V. B., 407, 414
Miller, D. R., 305, 306, 316
Miller, O. L., Jr., 336
Miller, W. H., 416
Mills, R. P., 255
Minagawa, M., 350
Misch, D. W., 202, 204
Miya, K., 49, 50, 51, 55, 57
Monsarrat, P., 528, 547, 548
Monson, J. M., 595
Montanelli, E., 204
Moor, H., 154
Morales, R., 461
Moran, D. T., 481, 489, 491, 492
Mordue, W., 275
Mori, T., 164
Morimoto, T., 327, 330
Mott, M. R., 81
Moulins, M., 491
Müller, B., 489, 502, 509
Mullins, D. E., 160
Mulnard, J., 21
Munk, R., 229
Murakami, A., 61
Murphey, R. K., 495

Nagashima, E., 49
Nagl, W., 9, 11, 21, 28, 30, 33, 38
Nakagaki, I., 327
Nakagawa, Y., 327
Nakahara, W., 324, 332
Nakajima, S., 327
Nappi, A. J., 580, 592
Nässel, D. R., 399
Natzle, J. E., 595
Neilson, M. M., 526
Neville, A. C., 494
Newman, D. J. S., 138
Nicholls, J. G., 436
Nickel, E., 405
Nicklaus, R., 487
Nicol, D., 425
Nishiitsutsuji-Uwo, J., 211
Nishimura, H., 351
Noble-Nesbitt, J., 289
Noda, H., 325, 334
Noirot, C., 145, 202, 208, 228, 229, 232, 247, 248, 254, 289, 306, 311, 458
Noirot-Timothée, C., 145, 202, 208, 228, 229, 248, 254, 289, 458

Nopanitaya, W., 202, 204
Nordheim, A., 83, 87
Nordmann, J. J., 277
Norman, T. C., 277, 279
Normann, I. C., 277
Nuccitelli, R., 19
Nunome, J., 325

Oba, H., 324, 325, 342
Oberlander, H., 291
O'Brien, R. D., 464, 465
Odhiambo, T. R., 274-277, 371, 377, 379, 382, 385, 390, 391
Odselius, R., 417
Oftedal, P., 592
Ohi, H., 355
Ohmachi, T., 331, 334
Ohshima, Y., 331
Ohtsuki, Y., 50, 51, 53, 57, 61, 65, 70, 71
Okabe, K., 330
Okada, M., 50, 57
Okada, T. A., 84, 87
Oka, H., 347
Okamoto, H., 343
Oliver, C., 379
O'Loughlin, G. T., 202
Ômura, S., 59, 61
Ono, M., 325, 326
Orkand, R. K., 436
Osborne, M. P., 199
Oschman, J. L., 233, 242, 248, 254, 483, 487
Osetō, C. Y., 13
Otsuki, E., 334

Painter, T. S., 76
Pak, W. L., 399
Palade, G. E., 131, 154
Palay, S. L., 77
Palevódy, C., 275, 276
Palka, J., 495
Panov, A. A., 274, 276
Papillon, M., 221, 223, 224
Paro, R., 86, 90
Paschke, J. D., 519
Paulian, R., 5
Paul, J., 87
Paulsen, R., 399
Paulus, H. F., 401
Pavan, O. H., 526, 527, 541
Payne, F., 11
Peachey, L. D., 135, 136, 137
Peacock, A., 248
Pearse, B. M. F., 375
Peeples, E., 587, 599
Pelling, C., 87

Pelttari, A., 153, 171
Pentreath, V. W., 436, 467
Pepe, I. M., 410
Percy, J., 307, 308, 318, 320
Perdrix-Gillot, S., 330
Perelet, A., 407, 410
Perotti, M. E., 365, 377, 381
Perrelet, A., 453
Perrin-Waldemer, C., 381
Peschke, K., 386
Pesson, P., 311, 313-318
Peters, A., 440
Peters, W., 479
Petre, A., 526
Philips, C. E., 485
Phillips, D., 240, 365
Phillips, J. E., 249
Philpott, D. E., 401
Pichon, Y., 464, 465
Piek, T., 305
Pinnock, D. E., 542, 545, 547
Pinto, L. H., 399
Pipa, R. L., 468, 469
Platzer-Schultz, I., 211, 219
Ploaie, P., 526
Poels, A., 387
Poinar, G. O., Jr., 528, 529, 550
Pollister, P. F., 318
Poodry, C. A., 289, 561
Pope, R. D., 305, 316, 320
Porte, A., 277
Porter, K., 131, 253, 411
Pratt, G., 275
Preusse, F., 3
Priesner, E., 501
Prillinger, L., 499, 507, 509
Pringle, J. W. S., 115, 117, 125, 129, 137, 139, 140, 141, 142
Prudhomme, J. C., 330, 334, 337

Quennedey, A., 306, 311
Quiot, J.-M., 537, 548

Raabe, M., 270
Rabinovitch, M., 338
Radojcic, T., 436, 467
Rae, P. M. M., 78
Raghow, R., 521, 523
Raina, A. K., 274, 276
Ramalingam, S., 383
Ramamurty, P. S., 24
Ramon, Y., 398
Ramon y Cajal, S., 436
Rasch, E. M., 330
Rasmussen, S. W., 55

Ratcliffe, N. A., 580, 592
Reedy, M. K., 119
Reese, T. S., 468
Reger, J. F., 137, 202
Reichardt, W., 399
Reimann, J. G., 377, 379, 382, 383, 385, 391
Reimann, K., 305
Reinecke, M., 468
Reinhardt, C., 580
Reinhardt, C. A., 205
Reisner, Y., 596
Reiss, F., 211
Rempel, J. G., 60
Rennie, J., 527
Retnakaran, A., 318
Rheuben, M. B., 468
Rhodin, J. A. G., 285
Ribbert, D., 485, 493
Ribi, W. A., 399, 407, 416, 422, 427, 455
Rice, M. J., 121, 491, 492
Riddiford, L. M., 293, 503, 506
Richard-Mercier, N., 30
Richards, A., 219, 220
Richards, A. G., 169, 303, 305, 499
Richards, G., 282
Richards, O. W., 306
Richards, P., 219, 220
Ris, H., 78, 81
Ritter, F. J., 499
Ritter, K. S., 533, 535
Rizki, R. M., 567, 579, 580, 581, 583, 585, 587, 588, 589, 592, 593, 595, 596, 597, 599, 600
Rizki, T. M., 579, 580, 581, 583, 585, 587, 588, 589, 592, 593, 595, 596, 597, 599, 600
Robert, A., 33
Robertson, C. W., 581
Robertson, H. A., 234
Robertson, J. S., 519
Roeder, K. D., 464
Rogojanu, P., 305
Röhrborn, G., 580
Romer, F., 282, 285, 483
Roots, B. I., 436, 440
Rosenbluth, J., 135
Rotheram, S., 529
Rowley, A. F., 580, 592
Rubin, G. M., 75, 76, 89
Rudin, W., 222
Rudkin, G. T., 90
Ruegg, R. P., 27
Ruska, H., 139
Ryerse, J. S., 191, 242, 291, 570, 577

Sacktor, B., 139
Safranek, L., 271
Sahai, Y. N., 11
Sainsbury, G. M., 120, 129
Saint Marie, R. L., 397, 417, 419, 421, 422, 425, 427, 435-439, 445, 447, 451, 457, 459, 461, 462, 463, 466, 468, 469
Sakaguchi, B., 50, 51
Saleuddin, A. S. M., 276
Salkeld, E. H., 51
Salpeter, M. M., 241
Salt, G., 579, 592
Sanchez, D., 436
Sander, K., 53
Sanes, J. R., 479, 480, 481
Sanford, M. T., 308
Sang, J. H., 592
Sanger, J. W., 121
Sasaki, S., 327, 338, 339
Sasaki, T., 325, 334
Sass, H., 81, 506
Satir, P., 465
Saura, A. O., 75, 78, 79, 80, 89, 93
Scali, V., 206
Schaller, F., 385
Scharrer, B., 274-277, 281, 390
Schinz, R. H., 405
Schlottman, L., 8, 11
Schmidt, K., 480, 485, 490, 495, 507
Schmidt, O., 9
Schneider, D., 501, 506
Schneiderman, H. A., 289, 291, 565, 567, 576
Schoffeniels, E., 436
Schofield, P. K., 465
Scholes, J. H., 399, 427
Schooneveld, H., 274
Schrader, F., 11
Schreiner, B., 4, 5, 8, 11, 13, 17, 23, 24, 27
Schultz, R. M., 25
Schwemer, J., 399
Scott, M. T., 579, 592
Scudder, G. E., 241, 249
Sears, T. A., 436
Sedlak, B. J., 269, 272, 273, 275, 276, 279, 281, 287, 289 290, 291, 293, 295, 296
Sehnal, F., 271
Sekhon, S. S., 481, 487, 506
Semeshin, V. F., 81
Seureau, C., 242
Shafiq, S. A., 138
Shalev, A., 601
Shankland, D. L., 240
Shaw, J., 245
Shaw, S. R., 399, 407, 417, 422, 425, 427, 459, 461, 469, 470
Shelton, P. M., 399
Shibukawa, A., 324, 325, 342, 345, 349

Shigematsu, H., 325, 330, 339
Shimizu, M., 342
Shimizu, S., 325
Shimura, K., 325, 331, 332, 334
Shinji, G. O., 13
Shrestha, R., 559, 567, 568, 570, 572, 573, 575, 576, 585
Siew, Y. C., 270, 274
Skaer, H. B., 371, 373, 399, 411, 436, 447, 449, 457, 458, 459, 463, 464
Skaer, R. J., 81
Slifer, E. H., 481, 487, 499, 502, 506
Slizynski, B. M., 76
Smith, D. S., 111, 112, 115, 119, 125, 127, 129, 131, 133, 136, 137, 139, 140, 141, 142, 143, 206, 208, 219, 244, 255, 274-277, 487, 491, 492, 494
Smith, K. M., 519, 541, 547
Smith, O. J., 527
Smith, R. E., 281
Smola, V., 407
Snyder, A. W., 399
Sohal, R. S., 214, 242
Somjen, G. G., 435, 436, 467
Somlyo, A. V., 135
Sonnenblick, B. P., 61
Sorsa, M., 77, 78, 83, 84
Sorsa, V., 75, 77-81, 83-87, 89, 90, 91, 93
Sotavalta, O., 141
Sparrow, J. C., 592
Speiser, C., 81
Spradling, A. C., 75, 76, 89
Sprague, K. U., 325, 332, 334, 342
Squire, J., 117, 119, 125
Srdic, Z., 580
Srivastava, U. S., 281
Stark, M. B., 567
Stark, W. S., 399, 414, 467, 468
Staübli, W., 223
Stavenga, D., 410
Stay, B., 277
Stebbings, H., 9, 12, 13, 16, 17, 19
Steel, C. G. H., 291
Steele, J. E., 176
Steffensen, D. M., 81
Steinbrecht, R. A., 477, 481, 487, 489, 490, 499, 501, 503, 506, 507, 509
Steinman, R. M., 373
Steitz, J. A., 332
Stepper, J., 507
Stobbart, R. H., 245
Stolarz, G., 271
Stoltz, D. B., 528, 529, 531, 541, 550
Stowe, S., 422, 425, 427
Strambi, C., 249

Straus-Durckheim, H., 111, 143
Strausfeld, N. J., 399, 422, 436, 437, 438, 470
Streams, F. A., 580, 592
Struve, G., 411
Sumimoto, K., 57, 164, 167, 209, 213, 214, 348, 349
Summers, M. D., 519, 521, 525, 529, 531, 535, 537, 542
Suzuki, E., 331, 354, 355
Suzuki, K., 55
Suzuki, T., 325
Suzuki, Y., 324, 325, 331, 332, 334, 352, 354, 355
Swales, L. S., 449, 459, 465
Swift, H., 77, 191, 240
Sykes, A. K., 173, 177, 180
Szöllösi, A., 390

Tagawa, T., 350
Takami, T., 49, 51, 61, 70
Takeda, N., 274, 276
Takei, R., 49, 50
Takeshita, H., 325, 339
Takesue, S., 50, 57, 63, 65
Tamaki, Y., 304, 315, 316
Tamura, T., 331, 356
Tanada, Y., 519, 521, 526, 529, 531, 533, 535, 537, 542, 545, 547
Tanaka, K., 255
Tanaka, Y., 324
Tandler, B., 381
Tanimura, I., 71
Tashiro, Y., 327, 330, 332, 333, 334, 338, 342
Tautz, J., 489, 491, 492, 493, 495
Taylor, H. H., 241
Tazima, Y., 49, 61, 323, 324
Teigler, D. J., 242
Telfer, W. H., 3, 4, 9, 11, 12, 13, 16, 19, 20, 21, 23, 25, 27, 28, 29, 30, 38, 365, 369
Terzakis, J. A., 205
Theiss, J., 491, 493, 494
Thomas, D., 215, 217, 218
Thomas, M. V., 465
Thomsen, E., 274, 276, 391
Thomsen, M., 274, 276
Thorell, B., 399
Thorson, B. J., 377, 379, 382, 383, 385, 391
Thorson, J., 489, 490
Throckmorton, L. H., 587
Thurm, U., 481, 483, 486, 487, 488, 489, 490, 491, 493, 494, 495, 497
Tichomiroff, A., 347
Tiegs, O. W., 112, 133, 139, 142
Tilney, L. G., 113

Author Index

Tinsley, T. W., 519
Tobe, S. S., 275, 276, 277
Tobias, M., 495
Toh, Y., 506
Tojo, S., 164, 167, 168
Tombes, A. S., 274, 275
Tongu, Y., 383
Torpier, G., 202
Toth, L., 13
Toyama, K., 49
Tregear, R. T., 117, 133
Treherne, J., 436, 464, 465, 466, 489
Trepte, H.-H., 485, 493
Trujillo-Cenoz, O., 399, 403, 409, 435, 447
Truman, J. W., 269
Tsacas, L., 585
Tsacopoulos, M., 453, 465, 467
Tscharntke, H., 407
Tsuda, M., 332
Tsujimoto, Y., 331, 332
Tucker, J. B., 27
Tung, A. S.-C., 468
Tuzet, O., 365, 385
Twarog, B. M., 464

Ullmann, S. L., 9, 11, 21, 24, 27, 28
Unnithan, G. C., 275, 277

Vande Berg, J. S., 485
Vanderberg, J. P., 9, 12, 21
van der Molen, L. G., 160, 481
van der Starre, 226
van der Wolk, F. M., 506
Varela, F. G., 403, 492
Varon, S. S., 435, 436, 467
Vernier, J. M., 35
Vince, R. K., 281
Vinson, S. B., 528, 529, 550
Virrankoski-Castrodeza, V., 81, 83
Vogt, R. G., 503, 506
Voigt, W.-H., 324, 327
Voinov, V., 158
Völker, W., 492, 493
Volkman, L. E., 525, 531
von Zwehl, V., 399
Vugman, I., 338

Waku, Y., 57, 164, 167, 209, 213, 214, 242, 276, 303, 309, 311, 313-316, 318, 348, 349
Wald, G., 403
Waldow, U., 509

Walker, F. D., 461
Walker, I., 580, 592
Walker, S., 521, 542, 545, 547
Wall, B. J., 241, 248, 254
Waterhouse, D. F., 202
Wattiaux, R., 191
Weber, K. M., 489
Webster, D., 167, 173
Wehner, R., 399, 407
Weih, M. A., 494
Weiler, R., 399
Weis-Fogh, T., 141, 143, 494
Wellman, S. E., 24
Welsch, U., 219
Went, D. F., 25, 27
Wessel, G., 488
Wessing, A., 240, 242, 249
West, A. S., 167, 173
Whalen, M. M., 526, 539
Wick, J. R., 5, 13, 28, 30, 33
Wielgus, J. J., 293
Wielowiejski, H. R. V., 3, 5
Wiemerslage, L. J., 12
Wigglesworth, V. B., 143, 154, 158, 186, 241, 305, 390, 466, 467, 479, 480
Wightman, J. A., 8
Wiitanen, W., 403
Wilcox, T. A., 521, 537
Wilson, L. P., 592
Winter, H., 9, 21
Winter, W. T., 489
Wolf, R., 63, 65
Wolken, J. J., 401
Wolstenholme, D. R., 81
Woodruff, R. I., 9, 19, 20
Woolever, P. S., 468
Wootton, R. J., 139
Wright, K. A., 234
Wright, M., 202
Wright, R. H., 499
Wurzelmann, S., 277
Wyatt, G. R., 152, 191

Xeros, N., 521, 541

Yamada, M., 325, 342
Yamamoto, T., 519
Yamanouchi, M., 324, 325, 342
Yamauchi, H., 31
Yasuzumi, G., 77
Yin, C. M., 276
Yokohari, F., 509
Yokoyama, T., 350

Yoshitake, N., 31, 50, 51, 57
Young, D., 142

Zacharuk, R. Y., 499, 507
Zelazny, B., 528
Zerbst-Boroffka, I., 249
Zettler, F., 399, 469, 470
Zhimulev, I. F., 79, 81, 86, 87, 89
Zierold, K., 489
Zimmerman, R. P., 422, 459, 461, 466, 469
Zinsmeister, P. P., 9
Zissler, D., 53
Zuniga, M., 332
Zylberberg, I., 249

Subject Index

A band, 119, 120, 133
Acanthoscelides obtectus, 375, 383, 389-391
Accessory glands, 257, 365-392
Acheta domesticus, 382, 387, 389-391, 485, 487, 491, 492, 494, 498, 500, 501
Acid mucopolysaccharides, 202, 382, 489
Acid phosphatases, 204, 467, 468
Acid phosphatase mutation of *Drosophila*, 467
Acinar salivary glands, 235-239
Acrosome, 60
Actin, 115, 120, 129
α-actinin, 120, 129
Acyrthosiphon pisum, 174
Adelges piceae, 318
Adipose tissue, 151-194, 527
Aedes aegypti, 158, 205, 220, 222, 234, 248, 251
  A. albopictus, 548
  A. dorsalis, 205
Aeshna cyaneae, 206, 223, 227
Aging, 593
Albuminoid spheres, 547
Alkaline phosphatase, 592
Allantoate, 190
Allatectomized larvae, 351
Amathes c-nigrum, 51
Amnion, 67, 70
Amoebocytes, 583
Amylase, 234, 239
Anagasta kühniella, 377, 379, 382, 383, 391
Annulate lamellae, 545, 548
Anomoneura mori, 304, 308, 318, 319
A. stephensi, 223
Antennal olfactory hair, 487
Anterior silk gland, 327, 345-347
Antheraea, 490
A. pernyi, 282, 283, 304, 331, 487
A. polyphemus, 487, 490, 502, 505
A. yamamai, 331
Antibody, 593

Anticarsia gemmatalis, 526, 537
Aonidiella aurantii, 305, 311, 313, 314, 315, 317, 318, 320
Apis mellifera, 120, 125, 131, 143, 145, 247, 248, 405
Apocrine secretion, 208, 379
Apolysis, 293
Apyrene sperm, 59
Arachnocampa luminosa, 242
Archips argyrospila, 542, 544, 545, 548
Arenivaga, 507, 509
Asynchronous muscle, 124, 129, 137
Atelotypic tumorous state, 561, 562
ATPase, 338
Autolysosome, 342
Autophagic vacuole, 156, 167, 177, 178, 184, 193, 206, 223, 291
Autophagosome, 342, 351, 353, 354
Autophagy, 164, 180, 190, 191
Autotypic tumorous state, 562
Axon terminals, 470

Bacterial symbionts, 232
Baculoviridae, 517
Baculovirus infection, 517-552
Basal infolds, 156, 206
Basal lamina, 156, 172, 186, 192, 205
Basement membrane, 202, 204, 373, 416, 419, 542, 562, 592, 593, 595
Bean-shaped accessory glands (BAG), 367, 369, 371, 375, 385, 389, 390
Black cell mutation of *Drosophila*, 581
Blastoderm, 63, 65, 67
Blatella germanica, 173, 241, 244, 245, 247, 248, 253, 255, 257, 260
Blood-brain barrier, 463, 464, 465
Blood forming organs, 580
Blood sucking insects, 222
Bombykal, 506

617

Bombykol, 503, 506
Bombyx mori, 49-71, 164, 242, 244, 245, 256, 282, 283, 323-356, 403, 482, 483, 487, 488, 490, 501, 502, 504, 505, 506, 508
  B. *mandarina*, 323
Bristle mother cell, 479, 480
Brochosomes, 244, 245
Bruchidius obtectus, 37

Calcium ions, 214
Calcium oxalate, 245
Calcium phosphate, 244
Calliphora, 122, 467, 468
  C. *erythrocephala*, 115, 119, 121, 158, 207, 219, 233, 242, 244, 248, 251, 256, 257, 274, 276, 277, 281, 283, 391, 465, 485-487, 490, 492-497, 506
  C. *vicina*, 485
Calpodes ethlius, 154, 160, 164, 167, 168, 171, 176, 177, 182, 184, 189, 190, 242, 243, 244, 282, 285, 288, 289, 295, 305, 308, 309, 316, 317, 377, 379, 382, 386, 391
Calyx cell, 550
Campaniform sensilla, 490, 491, 495
Capitate projections, 432, 447
Capsid, 529
Capsule formation, 592, 599
Carausius morosus, 217, 241, 244
Catalase, 190
Celithemis eponina, 131, 135, 136, 139
Cell junctions, 233
Cellular defense system of *Drosophila*, 579-602
Cenocorixa bifida, 241
Cercus, 485, 498, 501
Ceroplastes pseudoceriferus, 304, 315, 316
  C. *sinensis*, 311, 315, 316
Chemical synapses, 417, 422, 425
Chemoreceptors, 413, 477
Chironomus, 77, 219
Chitin-protein complexes, 219, 289
Cholesterol, 174, 282
Chordotonal receptors, 490
Chorion, 24, 50, 61, 65
Chromatin spreading technique, 332
Chromomere, 83-87, 103
Chromophore, 403
Chromosomal banding patterns, 75-103
Chromosomal puffs, 83, 541
Chromosome maps, 93-102
Cicada timbal muscle, 141
Ciliary region of sensillum, 481, 487, 494
Cimex lectularius, 509
Clavate hair, 494, 495
Cleavage nuclei, 65, 70
Coated vesicles, 167, 237, 256, 375, 449
Coccus hesperidum, 309, 311, 312, 315

Coeloconic sensilla, 507, 509
Coleoptera, 186
Collagen, 205, 595
Colonic cells, 229
Columnar cells of midgut, 205-208, 211-219, 222, 519
Companion photopigment, 405
Compound eye, 398, 453
Compound glands, 306
Concretions, 156, 190, 213-215
Condensing vacuole, 156, 173
Confronting cisternae, 156
Contractile system, 114
Copper-accumulating cells, 215
Corneal lens, 414, 415
Corneal pigment cells, 415, 455, 461
Corpora allata, 270-281
Corpora cardiaca, 277, 279
Creophilus maxillosus, 12, 21, 37, 38
Crystal cell, 567, 570, 581, 585-589, 599
Crystalline domain of fibroin, 337
Cryptoglossa verrucosa, 306
Culex tarsalis, 205, 222
Culicoides melleus, 385, 387
Cuticular intima, 337, 345, 348
Cuticular sensillae, 413
Cuticular wax, 304, 305
Cuticulin, 288, 291
Cytokinesis, 31
Cytoskeleton, 27, 192

Dacus tryoni, 309
Dark adaptation, 401, 410
Dauer larvae, 352, 354
Dendrite, 480-483, 485, 487, 491, 493, 504, 505
Dendritic cytoskeleton, 493, 497
Dendritic membrane, 492, 497
Dendritic sheath, 485, 493, 499
Dense body, 347, 348
Desmosomes, 151, 379, 391, 449, 458, 545
Detoxification, 215, 256
Diapause eggs, 50
Digestive juice, 519
Digestive organs, 199-240
Diparopsis watersi, 523
Diploptera punctata, 271, 275
Diptera, 398
Distal retinula, 400, 402
DNA amplification, 11, 12, 37
DNA content of silk gland cell, 330
DNA synthesis, 293, 330, 352
Double bands, 85, 86
Drosophila melanogaster, 61, 65, 70, 75-103, 140, 215, 239, 277, 281, 282, 283, 323, 366, 377, 381, 402, 467, 468, 559-577, 579-602
Drosophila polytene chromosomes, 75-103

Subject Index

Ecdysial canal, 497
Ecdysial droplets, 289, 291
Ecdysone, 27, 35, 180, 283, 353, 354
Ecdysteroid, 272, 287, 291, 295
Eclipse period, 533
Ectadenia, 366
Ectomicropyle, 51
E-face grooves, particles, 451, 455
Ejaculatory ducts, 366, 382, 387
Electrical compartmentalization, 461
Electrical synapses, 417, 427, 461
Elementary fibroin fiber, 334, 335, 337, 338, 351
Embryonic development, 30, 49-70
Embryonic envelope, 69
Encapsulation, 592, 601
Endobodies, 21
Endochorion, 28
Endocrine cells, 211, 212, 269-296
Endocuticle, 288, 291
Endocytosis, 152, 192
Endocytotic vesicles, 583
Endomitosis, 12, 330, 356, 479
Endoplasm, 53
Endoplasmic reticulum (ER), 53, 152, 156, 180, 190, 192, 373, 409, 410
Endopolyploidization, 11, 38, 330
Enteroglucagon, 211
Entomicropyle, 51
Epicuticular wax filaments, 316, 318
Epicuticle, 288
Epidermal cells, 288
Epidermal glands, 306
Epipyrops anomala, 308, 309, 316, 317, 318
Epithelial glia, 415, 437, 438, 439, 449
Eriocampa ovata, 306, 307, 309, 318
Eriococcus lagerstraemiae, 311, 313-318
Estigmene acrea, 525
Eupyrene sperm, 59, 60
Evacuating canal, 314, 317, 320
Excretory organs, 240-258
Exochorion, 28
Exocuticle, 289
Exocytosis, 152, 162, 192, 193, 277, 279, 379
Extrachromosomal DNA body, 37
Extraembryonic blastoderm, 69, 70
Eye, 346, 451, 461

Facet, 414
Fat body, 151-194, 525, 545, 548
Fat body vacuolar system, 152-194
Fat inclusions, 211
Fermentation chambers, 228-230
Ferritin, 215, 596
Fertilization, 51, 59, 60
Fertilization membrane, 61
Fiber-orienting mechanisms, 220

Fibrillogenesis, 115
Fibroin, 324, 325, 331-342, 356
Fibroin gene, 331, 332, 336, 337
Fibroin mRNA, 331, 336, 356, 364
Filiform hairs, 491, 494, 495
Filiform sensillum, 485
Filippi's gland, 347, 349
Filmsy cocoon mutation of *Bombyx*, 355
Filter chambers, 229-232
Flight muscle, 111, 114, 115, 127, 133, 135, 145, 146
Follicle cells, 8, 21, 24, 25
Formica polyctena, 202, 213, 214, 224, 226, 241, 244, 245, 247
Fulgora candelaria, 202

Galleria mellonella, 270, 271, 274, 281, 282, 285, 287, 468
Ganglion mother cells, 564
Gap junctions, 25, 151, 220, 427, 447, 461, 465, 486, 545
Gastric caeca, 540, 541
Gelatinous pseudocone, 415
Geneless DNA, 89
Germband, 67, 69
Germ cells, 16, 24, 29, 33, 35, 36
Glial barrier, 463, 470
Glial cells, 419, 420, 435-471, 478, 537, 564
Glucosides, 176
Glutamic acid, 466
Glycogen, 12, 155, 156, 176, 177, 455, 466, 467
Glycophospholipids, 204
Glycoproteins, 397
Goblet cells of midgut, 202, 208, 209
Golgi complexes, 152, 156, 171, 173, 176, 178, 180, 191, 206, 209, 224, 256, 281, 332, 334, 340, 356, 373, 375, 568, 583
Golgi complex beads, 156, 191, 192
Golgi complex organizer, 192
Golgi vacuoles, 334, 337
Gonads, 30
Gonomery, 63
Graded receptor potentials, 417
Granular cells, 377
Granulosis viruses (GV), 517, 542-547
Gustatory sensilla, 509
Gryllotalpa gryllotalpa, 241
Gnyllus bimaculatus, 241, 293
Gyrinus natator, 548

Hair components, 483-485
Harrisinia brillians, 527
Hemerocampa pseudotsugata, 526
Hemidesmosomes, 151, 156, 184, 186, 220, 449, 545
Hemiptera, 4, 13, 29, 173

Hemocytes, 526, 579-602
Hemolymph, 151
Hemopoietic malignancies, 570
Hemopoietic organs, 568, 570, 572
Heterophagic vacuoles, 156, 206
Hormonal controls, 269-296, 351-352
Hormonal target tissues, 288-296
House fly (see *Musca domestica*)
Hunger spherules, 352, 355
Hyalophora cecropia, 121, 270, 276, 281, 289, 290, 291, 295, 296
Hyaluronic acid, 489
Hydrogen peroxide, 190
20-hydroxyecdysone, 270, 271, 277
Hygroreceptors, 413, 509
Hyphantria junea, 539
Hypodermis, 525

Ileum, 228, 245, 253
Imaginal disc tumor, 561, 562
Incomplete cytokinesis, 16, 36
Interbands, 79, 81, 82, 83, 87, 89, 90
Intercellular bridges, 15, 16, 29, 31, 33, 35, 36
Intercellular canaliculus, 237
Intercellular cohesion, 417
Intercellular communication, 466
Intercellular junctions, 457
Interchromeric fibers, 89
Interchromomeres, 87
Interfollicular plug cells, 8
Interglial junctions, 447
Interloop fibers, 81
Intermediate retina, 438, 439, 440, 441, 445, 451, 458
Intestitial cells, 8, 9, 318, 319
Intracellular transport, 206
Intragenic recombination, 89
Intramembranous particles, 427
Intranuclear inclusions, 215
Intranuclear protein spherules, 217, 218
Intron, 89
Ion-barriers, 461
Ion channels, 405
Ionic homeostasis, 463
Ion transport, 232, 239, 417, 455, 459
Ips typographus, 509
Isolation body, 156

Joint membrane, 491, 494
Jumping muscle, 136
Junctional cells, 254
Junctional complexes, 62
Juvenile hormone (JH), 35, 175, 270-277, 281, 287, 288, 290, 291, 295, 296, 351, 356, 391
Juvenile hormone analogues, 271

Karyosome, 21
Kerria lacca, 304, 306, 309

Lamellar body, 156, 184, 185, 583
Lamellocytes, 567, 568, 572, 583, 584, 589, 593, 597, 601
Lamina depolarization, 469
Lamina ganglionaris, 421, 437, 449, 459
Lampbrush chromosomal organization, 81, 83
Lanthanum infiltration, 419
Laspeyresia pomonella, 519, 521
Lecanodiaspisis sardoa, 311, 315
Lectins, 596, 601
Lepidoptera, 186
Lepismodes inquilinus, 248, 252, 254
Leptinotarsa decemlineata, 23, 28, 30, 154, 274, 379
Lethal-benign imaginal disc neoplasms, 559
Lethal (1) L68 mutation of *Drosophila*, 89
Lethal malignant blood neoplasm mutations of *Drosophila*, 570-575, 576
Lethal (2) giant larvae mutation of *Drosophila*, 282, 561-566, 576
Lethocerus, 117, 119, 124, 125, 127, 141, 142
Leucophaea maderae, 274, 276, 277, 281, 390
Light adaptation, 401, 410, 413
Lipophorin, 167, 173, 182
Liquid silk, 348
Locusta migratoria, 142, 168, 221, 223, 224, 235, 236, 237, 239, 241, 245, 256, 259, 260, 275, 277, 375, 377, 378, 391
Loop organization of chromomere, 83-87
Lucilia cuprina, 215, 289, 309, 425
Lymantria dispar, 526
Lymph, 157
Lymph glands, 567, 580, 581, 586, 587
Lysosomes, 168, 171, 180, 184, 192, 215, 583

Macrochaetae, 493, 494
Macrophages, 583
Magnesium phosphates, 244
Male accessory glands, 365-392
Malignant neoplasms, 528, 559, 562, 564
Malpighian tubules, 155, 190, 229, 231, 240-245, 260
Mamestra brassicae, 270, 271
Manduca sexta, 234, 269, 271-296, 304
Margarodidae, 309, 315, 318
Mating refusal substances, 366
Mechano-electric transduction, 492, 493
Mechanoreceptive dendrites, 492, 493
Mechanoreceptors, 413, 477-479, 490
Meiotic divisions, 55, 61
Melanization, 585, 599
Melanized crystal cell, 587

Melanoplus, 256, 467, 506
Melanophila acuminata, 305
Melanotic tumor mutations of *Drosophila*, 592, 593, 600
Membrane-integrated cones, 492
Merocrine secretion, 377
Meroistic ovaries, 38
Mevalonate, 174
Microbody, 190
Micropylar apparatus, 51, 61
Microsprodian, 414
Microtrabecular lattice, 411
Microtubules, 15, 17, 19, 63, 127, 440, 463, 466, 467, 481, 487, 491, 494, 497, 519, 537, 599
Microvilli, 202, 206, 225, 232, 308, 311, 407, 483, 531
Middle layer sericin, 325, 342, 344, 345
Middle silk gland, 327, 342-345
Midgut, 201-228, 523, 548
Midgut filter chambers, 229
Mineralized concretions, 213-215
Mitochondria, 12, 28, 137, 202, 206, 220, 244, 282, 283, 411, 442, 459, 531, 545, 583
Mitochondria-scalariform junction complexes, 254
Monoclonal antibody, 593
Monopolar neurons, 419, 423, 442, 447, 469
mRNA, 21, 330
Mucopolysaccharides, 205, 209, 220
Mucoproteins, 230
Mucous secretions, 245
Multivesicular body, 55, 157, 162, 164, 168, 182, 192, 562
Musca domestica, 21, 51, 242, 244, 247, 377, 401, 407, 409, 410, 413, 419, 420, 427, 466, 467, 468, 488
Muscle cells, 111-146
Mushroom glands, 366
Mutant hemocytes, 572, 592
Mutations influencing silk production, 352, 354-356
Mycetocytes, 173, 174, 176
Myofibrils, 115, 133, 135
Myosin, 114, 115, 117, 119

Naked pupa (Nd) mutation of *Bombyx*, 332, 354, 355
Necrophorus, 503, 506
Nepa cinerea, 203, 204
Neural lamella, 437, 464
Neurilemmal cell, 479
Neuroblasts, 564
Neuroglia, 435-471
Neuropil, 437, 462, 469, 564
Neurosecretory cells, 270, 271, 274, 277, 279, 288, 296, 564

Neurotransmitter, 427, 470
Nondiapause eggs, 50
Notonecta, 19
Nuclear DNA body, 37
Nuclear envelope, 157
Nuclear fusion, 11
Nuclear polyhedroses viruses (NPV), 517-533, 547-550
Nuclear pore, 11
Nucleocapsid, 521, 523, 525, 529, 531, 533, 539, 548
Nucleolar organizer region, 37
Nucleoli, 11, 21, 36, 67
Nucleolonermata, 328, 351, 352
Nurse cell, 11, 15, 33, 35, 36, 39
Nurse cell-oocyte electrical gradients, 38
Nurse cell-oocyte interactions, 9, 19

Occlusion bodies, 523
Ocelli, 453
Odor molecules, 501
Oenocytes, 282, 304, 305
Olfactory sensilla, 477, 478, 481, 483, 488, 490, 499-501, 504-507
Ommatidia, 399, 407, 409, 416
Oncopeltus fasciatus, 5, 7, 8, 11, 21, 24, 27, 30, 31, 275, 276, 277
Oocyte, 15, 20, 25, 38, 51
Oocyte determination, 33, 38, 39
Oogenesis, 3-40, 50
Oolemma, 17, 21, 23, 25, 61, 63
Ooplasm, 53, 60, 63
Opsonization, 589
Optic cartridges, 422, 423, 427, 469
Optic neuroblasts, 567
Optic neuropil, 419, 427, 469
Orthoptera, 173, 489, 490
Oryctes nasicornis, 205, 221, 228, 229, 230
O. rhinoceros, 525, 528, 547, 548
Osmotic gradients, 254
Ovariole sheath, 5
Ovary, 3-40

Paracrystalline inclusions, 581, 585, 586
Paragonia, 366
Paramyosin, 124
Parietal cells, 235
Pectinophora gossypiella, 67
Pedicel cells, 8
Peplomers, 523
Pericardial cells, 255, 256
Perineurial cells, 437, 439, 445, 449, 465, 466
Periplaneta americana, 154, 162, 168, 173, 174, 202, 212, 228, 229, 235, 239, 241, 244, 248, 274, 281, 303, 373, 466, 468, 506, 509
Periplasm, 53, 55, 60, 70

Peritrophic membrane, 201, 204, 205, 219
Peroxidase, 190, 596
Peroxisomes, 155, 157, 178, 180, 190
Petrobius maritimus, 203, 209, 210, 213, 214, 221, 223, 225, 248, 250
P-face particles, 405, 425, 458, 459
Phagocytosis, 185, 467, 588, 590, 600, 601
Phanerotoma flavitestacea, 550
Phenol oxidases, 587, 599
Pheromone-binding protein, 503
Pheromone-sensitive sensilla, 501, 502
Philosamia cynthia, 270, 282, 283
Photopigment-containing organelle, 397
Photoreceptor cells, 397-428, 437, 445, 459, 470
Photostimulation, 465
Picornavirus, 575
Pieris brassicae, 141, 211, 245, 282
Pigment cell, 414, 415
Pigmented glia, 453-457
Pigment granules, 403, 410, 411
Pinocytosis, 23, 164, 167, 168, 192, 373, 445
Pinocytotic vesicle, 157, 182, 257
Planococcus, 315
Plasmalemma-mitochondria complex, 228
Plasma membrane recycling, 160
Plasma membrane reticular system, 157
Plasmatocytes, 567, 570, 572, 573, 581, 583, 584, 601
Plexiform layer, 443, 445
Plexiform surface coat, 202
Podocytes, 257, 258, 572, 583, 584, 601
Polar plasm, 55, 61
Polyhedra, 502, 523, 527, 535, 537
Polysaccharide matrix, 205
Polytene chromosomes, 75-103, 239, 485, 527, 541
Polytrophic ovary, 9, 19, 21, 24, 25, 31
Pore canal, 315
Pore tubule-dendrite, 501, 503
Porphyrophora crithmi, 309, 311, 312, 315
Posterior silk gland, 327-342
Postsynaptic processes, 425
Posttransition of fat body compartments, 153, 158-186, 193
Potassium, 234, 453, 465
Potassium transport, 465, 489
Potosia, 508
Precocenes, 275
Precocious pupae, 351
Prefollicular cells, 21
Prefollicular follicles, 38
Posttransition compartments of fat body, 153, 186-190, 193
Pretransition region, 155, 157
Previtellogenic follicle, 7, 20
Previtellogenic oocyte, 17, 23

Primary lysosomes, 153, 157, 182, 568, 570
Procrystal cells, 568, 570
Proctodeal fermentation chambers, 228-230
Proctodeum, 245, 248
Prohemocytes, 568
Promoter sequence, 332
Proplasmatocytes, 568, 572
Protein crystals, 215
Protein granules, 153, 155, 157, 168, 192
Prothoracicotropic hormone (PTTH), 270, 271, 288, 296
Prothoracic glands, 270, 271, 279, 281, 288
Provacuoles, 157, 158
Pseudaletia unipuncta, 519, 526, 531
Pseudoapolysis, 293
Pseudocartridge glia, 421, 437, 438, 441, 445, 449, 453, 459, 469
Pseudococcus maritimus, 318
Pseudocone, 457
Puffs, 83, 541
Purinic crystals, 245

R-cell, 411, 413, 419
rDNA amplification, 11
rDNA transcription, 333
Reabsorptive segment of salivary gland, 233
Receptorlymph, 485, 487, 489, 493, 503
Receptor membrane renewal, 410
Receptor-mediated endocytosis, 373
Recognition of self, 593
Rectal pads, 251, 254
Rectum, 245, 247
Reflexive gap junctions, 449
Regenerative cells, 217, 541
Reovirus, 575
Repetitive DNA sequences, 87
Replacement cells, 201, 223
Replication cycles, 90
Reproductive tract, 366
Reprogramming ecdysteroid peak, 293
Residual bodies, 157, 583
Retinal axons, 425, 468
Retinula, 399, 411, 437, 467, 469
Rhabdom, 399, 401, 403, 405
Rhabdomere, 397, 399, 403
Rhabdomeric microvilli, 407, 409
Rhodnius prolixus, 5, 7, 8, 10-37, 158, 160, 204, 241, 245, 276, 281, 291, 295, 371, 391
Rhodopsin, 397, 405, 411
Rhynchosciara angelae, 527, 540
Rossette organization of chromomeric fiber, 84
Rough endoplasmic reticulum (RER), 28, 61, 168, 171, 178, 189, 190, 205, 209, 220, 256, 282, 295, 308, 311, 318, 320, 371, 545, 568, 583

Salivary gland chromosomes, 75-103
Salivary glands, 232-240
Sarcomere, 117, 120, 129, 137
Sarcophaga, 122, 202, 485, 506
Sarcoplasmic reticulum (SR), 129, 135, 145
Satellite glia, 421, 437, 438, 439, 442, 445, 447, 449, 468
Scalariform junctions, 254, 421, 459
Scale insects, 304, 309, 311, 320
Schistocerca, 235, 237, 239, 274, 275, 276, 277, 377, 379, 382, 390, 391, 468
Scolopidial receptors, 490
Screening pigment granules, 410
Secondary lysosomes, 157, 572
Secondary yolk membrane, 67
Secretory vesicles of fat body, 155, 157, 171
Selective transport, 254
Seminal vesicles, 366, 391
Semper cells, 415, 416, 449, 455, 461
Sensilla, 477-510
Sensilla trichodea, 501, 504, 506, 507
Sensillar basiconica, 505, 506, 507
Septate desmosomes, 539
Septate junctions, 201, 220, 391, 421, 449, 458, 461, 484, 489, 490
Sericin, 342-346, 356
Serosa, 69, 70
Sex chromatin body, 330
Shellac, 304
Sibling cystocyte cluster, 15, 16
Sibling hemocytes, 582
Silk fibroin, 334
Silk fibroin DNA, 333
Silk gland, 323-356
Smooth endoplasmic reticulum (SER), 153, 206, 274, 277, 282, 304, 308, 309, 315, 318, 320, 410, 481
Sodium pump, 465
Somatostatin, 211
Sperm, 56, 59-61, 70
Spermatheca, 59, 386, 548
Spermatophore, 381, 385, 386, 387
Sperm transfer, 391
Spherocrystals, 213, 257
Spinneret, 350
Spiracular invaginations, 30
spk$^+$ lamellocyte, 596
Spodoptera, 271, 519, 521, 545, 548
Spoke channels, 507, 509
Spot desmosomes, 487
Stimulus-transmitting apparatus, 498
Striated border, 202
Sulfated mucopolysaccharides, 220
Supercontraction, 119, 120
Symboints, 155, 157, 173, 174
Synapse, 417, 422, 425, 427, 461

Synaptic neuropil, 459
Synaptic vesicles, 417, 422
Synaptonemal complexes, 20, 23
Synchronous muscle, 131, 133
Syncytium, 15, 16, 36, 37

Target cells of hormones, 269, 288-296
Taste hairs, 490
Telotrophic meroistic ovary, 3-40
Tenebrio molitor, 27, 218, 289, 367-391
Terminal filaments, 5, 7, 33
Terminal tracheoles, 142
Thecogen cells, 478, 480, 481, 485, 487
Thoracic glands, 35
Thoracic macrochaeta, 492
Thoracic muscle, 117
Thread press, 350
Thymidine incorporation, 330
Thysanurans, 219
Tibial thread hairs, 491
Tight junctions, 451, 561, 464
Timbal muscle, 141
Tipula paludosa, 527, 541
Tonofilaments, 127
Tormogen cells, 478, 479, 480, 483, 492
Tracheal air sac, 145
Tracheation, 142
Tracheoblast cell, 458
Tracheolar end cells, 369
Transdetermination, 561
Transition vesicle, 157
Trialeurodes vaporaviorum, 117, 127, 129, 135, 141
Triatoma infestans, 202
Trichogen cells, 478, 479, 480, 485, 486, 493
tRNA, 330, 332
Tropharium, 7, 9-16, 20
Trophic cord, 16-19
Trophocyte, 151, 193
Trophospongia, 436, 466, 489
T-tubule system, 133, 137, 141, 142
Tubular accessory gland (TAG), 367, 369, 389, 390
Tubular salivary glands, 239
Tumors of viral origin, 526
Tumorous blood cells, 567, 570, 575
Tumorous imaginal discs, 559
Tumorous neuroblasts, 567
Tunica propria, 5, 7, 31
tu-Sz^{ts} mutation of *Drosophila*, 593, 595
tu-W mutation of *Drosophila*, 592
Tyrosine storage vacuole, 153, 155, 157, 184, 185, 192

Urate cell, 259
Urate granules and vacuoles, 153, 155, 157, 162, 164, 167, 193

Urate oxidase, 164, 190
Uridine incorporation, 331
Urocytes, 160, 162
Urospherite, 214
UV cells, 407
UV receptors, 416

Vacuole, 156, 157, 186, 192
Vertebrate pancreas, 173
Vesiculation of RER, 223
Viral DNA, 529
Viral envelope, 529, 541
Virions, 521, 535, 545
Virogenesis, 523, 531, 541, 542, 547
Virogenic stroma, 523, 529, 533, 535, 541, 548, 550
Viropexis, 523
Viruses, 541, 542, 550
Viruslike particles, 218, 562, 567, 570, 572, 575, 576
Virus receptor site, 519
Virus replication, 539, 545, 547
Visual pigment molecules, 405
Vitamin A aldehyde, 403
Vitellarium, 7
Vitellin membrane, 50, 51, 53, 60, 61, 70
Vitellogenesis, 7, 17, 19, 20, 23, 25
Vitellophage, 67

Wachtiella kersicariae, 63, 65
Water transport, 232
Wax, 303
Wax blooms, 307, 316, 318, 320
Wax canals, 309
Wax filaments, 309, 317
Wax glands, 303–320
Wax-secreting cuticle, 319, 320
Wheat germ agglutinin, 596
White mutation of *Drosophila*, 89
Wound healing, 600

X chromosome, 78, 79, 89

Yolk, 55, 57, 67
Yolk protein, 23

Z bands, 117, 119, 120, 121, 125, 129
Z-DNA sequences, 87
Zonulae occludentes, 451
Zygogramma exclamationis, 15
Zymogen granules, 173
Zymogen-secreting cells, 235

If you have any concerns about our products,
you can contact us on
ProductSafety@springernature.com

In case Publisher is established outside the EU,
the EU authorized representative is:
**Springer Nature Customer Service Center GmbH
Europaplatz 3, 69115 Heidelberg, Germany**

Printed by Libri Plureos GmbH
in Hamburg, Germany